Numerical Computation of Internal and External Flows

Volume 1
Fundamentals of Computational Fluid Dynamics

Numerical Computation of Internal and External Flows

Volume 1
Fundamentals of Computational Fluid Dynamics

Second edition

Charles Hirsch

ELSEVIER

AMSTERDAM • BOSTON • HEIDELBERG • LONDON • NEW YORK • OXFORD
PARIS • SAN DIEGO • SAN FRANCISCO • SINGAPORE • SYDNEY • TOKYO
Butterworth-Heinemann is an imprint of Elsevier

Butterworth-Heinemann is an imprint of Elsevier
Linacre House, Jordan Hill, Oxford OX2 8DP
30 Corporate Drive, Suite 400, Burlington, MA 01803, USA

First published by John Wiley & Sons, Ltd
Second edition 2007

British Library Cataloguing in Publication Data
A catalogue record for this book is available from the British Library

Library of Congress Cataloguing in Publication Data
A catalog record for this book is available from the Library of Congress

ISBN: 978-0-7506-6594-0

For information on all Butterworth-Heinemann publications visit
our web site at http://books.elsevier.com

Printed and bound in the United Kingdom

Transferred to Digital Print 2010

To the Memory of

Leon Hirsch

and

Czypa Zugman

my parents,
struck by destiny

Contents

Preface to the second edition

This second, long (over)due, edition presents a major extension and restructuring of the initial two volumes edition, based on objective as well as subjective elements.

The first group of arguments is related to numerous requests we have received over the years after the initial publication, for enhancing the didactic structure of the two volumes in order to respond to the development of CFD courses, starting often now at an advanced undergraduate level.

We decided therefore to adapt the first volume, which was oriented at the fundamentals of numerical discretizations, toward a more self-contained and student-oriented first course material for an introduction to CFD. This has led to the following changes in this second edition:

- We have focused on a presentation of the essential components of a simulation system, at an *introductory* level to CFD, having in mind students who come in contact with the world of CFD for the first time. The objective being to make the student aware of the main steps required by setting up a numerical simulation, and the various implications as well as the variety of options available. This will cover Chapters 1–10, while Chapters 11 and 12 are dedicated to the first applications of the general methodology to inviscid simple flows in Chapter 11 and to 2D incompressible, viscous flows in Chapter 12.
- Several chapters are subdivided into two parts: an introductory level written for a first introductory course to CFD and a second, more advanced part, which is more suitable for a graduate and more advanced CFD course. We hope that by putting together the introductory presentation and the more advanced topics, the student will be stimulated by the first approach and his/her curiosity for the more advanced level, which is closer to the practical world of CFD, will be aroused. We also hope by this way to avoid frightening off the student who would be totally new to CFD, by a too 'brutal' contact with an approach that might appear as too abstract and mathematical.
- Each chapter is introduced by a section describing the 'Objectives and guidelines to this Chapter', and terminates by a section on 'Conclusions and main topics to remember', allowing the instructor or the student to establish his or her guide through the selected source material.
- The chapter on finite differences has been extended with additional considerations given to discretizations formulas on non-uniform grids.
- The chapters on finite element and finite volume methods have been merged, shifting the finite element description to the 'advanced' level, into Chapter 5 of this volume.
- A new Chapter 6 has been added devoted to an overview of various grids used in practice, including some recommendations related to grid quality.

- Chapters 7 and 8 of the first edition, devoted to the analysis of numerical schemes for consistency and stability have been merged and simplified, forming the new Chapter 7.
- Chapter 9 of the first edition has been largely reorganized, simplified and extended with new material related to general scheme properties, in particular the extremely important concept of monotonicity and the methodologies required to suppress numerical oscillations with higher order schemes, with the introduction of limiters. This is found in Chapter 8 of this volume.
- The former Chapters 10 and 11 have been merged in the new Chapter 9, devoted to the time integration schemes and to the general methodologies resulting from the combination of a selected space discretization with a separate time integration method.
- Parts of the second volume have been transferred to the first volume; in particular sections on potential flows (presented in Chapter 11) and two-dimensional viscous flows in Chapter 12. This should allow the student already to come in contact, at this introductory CFD level, with initial applications of fluid flow simulations.
- The number of problems has been increased and complete solution manuals will be made available to the instructors. Also a computer program for the numerical solutions of simple 1D convection and convection–diffusion equations, with a large variety of schemes and test cases can be made available to the instructors, for use in classes and exercises sessions. The objective of this option is to provide a tool allowing the students to develop their own 'feeling' and experience with various schemes, including assessment of the different types and level of errors generated by the combination of schemes and test cases. Many of the figures in the two volumes have been generated with these programs.

The second group of elements is connected to the considerable evolution and extension of Computational Fluid Dynamics (CFD) since the first publication of these books. CFD is now an integral part of any fluid-related research and industrial application, and is progressively reaching a mature stage. Its evolution, since the initial publication of this book, has been marked by significant advancements, which we feel have to be covered, at least partly, in order to provide the reader with a reliable and up-to-date introduction and account of modern CFD. This relates in particular to:

- Major developments of schemes and codes based on unstructured grids, which are today the 'standard', particularly with most of the commercial CFD packages, as unstructured codes take advantage of the availability of nearly automatic grid generation tools for complex geometries.
- Advances in high-resolution algorithms, which have provided a deep insight in the general properties of numerical schemes, leading to a unified and elegant approach, where concepts of accuracy, stability, monotonicity can be defined and applied to any type of equation.
- Major developments in turbulence modeling, including Direct Numerical Simulations (DNS) and Large Eddy Simulations (LES).
- Applications of full 3D Navier–Stokes simulations to an extreme variety of complex industrial, environmental, bio-medical and other disciplines, where fluids

play a role in their properties and evolution. This has led to a considerable overall experience accumulated over the last decade, on schemes and models.

- The awareness of the importance of verification and validation of CFD codes and the development of related methodologies. This has given rise to the definition and evaluation of families of test cases including the related quality assessment issues.
- The wide availability of commercial CFD codes, which are increasingly being used as teaching tools, to support the understanding of fluid mechanics and/or to generate simple flow simulations. This puts a strong emphasis on the need for educating students in the use of codes and providing them with an awareness of possible inaccuracies, sources of errors, grid and modeling effects and, more generally, with some global Best Practice Guidelines.

Many of these topics will be found in the second edition of Volume II.

I have benefited from the spontaneous input from many colleagues and students, who have been kind enough to send me notices about misprints in text and in formulas, helping hereby in improving the quality of the books and correcting errors. I am very grateful to all of them.

I also have to thank many of my students and researchers, who have contributed at various levels; in particular: Dr. Zhu Zong–Wen for the many problem solutions; Cristian Dinescu for various corrections. Benoit Tartinville and Dr. Sergey Smirnov have contributed largely to the calculations and derivations in Chapters 11 and 12.

Brussels, December 2006

Nomenclature

a	convection velocity or wave speed
A	Jacobian of flux function
c	speed of sound
c_p	specific heat at constant pressure
c_v	specific heat at constant volume
D	first derivative operator
e	internal energy per unit mass
e	vector (column matrix) of solution errors
$\vec{e}_x, \vec{e}_y, \vec{e}_z$	unit vectors along the x, y, z directions
E	total energy per unit volume
E	finite difference displacement (shift) operator
f	flux function
$\vec{f}_e$	external force vector
$\vec{F}(f, g, h)$	flux vector with components f, g, h
g	gravity acceleration
G	amplification factor/matrix
h	enthalpy per unit mass
H	total enthalpy
I	rothalpy
J	Jacobian
k	coefficient of thermal conductivity
k	wave number
M	Mach number
n	normal distance
$\vec{n}$	normal vector
p	pressure
P	convergence or conditioning operator
Pr	Prandtl number
q	non homogeneous term
q_H	heat source
Q	source term; matrix of non homogeneous terms
r	gas constant per unit mass
R	residual of iterative scheme
Re	Reynolds number
s	entropy per unit mass
S	space discretization operator
$\vec{S}$	surface vector
t	time
T	temperature
u	dependent variable
U	vector (column matrix) of dependent variables

U	vector of conservative variables; velocity
$\vec{v}\,(u, v, w)$	velocity vector with components u, v, w
V	eigenvectors of space discretization matrix
$\vec{w}$	relative velocity
W	weight function
x, y, z	cartesian coordinates
z	amplification factor of time integration scheme
α	diffusivity coefficient
β	dimensionless diffusion coefficient $\beta = \alpha \Delta t / \Delta x$, also called Von Neumann number
γ	specific heat ratio
Γ	circulation; boundary of domain Ω
δ	central-difference operator
δ^+, δ^-	forward and backward difference operators
Δ	Laplace operator
Δt	time step
ΔU	variation of solution U between levels $n + 1$ and n
$\Delta x, \Delta y$	spatial mesh size in x and y directions
ε	error of numerical solution
ε_v	turbulence dissipation rate
ε_D	dissipation or diffusion error
ε_ϕ	dispersion error
$\vec{\zeta}$	vorticity vector
θ	parameter controlling type of difference scheme
$\vec{\kappa}$	wave-number vector; wave propagation direction
λ	eigenvalue of amplification matrix
μ	coefficient of dynamic viscosity
μ	averaging difference operator
ξ	real part of amplification matrix
η	imaginary part of amplification matrix
ρ	density; spectral radius
σ	Courant number
$\overline{\overline{\sigma}}$	shear stress tensor
$\overline{\overline{\tau}}$	stress tensor
υ	kinematic viscosity
ϕ	velocity potential; phase angle in Von Neumann analysis
Φ	phase angle of amplification factor
ω	time frequency of plane wave; overrelaxation parameters
Ω	eigenvalue of space discretization matrix; volume

Subscripts

e	external variable
i, j	mesh point locations in x, y directions
I, J	nodal point index
J	eigenvalue number

min	minimum
max	maximum
n	normal or normal component
o	stagnation values
v	viscous term
x, y, z	components in x, y, z directions; partial differentiation with respect to x, y, z
∞	freestream value

Superscripts

n	iteration level; time level

Introduction: An Initial Guide to CFD and to this Volume

Computational Fluid Dynamics, known today as **CFD**, is defined as the set of methodologies that enable the computer to provide us with a numerical *simulation* of fluid flows.

We use the word '*simulation*' to indicate that we use the computer to solve numerically the laws that govern the movement of fluids, in or around a material system, where its geometry is also modeled on the computer. Hence, the whole system is transformed into a 'virtual' environment or *virtual product*. This can be opposed to an experimental investigation, characterized by a *material* model or prototype of the system, such as an aircraft or car model in a wind tunnel, or when measuring the flow properties in a prototype of an engine.

This terminology is also referring to the fact that we can visualize the whole system and its behavior, through computer visualization tools, with amazing levels of realism, as you certainly have experienced through the powerful computer games and/or movie animations, that provide a fascinating level of high-fidelity rendering. Hence the complete system, such as a car, an airplane, a block of buildings, etc. can be 'seen' on a computer, before any part is ever constructed.

I.1 THE POSITION OF CFD IN THE WORLD OF VIRTUAL PROTOTYPING

To situate the role and importance of CFD in our contemporary technological world, it might be of interest to take you down the road to the global world of *Computer-Assisted Engineering* or **CAE**. CAE refers to the ensemble of simulation tools that support the work of the engineer between the initial design phase and the final definition of the manufacturing process. The industrial production process is indeed subjected to an accelerated evolution toward the *computerization of the whole production cycle*, using various software tools.

The most important of them are: Computer-Assisted Design (**CAD**), Computer-Assisted Engineering (**CAE**) and Computer-Assisted Manufacturing (**CAM**) software. The CAD/CAE/CAM software systems form the basis for the different phases of the *virtual prototyping environment* as shown in Figure I.1.1.

This chart presents the different components of a computer-oriented environment, as used in industry to create, or modify toward better properties, a product. This product can be a single component such as a cooling jacket in a car engine, formed by a certain number of circular curved pipes, down to a complete car. In all cases the succession of steps and the related software tools are used in very much similar ways, the difference being the degree of complexity to which these tools are applied.

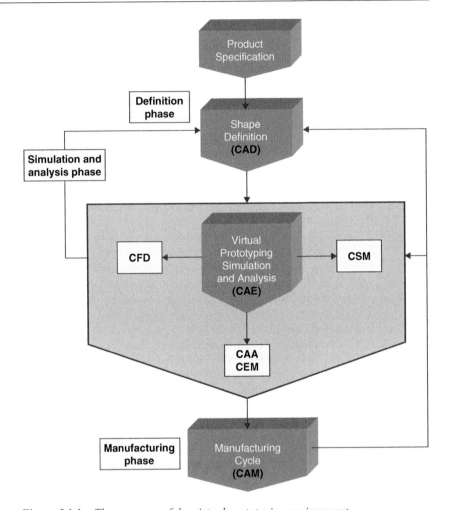

Figure I.1.1   *The structure of the virtual prototyping environment.*

I.1.1   The Definition Phase

The first step in the creation of the product is the **definition phase**, which covers the specification and geometrical definition. It is based on CAD software, which allows creating and defining the geometry of the system, in all its details. Typically, large industries can employ up to thousands of designers, working full time on CAD software. Their day-to-day task is to build the geometrical model on the computer screen, in interaction with the engineers of the simulation and analysis departments.

This CAD definition of the geometry is the required and unavoidable input to the CFD simulation task.

Figure I.1.2 shows several examples of CAD definitions of different models, for which we will see later results of CFD simulations. These examples cover a very wide range of applications, industrial, environmental and bio-medical.

Figure I.1.2a, is connected to environmental studies of wind effects around a block of buildings, with the main objective to improve the wind comfort of the people walking close to the main buildings. To analyze the problem we will have to look at the wind distribution at around 1.5 m above the ground and try to keep these wind velocities below a range of 0.5–1.0 m/s. Figure I.1.2b shows a CAD definition of an aircraft, in order to set up a CFD study of the flow around it.

Figure I.1.2c is a multistage axial compressor, one of the components of a gas turbine engine. The objective here is to calculate the 3D flow in all the blade rows, rotors and stators of this 3.5 stage compressor, simultaneously in order to predict the performance, identify flow regions generating higher losses and subsequently modify the blading in order to reduce or minimize these loss regions.

Figure I.1.2d, from Van Ertbruggen et al. (2005), is a section of several branches of the lung and the CFD analysis has as objective to determine the airflow configuration during inspiration and to determine the path of inhaled aerosols, typical of medical sprays, in function of the size of the particles. It is of considerable importance for the medical and pharmaceutical sector to make sure that the inhaled medication will penetrate deep enough in the lungs as to provide the maximal healing effect. Finally, Figure I.1.2e and f show, respectively, the complex liquid hydrogen pump of the VUL-CAIN engine of the European launcher ARIANE 5 and an industrial valve system, also used on the engines of the ARIANE 5 launcher. A CFD analysis is applied in both cases to improve the operating characteristics of these components and define appropriate geometrical changes.

I.1.2   The Simulation and Analysis Phase

The next phase is the **simulation and analysis phase**, which applies software tools to calculate, on the computer, the physical behavior of the system. This is called *virtual prototyping*. This phase is based on CAE software (eventually supported by experimental tests at a later stage), with several sub-branches related to the different physical effects that have to be modeled and simulated during the design process. The most important of these are:

- **Computational Solid Mechanics (CSM)**: The software tools able to evaluate the mechanical stresses, deformations, vibrations of the solid parts of a system, including fatigue and eventually life estimations. Generally, CSM software will also contain modules for the thermal analysis of materials, including heat conduction, thermal stresses and thermal dilation effects. Advanced software tools also exist for simulation of complex phenomena, such as crash, largely used in the automotive sector and allowing considerable savings, when compared with the cost of real crash experiments of cars being driven into walls.
- **Computational Fluid Dynamics (CFD)**: It forms the subject of this book, and as already mentioned designates the software tools that allow the analysis of the fluid flow, including the thermal heat transfer and heat conduction effects in the fluid and through the solid boundaries of the flow domain. For instance, in the case of an aircraft engine, CFD software will be used to analyze the flow in the multistage combination of rotating and fixed blade rows of the compressor and turbine; predict their performance; analyze the combustor behavior, analyze

(a) Computer (CAD) model of an urban environment.

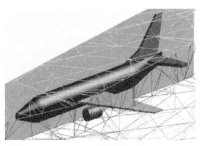

(b) Computer model (CAD) of an airplane.

(c) Computer model of a multistage compressor.

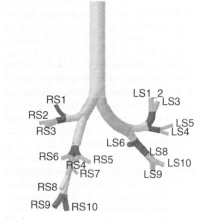

(d) Computer model of a section of pulmonary branches in the lung. From Van Ertbruggen et al. (2005).

(e) Computer model of the liquid hydrogen pump of the VULCAIN engine of the European launcher ARIANE 5.

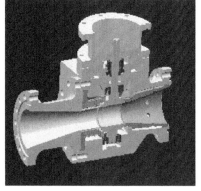

(f) Computer model (CAD) of an industrial valve system.

Figure I.1.2 *Examples of computer (CAD) models to initiate the steps toward a CFD simulation (for color image refer Plate I.1.2).*

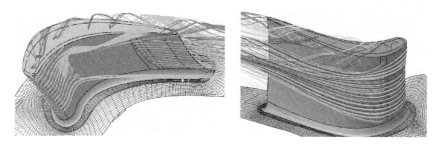

Figure I.1.3 *Simulation of the interaction between the cooling flow and the main external gas flow around a cooled turbine blade (for color image refer Plate I.1.3). Courtesy NUMECA Int. and KHI.*

the thermal parts to optimize the cooling passages, cavities, labyrinths, seals and similar sub-components. A growing number of sub-components are currently being investigated with CFD tools; while the ultimate objective is to be able to simulate the complete engine, from compressor entry to nozzle exit. An example of a complex simulation of a cooled gas turbine blade is shown in Figure I.1.3. In this simulation, the external flow around the cooled turbine interacts with the cooling flow ejected from the internal cooling passages. You can observe the very complex three-dimensional flow, which is affected by the secondary vortices, connected to the presence of the end-walls and by the tip clearance flow at the upper blade end.

- Other simulation areas related to specialized physical phenomena are also currently applied and/or in development, such as *Computational Aero-Acoustics* (CAA) and *Computational electromagnetics* (CEM). They play an important role when effects such as reduction of noise or electromagnetic interferences and signatures are important design objectives.

I.1.3 The Manufacturing Cycle Phase

In the last stage of the process, once the analysis has been considered satisfactory and the design objectives reached, **the manufacturing cycle** can start. This phase will attempt to simulate the fabrication process and verify if the shapes obtained from the previous phases can be manufactured within acceptable tolerances. This is based on the use of CAM software. This area is in strong development, as a growing number of processes are being simulated on computer, such as Forging, Stamping, Molding, Welding, for which appropriate software tools can indeed be found.

With the exploding growth of the computer hardware performance, both in terms of memory and speed, industrial manufacturers expect to simulate, in the near future, a growing number of design and fabrication processes on computer, prior to any prototype construction. This concept of *virtual product* associated to *virtual prototyping* is a major component of the technological progress, and it has already a considerable impact in all areas of industry. This impact is prone to grow further and to become a key-driving factor to all aspects of industrial analysis and design. In the automotive industry for instance, the time required for the design and production of a new car model has been reduced from 6 to 8 years in the 1970s to roughly 36 months in 2005,

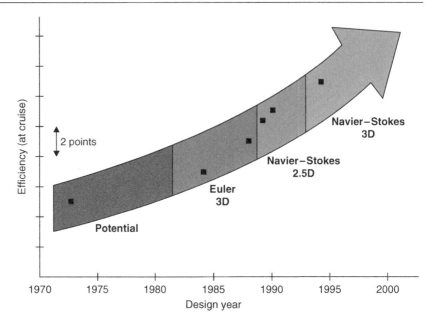

Figure I.1.4 *Impact of CFD on SNECMA fan performance, over a period of 30 years (for color image refer Plate I.1.4). From Escuret et al. (1998).*

with the announced objective of 24–18 months in the near future. A similar trend is observed in aerospace, as well as in many other highly competitive branches of industry.

It is important therefore that you realize that the major driving force behind this evolution is the wide use of computer simulations.

Coming back to the specific importance of CFD in this progress, the example of the propulsion industry is very instructive. The application of CFD has considerably improved the performance of the engines over the last 20 years, while reducing simultaneously the design cycle time. Figure I.1.4 shows the impact of the CFD tools, over a period of nearly 30 years, on the performance improvements of aircraft engines, as reported by the French engine manufacturer SNECMA. The evolution, from the initial use of simple 2D potential flow models in the early 1970s to the current applications of full 3D Navier–Stokes codes, has led to an overall gain in performance close to 10 points in efficiency. This figure also provides an interesting indication as to the period in time when the mentioned models were introduced in industry in the main design process. You will notice that 3D inviscid Euler CFD models were introduced around the mid-1980s, while the full 3D Navier–Stokes, turbulent CFD models entered the main design cycle by end of the 1990s. This evolution is due to the combination of growing computer hardware power and maturing CFD methodologies and algorithms.

A very similar impact of CFD is reported by the Boeing Company; the following statement by Boeing staff, Tinoco and Su (2004), is totally along the same line:

Effective use of Computational Fluid Dynamics (CFD) is a key ingredient in successful design of modern commercial aircraft. The application of CFD to the design of commercial transport aircraft has revolutionized the process of aerodynamic design.

Citing further from Boeing, you can find a very interesting account of 30 years of history of CFD development at this Company in Johnson et al. (2003). We highly recommend you to read this paper, as a fascinating account of how CFD evolved from an initial tool to a strategic factor in the Company's product development:

In 1973, an estimated 100 to 200 computer runs simulating flows about vehicles were made at Boeing Commercial Airplanes, Seattle. In 2002, more than 20,000 CFD cases were run to completion. Moreover, these cases involved physics and geometries of far greater complexity. Many factors were responsible for such a dramatic increase: (1) CFD is now acknowledged to provide substantial value and has created a paradigm shift in the vehicle design, analysis and support processes; ... (5) computing power and affordability improved by three to four orders of magnitude ...

Effective use of CFD is a key ingredient in the successful design of modern commercial aircraft. The combined pressures of market competitiveness, dedication to the highest of safety standards and desire to remain a profitable business enterprise all contribute to make intelligent, extensive and careful use of CFD a major strategy for product development at Boeing. Experience to date at Boeing Commercial Airplanes has shown that CFD has had its greatest effect in the aerodynamic design of the high-speed cruise configuration of a transport aircraft. The advances in computing technology over the years have allowed CFD methods to affect the solution of problems of greater and greater relevance to aircraft design, as illustrated in Figure 1.[1] Use of these methods allowed a more thorough aerodynamic design earlier in the development process, permitting greater concentration on operational and safety-related features.

The 777, being a new design, allowed designers substantial freedom to exploit the advances in CFD and aerodynamics. High-speed cruise wing design and propulsion/airframe integration consumed the bulk of the CFD applications. Many other features of the aircraft design were influenced by CFD. For example, CFD was instrumental in design of the fuselage. Once the body diameter was settled, CFD was used to design the cab. No further changes were necessary as a result of wind tunnel testing. In fact, the need for wind tunnel testing in future cab design was eliminated ... As a result of the use of CFD tools, the number of wings designed and wind tunnel tested for high-speed cruise lines definition during an airplane development program has steadily decreased (Figure 3).[2]

These advances in developing and using CFD tools for commercial airplane development have saved Boeing tens of millions of dollars over the past 20 years.

---

[1] See Figure I.1.5.

[2] See Figure I.1.6a. This figure shows information similar to Figure I.1.4. Figure I.1.6b shows the analogous evolution, seen from the European AIRBUS industry. We will come back to the various models mentioned in these figures in Chapter 2.

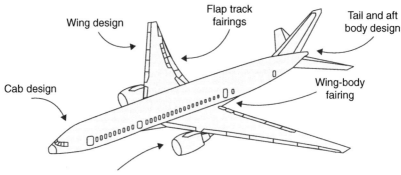

Figure I.1.5 *Role of CFD in the design of the Boeing 777. The arrows indicate the parts that were designed by CFD. From Johnson et al. (2003). Reproduced by permission of AIAA.*

However, significant as these savings are, they are only a small fraction of the value CFD delivered to the company.

The following general considerations, from the same Boeing paper, confirm the strategic impact of CFD:

A much greater value of CFD in the commercial arena is the added value of the product (the airplane) due to the use of CFD. Value is added to the airplane product by achieving design solutions that are otherwise unreachable during the fast-paced development of a new airplane. Value is added by shortening the design development process. Time to market is critical and very important in the commercial world is getting it right the first time. No prototypes are built. From first flight to revenue service is frequently less than one year! Any deficiencies discovered during flight test must be rectified sufficiently for government certification and acceptance by the airline customer based on a schedule set years before. Any delays in meeting this schedule may result in substantial penalties and jeopardize future market success. CFD is now becoming more interdisciplinary, helping provide closer ties between aerodynamics, structures, propulsion and flight controls. This will be the key to more concurrent engineering, in which various disciplines will be able to work more in parallel rather than in the sequential manner, as is today's practice. The savings due to reduced development flow time can be enormous!

To be able to use CFD in these multidisciplinary roles, considerable progress in algorithm and hardware technology is still necessary. Flight conditions of interest are frequently characterized by large regions of separated flows. For example, such flows are encountered on transports at low speed with deployed high-lift devices, at their structural design load conditions or when transports are subjected to in-flight upsets that expose them to speed and/or angle of attack

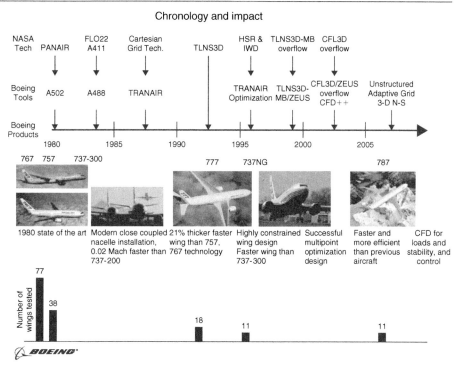

Figure I.1.6a *Evolution of the CFD tools over the last 40 years at Boeing, with an indication of the influence of CFD on the reduction of the number of wing tests (for color image refer Plate I.1.6a). Courtesy Enabling Technology and Research Organization, Boeing Commercial Airplanes.*

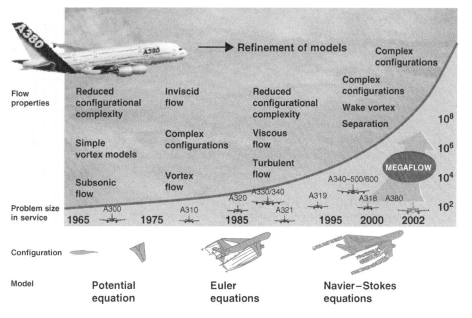

Figure I.1.6b *Evolution of the CFD tools over the last 40 years at Airbus, with an indication of the evolution of the applied models (for color image refer Plate I.1.6b). From Becker (2003).*

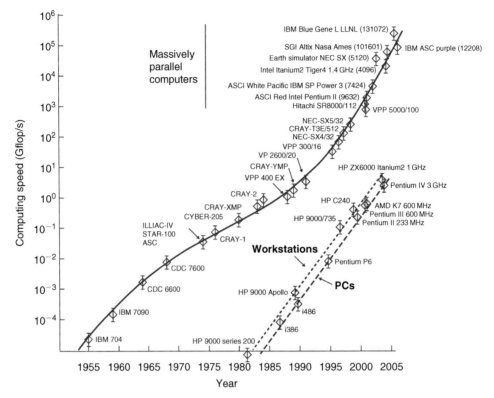

Figure I.1.7 *Evolution of Computer performance over the last 50 years, expressed in GfLOP/s, on a logarithmic scale. Courtesy Ch. Hinterberger and W. Rodi, University of Karlsruhe, Germany.*

conditions outside the envelope of normal flight conditions. Such flows can only be simulated using the Navier–Stokes equations. Routine use of CFD based on Navier–Stokes formulations will require further improvements in turbulence models, algorithm and hardware performance. Improvements in geometry and grid generation to handle complexity such as high-lift slats and flaps, deployed spoilers, deflected control surfaces and so on, will also be necessary. However, improvements in CFD alone will not be enough. The process of aircraft development, itself, will have to change to take advantage of the new CFD capabilities.

Another interesting section in this paper deals with the very important interaction between CFD and wind tunnel tests of components. We recommend you to read this section as a testimony of how CFD is contributing to raise the quality of experimental investigations.

In the previous paragraphs, we referred several times to the extraordinary growth of computing power over the last 50 years. This is summarized in Figure I.1.7, where the various computer systems are positioned by their CPU performance in function of their year of appearance. The CPU performance is measured in

GigaFlops: i.e. Billions (10^9) of floating point operations per second (Flop/s); a quite impressive number, a *Flop* being typically an addition or subtraction on the computer. The first computers in 1955 had a processor speed of 10^{-5} Gflop/s, that is of the order of 10,000 Flop/s; while the first PC with a 386 processor reached 100,000 Flop/s. Note that the level of 1000 Gflop/s, called TeraFlop/s, has been reached around the year 2000. The fastest computers shown on this figure turn around 200 TeraFlop/s, obtained through massively parallel computers over 100,000 processors. On the other hand, current high-end PCs, which are scalar computers, have a remarkable speed of the order of 5 Gflop/s.

I.2 THE COMPONENTS OF A CFD SIMULATION SYSTEM

Having positioned CFD, and its importance, in the global technological world of virtual prototyping, we should now look at the main components of a CFD system.

We wish to answer the following question: *What are the steps you have to define in order to develop, or to apply, a CFD simulation?* We make no difference at this stage between these two options, as it is similarly essential for the 'user' of a CFD code to understand clearly the different options available and to be able to exercise a critical judgment on all the steps involved.

Refer to Figure I.2.1 for a synthetic chart and guide to this section and the structure of this book. The CFD components are defined as follows:

- *Step 1*: It selects the mathematical model, defining the level of the approximation to reality that will be simulated (forms the content of Part I of this volume).
- *Step 2*: It covers the discretization phase, which has two main components, namely the space discretization, defined by the grid generation followed by the discretization of the equations, defining the numerical scheme (forms the content of Part II of this volume).
- *Step 3*: The numerical scheme must be analyzed and its properties of stability and accuracy have to be established (forms the content of Part III of this volume).
- *Step 4*: The solution of the numerical scheme has to be obtained, by selecting the most appropriate time integration methods, as well as the subsequent resolution method of the algebraic systems, including convergence acceleration techniques (forms the content of Part IV of this volume).
- *Step 5*: Graphic post-processing of the numerical data to understand and interpret the physical properties of the obtained simulation results. This is made possible by the existence of powerful visualization software.

Let us look at this in more details step by step.

I.2.1 Step 1: Defining the Mathematical Model

The first step in setting up a simulation is to define the physics you intend to simulate. Although we know the full equations of fluid mechanics since the second half of the 19th century, from the work of Navier and Stokes in particular, these equations are

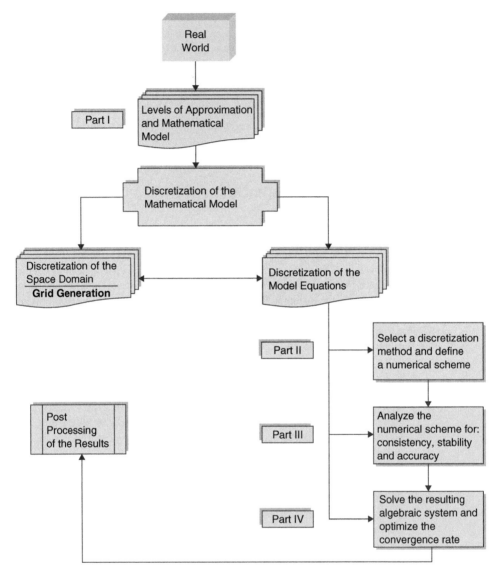

Figure I.2.1   *Structure of a CFD simulation system.*

extremely complicated. They form a system of *nonlinear* partial differential equations, with major consequences of this nonlinearity being the existence of turbulence, shock waves, spontaneous unsteadiness of flows, such as the vortex shedding behind a cylinder, possible multiple solutions and bifurcations. See Chapter 2 for some typical examples.

If we add to the basic flow more complex phenomena such as combustion, multiphase and multi-species flows with eventual effects of condensation, evaporation, bursting or agglomeration of gas bubbles or liquid drops, chemical reactions as in fire simulations, free surface flows, we need to model the physical laws describing these phenomena and provide the best possible approximations.

The essential fact to remember at this stage is that within the world of continua, as currently applied to describe the macroscopic behavior of fluids, there is **always** *an unavoidable level of empiricism in the models. It is therefore important that you take notice already that any modeling assumption will be associated with a generally undefined level of error when compared to the real world.*

Therefore, keep in mind that a good understanding of the physical properties and limitations of the accepted models is very important, as it is not unusual to discover that discrepancies between CFD predictions and experiments are not due to errors in experimental or numerical data, but are due to the fact that the theoretical model assumed in the computations might not be an adequate description of the real physics.

Consequently, with the exception of Direct Numerical Simulation (DNS) of the Navier–Stokes equations, we need to define appropriate modeling assumptions and simplifications. They will be translated into a mathematical model, formed generally by a set of partial differential equations and additional laws defining the type of fluid, the eventual dependence of key parameters, such as viscosity and heat conductivity in function of other flow quantities, such as temperature and pressure; as well as various quantities associated to the description of additional physics and other reactions, when present.

The establishment of adequate mathematical models for the physics to be described form the content of Part I of this volume. It is subdivided into three chapters dealing with:

- the basic flow equations (Chapter 1);
- an illustrated description of the different approximation levels that can be selected to describe a fluid flow (Chapter 2);
- the mathematical properties of the selected mathematical models (Chapter 3).

I.2.2   Step 2: Defining the Discretization Process

Once a mathematical model is selected, we can start with the major process of a simulation, namely the *discretization* process.

Since the computer recognizes only numbers, we have to translate our geometrical and mathematical models into numbers. This process is called *discretization*.

The first action is to discretize the space, including the geometries and solid bodies present in the flow field or enclosing the flow domain. The solid surfaces in the domain are supposed to be available from a CAD system in a suitable digital form, around which we can start the process of distributing points in the flow domain and on the solid surfaces. This set of points, which replaces the continuity of the real space by a finite number of isolated points in space, is called a *grid* or a *mesh*.

The process of grid generation is in general extremely complex and requires dedicated software tools to help in defining grids that follow the solid surfaces (this is called 'body-fitted' grids) and have a minimum level of regularity.

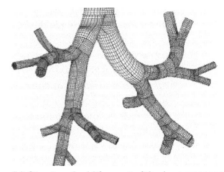

Surface grid
(a) Structured grid of a landing gear.
From Lockard et al. (2004).
Reproduced by permission from AIAA.

(b) Structured grid for part of the lung passages
shown in Plate I.1.2. From Van Ertbruggen
et al. (2005).

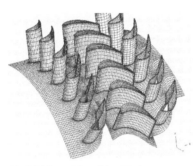

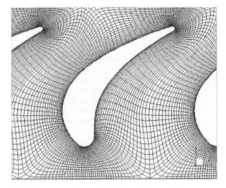

(c) Grid for a 3D turbine blade passage.

(d) Close-up view of the turbine grid.

Figure I.2.2   *Examples of structured grids.*

We will deal with the grid-related issues in Chapter 6, but we wish already here to draw your attention to the fact that, when dealing with complex geometries, the grid generation process can be very delicate and time consuming.

Grid generation is a major step in setting up a CFD analysis, since, as we will see later on, in particular in Chapters 4, 5 and 6, the outcome of a CFD simulation and its accuracy can be extremely dependent on the grid properties and quality.

Please notice here that the whole object of the simulation is for the computer to provide the numerical values of all the relevant flow variables, such as velocity, pressure, temperature, ..., at the positions of the mesh points.

Hence, this first step of grid generation is essential and cannot be omitted. Without a grid there is no possibility to start a CFD simulation.

Figure I.2.2 shows examples of 2D and 3D *structured* grids, while Figure I.2.3 shows some examples of *unstructured* grids. These concepts will be detailed further in Chapter 6.

So, once a grid is available, we can initiate the second branch of the discretization process, namely the discretization of the mathematical model equations, as shown in the chart of Figure I.2.1.

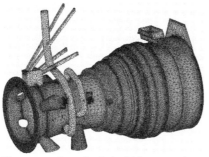

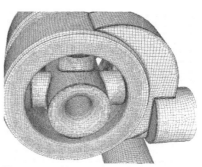

Unstructured tetrahedral grid for an engine. From ICEM-CFD.

Unstructured hexahedral grid for an oil valve. HEXPRESS mesh. Courtesy NUMECA Int.

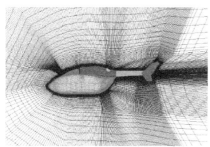

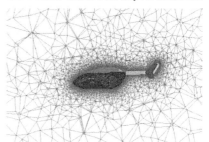

Figure 3: *ICEM-Hexa structured multiblock N.-S.mesh around the EC145 isolated fuselage: middle plane.*

Figure 4: *CENTAUR hybrid N.-S.mesh around the EC145 isolated fuselage: middle plane.*

From D´Alascio et al. (2004).
A middle plane section of an helicopter fuselage with structured and unstructured grids.

Figure I.2.3   *Examples of unstructured grids (for color image refer Plate I.2.3).*

*As the mesh point values are the **sole** quantities available to the computer, all mathematical operators, such as partial derivatives of the various quantities, will have to be transformed, by the discretization process, into arithmetic operations on the mesh point values.*

This forms the content of Part II, where the different methods available to perform this conversion from derivatives to arithmetic operations on the mesh point values will be introduced. In particular, we will cover the:

- finite difference method in Chapter 4,
- finite volume and finite element methods in Chapter 5,
- grid properties and guidelines in Chapter 6.

I.2.3   Step 3: Performing the Analysis Phase

After the discretization step, a set of algebraic relations between neighboring mesh point values is obtained, one relation for each mesh point. These relations are called a *numerical scheme*.

The numerical scheme must satisfy a certain number of rules and conditions to be accepted and subsequently it must be analyzed to establish the associated level of accuracy, as any discretization will automatically generate errors, consequence of the replacement of the continuum model by its discrete representation.

This analysis phase is critical; it should help you to select the most appropriate scheme for the envisaged application, while attempting at the same time to minimize the numerical errors. This will be introduced and discussed in Part III.

Part III covers many subjects and should be studied with great attention. The following topics will be dealt with:

- The concepts of consistency, stability and convergence of a numerical scheme and a method for the analysis of stability in Chapter 7, including the quantitative evaluation of the errors associated to a selected scheme.
- A general approach to properties of numerical schemes will be presented in Chapter 8, together with a methodology to generate schemes with prescribed accuracy. In addition this chapter will introduce the property of monotonicity leading to nonlinear high-resolution scheme.

I.2.4 Step 4: Defining the Resolution Phase

The last step in the CFD discretization process is solving the numerical scheme to obtain the mesh point values of the main flow variables. The solution algorithms depend on the type of problem we are simulating, i.e. time-dependent or steady flows. This will require techniques either to solve a set of ordinary differential equations in time, or to solve an algebraic system.

For time-dependent numerical formulations, a particular attention has to be given to the time integration, as we will see that for a given space discretization, not all the time integration schemes are acceptable.

It is essential at this stage to realize that at the end of the discretization process, all numerical schemes finally result in an *algebraic system of equations*, with as many equations as unknowns. This number can be quite considerable, as the present capacity of computer memory storage allows large grids to be used to enhance the accuracy of the CFD predictions. The flow around an aircraft, such as shown in Figure I.1.2, might require a grid close to 50 million points for a minimal acceptable accuracy. This number is substantiated by the outcome of a recent 'Drag Prediction' workshop, run in 2003 by AIAA[3] and NASA.[4]

The objective of the workshop was to assess the state-of-the-art of CFD for aircraft drag and lift prediction (see the review by Hemsch and Morrison, 2004). The main outcome of this workshop was that a grid of the order of 10–15 million points was required for acceptable accuracy of current CFD codes, on a wing–body–nacelle–pylon (WBNP) combination. The enhanced complexity of a full aircraft, compared

---

[3] American Institute of Aeronautics and Astronautics (USA).
[4] National Aeronautics and Space Administration (USA).

with this simplified WBNP combination, leads to a minimal estimate of the order of 50 million points for the full aircraft. With at least 5 unknowns per point (the three velocity components, pressure, and temperature) we wind up with an algebraic system of 250 million equations for 250 million unknowns; system that has to be solved many times during the iterative process toward convergence. You can understand on this example why the availability of very fast methods for the solution of these huge algebraic systems is crucial for an effective CFD simulation.

An introduction to the most important methods will be dealt with in Part IV, including also techniques for convergence acceleration, such as the important multigrid methods. Part IV is subdivided into:

- methods for ordinary differential equations, referring to the time integration methods, in Chapter 9;
- methods for the iterative solution of algebraic systems in Chapter 10.

Once the solution is obtained, we have to manipulate this considerable amount of numbers to analyze and understand the computed flow field. This can only be achieved through powerful *visualization systems*, which provide various software tools to study, qualitatively and quantitatively, the obtained results. Typical examples of outputs that can be generated are shown in Figure I.2.4:

- Cartesian plots for the distribution of a selected quantity in function of a coordinate direction or along a solid wall surface (Figure I.2.4a).
- Color plots of a given quantity on the solid surface or in the flow field (Figure I.2.4b and c).
- Visualization of streamlines, see Figure I.1.3 and of velocity vectors (Figure I.2.4d).
- Local values of a quantity in an arbitrary point, obtained by clicking the mouse on that point.
- Various types of animations.

Many other examples of visualizations will be shown in the following chapters.

The last part of Volume I, Part V, is devoted to several basic applications of the developed methodology, in order to guide you toward your first attempts in working out a CFD simulation. We will consider one-dimensional models for scalar variables, up to the Euler equations for nozzle flows, as well as two-dimensional potential and laminar flow models and present different numerical schemes in sufficient detail for you to program and solve these applications:

- Chapter 11 will deal with 2D potential flows and 2D inviscid flows governed by the system of Euler equations.
- Chapter 12 will deal with the 2D Navier–Stokes equations.

A particular section will be also devoted to some general Best Practice Guidelines to follow when applying existing, commercial or other, CFD tools. This will be based on the awareness of all possible sources of errors and uncertainties that can affect the quality and the validity of the obtained CFD results.

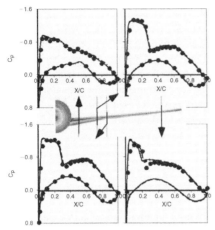

(a) Cartesian plot of pressure distribution at various positions along a wing–body–nacelle model, compared to experimental data.
From Tinoco and Su (2004),
Reproduced by permission from AIAA.

(b) Instantaneous iso-surfaces of vorticity colored by the span-wise component of vorticity of a 70° delta wing.
From: Morton (2004)

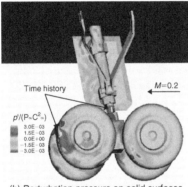

(b) Perturbation pressure on solid surfaces

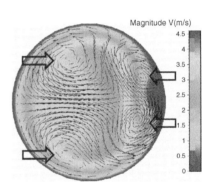

(c) Perturbation pressure distribution for an aero-acoustic simulation of the noise generated by a landing gear.
From Lockard et al. (2004).
Reproduced by permission from AIAA.

(d) Color plot and velocity vectors in one cross-section of the lung bifurcations shown in Figures I.1.2 and I.2.2. From Van Ertbruggen et al. (2005).

Figure I.2.4   *Examples of visual results from CFD simulations (for color image refer Plate I.2.4).*

I.3   THE STRUCTURE OF THIS VOLUME

The guideline to the overall organization of this volume is summarized on the following chart (Figure I.3.1), where each chapter is positioned. This will help you to situate at any time the topics you are studying within the global context.

As mentioned earlier, the structure and the presentation of this second edition of Volume I has been re-organized and focused in the first instance toward beginners and newcomers to CFD. We have attempted to guide the student and reader to progressively

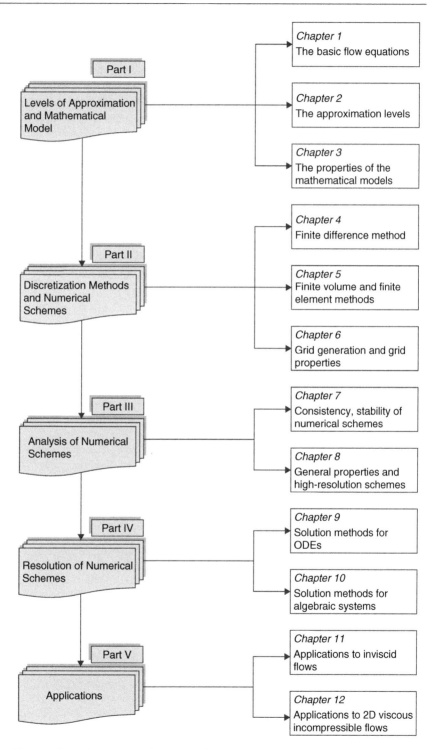

Figure I.3.1   *Structure and content of this volume.*

become familiar with the essential steps leading to a CFD application, either as a starting developer of CFD applications, or as a user of existing CFD tools, such as commercial software packages. In both cases, it is essential to acquire a deep understanding of all the components entering a CFD simulation, and in particular to develop a strong knowledge of the possible sources of errors and uncertainties.

On the other hand, we wish to give the opportunity to more advanced readers and students to also find material that would meet their objectives of accessing more advanced topics, while having at the same time a direct access to all the fundamentals.

Hence, we have identified in many chapters, topics and sections, indicated by A for Advanced, that we consider outside the introductory level and that can form the basis for a more advanced course. The relevant A-sections will be identified at the level of each chapter.

It goes without saying that any combination of 'A' sections with the other sections can be offered as course material at the discretion of the instructors.

On the other hand, we also hope that here and there, through the chapters, the newcomer to CFD will have his/her intellectual curiosity aroused by the subject and tempted to make an incursion in some of these more advanced subsections.

REFERENCES

Becker, K. (2003). *Perspectives for CFD*. DGLR-2002-013, DGLR Jahrbuch 2002, Band III, Germany.

D'Alascio, A., Pahlke, K. and Le Chuiton, K. (2004). Application of a structured and an unstructured CFD method to the fuselage aerodynamics of the EC145 helicopter. Prediction of the time averaged influence of the main rotor. *Proceedings ECCOMAS 2004 Conference*, P. Neittaanmäki, T. Rossi, S. Korotov, E. Oñate, J. Périaux and D. Knörzer (Eds.), Jyväskylä, Finland.

Escuret, J.F., Nicoud, D. and Veysseyre, Ph. (1998). Recent advances in compressor aerodynamic design and analysis. In *Integrated Multidisciplinary Design of High Pressure Multistage Compressor Systems*, RTO-LS-111, AC/323 (AVT) TP/1, ISBN 92-837-1000-2, RTO/NATO Paris, France.

Hemsch, M.J. and Morrison, J.H. (2004). Statistical analysis of CFD solutions from 2nd drag prediction workshop, *42nd AIAA Aerospace Sciences Meeting*, Reno, AIAA Paper 2004-556.

Johnson, F.T., Tinoco, E.N. and Yu, N.J. (2003). Thirty years of development and application of CFD at Boeing commercial airplanes, Seattle, *16th AIAA Computational Fluid Dynamics Conference*, Orlando, AIAA Paper 2003-3439.

Lockard, D.P., Khorrami, M.R. and Li, F. (2004). Aaeroacoustic analysis of a simplified landing gear, AIAA Paper 2004-2887. 10th AIAA/CEHS Aeroacoustics Conference, Manchester, UK.

Morton, S.A. (2004). Detached-Eddy simulations of a 70 degree delta wing in the ONERA F2 wind tunnel. *Proceedings ECCOMAS 2004 Conference*, P. Neittaanmäki, T. Rossi, S. Korotov, E. Oñate, J. Périaux and D. Knörzer (Eds.), Jyväskylä, Finland.

Tinoco, E.N. and Su, B. (2004). Drag prediction with the Zeus/CFL3D system, 42nd *AIAA Aerospace Sciences Meeting*. AIAA Paper 2004-552.

Van Ertbruggen, C., Hirsch, Ch. and Paiva, M. (2005). Anatomically based three-dimensional model of airways to simulate flow and particle transport using Computational Fluid Dynamics. *J. Appl. Physiol.*, 98, 970–980.

Part I

The Mathematical Models for Fluid Flow Simulations at Various Levels of Approximation

INTRODUCTION

The invention of the digital computer and its introduction in the world of science and technology has led to the development, and increased awareness, of the concept of approximation. This concerns the theory of the numerical approximation of a set of equations, taken as a mathematical model of a physical system. But it concerns also the notion of the approximation involved in the definition of this mathematical model with respect to the complexity of the physical world.

We are concerned here with physical systems for which it is assumed that the basic equations describing their behavior are known theoretically, but for which no analytical solutions exist, and consequently an approximate numerical solution will be sought instead.

For various reasons, the first of these being the great complexity, it is often not practically possible to describe completely the evolution of the system in its full complexity. Of course, the definition of these limits is relative to a given time and environment and these are being extended with the evolution of the computer technology. But taken at a given period, it is necessary to define mathematical models that will reduce the complexity of the original basic equations and make them tractable within fixed limits. Actually, the first level to be defined is the '*scale of reality*' level. Physicists propose various levels of description of our physical world, ranging from subatomic, atomic or molecular, microscopic, macroscopic (defined roughly as the scale of classical mechanics) up to the astronomical scale. As is well known, in the statistical description of a gas, the motion of the individual atoms or molecules are taken into consideration and the behavior is ruled by the Boltzmann equation. This description leads for instance to the definition of temperature as a measure of the mean kinetic energy of the gas molecules; to a definition of pressure as a result of the impulse of molecules on the walls of the body containing the gas; to a definition of viscosity connected to the momentum exchange due to the thermal molecular motion, and so on.

At this molecular level of description the fundamental variables are molecule velocities, number of particles per volume and other variables defining the motion of the individual molecules, while pressure, temperature, viscosity e.g. are mean properties which are deduced from other variables, more basic at this level of reality. Hence, we may consider that each level of reality can be associated with a set of fundamental variables, from which other variables can be defined as measures of certain mean properties. Continuing in the line of this example we have, beyond the molecular level of statistical mechanics, the atomic level, the nuclear and the subnuclear level that we do not plan to discuss here, since they are fully outside the domain of definition of a fluid.

Actually, fluid dynamics starts to exist as soon as the interaction between a sufficiently high number of particles affects and dominates, at least partly, the motion of each individual particle. Hence, fluid dynamics is essentially the study of the interactive motion and behavior of large number of individual elements.

The limit between individual motions of isolated particles or elements and their interactive motion is of significance in the study of rarefied gases. It is known that the interaction between the particles becomes negligible if the mean free path length attains a magnitude of the order of the length scale of the considered system. The ratio of the mean free path length to the reference length scale is called the *Knudsen number*.

For higher values of the mean free path length, or of the Knudsen number, the particles behave essentially as individual elements. These limit situations will not be considered here since they are outside the field of classical fluid dynamics. Note however that the intermediate range between the continuum and the rarefied gas approximations is of practical significance for the prediction of the re-entry phase of a Space Shuttle. When re-entering the earth atmosphere from space, the Space Shuttle crosses the atmosphere from very high altitudes, where it cannot be considered as a continuum, through an intermediate range that evolutes with reducing altitude to a continuum fluid. We need therefore special models, intermediate between the Navier–Stokes and the Boltzmann equations to handle these situations, which are extremely critical for the safety of the return phase of the Shuttle.

We will focus, in the following, on the level of reality in which the density of elements is high enough, so that we can make the approximation of considering the system of interacting elements as a ***continuum***. This expresses that continuity or closeness exists between the elements such that their mutual interaction dominates over the individual motions, although these are not suppressed. What actually happens, is that a collective motion is superimposed on the motion of the isolated elements as a consequence of the large number of these elements coexisting within the same domain.

From this point of view, we understand easily why the concepts of fluid mechanics can be applied to a variety of systems consisting of a large number of interacting individual elements.

This is the case for the current fluids and gases where the individual 'element', or fluid particle is actually not a single molecule, but consists of a large number of molecules occupying a small region with respect to the scale of the considered domain, but still sufficiently large in order to be able to define a meaningful and non-ambiguous average of the velocities and others properties of the individual molecules and atoms occupying this volume. It implies that this elementary volume contains a sufficiently high number of molecules, with for instance a well defined mean velocity, mean kinetic energy, allowing to define velocity, temperature, pressure, entropy and so on, at each point. Hence, associated fields, which will become basic variables for the description of the system, can be defined although the temperature, or pressure, or entropy of an individual atom or molecule is not defined and generally meaningless.

In the classical interpretation of turbulence, each fluid particle as defined above enters into a stochastic motion and in defining mean turbulent variables, such as a mean turbulent velocity field, an average is performed, in this case an average in time, over the motion of the fluid particles themselves.

A still higher level of averaging occurs in the description of flows through porous media such as soils. In the description of groundwater flows an 'element' is the set of fluid particles, as defined above, contained in a volume large enough as to contain a great number of soil particles and fluid particles such that a meaningful average can be performed, but still small with respect to the dimensions of the region to be analyzed. Such a volume is considered as a 'point' at this level of description, and the fields are attached to these points, implying that groundwater flow theories do study the behavior of collection of fluid particles.

Following this line, the movement or overall displacement of crowds at exit of railway stations during rush hours, or of a football stadium, can be analyzed with fluid mechanical concepts. In this case, an 'element' is the set of persons contained in a region small with regard to the dimensions of the station for instance, but still containing a sufficiently high number of individuals in order to define non-ambiguous average values, such as velocity and other variables. In this description, the displacement of an individual is not considered, but only the motion of groups of individuals.

A similar analysis can be defined for heavy traffic studies, where an 'element' is defined as a set of cars (in the one-dimensional space formed by the road). Obviously in a light traffic, the isolated car behaves as a single particle but collective motion comes in when a certain intensity of traffic has been reached such that the speed of an individual car is influenced by the presence of the other cars. This is actually to be considered as the onset of a 'fluid mechanical' description.

Finally, at a still larger scale, astrophysical fluid dynamics can be defined for the study of the interstellar medium or for the study of the formation and evolution of galaxies. In this latter case for instance, an 'element' consists of a set of stellar objects, including one or several solar systems and the dimensions of a 'point' can be of the order of light years.

In conclusion of these considerations, we can say that fluid mechanics is essentially the study of the behavior of averaged quantities and properties of a large number of interacting elements. The same is true for another domain of scientific knowledge, namely thermodynamics, which is also the study of systems of large numbers of interacting elements. It is therefore no wonder that thermodynamics is, with the exception of incompressible isothermal media, tightly interconnected with fluid mechanics and plays an important role in the description of the evolution of 'fluid mechanical' systems as mentioned above.

An essential step in fluid dynamics is therefore the averaging process. We have to decide, in front of a given system, which level of averaging will be performed in function of the quantities to be predicted, in function of the significant variables which can be defined in a meaningful way, in function of the precision and degree of accuracy to be achieved in the description of the system's behavior. This is a basic task for the scientists in charge of the analysis that requires a great understanding of the physics of the system, a judgment and sense of compromise between required level of accuracy and degree of sophistication of the chosen mathematical model.

The next step in the definition of the levels of approximation is to define a time or '*steadiness level*'. This implies an estimation of the various time constants of the considered flow situation and the choice of the lowest time constant to be taken into consideration in the modelization of this flow system. Then a time averaging will be performed with regard to the time constants lower than the chosen minimal value. The best-known example of this procedure is the system of time averaged Navier–Stokes equations for the mean turbulent flow variables. An averaging is performed over the turbulent fluctuations, since we are concerned in that case with variations of the flow slower that the turbulent fluctuations and hence, with time constants much larger than the time constant of these fluctuations. Through this procedure, extra terms appear in the equation, the Reynolds stresses, which are averaged products of fluctuations, and for which external information will have to be provided.

Along similar lines, in Large Eddy Simulations, known as the LES approximation, the turbulent fluctuations are averaged only over part of their spectrum, namely the small scales are modeled while the larger turbulent motion, associated with the lower frequencies, is directly simulated.

The *spatial level of approximation* defines the number of space variables used in the model. We have to decide in function of certain assumptions concerning the physical behavior of the system, if a one- or two-dimensional description will provide sufficiently accurate information about the behavior of the flow. It is of importance to note that the basic flow equations being three-dimensional any description with less than three space variables will be obtained by disregarding the flow variations with respect to the corresponding space coordinate and this can be formulated mathematically by averaging out the equations over that space variable.

Therefore, the averaging process, here over space, is again essential. In this space averaging, we will obtain equations in a two- or one-dimensional region, which contain terms describing the averaged influence of the full three-dimensional motion. These terms, analogous to the Reynolds stresses, will generally be neglected due to the lack of information to estimate them, although they can, or could, be estimated in certain cases.

Since the averaging procedure implies a loss of information in the averaged space variables, this information will have, in many cases, to be provided from 'outside' the model, for instance, through empirical data. It is also clear therefore, that simple models like one-dimensional flow descriptions, may require more empirical or external input than a viscous three-dimensional description, if some contributions from the three-dimensionality are to be taken into account.

The next level of approximation, the '*dynamical level*', is tied to an estimation of the relative influence of the various forces and their components on the system's behavior. The dynamical evolution of a flow system is determined by the equilibrium of the different forces acting on it, but it seldom occurs that all the force components are equally important. Therefore, a very basic step in setting up a mathematical model for the description of a system is an estimation of the dominant force components in order to simplify the model as strongly a possible. For instance, although gravity forces are always present on earth, in many cases these forces have only a negligible influence on the flow behavior. The detailed study of the influence of viscosity by Prandtl, which led to the boundary layer concept, is maybe the most fascinating example of the consequences of a deep analysis of the relative influence of forces. As is well known, the considerable simplification of the Navier–Stokes equations introduced through this analysis, allowed the practical calculation of many flow situations, which were largely intractable by the full Navier–Stokes equations.

This boundary layer concept led to the definition of the regions of validity of inviscid flows, in which the viscosity forces could be neglected. Therefore, inviscid flow approximations play an important role in fluid mechanics although their range of validity in internal flows is more restricted than in external aerodynamics.

It is to be noted that the different levels of approximation considered here can strongly interact with each other. For instance on a rotating blade of a turbomachine, the centrifugal forces will create a radial migration of the boundary layer fluid along the blade, leading to an increased spanwise mixing of the flow and hence will limit the validity of a purely two-dimensional description of the blade-to-blade flow. But in all cases, the final word with regard to the validity of a given model is the comparison to experimental data, or to computations at a higher level of approximation.

These remarks are presented here to introduce the methodology to be followed in the next chapters. After having summarized the basic flow equations in Chapter 1, a systematic presentation of various mathematical models describing the most current approximations will be given in Chapter 2. Finally, Chapter 3 will introduce the analysis of the properties of the system of equations describing the selected model; see the guideline chart (Figure I.3.1) for the global overview.

Chapter 1

The Basic Equations of Fluid Dynamics

OBJECTIVES AND GUIDELINES

This first chapter will introduce the initial step in defining a CFD application, namely the selection of the model to be discretized. In order to guide you through the complexities of fluid mechanics, which we alluded to in the introduction to this Volume, we need to establish the basic laws governing fluid flows.

You certainly have seen many flow situations and you have certainly recognized that they can be very complex, with phenomena such as turbulence, which is a global instability of a flow, as a dominant element of most of the flows encountered in nature and in technology.

In addition, applications to CFD have led to a new approach and a new way of looking at the laws of fluid mechanics. Although they can be written in many different mathematical forms, CFD has led us to put forward a specific form of these laws, through the concept of *conservation* and of *conservation laws*. This concept will be central to most of this chapter.

In Section 1.1 we develop and present the most general form of a conservation law, without specifying the nature of the 'conserved' quantity. To achieve this, we have to define first what *conservation* means and how we recognize an equation written in *conservative* form. We will see that this is a fundamental concept for CFD in many chapters later on, but the main reason for the privileged conservative form is connected to the requirement that, after the equations are discretized, essential quantities such as mass or energy will be conserved *at the discrete level*. This is certainly essential, as you can imagine, since a numerical simulation wherein mass or energy would be lost because of numerical artifacts, would be totally useless and not reliable.

A conservation law is strongly associated to the concept of fluxes and we will introduce in Section 1.1.2 the extremely important distinction between convective and diffusive fluxes. This distinction is central to the whole of fluid mechanics and of CFD.

With the basis obtained in Section 1.1, we are ready to apply the general conservation laws to the three quantities that define uniquely the laws of fluid mechanics; mass, momentum and energy, described and developed in detail in Sections 1.2, 1.3 and 1.4.

The flow chart in Figure 1.0.1 illustrates the links and the structure of this chapter. We strongly suggest that you refer regularly, while progressing through the material, to this chart as a guide for the order and relative importance of the various topics.

In addition, the part in gray indicates the sections containing more advanced material that can form the basis of a more advanced CFD course. Of course, any instructor can make his/her own 'cocktail' between the various topics, according to the level of the students.

These Advanced sections cover a few important topics, when dealing with CFD applications to rotating systems or moving grids, which often occur in practice,

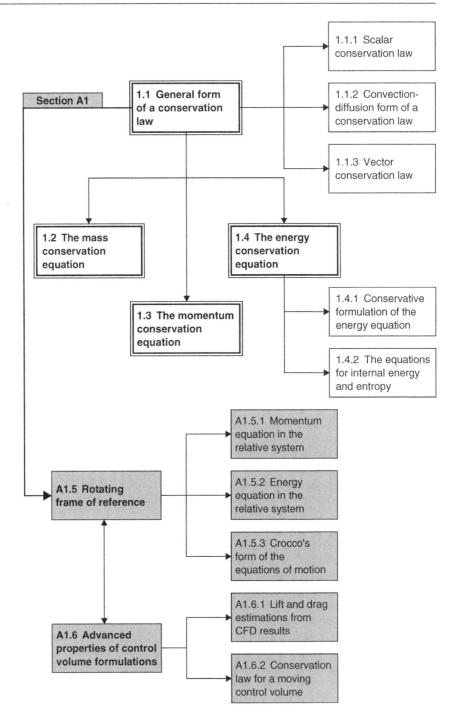

Figure 1.0.1 *Guide to this chapter.*

in large areas of industry, such as rotating machines, flow around helicopters, or with moving grids as encountered in fluid–structure interactions with vibrating surfaces.

We have also added a short section, which is a straightforward extension of the general integral form of the momentum equation, in presence of solid bodies, providing the formulation to post-process CFD data in order to extract the forces exerted on a body by the flow, such as lift and drag.

1.1 GENERAL FORM OF A CONSERVATION LAW

As mentioned in the introduction, the conservation law is the fundamental concept behind the laws of fluid mechanics.

But what is a conservation law?
It is altogether very simple in its basic logic, but can become complicated by its internal content. Conservation means that the variation of a conserved (intensive) flow quantity within a given volume is due to the net effect of some internal sources and of the amount of that quantity which is crossing the boundary surface. This amount is called the *flux* and its expression results from the mechanical and thermodynamic properties of the fluid. It will be defined more precisely in the next section. Similarly, the sources attached to a given flow quantity are also assumed to be known from basic studies. The fluxes and the sources are in general dependent on the space–time coordinates, as well as on the fluid motion. The associated fluxes are vectors for a scalar quantity and tensors for a vector quantity like momentum.

We can state the conservation law for a quantity U as the following logical consistency rule:

> *The variation of the total amount of a quantity U inside a given domain is equal to the balance between the amount of that quantity entering and leaving the considered domain, plus the contributions from eventual sources generating that quantity.*

Hence, we are looking at the rate of change of the quantity U during the flow evolution, as a flow is a moving and continuously changing system.

Although we will write the conservation law for an undefined quantity U, it should be mentioned at this stage that not all flow quantities obey a conservation law. The identification of the quantities that obey a conservation law is defined by the study of the physical properties of a fluid flow system. It is known today that the laws describing the evolution of fluid flows (this is what we call fluid dynamics) are totally defined by the conservation of the following three quantities:

1. Mass
2. Momentum
3. Energy.

This represents in total five equations, as the momentum, defined as the product of density and velocity, is a vector with three components in space.

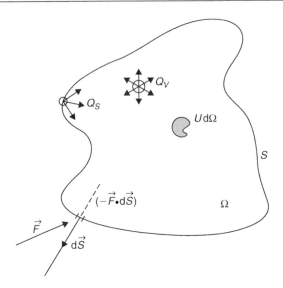

Figure 1.1.1 *General form of a conservation law for a scalar quantity.*

On the other hand, it is essential to keep in mind that other quantities, such as pressure, temperature, entropy, for instance, do **not** satisfy a conservation law. This does not mean that we cannot write an equation for these quantities, it just means that this equation will not be in the form of a conservation law.

1.1.1 Scalar Conservation Law

Let us consider a scalar quantity per unit volume U, defined as a flow related property.

We now consider an arbitrary volume Ω, *fixed in space*, bounded by a closed surface S (see Figure 1.1.1) crossed by the fluid flow.

The surface S is *arbitrary* and is called a *control surface*, while the volume Ω is called a *control volume*.

Our goal here is to write the fundamental law in its most general form, by expressing the balance of the variation of U, for a totally arbitrary domain Ω. *This control volume can be anywhere in the flow domain and can be of arbitrary shape and size.*

To apply the conservation law as defined above, we have to translate mathematically the quantities involved. The first one is the '*total amount of a quantity U inside a given domain*'. If we consider the domain of volume Ω, the total amount of U in Ω is given by

$$\int_{\Omega} U \, d\Omega$$

and the variation per unit time of the quantity U within the volume Ω is given by

$$\frac{\partial}{\partial t} \int_{\Omega} U \, d\Omega$$

Remark

I wish here to draw your attention to the interpretation of this mathematical expression. The above relation would be read by a mathematician as 'the partial derivative with respect to time of the volume integral of U over Ω'. However, I would like you to 'translate' this mathematical language in its physical meaning, by reading this relation instead as: 'the variation (∂) per unit of time (./∂t) of the total amount of U in Ω'. Try therefore, whenever appropriate, to always 'read' a mathematical expression by its translation of a physical property.

Coming back to the conservation law, we have now to translate mathematically *'the amount of that quantity U entering and leaving the considered domain'*.

This is where the physics comes in: we know from the study of the laws of physics that the local intensity of U will vary through the effect of quantities called **fluxes**, which express the contribution from the surrounding points to the local value of U, describing how the quantity U is transported by the flow.

The **flux** is a fundamental quantity associated to a conserved flow variable U, and is defined as **the amount of U crossing the unit of surface per unit of time**. It is therefore a directional quantity, with a direction and an amplitude, so that it can be represented as a vector. If this vector is locally parallel to the surface, then nothing will enter the domain. Consequently, only the component of the flux in the direction of the normal to the surface will enter the domain and contribute to the rate of change of U. So, *the amount of U crossing the surface element* $\mathrm{d}\vec{S}$ *per unit of time* is defined by the scalar product of the flux and the local surface element (see Figure 1.1.1),

$$F_n \, \mathrm{d}S = \vec{F} \cdot \mathrm{d}\vec{S}$$

with the surface element vector $\mathrm{d}\vec{S}$ pointing along the **outward normal**.

The net total contribution from the incoming fluxes is the sum over all surface elements $\mathrm{d}\vec{S}$ of the closed surface S, and is given by

$$-\oint_S \vec{F} \cdot \mathrm{d}\vec{S}$$

The minus sign is introduced because we consider the flux contribution as positive when it enters the domain. With the outward normal as positive, the scalar product will be negative for an entering flux, as seen from Figure 1.1.1. Hence the need to add the minus sign. If we had defined the inward normal as positive, we would not have added the minus sign. However, the generally accepted convention is to define as positive the outward normal, so that the minus sign is of current acceptance.

To finalize the balance accounts, we have to add contributions from the sources of the quantity U.

These sources can be divided into volume and surface sources, Q_V and $\vec{Q}_S$ and the total contribution is

$$\int_\Omega Q_V \, \mathrm{d}\Omega + \oint_S \vec{Q}_S \cdot \mathrm{d}\vec{S}$$

Hence, the general form of the conservation law for the quantity U is

$$\frac{\partial}{\partial t} \int_{\Omega} U \, d\Omega = -\oint_{S} \vec{F} \cdot d\vec{S} + \int_{\Omega} Q_{V} \, d\Omega + \oint_{S} \vec{Q}_{S} \cdot d\vec{S}$$

which is generally written as

$$\frac{\partial}{\partial t} \int_{\Omega} U \, d\Omega + \oint_{S} \vec{F} \cdot d\vec{S} = \int_{\Omega} Q_{V} \, d\Omega + \oint_{S} \vec{Q}_{S} \cdot d\vec{S} \tag{1.1.1}$$

This is called the integral conservation form and is the most general expression of a conservation law.

This form has some remarkable properties:

- Equation (1.1.1) is valid for any fixed surface S and volume Ω.
- The internal variation of U, in absence of volume sources, depends only on the flux contributions *through* the surface S and *not on the flux values inside the volume* Ω.
- The fluxes do not appear under a derivative or gradient operator and may therefore be discontinuous, as will be the case in the presence of shock waves.

Why are these properties so important, in particular the second one? The importance arises from the fact that we will require this property to remain valid also *after* discretization, to ensure hereby that we satisfy the conservation law at the discrete level. When this is the case, we will speak of a '*conservative numerical scheme*'.

For instance, in an internal flow calculation it is essential to ensure mass conservation, that is constancy of the mass flow in all sections, for any grid resolution. Basically, we see from equation (1.1.1) that if the discretization leads to values of fluxes inside the domain, they will not be distinguishable from the volume sources and will therefore act as such. These 'numerical' sources will then destroy the conservation property of the relevant quantity. For mass conservation, eventual numerical sources will create or destroy mass and hence the mass flow rate will not remain constant. As we will see in Chapter 5, this property can easily be satisfied on arbitrary grids, in particular through application of the finite volume method.

Differential form of a conservation law

An alternative, local differential form of the conservation law can be derived by applying Gauss' theorem to the surface integral term of the fluxes and the surface sources, assuming that these fluxes and surface sources are continuous.

Gauss' theorem states that the surface integral of the flux is equal to the volume integral of the divergence of this flux:

$$\oint_{S} \vec{F} \cdot d\vec{S} = \int_{\Omega} \vec{\nabla} \cdot \vec{F} \, d\Omega$$

for any volume Ω, enclosed by the surface S, where the gradient or divergence operator $\vec{\nabla}$ is introduced. The explicit expression of this gradient operator is defined later in Cartesian coordinates, see equation (1.2.7).

Introducing this relation in the integral conservation law (1.1.1), we obtain

$$\int_\Omega \frac{\partial U}{\partial t} \, d\Omega + \int_\Omega \vec{\nabla} \cdot \vec{F} \, d\Omega = \int_\Omega Q_V \, d\Omega + \int_\Omega \vec{\nabla} \cdot \vec{Q}_S \, d\Omega \qquad (1.1.2)$$

Since equation (1.1.2) is written for an arbitrary volume Ω, it must be valid locally in any point of the flow domain. This leads to the **differential form of the conservation law**,

$$\frac{\partial U}{\partial t} + \vec{\nabla} \cdot \vec{F} = Q_V + \vec{\nabla} . \vec{Q}_S \qquad (1.1.3)$$

or

$$\frac{\partial U}{\partial t} + \vec{\nabla} \cdot (\vec{F} - \vec{Q}_S) = Q_V \qquad (1.1.4)$$

It is seen from these equations that surface sources have the same effect on the system as a flux term and therefore we might as well consider them from the start as an additional flux. However, we favor the present classification in fluxes and sources, since it allows a clear physical interpretation of all the contributions to the evolution of the quantity U. In any case, the term $(\vec{F} - \vec{Q}_S)$ can be considered as an **effective flux**. This will be considered for the momentum conservation law, Section 1.3, where the pressure and shear stresses are indeed acting as surface sources, but they are currently added to the other flux terms to form one 'effective' flux for momentum conservation.
Note that:

- The fluxes (and surface sources) appear *exclusively* under the gradient operator, which is the only space derivative term. This is the direct translation of the surface integral of the fluxes in the integral form (1.1.1).
- **This indicates the way to recognize a conservation law in differential form. Look at all the space derivative terms: if they can be grouped as a divergence operator, then the equation is in conservation form**. If not, the equation is said to be in 'non-conservative' form, or in 'quasi-linear' form.
- This differential form is more restrictive than the integral form, as it requires the fluxes to be differentiable, i.e. having at least C1 continuity, which is not the case in presence of shock waves, for instance.
- For any quantity U, physical assumptions must provide definitions for the fluxes and the source terms, in function of other computed variables.

1.1.2 Convection–Diffusion Form of a Conservation Law

In Section 1.1.1, we have not provided any specific information concerning the fluxes, except for the fact that they do exist for any conserved quantity U. However, we can now be more specific and look closer to the physics of transport of a quantity U in a fluid flow.
The fluxes are generated from two contributions: a contribution due to the convective transport of the fluid and a contribution due to the molecular agitation, which can be present even when the fluid is at rest.

The first component, *which is always present*, is the **convective** flux $\vec{F}_C$, attached to the quantity U in a flow of velocity $\vec{v}$. *It represents the amount of U that is carried away or transported by the flow* and is defined as

$$\vec{F}_C = U\vec{v} \tag{1.1.5}$$

The local contribution of the convective flux through a surface element $\mathrm{d}\vec{S}$, $(\vec{F}_C.\mathrm{d}\vec{S})$ has an important physical significance.

For $U = \rho$ the fluid density, the corresponding convective flux through the surface $\mathrm{d}\vec{S}$ is equal to the local ***mass flow rate***, where we designate the mass flow rate by $\dot{m}$

$$\rho\vec{v} \cdot \mathrm{d}\vec{S} = \mathrm{d}\dot{m} \tag{1.1.6}$$

This quantity represents the amount of mass flowing through the surface dS, per unit of time, and is expressed in kg/s.

For a different conserved quantity $U = \rho u$, **where u is the quantity per unit mass**, the contribution of the convective flux is equal to

$$\vec{F}_C \cdot \mathrm{d}\vec{S} = \rho u\vec{v} \cdot \mathrm{d}\vec{S} = u\,\mathrm{d}\dot{m} \tag{1.1.7}$$

clearly showing the physical meaning of the convective flux as defined by the quantity u entrained by the local mass flow rate.

The second component is a ***diffusive*** flux $\vec{F}_D$, defined as the contribution present in fluids at rest, due to the macroscopic effect of the molecular thermal agitation. The effect of the molecular motion translates in the tendency of a fluid toward equilibrium and uniformity, since differences in the intensity of the considered quantity create a transfer in space such as to reduce the non-homogeneity. This contribution to the total flux is proportional to the gradient of the corresponding quantity, since it has to vanish for a homogeneous distribution.

Diffusive fluxes do not always exist; for instance, from an analysis of the physical properties of fluid, it is known that in a single-phase fluid at rest, no diffusion of specific mass is possible since any displacement of specific mass implies a macroscopic displacement of fluid particles. Therefore, there will be no diffusive flux contribution to the mass conservation equation.

The phenomenon of diffusion is indeed totally different from convection. We can best understand the physics of diffusion by the following experiment, which establishes the basics of diffusion.

Consider a reservoir of water, at rest, and inject a drop of a colored (black) dye, supposed having the same density as water.

Look at Figure 1.1.2. What is going to happen? Will the drop stay in its position? As you probably know from basics physics, we observe that after a certain time the whole of the reservoir will become colored. What has happened?

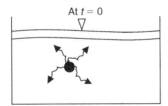
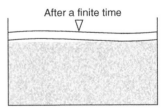

Figure 1.1.2   *A colored dye diffusing in a water reservoir.*

Due to the molecular agitation, the dye molecules are constantly in collision with the water molecules and will be 'hit' many times in all directions to wind up in arbitrary positions. As a result, after a certain time, we will find dye molecules everywhere and the whole reservoir will be colored. For us, as observer, we do not see the molecules, but we see the macroscopic results of these collisions and we notice the following *empirical* facts:

- The process appears as a diffusion, since a local concentration peak diffuses from a high local value to a lower concentration, in all directions.
- The process stops when there is no dye concentration differences anymore between two points, that is when uniformity is reached, which corresponds to a statistically homogenous distribution of the dye molecules between the water molecules.
- The diffusion process between two points is proportional to the concentration difference between these points and will tend to reduce these differences.
- An evolution whereby the concentration difference between two points increases (called anti-diffusion) has never been observed in this experiment.

The 'transport' of the dye molecules can therefore be described, from the point of view of continuum mechanics (since we view the fluid as a continuous media, instead of seen as constituted by molecules), by expressing the macroscopic observations by the existence of a diffusive flux, with the following properties:

- The diffusive flux is proportional to the gradient of the concentration, the gradient being the mathematical expression for the concentration difference between neighboring points, in their direction.
- It has to be opposite to the gradient, to express the tendency toward uniformity.
- It will be proportional to a diffusivity factor, which expresses its 'intensity', depending on the nature of the considered quantity and its environment.

This is summarized by the mathematical gradient *law of Fick*, where κ is the diffusivity coefficient:

$$\vec{F}_{\mathrm{D}} = -\kappa\rho\vec{\nabla}u \tag{1.1.8}$$

Observe that the diffusivity constant κ has units of m²/s for any quantity U.
Equation (1.1.3) then becomes

$$\frac{\partial \rho u}{\partial t} + \vec{\nabla}\cdot(\rho\vec{v}u) = \vec{\nabla}\cdot(\kappa\rho\vec{\nabla}u) + Q_V + \vec{\nabla}\cdot\vec{Q}_S \tag{1.1.9}$$

This equation is the general conservative form of a *transport equation* for the quantity $U = \rho u$ and is also referred to as a *convection–diffusion equation*.
The structure of this equation is of utmost importance, both from physical as well as mathematical point of view and you should study this very carefully, as it forms the backbone of all mathematical modeling of fluid flow phenomena.
The convection–diffusion equation takes its name from the physical properties of the two flux contributions and from their specific mathematical expressions,

which reflect very different physical properties. *Therefore, the following proper-ties will provide significant and fundamental guidelines when setting up a CFD discretization.*

Convective and diffusive fluxes have indeed totally different physical properties:

- Convective fluxes describe the 'passive' transport of the conserved variable by the flow, which you can visualize as a piece of wood carried away by a river flow.
- Consequently the convective flux describes a phenomenon that has directional properties, as it is proportional to the velocity. A convective flux cannot provide a contribution in a direction transverse or opposite to the flow direction. It has therefore properties very similar to wave propagation phenomena, which are also essentially directional, for a selected propagation direction. *We will see in Chapter 3 that the relation between convective transport and wave propagation is indeed very close.*
- Observe that the convective flux appears in the conservation law (1.1.9) as a ***first order partial derivative term***, through the gradient term on the left-hand side.
- Another very important property of the convective flux, is that it is mostly ***non-linear*** as the velocity field will generally depend on the transported variable. This nonlinearity is an essential property of fluid dynamics, the most notorious being turbulence, which is a direct consequence of the nonlinearity of the momentum conservation equation (see Section 1.3).
- The diffusion effects appear in the conservation law (1.1.9) as a ***second order partial derivative term***. In particular, for constant values of the product ($\kappa\rho$), the diffusion term is the Laplace operator. This gives us the physical interpretation of the Laplace operator, as describing an isotropic diffusion, in all directions x, y, z, of the three-dimensional space

$$\Delta u \triangleq \frac{\partial^2 u}{\partial x^2} + \frac{\partial^2 u}{\partial y^2} + \frac{\partial^2 u}{\partial z^2}$$

- *Remember therefore, that each time you come across a Laplace equation or a Laplace operator, it describes a physical phenomenon corresponding to an isotropic diffusion.*

We can summarize now the essential differences between convection and diffusion in Table 1.1.1.

These differences are crucial to the understanding of the physics of flows, but also to the rules for discretization and to essential properties of CFD numerical schemes.

We can already mention here a most fundamental rule of numerical discretizations, which will be elaborated further in the following chapters, namely:

The properties of a numerical discretization scheme may NEVER be in contradiction with the physics it aims to describe.

It is therefore of uttermost importance to clearly understand the physical properties of the equations to be discretized and the mathematical translation of these properties. This particular issue, namely the one-to-one relation between physical interpretation and mathematical properties of the equations will be treated in Chapter 3.

Table 1.1.1   *Differences between convection and diffusion.*

Convection	Diffusion
Expresses the transport of the considered quantity by the flow	Translates the effects of molecular collisions
Does not exist in a fluid at rest	Does exist in a fluid at rest
All quantities are convected by the flow	Not all quantities are subjected to diffusion
Directional behavior	Isotropic behavior
Leads to first order space derivatives in the conservation law	Leads to second order space derivatives in the conservation law
Is generally nonlinear, when the flow velocity depends on the transported variable	Is generally linear for constant fluid properties

For instance, a numerical scheme tuned to handle a diffusion equation, such as a Laplace equation ($\Delta u = 0$) or a Poisson equation ($\Delta u = q$), will not work when applied to a pure convection dominated equation. See Section 11.3.2.2 in Chapter 11 for an example of great historical significance, related to the first attempts to treat transonic potential flows numerically, during the 1960s, in the early years of CFD development.

The Peclet number

The solution of convection–diffusion equations will strongly depend on the relative strength of the two conflicting phenomena, which can range from pure convection to pure diffusion.

It is therefore important in many applications to be able to judge this relative strength by an appropriate indicator, which should be a non-dimensional number. If we compare the convective and diffusive fluxes, given respectively by equations (1.1.5) and (1.1.8), we can define a measure of their ratio, as follows:

$$\frac{|\vec{F}_C|}{|\vec{F}_D|} \approx \frac{\rho u V}{\rho \kappa u / L} = \frac{VL}{\kappa} \tag{1.1.10}$$

where V is a reference velocity and L a reference length, such that V/L is a measure of the gradient of u. The ratio in the right-hand side is the non-dimensional **Peclet number**, measuring the relative strength between convection and diffusion:

$$Pe \equiv \frac{VL}{\kappa} \tag{1.1.11}$$

Hence, if this ratio is much larger than one, the evolution of the quantity U will be dominated by convection, while it will be dominated by diffusion when the Peclet number is lower than 1. For values in the intermediate range, the solution U will have a mixed behavior, influenced both by convection and diffusion.

We will come back to these important properties in several chapters of this Volume.

1.1.3 Vector Conservation Law

If the conserved property is described by a vector quantity $\vec{U}$, then the flux becomes a tensor $\bar{\bar{F}}$, the volume source term a vector $\vec{Q}_V$ and the conservation equation (1.1.1) becomes

$$\frac{\partial}{\partial t} \int_\Omega \vec{U} \, d\Omega + \oint_S \bar{\bar{F}} \cdot d\vec{S} = \int_\Omega \vec{Q}_V \, d\Omega + \oint_S \bar{\bar{Q}}_S \cdot d\vec{S} \tag{1.1.12}$$

where the surface source term $\bar{\bar{Q}}_S$ can be written also as a tensor.

Applying Gauss's theorem, if the fluxes and the surface sources are continuous, we obtain

$$\frac{\partial}{\partial t} \int_\Omega \vec{U} \, d\Omega + \int_\Omega \vec{\nabla} \cdot \bar{\bar{F}} \, d\Omega = \int_\Omega \vec{Q}_V \, d\Omega + \int_\Omega \vec{\nabla}.\bar{\bar{Q}}_S \, d\Omega \tag{1.1.13}$$

and the equivalent differential form:

$$\frac{\partial \vec{U}}{\partial t} + \vec{\nabla} \cdot (\bar{\bar{F}} - \bar{\bar{Q}}_S) = \vec{Q}_V \tag{1.1.14}$$

Here again, the surface sources have the same effect as the flux term. Note here that the gradient of the flux tensor $\vec{\nabla} \cdot \bar{\bar{F}}$ is a vector.

The convective component of the flux tensor is given by

$$\bar{\bar{F}}_C = \vec{U} \otimes \vec{v} \tag{1.1.15}$$

where $\otimes$ denotes the tensor product of the vectors $\vec{v}$ and $\vec{U}$. In tensor notation, equation (1.1.15) becomes

$$(\bar{\bar{F}}_C)_{ij} = U_i v_j \tag{1.1.16}$$

and the diffusive component of the flux tensor takes the following form, for an homogeneous system

$$(\bar{\bar{F}}_D)_{ij} = -\kappa \rho \frac{\partial u_i}{\partial x_j} \tag{1.1.17}$$

with

$$U_i = \rho u_i \tag{1.1.18}$$

The general form (1.1.12) is the **integral formulation of a vector conservation law** and its most general expression, since it remains valid in presence of discontinuous variations of the flow properties such as inviscid shock waves or contact discontinuities. Only if continuity of the flow properties can be assumed, will equation (1.1.13) and its fully equivalent differential form (1.1.14) be valid.

Note that the differential form (1.1.14) is said to be in conservation form, recognizable, as for the scalar equation, by the fact that all spatial flux terms, with the exception of the volume sources, appear under the form of the divergence of a tensor quantity.

The Equations of Fluid Mechanics

As already mentioned above, the motion of a fluid is completely described by the conservation laws for the three basic properties: mass, momentum and energy.

The awareness of this fact has been one of the greatest achievements of modern science, due to the high level of generality and degree of abstraction involved. Indeed, how complicated the detailed evolution of a system might be, not only are the basic properties mass, momentum and energy conserved during the whole process at all times (in the sense to be defined later) but more than that, these three conditions *completely* determine the behavior of the system without any additional dynamical law. This is a very remarkable property, indeed. The only additional information concerns the specification of the nature of the fluid (e.g. incompressible fluid, perfect gas, condensable fluid, viscoelastic material, etc.).

Of course, an important level of knowledge implied in these statements has to be defined before the mathematical expression of these laws can be written and used to predict and describe the behavior of the system.

A fluid flow is considered as known if, at any instant of time, the velocity field and a minimum number of static properties are known at every point. The number of static properties to be known is dependent on the nature of the fluid. This number will be equal to one for an isothermal incompressible fluid (e.g. the pressure), two (e.g. pressure and density) for a perfect gas or any real compressible fluid in thermodynamic equilibrium.

We will consider that a separate analysis has provided the necessary knowledge enabling to define the nature of the fluid. This is obtained from the study of the behavior of the various types of continua and the corresponding information is summarized in the *constitutive laws* and in some other parameters such as viscosity and heat conduction coefficients. This study also provides the information on the nature and properties of the internal forces acting on the fluid since, by definition, a deformable continuum such as a fluid, requires the existence of internal forces connected to the nature of the constitutive law.

Besides, separate studies are needed in order to distinguish the various external forces that influence the motion of the system in addition to the internal ones. These external forces could be, e.g. gravity, buoyancy, Coriolis and centrifugal forces in rotating systems, electromagnetic forces in electrical conducting fluids.

Let us now move to the derivation of these basic fluid dynamic equations, by applying the general expressions derived in this section, to the specific quantities mass, momentum and energy.

The equation for mass conservation is also called the *continuity equation*, while the momentum conservation law is the expression of the generalized Newton law, defining the *equation of motion* of a fluid. The energy conservation law is also referred to as the expression of the *first principle of Thermodynamics*.

When applied to a viscous fluid, the set of these equations are known as the *Navier–Stokes equations*, while they are known as the *Euler equations* when applied to a prefect, inviscid fluid.[1]

1.2 THE MASS CONSERVATION EQUATION

The law of mass conservation is a general statement of kinematic nature, i.e. independent of the nature of the fluid or of the forces acting on it. It expresses the empirical fact that in a fluid system, mass cannot disappear from the system, nor be created. The quantity U is, in this case, the specific mass, $U = \rho$ in kg/m^3.

As noted above, no diffusive flux exists for the mass transport, which means that mass can only be transported through convection. With the convective flux defined by $\vec{F}_C = \rho\vec{v}$ and in absence of external mass sources, the general integral mass conservation equation then becomes

$$\frac{\partial}{\partial t} \int_{\Omega} \rho \, d\Omega + \oint_{S} \rho\vec{v} \cdot d\vec{S} = 0 \tag{1.2.1}$$

and in differential form following (1.1.3):

$$\frac{\partial \rho}{\partial t} + \vec{\nabla} \cdot (\rho\vec{v}) = 0 \tag{1.2.2}$$

This equation is also called the *continuity equation*.

An equivalent form to (1.2.2) is obtained by working out the divergence operator, leading to

$$\frac{\partial \rho}{\partial t} + (\vec{v} \cdot \vec{\nabla})\rho + \rho\vec{\nabla} \cdot \vec{v} = 0 \tag{1.2.3}$$

and introducing the *material or total derivative*:

$$\frac{d}{dt} \overset{\Delta}{=} \frac{\partial}{\partial t} + \vec{v} \cdot \vec{\nabla} \tag{1.2.4}$$

leads to the following form for the mass conservation law

$$\frac{d\rho}{dt} + \rho\vec{\nabla} \cdot \vec{v} = 0 \tag{1.2.5}$$

Although both forms (1.2.2) and (1.2.5) are fully equivalent from mathematical point of view, it will not necessarily remain so when a numerical discretization is performed. Equation (1.2.2) corresponds to the general form of a conservation law since it is

---

[1] Do not confuse a perfect fluid with the notion of a perfect gas: a perfect fluid is a fluid without viscosity, while a perfect gas is defined by the gas law $p = \rho r T$ and can be considered as viscous or not. On the other hand a perfect fluid can be a liquid or a gas.

written in *conservation form* or in *divergence* form; while equation (1.2.3) is said to be in *quasi-linear* or *non-conservative* form.

The conservative form in a numerical scheme is important, since if not properly taken into account, a discretization of equation (1.2.5) will lead to a numerical scheme in which all the mass fluxes through the mesh-cell boundaries will not cancel and hence the numerical scheme will not keep the total mass constant. The importance of a conservative discretization of the flow equations has also been stressed by Lax (1954) who demonstrated that this condition is necessary in order to obtain correct jump relations through a discontinuity in the numerical scheme. We will come back to this crucial point in Chapter 5 where the finite volume method is presented and in later chapters when discussing the discretization of Euler and Navier–Stokes equations.

Physical interpretation of the material derivative

The material derivative plays an important role in fluid mechanics and it is of interest to deeply catch its physical significance.

If you look at a log of wood entrained by a river flow, you can view the change in time of the log's position due to the flow in two ways. If you keep looking at a given fixed point, you will see the log pass in front of you and the change you will notice after a short time step, is described by $\partial/\partial t$; which by definition gives the variation per unit of time at a fixed point, with all space coordinates being kept fixed (see Figure 1.2.1).

However, if you follow the log in its movement, you will see an additional variation due to the motion of the fluid. Indeed, over a time interval Δt, point P will move to point Q, at a distance $\overrightarrow{PQ} = \vec{v}_P \Delta t$ and the value of an arbitrary quantity U has changed by the difference $(U_Q - U_P)$.

From a Taylor series expansion, this difference is equal to

$$U_Q - U_P = (\overrightarrow{PQ} \cdot \vec{\nabla})U_P + \cdots = \Delta t (\vec{v}_P \cdot \vec{\nabla})U_P + \cdots \tag{1.2.6}$$

and the corresponding variation per unit of time is equal to $(\vec{v} \cdot \vec{\nabla})U$.

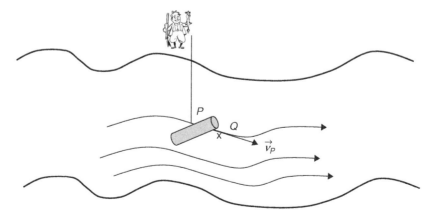

Figure 1.2.1 *Contribution to the material derivative.*

The total variation that you now can observe, when you follow the log in its motion is obtained by adding this contribution to the local one $\partial/\partial t$. This sum is called the **material or total derivative** and is defined by equation (1.2.4).

As can be seen form equation (1.2.3) above, it is derived directly from the convective flux and hence, as you also can notice from the arguments leading to equation (1.2.6), the material derivative is another expression for the *convection* effects. We also use the terminology **convective derivative** for the right-hand side of equation (1.2.6).

Some comment on the notations

For reasons of clarity and compactness of the equations, we use a vector notation for the space gradients, and we recommend you to become familiar with it, as it also allows a clear and direct physical interpretation. Indeed the operator $(\vec{v} \cdot \vec{\nabla})$ is a scalar product between the velocity vector and the gradient operator, defined as

$$\vec{\nabla} \stackrel{\Delta}{=} \vec{e}_x \frac{\partial}{\partial x} + \vec{e}_y \frac{\partial}{\partial y} + \vec{e}_z \frac{\partial}{\partial z} \qquad (1.2.7)$$

where $\vec{e}_i$ is the unit vector in the i direction. *The operator $\vec{v} \cdot \vec{\nabla}$ is therefore equal to the modulus of the velocity times the derivative in the direction of this velocity:*

$$\vec{v} \cdot \vec{\nabla} = |\vec{v}| \cdot \frac{\partial}{\partial l}$$

where ∂l is the differential arc length along the velocity direction, that is along the streamline.

Algebraically, this operator is defined in a Cartesian (x, y, z) coordinate system, with velocity components (u, v, w), as

$$\vec{v} \cdot \vec{\nabla} \stackrel{\Delta}{=} u \frac{\partial}{\partial x} + v \frac{\partial}{\partial y} + w \frac{\partial}{\partial z} \qquad (1.2.8)$$

On the other hand the divergence of the velocity, appearing in the third term of equation (1.2.3), is directly obtained by the scalar product of the two vectors, in the right order, leading to

$$\vec{\nabla} \cdot \vec{v} \stackrel{\Delta}{=} \frac{\partial u}{\partial x} + \frac{\partial v}{\partial y} + \frac{\partial w}{\partial z} \qquad (1.2.9)$$

Alternative form of a general conservation equation

The differential form of the general convection–diffusion conservation equation (1.1.9) can be written in another way. If equation (1.2.2) multiplied by u is subtracted from the left-hand side of equation (1.1.9), we obtain

$$\rho \frac{\partial u}{\partial t} + \rho \vec{v} \cdot \vec{\nabla} u = \vec{\nabla} \cdot (\kappa \rho \vec{\nabla} u) + Q_V + \vec{\nabla} \cdot \vec{Q}_S \qquad (1.2.10)$$

or

$$\rho \frac{du}{dt} = -\vec{\nabla} \cdot \vec{F}_D + Q_V + \vec{\nabla} \cdot \vec{Q}_S \qquad (1.2.11)$$

where $\vec{F}_D$ is the diffusive component of the flux vector. Again, the difference between equations (1.2.10) or (1.2.11) and (1.1.9) lies in the conservative form of the equations. Clearly equation (1.2.10) is *not* in conservation form and a straightforward discretization of this equation will generally not conserve the property u in the numerical simulation. Equation (1.2.10) is said to be in *quasi-linear* or *non-conservative* form.

It is also important to note that this conservation property is linked to the convective term and that, in a fluid at rest, there is no difference between the conservative form (1.1.9) and the non-conservative form (1.2.10).

Incompressible fluid

For an incompressible fluid, the density is constant and the continuity equation (1.2.2) or (1.2.5) reduces to the divergence free condition for the velocity

$$\vec{\nabla} \cdot \vec{v} = 0 \tag{1.2.12}$$

1.3   THE MOMENTUM CONSERVATION LAW OR EQUATION OF MOTION

Momentum is a vector quantity defined as the product of mass and velocity, which becomes when expressed per unit of volume, the product of density and velocity, i.e.

$$\vec{U} \stackrel{\Delta}{=} \rho\vec{v} \tag{1.3.1}$$

and therefore the conservation law will have the general form given by equations (1.1.12) and (1.1.14).

The convective flux tensor is defined by equation (1.1.15) applied to the momentum and becomes

$$\overline{\overline{F}}_C = \rho\vec{v} \otimes \vec{v} \tag{1.3.2}$$

and the flux contribution through the surface dS, takes the form:

$$\overline{\overline{F}}_C \cdot \mathrm{d}\vec{S} = \rho\vec{v}(\vec{v} \cdot \mathrm{d}\vec{S}) = \vec{v}\,\mathrm{d}\dot{m} \tag{1.3.3}$$

where d$\dot{m}$ is the mass flow rate through dS, as defined by equation (1.1.6).

As with the mass conservation equations, it is assumed that no diffusion of momentum is possible in a fluid at rest, and hence there is no diffusive contribution to the flux tensor $\overline{\overline{F}}$.

In order to determine all the terms of the conservation equations, it is necessary to define the sources influencing the variation of momentum. It is known, from Newton's law, that the sources for the variation of momentum in a physical system are the forces acting on it. These forces consist of the *external volume forces* $\vec{f}_e$ and the *internal forces* $\vec{f}_i$, *defined per unit mass*.

Hence, the source term $\vec{Q}_V$ of the conservation equation (1.1.12) consists of the sum of the external volume forces per unit volume $\rho\vec{f}_e$ and the sum of all the internal forces $\rho\vec{f}_i$.

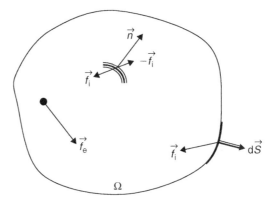

Figure 1.3.1 *Internal and external forces in the flow domain Ω. The internal forces cancel at all internal points due to the action = reaction rule and only the surface points remain for the internal forces contributions.*

The latter are dependent on the nature of the fluid considered and result from the assumptions made about the properties of the internal deformations within the fluid and their relation to the internal stresses.

Note that internal forces are the expression of the deformability of a fluid medium, as opposed to a solid rigid body, in which the distance between two internal points remains fixed during the body motion. In a fluid, two points initially very close to each other can be found some time later far apart due to the deformations of the fluid in its motion. This justifies the existence of internal forces, which are not present in a rigid body.

If you think about it, nobody has ever 'seen' a force, which is actually a very abstract concept. What we only can see are the *effects* of forces, for instance the displacement of an object due to gravitational forces, or the displacement of a pointer on a measuring instrument under an electric potential difference indicating the presence of an electrostatic or electromagnetic force. *In fact one of the more fundamental assumptions of modern physics is to consider that when we observe a certain effect, we assume the existence of a force behind it, as its cause.*

This is exactly what is considered in fluid mechanics: since a fluid can sustain internal deformations, a force, which is called the **internal force** of the fluid, must cause these deformations (see Figure 1.3.1).

We refer you to your fluid mechanics courses for more details, and we will summarize here only the main properties. The most important is the definition of the internal force, acting on a surface element dS. In the general case, the internal force acting on this surface element depends both on its position and on its orientation, defined by the normal. Therefore it should be described mathematically by a tensor $\bar{\bar{\sigma}}$, such that the local internal force vector is written as

$$\vec{f}_i = \bar{\bar{\sigma}} \cdot \vec{n} \tag{1.3.4}$$

where $\vec{n}$ denotes the unit normal vector to the surface element. The internal stress tensor $\bar{\bar{\sigma}}$ is a local property of the fluid and the internal force is, in addition, dependent on the orientation through the definition (1.3.4).

We will assume that the fluid is *Newtonian* and therefore, the total internal stress tensor $\bar{\bar{\sigma}}$ is taken to be

$$\bar{\bar{\sigma}} = -p\bar{\bar{I}} + \bar{\bar{\tau}} \tag{1.3.5}$$

where $\bar{\bar{I}}$ is the unit tensor. Here, the existence of the isotropic pressure component $p\bar{\bar{I}}$ is introduced and $\bar{\bar{\tau}}$ is the viscous shear stress tensor, equal to

$$\tau_{ij} = \mu\left[\left(\frac{\partial v_j}{\partial x_i} + \frac{\partial v_i}{\partial x_j}\right) - \frac{2}{3}(\vec{\nabla}\cdot\vec{v})\delta_{ij}\right] \tag{1.3.6}$$

where μ is the dynamic viscosity of the fluid, see for instance Batchelor (1970). A *kinematic viscosity* coefficient ν is also defined by $\nu = \mu/\rho$.

This relation is valid for a Newtonian fluid in local thermodynamic equilibrium. Otherwise, the most general form for the viscous stress tensor is

$$\tau_{ij} = \left[\mu\left(\frac{\partial v_j}{\partial x_i} + \frac{\partial v_i}{\partial x_j}\right) + \lambda(\vec{\nabla}\cdot\vec{v})\delta_{ij}\right] \tag{1.3.7}$$

Up to now, with the exception of very high temperature or pressure ranges, there is no experimental evidence that the Stokes relation:

$$2\mu + 3\lambda = 0 \tag{1.3.8}$$

leading to equation (1.3.6), is not satisfied. Therefore we will not consider in the following, the second viscosity coefficient λ as independent from μ.

It is very important that you remember here that the viscous shear stresses represent the internal friction force of fluid layers against each other.

By definition, internal forces cancel two per two in every point inside the volume. Therefore, after summation over all the volume points, the remaining internal forces within the volume Ω are those acting on the points of the boundary surface S, since they have no opposite counterpart within the considered volume.

Hence, *the internal forces act as surface sources* with the local intensity $\bar{\bar{\sigma}}\cdot d\vec{S}$ and the momentum conservation equation becomes, after taking the sum over the whole surface:

$$\frac{\partial}{\partial t}\int_\Omega \rho\vec{v}\,d\Omega + \oint_S \rho\vec{v}(\vec{v}\cdot d\vec{S}) = \int_\Omega \rho\vec{f_e}\,d\Omega + \oint_S \bar{\bar{\sigma}}\cdot d\vec{S}$$

$$= \int_\Omega \rho\vec{f_e}\,d\Omega - \oint_S p\cdot d\vec{S} + \oint_S \bar{\bar{\tau}}\cdot d\vec{S} \tag{1.3.9}$$

Note here that we could have considered from the start the internal forces as surface sources, but we have chosen for this presentation to illustrate its physical background.

Applying Gauss' theorem, we obtain

$$\frac{\partial}{\partial t}\int_\Omega \rho\vec{v}\,d\Omega + \int_\Omega \vec{\nabla}\cdot(\rho\vec{v}\otimes\vec{v})d\Omega = \int_\Omega \rho\vec{f_e}\,d\Omega + \oint_S \bar{\bar{\sigma}}\cdot d\vec{S} \tag{1.3.10}$$

which leads to the differential form of the equation of motion:

$$\frac{\partial \rho \vec{v}}{\partial t} + \vec{\nabla} \cdot (\rho \vec{v} \otimes \vec{v} + p\overline{\overline{I}} - \overline{\overline{\tau}}) = \rho \vec{f_e} \tag{1.3.11}$$

An equivalent non-conservative form is obtained after subtracting from the left-hand side the continuity equation multiplied by $\vec{v}$:

$$\rho \frac{d\vec{v}}{dt} \equiv \rho \frac{\partial \vec{v}}{\partial t} + \rho (\vec{v} \cdot \vec{\nabla}) \vec{v} = -\vec{\nabla} p + \vec{\nabla} \cdot \overline{\overline{\tau}} + \rho \vec{f_e} \tag{1.3.12}$$

where the material derivative d/dt has been introduced.

When the form (1.3.6) of the shear stress tensor for a Newtonian viscous fluid is introduced in equations (1.3.11) or (1.3.12), we obtain the **Navier–Stokes equations of motion**. For constant viscosity coefficients, it reduces to

$$\rho \frac{\partial \vec{v}}{\partial t} + \rho (\vec{v} \cdot \vec{\nabla}) \vec{v} = -\vec{\nabla} p + \mu \left[\Delta \vec{v} + \frac{1}{3} \vec{\nabla} (\vec{\nabla} \cdot \vec{v}) \right] + \rho \vec{f_e} \tag{1.3.13}$$

For an incompressible fluid, satisfying the divergence free velocity condition (1.2.12), the Navier–Stokes equation reduces to

$$\rho \frac{\partial \vec{v}}{\partial t} + \rho (\vec{v} \cdot \vec{\nabla}) \vec{v} = -\vec{\nabla} p + \mu \Delta \vec{v} + \rho \vec{f_e} \tag{1.3.14}$$

For an ideal fluid without internal shear stresses (i.e. for a perfect or inviscid fluid), the momentum equation reduces to the **Euler equation of motion**:

$$\rho \frac{d\vec{v}}{dt} \equiv \rho \frac{\partial \vec{v}}{\partial t} + \rho (\vec{v} \cdot \vec{\nabla}) \vec{v} = -\vec{\nabla} p + \rho \vec{f_e} \tag{1.3.15}$$

Remark

Observe that the convection term, under either the form of the second term of equations (1.3.11) or (1.3.13), is **nonlinear** even for incompressible flows. This is a crucially important property, as this term is in particular responsible for the appearance of turbulence. See Chapter 2 for an introductory presentation of turbulent flow properties.

The vorticity equation

The equations of motion can be written in many equivalent forms, one of them being obtained through the introduction of the vorticity vector $\vec{\zeta}$

$$\vec{\zeta} = \vec{\nabla} \times \vec{v} \tag{1.3.16}$$

and the vector identity

$$(\vec{v} \cdot \vec{\nabla}) \vec{v} = \vec{\nabla} \left(\frac{\vec{v}^2}{2} \right) - \vec{v} \times (\vec{\nabla} \times \vec{v}) \tag{1.3.17}$$

in the inertia term $d\vec{v}/dt$. Equation (1.3.12) becomes

$$\frac{\partial \vec{v}}{\partial t} - (\vec{v} \times \vec{\zeta}) = -\frac{1}{\rho}\vec{\nabla}p - \vec{\nabla}\left(\frac{\vec{v}^2}{2}\right) + \frac{1}{\rho}\vec{\nabla}\cdot\overline{\overline{\tau}} + \vec{f}_e \tag{1.3.18}$$

This equation will be transformed further by introduction of thermodynamical relations after having discussed the conservation law for energy.

An important equation for the vorticity $\vec{\zeta}$ can be obtained by taking the curl of the momentum equation (1.3.13). This leads to the **Helmholtz** equation:

$$\frac{\partial \vec{\zeta}}{\partial t} + (\vec{v}\cdot\vec{\nabla})\vec{\zeta} = (\vec{\zeta}\cdot\vec{\nabla})\vec{v} - \vec{\zeta}(\vec{\nabla}\cdot\vec{v}) + \vec{\nabla}p\times\vec{\nabla}\frac{1}{\rho} + \vec{\nabla}\times\left(\frac{1}{\rho}\vec{\nabla}\cdot\overline{\overline{\tau}}\right) + \vec{\nabla}\times\vec{f}_e \tag{1.3.19}$$

For a Newtonian fluid with constant kinematic viscosity coefficient ν, the shear stress term reduces to the Laplacian of the vorticity

$$\vec{\nabla}\times\left(\frac{1}{\rho}\vec{\nabla}\cdot\overline{\overline{\tau}}\right) = \nu\Delta\vec{\zeta} \tag{1.3.20}$$

The Reynolds number and viscosity as diffusion

Observe that, although derived from the contribution of the internal forces, the viscous shear stress term $\vec{\nabla}\cdot\overline{\overline{\tau}}$ has all the features of a diffusion flux. Indeed, it satisfies all the properties listed in Table 1.1.1 associated to diffusion, in particular the viscous terms appear as second order derivatives, reducing to a Laplacian for an incompressible fluid, as seen from equation (1.3.14). This confirms that the viscous stresses act as a diffusion, with the kinematic viscosity as the diffusion coefficient, with dimensions m^2/s.

The ratio between momentum convection and diffusion is given by the **Reynolds number**. It is defined as the particular form of the Peclet number (1.1.11), with the kinematic viscosity as diffusion coefficient:

$$Re = \frac{VL}{\nu} \tag{1.3.21}$$

The Reynolds number plays a most important role in fluid mechanics.

1.4   THE ENERGY CONSERVATION EQUATION

It is known, from the thermodynamic analysis of continua, that the energy content of a system is measured by its internal energy per unit mass e. This internal energy is a state variable of a system and hence its variation during a thermodynamic transformation depends only on the final and initial states.

In a fluid, the conserved quantity is the **total energy** defined as the sum of its internal energy and its kinetic energy per unit mass $\vec{v}^2/2$. We will indicate by E this total energy per unit mass and ρE the total energy per unit of volume, with

$$E = e + \frac{\vec{v}^2}{2} \tag{1.4.1}$$

The first law of thermodynamic states that the *sources* for the variation of the total energy are the work of the forces acting on the system plus the heat transmitted to this system.

Considering the general form of the conservation law for the quantity ρE, we have a convective flux of energy $\vec{F}_C$

$$\vec{F}_C = \rho \vec{v} \left(e + \frac{\vec{v}^2}{2} \right) \tag{1.4.2}$$

and a diffusive flux $\vec{F}_D$, written as

$$\vec{F}_D = -\gamma \rho \kappa \vec{\nabla} e \tag{1.4.3}$$

since, by definition, there is no diffusive flux associated with the motion. The coefficient κ is the thermal diffusivity coefficient and has to be defined empirically, together with the dynamic viscosity μ. The coefficient γ is the ratio of specific heat coefficients under constant pressure and constant volume $\gamma = c_p / c_v$.

Actually, this diffusive term (1.4.3) describes the diffusion of heat in a medium at rest due to molecular thermal conduction. It is generally written in a slightly different form, namely under the form of *Fourier's law of heat conduction*:

$$\vec{F}_D = -k \vec{\nabla} T \tag{1.4.4}$$

where T is the absolute temperature, and k is the thermal conductivity coefficient.

We have the relation

$$k = \rho c_p \kappa = \mu c_p / \mathrm{Pr} \tag{1.4.5}$$

where Pr is the **Prandtl number**:

$$\mathrm{Pr} = v/\kappa = \mu c_p / k \tag{1.4.6}$$

With regard to the sources of energy variations in a fluid system, a distinction has to made between the surface and the volume sources. The volume sources are the sum of the work of the volume forces $\vec{f}_e$ and of the heat sources other than conduction, such as radiation, heat released by chemical reactions, designated by q_H.

Hence we have, per unit volume, $Q_V = \rho \vec{f}_e \cdot \vec{v} + q_H$.

The surface sources $\vec{Q}_S$ are the result of the work done on the fluid by the internal shear stresses acting on the surface of the volume considering that there are no external surface heat sources,

$$\vec{Q}_S = \bar{\bar{\sigma}} \cdot \vec{v} = -p\vec{v} + \bar{\bar{\tau}} \cdot \vec{v} \tag{1.4.7}$$

1.4.1   Conservative Formulation of the Energy Equation

Grouping all the contributions, the energy conservation equation in integral form, becomes

$$\frac{\partial}{\partial t}\int_{\Omega} \rho E \, d\Omega + \oint_{S} \rho E \vec{v} \cdot d\vec{S}$$
$$= \oint_{S} k\vec{\nabla}T \cdot d\vec{S} + \int_{\Omega}(\rho\vec{f}_{e}\cdot\vec{v} + q_{H})d\Omega + \oint_{S}(\bar{\bar{\sigma}}\cdot\vec{v})\cdot d\vec{S} \quad (1.4.8)$$

After transformation to volume integrals, the differential form of the conservation equation for energy becomes

$$\frac{\partial\rho E}{\partial t} + \vec{\nabla}\cdot(\rho\vec{v}E) = \vec{\nabla}\cdot(k\vec{\nabla}T) + \vec{\nabla}\cdot(\bar{\bar{\sigma}}\cdot\vec{v}) + W_{f} + q_{H} \quad (1.4.9)$$

where W_f is the work of the external volume forces

$$W_f = \rho\vec{f}_e \cdot \vec{v} \quad (1.4.10)$$

Clarifying the term $\vec{\nabla}\cdot(\bar{\bar{\sigma}}\cdot\vec{v})$, and introducing the enthalpy of the fluid $h=(e+p/\rho)$, leads to the following alternative expression in differential form:

$$\frac{\partial\rho E}{\partial t} + \vec{\nabla}\cdot(\rho\vec{v}H - k\vec{\nabla}T - \bar{\bar{\tau}}\cdot\vec{v}) = W_f + q_H \quad (1.4.11)$$

where the **stagnation, or total, enthalpy H** is introduced

$$H = e + \frac{p}{\rho} + \frac{\vec{v}^2}{2} = h + \frac{\vec{v}^2}{2} = E + \frac{p}{\rho} \quad (1.4.12)$$

1.4.2   The Equations for Internal Energy and Entropy

An equation for the variation of the internal energy e, can be obtained after some manipulations (see Problem 1.2) and the introduction of the **viscous dissipation rate** ε_V:

$$\varepsilon_V = (\bar{\bar{\tau}}\cdot\vec{\nabla})\cdot\vec{v} = \frac{1}{2\mu}(\bar{\bar{\tau}}\otimes\bar{\bar{\tau}}^{\mathrm{T}})$$
$$= \tau_{ij}\frac{\partial v_i}{\partial x_j} \quad (1.4.13)$$

This leads to

$$\frac{\partial\rho e}{\partial t} + \vec{\nabla}\cdot(\rho\vec{v}h) = (\vec{v}\cdot\vec{\nabla})p + \vec{\nabla}\cdot(k\vec{\nabla}T) + \varepsilon_V + q_H \quad (1.4.14)$$

Observe that the term W_f, representing the work of the external forces, does *not* contribute to the internal energy balance. Note also that this equation is *not* in conservation from, since the pressure term is not under the form of a divergence.

An alternative form is obtained after introduction of the continuity equation:

$$\rho \frac{de}{dt} = -p(\vec{\nabla} \cdot \vec{v}) + \vec{\nabla} \cdot (k\vec{\nabla}T) + \varepsilon_V + q_H \qquad (1.4.15)$$

The first term is the reversible work of the pressure forces (and vanishes in an incompressible flow), while the other terms are being considered as heat additions, with the dissipation term ε_V acting as an irreversible heat source. This appears clearly by introducing the **entropy per unit mass s** of the fluid, through the thermodynamic relation

$$T \, ds = de + p d \left(\frac{1}{\rho} \right) = dh - \frac{dp}{\rho} \qquad (1.4.16)$$

The separation between reversible and irreversible heat additions is defined by

$$T \, ds = dq + dq' \qquad (1.4.17)$$

where dq is a reversible heat transfer to the fluid, while dq' is an irreversible heat addition. As is known from the second principle of thermodynamics, dq' is always non-negative and hence *in an adiabatic flow ($dq = 0$), with irreversible transformations, the entropy will always increase.*

Introducing the definition (1.4.16) in equation (1.4.15), we obtain

$$\rho T \frac{ds}{dt} = \varepsilon_V + \vec{\nabla} \cdot (k\vec{\nabla}T) + q_H \qquad (1.4.18)$$

where the last two terms can be considered as reversible head additions by conduction and by other sources. Therefore, in an adiabatic flow, $q_H = 0$, without heat conduction ($k = 0$) the non-negative dissipation term ε_V behaves as a non-reversible heat source. Equation (1.4.18) is the entropy equation of the flow. Although this equation plays an important role, it is not independent from the energy equation. Only one of these has to be added to the conservation laws for mass and momentum. Note also that the entropy is not a 'conserved' quantity in the sense of the previously derived conservation equations.

1.4.3   Perfect Gas Model

The system of Navier–Stokes equations has still to be supplemented by the constitutive laws and by the definition of the shear stress tensor in function of the other flow variables. We will consider here only Newtonian fluids for which the shear stress tensor is defined by equation (1.3.6). The thermodynamic laws define the internal energy e, or the enthalpy h as a function of only two other thermodynamic variables chosen between pressure p, specific mass ρ, temperature T, entropy s or any other intensive variable. For instance,

$$e = e(p, T) \qquad (1.4.19)$$

or

$$h = h(p, T) \tag{1.4.20}$$

In addition, the laws of dependence of the two fluid properties, the dynamic viscosity coefficient μ and the coefficient of thermal conductivity k are to be given in function of the fluid state, for instance in function of temperature and eventually pressure. In particular, the viscosity coefficient μ is strongly influenced by temperature. For gases, a widely used relation is given by Sutherland's formula, for instance for air, in the standard international, metric system

$$\mu = \frac{1.45 \, T^{3/2}}{T + 110} 10^{-6} \tag{1.4.21}$$

where T is in degrees Kelvin. Note that for liquids, the dynamic viscosity decreases strongly with temperature, and that the pressure dependence of μ, for both gases and liquids is small. The temperature dependence of k is similar to that of μ for gases while for liquids, k is nearly constant. ***In any case the temperature and pressure dependence of μ and k can only be obtained, within the framework of continuum mechanics, by experimental observation.***

In many instances a compressible fluid can be considered as a perfect gas, even if viscous effects are taken into account, and the equation of state is written as

$$p = \rho r T \tag{1.4.22}$$

where r is the gas constant per unit of mass, and is equal to the universal gas constant divided by the molecular mass of the fluid. The internal energy e and the enthalpy h are only function of temperature and we have the following relations, taking into account that

$$c_p = \frac{\gamma}{\gamma - 1} r \tag{1.4.23}$$

where

$$\gamma = \frac{c_p}{c_v} \tag{1.4.24}$$

is the ratio of specific heat coefficients under constant pressure and constant volume:

$$e = c_v T = \frac{1}{\gamma - 1} \frac{p}{\rho}$$
$$h = c_p T = \frac{\gamma}{\gamma - 1} \frac{p}{\rho} \tag{1.4.25}$$

The entropy variation from a reference state indicated by the subscript A, is obtained from equation (1.4.16) as

$$s - s_A = c_p \ln \frac{T}{T_A} - r \ln \frac{p}{p_A} \tag{1.4.26}$$

or

$$s - s_A = -r \ln \frac{p/p_A}{(T/T_A)^{\gamma/(\gamma-1)}}$$ (1.4.27)

Introducing the equation of state, we also obtain

$$s - s_A = c_v \ln \frac{p/p_A}{(\rho/\rho_A)^{\gamma}}$$ (1.4.28)

The stagnation variables can be derived from the total enthalpy H

$$H = E + \frac{p}{\rho} = h + \frac{\vec{v}^2}{2} = c_p T_0$$ (1.4.29)

where the total or stagnation temperature T_0 is defined by

$$T_0 = T + \frac{\vec{v}^2}{2c_p} = T\left(1 + \frac{\gamma - 1}{2}M^2\right)$$ (1.4.30)

The Mach number M has been introduced by

$$M = \frac{|\vec{v}|}{c}$$ (1.4.31)

with

$$c^2 = \left(\frac{\partial p}{\partial \rho}\right)_s = \gamma r T = \gamma \frac{p}{\rho}$$ (1.4.32)

being the square of the speed of sound. Similarly we have

$$E = c_v T_0$$ (1.4.33)

Considering that the transition of the fluid from static to stagnation state is isentropic, we have for the stagnation pressure p_0

$$\frac{p_0}{p} = \left(\frac{T_0}{T}\right)^{\gamma/(\gamma-1)} = \left(1 + \frac{\gamma - 1}{2}M^2\right)^{\gamma/(\gamma-1)}$$ (1.4.34)

and hence the relations (1.4.27) and (1.4.28) remain unchanged if the static variables are replaced by the stagnation variables. For instance, we have

$$s - s_A = -r \ln \frac{p_0/p_{0A}}{(T_0/T_{0A})^{\gamma/(\gamma-1)}}$$ (1.4.35)

or

$$s - s_A = -r \ln \frac{p_0/p_{0A}}{(H/H_A)^{\gamma/(\gamma-1)}}$$ (1.4.36)

Various other forms of the relations between the thermodynamic variables p, ρ, T, s, e, h can be obtained according to the choice of the independent variables. As a function of h and s we have

$$\frac{p}{p_A} = \left(\frac{h}{h_A}\right)^{\gamma/(\gamma-1)} e^{-(s-s_A)/r} \tag{1.4.37}$$

or from (1.4.27)

$$\frac{\rho}{\rho_A} = \left(\frac{h}{h_A}\right)^{1/(\gamma-1)} e^{-(s-s_A)/r} \tag{1.4.38}$$

Many other relations can be derived by selecting other combinations of variables.

1.4.4 Incompressible Fluid Model

The Navier–Stokes equations simplify considerably for incompressible fluids for which the specific mass may be considered as constant. This leads generally to a decoupling of the energy equation from the other conservation laws if the flow remains isothermal. This is the case for many applications that do not involve heat transfer.

For flows involving temperature variations, the coupling between the temperature field and the fluid motion can occur through various effects, such as variations of viscosity or heat conductivity with temperature; influence of external forces function of temperature, such as buoyancy forces in atmospheric flows; electrically, mechanically or chemically generated heat sources.

The mass conservation equation reduces in the case of incompressible flows to

$$\vec{\nabla} \cdot \vec{v} = 0 \tag{1.4.39}$$

which appears as a kind of constraint to the general time-dependent equation of motion, written here in non-conservative form:

$$\rho \frac{\partial \vec{v}}{\partial t} + \rho(\vec{v} \cdot \vec{\nabla})\vec{v} = -\vec{\nabla}p + \mu \Delta \vec{v} + \rho \vec{f}_e \tag{1.4.40}$$

The system of equations for incompressible flow presents a particular situation in which one of the five unknowns, namely the pressure, does not appear under a time dependence form due to the non-evolutionary character of the continuity equation. This creates actually a difficult situation for the numerical schemes and special techniques have to be adapted in order to treat the continuity equation. This will be introduced in Chapter 12.

An equation for the pressure can be obtained by taking the divergence of the momentum equation (1.4.40), and introducing the divergence free velocity condition (1.4.39), leading to

$$\frac{1}{\rho}\Delta p = -\vec{\nabla} \cdot (\vec{v} \cdot \vec{\nabla})\vec{v} + \vec{\nabla} \cdot \vec{f}_e \tag{1.4.41}$$

which can be considered as a Poisson equation for the pressure for a given velocity field. Note that the right-hand side contains only products of first order velocity derivatives, because of the incompressibility condition (1.4.39). Indeed, in tensor notations, the velocity term in the right-hand side term is equal to $(\partial_j v_i).(\partial_i v_j)$.

A1.5 ROTATING FRAME OF REFERENCE

In many applications such as geophysical flows, turbomachinery problems, flows around helicopter blades, propellers, windmills, we have to deal with rotating systems and it is necessary to be able to describe the flow behavior relatively to a rotating frame of reference. We will assume that the moving system is *rotating steadily* with angular velocity $\vec{\omega}$ around an axis along which a coordinate z is aligned.

A1.5.1 Equation of Motion in the Relative System

Defining $\vec{w}$ as the velocity field relative to the rotating system and $\vec{u} = \vec{\omega} \times \vec{r}$ as the entrainment velocity, the composition law holds

$$\vec{v} = \vec{w} + \vec{u} = \vec{w} + \vec{\omega} \times \vec{r} \tag{1.5.1}$$

Since the entrainment velocity does not contribute to the mass balance, the continuity equation remains invariant and can be written in the relative system:

$$\frac{\partial \rho}{\partial t} + \vec{\nabla}.(\rho \vec{w}) = 0 \tag{1.5.2}$$

With regard to the momentum conservation law, observers in the two systems of reference will not see the same field of forces since the inertia term $d\vec{v}/dt$ is not invariant when passing from one system to the other. It is known that we have to add in the rotating frame of reference two forces, the **Coriolis force** per unit mass $\vec{f}_C$

$$\vec{f}_C = -2(\vec{\omega} \times \vec{w}) \tag{1.5.3}$$

and the centrifugal force per unit mass $\vec{f}_c$

$$\vec{f}_c = -\vec{\omega} \times (\vec{\omega} \times \vec{r}) = \omega^2 \vec{R} \tag{1.5.4}$$

where $\vec{R}$ is the component of the position vector perpendicular to the axis of rotation. Hence, additional force terms appear in the right-hand side of the conservation law (1.3.9) if this equation is written directly in the rotating frame of reference. These two forces, acting on a fluid particle in the rotating system, play a very important role in rotating flows, especially when the relative velocity vector $\vec{w}$ has large components in the direction perpendicular to $\vec{\omega}$.

The conservation law for momentum in the relative system then becomes

$$\frac{\partial}{\partial t} \int_\Omega \rho \vec{w} \, d\Omega + \oint_S \rho \vec{w} (\vec{w} \cdot d\vec{S}) = \int_\Omega \rho \vec{f}_e \, d\Omega + \int_\Omega \rho \vec{f}_C \, d\Omega$$

$$+ \int_\Omega \rho \vec{f}_c \, d\Omega - \oint_S p \cdot d\vec{S} + \oint_S \bar{\bar{\tau}} \cdot d\vec{S} \quad (1.5.5)$$

and the transformation of the surface integrals into volume integrals, leads to the differential form:

$$\frac{\partial \rho \vec{w}}{\partial t} + \vec{\nabla} \cdot (\rho \vec{w} \otimes \vec{w}) = \rho \vec{f}_e - \rho \vec{\omega} \times (\vec{\omega} \times \vec{r}) - 2\rho(\vec{\omega} \times \vec{w}) - \vec{\nabla} p + \vec{\nabla} \cdot \bar{\bar{\tau}} \quad (1.5.6)$$

The shear stress tensor $\bar{\bar{\tau}}$ is to be expressed in function of the relative velocities. It is considered indeed that the rotation of the relative system has no effect on the internal forces within the fluid, since these internal forces cannot, by definition, be influenced by solid body motions of one system of reference with respect to the other. A non-conservative form of the relative momentum equation similar to equation (1.3.18) can be obtained as

$$\frac{\partial \vec{w}}{\partial t} - (\vec{w} \times \vec{\zeta}) = -\frac{1}{\rho} \vec{\nabla} p - \vec{\nabla} \left(\frac{\vec{w}^2}{2} - \frac{\vec{u}^2}{2} \right) + \frac{1}{\rho} \vec{\nabla} \cdot \bar{\bar{\tau}} + \vec{f}_e \quad (1.5.7)$$

where the presence of the absolute vorticity vector is to be noticed.

A1.5.2 Energy Equation in the Relative System

The energy conservation equation in a relative system *with steady rotation* is obtained by adding the work of the centrifugal forces, since the Coriolis forces do not contribute to the energy balance of the flow.

In differential form, one obtains the following full conservative form of the equation corresponding to equation (1.4.11):

$$\frac{\partial}{\partial t} \rho \left(e + \frac{\vec{w}^2}{2} - \frac{\vec{u}^2}{2} \right) + \vec{\nabla} \cdot \left[\rho \vec{w} \left(h + \frac{\vec{w}^2}{2} - \frac{\vec{u}^2}{2} \right) - k \vec{\nabla} T - \bar{\bar{\tau}} \cdot \vec{w} \right] = W_f^{\text{rel}} + q_H$$

$$(1.5.8)$$

where

$$W_f^{\text{rel}} = \rho \vec{f}_e \cdot \vec{w} \quad (1.5.9)$$

is the work performed by the external forces *in the relative system*.

In non-conservative form, equation (1.5.8) becomes, where time derivatives d/dt and $\partial/\partial t$ are considered in *the relative system*:

$$\rho \frac{d}{dt} \left(h + \frac{\vec{w}^2}{2} - \frac{\vec{u}^2}{2} \right) = \frac{\partial p}{\partial t} + \vec{\nabla} \cdot (k \vec{\nabla} T) + \vec{\nabla} \cdot (\bar{\bar{\tau}} \cdot \vec{w}) + W_f^{\text{rel}} + q_H \quad (1.5.10)$$

The quantity

$$I = h + \frac{\vec{w}^2}{2} - \frac{\vec{u}^2}{2} = H - \vec{u} \cdot \vec{v} \tag{1.5.11}$$

appearing in the left-hand side of the above equations, plays an important role, since it appears as a stagnation enthalpy term for the rotating system. This term has been called the ***rothalpy*** and it measures the total energy content in a steadily rotating frame of reference.

A1.5.3 Crocco's Form of the Equations of Motion

The pressure gradient term in the equation of motion can be eliminated by making use of the entropy equation (1.4.16) written for arbitrary variations of the state of the fluid. In particular, if the flow is followed in its displacement along its (absolute) velocity line,

$$T\vec{\nabla}s = \vec{\nabla}h - \frac{\vec{\nabla}p}{\rho} \tag{1.5.12}$$

and introducing this relation in equation (1.3.18), we obtain

$$\frac{\partial \vec{v}}{\partial t} - (\vec{v} \times \vec{\zeta}) = T\vec{\nabla}s - \vec{\nabla}H + \frac{1}{\rho}\vec{\nabla} \cdot \bar{\bar{\tau}} + \vec{f}_e \tag{1.5.13}$$

where the stagnation enthalpy H has been introduced. Similarly, in the relative system, we obtain from equation (1.5.7):

$$\frac{\partial \vec{w}}{\partial t} - (\vec{w} \times \vec{\zeta}) = T\vec{\nabla}s - \vec{\nabla}I + \frac{1}{\rho}\vec{\nabla} \cdot \bar{\bar{\tau}} + \vec{f}_e \tag{1.5.14}$$

where the rothalpy I appears, as well as the *absolute* vorticity $\vec{\zeta}$.

The introduction of entropy and stagnation enthalpy gradients in the equation of motions is due to Crocco (1937), and equations (1.5.13) and (1.5.14) reveal important properties. A first observation is that, even in steady flow conditions, the flow will be rotational, except in very special circumstances, namely frictionless, isentropic and isoenergetic flow conditions, without external forces or with forces that can be derived from a potential function where the corresponding potential energy is added to the total energy H. An analogous statement can be made for the equation in the relative system where the total energy is measured by I. However, since the absolute vorticity appears in the relative equation of motion, even under steady relative conditions, with constant energy I and inviscid flow conditions without body forces, the relative vorticity will not be zero, but equal to $(-2\vec{\omega})$. The relative motion is therefore never irrotational but will have at least a vorticity component equal to minus twice the solid body angular velocity. This shows that under the above-mentioned conditions of absolute vorticity equal to zero, the relative flow undergoes a solid body rotation equal to $2\vec{\omega}$ in opposite direction to the rotation of the relative system.

A1.6 ADVANCED APPLICATIONS OF CONTROL VOLUME FORMULATIONS

In this section we introduce some advanced applications of the control volume concept, which are of great importance in practical applications of CFD.

In external flows, CFD simulations are applied to predict the forces on the bodies, in particular lift and drag for aircrafts and cars, particularly for racing cars, where these quantities are crucial for the performance improvements. Reduction of drag by very small amounts in aircrafts and racing cars, can make the difference between acceptance or not, between winning and losing a race. Section 1.6.1 presents two complementary methods to evaluate these forces from the CFD results.

Another important application is connected to cases with moving grids or moving control volumes. This is presented in Section 1.6.2.

A1.6.1 Lift and Drag Estimations from CFD Results

Solid bodies inside control volume
If the volume Ω contains solid bodies, then an additional force $(-\vec{R})$ has to be added to the right-hand side of equation (1.3.9), where $\vec{R}$ is the total force exerted by the fluid on the body.

$$\frac{\partial}{\partial t} \int_\Omega \rho \vec{v} \, d\Omega + \oint_S \rho \vec{v}(\vec{v} \cdot d\vec{S}) = -\oint_S p \cdot d\vec{S} + \oint_S \bar{\bar{\tau}} \cdot d\vec{S} + \int_\Omega \rho \vec{f_e} \, d\Omega + (-\vec{R}) \quad (1.6.1)$$

This equation is currently applied for the determination of lift force $\vec{L}$ and drag force $\vec{D}$ on solid bodies, particularly for stationary flows.

For a surface S located in the steady far field, where the viscous shear stresses can be considered as negligible, the sum of the stationary lift and drag forces are given by the following relation, in absence of external forces:

$$\oint_S \rho \vec{v}(\vec{v} \cdot d\vec{S}) + \oint_S p \cdot d\vec{S} = (-\vec{R}) = -\vec{L} - \vec{D} \quad (1.6.2)$$

The lift and drag forces are obtained from the calculated left-hand side vector after projection respectively along the direction perpendicular to the far-field incoming velocity $\vec{U}_\infty$ and in its direction (Figure 1.6.1).

If the control surface S is taken along the solid body surface S_b, where the velocity field is zero due to the non-slip condition of viscous flows, then the lift and drag forces are also defined by

$$\vec{R} = \vec{L} + \vec{D} = \oint_{S_b} p \cdot d\vec{S} + \oint_{S_b} \bar{\bar{\tau}} \cdot d\vec{S} \quad (1.6.3)$$

These formulas are currently applied to determine these forces from computed flow fields, by either integrating pressure and shear stresses along the solid walls, following equation (1.6.3) or alternatively by integrating momentum and pressure along the far-field enclosing surface, following equation (1.6.2).

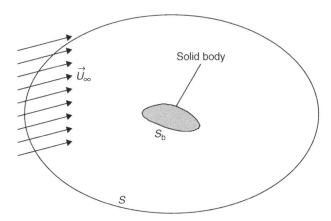

Figure 1.6.1 *Far-field control surface for lift and drag determination on an enclosed solid body.*

A1.6.2 Conservation Law for a Moving Control Volume

The general conservation law (1.1.1) was derived for a fixed control surface. However, many flow situations involve the simultaneous presence of moving and fixed parts, for which moving grids are required. In these cases, attaching a control volume to a moving grid or a moving body requires to take into account the displacement of the control surfaces. Representative examples are:

- flow around a train entering a tunnel, or between two crossing trains in opposite directions;
- flow around an oscillating wing, where the oscillations can be forced or flow induced in an aero-elastic interaction;
- the flow between an aircraft and a separating store;
- the flow of gases in internal combustion engines where the piston head has a periodic motion with respect to the cylinder walls, modifying the available flow volume accordingly;
- hydrodynamic of moving ships with free sea surface and wind induced waves.

In all these examples, two systems of reference are present simultaneously and depending on the particular situation, fixed or deformable grids have to be attached to the moving system and the conservation laws of mass, momentum and energy have to take into account the effects of the relative motion between the two systems.

Scalar conservation law
Referring to Section 1.1.1, we consider now the volume Ω where each point of the bounding surface S has a velocity $\vec{v}_s$ in the inertial reference system (see Figure 1.6.2).

Over a time interval Δt, the observer moving with the surface S will see a variation of the content of U inside the volume determined by the sum of the fluxes and the source terms, plus an additional contribution from the volume deformation caused by the surface motion.

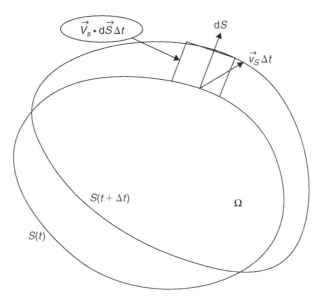

Figure 1.6.2 *Moving control volume with surface velocity.*

Each surface element $d\vec{S}$ gives rise to a volume variation during the time interval Δt equal to $\vec{v}_s.d\vec{S}\,\Delta t$ (see Figure 1.6.2) and therefore the content of U in Ω will increase by an additional amount given by the surface integral $\oint_S U\vec{v}_s.d\vec{S}\,\Delta t$. Summing up all contributions, we have

$$\frac{\partial}{\partial t}\int_\Omega U\,d\Omega = \int_\Omega \frac{\partial U}{\partial t}\,d\Omega + \int_\Omega U\frac{\partial d\Omega}{\partial t}$$

$$= \int_\Omega \frac{\partial U}{\partial t}\,d\Omega + \oint_S U\vec{v}_s \cdot d\vec{S}$$

$$= -\oint_S \vec{F}\cdot d\vec{S} + \int_\Omega Q_V\,d\Omega + \oint_S \vec{Q}_S \cdot d\vec{S} \qquad (1.6.4)$$

or, generalizing in this way equation (1.1.1) to a moving control volume:

$$\int_\Omega \frac{\partial U}{\partial t}\,d\Omega + \oint_S (\vec{F} - U\vec{v}_s - \vec{Q}_S) \cdot d\vec{S} = \int_\Omega Q_V\,d\Omega \qquad (1.6.5)$$

The time derivative is to be considered here in the moving system, that is for fixed space coordinates attached to the volume Ω.

The quantity $\vec{F}^{\mathrm{rel}} \overset{\Delta}{=} (\vec{F} - U\vec{v}_s)$ can be considered as a ***relative flux*** vector, since it contributes to the convective flux in (1.6.5).

Indeed, introducing the ***relative velocity***

$$\vec{w} = \vec{v} - \vec{v}_s \qquad (1.6.6)$$

a *relative convective flux* can be defined as

$$\vec{F}_C^{\text{rel}} \overset{\Delta}{=} (\vec{F}_C - U\vec{v}_s) = U\,\vec{w} \tag{1.6.7}$$

applying the definition (1.1.5) for the reference convective flux.

Vector conservation law

The conservation law for a vector quantity, such as momentum, for a moving control surface is derived in the same way as above, introducing *the relative convective flux tensor*, which generalizes equation (1.1.15):

$$\overline{\overline{F}}_C^{\text{rel}} \overset{\Delta}{=} (\overline{\overline{F}}_C - \vec{U} \otimes \vec{v}_s) = \vec{U} \otimes \vec{w} \tag{1.6.8}$$

leading to the following extension of equation (1.1.12):

$$\int_\Omega \frac{\partial \vec{U}}{\partial t}\, d\Omega + \oint_S (\overline{\overline{F}} - \vec{U} \otimes \vec{v}_s - \overline{\overline{Q}}_S) \cdot d\vec{S} = \int_\Omega \vec{Q}_V\, d\Omega \tag{1.6.9}$$

The quantity $\vec{U}$ is defined in the inertial system, but the partial time derivative is considered for fixed coordinates in the moving system.

   If $\vec{X}$ is the local coordinate attached to a point in the reference system moving with the velocity $\vec{v}_s$ and $\vec{x}$ the local coordinate in the reference inertial system, the relation

$$d\vec{x} = d\vec{X} + \vec{v}_s\, dt \tag{1.6.10}$$

connects the two reference frames and defines also the relation between the local time variations in the absolute and moving systems, as follows

$$\left[\frac{\partial}{\partial t} \right]_{\text{rel}} = \left[\frac{\partial}{\partial t} \right]_{\text{abs}} + \vec{v}_s \cdot \vec{\nabla} \tag{1.6.11}$$

This relation is obtained by applying the chain rule for differentials to an arbitrary function $U[\vec{x}(\vec{X}, t), t]$ with the definitions

$$\left[\frac{\partial \vec{x}}{\partial t} \right]_X = \vec{v}_s \quad \text{and} \quad \left[\frac{\partial \vec{X}}{\partial t} \right]_x = -\vec{v}_s \tag{1.6.12}$$

based on equation (1.6.10).

   These relations can be very useful in order to connect the time variation of the two systems.

SUMMARY OF THE BASIC FLOW EQUATIONS

The equations derived in the previous sections are valid in all generality for any Newtonian compressible fluid in an absolute or a relative frame of reference with constant rotation. The various forms of these equations can be summarized in the following tables. Table 1 corresponds to the equations in the absolute system, while Table 2 contains the equations written in the steadily rotating, relative frame of reference.

Table 1 The system of flow equations in an absolute frame of reference.

Equation form	Integral	Differential Conservative form	Non-conservative form
Conservation of mass	$$\frac{\partial}{\partial t}\int_\Omega \rho\,d\Omega + \oint_S \rho\vec{v}\cdot d\vec{S} = 0$$	$$\frac{\partial \rho}{\partial t} + \vec{\nabla}\cdot(\rho\vec{v}) = 0$$	$$\frac{d\rho}{dt} + \rho\vec{\nabla}\cdot\vec{v} = 0$$
Conservation of momentum	$$\frac{\partial}{\partial t}\int_\Omega \rho\vec{v}\,d\Omega + \oint_S \rho\vec{v}(\vec{v}\cdot d\vec{S})$$ $$= -\oint_S p\cdot d\vec{S} + \oint_S \bar{\bar{\tau}}\cdot d\vec{S} + \int_\Omega \rho\vec{f}_e\,d\Omega$$	$$\frac{\partial \rho\vec{v}}{\partial t} + \vec{\nabla}\cdot(\rho\vec{v}\otimes\vec{v} + p\bar{\bar{I}} - \bar{\bar{\tau}}) = \rho\vec{f}_e$$	$$\rho\frac{d\vec{v}}{dt} = -\vec{\nabla}p + \vec{\nabla}\cdot\bar{\bar{\tau}} + \rho\vec{f}_e$$ **Crocco's form** $$\frac{\partial\vec{v}}{\partial t} - (\vec{v}\times\vec{\zeta}) = T\vec{\nabla}s - \vec{\nabla}H + \frac{1}{\rho}\vec{\nabla}\cdot\bar{\bar{\tau}} + \vec{f}_e$$
Conservation of energy	$$\frac{\partial}{\partial t}\int_\Omega \rho E\,d\Omega + \oint_S (\rho\vec{v}H - k\vec{\nabla}T - \bar{\bar{\tau}}\cdot\vec{v})\cdot d\vec{S}$$ $$= \int_\Omega (\rho\vec{f}_e\cdot\vec{v} + q_H)\,d\Omega$$ $W_f = \rho\vec{f}_e\cdot\vec{v}$ work of external forces q_H external heat sources	$$\frac{\partial\rho E}{\partial t} + \vec{\nabla}\cdot(\rho\vec{v}H - k\vec{\nabla}T - \bar{\bar{\tau}}\cdot\vec{v})$$ $$= W_f + q_H$$ $H = h + \vec{v}^2/2$ stagnation enthalpy $E = e + \vec{v}^2/2$ total energy	$$\rho\frac{de}{dt} = -p(\vec{\nabla}\cdot\vec{v}) + \vec{\nabla}\cdot(k\vec{\nabla}T) + (\bar{\bar{\tau}}\cdot\vec{\nabla})\cdot\vec{v} + q_H$$ **Alternative form** $$\rho\frac{dH}{dt} = \frac{\partial p}{\partial t} + \vec{\nabla}\cdot(k\vec{\nabla}T + \bar{\bar{\tau}}\cdot\vec{v}) + W_f + q_H$$ **Entropy equation** $$\rho T\frac{ds}{dt} = \varepsilon_V + \vec{\nabla}\cdot(k\vec{\nabla}T) + q_H$$
Definitions	$$\tau_{ij} = \mu\left[(\partial_i v_j + \partial_j v_i) - \frac{2}{3}(\vec{\nabla}\cdot\vec{v})\delta_{ij}\right]\qquad \vec{\nabla}\cdot\bar{\bar{\tau}} = \mu\left[\Delta\vec{v} + \frac{1}{3}\vec{\nabla}(\vec{\nabla}\cdot\vec{v})\right]\quad \text{for constant viscosity coefficients}$$ Viscous dissipation rate: $\varepsilon_V = (\bar{\bar{\tau}}\cdot\vec{\nabla})\cdot\vec{v} = \frac{1}{2\mu}(\bar{\bar{\tau}}\otimes\bar{\bar{\tau}}^T)$ vorticity vector: $\vec{\zeta} = \vec{\nabla}\times\vec{v}$		

Table 2 *The system of flow equations in a relative rotating frame of reference.*

Equation form	Integral	Differential — Conservative form	Differential — Non-conservative form
Conservation of mass	$\dfrac{\partial}{\partial t}\displaystyle\int_\Omega \rho\,\mathrm{d}\Omega + \oint_S \rho\vec{w}\cdot\mathrm{d}\vec{S} = 0$	$\dfrac{\partial\rho}{\partial t} + \vec{\nabla}\cdot(\rho\vec{w}) = 0$	$\dfrac{\mathrm{d}\rho}{\mathrm{d}t} + \rho\vec{\nabla}\cdot\vec{w} = 0$ with $\dfrac{\mathrm{d}}{\mathrm{d}t} = \dfrac{\partial}{\partial t} + \vec{w}\cdot\vec{\nabla}$
Conservation of momentum	$\dfrac{\partial}{\partial t}\displaystyle\int_\Omega \rho\vec{w}\,\mathrm{d}\Omega + \oint_S \rho(\vec{w}\otimes\vec{w}+p\bar{\bar{I}}-\bar{\bar{\tau}})\cdot\mathrm{d}\vec{S}$ $= \displaystyle\int_\Omega \rho[\vec{f}_e - 2(\vec{\omega}\times\vec{w}) - \vec{\omega}\times(\vec{\omega}\times\vec{r})]\mathrm{d}\Omega$ $\vec{u}=\vec{\omega}\times\vec{r}$ $\vec{w}$ relative velocity $\vec{v}=\vec{u}+\vec{w}$ absolute velocity	$\dfrac{\partial\rho\vec{w}}{\partial t} + \vec{\nabla}\cdot(\rho\vec{w}\otimes\vec{w}+p\bar{\bar{I}}-\bar{\bar{\tau}})$ $= \rho[\vec{f}_e - 2(\vec{\omega}\times\vec{w}) - \vec{\omega}\times(\vec{\omega}\times\vec{r})]$	$\rho\dfrac{\mathrm{d}\vec{w}}{\mathrm{d}t} = -\vec{\nabla}p + \vec{\nabla}\cdot\bar{\bar{\tau}} +$ $\rho[\vec{f}_e - 2(\vec{\omega}\times\vec{w}) - \vec{\omega}\times(\vec{\omega}\times\vec{r})]$ *Crocco's form* $\dfrac{\partial\vec{w}}{\partial t} - (\vec{w}\times\vec{\zeta}) = T\vec{\nabla}s - \vec{\nabla}I + \dfrac{1}{\rho}\vec{\nabla}\cdot\bar{\bar{\tau}} + \vec{f}_e$
Conservation of energy	$\dfrac{\partial}{\partial t}\displaystyle\int_\Omega \rho E^*\,\mathrm{d}\Omega + \oint_S (\rho\vec{w}I - k\vec{\nabla}T - \bar{\bar{\tau}}\cdot\vec{w})\cdot\mathrm{d}\vec{S}$ $= \displaystyle\int_\Omega (\rho\vec{f}_e\cdot\vec{w}+q_H)\mathrm{d}\Omega$ $W_f^{\mathrm{rel}} = \rho\vec{f}_e\cdot\vec{w}$ work of external forces in relative system q_H external heat sources	$\dfrac{\partial\rho E^*}{\partial t} + \vec{\nabla}\cdot(\rho\vec{w}I - k\vec{\nabla}T - \bar{\bar{\tau}}\cdot\vec{w})$ $= W_f^{\mathrm{rel}} + q_H$ $E^* = e + \vec{w}^2/2 - \vec{u}^2/2 = E - \vec{u}\cdot\vec{v}$ $I = h + \vec{w}^2/2 - \vec{u}^2/2 = H - \vec{u}\cdot\vec{v}$	$\rho\dfrac{\mathrm{d}e}{\mathrm{d}t} = -p(\vec{\nabla}\cdot\vec{W}) + \vec{\nabla}\cdot(k\vec{\nabla}T) + (\bar{\bar{\tau}}\cdot\vec{\nabla})\cdot\vec{W} + q_H$ *Alternative form* $\rho\dfrac{\mathrm{d}I}{\mathrm{d}t} = \dfrac{\partial p}{\partial t} + \vec{\nabla}\cdot(k\vec{\nabla}T + \bar{\bar{\tau}}\cdot\vec{w}) + W_f^{\mathrm{rel}} + q_H$ *Entropy equation* $\rho T\dfrac{\mathrm{d}s}{\mathrm{d}t} = \varepsilon_V + \vec{\nabla}\cdot(k\vec{\nabla}T) + q_H$
Definitions	$\tau_{ij} = \mu\left[(\partial_i w_j + \partial_j w_i) - \dfrac{2}{3}(\vec{\nabla}\cdot\vec{w})\delta_{ij}\right]$ $\vec{\nabla}\cdot\bar{\bar{\tau}} = \mu\left[\Delta\vec{w} + \dfrac{1}{3}\vec{\nabla}(\vec{\nabla}\cdot\vec{w})\right]$ for constant viscosity coefficients Viscous dissipation rate: $\varepsilon_V = (\bar{\bar{\tau}}\cdot\vec{\nabla})\cdot\vec{W} = \dfrac{1}{2\mu}(\bar{\bar{\tau}}\otimes\bar{\bar{\tau}}^T)$ absolute vorticity vector: $\vec{\zeta} = \vec{\nabla}\times\vec{v}$		

CONCLUSIONS AND MAIN TOPICS TO REMEMBER

- The fundamental properties of fluid mechanics are contained in the conservation laws, which can be written in the very general form given by equation (1.1.1) for a scalar quantity and by equation (1.1.12) for a vector quantity.
- To any conserved quantity, we can associate a flux (vector or tensor) describing how that quantity is changed by the flow.
- You recognize an integral conservation law by the presence of the surface integral of the fluxes, which is the only place where the fluxes appear. This fundamental property is the key to any integral conservation law: fluxes may never appear inside the volume, as they will not be distinguishable from volume sources.
- For the differential form of the conservation law, equations (1.1.4) and (1.1.14), fluxes appear exclusively under a divergence operator. This is how a differential conservation law is recognized.
- In general the flux associated to a conserved quantity will contain a convective component, which is *always* present in a fluid in motion, and a diffusive component, which is present in a fluid at rest, but may not always exist.
- The convective flux appears as a first order derivative in space, while the diffusion terms are always expressed by second order spatial derivative terms, reducing to a Laplacian for constant diffusivity properties.
- The distinction between convective and diffusive fluxes is of crucial importance and translates the fundamental differences in their physical interpretation, as described in Section 1.1.2. Please go back to this section if you feel that these differences are not fully clear to you.
- The laws of fluid mechanics are governed by the conservation equations for the three basic quantities: mass, momentum and energy. Make sure that you have a good insight into the form of these equations and of the significance of the various contributions to the conservation of momentum (the equation of motion) and of energy.
- Although equations can be written for other quantities, such as pressure, internal energy, temperature, entropy, they nevertheless do *not* obey a conservation law, as they have no associated flux.

REFERENCES

Batchelor, G.K. (1970). *An Introduction to Fluid Dynamics*. Cambridge University Press, UK.

Crocco, L. (1937). Eine neue stromfunktion fur die Erforschung der Bewegung der Gase mit Rotation. *Z. Angew. Math. Mech.*, 17, 1–7.

Lax, P.D. (1954). Weak solutions of nonlinear hyperbolic equations and their numerical computation. *Comm. Pure Appl. Math.*, 7, 159–193.

PROBLEMS

P.1.1 Develop the algebraic form for the gradient of the flux tensor appearing in the vector conservation law (1.1.14), in Cartesian coordinates.

With the flux components denoted by F_{ij} with $i, j = x, y, z$, show that

$$(\vec{\nabla}.\overline{\overline{F}})_x = \frac{\partial F_{xx}}{\partial x} + \frac{\partial F_{yx}}{\partial y} + \frac{\partial F_{zx}}{\partial z}$$

Write out the two other components y and z.

P.1.2 Write out the Navier–Stokes equation of motion (1.3.11) explicitly for the three components u, v, w in Cartesian coordinates.

Repeat it for the non-conservative form (1.3.12).

P.1.3 Obtain equation (1.3.13).

P.1.4 Prove equation (1.3.20).

P.1.5 Obtain the energy equations (1.4.14) and (1.4.15) for the internal energy e.

Hint: Introduce the momentum equation multiplied by the velocity vector into equation (1.4.9).

P.1.6 Show that in an incompressible fluid at rest the energy equation (1.4.14) reduces to the temperature conduction equation:

$$\rho c_V \frac{\partial T}{\partial t} = \vec{\nabla}.(k\vec{\nabla}T) + q_H$$

P.1.7 Show that the energy equation (1.4.11) reduces to a convection–diffusion balance of the stagnation enthalpy H when the Prandtl number is equal to one and when only the contribution from the work of the shear stresses related to the viscous diffusion of the kinetic energy is taken into account.

Hint: Assume constant flow properties, setting $k = \mu c_p$ in absence of external sources, and separate the contributions to the term $\vec{\nabla}.(\overline{\overline{\tau}}.\vec{v})$ according to the following relations, valid for incompressible flows:

$$\vec{\nabla}.(\overline{\overline{\tau}}.\vec{v}) = \partial_i \left[\mu(\partial_i v_j + \partial_j v_i)v_j \right] = \vec{\nabla}. \left[\mu\vec{\nabla}(\vec{v}^2/2) \right] + \vec{\nabla}. \left[\mu(\vec{v}.\vec{\nabla}).\vec{v} \right]$$

Neglecting the second term, setting $H = c_p T + (\vec{v}^2/2)$ leads to

$$\frac{\partial \rho E}{\partial t} + \vec{\nabla}.(\rho\vec{v}H) = \vec{\nabla}.(\mu\vec{\nabla}H)$$

P.1.8 Obtain the entropy equation (1.4.18).

P.1.9 Derive equation (1.5.7).

P.1.10 Consider the integral mass conservation equation (1.2.1) for a permanent (steady) flow, defined as a time-independent flow for which all partial time derivatives vanish. Consider a channel of arbitrary varying section and two cross-sections at an arbitrary distance apart. Show that the application of the steady form of the integral mass conservation law leads to the constancy of the mass flow rate through each cross-section.

Hint: select a control surface formed by the two cross-sections and the channel walls in-between and apply the steady integral mass conservation law together with the definition (1.1.6) of the element of mass flow rate.

Chapter 2

The Dynamical Levels of Approximation

OBJECTIVES AND GUIDELINES

The main objective of this chapter is to guide you through the different approximations that can be defined to reduce the complexity of the system of flow equations. This process of simplification is based on *physical* considerations, connected to the dynamical properties of fluid flows, hence the denomination 'dynamical' in the title of this chapter.

In Chapter 1, we derived and discussed the fundamental form of a conservation law and applied it to obtain the basic equations of fluid mechanics, known as the system of Navier–Stokes equations, expressing the conservation of the three fundamental quantities, mass, momentum and energy.

These equations contain many levels of complexity; the most significant being the following:

- They form a system of five (in three-dimensional space) *fully coupled* time-dependent partial differential equations for the five unknowns, velocity vector (three unknowns), and two thermodynamic quantities, such as for instance pressure and density, or pressure and temperature. The coupling occurs through the velocity and density fields, possibly also through the temperature field, when thermal effects are significant.
- Each of these equations is *nonlinear*. The nonlinearity of the flow equations is not just a mathematical observation, as it has *major consequences on the whole of fluid mechanics*:
  - The dominant nonlinearity is provided by the convection term $\rho(\vec{v}.\vec{\nabla})\vec{v}$, see, e.g., the momentum equation under the form (1.3.13). This term is responsible for the appearance of *turbulence*, which is a spontaneous *instability* of the flow, whereby all quantities take up a *statistical* (chaotic) behavior.
  - For compressible flows, the products of density and velocity represent another nonlinearity, leading to the existence of *shock waves* in supersonic flows. Through a shock, velocities, pressure, temperature, undergo a *discontinuous jump* and, as we will see later on, these discontinuities are indeed *exact solutions* of the nonlinear inviscid Euler equations.
  - With non-uniform temperature fields, flow-thermal nonlinearities appear, such as Bénard cells in shallow heated fluid layers, representative of complex thermal convection phenomena, which are crucial, among others, in weather forecasting.
  - Other nonlinearities can appear in flows with free surfaces, such as the breaking of waves that you can observe on the seashores, or during heavy sloshing of a liquid in a tank. Also in two phase flows, the coalescence or

breaking up of droplets or bubbles result from the nonlinear properties of flows and nonlinearities of thermodynamic origin lead to phase changes, such as evaporation, condensation.

– Nonlinearities lead to ***non-unique solutions***. This has also major consequences, under the form of the existence of multiple flow configurations for the same initial and boundary conditions, resulting from ***bifurcations*** from one flow state to another. Numerous examples of flow non-uniqueness and of bifurcations have been observed, experimentally and numerically, and we will show some examples in this chapter.

These complexities in fluid dynamics pose considerable challenges for CFD, in particular for turbulent flows.

The link with the available computing power at a given time is illustrated in Figures I.1.4 and I.1.6, showing the evolution of the models used in industry, progressing toward increased complexity, and hence reliability, due to the progress in available computing power shown in Figure I.1.7.

We will present in this chapter an overview of the most significant and most widely used approximation levels. All these models, ranging from Direct Numerical Simulation (DNS), Large Eddy Simulation (LES) and Reynolds Averaged Navier–Stokes (RANS), to various forms of simplified treatment of the viscous terms, including boundary layer approximations, down to inviscid models of the Euler equations and the most simplified forms of potential flow, have been used, and several of them are still in use, in various sectors of industry.

The Issue of the Time and Length Scales

The issue of the time and length scales of the description of the physical flow features is critical to the world of simulation. The same flow will appear very different when we reduce the scale at which we look at it. This is illustrated by the following example of the experimental observation of a double annular jet, Figures 2.0.1 and 2.0.2. When measurements are taken with a standard Laser Doppler Velocimeter (LDV) instrumentation, which averages the flow over a certain time scale, a steady averaged flow, with three backflow regions appear, as seen on Figure 2.0.1. On the other hand, the same flow seen with a Particle Image Velocimetry (PIV) laser sheet technique, which takes an instantaneous snapshot of the flow in the plane of the laser sheet, shows a highly unsteady flow with large-scale fluctuations, although the inlet conditions are constant, Figure 2.0.2.

Look carefully at these figures, as they illustrate a fundamental issue in CFD simulations. Before you consider a CFD application, you should evaluate and define the time and length scales at which you want to model your flow system, in the same way experimentalists choose their instrumentation in function of the level of details they require.

This is the main objective of this chapter. It should help you develop a knowledge and awareness of the best suited model for a given flow problem, by making a proper balance between the acceptable approximations with respect to the reality, and compatible with the computer capacity you have available.

We have chosen to illustrate the various approximations by typical results and examples from CFD calculations performed at each level of approximation, as an illustration of the type of computations achievable with the model being considered.

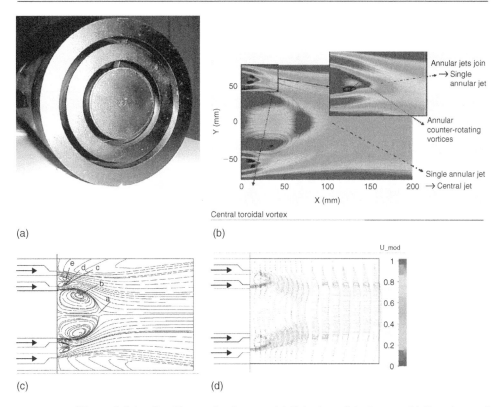

(a)   (b)

(c)   (d)

Figure 2.0.1   *Double annular burner: (a) Exit view of the burner. (b) Experimental color plot of axial velocity in the symmetry plane, obtained from LDV. (c) Experimental streamlines of the flow with designation of specific position points, related to the vortex structure, obtained from LDV. (d) Experimental vector plot of the velocity field obtained from LDV data (for color image refer Plate 2.0.1). From S. Geerts et al. (2005). Courtesy Vrije Universiteit Brussel (VUB).*

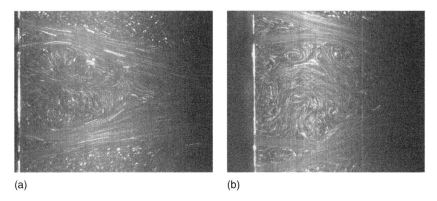

(a)   (b)

Figure 2.0.2   *Double annular burner: two successive snapshots taken with a PIV laser sheet technique. Courtesy Vrije Universiteit Brussel.*

Table 2.0.1 *The simplification process leading to the scalar, linearized model equations.*

Degrees of complexity	Full system of flow equations	Reduction of complexity
Three-dimensional space	Vector of minimum five unknowns	Reduce to one-dimensional space, with three unknowns
Coupled system	System of minimum five equations	Decouple the equations and reduce to a single scalar equation for one unknown quantity
Nonlinearities	Full nonlinear system	Linearize the scalar equations

Simplified Model Equations

Most of the approximate models retain, at various degrees, the basic complexities of the Navier–Stokes equations, namely the three dimensions of space, the coupled nature of the equations and the nonlinearities.

As you easily can imagine, it is hardly possible to develop and study the basics of numerical discretization methods on such complex models. Hence, we need a second level of simplification, reducing these three degrees of complexity to their essentials, namely the fundamental effects of convection and diffusion, as seen in Section 1.1.2. This leads us to simplified ***model equations***, obtained by the following simplification process, summarized in Table 2.0.1, which can be followed by going down along the third column:

- Three-dimensional space, to be simplified to one-dimensional space.
- The coupled equations, to be decoupled to a single scalar equation.
- The nonlinearities, to be removed by a linearization assumption.

These mathematically based simplifications lead to sufficiently simple model equations to guide the development of the basic numerical schemes for CFD.

They will form the basis for all of Chapters 4–10, while Chapters 11 and 12, will focus on applying the methods developed on the model equations to 'real' approximate flow models formed by coupled and nonlinear equations.

Note that the numerical schemes developed and analyzed on these simplified model equations will ultimately be applied for the discretization of the full Navier–Stokes equations.

The simplified models will be introduced in Chapter 3.

Content of this chapter

It is considered that the system of Navier–Stokes equations, supplemented by empirical laws for the dependence of viscosity and thermal conductivity with other flow variables and by a constitutive law defining the nature of the fluid, completely describes all flow phenomena (Section 2.1).

For laminar flows, no additional information is required and one can consider that any experiment in laminar flow regimes can be accurately duplicated by computations. However, and we could say unfortunately from the point of view of computational fluid dynamics, most of the flow situations occurring in nature and in technology enter into a particular form of instability called turbulence. Turbulence occurs in all flow situations when the velocity, or more precisely, the Reynolds number, defined as the product of representative scales of velocity and length divided by the kinematic viscosity, exceeds a certain critical value. The particular form of instability generated in the turbulent flow regime, is characterized by the presence of statistical fluctuations of all the flow quantities. These fluctuations can be considered as superimposed on mean or averaged values and can attain, in many situations, the order of 10% of the mean values, although certain flow regions, such as separated zones, can attain much higher levels of turbulent fluctuations.

Clearly, the numerical description of the turbulent fluctuations is a formidable task, which puts extremely high demands on computer resources. With increasing computer power, in both speed and memory, we are progressively able to simulate the large-scale turbulent fluctuations, or even the small-scale turbulent motion, directly on the computer from the time-dependent Navier–Stokes equations. This forms the basis of the growing development of ***Direct Numerical Simulation (DNS)***. An estimate of the computer requirements connected to this level of approximation can be found already in Chapman (1979) and in the recent books of Sagaut (2001) and Geurts (2004).

The computer requirements for DNS simulations of turbulent flows are out of reach in the foreseeable future for industrial applications and it is therefore essential to resort to approximations enabling the numerical description of turbulent flows in acceptable computer CPU times. The highest approximation, with good prospects for reaching the industrial stage in the near future is the approximation known as ***Large Eddy Simulation (LES)***. This LES approach is similar to DNS, in its objective to simulate directly the turbulent fluctuations, but restricted to the larger scales with the smaller scales being modeled.

The next highest level of approximation is the ***Reynolds Averaged Navier–Stokes (RANS)*** model, which is restricted to the computation of the averaged turbulent flow. This requires the RANS equations to be supplemented by models for the Reynolds stresses. These models can range from simple eddy viscosity or mixing length models to transport equations for the turbulent kinetic energy and dissipation rates, or their many two equation variants, or to still more complicated models solving directly the transport equations for the Reynolds stresses. LES and RANS models are introduced in Section 2.2.

Considering the various stages within the dynamical level of approximation, a first reduction in complexity can be introduced for flows with small amount of separation or back-flow and with a predominant mainstream direction at high Reynolds numbers. This allows neglecting viscous and turbulent diffusion in the mainstream direction and hence to reduce the number of shear stress terms to be computed, considering that they have a negligible action on the flow behavior. This is the '***Thin Shear Layer Approximation***' discussed in Section 2.3.

Within the same level, we can situate the ***parabolic approximations*** for the steady state Navier–Stokes equations. In these approximations, the elliptic character of the flow is put forward through the pressure field, while all other variables are considered as transported or as having a parabolic behavior (Section 2.4).

The next level to be considered is the ***Boundary Layer Approximation*** referred to in Section 2.5. As is well known, this analysis of the effects of viscosity by L. Prandtl, is a most spectacular example of the impact of a careful investigation of the magnitude of force components on the description of a flow system. For flows with no separation and thin viscous layers, that is at high Reynolds numbers, a decoupling of the viscous and inviscid parts of the flow can be introduced. The pressure field is hereby decoupled from the viscous effects, showing that the influence of the viscous and turbulent shear stresses is confined to small regions close to the walls and that outside these layers the flow behaves as inviscid. This analysis, which was perhaps the greatest breakthrough in fluid mechanics since the discovery of the Navier–Stokes equations, showed that many of the flow properties could be described by the inviscid approximation, e.g. determination of the pressure distributions, and that a simplified boundary layer approximation allows for the determination of the viscous effects.

When this influence or interaction is neglected, we enter the field of the inviscid approximations, which allow generally a good approximation of the pressure field, and hence of lift coefficient for attached external flows.

An intermediate level between the partially or fully viscous flow descriptions and the inviscid approximation is the ***distributed loss model***, used in confined flow problems, in particular in the simulation of multistage turbomachinery flows and shallow water models in ocean dynamics. The overall effect of boundary layers and wakes is expressed as a distributed friction force and the implications of this approximation are presented in Section 2.6. At the same level, we can consider various ***viscid–inviscid interaction*** models, which couple a boundary layer calculation, as a correction to an inviscid simulation, in order to obtain an approximation of viscous effects, including friction losses. Within the inviscid approximations, the model of the time-dependent Euler equation is summarized in Section 2.7.

The potential flow model, restricted to non-rotational flows, is at a lower level of approximation, due to the associated assumption of isentropic flow (Section 2.8). This leads to a description of shock discontinuities which deviate from the Rankine–Hugoniot relations and occasionally to problems of non-uniqueness. However, the potential flow model is equivalent to the Euler equations for subsonic, non-rotational flows.

The content of Chapter 2 is summarized in the chart of Figure 2.0.3.

2.1 THE NAVIER–STOKES EQUATIONS

The most general description of a fluid flow is obtained from the full system of Navier–Stokes equations. Referring to Chapter 1, the conservation laws for the three basic flow quantities $(\rho, \rho\vec{v}, \rho E)$ can be written in a compact form, expressing the coupled nature of the equations:

$$
\frac{\partial}{\partial t}
\begin{vmatrix} \rho \\ \rho\vec{v} \\ \rho E \end{vmatrix}
+ \vec{\nabla} \cdot
\begin{vmatrix} \rho\vec{v} \\ \rho\vec{v} \otimes \vec{v} + p\bar{\bar{I}} - \bar{\bar{\tau}} \\ \rho\vec{v}H - \bar{\bar{\tau}} \cdot \vec{v} - k\vec{\nabla}T \end{vmatrix}
=
\begin{vmatrix} 0 \\ \vec{f}_e \\ W_f + q_H \end{vmatrix}
\tag{2.1.1}
$$

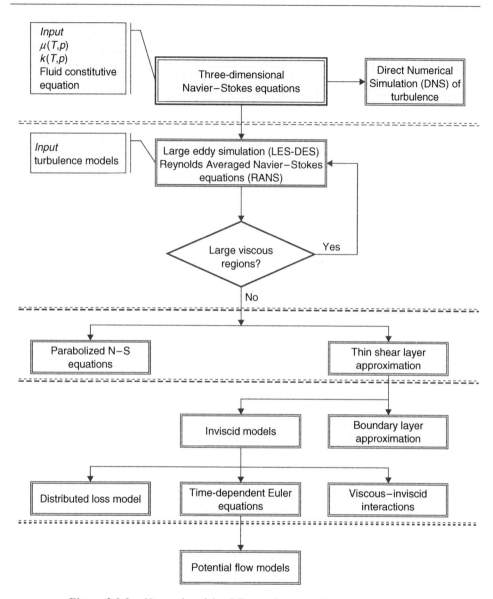

Figure 2.0.3 *Hierarchy of the different dynamical levels of approximation.*

The above equation defines a (5×1) column vector U of the conservative variables:

$$U = \begin{vmatrix} \rho \\ \rho\vec{v} \\ \rho E \end{vmatrix} = \begin{vmatrix} \rho \\ \rho u \\ \rho v \\ \rho w \\ \rho E \end{vmatrix} \qquad (2.1.2)$$

and a generalized (5×3) flux vector $\vec{F}$:

$$\vec{F} = \begin{vmatrix} \rho \vec{v} \\ \rho \vec{v} \otimes \vec{v} + p\overline{\overline{I}} - \overline{\overline{\tau}} \\ \rho \vec{v} H - \overline{\overline{\tau}} \cdot \vec{v} - k \vec{\nabla} T \end{vmatrix} \tag{2.1.3}$$

with Cartesian coordinates f, g, h each of these components being a (5×1) column vector. The right-hand side contains the source terms and these can be grouped into a (5×1) column vector Q, defined by

$$Q = \begin{vmatrix} 0 \\ \vec{f}_e \\ W_f + q_H \end{vmatrix} \tag{2.1.4}$$

The source terms express the effects of the external forces $\vec{f}_e$, of the heat sources q_H and of the work performed by the external forces $W_f = \rho \vec{f}_e \cdot \vec{v}$. The group of equations (2.1.1) takes then the following condensed form:

$$\frac{\partial U}{\partial t} + \vec{\nabla} \cdot \vec{F} = Q \tag{2.1.5}$$

Expressed in Cartesian coordinates, we obtain the more explicit algebraic form:

$$\frac{\partial U}{\partial t} + \frac{\partial f}{\partial x} + \frac{\partial g}{\partial y} + \frac{\partial h}{\partial z} = Q \tag{2.1.6}$$

or in an alternative condensed notation:

$$\partial_t U + \partial_x f + \partial_y g + \partial_z h = Q$$

where u, v, w are the x, y, z components of the velocity vector and the flux vector (2.1.3) is defined by its components f, g, h (subscripts indicate the corresponding Cartesian components):

$$f = \begin{vmatrix} \rho u \\ \rho u^2 + p - \tau_{xx} \\ \rho uv - \tau_{xy} \\ \rho uw - \tau_{xz} \\ \rho uH - (\overline{\overline{\tau}} \cdot \vec{v})_x - k\partial_x T \end{vmatrix} \qquad g = \begin{vmatrix} \rho v \\ \rho vu - \tau_{yx} \\ \rho v^2 + p - \tau_{yy} \\ \rho vw - \tau_{yz} \\ \rho uH - (\overline{\overline{\tau}} \cdot \vec{v})_y - k\partial_y T \end{vmatrix}$$

$$h = \begin{vmatrix} \rho w \\ \rho wu - \tau_{zx} \\ \rho wv - \tau_{yz} \\ \rho w^2 + p - \tau_{zz} \\ \rho wH - (\overline{\overline{\tau}} \cdot \vec{v})_z - k\partial_z T \end{vmatrix} \tag{2.1.7}$$

Refer to Chapter 1 for a more detailed discussion, in particular for the association with the Newtonian fluid model, the perfect gas model, or for the particular formulation related to incompressible fluids.

This system of Navier–Stokes equations is valid for any laminar or turbulent flow of any fluid, defined by its constitutive equation relating the shear stresses to the other flow variables.

2.1.1 Non-uniqueness in Viscous Flows

Non-unique solutions of the Navier–Stokes equations are known to exist for many flow situations when some non-dimensional number, representing a measure of the balance between various forces reach a critical value. For instance, the Bénard problem of a fluid heated from below or the Taylor problem of the flow between concentric cylinders, of which the inner one is rotating are known to generate more than one physical state for the same physical conditions. It is interesting to observe here that the non-uniqueness of the stationary Navier–Stokes equations has been proven theoretically for these flow cases (see for instance Temam, 1977).

However, many other flow systems show this non-unique behavior, which is often associated with additional complexities resulting from spontaneous unsteadiness, such as bifurcations, symmetry breaking and route to chaos. The latter step is often the road to turbulence.

We will illustrate this with two examples. The first one is less familiar as it describes the flow induced by the temperature gradient of surface tension on a cylinder of liquid bound by above and below by two surfaces at different temperatures. It is known as a *liquid bridge* and the flow phenomenon is known as the ***Marangoni effect***.

The second example is the well-known flow over a cylinder with a uniform steady incident velocity. The resulting flow is subject to the appearance of vortex shedding, known as the Von Karman street, but many levels of additional complexities appear when either compressibility and/or 3D effects are taken into consideration.

The non-uniqueness property of the viscous flows, connected to the spontaneously generated unsteadiness, pose considerable problems to the numerical simulation and represent one of the main challenges in CFD. The flow can undergo sudden changes in its unsteady behavior, or be subjected to bifurcations, route to chaos and other effects typical of nonlinear systems. Very high accuracy, at the level of the discretization schemes, as well as in the treatment of the boundary conditions are required in order to be able to recover numerically multiple solutions, when they exist; while avoiding spurious states of numerical origin.

2.1.1.1 *Marangoni thermo-capillary flow in a liquid bridge*

A liquid bridge is a small cylinder of liquid, of height h and diameter d, contained between two endplates at different temperatures, but with a free external surface, held together by surface tension. The Marangoni thermo-capillary effect designates the flow system that appears induced by a temperature gradient of the surface tension and is of interest in float-zone crystal growth technology.

In the above equations, the external surface forces are defined by

$$\vec{f}_{eS} = \sigma \kappa \vec{n} + \vec{\nabla}_t \sigma \qquad (2.1.8)$$

The first term on the right-hand side represents the normal component of the surface tension σ, function of temperature, for a surface curvature κ, directed along the normal direction n. The second term is the tangential component of the surface tension gradient, which appears if the surface tension is not uniform, in particular due to temperature variations.

An associated heat source is generated by the work of these capillary forces, defined by

$$W_{fS} = \rho \vec{f}_{eS} \cdot \vec{v}_S \tag{2.1.9}$$

where the subscript S indicates values defined on the peripheral surface.

The surface tension forces hold the column of fluid together between two disks maintained at different temperatures, and a convection pattern is generated by the surface tension gradients induced by the temperature gradient on the interfacial surface.

A non-dimensional Reynolds number is defined as

$$\mathrm{Re} = \frac{|\sigma_T| h \Delta T}{\rho v^2} \tag{2.1.10}$$

h is the height of the liquid bridge, ΔT is the imposed temperature difference between the endplates, $\sigma_T = -d\sigma/dT$ is the negative rate of change of surface tension with temperature.

The Marangoni number is defined by multiplication with the Prandtl number $\mathrm{Pr} = v/\kappa$:

$$\mathrm{Ma} = \mathrm{Re}.\,\mathrm{Pr} = \frac{|\sigma_T| h \Delta T}{\rho v^2} \cdot \frac{v}{\kappa} = \frac{|\sigma_T| h \Delta T}{\rho v \kappa} \tag{2.1.11}$$

The flow structure depends on three non-dimensional parameters, Re, Pr and the aspect ratio of the liquid bridge, $A = 2h/d$.

For small temperature differences ΔT between the upper (hot) and lower (cold) endplate, that is for low values of the Marangoni number Ma, a steady axisymmetric flow is being generated by the temperature-dependent surface tension forces. This steady flow is characterized by two toroidal vortices and by a temperature field with a cold point at the lower corner. Figure 2.1.1 shows the liquid bridge configuration and the steady flow for half the cylinder, where the central limit of the velocity and temperature plots is the vertical symmetry axis of the liquid bridge.

When the temperature difference, that is the Marangoni number increases, the flow structure changes completely as the velocity and temperature fields become unsteady, with increasingly complex flow configurations.

The following flow structures are observed; Dinescu and Hirsch (2001), Hirsch and Dinescu (2003):[1]

- A first spontaneous unsteadiness appears under the form of a standing pulsating wave, with 2, 3 or more modes, depending on the Prandtl number and the aspect

---

[1] The another would like to thank Cristian Dinescu for putting together these figures extracted from the CFD animation.

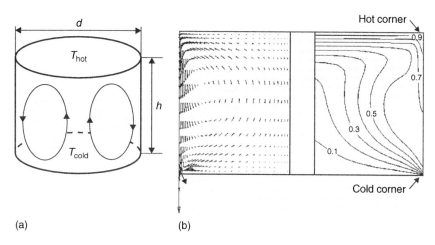

(a)   (b)

Figure 2.1.1   *(a) Liquid bridge configuration and (b) Marangoni effect. Steady axisymmetric flow and temperature pattern shown on half of the cylinder, the axis of symmetry is the internal vertical line in each of these figures. From Dinescu and Hirsch (2001).*

ratio. Figure 2.1.2a shows four snapshots of the associated temperature and velocity fields, for $m = 2$ mode.

- This pulsating mode is not permanent and after a certain time transforms itself to a traveling mode, which remains unchanged over time (Figure 2.1.2b).

  With further increasing Marangoni numbers, a succession of nonlinear generated phenomena appear:

  - Symmetry breaking of the flow configuration first via a period doubling as shown on (Figure 2.1.2c).
  - At different conditions, another route to chaos is identified, showing a quasi-periodic state with frequency doubling. Inspecting Figure 2.1.2d, we can see the pattern of the temperature disturbance field with the three cold (blue) spots placed near the free surface and the three hot spots located near the vertical axis of the liquid bridge. The loss of symmetry and the two frequencies of the traveling waves captured by the numerical simulation are confirmed by the experiments of Schwabe (2001), identified as quasi-periodic states.
  - Route to chaos: with increasing Marangoni numbers an increasingly chaotic motion starts to appear.

These results are confirmed by experimental observations and other simulations. The interested reader can consult the following additional references to learn more about these fascinating phenomena: Velten et al. (1991), Frank and Schwabe (1997), Shevtsova and Legros (1998), Kuhlmann (1999), Leypoldt et al. (2000), and Zeng et al. (2004).

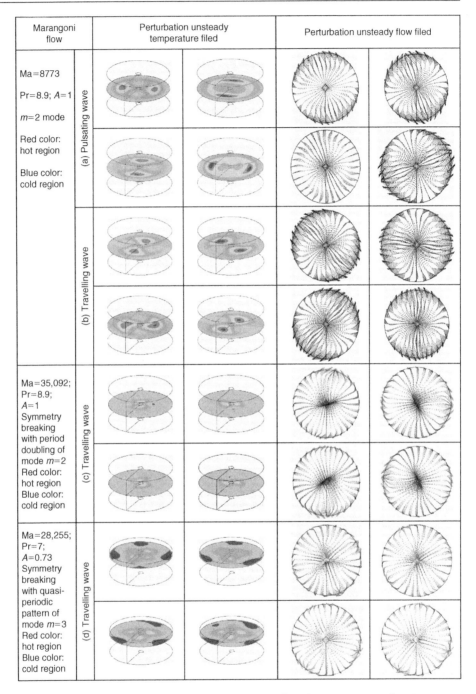

Figure 2.1.2 *Liquid bridge and Marangoni effect. Representation of various unsteady and symmetry broken solutions for velocity and temperature Perturbation fields. Each group of four figures represents four snapshots of the corresponding unsteady peturbation field (for color image refer Plate 2.1.2). From Dinescu and Hirsch (2001), Hirsch and Dinescu (2003).*

2.1.1.2   *Flow around a circular cylinder*

The flow around a circular cylinder generated by a horizontal incident uniform steady velocity field U_∞ is one of the simplest possible setups. Nevertheless, the resulting flow is of great complexity, known to generate a spontaneous unsteadiness under the form of a periodic vortex shedding.

Two-dimensional configuration

For Reynolds numbers $Re = U_\infty D / \nu$ below 40 the flow is steady with a symmetrical backflow region behind the cylinder (Figure 2.1.3).

However, above this critical value of 40, a spontaneous unsteadiness appears, under the form of a periodic vortex shedding. This can be explained as follows: the backflow regions result from the separation of the flow at positions around 90° and form a symmetrical pattern. The viscous boundary layers are regions with high vorticity and it is known that the intensity of the vortices increases with Reynolds number. At the separation points of the upper and lower parts of the cylinder, the vortices are equal and have opposite signs, such that a symmetrical flow pattern arises. When the vortex intensities grow, a small perturbation, which would give to the upper vortex for instance a slightly larger value than the lower one, would influence the flow on the lower part and attract the lower vortex. This breaks the symmetry with the consequence that the upper vortex is not balanced anymore by the lower one of opposite sign. This vortex is then convected by the flow away from the cylinder surface, leaving the lower vortex as the dominating one. This lower vortex attracts the flow to the lower side and after being at his turn convected by the flow away from the surface, handles back the dominating role to the upper vortex. This results then in a periodic motion, known as the periodic ***Von Karman street of shed vortices***. The frequency of this effect increases with Reynolds number.

This is illustrated in the series of Figure 2.1.4 displaying pictures at $Re = 100$, at different time instants of the progressive generation of the vortex shedding, from http://www.idi.ntnu.no/~zoran/NS-imgs/lics.html

 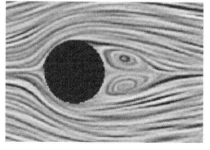

Circular cylinder at Re=26–experimental image
(Van Dyke–An Album of Fluid Motion)

Figure 2.1.3 *Visualizations of the 2D flow around a cylinder at Re = 26, showing the symmetric pattern of the separated regions; from Van Dyke (1982); compared with a numerical simulation from http://www.idi.ntnu.no/~zoran/NS-imgs/lics.html*

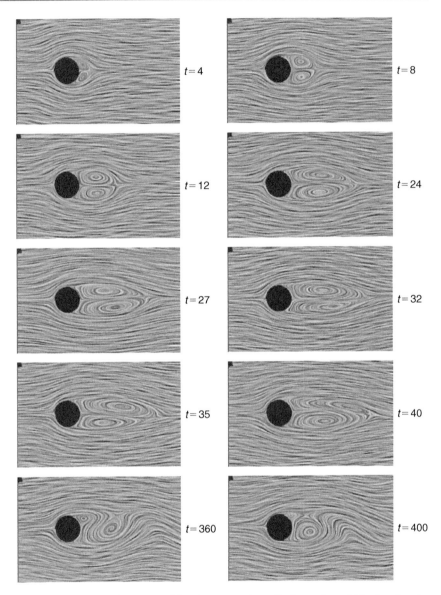

Figure 2.1.4 *Visualizations of the flow around a cylinder at Re = 100, showing the progressive generation of the vortex shedding, at different time instants. From http://www.idi.ntnu.no/~zoran/NS-imgs/lics.html*

What this actually means is that above the critical value of the Reynolds number, *there is no steady flow configuration anymore*, although the cylinder and the incident conditions remain fixed and constant.

If we consider in addition a high Mach number flow, an additional complexity appears, due to the interaction with compressibility effects.

Compressibility effects and non-uniqueness

The influence of the compressibility on the flow around the cylinder is summarized in the beautiful series of pictures shown on Figure 2.1.5 taken at the Institut de Mécanique des Fluides de Lille, France, at different Mach numbers and at a Reynolds number close to 10^5; see Dyment (1982). With increasing Mach number and intensity of the acoustic waves, the interaction and the coupling between compressibility effects, vortex shedding and separation on the cylinder becomes more pronounced. A strong shock is gradually generated downstream of the cylinder, and a steady wake of increasing length appears for Mach numbers from 0.70 to 0.90, with a periodic vortex shedding downstream of the shock. Above a certain value of Mach number lambda shocks appear and when they join, no disturbances can travel upstream preventing the coupling between the wake and the vortex street. This can lead to a stationary regime such as observed in certain circumstances at $M = 0.98$. The flow visualizations show another important phenomena, namely non-uniqueness under the form of the appearance of more than one flow regime at certain values of Mach number and Reynolds number. Two unsteady flow configurations can be distinguished at $M = 0.8$ while at $M = 0.98$ both an unsteady and a steady flow regime can occur.

Three-dimensional effects

When the flow around the cylinder is analyzed, experimentally or numerically, taking into account its spanwise dimension and length, new phenomena appear, characterized by the presence of streamwise vortices, with non-unique properties and additional nonlinear interactions leading to frequency doubling effects and route to chaos with increasing Reynolds numbers. These effects were initially found experimentally by Williamson (1998a, b, 1989, 1992). See also Williamson (1996a, b) for a general overview. They have been confirmed and analyzed in depth in a series of numerical simulations by M. Braza and her coworkers, where the detailed mechanisms of the two modes have been clearly identified (Persillon and Braza (1998) and Braza et al. (2001)).[2]

Figure 2.1.6 shows the experimental data for the variation of the Strouhal number $St = fD/U_\infty$, where f is the frequency of the vortex shedding, with Reynolds number. This figure shows the presence of a first bifurcation, with two possible states between $Re = 180$ and 190 and a second bifurcation, with a region of multiple solutions between $Re = 230$ and 280. The upper curve of points corresponds to two-dimensional vortex shedding and there is a marked difference in the Strouhal number variation between the two- and three-dimensional cases, due to the strictly three-dimensional character of the two discontinuities and of their intermediate region. These multiple solutions correspond to two different 3D vortex structures, referred to as modes A and B in Figure 2.1.6 from Williamson (1996a, b).

The numerical simulations, supported by the experimental evidence, allows an in-depth analysis of the bifurcation mechanisms between the modes A and B.

---

[2] The author gratefully acknowledges the friendly support of Prof. Marianna Braza in the analysis of this section and for providing the quoted figures.

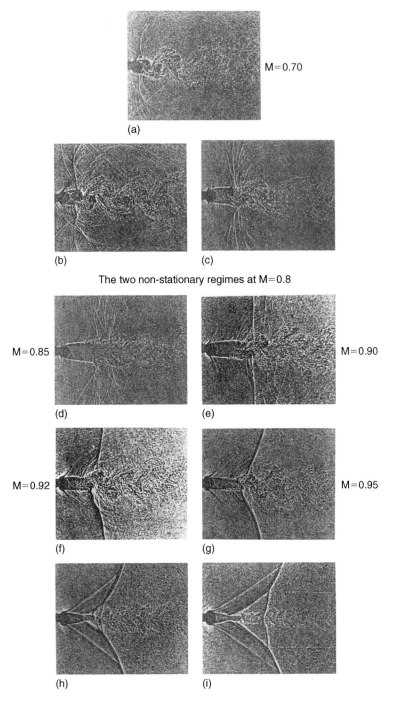

M=0.70

(a)

(b)　　　　　　　　　(c)

The two non-stationary regimes at M=0.8

M=0.85　　　　　　　　　M=0.90

(d)　　　　　　　　　(e)

M=0.92　　　　　　　　　M=0.95

(f)　　　　　　　　　(g)

(h)　　　　　　　　　(i)

The two regimes at M=0.98; one non-stationary, the other stationary

Figure 2.1.5 *Visualizations of the flow around a cylinder for various Mach numbers at a Reynolds number of 10^5. Cylinder $H = 8\,mm$; $UH/v = 10^5$ shadowgraphs 0.3 ms. Courtesy A. Dyment and M. Pianko, Institut de Mécanique des Fluides de Lille, France.*

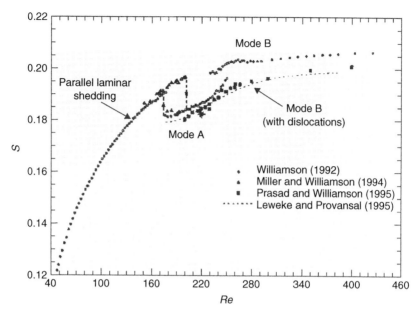

Figure 2.1.6 *Strouhal number versus Reynolds number relation, showing the different modes and a lower path with dislocations. Experiments reported by Williamson (1996a, b) (detailed in the legend); * direct simulation by the present study. From Braza et al. (2001).*

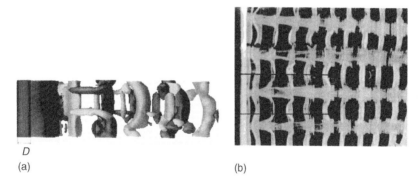

Figure 2.1.7 *Spanwise undulation of the main vortex rows and streamwise vortices, at Re = 220; shown by iso-contours of vorticity components. (b) Spanwise experimental flow visualization, provided by Williamson (1992). The frame shows correspondence to the computational region explored (for color image refer Plate 2.1.7). From Persillon and Braza (1998). Courtesy M. Braza. IMFT, Toulouse.*

Figure 2.1.7, from Persillon and Braza (1998) shows the A configuration at a Reynolds number of 220. The figure displays the three-dimensional vortex structure through the iso-contours of the transverse and streamwise vorticity components. A distortion of the main vortex filaments is seen along the span, with the presence of

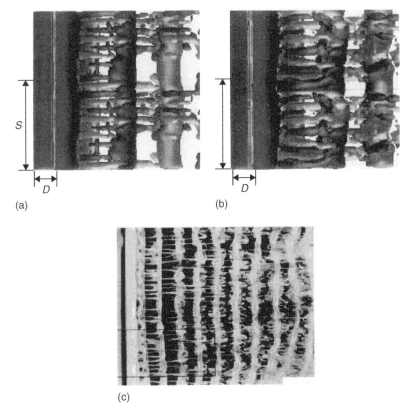

(a) (b)

(c)

Figure 2.1.8 *Modification of the spanwise vortex structures as Reynolds number increases; passage to mode B; (a) Re = 270; (b) Re = 300. (c) Spanwise experimental flow visualization, provided by Williamson (1992, 1996 a, b). The frame shows correspondence to the computational region explored (for color image refer Plate 2.1.8). From Persillon and Braza (1998). Courtesy M. Braza. IMFT, Toulouse.*

streamwise vortex structures. This pattern is compared with experimental visualizations by Williamson (1992). At higher Reynolds numbers, we observe a modification of the spanwise vortex structures and a transition to mode B, as seen from Figure 2.1.8 where Figure 2.1.8a corresponds to Re = 270 and Figure 2.1.8b to Re = 300. Figure 2.1.8c shows the spanwise experimental flow visualization, from Williamson (1992, 1996a, b).

The mechanism behind the bifurcation between modes A and B is related to a dislocation mechanism of the transverse vortices, as shown by the simulations performed by Braza et al. (2001). Details of this mechanism, involving a fundamental frequency reduction and route to chaos are summarized in Figure 2.1.9, showing the transition steps from mode A to mode B. The figure displays the modification of the spanwise wavelengths, as well as the junction between two adjacent Von Karman vortex rows, that illustrates the vortex dislocation pattern with an increase of the chaotic components.

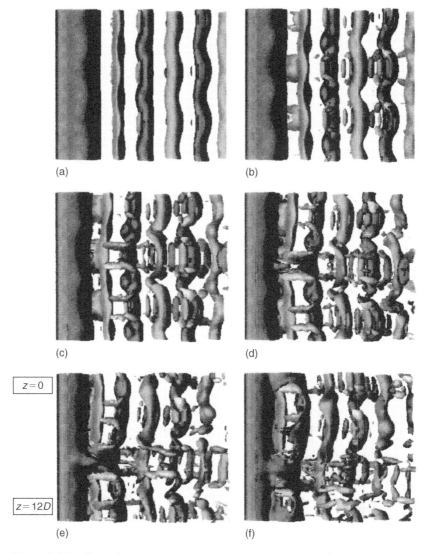

Figure 2.1.9 *Spanwise and streamwise iso-vorticity contours showing mode A formation and the transition to the vortex dislocations pattern at Re = 220 (for color image refer Plate 2.1.9). From Braza et al. (2001). Courtesy M. Braza. IMFT, Toulouse.*

2.1.2 Direct Numerical Simulation of Turbulent Flows (DNS)

A fundamental property of fluid mechanics is the appearance of ***turbulence***.

Any flow system will remain laminar up to a certain critical value of the Reynolds number $V \cdot L/v$, where V and L are representative values of velocity and length scales for the considered flow system and v is the kinematical viscosity (expressed in m^2/s). *Above a critical value of the Reynolds number, all flows become turbulent.* They are then characterized by the appearance of statistical fluctuations of all the variables (velocity, pressure, density, temperature, etc.) around mean values. These fluctuations are a form of instability of the flow system, as a consequence of the nonlinear convection terms. Hence, they cannot be described anymore in a deterministic way.

However, they could be computed numerically in direct simulations of turbulence, DNS or at a lower level of approximation by the 'large eddy simulation' (LES) approach whereby only the small-scale turbulent fluctuations are modeled and the larger-scale fluctuations are computed directly.

The reader can find a review of the state of art of direct numerical simulation of turbulence in Jimenez (2003), B Geurts (2004) and in Rogallo and Moin (1984) for a historical perspective. Although this approach requires considerable computer resources, it has already led to very informative results on the fundamental physics of turbulence.

Direct Numerical Simulation of Turbulent Flows (DNS) has as objective to simulate on computer the whole range of the turbulent statistical fluctuations at all relevant physical scales. This is a formidable challenge, which grows with increasing Reynolds numbers, since the size of the smallest turbulent eddies is inversely proportional to $Re^{3/4}$, the well-known Kolmogorov scale related to the turbulent dissipation. If we wish a resolution of n points per unit length of the smallest eddy, the total number of mesh points required, and the number of arithmetic operations, will scale with $n^3 \cdot Re^{9/4}$. As the Navier–Stokes equations have to be integrated in time, with a time step determined by the smallest turbulent time scales, which are proportional to $Re^{3/4}$, the total computational effort for DNS simulations is proportional to Re^3 for homogeneous turbulence! Wall flows and other inhomogeneous cases, are even more expensive, since the mesh should adapt to the resolution scales of the near-wall structures.

This means that increasing the Reynolds number by a factor 10, requires an increase in the computational power of at least a factor 1000, and by a factor $10^{9/4} = 178$ for the memory requirements.

Therefore, DNS simulations for realistic Reynolds numbers of the order of $10^5 - 10^7$, as found in many industrial external flows around aircrafts, cars, buildings, or internal flows in engines, pumps, compressors, turbines, etc. are out of reach for a long time, based on the current and projected computer capacities.

Nevertheless, DNS is widely applied as a basic research tool to better understand the fundamental mechanisms of turbulence, with the objective to establish a database of information to be used to improve lower level approximations such as LES or turbulence models for RANS simulations (Section 2.2).

Some of the more advanced DNS simulations are being performed by J. Jimenez and his coworkers at the University of Madrid and University of Illinois (in particular R.D. Moser). See for instance Jimenez (2004), Jimenez et al. (2004), Del Alamo et al. (2004, 2006), Hoyas and Jimenez (2006) and Flores and Jimenez (2006).

Their fundamental DNS simulations of the turbulent flow in a simple channel provide a wealth of information on the basic turbulent properties at all scales. It is not the place here to enter into a detailed discussion on these properties, but Figures 2.1.10 and 2.1.11 provide representative examples of DNS simulations, under the form of

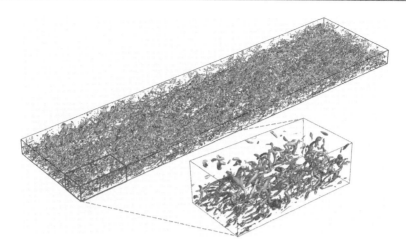

Figure 2.1.10 *Instantaneous view of the turbulent vortices colored with their distance to the wall (red is closest to the wall and yellow is at the center of the channel. Only 1/4 of the channel is shown (full length, half-width and half-height), and the flow direction is from bottom-left to top-right (for color image refer Plate 2.1.10). From del Alamo et al. (2004). Courtesy J. Jimenez and coworkers.*

snapshots of the instantaneous turbulent fluctuation field and its underlying vortex structure.[3]

The two figures correspond to simulations performed at two different Reynolds numbers. Figure 2.1.10 has been obtained at a Reynolds number based on the channel width and the center line velocity of 47,500 for a friction Reynolds numbers Re_{tau} (based on the channel half-width and the friction velocity) $= 950$. The simulations have been performed on a grid of $NX \times NY \times NZ = 1048 \times 385 \times 1556 = 1{,}226{,}874{,}880$ mesh points. The case was run on several computers belonging to DoE in the US or at San Diego, mostly by the group of Prof. R.D. Moser, then at University of Illinois in Urbana. It took about 10^6 processors hours, usually on 384 SP2/SP3 processors. The insert in the figure allows observing the structure of single vortices, while the complexity of the flow is highlighted in the main part of the figure.

Figure 2.1.11 shows a similar view of the instantaneous vorticity field at a Reynolds number based on the channel width and the center line velocity of 100,000 (Re_{tau} based on the channel half-width and the friction velocity $= 2000$). The simulations have been performed on a grid of $NX \times NY \times NZ = 4096 \times 633 \times 3072 = 7{,}964{,}983{,}296$ mesh points, with typical computation times of 6.10^6 CPU hours on the 'Mare Nostrum' supercomputer in Barcelona on 2100 processors for about 6 months. This is the highest Reynolds number up to date applied for DNS simulations, and we

---

[3] The author is grateful to Prof. Javier Jimenez, for the information on his work and on the shown figures, as well as to J.C. Del Alamo and O. Flores, and for permission to publish them.

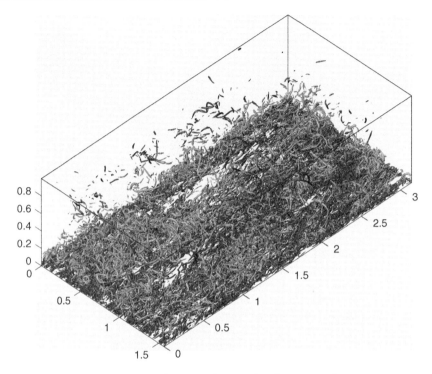

Figure 2.1.11 *Instaneous realization of a complex clustering of vortices in a turbulent channel flow at a Reynolds number of 100,000 (Retau = 2000). The flow is from left to right and the vortices are colored with their distance to the wall (blue is near the wall while red is far from the wall) (for color image refer Plate 2.1.11). From Hoyas and Jimenez (2006). Courtesy J. Jimenez and coworkers.*

recommend the cited papers to the interested reader for detailed analysis of the turbulence properties that emerge from these unique simulations.

Another important domain of application of DNS is the simulation of laminar–turbulent transition, of which many fundamental features are still unknown. DNS offers a significant way of investigating the complexity of transition phenomena and Figure 2.1.12 shows a snapshot of a DNS simulation performed by J. Wissink and W. Rodi of the University of Karlsruhe, with a mesh of 56 million points, at a Reynolds number of 60,000. The simulations investigate the effects of an external turbulence on the transition, comparing a laminar incoming separation bubble with no turbulence with an incoming 7% turbulence intensity.

2.2 APPROXIMATIONS OF TURBULENT FLOWS

The applications of CFD to real life flow systems, in nature or in technology, require the ability to handle turbulent flows, as these are the most widely encountered situation. Hence we need to take into account the effects of turbulence on the mean flow and this requires approximate models, as DNS is not a short-term option.

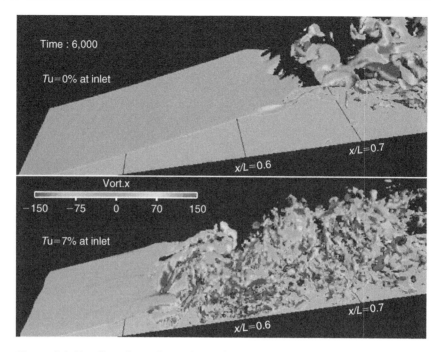

Figure 2.1.12   *Snapshots of a DNS simulation at a Reynolds number of 60,000, showing the effects of an external turbulence on the transition, comparing the vorticity field of a laminar incoming separation bubble with no turbulence and with 7% turbulence intensity (for color image refer Plate 2.1.12). Courtesy J. Wissink and W. Rodi, University of Karlsruhe.*

Two families of models are presently available: one family, called **Large Eddy Simulation (LES)** is of the same category as DNS, in that it computes directly the turbulent fluctuations in space and time, but only above a certain length scale. Below that scale, called the **subgrid scale**, the turbulence is modeled by semi-empirical laws.

The other family, called the **Reynolds Averaged Navier–Stokes (RANS)** model, ignores the turbulent fluctuations and aims at calculating only the turbulent-averaged flow. This is currently the most widely applied approximation in the CFD practice.

The hierarchy between these three levels of turbulence modeling is summarized in Figure 2.2.1, which shows the turbulent energy spectrum in function of wave number k, and the limits of the range of application of LES and RANS models. Remember that the wave number is defined as $k = 2\pi/\lambda$, where λ is the wavelength.

2.2.1   Large Eddy Simulation (LES) of Turbulent Flows

The equations describing LES models are obtained from the Navier–Stokes conservation laws by a **filtering** operation whereby the equations are averaged over the part of the spectrum that is not computed, that is over the smaller length scales (that is the

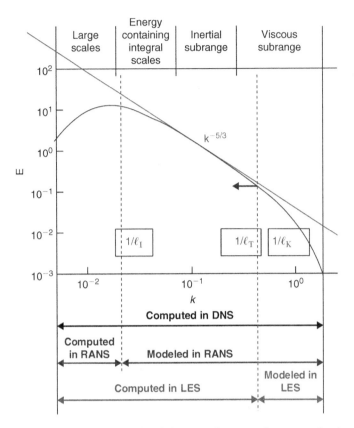

Figure 2.2.1   *Energy spectrum of turbulence in function of wave number k, with indication of the range of application of the DNS, LES and RANS models. The length scales l_T and l_I are associated with the LES and RANS approximations, respectively (for color image refer Plate 2.2.1). Courtesy C. Fureby (FOI, Sweden).*

high wave number region). In practice, the lowest identified scales are related to the mesh size and therefore the LES models are often referred to as **subgrid scale models**.

Since the remaining larger-scale turbulent fluctuations are directly simulated, the computational requirements on LES simulations are still very high. It can be shown that for a resolution of n points per unit length of the simulated eddies, the number of arithmetic operations will scale with $n^3 \cdot Re^{3/2}$ and taking into account the time integration, the total computational effort for LES simulations is proportional to $Re^{9/4}$. This is significantly lower than the DNS requirements, but still excessively high for large Reynolds number applications, particularly for wall-bounded flows.

A domain where LES is clearly coming close to practical industrial applications is the modeling of combustion phenomena.

For many applications involving wall-bounded flows and attached boundary layers, various hybrid combinations of LES and RANS are being considered, whereby the RANS approximation is kept in the regions where the boundary layers are attached to the solid walls. A recent account can be found in Haase et al. (2006).

2.2.2 Reynolds Averaged Navier–Stokes Equations (RANS)

The most widely applied approximation for industrial applications of CFD is the approximation whereby the turbulent equations are averaged out, in time, over the whole spectrum of turbulent fluctuations. This leads to the so-called '*Reynolds Averaged Navier–Stokes equations*' which require, in addition, empirical or at least semi-empirical information on the turbulence structure and its relation to the averaged flow.

This approach goes back to O. Reynolds himself.

Biographical note
Osborne Reynolds (1842–1912) was born in Belfast and went to school at Dedham (Essex) where his father (a priest in the Anglican church, but having an academic degree from Cambridge university) was headmaster. He studied mathematics at Cambridge, graduating in 1867. In 1868, Osborne Reynolds became the first professor of engineering at Owens College in Manchester. That College is a predecessor of the Victoria University of Manchester, merged with UMIST in 2004 and now the University of Manchester.

*Before and after his university studies, O. Reynolds was shortly employed by engineering firms, which marked his interest for phenomena encountered in practice. He wrote about himself: '**From my earliest recollection I have had an irresistible liking for mechanics and the physical laws on which mechanics as a science is based**', as reported in Lamb (1912–1913).*

He initially worked on a wide range of phenomena: condensation and heat transfer between solids and fluids, the effect of rain and oil in calming waves at sea, the refraction of sound by the atmosphere, as well as various engineering works: the first multi-stage turbine, a laboratory-scale model of the Mersey estuary that mimicked tidal effects.

*By 1880 O. Reynolds became interested by the detailed mechanics of fluid motion, especially the sudden transition between direct and sinuous flow in circular pipes which he found occurred when $UD/\nu \approx 2000$. He published his first experimental observations of turbulent flows and laminar–turbulent transition in 1883, obtained by injecting an ink tracer into water flowing in a circular glass pipe, and varying the diameter and the velocity. His paper was called: '**An experimental investigation of the circumstances which determine whether the motion of water in parallel channels shall be direct or sinuous and of the law of resistance in parallel channels**'. What is known today as the 'Reynolds number', namely the combination UD/ν, appears in this work. He further developed the theoretical basis of the RANS models and presented his theoretical ideas to the Royal Society in 1894, which included 'Reynolds averaging', Reynolds stresses and the first derivation of the turbulence energy equation.*

O. Reynolds made also significant contributions to of the theory of lubrication, and he is widely recognized as the founder of the science of tribology (friction, lubrication and wear).

H. Lamb, who new him well, wrote about the personality of Osborne Reynolds: 'The character of Reynolds was like his writings, strongly individual. He was conscious of the value of his work, but was content to leave it to the mature judgment of the scientific world. For advertisement, he had no taste, and undue pretension on the part of others only elicited a tolerant smile. To his pupils he was most generous in the opportunities for valuable work, which he put in their way, and in the share of cooperation. Somewhat reserved in serious or personal matters and occasionally combative and tenacious in debate, he was in the ordinary relations of life the most kindly and genial of companions'.

A lively account of his particular teaching habits is cited by R.A. Smith (http://www.queens.cam.ac.uk/Queens/Record/1997/History/Reynolds.html): Reynolds had a characteristically uncompromising style of both written and oral communication, the latter well illustrated by this account of one of his lectures, given by his most famous pupil, Sir J.J. Thompson, later Nobel Laureate, President of the Royal Society and Master of Trinity; 'He was one of the most original and independent of men and never did anything or expressed himself like anybody else. The result was that it was very difficult to take notes at his lectures so that we had to trust mainly to Rankine's textbooks. Occasionally in the higher classes he would forget all about having to lecture and, after waiting for ten minutes or so, we sent the janitor to tell him that the class was waiting. He would come rushing into the door, taking a volume of Rankine from the table, open it apparently at random, see some formula or other and say it was wrong. He then went up to the blackboard to prove this. He wrote on the board with his back to us, talking to himself, and every now and then rubbed it all out and said it was wrong. He would then start afresh on a new line, and so on. Generally, towards the end of the lecture he would finish one which he did not rub out and say that this proved Rankine was right after all'.

Reynolds became a Fellow of the Royal Society in 1877 and won the Royal Medal in 1888. By the beginning of the 1900s, Reynolds health began to decline and he retired in 1905. He died in 1912.

The original experimental equipment of Osborne Reynolds is still operational and can be seen at the University of Manchester. A permanent exhibition of the life and achievements of O. Reynolds is visible in the Simon Engineering Laboratories of the University of Manchester.

References for additional reading:

Lamb, H. (1912–1913). Osborne Reynolds. Proc. Roy. Soc., 88A.

Rott, N. (1990). Note on the history of the Reynolds number. Annu. Rev. Fluid Mech., 22 pp. 1–11.

http://www-history.mcs.st-andrews.ac.uk/history/: The MacTutor History of Mathematics Archive is a website maintained by John J. O'Connor and Edmund F. Robertson and hosted by the University of St. Andrews in Scotland. It contains detailed biographies of many notable mathematicians.

The turbulent averaging process is introduced in order to obtain the laws of motion for the 'mean', time-averaged, turbulent quantities. This time averaging is to be defined in such a way as to remove the influence of the turbulent fluctuations, while not destroying the time dependence connected with other time-dependent phenomena with time scales distinct from those of turbulence.

Turbulent averaged quantities

For any turbulent quantity A, the separation

$$A = \overline{A} + A'$$ (2.2.1)

is introduced with

$$\overline{A}(\vec{x}, t) = \frac{1}{T} \int_{-T/2}^{T/2} A(\vec{x}, t + \tau) \mathrm{d}\tau$$ (2.2.2)

$\overline{A}$ represents a time-averaged turbulent quantity, where T is to be chosen large enough compared to the time scale of the turbulence but still small compared to the time scales of all other unsteady phenomena. Obviously, this might not be always possible: if unsteady phenomena occur with time scales of the same order as those of the turbulent fluctuations, the Reynolds averaged equations will not allow to model these phenomena. However, it can be considered that most of the unsteady phenomena in fluid dynamics have frequency ranges outside the frequency range of turbulence, Chapman (1979). The remaining term A' represents the turbulent fluctuating part, which is of **stochastic nature**.

For compressible flows, the averaging process leads to products of fluctuations between density and other variables such as velocity or internal energy. In order to avoid their explicit occurrence, a density-weighted average can be introduced, called **Favre-averaging**, through

$$\tilde{A} = \frac{\overline{\rho A}}{\overline{\rho}}$$ (2.2.3)

with

$$A = \tilde{A} + A''$$ (2.2.4)

and

$$\overline{\rho A''} = 0$$ (2.2.5)

This way of defining mean turbulent variables will remove all extra products of density fluctuations with other fluctuating quantities. This is easily seen by performing the averaging process defined by equation (2.2.3) on the continuity equation, leading to

$$\frac{\partial \overline{\rho}}{\partial t} + \vec{\nabla} \cdot (\overline{\rho} \tilde{\vec{v}}) = 0 \tag{2.2.6}$$

Applied to the momentum equations, we obtain the following equation for the turbulent mean momentum, in absence of body forces:

$$\frac{\partial \overline{\rho} \tilde{\vec{v}}}{\partial t} + \vec{\nabla} \cdot (\overline{\rho} \tilde{\vec{v}} \otimes \tilde{\vec{v}} + \overline{p} \overline{\overline{I}} - \overline{\overline{\tilde{\tau}}}^V - \overline{\overline{\tau}}^R) = \rho f_e \tag{2.2.7}$$

where the **Reynolds stresses** $\overline{\overline{\tau}}^R$, defined by

$$\overline{\overline{\tau}}^R = -\overline{\rho \vec{v}'' \otimes \vec{v}''} \tag{2.2.8}$$

are added to the *averaged* viscous shear stresses $\overline{\overline{\tilde{\tau}}}^V$. In Cartesian coordinates we have

$$\overline{\overline{\tau}}^R_{ij} = -\overline{\rho v_i'' v_j''}$$

The relations between the Reynolds stresses and the mean flow quantities are unknown. Therefore, the application of the Reynolds averaged equations to the computation of turbulent flows, requires the introduction of models for these unknown relations, based on theoretical considerations coupled to unavoidable empirical information. A wide variety of models, from simple algebraic relations to transport equations for turbulent quantities, such as the turbulent kinetic energy, the turbulent dissipation or even transport equations for the Reynolds stress components have been developed and applied with varying degrees of success.

Reviews of turbulence models can be found in the books of Wilcox (1998), Pope (2000), Haase et al. (2006) and in the scientific literature; see for instance Leschziner and Drikakis (2002) for an excellent review; and various conference proceedings devoted to turbulent flows.

It is to be mentioned that none of the available turbulence models offers today a totally accurate description of turbulent flows and although the RANS approximations is the most widely used in practice, the turbulent model components are their weakest link.

Practical example: The OBI diffuser

An asymmetric plane diffuser, known as the OBI diffuser, was computed on a grid made of 39,000 cells with the models of Spalart–Allmaras (SA), Spalart–Allmaras with curvature corrections (SARC), the k–ε model of Yang–Shih, Wilcox k–ω model, Menter's SST variant and the v_2-f model. The description of these various models as well as details on this test case and an extended set of results can be found in Haase et al. (2006).

Figure 2.2.2 shows the influence of the choice of the turbulence model on the length of the recirculation zone, as well as comparisons between calculated and measured

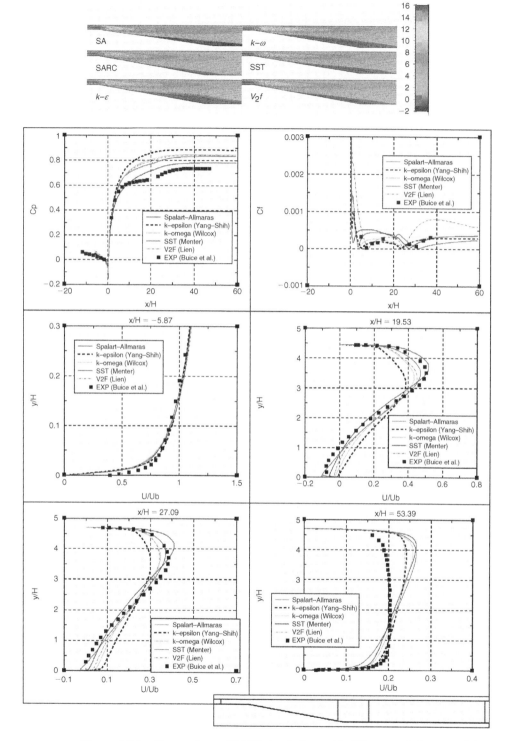

Figure 2.2.2 *Comparison of RANS simulations with different turbulence models for the OBI axisymmetric diffuser. The top figure shows the position and extends of the separation region, while the bottom figure compares calculated and measured pressure distribution, wall shear stress at the bottom wall and velocity profiles at the four positions indicated in the lower insert (for color image refer Plate 2.2.2). Courtesy NUMECA International and Haase et al. (2006).*

pressure distribution and wall shear stress at the bottom wall, velocity profiles at the positions indicated in the lower insert. The last figure is noteworthy, as it demonstrates a clear weakness of all the tested turbulence models, in that the velocity profile in the downstream duct of the diffuser is experimentally fully recovered, while the calculated profiles still show remaining effects of their earlier separation.

We can hope that the gained knowledge on turbulence from advanced DNS and LES simulations will contribute to the improvement of current turbulence models.

2.3 THIN SHEAR LAYER APPROXIMATION (TSL)

At high Reynolds numbers, wall shear layers, wakes or free shear layers will be of limited size and if the extension of the viscous region remains limited during the flow evolution, then the dominating influence of the shear stresses will come essentially from the gradients transverse to the main flow direction.

If we consider an arbitrary curvilinear system of coordinates with ξ^1 and ξ^2 along the surface and $\xi^3 = n$ directed toward the normal, then the ***Thin Shear Layer (TSL) approximation*** of the Navier–Stokes equations consists in neglecting all ξ^1 and ξ^2 derivatives occurring in the turbulent and viscous shear stress terms, Steger (1978), Pulliam and Steger (1978, 1985). This approximation is also supported by the fact that generally, at high Reynolds numbers (typically $Re > 10^4$) the mesh is dense in the direction normal to the shear layer and therefore the neglected terms are computed with a lower accuracy than the normal derivatives.

This approximation is actually close to a boundary layer approximation, since viscous terms, which are neglected in the boundary layer approximation, are also neglected here. However, the momentum equation in the directions normal to the shear layer is retained, instead of the constant pressure rule over the boundary layer thickness along a normal to the wall. Therefore the transition from viscous dominated regions to the inviscid region outside the wall layer is integrally part of the calculation, and one has here a form of 'higher order' boundary layer approximation. The classical boundary layer approximation is obtained, when the momentum equation in the direction normal to the wall is replaced by the condition:

$$\frac{\partial p}{\partial n} \simeq 0$$

The thin shear layer approximation amounts to neglect the viscous diffusion in the direction parallel to the shear surface and keep only the contributions from the diffusion in the normal direction.

The main motivation for the TSL approximation is historical, as it allowed in the 1980s some saving in computer times, compared to a full RANS approximation.

Today, this approximation is hardly justified, as it has to rely on the same turbulence models as current RANS simulations.

2.4 PARABOLIZED NAVIER–STOKES EQUATIONS

The Parabolized Navier–Stokes (PNS) approximation is based on considerations similar to the thin shear layer approximation, but applies only to the *steady state formulation* of the Navier–Stokes equations. This approximation is directed toward flow situations with a predominant main flow direction, as would be the case in a channel flow, whereby the cross-flow components are of a lower order of magnitude. In addition, along the solid boundaries, the viscous regions are assumed to be dominated by the normal gradients and hence, the streamwise diffusion of momentum and energy can be neglected.

If x is the streamwise coordinate, the x-derivatives in the shear stress terms are all neglected compared to the derivatives in the two transverse directions y and z. A similar approximation is introduced in the energy diffusion terms. This approximation is therefore valid as long as the mainstream flow of velocity u is dominant, that is as long as the positive x-direction corresponds to the forward flow direction. This will not be the case anymore if there is a region of reverse flow of the streamwise velocity component. In this case, the streamwise derivatives of u will become of the same order as the transverse derivatives and the whole approximation breaks down.

Note that this approximation can also be applied to supersonic steady flows, where the streamwise direction appears as the 'timelike' direction; see Chapter 3 for a definition of his concept.

2.5   BOUNDARY LAYER APPROXIMATION

It was the great achievement of L. Prandtl to recognize that at high Reynolds numbers, the viscous regions remain of limited extension δ (of the order of $\delta/L \approx \sqrt{\nu/UL}$, for a body of length L) along the surfaces of the solid bodies immersed in, or limiting the flow. In practical terms, this means that for a Reynolds number of 10^6, on a body with a chord length of 1 m, the boundary layer will have a thickness of the order of a few mm! If you consider an aircraft wing, flying at a speed of say 800 km/h, the velocity over the wing will change from zero on the wall to velocities in the range of 800 km/h over a distance of a few mm. Hence, the velocity gradients in the normal direction are much larger than the corresponding gradients in the streamwise direction.

In all cases where these viscous regions remain close to the body surfaces, that is in absence of separation, the calculation of the pressure field may be decoupled from the calculation of the viscous velocity field. A detailed discussion of the conditions for the derivation of the boundary layer equations can be found in the books of Batchelor (1970), Schlichting (1971) and Cebeci and Bradshaw (1984).

With the assumption that the vertical velocity component in the boundary layer is very small compared to the mainstream velocity, the momentum equation in the normal direction reduces to the condition of vanishing normal pressure gradient:

$$\frac{\partial p}{\partial z} \equiv \frac{\partial p}{\partial n} \cong 0 \qquad (2.5.1)$$

As a consequence, the pressure $p(x,y,z)$ inside the viscous boundary layer may be taken equal to the pressure outside of this layer and therefore equal to the value of the pressure $p_e(x,y)$, obtained from an inviscid computation. The pressure $p_e(x,y)$ is the value taken by the inviscid pressure field at the edge of the boundary layer of surface point (x,y).

Hence, the boundary layer equations are obtained from the streamwise and cross-flow momentum equations with the replacement of $p(x, y, z)$ by $p_e(x, y)$:

$$\frac{\partial}{\partial t}(\rho u) + \frac{\partial}{\partial x}(\rho u^2) + \frac{\partial}{\partial y}(\rho uv) + \frac{\partial}{\partial z}(\rho uw) = -\frac{\partial p_e}{\partial x} + \frac{\partial}{\partial z}\left(\mu \frac{\partial u}{\partial z}\right) \qquad (2.5.2)$$

$$\frac{\partial}{\partial t}(\rho v) + \frac{\partial}{\partial x}(\rho uv) + \frac{\partial}{\partial y}(\rho v^2) + \frac{\partial}{\partial z}(\rho vw) = -\frac{\partial p_e}{\partial y} + \frac{\partial}{\partial z}\left(\mu \frac{\partial v}{\partial z}\right) \qquad (2.5.3)$$

The inviscid pressure gradient, which is obtained from an inviscid calculation prior to the resolution of the boundary layer equations, acts as an external force on the viscous region. The same inviscid computation also provides the velocities $u_e(x, y)$ and $v_e(x, y)$ at the edge of the boundary layer, connected to the pressure field p_e by the inviscid equation:

$$\rho \frac{\partial \vec{v}_e}{\partial t} + \rho(\vec{v}_e \cdot \vec{\nabla})\vec{v}_e = -\vec{\nabla}p_e \qquad (2.5.4)$$

where $\vec{v}_e$ is the velocity vector parallel to the body surface with components $u_e(x, y)$ and $v_e(x, y)$. Equations (2.5.2) and (2.5.3) are to be solved with the additional boundary conditions:

$$u = u_e(x, y) \quad \text{and} \quad v = v_e(x, y) \quad \text{at } z = \delta \qquad (2.5.5)$$

at the edge of the boundary layer.

The system of equations obtained in this way has only the velocities as unknowns and this represents a significant simplification of the Navier–Stokes equations. Therefore, the boundary layer equations are much easier to solve, being close to standard parabolic second order partial differential equation and many excellent numerical methods have been developed (e.g. Kline, 1968; Cebeci and Bradshaw, 1977, 1984).

The inviscid region is limited by the edge of the boundary layer, which is initially unknown since the computational process has to start by the calculation of the pressure field. In the classical boundary layer approximation, the limits of the inviscid region are taken on the surface, which is justified for small boundary layer thicknesses. This leads to a complete separation between the pressure field and the velocity field since the pressure, in the remaining momentum equations (2.5.2) and (2.5.3), is equal to the values of the inviscid pressure field at the wall and is known when these equations are to be solved.

When the influence of the boundary layers on the inviscid flow field is considered as non-negligible, this interaction can be taken into account in an iterative way, by recalculating the inviscid pressure field with the limits of the inviscid region located at the edge of the boundary layer obtained at the previous iteration. This procedure is applied for thick boundary layers up to small separated regions and is known as the *viscid–inviscid* interaction approximation; see Le Balleur (1983) for a recent review on the subject.

2.6   THE DISTRIBUTED LOSS MODEL

The distributed loss model is an approximation applied essentially in internal and channel flows more particularly in the fields of turbomachinery, river hydraulics and oceanography.

This model is defined by the assumption that the effect of the shear stresses on the motion is equivalent to a ***distributed friction force***, defined by semi-empirical data. Obviously, a certain number of three-dimensional flow details will be lost in such an approximation, in particular, all flow aspects that can be attributed to, or are strongly influenced, by viscous effects.

Since the details of the loss mechanism, that is of the shear stresses, are not considered, these equations are to be taken as describing an inviscid model, however with an entropy producing term. The boundary conditions for the velocity field are therefore the inviscid conditions of vanishing normal velocity components at the walls, with a non-vanishing tangential velocity along these boundaries. The resulting approximation is then of inviscid nature but not isentropic since the entropy variation along the path of a fluid particle will be connected to the energy dissipation along this path. See for instance Hirsch and Deconinck (1985) for an application to internal turbomachinery flows.

A similar approximation is introduced in river hydraulics where the effects of the wall friction are represented by an empirical resistance force. The distributed loss model therefore consists in replacing the shear stress terms by an external friction force, function of velocity or other flow variables, but not directly expressed as second order derivatives of the velocity field.

2.7   INVISCID FLOW MODEL: EULER EQUATIONS

The most general flow configuration for a non-viscous, non-heat conducting fluid is described by the set of Euler equations, obtained from the Navier–Stokes equations (2.1.1) by neglecting all shear stresses and heat conduction terms. As is known from Prandtl's boundary layer analysis, this is a valid approximation for flows at high Reynolds numbers, outside viscous regions developing near solid surfaces.

This approximation introduces a drastic change in the mathematical formulation with respect to all the previous models containing viscosity terms since the system of partial differential equations describing the inviscid flow model reduces from second order to first order. This is of paramount importance since it will determine the numerical and physical approach to the computation of these flows. In addition, the number of allowable boundary conditions is modified by passing from the second order viscous equations to the first order inviscid system.

The time-dependent Euler equations, in conservation form and in an absolute frame of reference, for the conservative variables U defined by equation (2.1.2):

$$\frac{\partial U}{\partial t} + \vec{\nabla} \cdot \vec{F} = Q \tag{2.7.1}$$

form a system of first order partial differential equations hyperbolic in time (as will be shown later), where the flux vector F has the Cartesian components (f, g, h) given by

$$f = \begin{vmatrix} \rho u \\ \rho u^2 + p \\ \rho uv \\ \rho uw \\ \rho uH \end{vmatrix} \quad g = \begin{vmatrix} \rho v \\ \rho vu \\ \rho v^2 + p \\ \rho vw \\ \rho uH \end{vmatrix} \quad h = \begin{vmatrix} \rho w \\ \rho wu \\ \rho wv \\ \rho w^2 + p \\ \rho wH \end{vmatrix} \tag{2.7.2}$$

and the source term Q is given by equation (2.1.4). Generally, heat sources will not be considered since heat conduction effects are neglected in the system of Euler equations.

It is important to notice the properties of the entropy variations in an inviscid flow. From equation (1.4.18) and in absence of heat sources, the entropy equation for continuous flow variations reduces to

$$T \left(\frac{\partial s}{\partial t} + \vec{v} \cdot \vec{\nabla} s \right) = 0 \tag{2.7.3}$$

expressing that entropy is constant along a flow path. Hence, the Euler equations describe isentropic flows, in absence of discontinuities.

As is known, the set of Euler equations allows also discontinuous solutions in certain cases, namely, vortex sheets, contact discontinuities or shock waves occurring in supersonic flows. The properties of these discontinuous solutions can only be obtained from the integral form of the conservation equations, since the gradients of the fluxes are not defined at discontinuity surfaces.

2.8 POTENTIAL FLOW MODEL

The most impressive simplification of the mathematical description of a flow system is obtained with the approximation of a non-viscous, *irrotational* flow.

Setting the vorticity to zero, by

$$\vec{\zeta} = \vec{\nabla} \times \vec{v} = 0 \tag{2.8.1}$$

the three-dimensional velocity field can be described by a single scalar function ϕ, the potential function defined by

$$\vec{v} = \vec{\nabla} \phi \tag{2.8.2}$$

reducing the knowledge of the three velocity components to the determination of a single potential function ϕ.

As seen from the preceding section, if the initial–conditions are compatible with uniform entropy, than for continuous flows, equation (2.8.1) implies that the entropy is constant over the whole flow field. Hence, for isentropic flows, the momentum equation under Crocco's form (1.5.13) becomes

$$\frac{\partial}{\partial t}(\vec{\nabla}\phi) + \vec{\nabla} H = 0 \tag{2.8.3}$$

or

$$\frac{\partial \phi}{\partial t} + H = H_0 \tag{2.8.4}$$

the constant H_0 having the same value along all the streamlines.

This equation shows that the energy equation is not independent anymore from the momentum equation, and therefore the flow will be completely determined by *initial and boundary conditions on one hand and by the knowledge of the single function ϕ on the other hand. This is a very considerable simplification indeed.*

The equation for the potential function is obtained from the continuity equation, taking into account the isentropic conditions to express the density in function of velocity and hence in function of the gradient of the potential function. We obtain the basic potential equation in conservation form:

$$\frac{\partial \rho}{\partial t} + \vec{\nabla} \cdot (\rho \vec{\nabla} \phi) = 0 \qquad (2.8.5)$$

and the relation between density and potential function obtained by introducing the definition of stagnation enthalpy in function of velocity and static enthalpy h, for a perfect gas:

$$\frac{\rho}{\rho_A} = \left(\frac{h}{h_A} \right)^{1/(\gamma-1)} = \left[\left(H_0 - \frac{\vec{v}^2}{2} - \frac{\partial \phi}{\partial t} \right) / h_A \right]^{1/(\gamma-1)} \qquad (2.8.6)$$

The subscript A refers to an arbitrary reference state, for instance the stagnation conditions $\rho_A = \rho_0$ and $h_A = H_0$.

Steady potential flows

A further simplification for steady potential flows is obtained since the potential equation reduces to, with $H = H_0 = $ constant

$$\vec{\nabla} \cdot (\rho \vec{\nabla} \phi) = 0 \qquad (2.8.7)$$

with the density given by equation (2.8.6) where h_A can be chosen equal to H_0. Hence, we have

$$\frac{\rho}{\rho_0} = \left[1 - \frac{(\vec{\nabla} \phi)^2}{2H_0} \right]^{1/(\gamma-1)} \qquad (2.8.8)$$

where ρ_0 is the stagnation density, constant throughout the whole flow field.

Both for steady and unsteady flows, the boundary condition along a solid boundary is the condition of vanishing relative velocity between flow and solid boundary in the direction n normal to the solid wall

$$v_n = \frac{\partial \phi}{\partial n} = \vec{u}_w . \vec{e}_n \qquad (2.8.9)$$

where $\vec{u}_w$ is the velocity of the solid boundary with respect to the considered system of reference.

Kutta Joukowski condition

Although the local vorticity in the flow is zero it may occur for a potential flow in non-simply connected domains that the circulation around a closed curve C becomes non-zero. This is essentially the case for lifting airfoils. To achieve a non-zero lift on the body, a circulation Γ around the airfoil is imposed. This circulation is represented by a free vortex singularity, although it originates from a vorticity production physically generated in the boundary layer. It follows that the value of Γ cannot be determined from irrotational theory and is an externally given value for a potential flow. It is also to be remembered that, with the addition of the free vortex singularity Γ, an infinity of different potential flows can be obtained, for the same incident flow conditions, each one of these solutions having another value of Γ. However, for aerodynamically shaped bodies such as airfoil profiles, a fairly good approximation of the circulation and hence the lift, may be obtained by the **Kutta–Joukowski condition**, provided that no boundary layer separation occurs in the physical flow. The Kutta–Joukowski condition states that the value of the circulation, which approximates best the real (viscous) attached flow, is obtained if the stagnation point at the downstream end of the body, is located at the trailing edge.

Supercritical airfoils

The development of supercritical airfoils, defined as having a shock-free transition from supersonic to subsonic surface velocities, is one of the most spectacular outcomes of the early developments of computational fluid dynamics. These airfoils are now of general use on civil aircrafts, allowing important savings on fuel costs due to the absence of the pressure drag produced by a shock.

Subsonic potential flows

In the subsonic range, the potential model has the same validity as the Euler model for uniform inflow conditions on a body, since the flow remains irrotational in this case.

The small disturbance approximation of the potential equation

In steady or unsteady transonic flow around wings and airfoils with thickness to chord ratios of a few percent, we can generally consider that the flow is predominantly directed along the chordwise direction, taken as the x-direction. In this case, the velocities in the transverse direction can be neglected and the potential equation reduces to the so-called **small disturbance potential equation**:

$$(1 - M_\infty^2)\frac{\partial^2 \phi}{\partial x^2} + \frac{\partial^2 \phi}{\partial x \, \partial y} + \frac{\partial^2 \phi}{\partial z^2} = \frac{1}{a^2}\left(\frac{\partial^2 \phi}{\partial t^2} + 2\frac{\partial \phi}{\partial x}\frac{\partial^2 \phi}{\partial t \, \partial x}\right) \qquad (2.8.10)$$

Historically, the steady state, two-dimensional form of this equation was used by Murman and Cole (1961) to obtain the first numerical solution for a transonic flow around an airfoil with shocks.

Linearized potential flows: singularity methods

If the flow can be considered as incompressible, the potential equation becomes a linear Laplace equation for which many standard solution techniques exist. One of these, based on the linearity of the equation is the singularity method whereby a linear superposition of known elementary flow fields such as vortex and source singularities are defined. The unknown coefficients of this linear superposition are obtained by imposing that the resultant velocity field satisfies the condition of vanishing normal velocity along solid body surfaces (in absence of wall suction or blowing).

The three-dimensional extension of the singularity method, the Panel method, has been widely used in the aeronautical industry in order to compute the three-dimensional flow field around complex configurations. The method is still in use and although extensions to handle compressibility and transonic regimes can be developed, these methods are best replaced, for high speed flows, by higher approximations such as the nonlinear potential model and the Euler equations for the inviscid flow description. We will therefore omit any detailed discussion of this approach and the interested reader will find detailed information in the specialized literature.

2.9   SUMMARY

Different flow models, involving various degrees of approximation, have been defined and illustrated by a variety of examples. With the exception of laminar flows, which can be resolved by the Navier–Stokes model with the addition of empirical information on the dependence of viscosity and heat conductivity coefficients all other models are limited by either empirical knowledge about turbulence, as for the Reynolds averaged Navier–Stokes equations, or by some approximations.

Thin shear layer models are valid if no severe viscous separated regions exist and similarly, the parabolized Navier–Stokes models for stationary formulations are limited by the presence of streamwise separation.

Inviscid flow models provide a valid approximation far from solid walls or when the influence of boundary layers can be neglected and, although the isentropic potential flow model is of questionable accuracy in transonic flows with shocks, it remains a valid and economical model for subsonic flows and for shock-free supercritical flows.

This should be kept in mind in the selection of a flow model, and the limits of validity have to be established for each family of applications, by comparison with experimental data or with computations from a higher level model.

REFERENCES

Batchelor, G.K. (1970). *An Introduction to Fluid Dynamics*. Cambridge University Press, UK.

Braza, M., Faghani, D., Persillon, H. (2001). Successive stages and the role of natural vortex dislocations in three-dimensional wake transition. *J. Fluid Mech.*, 439, 1–41.

Cebeci, T. and Bradshaw, P. (1977). *Momentum Transfer in Boundary Layers*. Hemisphere, Washington, DC.

Cebeci, T. and Bradshaw, P. (1984). *Physical and Computational Aspects of Convective Heat Transfer*. Springer Verlag, New York.

Chapman, D.R. (1979). Computational aerodynamics development and outlook. *AIAA J.*, 17, 1293–1313.

del Alamo, J.C., Jimenez, J., Zandonade, P. and Moser, R.D. (2004). Scaling of the energy spectra of turbulent channels. *J. Fluid Mech.*, 500, 135–144.

del Alamo, J.C., Jimenez, J., Zandonade, P. and Moser, R.D. (2006). Self-similar vortex clusters in the logarithmic region. *J. Fluid Mech.*, 561, 329–358.

Dinescu, C. and Hirsch, Ch. (2001). Numerically computed structure of the thermocapillary flow in cylindrical liquid bridges. *Proceedings of the ECCOMAS Computational Fluid Dynamics Conference 2001*, Swansea, Wales, UK.

Dyment, A. (1982). Vortices following two dimensional separation. In *Vortex Motion*, H.G. Hornung and E.A. Muller (Eds). Vieweg & Sohn, Braunschweig, pp. 18–30.

Flores, O. and Jimenez, J. (2006). Effect of wall-boundary disturbances on turbulent channel flows. *J. Fluid Mech.*, 566, 357–376.

Frank, S. and Schwabe, D. (1997). Temporal and spatial elements of thermocapillary convection in floating zones. *Exp. Fluids*, 23, 234–251.

Geerts, S., Hirsch, Ch., Broeckhoven, T. and Lacor, C. (2005). Validation of CFD and turbulence models for a confined double annular jet. AIAA 38th Fluid Dynamics, Plasmadynamics and Lasers Conference, AIAA Paper 2005-5894

Geurts, B.J. (2004). *Elements of Direct and Large Eddy Simulation*. R.T. Edwards, Philadelphia, PA.

Haase, W., Aupoix, B., Bunge, U. and Schwamborn, D. (Eds.) (2006). *FLOMANIA – A European Initiative on Flow Physics Modelling. Notes on Numerical Fluid Mechanics and Multidisciplinary Design (NNFM)*, Vol. 94, XI. Springer Verlag, Berlin.

Hirsch, Ch. and Deconinck, H. (1985). Through flow models in turbomachines: stream surface and passage averaged representations. In *Thermodynamics and Fluid Mechanics of Turbomachinery*, Vol. I, S.A. Ucer, P. Stow, and Ch. Hirsch, (Eds.) NATO ASI Series, Series E: Applied Sciences-No.97A, Martinus Nijhoff Publishers, Dordrechtt.

Hirsch, Ch. and Dinescu, C. (2003). A three dimensional finite volume approach of thermocapillary flows in liquid bridges at medium Prandtl numbers. In *Interfacial Fluid Dynamics and Transport Processes. Lecture Notes in Physics*, Vol. 628. Springer Verlag, Berlin. pp. 369–385.

Hoyas, S. and Jimenez, J. (2006). Scaling of velocity fluctuations in turbulent channels up to Retau = 2000. *Phys. Fluids*, 18, 117.

Jimenez, J. (2003). Computing high-Reynolds number flows: will simulations ever replace experiments. *J. Turbul.*, 4, 22.

Jimenez, J. (2004). Turbulent flows over rough walls. *Annu. Rev. Fluid Mech.*, 36, 176–196.

Jimenez, J., del Alamo, J.C. and Flores, O. (2004). The large-scale dynamics of near-wall turbulence. *J. Fluid Mech.*, 505, 179–199.

Kline, S.J. (1968). *Proceedings of the 1968 AFOSR-HTTM-STANFORD Conference on Complex Turbulent Flows*, Vol. I, II. University of Stanford, Thermosciences Division, Stanford, CA.

Kuhlmann, H.C. (1999). *Thermocapillary Convection in Models of Crystal Growth. Springer Tracts in Modern Physics, Vol. 152.* Springer Verlag, Berlin.

Le Balleur, J.C. (1983). Progres dans le calcul de l'interacton fluide parfait-fluide visqueux. *AGARD Conference Proceedings CP-351 on Viscous Effects in Turbomachines*. NATO-RTO Publications, Neuilly-sur-Seine, France.

Leschziner, M. and Drikakis, D. (2002). Turbulence modelling and turbulent-flow computation in aeronautics. *The Aeronaut. J.*, 106, 349–384.

Leypoldt, J., Kulhmann, C.H. and Rath, J.H. (2000). Three-dimensional numerical simulation of thermocapillary flows in cylindrical liquid bridges. *J. Fluid Mech.*, 414, 285–314.

Murman E.M. and Cole J.D. (1971). Calculation of Plane Steady Transonic Flows. *AIAA Journal*, 9, 114–121.

Persillon, H. and Braza, M. (1998). Physical analysis of the transition to turbulence in the wake of a circular cylinder by three-dimensional Navier-Stokes simulation. *J. Fluid Mech.*, 365, 23–88.

Pope, S. (2000). *Turbulent Flows*. Cambridge University Press, Cambridge, UK.

Pulliam, T.H. and Steger, J.L. (1978). Implicit finite difference simulations of three-dimensional flows. *16th Aerospace Sciences Meeting*, AIAA Paper 78-10. see also AIAA J., 18, 1980, 159–167.

Pulliam, T.H. and Steger, J.L. (1985). Recent Improvements in Efficiency, Accuracy and Convergence for Implicit Approximate Factorization Algorithms. *AIAA 23rd Aerospace Sciences Meeting*. AIAA Paper 85-0360.

Rogallo, R.S. and Moin, P. (1984). Numerical simulation of turbulent flows. *Ann. Rev. Fluid Mech.*, 16, 99–137.

Sagaut, P. (2001). *Large Eddy Simulation for Incompressible Flows*. Springer Verlag, Berlin.

Schlichting, H. (1971). *Boundary Layer Theory*. Mc Graw Hill, New York.

Schwabe, D. (2001). Standing waves of oscillatory thermocapillary convection in floating zones under microgravity observed in the experiment MAUS G141. *33rd COSPAR Conference*, Warzaw, Paper GO.1-0030.

Shevtsova, V. and Legros, J.C. (1998). Oscillatory convective motion in deformed liquid bridges. *Phys. Fluids*, 7(5), 1631–1634.

Steger, J.L. (1978). Implicit finite difference simulation of flows about arbitrary geometries. *AIAA J.*, 16, 679–686.

Temam, R. (1977). *Navier–Stokes Equations*. North-Holland Co, Amsterdam.

Van Dyke, M. (1982). *An Album of Fluid Motion*. The Parabolic Press, Stanford, USA.

Velten, R., Schwabe, D. and Scharmann, A. (1991). The periodic instability of thermocapillary convection in cylindrical liquid bridges. *Phys. Fluids A*, 3(2), 267–279.

Wilcox, D.C. (1998). *Turbulence Modeling for CFD*, 2nd edn. DCW Industries, California, USA. https://www.dcwindustries.com/

Williamson, C.H.K. (1988a). Defining a universal and continuous Strouhal–Reynolds number relationship for the laminar vortex shedding of a circular cylinder. *Phys. Fluids*, 31, 2742.

Williamson, C.H.K. (1988b). The existence of two stages in the transition to three-dimensionality of a cylinder wake. *Phys. Fluids*, 31, 3165–3168.

Williamson, C.H.K. (1989). Oblique and parallel mode of vortex shedding in the wake of a cylinder at low Reynolds numbers. *J. Fluid Mech.*, 206, 579–627.

Williamson, C.H.K. (1992). The natural and forced formation of spot-like vortex dislocations in the transition of a wake. *J. Fluid Mech.*, 243, 393–441.

Williamson, C.H.K. (1996a). Vortex dynamics in the cylinder wake. *Ann. Rev. Fluid Mech.*, 28, 477–539.

Williamson, C.H.K. (1996b). Three-dimensional wake transition. *J. Fluid Mech.*, 328, 345–407.

Zeng, Z., Mizuseki, H., Chen, J., Ichinoseki, K. and Kawazoe, Y. (2004). Oscillatory thermocapillary convection in liquid bridge under microgravity. *Mater. Trans.*, 45 (5), 1522–1527.

PROBLEMS

P.2.1 By developing explicitly the shear stress gradient and the momentum terms, derive the equations (2.1.7).

P.2.2 By using the definition of the shear stress tensor, equation (1.3.6), work out the full, explicit form of the Navier–Stokes equations, for non-constant viscosity coefficients, in function of velocity components, in Cartesian coordinates. Show also, that in the case of constant viscosity, the equations reduce to the projections of equation (1.3.13).

Hint: Applying equation (1.3.6), we have

$$\tau_{xx} = \frac{4}{3}\mu\frac{\partial u}{\partial x} - \frac{2}{3}\mu\left(\frac{\partial v}{\partial y} + \frac{\partial w}{\partial z}\right)$$

$$\tau_{xy} = \mu\left(\frac{\partial v}{\partial x} + \frac{\partial u}{\partial y}\right)$$

$$\tau_{xz} = \mu\left(\frac{\partial w}{\partial x} + \frac{\partial u}{\partial z}\right)$$

and the x-projection of the momentum equation becomes

$$\frac{\partial \rho u}{\partial t} + \frac{\partial}{\partial x}(\rho u^2 + p) + \frac{\partial}{\partial y}(\rho uv) + \frac{\partial}{\partial z}(\rho uw) = \frac{\partial \tau_{xx}}{\partial x} + \frac{\partial \tau_{yx}}{\partial y} + \frac{\partial \tau_{zx}}{\partial z}$$

P.2.3 Derive the energy conservation equation for a three-dimensional incompressible flow, in presence of gravity forces.

Hint: Apply equation (1.3.18) to the momentum equation (1.4.40) and multiply scalarly by v. Introducing the total energy ($H = p/\rho + v^2/2 + gz$) where z is the vertical coordinate, proof the Bernoulli equation:

$$\frac{\partial}{\partial t}\left(\frac{\vec{v}^2}{2}\right) + (\vec{v} \cdot \vec{\nabla})H = v\vec{v} \cdot \Delta\vec{v}$$

P.2.4 By working out explicitly the gradients of specific mass in function of the velocities, show that the potential equation (2.8.5) can be written in the quasilinear form, in function of the Mach numbers $M_i = v_i/c$:

$$(\delta_{ij} - M_iM_j)\frac{\partial^2 \phi}{\partial x_i \partial x_j} = \frac{1}{c^2}\left[\frac{\partial^2 \phi}{\partial t^2} + \frac{\partial}{\partial t}(\vec{\nabla}\phi)^2\right]$$

with a summation on the Cartesian subscripts i, j = 1,2,3 or x,y,z. Show that in two dimensions the steady state potential equation reduces to

$$\left(1 - \frac{u^2}{c^2}\right)\frac{\partial^2 \phi}{\partial x^2} - 2\frac{uv}{c^2}\frac{\partial^2 \phi}{\partial x \partial y} + \left(1 - \frac{v^2}{c^2}\right)\frac{\partial^2 \phi}{\partial y^2} = 0$$

Hint: Apply the isentropic laws and the energy equation (2.8.4) to derive the relations

$$dh = c^2 d\rho/\rho$$

$$\frac{c^2}{\rho}\frac{\partial \rho}{\partial t} = -\frac{\partial^2 \phi}{\partial t^2} - \frac{\partial}{\partial t}\frac{(\vec{\nabla}\phi)^2}{2}$$

$$\frac{c^2}{\rho}\vec{\nabla}\rho = -\frac{\partial(\vec{\nabla}\phi)}{\partial t} - \vec{\nabla}\frac{(\vec{\nabla}\phi)^2}{2}$$

where c is the speed of sound, and substitute into equation (2.8.5).

Chapter 3

The Mathematical Nature of the Flow Equations and Their Boundary Conditions

OBJECTIVES AND GUIDELINES

We have learned in Chapter 1, how to derive the basic equations of fluid mechanics and to recognize that any flow configuration is the outcome of a balance between the effects of convective fluxes, diffusive fluxes and external or internal sources.[1]

From the mathematical point of view, diffusive fluxes appear through second order derivative terms in space, as a consequence of the generalized Fick law, equation (1.1.8), which expresses the essence of the molecular diffusion phenomenon as a tendency to smooth out gradients. The convective fluxes, on the other hand, appear as first order derivative terms in space and express the transport properties of a flow system.

Next, we have seen in Chapter 2 how various approximation levels give rise to different mathematical models. The common property of all possible models describing flow behavior is that they constitute a *system of partial differential equations* (PDEs) in space or in space–time, which can take up various forms, but where the highest space derivatives do not exceed second order.

Because of this variety of mathematical flow models, we need tools to analyze their properties, independently of their appearance and to tell us something about the behavior of their solutions.

Let us illustrate this by the example of a stationary compressible potential flow. We have seen in Chapter 2 that we can describe this flow by two different models, namely the set of time-independent Euler equations (2.7.1), or by the potential equation (2.8.5). If we restrict ourselves to two dimensions, the steady Euler model would be described as

$$u\frac{\partial \rho}{\partial x} + v\frac{\partial \rho}{\partial y} + \rho\frac{\partial u}{\partial x} + \rho\frac{\partial v}{\partial y} = 0$$

$$u\frac{\partial u}{\partial x} + v\frac{\partial u}{\partial y} = -\frac{1}{\rho}\frac{\partial p}{\partial x} = -\frac{c^2}{\rho}\frac{\partial \rho}{\partial x}$$

---

[1] If these properties are not very clear to you at this stage, we recommend you to go back to Section 1.1.2, and to study it again with great care. Give also a special attention to Table 1.1 and to the associated discussion.

$$u\frac{\partial v}{\partial x} + v\frac{\partial v}{\partial y} = -\frac{1}{\rho}\frac{\partial p}{\partial y} = -\frac{c^2}{\rho}\frac{\partial \rho}{\partial y} \qquad (13.1)$$

with the isentropic relation between pressure and density defined by the speed of sound

$$c^2 = \left(\frac{\partial p}{\partial \rho}\right)_s \qquad (13.2)$$

On the other hand, the steady compressible potential equation model can be written as (see Problem P. 2.4)

$$\left(1 - \frac{u^2}{c^2}\right)\frac{\partial^2 \phi}{\partial x^2} - 2\frac{uv}{c^2}\frac{\partial^2 \phi}{\partial x\,\partial y} + \left(1 - \frac{v^2}{c^2}\right)\frac{\partial^2 \phi}{\partial y^2} = 0 \qquad (13.3)$$

where ϕ is the potential function defined by $u = \partial\phi/\partial x$, $v = \partial\phi/\partial y$.

For a uniform incoming flow, these two sets of PDEs describe exactly the same physics of steady irrotational flows, but their formulation appears very different. We have therefore to develop methods that allow us to analyze the mathematical properties of PDEs, independently of the different forms they might take.

Hence, we wish to answer the following basic questions:

- How do we recognize if, or when, a mathematical model describes a convection or diffusion phenomenon, taking into account that the models are generally expressed by a system of partial differential equations, where the presence of convection or diffusion terms might not be as obvious as in Section 1.1.2.
- What are the different types of physical situations, in addition to pure convection or pure diffusion that can occur and how do we recognize them.
- What are the associated initial and/or boundary conditions.

In the process of numerical discretization, it is essential indeed to be able to identify these differences, since it cannot be expected that a discretization compatible with the physics of diffusion will be valid for the physics associated to convection, since the physical properties of these two phenomena are fundamentally different, as discussed in Section 1.1.2.

In addition, there is another very important distinction we have to consider, namely the option between steady and unsteady flow descriptions. In the former case, no time derivatives will appear in the mathematical model, which will contain only space derivatives. In the latter case, time derivatives are present, next to the space derivatives, these derivatives being also assumed not to exceed second order. As we will see in the next sections, this is an important distinction since the mathematical properties of the considered equations can change significantly when introducing or removing the time dependence.

Moreover, since the laws of fluid mechanics are nonlinear (with the sole exception of incompressible potential flows), we have to consider the possibility for the mathematical nature of the equations to be flow dependent, with different properties in different regions of the flow domain. This will be the case for instance for steady transonic flows, as we will see in the following sections in more detail. But we can already have a hint as to what we are referring to hereby looking at equation (I3.3) for the simplified case of a unidirectional flow in the x-direction, with $v = 0$, reducing to $\left(1 - \frac{u^2}{c^2}\right) \frac{\partial^2 \phi}{\partial x^2} + \frac{\partial^2 \phi}{\partial y^2} = 0$. Observe that for subsonic flows, i.e. $u < c$, the two second derivative terms have the same positive sign, but they become of opposite sign for supersonic flows, when $u > c$.

This chapter will introduce you, therefore, to the analysis of a chosen mathematical model and to the derivation of the specific properties behind its set of partial differential equations. It is structured in the following way:

- We introduce first, in Section 3.1, the most simplified forms of the basic convection–diffusion equation formulation of a conservation law (1.1.9), in order to bring forward the essence of this type of equation, via reduction to two independent space–time variables. That is, to one space dimension for time-dependent problems and to two space dimensions for time-independent problems. A further simplification is obtained after a linearization, whereby the nonlinear properties are removed leading to linear convection or convection–diffusion equations.
- Section 3.2 is defining the basic methodology for the mathematical analysis of the properties of PDEs. We will introduce the fundamental concepts and distinction between **hyperbolic, parabolic and elliptic** PDEs. This will also lead us to a direct physical interpretation of these properties and to their fundamental link with the physical phenomena of convection and diffusion.
- An important extension of this analysis is developed in Section 3.3, where the concepts of characteristic surfaces and the associated very fundamental properties of **domain of dependence and domain of influence** are introduced. These are essential properties of hyperbolic and parabolic equations, and play a very critical role in many areas of CFD.
- Section 3.4 redefines the mathematical properties for time-dependent models and introduces the notion of time-like variable. In addition, this will allow us to establish the link with the conservation form of the PDEs.
- The distinction between hyperbolic, parabolic and elliptic equations is important as it will form the basis of the analysis of the number and type of initial and boundary conditions associated to the various properties of a system of PDEs. This is described and analyzed in Section 3.5.
- An Advanced Section A3.6 will focus on the introduction of **compatibility relations**, associated to characteristic surfaces. These relations play a very important role in the numerical treatment of boundary conditions for the Euler and Navier–Stokes equations in many CFD codes.

Figure 3.0.1 provides you with the guide to this chapter, for further reference, while going through the different sections.

Finally, in order to situate globally the important content of this chapter, we refer you to the general introduction and to the section entitled 'The components of a CFD simulation system', Figure I.3.1 in particular. You will notice that this chapter is the last step of Part I aimed at guiding you toward the definition of the approximation levels and the associated mathematical model to be considered as candidate for discretization.

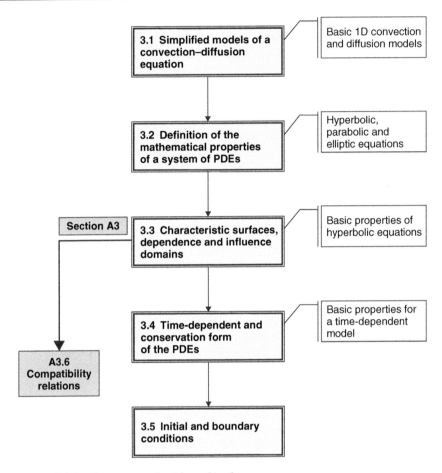

Figure 3.0.1 *Structure and guide to this chapter.*

3.1 SIMPLIFIED MODELS OF A CONVECTION–DIFFUSION EQUATION

We wish here to look at the simplest forms that can be taken by a convection–diffusion conservation equation, in order to focus on the basic mathematical properties, without being burdened by non-essential complexities, such as extra space dimensions, or specific source terms.

Hence, we limit ourselves in this section to two independent space–time variables. Consequently, for time-dependent models, we will consider only one space dimension (1D formulation); while two space variables (x and y) will be considered for time-independent models.

3.1.1 1D Convection–Diffusion Equation

Focusing first on *time-dependent* models, we consider the quasi-linear equation (1.2.10), assuming constant density and constant diffusivity coefficients, absence of source terms and one space dimension x. The resulting simplified *1D convection–diffusion equation* is written as follows:

$$\frac{\partial u}{\partial t} + a(u)\frac{\partial u}{\partial x} = \alpha \frac{\partial^2 u}{\partial x^2} \qquad (3.1.1)$$

where we have written $a(u)$ for the x-component of the convection velocity and α for the diffusivity coefficient. The convection velocity $a(u)$ can be an arbitrary function of u.

We do not specify the significance of u at this stage, but when the simplifications mentioned here are applied to the temperature equation, u will be replaced by the temperature T, while if one considers the x-projection of the momentum equation (1.3.14) in absence of pressure and external forces, u will represent the velocity and $a(u)$ will be equal to u, i.e. $a(u) = u$. In this case we obtain the well-known *Burgers equation*:

$$\frac{\partial u}{\partial t} + u\frac{\partial u}{\partial x} = \alpha \frac{\partial^2 u}{\partial x^2} \qquad (3.1.2)$$

This equation plays an important role since it contains the full convective nonlinearity of the flow equations and a large number of analytical solutions of this equation are known (see Whitham, 1974). The general importance of analytical solutions to simplified models cannot be enough underlined, as these solutions are essential to the verification and the validation process of numerical methods and solutions.

A further simplification is obtained by assuming $a(u)$ to be constant, leading to the *one-dimensional linear convection–diffusion equation*:

$$\frac{\partial u}{\partial t} + a\frac{\partial u}{\partial x} = \alpha \frac{\partial^2 u}{\partial x^2} \qquad (3.1.3)$$

Let us look now at specific particular cases.

3.1.2   Pure Convection

In absence of diffusion, $\alpha = 0$, equation (3.1.1) reduces to the *nonlinear convection equation*:

$$\frac{\partial u}{\partial t} + a(u)\frac{\partial u}{\partial x} = 0 \qquad (3.1.4)$$

A very important model is the *'inviscid' Burgers equation*:

$$\frac{\partial u}{\partial t} + u\frac{\partial u}{\partial x} = 0 \qquad (3.1.5)$$

This equation generates discontinuous solutions from an initial continuous field, in a process very similar to the process of shock creation in supersonic flows, with the advantage that this process can be fully analyzed from the corresponding analytical solution of the Burgers equation (3.1.5) (see Whitham, 1974).

The ultimate simplification is obtained by assuming the convection velocity to be a constant, leading to the ***one-dimensional linear convection equation***:

$$\frac{\partial u}{\partial t} + a\frac{\partial u}{\partial x} = 0 \qquad (3.1.6)$$

This equation describes the transport of the quantity u by the constant convection velocity a, whereby an arbitrary initial profile $u(x, t = 0) = u_0(x)$ is translated without change with the velocity a. Stated otherwise, after a time t, the quantity u is the initial profile displaced with distance $x = at$. Hence, the solution $u(x, t)$ is given by

$$u(x, t) = u_0(x - at) \qquad (3.1.7)$$

This equation can also be seen as a wave propagation equation, with the interpretation of u as a wave amplitude having a phase propagation speed equal to a. Indeed, considering a plane wave of amplitude $\hat{u}$, wavelength λ, pulsation $\omega = 2\pi f$, where f designates the wave frequency, defined by

$$u = \hat{u}e^{\mathrm{I}(kx-\omega t)} \qquad (3.1.8)$$

with wave number $k = 2\pi/\lambda$ and $\mathrm{I} = \sqrt{-1}$, this propagating wave will be solution of equation (3.1.6) if

$$\omega = ak \qquad (3.1.9)$$

This relation states that the solution (3.1.8) is a plane wave satisfying the basic relation between wavelength and frequency $f = a/\lambda$.

This equivalence between pure convection and wave propagation is critical to the understanding of the physics of convection and is central to the approach taken in this chapter.

Hence, we advise you to always remember that these two phenomena, convection and wave propagation, are two facets of the same physical properties.

3.1.3   Pure Diffusion in Time

In the absence of convection, we have a time-dependent diffusion, described by

$$\frac{\partial u}{\partial t} = \alpha\frac{\partial^2 u}{\partial x^2} \qquad (3.1.10)$$

Applied to the temperature, this equation is known as the ***heat diffusion equation of Fourier***, describing heat conduction in solids or fluids at rest.

For constant values of the diffusion constant, exact solutions to this equation are known. Looking for a solution of the type (3.1.8), we obtain after introduction in (3.1.10), the relation

$$\mathrm{I}\omega = \alpha k^2 \qquad (3.1.11)$$

leading to the solution

$$u = \hat{u}e^{Ikx}e^{-\alpha k^2 t} \tag{3.1.12}$$

This represents the behavior of a wave in space, with wave number k, exponentially damped in time, due to the diffusion coefficient α. Observe that the diffusion coefficient α has to be positive to describe a physical diffusion. A negative value of α, will represent an exponentially growing phenomenon, typical of an ***explosion***.

We recommend you to closely remember this property of the diffusion equation in time: it represents an exponentially damped spatial wave, if the diffusion coefficient is positive.

3.1.4 Pure Diffusion in Space

In two dimensions, and for a ***time-independent model***, the convection–diffusion equation (1.2.10) reduces to the Poisson equation:

$$\frac{\partial^2 u}{\partial x^2} + \frac{\partial^2 u}{\partial y^2} = q \tag{3.1.13}$$

Hence, as already pointed out in Section (1.1.2), the Laplace operator describes a pure spatial diffusion.

These various models form a basis for the development of numerical schemes and the investigation of their properties. We will extensively come back to these models in Part II of this book, where they will be used to analyze and compare a large variety of numerical schemes.

3.2 DEFINITION OF THE MATHEMATICAL PROPERTIES OF A SYSTEM OF PDES

Let us now move to the key part of this chapter, namely how do we define a general method to identify the mathematical properties of a system of PDEs?

Although we have extensively stressed in Chapter 1, the importance of the conservation form of the equations, as compared to the ***quasi-linear and non-conservative form***, it is obvious that these two formulations will describe the same physics, and hence will share the same mathematical properties. **Therefore, because of its explicit formulation, the quasi-linear form is more appropriate for the analysis of the mathematical properties and will be used in this section and throughout this chapter.**

Various approaches for the classification of PDEs can be found in the mathematical literature, connected to the possible existence of specific surfaces, called ***characteristic surfaces***, which can be defined as families of surfaces, or hypersurfaces for a general three-dimensional unsteady flow, along which certain properties remain constant or certain derivatives can become discontinuous. The discussion of these properties can be found in many textbooks, and we refer for instance to Courant and Hilbert (1962), Ames (1965) and Hildebrandt (1976) for a mathematical presentation.

However, we will give here the preference to a more 'physical' presentation of the structure of PDEs and of the associated concept of characteristic surfaces.

We offer the following definition:

A system of quasi-linear partial differential equations will be called of hyperbolic type, if its homogeneous part admits wave-like solutions.

This implies that a hyperbolic set of equations will be associated to propagating waves and that the behavior and properties of the physical system described by these equations will be dominated by wave-like phenomena.

In other words, a hyperbolic system describes convection phenomena and inversely, convection phenomena are described by hyperbolic equations.

On the other hand, if the equations admit solutions corresponding to damped waves the system will be called parabolic and if it does not admit wave-like solutions the equations are said to be elliptic. In this case, the behavior of the physical system considered is dominated by diffusion phenomena.

3.2.1 System of First Order PDEs

The systems of partial differential equations (PDE) describing the various levels of approximation discussed in the previous chapter are quasi-linear, and at most of second order. It can be shown however that any second order equation, or system of equations, can be transformed into a first order system. Although this transformation is not unique and could lead to an artificially degenerate system, it will be considered that an appropriate transformation has been defined such that the system of first order represents correctly the second order equations.

Hence, the following steps define the procedure to identify the properties of a mathematical model:

Step 1: Write the system of PDEs describing the mathematical model under the form of a system of first order PDEs

Suppose we have n unknown variables u^j, in an $(m + 1)$-dimensional space x^k, we can group all the variables u^j in an $(n \times 1)$ vector column U and write the system of first order PDEs under the general form:

$$A^k \frac{\partial U}{\partial x^k} = Q \quad k = 1, \ldots, m + 1$$

$$U \stackrel{\Delta}{=} \begin{vmatrix} u^1 \\ \cdot \\ \cdot \\ u^n \end{vmatrix} \tag{3.2.1}$$

where A^k are $(n \times n)$ matrices and Q is a column vector of the non-homogeneous source terms. The matrices A^k and Q can depend on x^k and U, but not on the derivatives of U.

Note that we always assume that a summation is performed on repeated indices. This is called the Einstein summation convention.

Step 2: Consider a plane wave solution of amplitude $\hat{U}$ in the space of the independent variables $\vec{x}$ with components x^k $(k = 1, \ldots, m + 1)$, defined by

$$U = \hat{U} e^{\mathrm{I}(\vec{n} \cdot \vec{x})} = \hat{U} e^{\mathrm{I}(n_k x^k)} \tag{3.2.2}$$

where $1 = \sqrt{-1}$ and $\vec{n}$ is a vector in the m-dimensional space of the independent variables x^k.

Step 3: Introduce this solution in the homogenous part of the system (3.2.1) and find the values of $\vec{n}$ satisfying the resulting equation

The homogenous part of equation (3.2.1) is written as

$$A^k \frac{\partial U}{\partial x^k} = 0 \quad k = 1, \ldots, m+1 \tag{3.2.3}$$

and the function (3.2.2) is a solution of this system of equations if the homogenous algebraic system of equations:

$$[A^k n_k] \hat{U} = 0 \tag{3.2.4}$$

has non-vanishing solutions for the amplitude $\hat{U}$. This will be the case if and only if the determinant of the matrix $A^k n_k$ vanishes.

Step 4: Find the n solutions of the equation

$$\det \left| A^k n_k \right| = 0 \tag{3.2.5}$$

Equation (3.2.5) defines a condition on the normals $\vec{n}$. This equation can have at most n solutions, and for each of these normals $\vec{n}^\alpha$, the system (3.2.5) has a non-trivial solution.

The system is said to be *hyperbolic in the space x^k* if all the n characteristic normals $\vec{n}^\alpha$ are real and if the solutions of the n associated systems of equations (3.2.5) are linearly independent. If all the characteristics are complex, the system is said to be *elliptic in the space x^k* and if some are real and other complex the system is considered as *hybrid*.

If the matrix $(A^k n_k)$ is not of rank n, i.e. there are less than n real characteristic normals then the system is said to be *parabolic in the space x^k*.

This will occur, for instance, when at least one of the variables, say u^1 has derivatives with respect to one coordinate, say x^1, missing. This implies that the components $A^1_{i1} = 0$ for all equations i.

Example E.3.2.1: System of two first order equations in two dimensions

The above-mentioned properties can be illustrated in a two-dimensional space x, y with the system:

$$a \frac{\partial u}{\partial x} + c \frac{\partial v}{\partial y} = f_1$$

$$b \frac{\partial v}{\partial x} + c \frac{\partial u}{\partial y} = f_2 \tag{E.3.2.1}$$

or in matrix form:

$$\begin{vmatrix} a & 0 \\ 0 & b \end{vmatrix} \frac{\partial}{\partial x} \begin{vmatrix} u \\ v \end{vmatrix} + \begin{vmatrix} 0 & c \\ d & 0 \end{vmatrix} \frac{\partial}{\partial y} \begin{vmatrix} u \\ v \end{vmatrix} = \begin{vmatrix} f_1 \\ f_2 \end{vmatrix} \tag{E.3.2.2}$$

Hence, with $x^1 = x$, $x^2 = y$:

$$U = \begin{vmatrix} u \\ v \end{vmatrix} \quad A^1 = \begin{vmatrix} a & 0 \\ 0 & b \end{vmatrix} \quad A^2 = \begin{vmatrix} 0 & c \\ d & 0 \end{vmatrix} \quad Q = \begin{vmatrix} f_1 \\ f_2 \end{vmatrix} \tag{E.3.2.3}$$

Equation (E.3.2.2) is written as

$$A^1 \frac{\partial U}{\partial x} + A^2 \frac{\partial U}{\partial y} = Q \tag{E.3.2.4}$$

The determinant equation (3.2.5) becomes, after division by n_y supposed to be different from zero

$$|A^1 n_x + A^2 n_y| = n_y \begin{vmatrix} an_x/n_y & c \\ d & bn_x/n_y \end{vmatrix} = 0 \tag{E.3.2.5}$$

leading to the conditions for the characteristic normals

$$\left(\frac{n_x}{n_y} \right)^2 = \frac{cd}{ab} \tag{E.3.2.6}$$

If $cd/ab > 0$, the solutions n_x/n_y are real and the system is **hyperbolic in the space** *(x, y)*, for instance $a = b = 1$; $c = d = 1$ with vanishing right-hand side leading to the well-known wave equations, obtained after elimination of the variable v:

$$\frac{\partial^2 u}{\partial x^2} - \frac{\partial^2 u}{\partial y^2} = 0 \tag{E.3.2.7}$$

If $cd/ab < 0$, equation (E.3.2.6) has no real solutions and the system is **elliptic in the space (x, y)**. For instance, $a = b = 1$; $c = -d = -1$ and vanishing right-hand side leading to the Laplace equation which is the standard form of elliptic equations and describes diffusion phenomena.

Finally, if $b = 0$, there is only one characteristic normal $n_y = 0$ and the system is **parabolic**. For instance, with $a = 1$, $b = 0$, $c = -d = -1$ and $f_1 = 0$, $f_2 = v$, we obtain the standard form for a parabolic equation:

$$\frac{\partial u}{\partial x} = \frac{\partial^2 u}{\partial y^2} \tag{E.3.2.8}$$

This is recognizable by the fact that the equation presents a combination of first and second order derivatives.

Example E.3.2.2: The stationary Euler equations in two dimensions

This system is defined by equations (13.1), which can be written in the matrix form (3.2.3) as follows:

$$\begin{vmatrix} u & \rho & 0 \\ c^2/\rho & u & 0 \\ 0 & 0 & u \end{vmatrix} \frac{\partial}{\partial x} \begin{vmatrix} \rho \\ u \\ v \end{vmatrix} + \begin{vmatrix} v & 0 & \rho \\ 0 & v & 0 \\ c^2/\rho & 0 & v \end{vmatrix} \frac{\partial}{\partial y} \begin{vmatrix} \rho \\ u \\ v \end{vmatrix} = 0 \tag{E.3.2.9}$$

Introducing the vector

$$U = \begin{vmatrix} \rho \\ u \\ v \end{vmatrix}$$

(E.3.2.10)

the system is written in the condensed form:

$$A^1 \frac{\partial U}{\partial x} + A^2 \frac{\partial U}{\partial y} = 0$$

(E.3.2.11)

The three characteristic normals $\vec{n}$, are obtained as the solutions of equation (3.2.5), with $\lambda = n_x/n_y$

$$\begin{vmatrix} u\lambda + v & \rho\lambda & \rho \\ \lambda c^2/\rho & u\lambda + v & 0 \\ c^2/\rho & 0 & u\lambda + v \end{vmatrix} = 0$$

(E.3.2.12)

Working out the determinant (E.3.2.12) leads to the solution

$$\lambda^{(1)} = -\frac{v}{u}$$

(E.3.2.13)

and the two solutions of the quadratic equation

$$(u^2 - c^2)\lambda^2 + 2\lambda uv + (v^2 - c^2) = 0$$

(E.3.2.14)

$$\lambda^{(2),(3)} = \frac{-uv \pm c\sqrt{u^2 + v^2 - c^2}}{u^2 - c^2}$$

(E.3.2.15)

The first solution is always real, and the two others are real if the flow is supersonic, since equation (E.3.2.15) can be written as follows, after introduction of the Mach number:

$$M = \frac{\sqrt{u^2 + v^2}}{c}$$

(E.3.2.16)

$$\lambda^{(2),(3)} = \frac{-uv \pm c^2\sqrt{M^2 - 1}}{u^2 - c^2}$$

(E.3.2.17)

Hence, the stationary Euler equations are **hyperbolic** in (x, y) for supersonic flows. For subsonic flows, the second and third solutions are complex conjugate, and hence the system is a hybrid mix **elliptic–hyperbolic**, since one solution is always real. At the sonic velocity $M = 1$, the two solutions $\lambda^{(2)} = \lambda^{(3)}$ and the system is **parabolic**.

3.2.2 Partial Differential Equation of Second Order

These different concepts can also be applied to the classical example of the quasi-linear partial differential equation of second order:

$$a\frac{\partial^2 \phi}{\partial x^2} + 2b\frac{\partial^2 \phi}{\partial x\, \partial y} + c\frac{\partial^2 \phi}{\partial y^2} = 0 \tag{3.2.6}$$

where a, b and c can depend on the coordinates x, y, the function ϕ and its first derivatives. This equation can be written as a system of first order equations, after introduction of the variables u and v defined by

$$u = \frac{\partial \phi}{\partial x} \qquad v = \frac{\partial \phi}{\partial y} \tag{3.2.7}$$

Equation (3.2.6) is then equivalent to the following system:

$$a\frac{\partial u}{\partial x} + 2b\frac{\partial u}{\partial y} + c\frac{\partial v}{\partial y} = 0$$
$$\frac{\partial v}{\partial x} - \frac{\partial u}{\partial y} = 0 \tag{3.2.8}$$

which can be written in matrix form:

$$\begin{vmatrix} a & 0 \\ 0 & 1 \end{vmatrix}\frac{\partial}{\partial x}\begin{vmatrix} u \\ v \end{vmatrix} + \begin{vmatrix} 2b & c \\ -1 & 0 \end{vmatrix}\frac{\partial}{\partial y}\begin{vmatrix} u \\ v \end{vmatrix} = 0 \tag{3.2.9}$$

Introducing the vector U and the matrices A^1 and A^2

$$U = \begin{vmatrix} u \\ v \end{vmatrix} \qquad A^1 = \begin{vmatrix} a & 0 \\ 0 & 1 \end{vmatrix} \qquad A^2 = \begin{vmatrix} 2b & c \\ -1 & 0 \end{vmatrix} \tag{3.2.10}$$

we obtain the form (3.2.3) and equation (3.2.5) leads to

$$\begin{vmatrix} an_x + 2bn_y & cn_y \\ -n_y & n_x \end{vmatrix} = 0 \tag{3.2.11}$$

Hence, from the roots of

$$a\left(\frac{n_x}{n_y}\right)^2 + 2b\left(\frac{n_x}{n_y}\right) + c = 0 \tag{3.2.12}$$

the well-known conditions defining the type of the second order quasi-linear partial differential equation (3.2.6) are obtained.

The solutions of equation (3.2.12) are given by

$$\frac{n_x}{n_y} = \frac{-b \pm \sqrt{b^2 - ac}}{a} \tag{3.2.13}$$

The solution (3.2.2) will represent a true wave if n_y is real for all real values of n_x. Therefore, if $(b^2 - ac)$ is positive there are two wave-like solutions, and the equation is **hyperbolic**, while for $(b^2 - ac) < 0$ the two solutions are complex conjugate and the equation is **elliptic**. When $(b^2 - ac) = 0$ the two solutions are reduced to one single direction $n_x/n_y = -b/2$ and the equation is **parabolic**.

Example E.3.2.3: Stationary potential equation

An interesting example is provided by the stationary potential flow equation in two dimensions x, y, defined by equation (I3.3) where c designates the speed of sound:

$$\left(1 - \frac{u^2}{c^2}\right)\frac{\partial^2\phi}{\partial x^2} - \frac{2uv}{c^2}\frac{\partial^2\phi}{\partial x\,\partial y} + \left(1 - \frac{v^2}{c^2}\right)\frac{\partial^2\phi}{\partial y^2} = 0 \tag{E.3.2.18}$$

with

$$a = \left(1 - \frac{u^2}{c^2}\right) \quad b = -\frac{uv}{c^2} \quad c = \left(1 - \frac{v^2}{c^2}\right) \tag{E.3.2.19}$$

we can write the potential equation under the form (3.2.6). In this particular case the discriminant $(b^2 - ac)$ becomes, introducing the Mach number M

$$b^2 - ac = \frac{u^2 + v^2}{c^2} - 1 = M^2 - 1 \tag{E.3.2.20}$$

and hence the stationary potential equation is **elliptic for subsonic flows and hyperbolic for supersonic flows**. Along the sonic line $M = 1$, the equation is **parabolic**.

The solution (3.2.13) takes the following form for the two-dimensional potential equation:

$$\frac{n_x}{n_y} = \frac{uv \pm c^2\sqrt{M^2 - 1}}{c^2 - u^2} \tag{E.3.2.21}$$

This mixed nature of the potential equation has been a great challenge for the numerical computation of transonic flows since the transition line between the subsonic and the supersonic regions is part of the solution. An additional complication arises from the presence of shock waves which are discontinuities of the potential derivatives and which can arise in the supersonic regions. The particular problems of transonic potential flow with shocks and their numerical treatment will be discussed in Volume II.

3.3   HYPERBOLIC AND PARABOLIC EQUATIONS: CHARACTERISTIC SURFACES AND DOMAIN OF DEPENDENCE

Parabolic and hyperbolic equations play an important role in CFD, due to their specific properties associated to propagation, or convection, phenomena. They are recognized by the existence of real characteristic normals, solutions of equation (3.2.5). Each of these normals $\vec{n}^\alpha$ defines therefore a surface, called ***characteristic surface***, to which it is orthogonal.

We will show here the very important consequences of these properties, as they have a significant effect on the whole process of discretization in CFD.

3.3.1   Characteristic Surfaces

Indeed, if we define a surface $S(x^k)=0$, in the $(m+1)$-dimensional space of the independent variables x^k, the normal to this surface is defined by the gradient of the function $S(x^k)$, as

$$\vec{n} = \vec{\nabla}S \tag{3.3.1}$$

See Figure 3.3.1 for a surface in the 3D space.

What is the significance of this characteristic surface in terms of wave propagation, referring to the plane wave solution (3.2.2)?

If equation (3.3.1) is introduced in the plane wave (3.2.2), a general representation is defined as

$$U = \hat{U}e^{I(\vec{x}\cdot\vec{\nabla}S)} = \hat{U}e^{I(x^k S_k)} \quad \text{with } S_k \overset{\Delta}{=} \frac{\partial S}{\partial x^k} \tag{3.3.2}$$

If we consider the tangent plane to the surface $S(x^k)=0$, defined by

$$S(x^k) = S(0) + \vec{x}\cdot\vec{\nabla}S \equiv S(0) + x^k\frac{\partial S}{\partial x^k} \equiv S(0) + x^k n_k \tag{3.3.3}$$

we observe that along the constant values of the phase of the wave $\Phi \overset{\Delta}{=} \vec{x}\cdot\vec{\nabla}S$, the quantity U is constant.

Hence, we can consider that following the direction of the normal $\vec{n}$ the quantity U is propagating at a constant value.

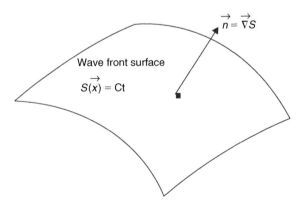

Figure 3.3.1   *Wavefront surface and associated normal.*

The surface S is called a **wavefront surface**, defined as the surface separating the space domain already influenced by the propagating quantity U from the points not yet reached by the wave.

Observe that in the general case of n unknown flow quantities u^j, we have n characteristic surfaces, for a pure hyperbolic problem.

In a two-dimensional space the characteristic surface reduces to a characteristic line. The properties U are transported along the line $S(x,y)=0$ and the vectors tangent to the characteristic line are obtained by expressing that along the wavefront:

$$dS = \vec{\nabla}S \cdot d\vec{x} = \frac{\partial S}{\partial x}dx + \frac{\partial S}{\partial y}dy = 0 \tag{3.3.4}$$

Hence, the direction of the characteristic line in two dimensions is given by

$$\frac{dy}{dx} = -\frac{S_x}{S_y} = -\frac{n_x}{n_y} \tag{3.3.5}$$

In two dimensions, there are two characteristic directions for a hyperbolic equation, such as the two solutions (3.2.13). Hence out of each point in the (x, y) domain, two characteristics can be defined, along which two quantities propagate. As we have as many unknowns, at each point the solution can be obtained from the characteristic-related quantities that have propagated from the boundary to the point P. See Figure 3.3.2 for an example.

A numerical method based on these properties has been applied in the past, particularly for two-dimensional problems, and is known as the **Method of Characteristics**.

Example E.3.3.1: The small disturbance potential equation

Referring to Example E.3.2.3, the small disturbance potential equation is obtained for a horizontal incoming flow of Mach number M_∞ in case the vertical velocity

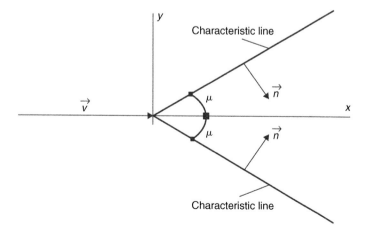

Figure E.3.3.1 *Characteristics for two-dimensional potential equation.*

component can be considered as negligible, for instance for a flow along a thin body. The stationary potential equation reduces to the form, assuming that the horizontal velocity component remains constant,

$$(1 - M_\infty^2)\frac{\partial^2 \phi}{\partial x^2} + \frac{\partial^2 \phi}{\partial y^2} = 0 \qquad\qquad\text{(E.3.3.1)}$$

where M_∞ is the upstream Mach number.

The solutions (E.3.2.21) reduce to the following, considering a supersonic flow, for which the solutions are real

$$\frac{n_y}{n_x} = \mp\sqrt{M_\infty^2 - 1} \qquad\qquad\text{(E.3.3.2)}$$

hereby defining the normals to the two characteristics for supersonic flows. Their directions are obtained from equation (3.3.5) as

$$\frac{dy}{dx} = \pm 1/\sqrt{M_\infty^2 - 1} = \pm\tan\mu \qquad\qquad\text{(E.3.3.3)}$$

Referring to Figure E.3.3.1, it can be seen that these characteristics are identical to the **Mach lines** at an angle μ to the direction of the velocity, with

$$\sin\mu = 1/M_\infty \qquad\qquad\text{(E.3.3.4)}$$

3.3.2 Domain of Dependence: Zone of Influence

The propagation property of hyperbolic problems has important consequences with regard to the way the information is transmitted through the flow region.

Considering Figure 3.3.2, where Γ is a boundary line distinct from a characteristic, the solution U along a segment AB of Γ will propagate in the flow domain along the characteristics issued from AB. For a two-dimensional problem in the variables x, y, there are two characteristics if the problem is hyperbolic.

Hence, the two characteristics out of A and B limit the region PAB, which determines the solution at point P. The region PAB is called the **region of dependence** of point P, since the characteristics out of any point C outside AB will never reach point P. On the other hand, the region downstream of P, and located between the characteristics, defines the zone where the solution is influenced by the function value in P. This region is called the **zone of influence** of P.

3.3.2.1 *Parabolic problems*

For parabolic problems the two characteristics are identical, (Figure 3.3.3) and the region of dependence of point P reduces to the segment BP. The zone of influence of P, on the other hand, is the whole region right of the characteristic BP.

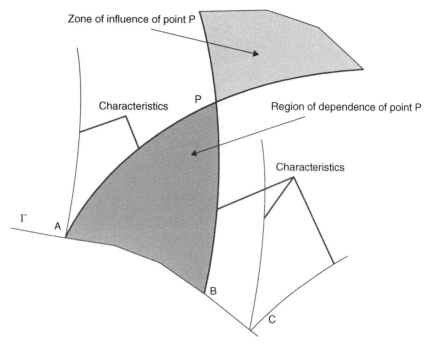

Figure 3.3.2   *Region of dependence and zone of influence of point P for a hyperbolic problem with two characteristics per point.*

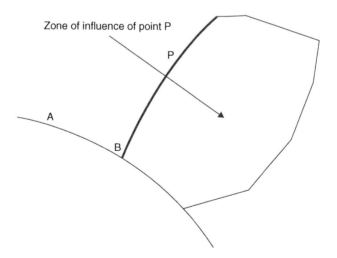

Figure 3.3.3   *Region of dependence and zone of influence of point P for a parabolic problem with one characteristic per point.*

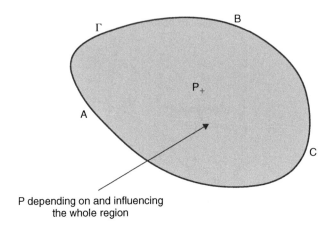

P depending on and influencing
the whole region

Figure 3.3.4 *Region surrounding P in an elliptic problem.*

3.3.2.2 *Elliptic problems*

In this case, there are no real characteristics and the solution in a point P depends on all the surrounding points, since the physical problem is of the diffusive type. Inversely, the whole boundary ACB surrounding P, is influenced by point P (Figure 3.3.4). Hence, one can consider that the dependence region is identical to the zone of influence, both of them being equal to the whole of the flow domain.

3.4 TIME-DEPENDENT AND CONSERVATION FORM OF THE PDES

Although we have not specified in the previous sections the exact definition of all the x^k coordinates, the examples in these sections all refer to pure space variables, implying a system of PDEs describing a steady flow model, that is not involving time.

However, many flow systems are time dependent, as we have seen from the derivation of the conservation laws in the first chapter, and therefore we need to analyze the properties of the PDEs, in presence of time derivatives.

In addition, an important alternative way to simulate steady flows is obtained by discretizing the time-dependent equations in time and space and letting the time evolution take us to the steady state solution. This is called a '*time-marching*' approach.

Although the time-marching approach will converge to the same solution as obtained from the stationary model, the properties of the time-dependent PDEs can be significantly different from the properties of their associated spatial parts.

Let us illustrate this more precisely, by considering the system of *n* first order partial differential equations *in time and space*, written in conservation form, following the derivations seen in Chapter 1, which we rewrite here for convenience

$$\frac{\partial U}{\partial t} + \vec{\nabla} \cdot \vec{F} = Q \tag{3.4.1}$$

where U is the column vector of the n unknown functions u^j, in the m-dimensional space x^k ($k = 1, \ldots, m$). Here m is the space dimension, i.e. $m = 1, 2$ or 3 according to the number of space coordinates considered.

The equivalent algebraic form (with a summation convention on repeated superscripts or subscripts), is written as the system of n equations, for line i:

$$\frac{\partial u^i}{\partial t} + \left(\frac{\partial F^k}{\partial x^k} \right)^i = Q^i \quad (i = 1, \ldots, n) \tag{3.4.2}$$

We consider here that this *system is of first order in space*; i.e., that the flux vector $\vec{F}$ does not contain space derivatives of U.

As stated in Section 3.2, the analysis of the properties of this system relies on the quasi-linear form, as opposed to the conservation form. This can be obtained after introduction of the **Jacobian matrices** A^k, where

$$A^k_{ij} = \frac{\partial F^k_i}{\partial u^j} \tag{3.4.3}$$

are the Jacobian matrix element of the flux F^k_i with respect to the variable u^j.

In a compact notation, the Jacobian matrix is defined by

$$\overline{\overline{A}}^k \stackrel{\wedge}{=} \frac{\partial F^k}{\partial U} \tag{3.4.4}$$

With this definition, the time-dependent quasi-linear form of PDEs, derived from the conservation form, becomes

$$\frac{\partial U}{\partial t} + A^k \frac{\partial U}{\partial x^k} = Q \tag{3.4.5}$$

We see here that the stationary model, obtained when the time derivative is equal to zero for a time-independent vector U, reduces to equation (3.2.1), with $k = 1, \ldots, m$.

3.4.1   Plane Wave Solutions with Time Variable

We have now to adapt the procedures defined in Section 3.2 to the time-dependent model (3.4.5).

To establish this link we observe that the general form (3.2.1) reduces to equation (3.4.6), if one variable, say x^{m+1}, is singled out and the corresponding Jacobian matrix A^{m+1} is taken as the unit matrix. We call this variable *time like* and take $x^{m+1} = t$, with the conditions:

$$x^{m+1} \equiv t \quad \text{and} \quad A^{m+1} = \overline{\overline{I}} \, (\text{unit matrix}) \tag{3.4.6}$$

With the introduction of the time variable, a plane wave solution can be written in a more conventional way:

$$U = \hat{U} e^{I(\vec{\kappa} \cdot \vec{x} - \omega t)} \tag{3.4.7}$$

The vector $\vec{\kappa}$ is called the **wave number vector** and its magnitude is the number of periods or wavelengths over a distance 2π in the direction of the vector $\vec{\kappa}$ and ω is the wave pulsation. This simple wave is a solution of the linear convection equation:

$$\frac{\partial U}{\partial t} + (\vec{a} \cdot \vec{\nabla})U = 0 \tag{3.4.8}$$

where

$$\vec{a} \cdot \vec{\kappa} = \omega \tag{3.4.9}$$

If the direction of $\vec{\kappa}$ is the propagation direction of the wave, then the phase velocity in the direction of propagation is given by:

$$a = \frac{\omega}{\kappa} \tag{3.4.10}$$

One has also the following relation between frequency ν, wavelength λ, and the variables κ and ω

$$\lambda = \frac{2\pi}{\kappa} \quad \omega = 2\pi\nu \quad \lambda\nu = a \tag{3.4.11}$$

typical of plane waves.

The general solution (3.3.2) becomes, with the explicit time variable $x^{m+1} = t$:

$$U = \hat{U}\, e^{I(\vec{\kappa}\cdot\vec{\nabla}S + tS_t)} \tag{3.4.12}$$

identifying the frequency ω of the wave as

$$\omega = -\frac{\partial S}{\partial t} = -n_t \tag{3.4.13}$$

and the wave number vector $\vec{\kappa}$ is defined by

$$\vec{\kappa} = \vec{\nabla}S = \vec{n} \tag{3.4.14}$$

In this notation $\vec{\nabla}S$ is the normal to the intersection of the characteristic or wavefront surface $S(\vec{x}, t)$ with the hypersurfaces $t = $ constant. Hence, the normals are defined here in the m-dimensional space of the space variables, while the normals $\vec{n}$ defined in Section 3.3.1 (equation (3.3.1)) are normals to the wavefront surfaces in the $(m + 1)$-dimensional space $x^1, \ldots, x^{m+1}$. Observe also that the wave number $\vec{\kappa}$, in the space $(x^1, \ldots, x^m)$ is normal to the characteristic subsurfaces $S(\vec{x}, t)$ at constant t.

In order for this wave to be a solution of the homogeneous system

$$\frac{\partial U}{\partial t} + A^k \frac{\partial U}{\partial x^k} = 0 \quad k = 1, \ldots, m \tag{3.4.15}$$

S has to satisfy the equation

$$\det\left|\frac{\partial S}{\partial t} + A^k \frac{\partial S}{\partial x^k}\right| = 0 \tag{3.4.16}$$

or

$$\det \left| -\omega \delta_{ij} + \kappa_k A_{ij}^k \right| = 0 \tag{3.4.17}$$

This important condition actually defines the wave pulsations ω, as the eigenvalues of the matrix $K_{ij} = \kappa_k A_{ij}^k$.

If we group the m matrices A^k in a vector $\vec{A}$ of dimensions m, $\vec{A}\,(A^1, \ldots, A^m)$, we can write the matrix K as a scalar product:

$$K = \vec{A} \cdot \vec{\kappa} \tag{3.4.18}$$

The system is said to be *hyperbolic in space and time* if all the n-eigenvalues of the matrix K *are real* and if the n associated solutions of equation (3.4.15) are linearly independent.

If all the eigenvalues are complex, the system is said to be *elliptic in space in time* and if some are real and other complex the system is considered as *hybrid*.

If the eigenvalues are purely imaginary, then the system is said to be *parabolic in space in time*.

If some eigenvalues are real and other purely imaginary, the system is said to be *parabolic–hyperbolic in space in time*.

Referring to equation (3.4.10), the propagation speed associated with a characteristic frequency ω is obtained as the eigenvalue of the matrix $\hat{K}$:

$$\hat{K} \equiv \frac{K}{\kappa} = \vec{A} \cdot \frac{\vec{\kappa}}{\kappa} = \vec{A} \cdot \vec{e}_\kappa \tag{3.4.19}$$

where $\vec{e}_\kappa$ is the unit vector in the direction $\vec{\kappa}$.

Hence, for each wave number vector $\vec{\kappa}$, a perturbation in the surface normal to $\vec{\kappa}$ propagates in the direction of $\vec{\kappa}$ with phase velocity $\vec{a}$ and frequency ω equal to the eigenvalue of the matrix $K = A^k \kappa_k$. The associated characteristic speeds $a_{(\alpha)}$ are obtained as solutions of the eigenvalue problem:

$$\det \left| -a_{(\alpha)} \delta_{ij} + \hat{K}_{ij} \right| = 0 \tag{3.4.20}$$

For an elliptic system, the eigenvalues are complex and the solution takes the form, for an eigenvalue $\omega_1 = \xi + I\eta$:

$$U_1 = \hat{U}_1 \, e^{-I\xi t} e^{I\vec{\kappa}\cdot\vec{x}} e^{+\eta t} \tag{3.4.21}$$

Since the coefficients of the Jacobian matrices A^k are considered to be real, each eigenvalue $\omega_1 = \xi + I\eta$ is associated to a complex conjugate eigenvalue $\omega_2 = \xi - I\eta$, leading to a solution of the form:

$$U_2 = \hat{U}_2 \, e^{-I\xi t} e^{I\vec{\kappa}\cdot\vec{x}} e^{-\eta t} \tag{3.4.22}$$

Hence, according to the sign of the imaginary part of the eigenvalue, one of the two solutions U_1 or U_2 will be damped in time, while the other will amplified.

Example E.3.4.1: Time-dependent shallow water equations in one dimension

The one-dimensional form of the time-dependent shallow water equations can be written as

$$\frac{\partial h}{\partial t} + u\frac{\partial h}{\partial x} + h\frac{\partial u}{\partial x} = 0$$

$$\frac{\partial u}{\partial t} + u\frac{\partial u}{\partial x} + g\frac{\partial u}{\partial x} = 0 \qquad\qquad (E.3.4.1)$$

where h represents the water height, g is the gravity acceleration and u the horizontal velocity.

In matrix form, we have

$$\frac{\partial}{\partial t}\begin{vmatrix} h \\ u \end{vmatrix} + \begin{vmatrix} u & h \\ g & u \end{vmatrix}\frac{\partial}{\partial x}\begin{vmatrix} h \\ u \end{vmatrix} = 0 \qquad\qquad (E.3.4.2)$$

The two characteristic velocities $a_{1,2}$ are obtained from equation (3.4.20) as solutions of

$$\det\begin{vmatrix} -a+u & h \\ g & -a+u \end{vmatrix} = 0 \qquad\qquad (E.3.4.3)$$

or

$$a_{1,2} = u \pm \sqrt{gh} \qquad\qquad (E.3.4.4)$$

Since these eigenvalues are always real, the system is always hyperbolic in (x, t).

Example E.3.4.2: Time-dependent Euler equations in one dimension

Let us consider, as a second example, the one-dimensional form of the time-dependent Euler equations, for isentropic flows.

This is obtained by adding a time derivative of density and x-velocity component to the first two equations of the system (I3.1), leading to

$$\frac{\partial \rho}{\partial t} + u\frac{\partial \rho}{\partial x} + \rho\frac{\partial u}{\partial x} = 0$$

$$\frac{\partial u}{\partial t} + u\frac{\partial u}{\partial x} + \frac{c^2}{\rho}\frac{\partial \rho}{\partial x} = 0 \qquad\qquad (E.3.4.5)$$

or in matrix form, defining the vector U and the matrix A:

$$U = \begin{vmatrix} \rho \\ u \end{vmatrix} \quad A = \begin{vmatrix} u & \rho \\ c^2/\rho & u \end{vmatrix} \qquad\qquad (E.3.4.6)$$

as

$$\frac{\partial U}{\partial t} + A\frac{\partial U}{\partial x} = 0 \tag{E.3.4.7}$$

Equation (3.4.7) becomes, in one space dimension x:

$$U = \hat{U}e^{I(\kappa x - \omega t)} \tag{E.3.4.8}$$

and the condition (3.4.17) reduces to the eigenvalue equation:

$$\det \begin{vmatrix} u\kappa - \omega & \rho\kappa \\ \kappa c^2/\rho & u\kappa - \omega \end{vmatrix} = 0 \tag{E.3.4.9}$$

The two eigenvalues

$$a_1 = \omega_1/\kappa = u + c \quad \text{and} \quad a_2 = \omega_2/\kappa = u - c \tag{E.3.4.10}$$

are real, for *all* values of the velocity u, and hence the system is always hyperbolic in space and time.

This is an extremely important property comparing to the analysis of Example E.3.2.2 (confirmed by Example E.3.2.3) where it is shown that the steady isentropic Euler equations are elliptic in the space (x, y) for subsonic velocities and hyperbolic in the space (x, y) for supersonic velocities.

Here, in space and time, the inviscid isentropic equations are always hyperbolic independently of the subsonic or supersonic state of the flow. As a consequence, the same numerical algorithms can be applied for all flow velocities.

On the other hand, dealing with the steady state equations, the numerical algorithms will have to adapt to the flow regime, as the mathematical nature of the system of equations is changing when passing from subsonic to supersonic, or inversely.

This is the main reason for the very widespread choice of the time-dependent form of the conservation laws as basis for the numerical discretization, even for the simulation of steady flows.

In this approach, we solve the flow equations in time until a numerical steady state is reached, while the numerical transient is defined in such a way as to reach the steady state as fast as possible, through different numerical acceleration techniques, such as local time steps, multigrid. These techniques will be introduced later on.

Take very carefully notice of the above fundamental property, as it conditions a wide area of CFD applications.

It is indeed considered as a major advantage to be in a position to develop numerical algorithms, without having to bother about the subsonic or the supersonic state of the flow.

The following example demonstrates that this property is not restricted to one space dimension. It is indeed valid for all dimensions, and is shown here for two space dimensions, considering the time-dependent form of the system (I3.1).

Example E.3.4.3: Time-dependent Euler equations in two space dimensions

The time-dependent form of system (I3.1) for the two-dimensional isentropic Euler equations is written as

$$\frac{\partial \rho}{\partial t} + u\frac{\partial \rho}{\partial x} + v\frac{\partial \rho}{\partial y} + \rho\frac{\partial u}{\partial x} + \rho\frac{\partial v}{\partial y} = 0$$

$$\frac{\partial u}{\partial t} + u\frac{\partial u}{\partial x} + v\frac{\partial u}{\partial y} + \frac{c^2}{\rho}\frac{\partial \rho}{\partial x} = 0 \qquad\qquad (E.3.4.11)$$

$$\frac{\partial v}{\partial t} + u\frac{\partial v}{\partial x} + v\frac{\partial v}{\partial y} + \frac{c^2}{\rho}\frac{\partial \rho}{\partial y} = 0$$

With the same definitions as in Example E.3.2.2, the system is written in the matrix form (3.4.15):

$$\frac{\partial U}{\partial t} + A^1\frac{\partial U}{\partial x} + A^2\frac{\partial U}{\partial y} = 0 \qquad\qquad (E.3.4.12)$$

and the eigenvalue equation (3.4.17) becomes

$$\det \begin{vmatrix} \vec{u}\cdot\vec{\kappa} - \omega & \rho\kappa_x & \rho\kappa_y \\ c^2\kappa_x/\rho & \vec{u}\cdot\vec{\kappa} - \omega & 0 \\ c^2\kappa_y/\rho & 0 & \vec{u}\cdot\vec{\kappa} - \omega \end{vmatrix} = 0 \qquad\qquad (E.3.4.13)$$

The three solutions are given by the following three pulsations:

$$\omega_1 = \vec{u}\cdot\vec{\kappa}$$

$$\omega_2 = \vec{u}\cdot\vec{\kappa} + c\kappa \qquad\qquad (E.3.4.14)$$

$$\omega_3 = \vec{u}\cdot\vec{\kappa} - c\kappa$$

and as they are always real, the system (E.3.4.11) is *always* hyperbolic in space and time.

3.4.2 Characteristics in a One-Dimensional Space

The particular case of the one-dimensional space allows obtaining an important property of hyperbolic systems, typical for time-dependent first order equations, such as the Euler equations.

For a general nonlinear equation of the form:

$$\frac{\partial u}{\partial t} + a(u)\frac{\partial u}{\partial x} = 0 \qquad\qquad (3.4.23)$$

the above formalism, see in particular equation (3.3.5), defines the characteristics in the space–time domain $(x - t)$, as

$$\frac{dt}{dx} = -\frac{S_x}{S_t} = -\frac{n_x}{n_t} = \frac{1}{a(u)} \qquad\qquad (3.4.24)$$

Hence, we can state that the general solution of the nonlinear convection equation (3.4.23) is given by the characteristic property:

$$u = \text{Const} \quad \text{along} \quad \frac{dx}{dt} = a(u) \tag{3.4.25}$$

3.4.3 Nonlinear Definitions

The n-eigenvalues of the matrix (3.4.18) define the n **dispersion relations** for the frequencies $\omega_{(\alpha)}$:

$$\omega_{(\alpha)} = \lambda_{(\alpha)}(\vec{\kappa}(\vec{x}, t)) \tag{3.4.26}$$

Note that a nonlinear wave can be only written under the form (3.4.7) with:

$$\vec{\kappa} = \vec{\kappa}(\vec{x}, t)$$

$$\omega = \omega(\vec{\kappa}(\vec{x}, t)) \tag{3.4.27}$$

and the same local definition of wave number and frequency, under certain conditions.

Indeed, when introduced in equation (3.4.7), the above nonlinear solution gives the following contributions:

$$\frac{\partial U}{\partial t} = IU\left[-\omega - t\frac{\partial \omega}{\partial t} + \vec{x} \cdot \frac{\partial \vec{\kappa}}{\partial t}\right] = IU\left[-\omega + (\vec{x} - t\vec{\nabla}_{\kappa}\omega) \cdot \frac{\partial \vec{\kappa}}{\partial t}\right]$$

$$\frac{\partial U}{\partial x^k} = IU\left[\kappa_k - t\frac{\partial \omega}{\partial x^k} + \vec{x} \cdot \frac{\partial \vec{\kappa}}{\partial x^k}\right] = IU\left[\kappa_k + (\vec{x} - t\vec{\nabla}_{\kappa}\omega) \cdot \frac{\partial \vec{\kappa}}{\partial x^k}\right] \tag{3.4.28}$$

The derivative of the frequency with respect to the wave number component κ_j is the j-component of the **group velocity** of the wave

$$v_j^{(G)} = \frac{\partial \omega}{\partial \kappa_j} \tag{3.4.29}$$

or, in condensed notation,

$$\vec{v}^{(G)} = \vec{\nabla}_{\kappa}\omega \tag{3.4.30}$$

The terms within the round brackets in equation (3.4.28) will vanish for an observer moving with the group velocity, i.e. for

$$\frac{\vec{x}}{t} = \vec{\nabla}_{\kappa}\omega \tag{3.4.31}$$

Note that the group velocity is the velocity at which the wave energy propagates. The interested reader will find an extensive discussion of nonlinear waves in Whitham (1974).

3.5 INITIAL AND BOUNDARY CONDITIONS

The information necessary for the initial and boundary conditions to be imposed with a given system of differential equations in order to have a well-posed problem can be gained from the preceding considerations. Well-posedness in the sense of Hadamard is established if the solution depends in continuous manner on the initial and boundary conditions. That is, a small perturbation of these conditions should give rise to a small variation of the solution at any point of the domain at a finite distance from the boundaries.

Two types of problems are considered with regard to the time variable t: an initial value problem or Cauchy problem where the solution is given in the subspace $t = 0$ as $U = U(\vec{x}, t = 0)$ and is to be determined at subsequent values of t. If the subspace $t = 0$ is bounded by some surface $\Omega(\vec{x})$ then additional conditions have to be imposed along that surface at all values of t and this defines an *initial–boundary value problem*.

A solution of the system of first order partial differential equations can be written as a superposition of wave-like solutions of the type corresponding to the n-eigenvalues of the matrix K:

$$U = \sum_{\alpha=1}^{n} \hat{U}_\alpha e^{I(\vec{\kappa}\cdot\vec{x} - \omega_{(\alpha)} t)} \tag{3.5.1}$$

where the summation extends over all the eigenvalues $\omega_{(\alpha)}$, U being the column containing the unknowns u^j.

If N_r and N_c denote, respectively, the number of real and complex eigenvalues, considered to be of multiplicity one, with $n = N_r + N_c$, it is seen from equation (3.4.21) that the complex eigenvalues will generate amplified modes for $\eta > 0$. If such a mode is allowed the problem will not be well-posed according to Hadamard. Therefore, the number of initial and boundary conditions to be imposed have to be selected as to make sure that such modes are neither generated nor allowed. *If the problem is hyperbolic, $N_c = 0$ and $N_r = n$ and since no amplified modes are generated, n initial conditions for the Cauchy problem have to be given in order to determine completely the solution.*

That is, as many conditions as unknowns have be given at $t = 0$. On the other hand, if the problem is elliptic or hybrid, there will be $N_c/2$ amplified modes, and hence only $N_r + N_c/2$ conditions are allowed. Since this number is lower than n, the pure initial value or Cauchy problem is not well posed for non-hyperbolic problems and only boundary value problems will be well-posed in this case. The inverse is also true, a pure boundary value problem is ill-posed for a hyperbolic problem.

For an elliptic system $N_r = 0$ and the number of boundary conditions to be imposed at every point of the boundary is equal to half the order of the system. For instance, for a second order hyperbolic equation two conditions will have to be fixed along the initial Cauchy line, while for a second order elliptic equation one condition will have to be given along the boundaries.

For hyperbolic problems, the n boundary conditions have to be distributed along the boundaries at all values of t, according to the direction of propagation of the corresponding waves. If a wave number $\vec{\kappa}$ is taken in the direction of the interior normal vector $\vec{n}$, then the corresponding wave, whose phase velocity is obtained as eigenvalue of the matrix $(A^k n_k)$, will propagate information inside the domain if this velocity is positive. Hence, the number of conditions to be imposed for the hyperbolic initial–boundary value problem at a given point of the boundary is equal to the number of positive eigenvalues of the matrix $A^k n_k$ at that point. Refer to Chapter 11, section 11.4.1.3 for a detailed application to the 2D system of time-dependent Euler equations. The total number of conditions remains obviously equal to the total number of eigenvalues that is to the order of the system.

For hybrid problems, the conclusions are the same for the real characteristics, but $N_c/2$ additional conditions have to be imposed everywhere along the boundary. Note also that, next to the number of boundary conditions to impose, the nature of these conditions can also be important in order to avoid ill-posed conditions along the boundaries.

Parabolic problems in t and $\bar{x}$ define initial–boundary value problems. Hence, the solution is to be defined at $t = 0$, that is for an order n, n conditions have to be given at $t = 0$. Along the boundaries for all times, $n/2$ boundary conditions have to be imposed. This is the case for the standard form of parabolic equations $\partial_t u = L(u)$, where $L(u)$ is a second order elliptic operator in space.

A more complex parabolic structure arises in boundary layer theory, where the equations are of the form $u_{yy} = L(u)$ where $L(u)$ is a hyperbolic first order operator in the space x, y, z, with y the coordinate normal to the wall. This leads to complex mixed parabolic–hyperbolic phenomena in space for three-dimensional boundary layer calculations. Some of these aspects are described in Krause (1973) and Dwyer (1981). The boundary conditions are of the initial value type for the hyperbolic components and of boundary value nature for the elliptic parts of the system.

The whole system of Navier–Stokes equations is essentially parabolic in time and space or parabolic–hyperbolic while the steady state part is elliptic–hyperbolic due to the hyperbolic character of the continuity equation considered for a known velocity field. On the other hand, in absence of viscosity and heat conduction effects, the system of time-dependent Euler equations is purely hyperbolic in space and time.

The various approximations to the Navier–Stokes equations discussed in this chapter have evidently different mathematical properties. For instance, in the thin shear layer approximation or the boundary layer approximations, the diffusive effect of viscosity is neglected in all directions except in the directions normal to the wall. Therefore, the resulting equations remain parabolic in time and in the direction normal to the surface, while the behavior of the system will be purely hyperbolic in the other two directions and time. The global property remains, however, parabolic although the local behavior of the system is modified compared to the full Navier–Stokes model. This leads to important consequences for the numerical simulation of three-dimensional boundary layers, see Dwyer (1981) for a review of these issues.

From mathematical point of view, no general, global existence theorems for the non-stationary compressible Navier–Stokes equation with a defined set of boundary and initial conditions can be defined. Some partial, local existence theorems have been obtained (see Temam, 1977; Solonnikov and Kazlikhov, 1981), for both the Cauchy problems, i.e. given distributions of density, velocity and temperature at time $t = 0$, and for initial–boundary value problems where the flow parameters are given at $t = 0$ and boundary conditions are imposed at all times for velocity and temperature on the boundary of the flow domain. These investigations do not lead presently to practical rules for the establishment of boundary conditions and, therefore, case-by-case considerations have to be used in function of the type of the equations and of physical properties of the system. In general, the elliptic time-independent problems will impose the values of the flow variables (Dirichlet conditions) or their derivatives (Neumann conditions) on the boundaries of the flow domain. From physical considerations, for fluid conditions far away from the molecular free motions (Knudsen numbers below 10^{-2}) the velocity should be continuous at the material boundaries. This leads to the well-known no-slip conditions for the velocity for the Navier–Stokes equations.

For the temperature one of the following three conditions can be used, T_w being the wall temperature:

$$T = T_w \qquad\qquad \text{fixed wall temperature} - \text{Dirichlet condition}$$

$$k\frac{\partial T}{\partial n} = q \qquad\qquad \text{fixed heat flux} - \text{Neumann condition}$$

$$k\frac{\partial T}{\partial n} = \alpha(T - T_w) \quad \text{heat flux proportional to local heat transfer} \\ \text{(mixed condition)}$$

For other elliptic equations such as the subsonic potential or stream function equations the choice will be made on the basis of the physical interpretation of these functions and this will be discussed in the appropriate chapters.

Inviscid flow equations, being first order, allow only one condition on the velocity, namely that the velocity component normal to the wall is fixed by the mass transfer through that wall, while the tangential component will have to be determined from the computation and will generally be different from the non-slip value, since slip velocities are allowed. For free surfaces, the physical conditions are chosen on the basis of continuity of the normal and tangential stresses and of the statement that the free boundary is a stream surface.

A.3.6 ALTERNATIVE DEFINITION: COMPATIBILITY RELATIONS

An alternative definition of characteristic surfaces and hyberbolicity can be obtained because wavefront surfaces carry certain properties and that a complete description of the physical system is obtained when all these properties are known. This implies that the original system of equations, *if hyperbolic*, can be reformulated as differential relations written along the wavefront or the characteristic surfaces only. Hence, the following definition can be given: A *characteristic surface* $S(x^1,\ldots,x^m) = 0$ will exist, if the first order system of equation (3.2.1) can be transformed, through a linear combination of the form, where the l^i are n arbitrary coefficients:

$$l^i A^k_{ij}\frac{\partial u^j}{\partial x^k} = l^i Q_i \quad i,j = 1,\ldots,n \ \ k = 1,\ldots,m \tag{3.6.1}$$

into an equivalent system containing only derivatives along the surface S. Along the surface $S(x^1,\ldots,x^m) = 0$ one of the coordinates can be eliminated, for instance x^m, by expressing

$$dS = \frac{\partial S}{\partial x^k}\,dx^k = 0 \tag{3.6.2}$$

or, along the surface S:

$$\frac{\partial x^m}{\partial x^k}\Big|_S = -\frac{\partial S/\partial x^k}{\partial S/\partial x^m} = -\frac{n_k}{n_m} \tag{3.6.3}$$

where the components of the normal vector $\vec{n} = \vec{\nabla}S$ are introduced. Hence, we can define derivatives $\overline{\partial/\partial x}^k$ along the surface S, in the following way. For any variable u^j the partial derivative $\overline{\partial/\partial x}^k$ along the surface S is given by

$$\frac{\overline{\partial}}{\partial x^k} = \frac{\partial}{\partial x^k} + \left(\frac{\partial x^m}{\partial x^k}\right)\Big|_S \frac{\partial}{\partial x^m} = \frac{\partial}{\partial x^k} - \left(\frac{n_k}{n_m}\right)\frac{\partial}{\partial x^m} \quad k = 1, ..., m \tag{3.6.4}$$

Note that for the variable x^m the surface derivative is zero, i.e. $\overline{\partial/\partial x}^m = 0$. Introducing this relation into the linear combination (3.6.1), leads to

$$l^i A^k_{ij}\left[\frac{\overline{\partial}}{\partial x^k} + \left(\frac{n_k}{n_m}\right)\frac{\partial}{\partial x^m}\right]u^j = l^i Q_i \tag{3.6.5}$$

The summation over k extends from $k = 1$ to $k = m$. A characteristic surface will exist for any u^j if the system is reduced to the form (3.6.1).

This is satisfied if the surface S obeys the relations, for any u^j

$$l^i A^k_{ij} n_k = 0 \tag{3.6.6}$$

$$l^i A^k_{ij}\frac{\partial S}{\partial x^k} = 0 \quad i,j = 1,\ldots,n \;\; k = 1,\ldots,m \tag{3.6.7}$$

The conditions for this homogeneous system, in the l^i unknowns, to be compatible are the vanishing of the determinant of the coefficients, leading to the condition (3.2.5). For each solution $\vec{n}^{(\alpha)}$ of equation (3.2.5), the system (3.6.7) has a non-trivial solution for the coefficients $l^{i(\alpha)}$, up to an arbitrary scale factor. The $l^{i(\alpha)}$ coefficients can be grouped into a $(n \times 1)$ line vector $\vec{l}^{(\alpha)}$. The system is said to be *hyperbolic* if all the n characteristic normals $\vec{n}^{(\alpha)} = \vec{\nabla}S^{(\alpha)}$ are real and if the n vectors $\vec{l}^{(\alpha)}$ $(\alpha = 1,\ldots,n)$, solutions of the n systems of equations (3.6.7) are linearly independent.

A.3.6.1   Compatibility Relations

The reduced form (3.6.1) expresses that the basic equations can be combined to a form containing only derivatives confined to a $(n-1)$-dimensional space. That is, the system of equations, if hyperbolic, can be considered as describing phenomena occurring on hypersurfaces $S^{(\alpha)}$. Indeed, defining a set of n vectors $\vec{Z}_j$ in the m-dimensional space with components $Z^k_j, j = 1,\ldots,n; k = 1,\ldots,m$ by the relations

$$Z^k_j = l^i A^k_{ij} \tag{3.6.8}$$

equation (3.6.1) can be rewritten as

$$Z_j^k \frac{\partial u^j}{\partial x^k} = l^i Q_i \tag{3.6.9}$$

The vectors $\vec{Z}_j$ define n *characteristic directions* of which $(n-1)$ are independent. The operators $(Z_j^k \partial_k)$ are the derivatives in the direction of the vector $\vec{Z}_j$. Hence defining

$$d_j \overset{\Delta}{=} Z_j^k \frac{\partial}{\partial x^k} = \vec{Z}_j.\vec{\nabla} \tag{3.6.10}$$

in the m-dimensional space, the transformed equation (3.6.1) can be written as a derivatives along the vectors $\vec{Z}_j$:

$$d_j u^j \equiv \vec{Z}_j \cdot \vec{\nabla} u^j = l^i Q_i \tag{3.6.11}$$

The above equation is known as the **compatibility relation**, and represent an alternative formulation to the system (3.2.1). Condition (3.6.7) expresses that all the $\vec{Z}_j$ vectors lie in the characteristic surface whose normal is $\vec{n}$. Indeed, equation (3.6.7) becomes, with the introduction of $\vec{Z}_j$

$$\vec{Z}_j \cdot \vec{n} = Z_j^k n_k = 0 \quad \text{for all } j = 1, \ldots, n \tag{3.6.12}$$

These phenomena correspond to propagating wavefronts, as seen earlier, and it can be shown (see for instance Whitham, 1974) that the characteristic surfaces can also contain discontinuities of the normal derivatives $\partial u^j / \partial n$, satisfying

$$A_{ij}^k n_k \left[\frac{\partial u^j}{\partial n} \right] = 0 \tag{3.6.13}$$

where

$$\left[\frac{\partial u^j}{\partial n} \right] = \left(\frac{\partial u^j}{\partial n} \right)_+ - \left(\frac{\partial u^j}{\partial n} \right)_- \tag{3.6.14}$$

is the jump, over the surface S, of the normal derivatives of the solution u^j. This relation can also be written:

$$n_k \left[\frac{\partial F_i^k}{\partial n} \right] = 0 \tag{3.6.15}$$

where $\partial/\partial n$ is the normal derivative to the surface S. The corresponding vectors $\vec{l}$ are obtained from equation (3.6.6), written as

$$l^i (\delta_{ij} n_t + A_{ij}^k n_k) = 0 \quad k = 1, \ldots, m-1 \tag{3.6.16}$$

or

$$l^i K_{ij} = \lambda_{(\alpha)} l^i \delta_{ij} \quad i,j = 1,\ldots,n \ \ \alpha = 1,\ldots,n \tag{3.6.17}$$

Hence, the vectors $\vec{l}^{(\alpha)}$ of components $l^{i(\alpha)}$ are the left eigenvectors of the matrix $\mathbf{K}$ corresponding to the eigenvalues $\lambda_{(\alpha)}$. If the n eigenvectors $\vec{l}^{(\alpha)}$ are linearly independent, the system will be hyperbolic. If the n eigenvectors $\vec{l}^{(\alpha)}$ are grouped in a matrix L^{-1}, where each row contains the components of an eigenvector $\vec{l}^{(\alpha)}$, i.e.

$$\left(L^{-1}\right)^{i\alpha} = l^{i(\alpha)} \tag{3.6.18}$$

we obtain from the eigenvector equation (3.4.13) that the matrix L diagonalizes the matrix K:

$$L^{-1} K L = \Lambda \tag{3.6.19}$$

where Λ is the diagonal matrix containing the eigenvalues $\lambda_{(\alpha)}$:

$$\Lambda = \begin{vmatrix} \lambda_{(1)} & & & 0 \\ & \lambda_{(2)} & & \\ & & \cdot & \\ & & & \cdot \\ 0 & & & \lambda_{(n)} \end{vmatrix} \tag{3.6.20}$$

It is also interesting to notice that, within the same assumptions, the intensities of the propagating disturbances are the right eigenvectors $r^{(\alpha)}$ of K, since equation (3.6.13) can be written as

$$K_{ij} r^j = \lambda_{(\alpha)} r^j \delta_{ij} \tag{3.6.21}$$

Example E.3.6.1: Small disturbance potential equation

The two vectors $\vec{l}$ associated to the two characteristic normals (E.3.3.2) are obtained from the system (3.6.6), which is written here as

$$(l^1, l^2) \begin{vmatrix} (1 - M_\infty^2)n_x & n_y \\ -n_y & n_x \end{vmatrix} = 0 \tag{E.3.6.1}$$

where the ratio $\lambda = n_y/n_x$ is defined as the solutions (E.3.3.2). Choosing $l^1 = 1$, the system (E.3.6.1) has the solution:

$$l^1 = 1$$

$$l^2 = \sqrt{(M_\infty^2 - 1)} \tag{E.3.6.2}$$

The two characteristic directions $\vec{Z}_j$ defined by equation (3.6.8), become here

$$\vec{Z}_1 = (1 - M_\infty^2, \lambda)$$

$$\vec{Z}_2 = (-\lambda, 1) \tag{E.3.6.3}$$

Observe that these two directions are parallel to each other, since $\lambda^2 = (M_\infty^2 - 1)$ and that their common direction is the characteristic line of Figure E.3.3.1, making the angle μ with the x-direction, since $\lambda = \cos\mu/\sin\mu$. This can also be seen from a direct verification of equation (3.6.12), with the vector of the characteristic normal defined by the components $\vec{n} \equiv (1, \lambda)$, which indicates that the Z-directions are orthogonal to the normals n. The compatibility relation (3.6.9) or (3.6.11) becomes here

$$\left[(1 - M_\infty^2)\frac{\partial}{\partial x} + \lambda\frac{\partial}{\partial y} \right] u + \left[-\lambda\frac{\partial}{\partial x} + \frac{\partial}{\partial y} \right] v = 0 \tag{E.3.6.4}$$

or

$$\cot\mu \left[\cos\mu\frac{\partial}{\partial x} \mp \sin\mu\frac{\partial}{\partial y} \right] u \pm \left[\cos\mu\frac{\partial}{\partial x} \mp \sin\mu\frac{\partial}{\partial y} \right] v = 0 \tag{E.3.6.5}$$

For constant Mach angles μ, the compatibility relation expresses the property that the velocity component along one characteristic ($u\cos\mu \pm v\sin\mu$) is conserved along the other characteristic.

CONCLUSIONS AND MAIN TOPICS TO REMEMBER

This chapter has introduced the general methodology for the determination of the mathematical properties of a system of PDEs, defining a selected approximation model of the flow system.

It is of crucial importance to identify and distinguish when a system of PDEs describes convection, or diffusion, or mixed phenomena.

The main topics to remember are:

- A system of quasi-linear partial differential equations is hyperbolic if its homogeneous part admits wave-like solutions. In other words, a hyperbolic system describes convection phenomena and inversely, convection phenomena are described by hyperbolic equations.
- If the equations admit damped wave solutions the system will be called parabolic.
- If the equations do not admit wave-like solutions, they are elliptic, and the behavior of the physical system considered is dominated by diffusion phenomena.
- Characteristic surfaces represent an important property of hyperbolic systems, as they are associated to the propagation phenomena.
- The notions of domain of dependence and zone of influence are very essential properties of PDEs and should be kept in mind in all cases.

- Steady inviscid flows have the property of being elliptic in the subsonic range and hyperbolic in the supersonic regions.
- Time-marching methods for steady state problems offer the great advantage that their properties in space and time are independent of the Mach number. In the inviscid case, they are always hyperbolic in space and time, for all values of the Mach number.

REFERENCES

Ames, W.F. (1965). *Nonlinear Partial Differential Equations in Engineering*. Academic Press, New York.

Courant, R. and Hilbert, D. (1962). *Methods of Mathematical Physics*. Interscience, New York.

Dwyer, H.A. (1981). Some aspects of three dimensional laminar boundary layers. Annu. Rev. Fluid Mech., 13, 217–229.

Hildebrand, F.B. (1976). *Advanced Calculus for Applications*. Prentice Hall, Englewood Cliffs, NJ.

Krause, E. (1973). Numerical treatment of boundary layer problems. In *Advances in Numerical Fluid Dynamics*, AGARD-LS-64.

Solonnikov, J.A. and Kazhikhov, A.V. (1981). Existence theorems for the equations of motion of a compressible viscous fluid. Annu. Rev. Fluid Mech., 13, 79–95.

Temam, R. (1977). *Navier–Stokes Equations*. North-Holland Publ. Co, Amsterdam.

Whitham, G.B. (1974). *Linear and Nonlinear Waves*. J. Wiley & Sons, New York.

PROBLEMS

P.3.1 Show that the system of Cauchy–Riemann equations

$$\frac{\partial u}{\partial x} + \frac{\partial v}{\partial y} = 0$$

$$\frac{\partial v}{\partial x} - \frac{\partial u}{\partial y} = 0$$

is of elliptic nature.

P.3.2 Consider the two-dimensional stationary shallow water equations describing the spatial distribution of the height h of the free water surface in a stream with velocity components u and v. They can be written in the following form, where g is the earth's gravity acceleration:

$$u\frac{\partial h}{\partial x} + v\frac{\partial h}{\partial y} + h\frac{\partial u}{\partial x} + h\frac{\partial v}{\partial y} = 0$$

$$u\frac{\partial u}{\partial x} + v\frac{\partial u}{\partial y} + g\frac{\partial h}{\partial x} = 0$$

$$u\frac{\partial v}{\partial x} + v\frac{\partial v}{\partial y} + g\frac{\partial h}{\partial y} = 0$$

(a) Introduce the vector

$$U = \begin{vmatrix} h \\ u \\ v \end{vmatrix}$$

and write the system (E.3.2.8) under the form (3.2.3).

(b) Obtain the three characteristic normals $\bar{n}$, as solutions of equation (3.2.5), by defining $\lambda = n_x/n_y$. Show that we obtain

$$\lambda^{(1)} = -\frac{v}{u}$$

and the two solutions of the quadratic equation:

$$(u^2 - gh)\lambda^2 + 2\lambda uv + (v^2 - gh) = 0$$

$$\lambda^{(2),(3)} = \frac{-uv \pm \sqrt{gh}\sqrt{u^2 + v^2 - gh}}{u^2 - gh}$$

Observe that $\sqrt{gh}$ plays the role of a sonic, critical, velocity and the system is hyperbolic for supercritical velocities $\bar{v}^2 = u^2 + v^2 > gh$.

P.3.3 Show that the one-dimensional Navier–Stokes equation without pressure gradient (known as the 'viscous' Burger's equation)

$$\frac{\partial u}{\partial t} + u\frac{\partial u}{\partial x} = \alpha\frac{\partial^2 u}{\partial x^2}$$

is parabolic in x, t.

Hint: Write the equation as a system, introducing $v = \partial u/\partial x$ as second variable and apply equation (3.2.5) to show that the matrix is not of rank 2.

P.3.4 Consider the one-dimensional Euler equations:

$$\frac{\partial \rho}{\partial t} + u\frac{\partial \rho}{\partial x} + \rho\frac{\partial u}{\partial x} = 0$$

$$\frac{\partial u}{\partial t} + u\frac{\partial u}{\partial x} + \frac{1}{\rho}\frac{\partial p}{\partial x} = 0$$

$$\frac{\partial H}{\partial t} + u\frac{\partial H}{\partial x} = -\frac{1}{\rho}\frac{\partial p}{\partial t}$$

Introduce the isentropic assumption $\partial p/\partial \rho = c^2$, with c the speed of sound and replace the third equation by an equation on the pressure, by applying the perfect gas laws and the definition of H. Obtain the equation:

$$\frac{\partial p}{\partial t} + u\frac{\partial p}{\partial x} + \rho c^2\frac{\partial u}{\partial x} = 0$$

Write the system in matrix form for the variable vector:

$$\begin{vmatrix} \rho \\ u \\ p \end{vmatrix}$$

Show that the system is hyperbolic and has the eigenvalues u, $u+c$ and $u-c$.
Obtain the left and right eigenvectors.

P.3.5 Consider the system

$$\frac{\partial u}{\partial t} + \frac{1}{2}\frac{\partial u}{\partial x} + \frac{\partial v}{\partial x} = 0$$

$$\frac{\partial v}{\partial t} + \frac{\partial u}{\partial x} + \frac{1}{2}\frac{\partial v}{\partial x} = 0$$

(a) Write the system in matrix form (3.4.5) and obtain the matrix A:

$$\frac{\partial U}{\partial t} + A\frac{\partial U}{\partial x} = 0 \quad \text{with } U = \begin{vmatrix} u \\ v \end{vmatrix}$$

(b) Find the eigenvalues of A and show that the system is hyperbolic.
(c) Derive the left and right eigenvectors and obtain the matrix L which diagonalizes A. Explain why the left and right eigenvectors are identical.
(d) Obtain the characteristic variables and the compatibility relations.
Hint: The eigenvalues of A are $3/2$ and $-1/2$. The matrix L has the form:

$$L = \frac{1}{\sqrt{2}}\begin{vmatrix} 1 & 1 \\ 1 & -1 \end{vmatrix}$$

The characteristic variables are

$$(u+v)\sqrt{2} \quad \text{and} \quad (u-v)\sqrt{2}$$

The compatibility relations are

$$\frac{\partial(u+v)}{\partial t} + \frac{3}{2}\frac{\partial(u+v)}{\partial x} = 0$$

$$\frac{\partial(u-v)}{\partial t} - \frac{1}{2}\frac{\partial(u-v)}{\partial x} = 0$$

P.3.6 Consider the steady potential equation (E.3.2.18) for supersonic flows, $M > 1$. From Example E.3.2.3, it is known that the equation is hyperbolic. Obtain the two vectors $\vec{l}^{(\alpha)}$, $\alpha = 1, 2$ associated to the two characteristic normal directions $\vec{n}^{(\alpha)}$, solutions of equation (3.2.11). Show that the two characteristics form an angle $\pm\mu$ with the velocity vector v, with $\sin\mu = 1/M$. The angle μ is called the Mach angle.
Hint: Define β as the angle of the velocity vector by $\cos\beta = u/|\vec{v}|$, $\sin\beta = v/|\vec{v}|$. Setting $n_x = 1$ shows that $n_y = -\cotan(\beta \pm \mu)$. Selecting $l_1 = 1$, obtain $l_2 = -cn_y$ from equation (3.2.2).

P.3.7 Show that for the transformation leading to equation (3.4.5) $A_{ij}^m = \delta_{ij}$, the characteristic directions (3.6.8) become

$$Z_j^m = l^i \delta_{ij}$$

$$Z_j^k = l^i A_{ij}^k \quad k = 1, \ldots, m-1$$

Form the $(n \times m)$ matrix Z and note that the last line (j being the column index) is formed by the vector $\vec{l}$.

Show that the orthogonality condition (3.6.12) is equivalent to equation (3.6.17) and that we have, for a wave number vector $\vec{\kappa}$:

$$Z_j^k \kappa_k = \lambda l^i \delta_{ij}$$

P.3.8 Referring to the one-dimensional shallow water equations treated in Example E.3.4.1, find the eigenvectors $\vec{l}^{(\alpha)}$, $\alpha = 1, 2$ as well as the characteristic vectors $\vec{Z}_1$ and $\vec{Z}_2$. Derive also the compatibility relations (3.6.11).
Hint: Show that the left eigenvectors are proportional to $\vec{l}[1, \pm \sqrt{h/g}]$ and that the characteristic vectors have the components:

$$\vec{Z}_1 = \begin{vmatrix} u \pm \sqrt{gh} \\ 1 \end{vmatrix} \quad \vec{Z}_2 = \pm \frac{\sqrt{h}}{g} \begin{vmatrix} u \pm \sqrt{gh} \\ 1 \end{vmatrix}$$

Show that we obtain

$$\frac{\partial h}{\partial t} + (u \pm \sqrt{gh})\frac{\partial h}{\partial x} \pm \frac{\sqrt{h}}{g} \left\{ \frac{\partial u}{\partial t} + (u \mp \sqrt{gh})\frac{\partial u}{\partial x} \right\} = 0$$

where the upper signs are to be taken together for the first and the lower signs for the second compatibility relation.

P.3.9 Consider equation (3.2.6) and write also the compatibility relation (3.6.11) after having defined the characteristic directions $\vec{Z}_j$ according to equation (3.6.8).
Hint: Obtain

$$\vec{Z}_1 = \begin{vmatrix} a \\ -2b + cn_y \end{vmatrix} \quad \vec{Z}_2 = \begin{vmatrix} -cn_y \\ c \end{vmatrix}$$

and verify equation (3.6.12). Note also that $\vec{Z}_1$ and $\vec{Z}_2$ are in the same direction, since they are both orthogonal to $\vec{n}$. Show by a direct calculation that the vector product of $\vec{Z}_1$ and $\vec{Z}_2$ is indeed zero. Referring to the general form of equation (3.2.6), obtain the compatibility relation

$$a\frac{\partial u}{\partial x} + (cn_y - 2b)\frac{\partial u}{\partial y} - cn_y\frac{\partial v}{\partial x} + c\frac{\partial v}{\partial y} = 0$$

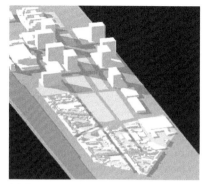

(a) Computer (CAD) model of an urban environment.

(b) Computer model (CAD) of an airplane.

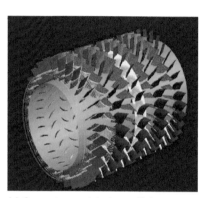

(c) Computer model of a multistage compressor.

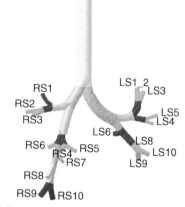

(d) Computer model of a section of pulmonary branches in the lung. From Van Ertbruggen et al. (2005).

(e) Computer model of the liquid hydrogen pump of the VULCAIN engine of the European launcher ARIANE 5.

(f) Computer model (CAD) of an industrial valve system.

Plate I.1.2 *Examples of computer (CAD) models to initiate the steps toward a CFD simulation.*

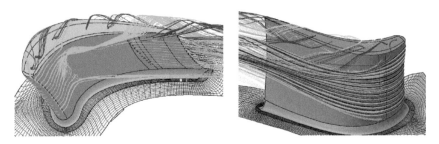

Plate I.1.3 *Simulation of the interaction between the cooling flow and the main external gas flow around a cooled turbine blade. Courtesy NUMECA Int. and KHI.*

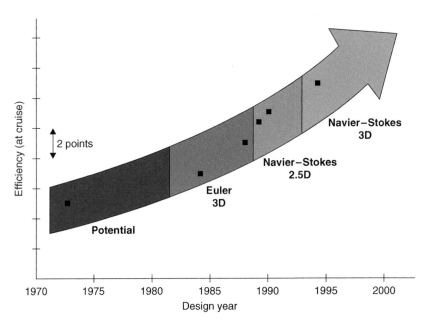

Plate I.1.4 *Impact of CFD on SNECMA fan performance, over a period of 30 years. From Escuret et al. (1998).*

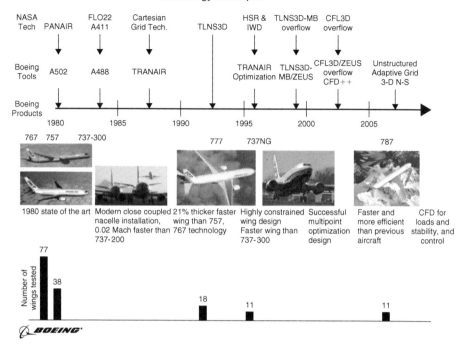

Plate I.1.6a *Evolution of the CFD tools over the last 40 years at Boeing, with an indication of the influence of CFD on the reduction of the number of wing tests. Courtesy Enabling Technology and Research Organization, Boeing Commercial Airplanes.*

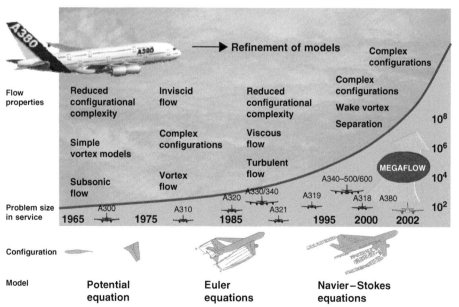

Plate I.1.6b *Evolution of the CFD tools over the last 40 years at Airbus, with an indication of the evolution of the applied models. From Becker (2003).*

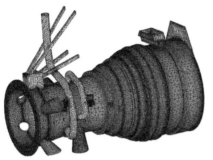

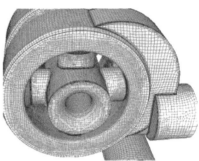

Unstructured tetrahedral grid for an engine.
From ICEM-CFD.

Unstructured hexahedral grid for an oil valve.
HEXPRESS mesh. Courtesy NUMECA Int.

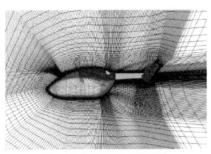

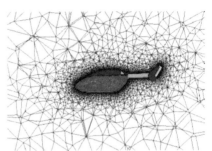

Figure 3: *ICEM-Hexa structured multiblock N.-S.mesh around the EC145 isolated fuselage: middle plane.*

Figure 4: *CENTAUR hybrid N.-S.mesh around the EC145 isolated fuselage: middle plane.*

From D´Alascio et al. (2004).
A middle plane section of an helicopter fuselage with structured and unstructured grids.

Plate I.2.3 *Examples of unstructured grids.*

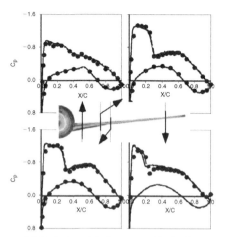

(a) Cartesian plot of pressure distribution at various positions along a wing–body–nacelle model, compared to experimental data. From Tinoco and Su (2004), Reproduced by permission from AIAA.

(b) Instantaneous iso-surfaces of vorticity colored by the span-wise component of vorticity of a 70° delta wing. From: Morton (2004)

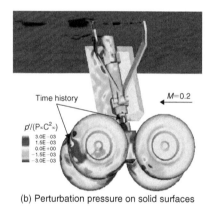

(b) Perturbation pressure on solid surfaces

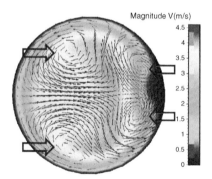

(c) Perturbation pressure distribution for an aero-acoustic simulation of the noise generated by a landing gear. From Lockard et al. (2004). Reproduced by permission from AIAA.

(d) Color plot and velocity vectors in one cross-section of the lung bifurcations shown in Figures I.1.2 and I.2.2. From Van Ertbruggen et al. (2005).

Plate I.2.4   *Examples of visual results from CFD simulations.*

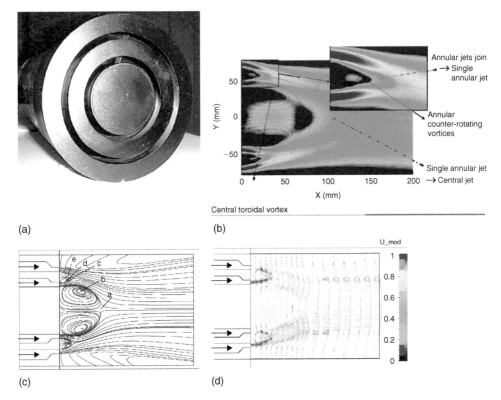

Plate 2.0.1 *Double annular burner: (a) Exit view of the burner. (b) Experimental color plot of axial velocity in the symmetry plane, obtained from LDV. (c) Experimental streamlines of the flow with designation of specific position points, related to the vortex structure, obtained from LDV. (d) Experimental vector plot of the velocity field obtained from LDV data. From S. Geerts et al. (2005). Courtesy Vrije Universiteit Brussel (VUB).*

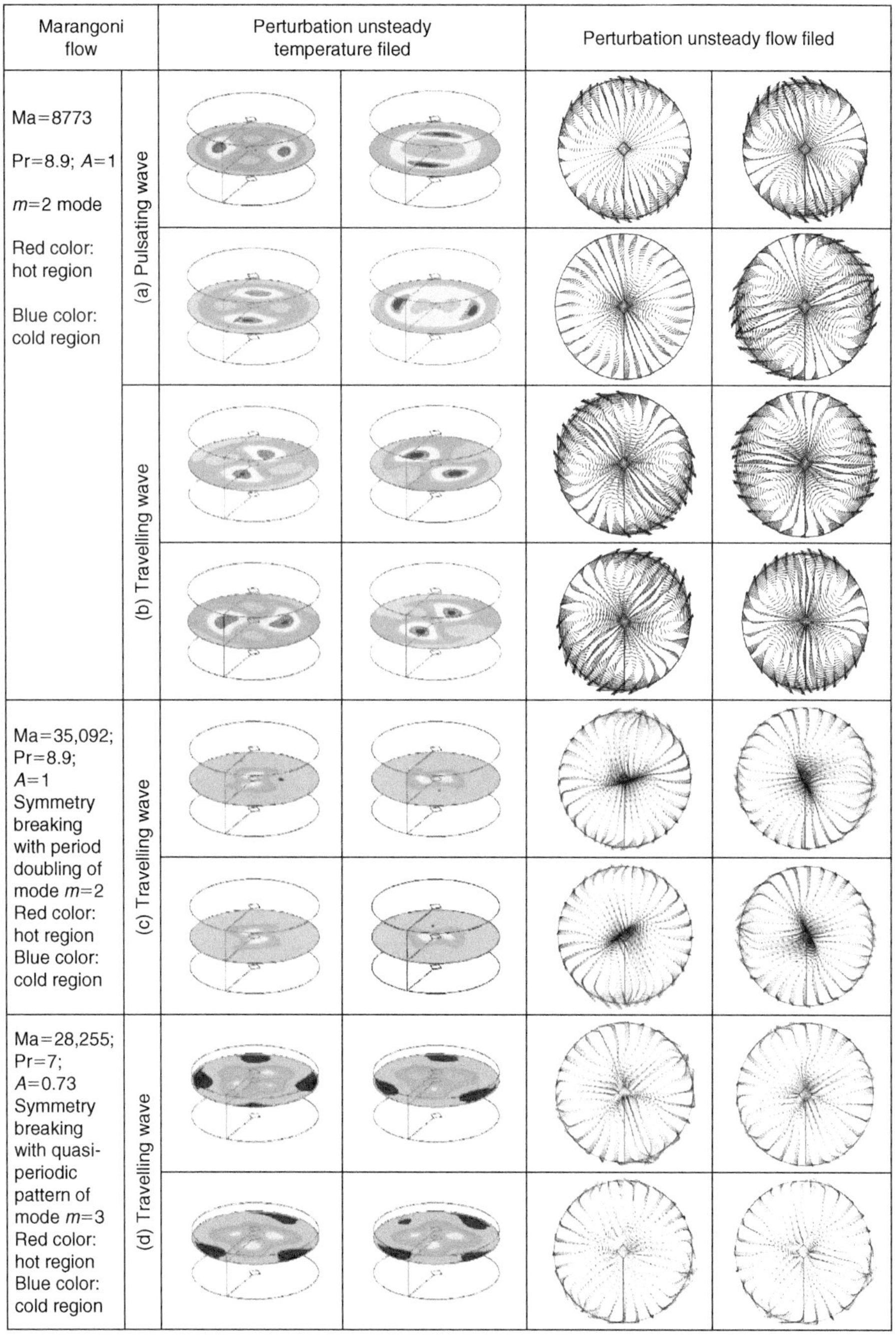

Marangoni flow		Perturbation unsteady temperature filed		Perturbation unsteady flow filed	

Ma=8773

Pr=8.9; A=1

m=2 mode

Red color: hot region

Blue color: cold region

(a) Pulsating wave

(b) Travelling wave

Ma=35,092; Pr=8.9; A=1 Symmetry breaking with period doubling of mode m=2 Red color: hot region Blue color: cold region

(c) Travelling wave

Ma=28,255; Pr=7; A=0.73 Symmetry breaking with quasi-periodic pattern of mode m=3 Red color: hot region Blue color: cold region

(d) Travelling wave

Plate 2.1.2 *Liquid bridge and Marangoni effect. Representation of various unsteady and symmetry broken solutions for velocity and temperature Perturbation fields. Each group of four figures represents four snapshots of the corresponding unsteady peturbation field. From Dinescu and Hirsch (2001), Hirsch and Dinescu (2003).*

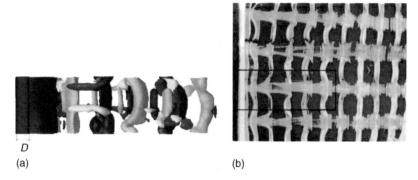

(a) (b)

Plate 2.1.7 *Spanwise undulation of the main vortex rows and streamwise vortices, at Re = 220; shown by iso-contours of vorticity components. (b) Spanwise experimental flow visualization, provided by Williamson (1992). The frame shows correspondence to the computational region explored. From Persillon and Braza (1998). Courtesy M. Braza. IMFT, Toulouse.*

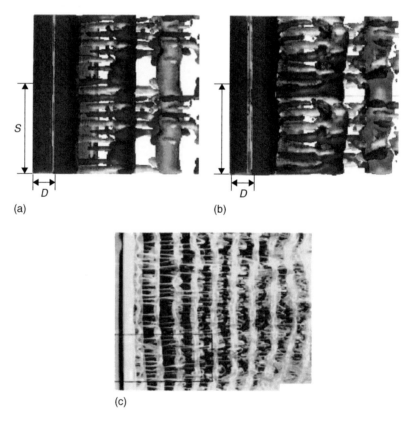

(a) (b)

(c)

Plate 2.1.8 *Modification of the spanwise vortex structures as Reynolds number increases; passage to mode B; (a) Re = 270; (b) Re = 300. (c) Spanwise experimental flow visualization, provided by Williamson (1992, 1996a, b). The frame shows correspondence to the computational region explored. From Persillon and Braza (1998). Courtesy M. Braza. IMFT, Toulouse.*

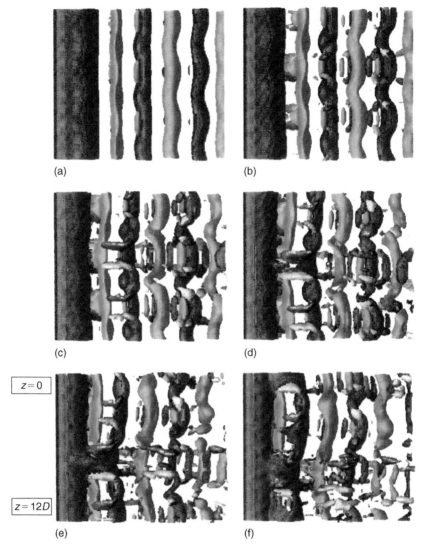

Plate 2.1.9 *Spanwise and streamwise iso-vorticity contours showing mode A formation and the transition to the vortex dislocations pattern at Re = 220. From Braza et al. (2001). Courtesy M. Braza. IMFT, Toulouse.*

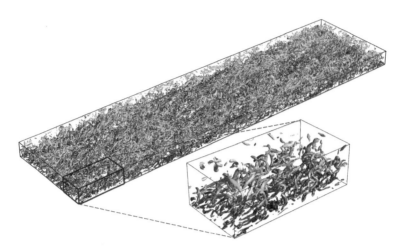

Plate 2.1.10 *Instantaneous view of the turbulent vortices colored with their distance to the wall (red is closest to the wall and yellow is at the center of the channel. Only 1/4 of the channel is shown (full length, half-width and half-height), and the flow direction is from bottom-left to top-right. From del Alamo et al. (2004). Courtesy J. Jimenez and coworkers.*

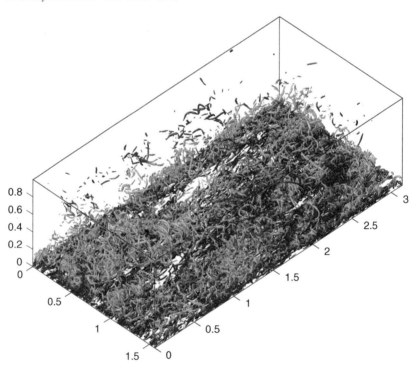

Plate 2.1.11 *Instaneous realization of a complex clustering of vortices in a turbulent channel flow at a Reynolds number of 100,000 (Retau = 2000). The flow is from left to right and the vortices are colored with their distance to the wall (blue is near the wall while red is far from the wall). From Hoyas and Jimenez (2006). Courtesy J. Jimenez and coworkers.*

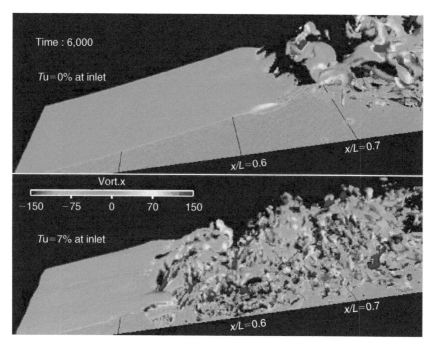

Plate 2.1.12 *Snapshots of a DNS simulation at a Reynolds number of 60,000, showing the effects of an external turbulence on the transition, comparing the vorticity field of a laminar incoming separation bubble with no turbulence and with 7% turbulence intensity. Courtesy J. Wissink and W. Rodi, University of Karlsruhe.*

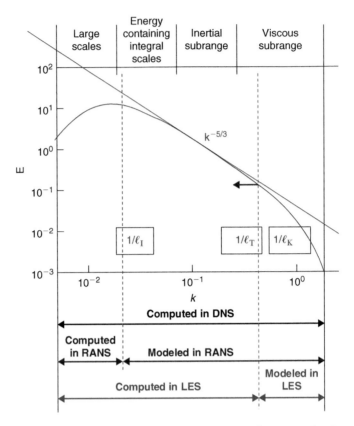

Plate 2.2.1 *Energy spectrum of turbulence in function of wave number k, with indication of the range of application of the DNS, LES and RANS models. The length scales l_T and l_I are associated with the LES and RANS approximations, respectively. Courtesy C. Fureby (FOI, Sweden).*

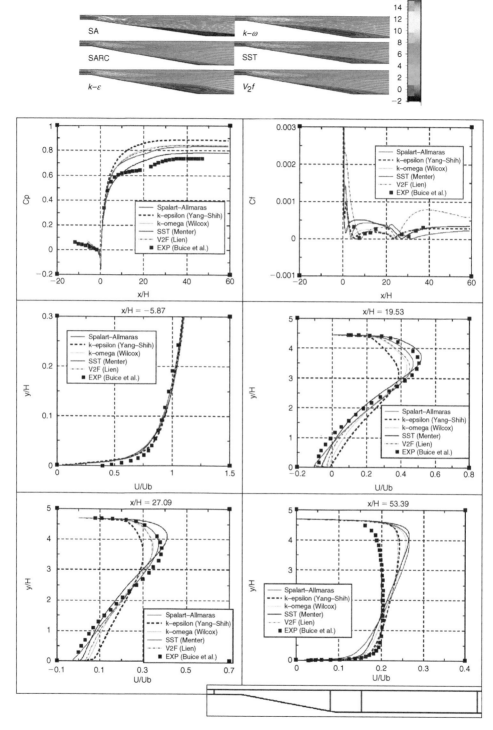

Plate 2.2.2 *Comparison of RANS simulations with different turbulence models for the OBI axisymmetric diffuser. The top figure shows the position and extends of the separation region, while the bottom figure compares calculated and measured pressure distribution, wall shear stress at the bottom wall and velocity profiles at the four positions indicated in the lower insert. Courtesy NUMECA International and Haase et al. (2006).*

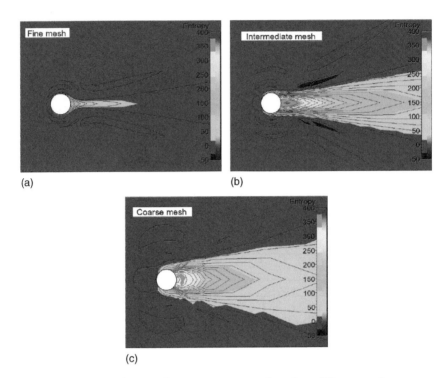

(a)

(b)

(c)

Plate 11.5.4 *Distribution of entropy as computed on three different meshes: (a) fine mesh, (b) intermediate mesh and (c) coarse mesh.*

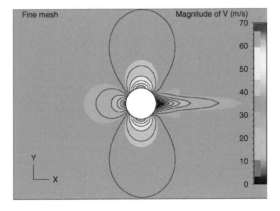

Plate 11.5.5 *Distribution of velocity magnitude and iso-velocity lines, as computed on the finest mesh.*

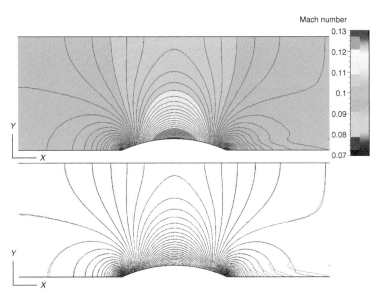

Plate 11.5.8 *Isolines of Mach number as computed using the (65 × 33) mesh superimposed on the color map of a reference solution on a (225 × 113) mesh. One isoline has been drawn every 0.001 ranging from 0.07 to 0.13. The bottom figure compares the Mach number isolines of the two solutions, where the darkest line is the reference solution.*

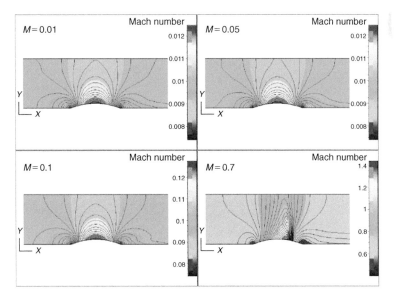

Plate 11.5.10 *Distribution of velocity as computed on the (65 × 33) mesh, for different values of the incident Mach number, from 0.01 to 0.7. Observe the shock appearing at M = 0.7.*

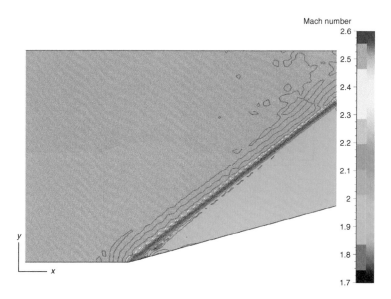

Plate 11.5.14 *Distribution of computed Mach number.*

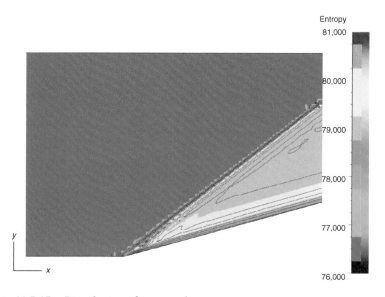

Plate 11.5.15 *Distribution of computed entropy.*

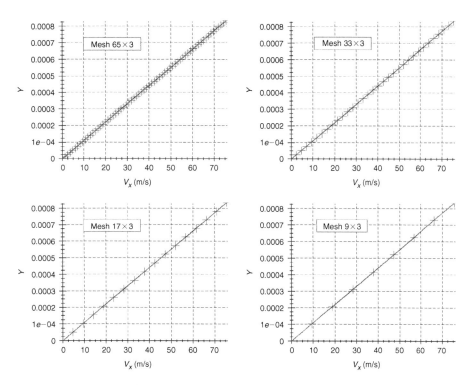

Plate 12.3.3 *Wall-to-wall distribution of axial velocity as deduced from analytical result (continuous line) and from the numerical result (plus signs).*

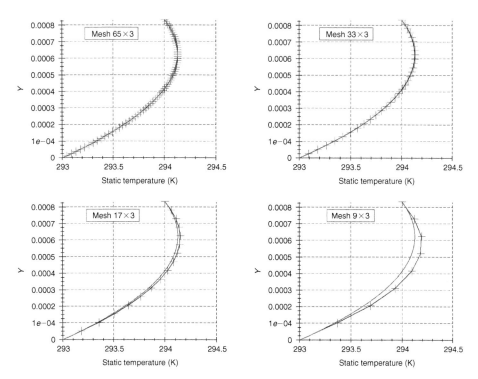

Plate 12.3.4 *Wall-to-wall distribution of static temperature (line with plus signs), compared to the analytical solution (continuous line), for the four different grids.*

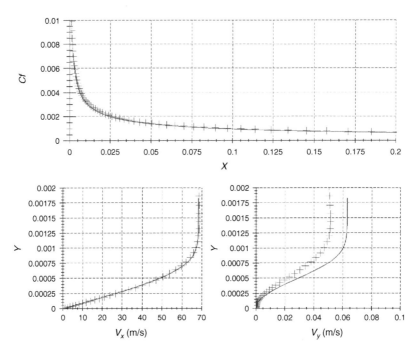

Plate 12.3.8 *Distribution of friction coefficient (upper panel), axial and wall normal velocities at x = 0.2 m (lower left and right panels) as obtained from analytical result (continuous line) and from the numerical result on the finest mesh (plus signs). The vertical axis is the distance in meters.*

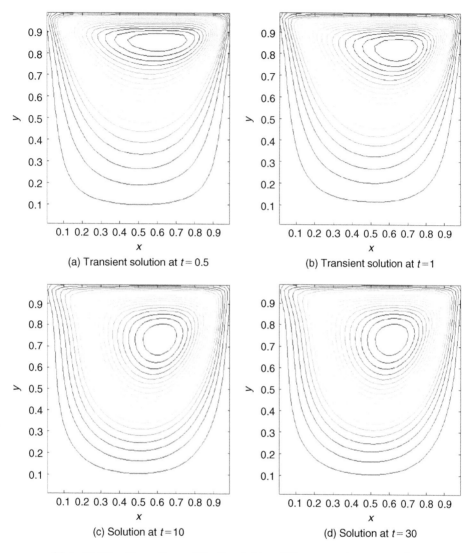

(a) Transient solution at $t=0.5$

(b) Transient solution at $t=1$

(c) Solution at $t=10$

(d) Solution at $t=30$

Plate 12.5.2 *Streamlines of the flow field at different transient stages, for $t=0.5$, 1, 10 and 30.*

Part II

Basic Discretization Techniques

Having defined in Part I, the mathematical models for fluid dynamics and a methodology for analyzing their fundamental properties in Chapter 3, we are now ready to move on to the next step toward setting up a CFD algorithm, namely *the discretization phase*.

As outlined in Figures I.2.1 and I.3.1, to which we refer you again, this second step in the definition of the computational approach deals with the choice of the discretization method of the selected mathematical model and involves two components, the *space discretization* and the *equation discretization*.

The *space discretization* consists in setting up a *mesh*, or a *grid*, by which the continuum of space is replaced by a finite number of points where the numerical values of the variables will have to be determined. It is intuitively obvious that the accuracy of a numerical approximation will be directly dependent on the size of the mesh, that is the closer the points, the better the discretized space approaches the continuum, the better the approximation of the numerical scheme. In other words, the error of a numerical simulation has to tend to zero when the mesh size tends to zero, and the pace of this variation will be characterized by the *order* of the numerical discretization.

For complex geometries the solution will also be dependent on the form of the mesh, since in these cases we will tend to develop meshes which are adapted to the geometrical complexities, as for flows along solid walls, and the mesh shape and size will vary through the flow field. Therefore, the generation of grids for complex geometries is a crucial problem, whose importance increases with the space dimension, making this step the most significant in three-dimensional CFD simulations.

As described already in the general introduction, we distinguish between structured and unstructured grids, the latter being of more general nature. Structured grids are formed by families of lines (one for each space dimension), each mesh point being at the intersection of one line of each family and correspond to Cartesian grids in the mathematical space of the curvilinear coordinates. In unstructured grids, the mesh point distribution is arbitrary since they are not localized on identified lines and they can be connected through various polynomials in 2D or polyhedrals in 3D.

Grid generation and grid quality are essential elements of the whole discretization process. Not only is grid generation today a most critical element in the cost of running CFD simulations, but more importantly, *the accuracy of the obtained numerical results is critically dependent on mesh quality*.

In recent years methods have been, and still are developed in order to generate efficiently and as automatically as possible meshes adapted to arbitrary geometries.

We will devote Chapter 6 of Part II to a presentation of different grid types and their properties, with some considerations and recommendations related to grid quality and grid-related errors.

Once a mesh has been defined, the equations can be discretized leading to the transformation of the differential or integral equations to discrete algebraic operations involving the values of the unknowns related to the mesh points. *The basis of all numerical methods consists in this transformation of the mathematical model into an algebraic, linear or nonlinear, system of equations for the mesh-related unknown quantities.*

A very important aspect in the definition of an algorithm is the choice to be made between a time-dependent or steady state model for the flow equations, as mentioned already in Chapter 3.

For physical time-dependent problems with a transient flow behavior or connected to time varying boundary conditions, there is obviously no alternative to the use of a time-dependent mathematical model as time accuracy of the numerical solution is required.

However, with stationary or steady state problems, an alternative exists and you can decide to work either with a time-independent mathematical model of your flow problem or to use an unsteady formulation and follow the numerical solution in time until the steady state is reached. This last family of methods is often called '*time marching*' or '*pseudo unsteady*' since the time accuracy is not required and we attempt to reach the steady state in the smallest possible number of time steps, without requiring the correct numerical simulation of the transient behavior. In this case, the numerical schemes will rely on the solution of systems of ODEs in time, while in the former case the numerical solution techniques will have to rely on the methods for solving algebraic systems of equations (in space).

This has been explained already in Chapter 3 and the advantages of the time-dependent formulation, whereby the properties of the equations do not change when passing from subsonic to supersonic flow conditions, have been stressed. Another very important advantage of the time dependent formulation is related to the fact that many flow configurations do not have a steady state behavior, even in presence of stationary boundary conditions. A most popular example is the flow over a cylinder, characterized by the presence of a periodic vortex shedding, for Reynolds numbers high enough (typically above $\sim$40), as shown in Section 2.1. This flow has no steady state solution and hence attempting to solve the time-independent flow equations, will lead to non-physical solutions, or to no solution at all.

It is essential to be aware that, seen at small enough length scales all turbulent flow configurations are unsteady and that steady state flow conditions are the exception and not the rule. Although in practice, as we do not always look at the fine scales, many flows appear as steady.

For instance, the flow along an aircraft wing, under constant upstream conditions, considered at the level of the Reynolds averaged turbulent approximation (RANS model), may appear as steady, provided there is no large-scale separation. Since nearly all separated flow regions tend to have an unsteady behavior when the grid is refined, it is required to work with a mathematical model that is capable of detecting large-scale unsteadiness, as they appear.

Therefore, we recommend working with the time-dependent equations, unless there is an assurance that the flow will remain steady. However, even in this case there is no significant advantage in working with the stationary flow models.

For all those reasons, we are focusing on the discretization properties of time-dependent algorithms as the most general approach.

We will show, however, in the later Chapters 9 and 10, that bridges can be established between the two approaches, and they can be viewed as belonging to a common family of schemes, when iterative methods are applied for the resolution of the algebraic systems resulting from the space discretization.

With time-dependent numerical formulations, we will distinguish two families of methods, *explicit or implicit*. In *explicit* methods, the matrix of the unknown variables at the new time is a diagonal matrix while the right-hand side of the system is being dependent only on the flow variables at the previous times. This leads therefore to a trivial matrix inversion and hence to a solution with a minimal number of arithmetic operations for each time step. However, this advantage is counter-balanced by the fact that stability and convergence conditions impose severe restrictions on the maximum admissible time step. While this might not be a limitation for physical unsteady problems, it leads to the necessity of a large number of time steps in order to reach the steady state solution corresponding to a physical time-independent problem.

In *implicit* methods, the matrix to be inverted is not diagonal since more than one set of variables are unknown at the same time level. In many cases however, the structure of the matrix will be rather simple, such as block pentadiagonal, block tridiagonal or block bidiagonal, allowing simple algorithms for the solution of the system at each time step, although the number of operations required will be higher when compared to the explicit methods. This is compensated by the fact that many implicit methods have, at least for linear problems, no limitation on the time step and hence a lesser number of iterations will be needed to reach the steady state.

In summary, the following steps have to be defined in the process of setting up a numerical scheme:

(i) Selection of a discretization method of the equations. This implies the selection between finite difference, finite volume or finite element methods as well as the selection of the order of accuracy of the spatial and eventually time discretization.

(ii) Analysis of the selected numerical algorithm. This step concerns the analysis of the 'qualities' of the scheme in terms of stability and convergence properties as well as the investigation of the generated errors.

(iii) Selection of a resolution method for the system of ordinary differential equations in time, for the algebraic system of equations and for the iterative treatment of eventual nonlinearities.

Step (i) will be discussed in Part II, Chapters 4–6; step (ii) will be addressed in Part III, Chapters 7 and 8 and step (iii) in Part IV, Chapters 9 and 10.

THE STRUCTURE OF PART II

In Part II, we will introduce you to the most important methods for the discretization of the *space derivatives* entering in the conservation laws. Three families of methods are available, with varying degrees of generality. The most traditional and oldest method is the *finite difference method (FDM)*, which remains the reference for all studies of numerical discretization, although it is only applicable in practice to structured grids.

Therefore, it is very important that you develop a strong understanding of the main properties of finite difference formulas.

By far the most widely applied method today in CFD is the *finite volume method* (*FVM*), which discretizes directly the integral form of the conservation laws. Its popularity is due to its generality, its conceptual simplicity and the relative ease of application to both structured as well as to any kind of unstructured grids. A large body of literature has been developed around the FVM and we will introduce its main properties. Although the FV discretization will lead to similar formulas as a FDM when applied to structured grids, it is equally important that you develop a good understanding of the main properties of FVMs. This will help you to follow the developments behind current CFD tools and to better interpret the results obtained by applying for instance commercial CFD codes to practical problems.

The third method is derived from the world of structural mechanics, where the *finite element method* (*FEM*) is most widely, if not exclusively, applied. Its application to CFD is of interest, but is not dominant and we will restrict ourselves to a short presentation, *at advanced level*, mainly oriented at its basic properties and its relation with the finite volume method.

Although the analysis of discretization methods is best performed on uniform grids, this is seldom the case in practice, where geometrical complexities can lead to highly irregular and distorted grids. Due to the strong influence of the grid properties on the quality of the CFD results and the associated loss of accuracy, it is important to develop an understanding of these effects and to extract possible guidelines on grid quality in order to reduce the associated numerical errors. This will be introduced in Chapter 4 on a one-dimensional basis, and further discussed in Chapter 6.

This second part is therefore organized in three chapters:

1. Chapter 4 deals with the basic method for the discretization of PDEs, namely the *finite difference method* (*FDM*), which can be applied to any structured mesh configuration.
2. Chapter 5 will cover the fundamentals of what is today the most widely applied discretization method, valid for both structured and unstructured grids, the *finite volume method* (*FVM*). Some advanced sections will introduce the essentials of the *finite element method* (*FEM*), which is the reference method in structural mechanics, but is also applicable to fluid mechanics.
3. Chapter 6 will introduce the important issue of mesh properties. As stated in the introduction, two families of grids can be selected: structured or unstructured, the latter being the most general option. This chapter is not oriented at the techniques for grid generation, which are quite specialized and outside the scope of this book. Several excellent publications can be consulted on this subject and references will be given in Chapter 6. Instead, this new chapter will present the *different grid types* as encountered in CFD and will focus on issues of grid generated errors, grid quality and provide some best practice recommendations on grid properties in order to minimize the grid-related error sources.

Chapter 4

The Finite Difference Method for Structured Grids

OBJECTIVES AND GUIDELINES

The Finite Difference Method (FDM) is based on the properties of Taylor expansions and on the straightforward application of the definition of derivatives. It is probably the simplest method to apply, particularly on uniform meshes, but it requires the mesh to be set up in a structured way, whereby the mesh points, in an n-dimensional space, are located at the intersections of n families of lines and each point must lie on one, and only one, line of each family.

In Section 4.1, we explain the basic steps toward the establishment of finite difference formulas and introduce the very important concept of ***order of accuracy***, as applied to first and second derivatives. It will be immediately clear from this section and further confirmed in the subsequent sections, that an unlimited number of FD formulas can be derived for any derivative. This implies that for any partial differential equation (PDE) of a mathematical model, we will have an infinite number of possible numerical schemes. This richness of numerical algorithms makes the selection process and the associated criteria a challenging and altogether exciting issue and we hope to guide you in this wonderful world and to share with you the sense of beauty it can provide. This will be directly illustrated on the simplest one-dimensional model equations for linear convection and linear diffusion.

In Section 4.2, we introduce the extensions of FD formulas for partial derivatives in two dimensions, on uniform structured grids. This extension can be applied to single or mixed derivatives of any order. Although Section 4.2 is restricted to two-dimensional space, its extension to three dimensions is straightforward. A particular attention is given to finite difference formulas for the very important Laplace equation, which is the standard equation describing diffusion phenomena.

In Section 4.3, we introduce some issues related to FD formulas on non-uniform grids, in one-dimensional space. This section is of great importance, since most of the grids used in practical CFD simulations are non-uniform. We will show that standard FD formulas can easily loose at least one order of accuracy, when applied to a non-uniform grid. In addition to the derivation of representative FD formulas, the presented analysis will provide some recommendations to be considered when dealing with non-uniform grids.

We consider Section 4.3 as one of the most important of this chapter, and recommend that you give a particular attention to its content and its conclusions.

The two following sections are of a more advanced level and we suggest considering this section for a more advanced course. They will be marked as 'Advanced' in the roadmap of Figure 4.0.1.

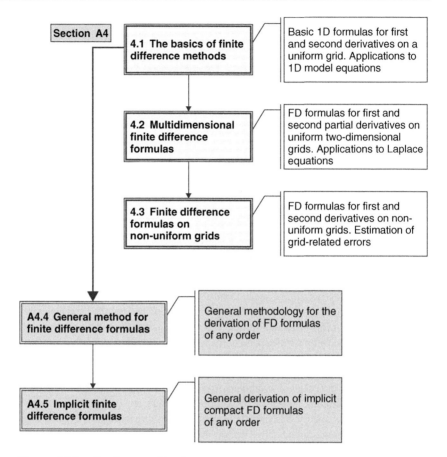

Figure 4.0.1 *Roadmap to this chapter.*

In Section 4.4, we introduce a general methodology to obtain **arbitrary FD formulas for any derivative**, with a prescribed order of accuracy. This section is based on general mathematical expressions for finite difference operators, linking them to the related differential operator, establishing at the same time the associated order of accuracy and the dominant truncation error. This applies to derivatives on a uniform one-dimensional mesh. The presented framework has a large range of application and is of particular interest for the derivation and properties of FD formulas for higher order derivatives.

Section 4.5 deals with the derivation of **implicit FD formulas**, defined as expressions where derivatives at different mesh points appear simultaneously. This approach is an alternative to the standard FD formulas, for high orders of accuracy. It is shown in Section 4.4 that in order to achieve high orders of accuracy, more mesh points have to be involved. Typically, an FD formula for a first derivative with order of accuracy n, will generally require contributions from at least $(n + 1)$ points. Hence, we can ask the question if we could generate high accuracy FD formulas, with a restricted number of points, defining hereby **compact** formulas. The answer to this question is positive, leading to compact, implicit schemes, which can be of interest when high order accuracy is required, for instance in simulations such as direct numerical simulation (DNS), large eddy simulation (LES) for turbulent flows, computational electromagnetic (CEM) or computational aero acoustic (CAA).

The roadmap of this chapter is summarized in Figure 4.0.1.

4.1 THE BASICS OF FINITE DIFFERENCE METHODS

The finite difference approximation is the oldest of the methods applied to obtain numerical solutions of differential equations, and the first application is attributed to Leonhard Euler (1707–1783) in 1768. The idea of finite difference methods is actually quite simple, since it corresponds to an estimation of a derivative by the ratio of two differences according to the theoretical definition of the derivative.

For a function $u(x)$, the derivative at point x is defined by

$$u_x = \frac{\partial u}{\partial x} = \lim_{\Delta x \to 0} \frac{u(x + \Delta x) - u(x)}{\Delta x} \tag{4.1.1}$$

If we remove the limit in the above equation, we obtain a *finite difference*, which explains the name given to this method.

If Δx is small but finite, the expression on the right-hand side is an approximation to the exact value of u_x. The approximation will be improved by reducing Δx, but for any finite value of Δx, an error is introduced, the **truncation error**, which goes to zero for Δx tending to zero.

The power of Δx with which this error tends to zero is called the **order of accuracy of the difference approximation**, and can be obtained from a Taylor series development of $u(x + \Delta x)$ around point x.

Actually, the whole concept of finite difference approximations is based on the properties of Taylor expansions. Developing $u(x + \Delta x)$ around $u(x)$ we have

$$u(x + \Delta x) = u(x) + \Delta x \frac{\partial u}{\partial x} + \frac{\Delta x^2}{2} \frac{\partial^2 u}{\partial x^2} + \frac{\Delta x^3}{3!} \frac{\partial^3 u}{\partial x^3} + \cdots \tag{4.1.2}$$

This relation can be written as follows:

$$\frac{u(x + \Delta x) - u(x)}{\Delta x} = u_x(x) + \underbrace{\frac{\Delta x}{2} u_{xx}(x) + \frac{\Delta x^2}{6} u_{xxx}(x) + \cdots}_{\text{Truncation error}} \tag{4.1.3}$$

showing that

- The right-hand side (r.h.s) of equation (4.1.1) is indeed an approximation to the first derivative u_x in point x.
- The remaining terms in the r.h.s represent the error associated to this formula.

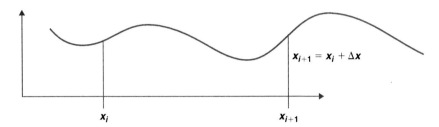

Figure 4.1.1 *Taylor expansion of the function u(x), around point x.*

If we restrict the truncation error to its dominant term, that is to the lower power in Δx, we see that this approximation for $u(x)$ goes to zero like the first power of Δx and is said to be *first order* in Δx and we write

$$\frac{u(x + \Delta x) - u(x)}{\Delta x} \cong u_x(x) + \frac{\Delta x}{2} u_{xx}(x) = u_x(x) + O(\Delta x) \tag{4.1.4}$$

indicating that the truncation error $O(\Delta x)$ goes to zero like the first power in Δx.

A very large number of finite difference approximations can be obtained for the derivatives of functions as shown next and a general procedure will be described in Section 4.4, based on formal difference operators and their manipulation.

A remark about the significance of Taylor expansions

The Taylor expansion (4.1.2) actually tells us something quite remarkable about the properties of continuous functions. The left-hand side (l.h.s) is the value of the function u at an arbitrary distance Δx from point x, with *no restriction* on this distance. In the right-hand side (r.h.s), all quantities are evaluated at point x. Hence, what the Taylor expansion tells us is that we can know the value of the function at an *arbitrary distance* far away from point x (say 5000 km), if we know 'everything', that is all the derivatives, at *this single point* x (Figure 4.1.1). In practice, for any finite value of Δx, the knowledge of a finite number of derivatives in point x, will suffice to evaluate the value of u at point $(x + \Delta x)$ with a preset accuracy.

Example E.4.1.1

This is illustrated by the following plot of the Taylor expansion of $(e^{\sqrt{x}} - 1)$ around $x = 0$, obtained by the Maple symbolic mathematical software (version Maple 10):

$$(e^{\sqrt{x}} - 1) = \sqrt{x} + \frac{1}{2}x + \frac{1}{6}x^{3/2} + \frac{1}{24}x^2 + \frac{1}{120}x^{5/2} + \frac{1}{720}x^3 + \frac{1}{5040}x^{7/2}$$

$$+ \frac{1}{40,320}x^4 + \frac{1}{3,62,880}x^{9/2} + O(x^5) \tag{E.4.1.1}$$

Figure E.4.1.1 illustrates that by increasing the number of terms of the Taylor expansion, we increase the distance over which the expansion provides a valid representation of the function. With four terms, that is the knowledge of up to four derivatives in point $x = 0$, the four term expansion covers nearly all the domain from 0 to 2.

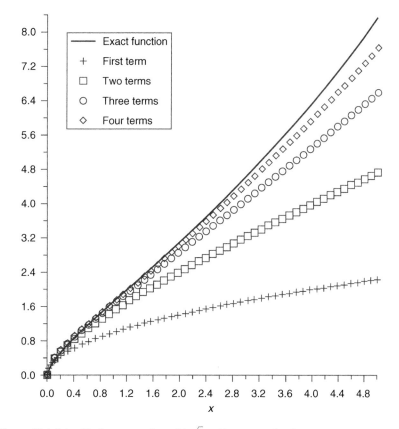

Figure E.4.1.1   *Taylor expansion of* $(e^{\sqrt{x}} - 1)$ *up to order 5.*

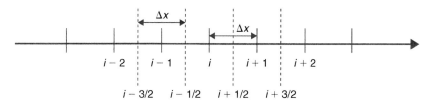

Figure 4.1.2   *One-dimensional uniform grid on the x-axis.*

4.1.1   Difference Formulas for First and Second Derivatives

To apply this general definition, we consider a one-dimensional space, the x-axis, where a space discretization has been performed such that the continuum is replaced by N discrete mesh points x_i, $i = 1, \ldots, N$ (Figure 4.1.2).

We will indicate by u_i the value of the function $u(x)$ at the points x_i, i.e. $u_i = u(x_i)$ and consider that the spacing between the discrete points is constant and equal to Δx. Without loss of generality, we can consider that $x_i = i \Delta x$ and this point will also be referred to as 'point x_i' or as 'point i'.

4.1.1.1 *Difference formula for first derivatives*

Applying the above relations (4.1.3) and (4.1.4) at point i, we obtain the following finite difference approximation for the first derivative $(u_x)_i = (\partial u/\partial x)_i$:

$$(u_x)_i = \left(\frac{\partial u}{\partial x}\right)_i = \frac{u_{i+1} - u_i}{\Delta x} \underbrace{- \frac{\Delta x}{2}(u_{xx})_i - \frac{\Delta x^2}{6}(u_{xxx})_i + \cdots}_{\text{Truncation error}} \tag{4.1.5}$$

$$= \frac{u_{i+1} - u_i}{\Delta x} + O(\Delta x)$$

As this formula involves the point $(i+1)$ to the right of point i, it is called the *first order forward difference* for the first derivative $(u_x)_i = (\partial u/\partial x)_i$.

This is certainly not the only formula we can think of, as we can re-apply the relations (4.1.1)–(4.1.4) by replacing everywhere Δx by $(-\Delta x)$. This leads to the relation:

$$(u_x)_i = \left(\frac{\partial u}{\partial x}\right)_i = \frac{u_i - u_{i-1}}{\Delta x} \underbrace{+ \frac{\Delta x}{2}(u_{xx})_i - \frac{\Delta x^2}{6}(u_{xxx})_i + \cdots}_{\text{Truncation error}} \tag{4.1.6}$$

$$= \frac{u_i - u_{i-1}}{\Delta x} + O(\Delta x)$$

With respect to the point $x = x_i$, this formula is the *first order backward difference* for the derivative $(u_x)_i$. Both formulas (4.1.5) and (4.1.6) are called *one-sided difference* formulas, since they involve points at one side of point i only.

Looking carefully at the two one-sided formulas, we observe that the dominant first order truncation errors are of opposite signs. Hence, if we add them up, we obtain a *second order approximation*

$$(u_x)_i = \frac{u_{i+1} - u_{i-1}}{2\Delta x} - \frac{\Delta x^2}{6}(u_{xxx})_i + \cdots = \frac{u_{i+1} - u_{i-1}}{2\Delta x} + O(\Delta x^2) \tag{4.1.7}$$

This formula, which involves the points to the left and to the right of point i, is called therefore a *central difference formula*.

These three approximations are represented geometrically on Figure 4.1.3 while the derivative is represented by the tangent to the curve $u(x)$, the forward, backward and central chords represent the approximations defined by the corresponding difference formulas. It is clear that the central chord is always a much better approximation than the one-sided chords and this is reflected by its second order accuracy.

The formula (4.1.7) is indeed particularly interesting, as it provides, on the same support as the one-sided differences, namely points $(i+1)$, i, $(i-1)$, a second order accuracy, compared to first order. To better realize what this means, consider a domain between 0 and 1 on the x-axis, with 11 mesh points and $\Delta x = 0.1$. A first order formula can be interpreted as generating an error $O(\Delta x)$, that is of the order of 10%, while applying the second order central formula on the same 11-grid point mesh, will give an accuracy $O(\Delta x^2)$ of the order of 1%. If we want an accuracy of 1% with the one-sided difference formulas, we will need to generate a mesh with 101 points and $\Delta x = 0.01$, which represents a significantly higher cost, as we need to apply the formulas 100 times instead of 10 times.

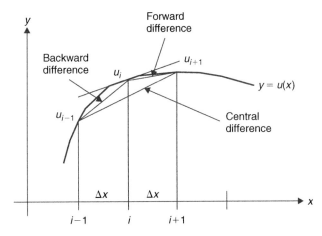

Figure 4.1.3   *Geometrical interpretation of difference formulas for first order derivatives.*

Important remarks and interpretations

- As the truncation error of a first order FD formula is proportional to the second derivative, as seen from the above equations, we can state that a first order FD formula is exact for a linear function. Similarly, a second order formula has its truncation error proportional to the third derivative; hence, this formula will be exact for a quadratic (parabolic) function.
- The first order forward difference formula for $(u_x)_i$ can be considered as a central difference with respect to the mid-point

$$x_{i+1/2} = \frac{(x_i + x_{i+1})}{2} \qquad (4.1.8)$$

leading to a second order approximation for the derivative $(u_x)_{i+1/2}$ in this point. This is an important property, which is often used in computations due to its compact character. The same formula (4.1.5) is either a first order forward difference for $(u_x)_i$ or a second order central approximation for $(u_x)_{i+1/2}$ but involving only the same two mesh points i and $(i + 1)$. We therefore have

$$(u_x)_{i+1/2} = \left(\frac{\partial u}{\partial x} \right)_{i+1/2} = \frac{u_{i+1} - u_i}{\Delta x} + O(\Delta x^2) \qquad (4.1.9)$$

and similarly at $(i - 1/2)$

$$(u_x)_{i-1/2} \equiv \left(\frac{\partial u}{\partial x} \right)_{i-1/2} = \frac{u_i - u_{i-1}}{\Delta x} + O(\Delta x^2) \qquad (4.1.10)$$

Compared to the formulas (4.1.5) and (4.1.6) for $(u_x)_i$, we have gained an order of accuracy by considering the same expressions as approximations for the mid-points $(i + 1/2)$ or $(i - 1/2)$, respectively.

Actually, difference formulas for the first derivative $(u_x)_i$ can be constructed involving any number of adjacent points with the order of the approximation increasing with the number of points. In any numerical scheme, a balance will have to be defined between the order of accuracy and the number of points simultaneously involved in the computation. The bandwidth of the algebraic system that has finally to be solved in order to obtain the solutions u_i is generally proportional to the number of simultaneous points involved.

Second order one-sided differences

For instance, a one-sided, second order difference formula for $(u_x)_i$, containing only the upstream points, $i - 2, i - 1, i$, can be obtained by an expression of the form

$$(u_x)_i = \frac{au_i + bu_{i-1} + cu_{i-2}}{\Delta x} + O(\Delta x^2) \tag{4.1.11}$$

The coefficients (a, b, c) are found from a Taylor expansion of u_{i-2} and u_{i-1} around u_i. Writing

$$u_{i-2} = u_i - 2\Delta x(u_x)_i + \frac{(2\Delta x)^2}{2}(u_{xx})_i - \frac{(2\Delta x)^3}{6}(u_{xxx})_i + \cdots \tag{4.1.12}$$

$$u_{i-1} = u_i - \Delta x(u_x)_i + \frac{\Delta x^2}{2}(u_{xx})_i - \frac{\Delta x^3}{6}(u_{xxx})_i + \cdots \tag{4.1.13}$$

and multiplying the first equation by c, the second by b and adding to au_i, leads to

$$au_i + bu_{i-1} + cu_{i-2} = (a + b + c)u_i - \Delta x(b + 2c)(u_x)_i$$
$$+ \frac{\Delta x^2}{2}(b + 4c)(u_{xx})_i + O(\Delta x^3) \tag{4.1.14}$$

Hence, identifying with equation (4.1.11), we obtain the three conditions

$$a + b + c = 0$$
$$(b + 2c) = -1 \tag{4.1.15}$$
$$b + 4c = 0$$

and the second order accurate one-sided formula:

$$(u_x)_i = \frac{3u_i - 4u_{i-1} + u_{i-2}}{2\Delta x} + O(\Delta x^2) \tag{4.1.16}$$

This is a general procedure for obtaining finite difference formulas with an arbitrary number of points and an adapted order of accuracy. In general, a first order derivative at mesh point i, can be made of order of accuracy p, by an explicit formula such as (4.1.11) involving $(p + 1)$ points. For instance, a formula involving the forward points $i + 2, i + 1, i$, is

$$(u_x)_i = \frac{-3u_i + 4u_{i+1} - u_{i+2}}{2\Delta x} + O(\Delta x^2) \tag{4.1.17}$$

The first of the three equations (4.1.15) is of great importance, since it states that the sum of the coefficients of a finite difference formula, in this case (4.1.11), has to be zero.

This is a general condition of consistency, ensuring that the numerical approximation to the derivative of a constant will always vanish, as can be seen by replacing all the u_i by the constant value of one.

Therefore, any difference formula, for any order of the derivative and in any number of dimensions, must always satisfy the condition that the sum of its coefficients is equal to zero.

Higher orders of accuracy can also be obtained with a reduced number of mesh points at the cost of introducing implicit formulas, as will be seen in Section A4.5.

4.1.1.2 *FD formulas for second derivatives*

Finite difference approximations of higher order derivatives can be obtained by repeated applications of first order formula. For instance, a second order approximation to the second derivative $(u_{xx})_i$ is obtained by

$$(u_{xx})_i \equiv \left(\frac{\partial^2 u}{\partial x^2}\right)_i = \frac{(u_x)_{i+1} - (u_x)_i}{\Delta x} \tag{4.1.18}$$

$$= \frac{u_{i+1} - 2u_i + u_{i-1}}{\Delta x^2} + O(\Delta x^2)$$

where backward approximations for $(u_x)_{i+1}$ and $(u_x)_i$ are selected.

This symmetrical, central difference formula is of second order accuracy as can be seen from a Taylor expansion. We obtain indeed

$$\frac{u_{i+1} - 2u_i + u_{i-1}}{\Delta x^2} = (u_{xx})_i + \frac{\Delta x^2}{12}\left(\frac{\partial^4 u}{\partial x^4}\right)_i + \cdots \tag{4.1.19}$$

As with equation (4.1.11) we can define formulas with an arbitrary number of points, around point i, by combination of Taylor series developments. For instance, an expression such as equation (4.1.11) for the second derivative $(u_{xx})_i$ will lead to the conditions

$$a + b + c = 0$$
$$b + 2c = 0 \tag{4.1.20}$$
$$b + 4c = 2$$

and to the one-sided, backward formula for the second derivative

$$(u_{xx})_i = \frac{u_i - 2u_{i-1} + u_{i-2}}{\Delta x^2} + \Delta x (u_{xxx})_i + \cdots \tag{4.1.21}$$

This one-sided formula is only first order accurate at point i. Note also that this same formula is a second order accurate approximation to the second derivative at point $(i-1)$, as can be seen by a comparison with the central formula (4.1.18).

The above procedure, with undetermined coefficients, can be put into a systematic framework in order to obtain finite difference approximations for any derivatives, with a pre-selected order of accuracy. In order to achieve this a formalization of the relations between differentials and difference approximations is to be defined, via the introduction of appropriate difference operators. The methodology behind this approach is developed in Section A4.4.

4.1.2   Difference Schemes for One-Dimensional Model Equations

We are now ready to derive the first *numerical finite difference schemes*, by applying different FD formulas to some of the model equations introduced in Section 3.1.

4.1.2.1   *Linear one-dimensional convection equation*

Let us consider first the fundamental linear one-dimensional convective, hyperbolic equation (3.1.6), written here in the following notation:

$$\frac{\partial u}{\partial t} + a\frac{\partial u}{\partial x} = 0 \tag{4.1.22}$$

where $u(x, t)$ is the unknown function of (x, t) and a the convection speed, or the wave speed according to the interpretation given to equation (3.1.6).

In the following, when no danger of ambiguity can arise, we will also use a short-hand notation, where the derivatives are indicated as subscripts. Hence, we will write equation (4.1.22) as follows:

$$u_t + au_x = 0 \tag{4.1.23}$$

Considering an initial, boundary value problem, this equation has to be substantiated by the following initial and boundary conditions, for $a > 0$:

$$
\begin{aligned}
\text{At } t &= 0 \quad u(x,0) = u^{(0)}(x) \quad 0 \le x \le L \\
\text{At } x &= 0 \quad u(0,t) = g(t) \qquad t \ge 0
\end{aligned}
\tag{4.1.24}
$$

In order to apply a finite difference method to this equation, we discretize the space and the time domains, with constant steps. That is, the x-axis is discretized with N constant mesh intervals Δx, and the time axis is subdivided in constant time intervals Δt, as in Figure 4.1.4, with

$$
\begin{aligned}
x_i &= i\Delta x \qquad\qquad t^n = n\Delta t \\
u_i^n &= u(i\Delta x, n\Delta t)
\end{aligned}
\tag{4.1.25}
$$

We indicate the time level n by a superscript and the space position is indicated by the subscript i.

In order to obtain a numerical scheme, we have to discretize separately the space and the time derivatives and let us concentrate first on the space discretization. We could select for instance, a central, second order difference formula for the discretization

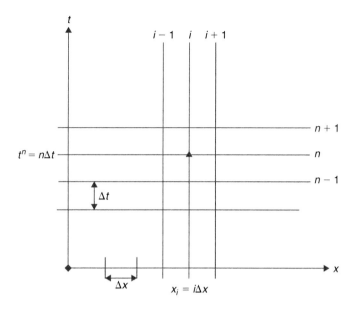

Figure 4.1.4   *Discretization of the time and the space axis.*

of the space derivative at mesh point i. This leads to the **semi-discrete** scheme, also called **method of lines**:

$$(u_t)_i = -\frac{a}{2\Delta x}(u_{i+1} - u_{i-1})$$ (4.1.26)

The left-hand side represents the time derivatives evaluated at point i, and the next step is to define a discretization in time. This implies the replacement of the time derivative by a discrete form but also a decision as to the time level at which the right-hand side will be evaluated.

In a convection (or propagation) problem, we know the solution at $t = 0$ and we look for the solution at later times. This is translated numerically, when we progress in time from time level n to time level $(n + 1)$, by considering that we know the solution at time level n and are looking for its evolution to time level $(n + 1)$. It is therefore logical to select a forward difference formula for $(u_t)_i$, leading to

$$\frac{u_i^{n+1} - u_i^n}{\Delta t} = -\frac{a}{2\Delta x}(u_{i+1} - u_{i-1})$$ (4.1.27)

The simplest scheme would be obtained with an evaluation of the right-hand side of equation (4.1.27) at time step n, for which all quantities are considered as known. This method is known as the **Euler method** for the time integration of ordinary differential equations, leading to the **explicit** numerical scheme:

$$\frac{u_i^{n+1} - u_i^n}{\Delta t} + \frac{a}{2\Delta x}(u_{i+1}^n - u_{i-1}^n) = 0$$ (4.1.28)

This is an explicit scheme, since the discretized equation contains only one unknown at level $(n + 1)$. It will be programmed by isolating the unknown value, as

$$u_i^{n+1} = u_i^n - \frac{a\Delta t}{2\Delta x}(u_{i+1}^n - u_{i-1}^n) \tag{4.1.29}$$

This form shows that the unknown value at the time $(n + 1)$ is obtained by a few arithmetic operations on known quantities. This is typical of explicit schemes, which are very economical in terms of number of arithmetic operations necessary for progressing in time.

As we will see in the following chapters (chapter 7) the price to pay is a severe restriction on the time step Δt, as a consequence of stability conditions, which requires many short time steps to advance in time.

Evaluating the right-hand side at level $(n + 1)$, leads to the *implicit* scheme:

$$\frac{u_i^{n+1} - u_i^n}{\Delta t} + \frac{a}{2\Delta x}(u_{i+1}^{n+1} - u_{i-1}^{n+1}) = 0 \tag{4.1.30}$$

known as the **backward or implicit Euler method**, since three unknowns appear simultaneously at time level $(n + 1)$.

Equation (4.1.30) leads to a system of equations with a tridiagonal matrix and we will present in an appendix to Chapter 10, algorithm leading to an efficient solution of tridiagonal systems, known as the Thomas algorithm.

Note that this equation could also be obtained from equation (4.1.26) by applying a backward difference in time for the discretization of u_t.

From the definitions of the order of accuracy of the finite difference formulas, we expect schemes (4.1.28) and (4.1.30) to be first order in time and second order in space at points i and time level n.

First order in time, first order in space

Another alternative, with a first order approximation for the space derivative, would be obtained with a backward difference in space, leading to the semi-discrete form:

$$(u_t)_i = -\frac{a}{\Delta x}(u_i - u_{i-1}) \tag{4.1.31}$$

With a forward difference in time, we obtain the following explicit scheme:

$$\frac{u_i^{n+1} - u_i^n}{\Delta t} + \frac{a}{\Delta x}(u_i^n - u_{i-1}^n) = 0 \tag{4.1.32}$$

The corresponding implicit version, evaluating the right-hand side at $(n + 1)$, would be

$$\frac{u_i^{n+1} - u_i^n}{\Delta t} + \frac{a}{\Delta x}(u_i^{n+1} - u_{i-1}^{n+1}) = 0 \tag{4.1.33}$$

We could also choose to apply a forward difference in space instead, leading to

$$(u_t)_i = -\frac{a}{\Delta x}(u_{i+1} - u_i) \tag{4.1.34}$$

and with the same two choices for explicit and implicit schemes

$$\frac{u_i^{n+1} - u_i^n}{\Delta t} + \frac{a}{\Delta x}(u_{i+1}^n - u_i^n) = 0 \tag{4.1.35}$$

$$\frac{u_i^{n+1} - u_i^n}{\Delta t} + \frac{a}{\Delta x}(u_{i+1}^{n+1} - u_i^{n+1}) = 0 \tag{4.1.36}$$

These schemes are first order in space and in time and are known as the **first order upwind schemes** for the convection equation.

The richness in the world of numerical schemes is unlimited, as we can select any combination of discretization formulas for the space and the time differences separately. Hence, we can write an unlimited number of possible schemes, even for the simplest model equation, such as the 1D linear convection equation. These schemes will have different properties, in terms of accuracy, stability and error properties and the analysis and prediction of these properties will form the subject of Part III.

To further illustrate this variety in possible schemes, let us look at some additional options.

First order in time, second order backward difference in space

$$\frac{u_i^{n+1} - u_i^n}{\Delta t} + a\frac{3u_i - 4u_{i-1} + u_{i-2}}{2\Delta x} = 0 \tag{4.1.37}$$

with the explicit

$$\frac{u_i^{n+1} - u_i^n}{\Delta t} + a\frac{3u_i^n - 4u_{i-1}^n + u_{i-2}^n}{2\Delta x} = 0 \tag{4.1.38}$$

or implicit options

$$\frac{u_i^{n+1} - u_i^n}{\Delta t} + a\frac{3u_i^{n+1} - 4u_{i-1}^{n+1} + u_{i-2}^{n+1}}{2\Delta x} = 0 \tag{4.1.39}$$

Second order in time, second order central difference in space

$$\frac{u_i^{n+1} - u_i^{n-1}}{2\Delta t} + a\frac{u_{i+1}^n - u_{i-1}^n}{2\Delta x} = 0 \tag{4.1.40}$$

This explicit scheme is called the **leapfrog scheme** and is second order in space and time. Its properties will also be discussed in Part III.

RECOMMENDATION FOR PRACTICAL TESTS

You can observe that the different schemes above differ by what we could consider as 'small' changes in the location of some points or in values of coefficients. However, their properties can differ significantly and we can already recommend that you program some of these schemes, even before studying the next chapters, where methods for predicting their properties will be presented.

We suggest that you program first the explicit central scheme (4.1.29), with the non-dimensional parameter $\sigma = a\Delta t/\Delta x$ equal to 0.8. Consider the initial solution of triangular shape

$$
\begin{aligned}
u^{(0)}(x) &= 0 & x \leq 0.9 \\
&= 10(x - 0.9) & 0.9 \leq x \leq 1.0 \\
&= 10(1.1 - x) & 1.0 \leq x \leq 1.1 \\
&= 0 & x \geq 1.1
\end{aligned}
$$

and calculate five consecutive time steps. Compare with the exact solution $u(x, t) = u^{(0)}(x - at)$, with $a = 1$, $\Delta x = 0.05$ and 41 mesh points over the initial domain between $x = 0$ and $x = 2$.

You will observe that the numerical solution grows erratically, showing that this scheme is **unstable**, and therefore useless.

Write now a program for the same initial solution, but applying the first order upwind scheme (4.1.32), with the parameter $\sigma = a\Delta t/\Delta x$ equal to 0.8 and calculate 5, 10, 15 and 30 time steps. When you compare now the graphical results with the exact solutions, you will observe that the numerical solution is acceptable, but that it is significantly diffused with increasing number of time steps.

Repeat now the same calculation with the parameter $\sigma = a\Delta t/\Delta x$ equal to 1.5 and observe that the solution is again erratic. This is typical for what we will define as **conditional stability**.

Here is what you should obtain, where the full line is the exact solution after, respectively 5 and 30 time steps (Figure 4.1.5).

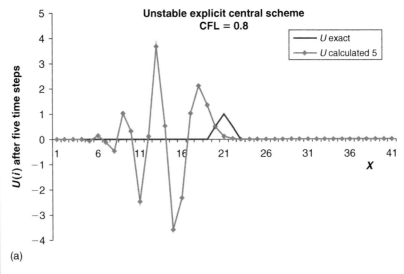

(a)

Figure 4.1.5a *Explicit central scheme for triangular initial profile and $\sigma = 0.8$, after 5 time steps.*

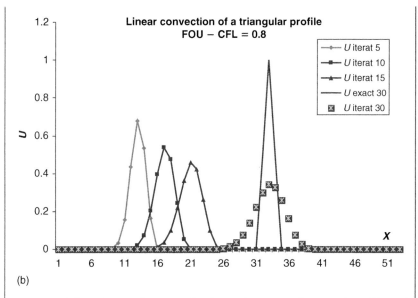

(b)

Figure 4.1.5b *First order upwind scheme for triangular initial profile and* $\sigma = 0.8$, *after 5, 10, 15 and 30 time steps.*

4.1.2.2 *Linear diffusion equation*

We consider now the time-dependent diffusion equation (3.1.10), describing a damped diffusion in time

$$\frac{\partial u}{\partial t} = \alpha \frac{\partial^2 u}{\partial x^2} \tag{4.1.41}$$

which we write also in condensed notation as

$$u_t = \alpha u_{xx} \tag{4.1.42}$$

The physics of diffusion is of isotropic nature and therefore the FD formula which correspond best to this property is a central difference, which does not distinguish between upstream and downstream directions. Hence, we select the second order central difference (4.1.18) for the space derivative and a forward difference in time, leading to

$$u_i^{n+1} = u_i^n + \frac{\alpha \Delta t}{\Delta x^2}(u_{i+1} - 2u_i + u_{i-1}) \tag{4.1.43}$$

Here again we can choose an explicit scheme

$$u_i^{n+1} = u_i^n + \frac{\alpha \Delta t}{\Delta x^2}(u_{i+1} - 2u_i + u_{i-1})^n \tag{4.1.44}$$

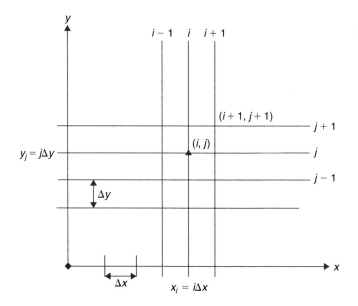

Figure 4.2.1 *Two-dimensional Cartesian mesh.*

or an implicit scheme

$$u_i^{n+1} = u_i^n + \frac{\alpha \Delta t}{\Delta x^2}(u_{i+1} - 2u_i + u_{i-1})^{n+1} \tag{4.1.45}$$

Both schemes are first order in time and second order in space.

A well-known scheme for this parabolic time-dependent diffusion equation is obtained by taking the average of the explicit and implicit schemes, leading to the scheme known as the ***Crank–Nicholson scheme***

$$u_i^{n+1} = u_i^n + \frac{1}{2}\frac{\alpha \Delta t}{\Delta x^2}(u_{i+1}^{n+1} - 2u_i^{n+1} + u_{i-1}^{n+1}) + \frac{1}{2}\frac{\alpha \Delta t}{\Delta x^2}(u_{i+1}^n - 2u_i^n + u_{i-1}^n) \tag{4.1.46}$$

This scheme is second order in time and in space.

4.2 MULTIDIMENSIONAL FINITE DIFFERENCE FORMULAS

It is easy to extend the one-dimensional FD formulas to the partial derivatives in two- or three-dimensional space, by applying the definition stating that a partial derivative with respect to x is the variation in the x-direction at constant y.

In a two-dimensional space, a rectangular mesh can be defined by the points of coordinates $x_i = x_0 + i\Delta x$ and $y_j = y_0 + j\Delta y$ (Figure 4.2.1).

Defining $u_{ij} = u(x_i, y_j)$, all the above formulas can be applied on either variable x, y, acting separately on the i and j subscripts, and all derivations are based on a two-dimensional Taylor expansion of mesh point value around u_{ij}. For instance, for point $(i+1, j+1)$

$$u(x_i + \Delta x, y_j + \Delta y) \equiv u_{i+1,j+1} = u_{ij} + \left(\Delta x \frac{\partial}{\partial x} + \Delta y \frac{\partial}{\partial y} \right) u \Big|_{ij}$$

$$+ \frac{1}{2} \left(\Delta x \frac{\partial}{\partial x} + \Delta y \frac{\partial}{\partial y} \right)^2 u \Big|_{ij}$$

$$+ \frac{1}{6} \left(\Delta x \frac{\partial}{\partial x} + \Delta y \frac{\partial}{\partial y} \right)^3 u \Big|_{ij} + \cdots \qquad (4.2.1)$$

For the first partial derivative in the x-direction a forward difference of first order accuracy is

$$(u_x)_{ij} = \left(\frac{\partial u}{\partial x} \right)_{ij} = \frac{u_{i+1,j} - u_{ij}}{\Delta x} + O(\Delta x) \qquad (4.2.2)$$

and similarly in the y-direction

$$(u_y)_{ij} = \left(\frac{\partial u}{\partial y} \right)_{ij} = \frac{u_{i,j+1} - u_{ij}}{\Delta y} + O(\Delta y) \qquad (4.2.3)$$

Backward partial differences can be defined in a similar way, also with first order accuracy

$$(u_x)_{ij} = \left(\frac{\partial u}{\partial x} \right)_{ij} = \frac{u_{ij} - u_{i-1,j}}{\Delta x} + O(\Delta x) \qquad (4.2.4)$$

$$(u_y)_{ij} = \left(\frac{\partial u}{\partial y} \right)_{ij} = \frac{u_{ij} - u_{i,j-1}}{\Delta y} + O(\Delta y) \qquad (4.2.5)$$

For central difference formulas, we have

$$(u_x)_{ij} = \left(\frac{\partial u}{\partial x} \right)_{ij} = \frac{u_{i+1,j} - u_{i-1,j}}{2\Delta x} + O(\Delta x^2) \qquad (4.2.6)$$

$$(u_y)_{ij} = \left(\frac{\partial u}{\partial y} \right)_{ij} = \frac{u_{i,j+1} - u_{i,j-1}}{2\Delta y} + O(\Delta y^2) \qquad (4.2.7)$$

Also, a second order, central difference formula for the second derivative will be, referring to formula (4.1.18):

$$(u_{xx})_{ij} = \left(\frac{\partial^2 u}{\partial x^2} \right)_{ij} = \frac{u_{i+1,j} - 2u_{ij} + u_{i-1,j}}{\Delta x^2} - \frac{\Delta x^2}{12} \left(\frac{\partial^4 u}{\partial x^4} \right)_{ij} \qquad (4.2.8)$$

and similar expressions can be derived for the y-derivatives:

$$(u_{yy})_{ij} = \left(\frac{\partial^2 u}{\partial y^2} \right)_{ij} = \frac{u_{i,j+1} - 2u_{ij} + u_{i,j-1}}{\Delta y^2} - \frac{\Delta y^2}{12} \left(\frac{\partial^4 u}{\partial y^4} \right)_{ij} \qquad (4.2.9)$$

Besides the straightforward application of the various formulas presented in the previous section additional forms can be defined by introducing an interaction between the two space directions for instance through a semi-implicit form on one of the two space coordinates. A representative example for a first x-derivative is a weighted average of the central difference formulas on the lines $j-1, j, j+1$, as

$$\left(\frac{\partial u}{\partial x}\right)_{ij} = \frac{1}{6}\left[\frac{u_{i+1,j+1} - u_{i-1,j+1}}{\Delta x} + 4\frac{u_{i+1,j} - u_{i-1,j}}{\Delta x}\right.$$

$$\left. + \frac{u_{i+1,j-1} - u_{i-1,j-1}}{\Delta x}\right] + O(\Delta x^2) \qquad (4.2.10)$$

which is also of second order accuracy.

4.2.1 Difference Schemes for the Laplace Operator

The Laplace equation plays an important role in CFD, as it appears in the Navier–Stokes equations, as well as in simple models such as heat conduction or potential flows. Therefore, its discretization is of major importance for many aspects of CFD.

In order to illustrate this point, let us consider the Laplace operator $\Delta u = u_{xx} + u_{yy}$ in two dimensions.

As the Laplace equation is typical for diffusion phenomena, it has to be discretized with central differences, in order for the discretization to be consistent with the physics it simulates.

This is a crucial element in the selection of an adequate numerical scheme, among all the possible options.

Application of second order central differencing in both directions leads to the well-known *five-point difference operator*

$$\Delta u_{ij} = \frac{u_{i+1,j} - 2u_{ij} + u_{i-1,j}}{\Delta x^2} + \frac{u_{i,j+1} - 2u_{ij} + u_{i,j-1}}{\Delta y^2} + O(\Delta x^2, \Delta y^2) \quad (4.2.11)$$

or for a uniform mesh, $\Delta x = \Delta y$

$$\frac{u_{i+1,j} + u_{i-1,j} + u_{i,j+1} + u_{i,j-1} - 4u_{ij}}{\Delta x^2} = \Delta u_{ij} + \frac{\Delta x^2}{12}\left(\frac{\partial^4 u}{\partial x^4} + \frac{\partial^4 u}{\partial y^4}\right)_{ij} \quad (4.2.12)$$

referring to the truncation errors given by equation (4.1.19).

This is the most widely applied difference scheme, of second order accuracy, for the Laplace operator.

This formula is illustrated by the computational molecule of Figure 4.2.2. The concept of the computational molecule is based on representing, in an (i, j) plane, only the points which contribute to the difference formula, with their coefficients. It provides a visual, easy to remember, representation of a two-dimensional difference formula.

Other combinations are possible whereby difference operators on the two space coordinates are mixed. For instance the following formula, for $\Delta x = \Delta y$, is also

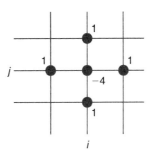

Figure 4.2.2   *Computational molecule for the five-point Laplace operator,*
equation (4.2.12).

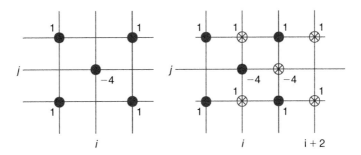

Figure 4.2.3   *Five-point molecules for Laplace operator of equation (4.2.13) for*
$\Delta x = \Delta y$.

a second order approximation of the Laplacian operator which is represented in
Figure 4.2.3:

$$\Delta u_{ij} = \frac{1}{4\Delta x^2}(u_{i+1,\,j+1} + u_{i+1,\,j-1} + u_{i-1,\,j-1} + u_{i-1,\,j+1} - 4u_{ij})$$
$$+ O(\Delta x^2, \Delta y^2) \tag{4.2.13}$$

Its truncation error can be obtained from Taylor expansions of all the points involved
around the point (i, j), leading to (see Problem P.4.5) where we represent the difference
operator of equation (4.2.13) by $\Delta^{(2)}u_{ij}$.

$$\Delta^{(2)}u_{ij} = \Delta u_{ij} + \frac{\Delta x^2}{12}\frac{\partial^4 u}{\partial x^4}\bigg|_{ij} + \frac{\Delta y^2}{12}\frac{\partial^4 u}{\partial y^4}\bigg|_{ij} + \left(\frac{\Delta x^2 + \Delta y^2}{4}\right)\frac{\partial^4 u}{\partial x^2 \partial y^2}\bigg|_{ij} \tag{4.2.14}$$

This other five-point scheme is interesting, as it is a rotated version of the scheme
(4.2.12), but actually, it is associated with a major problem. This can be seen by
looking at the link between the schemes of two neighboring points, for instance the
scheme written for point $(i + 1, j)$. By shifting the molecule horizontally by Δx,
represented by open symbols in Figure 4.2.3, we see that the molecules of the points
(i, j) and $(i + 1, j)$ have not a single common point. This is also the case for all the
molecules shifted by one cell in either x- or y-direction. If i, j are even numbers, all

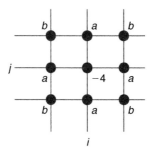

Figure 4.2.4   *Nine-point molecule for the Laplace operator (4.2.15) with $a + b = 1$.*

these shifted points will have either i or j uneven. Hence, this total decoupling of the computational molecules is called an ***odd–even decoupling***, leading to solutions on even points which will have different error levels than the solutions on the uneven points.

Therefore, this scheme is not recommended.

We can define a family of nine-point schemes for the Laplace operator on a uniform cartesian mesh $\Delta x = \Delta y$, by the combination

$$\Delta^{(3)} u_{ij} = (a\Delta^{(1)} + b\Delta^{(2)})u_{ij} \quad \text{with} \quad a + b = 1 \tag{4.2.15}$$

where we designate the standard five-point scheme of equation (4.2.12) by $\Delta^{(1)} u_{ij}$.

Combining these two operators, we obtain (see also Problem P.4.6)

$$\Delta^{(3)} u_{ij} = \Delta u_{ij} + \frac{\Delta x^2}{12} \left[\frac{\partial^4 u}{\partial x^4} + \frac{\partial^4 u}{\partial y^4} + 6b\frac{\partial^4 u}{\partial x^2 \partial y^2} \right]_{ij} \tag{4.2.16}$$

The computational molecule associated to this scheme is shown in Figure 4.2.4

The particular choice of $a = b = 1/2$ leads to the well-known scheme of Figure 4.2.5a, which is also obtained from a Galerkin finite element discretization of the Laplace operator on the same mesh, using bilinear quadrilateral elements, as shown in Chapter 5.

With $b = 1/3$, we obtain the computational molecule of Figure 4.2.5b, which is recommended by Dahlquist and Bjorck (1974), because the truncation error is equal to

$$-\frac{\Delta x^2}{12} \left(\frac{\partial^2}{\partial x^2} + \frac{\partial^2}{\partial y^2} \right)^2 u = -\frac{\Delta x^2}{12} \Delta^2 u$$

Hence, the equation $\Delta u = \lambda u$ can be discretized with this nine-point operator $\Delta^{(3)} = (\frac{2}{3}\Delta^{(1)} + \frac{1}{3}\Delta^{(2)})$ and will have a truncation error equal to $-\frac{\lambda^2 \Delta x^2}{12} u$

Therefore, the corrected difference scheme

$$\Delta^{(3)} u_{ij} = \left(\lambda + \frac{\lambda^2 \Delta x^2}{12} \right) u$$

will have a fourth order truncation error.

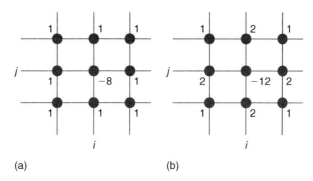

(a)   (b)

Figure 4.2.5   *Nine-point molecule for the Laplace operator (4.2.15) for b = 1/2 and b = 1/3. All coefficients are multiplied by (a) 1/2 and (b) 1/3.*

Example E.4.2.1: A program for solving the Laplace equation

The schemes for the Laplace equation allow us now to write a simple program for the Poisson equation:

$$\Delta u = f \tag{E.4.2.1}$$

on a Cartesian grid. If we select the five-point scheme (4.2.11), we can discretize the Poisson equation as

$$\frac{u_{i+1,j} - 2u_{ij} + u_{i-1,j}}{\Delta x^2} + \frac{u_{i,j+1} - 2u_{ij} + u_{i,j-1}}{\Delta y^2} = f_{ij} \tag{E.4.2.2}$$

or for $\Delta x = \Delta y$

$$u_{i+1,j} + u_{i-1,j} + u_{i,j+1} + u_{i,j-1} - 4u_{ij} = f_{ij}\Delta x^2 \tag{E.4.2.3}$$

We can now write a five-line FORTRAN program to solve this numerical scheme by an iterative method:

```
do N = 1, NTmax
do i = 1, Imax
do j = 1, Jmax
u(i,j) = 0.25* (u(i+1,j)+u(i-1,j)+u(i,j+1)+
        u(i,j-1)-f(i,j)*Δx**2)
continue
```

where NTmax is the maximum number of iterations, Imax and Jmax being the number of points in the *x*- and *y*-directions, respectively.

The iterative method behind this algorithm is the Gauss–Seidel method, to be introduced in Part IV, Chapter 10.

Of course, in a practical code for the Poisson equation, we will have to add instructions for the treatment of the boundary conditions, as well as appropriate routines for reading in the geometrical data and post-processing the results. These additions can require many thousands of lines, depending on the level of generality of the Poisson solver.

But the point we want to make here is that, once the theoretical developments are done, and this can take many pages of theory, ultimately the code implementing the selected numerical scheme can be very short and simple.

We expect that you will have the opportunity to verify this in the following when programming some of the proposed schemes in the later chapters.

Nonlinear diffusion terms

In the Navier–Stokes equations, the diffusion terms have a more general form compared to the Laplace equation when the diffusion coefficients are not constant. They appear under the form of the operator $\vec{\nabla}(\kappa\vec{\nabla}u)$ as shown in Chapter 1. A second order discretization of this operator can be written as follows, introducing the values at mid-points $(i \pm 1/2, j)$ and $(i, j \pm 1/2)$:

$$\vec{\nabla}(\kappa\vec{\nabla}u)_{ij} = \frac{1}{\Delta x^2}(\kappa_{i+1/2,\, j}(u_{i+1,\, j} - u_{i,\, j}) - \kappa_{i-1/2,\, j}(u_{i,\, j} - u_{i-1,\, j}))$$

$$+ \frac{1}{\Delta y^2}(\kappa_{i,\, j+1/2}(u_{i,\, j+1} - u_{i,\, j}) - \kappa_{i,\, j-1/2}(u_{i,\, j} - u_{i,\, j-1})) \quad (4.2.17)$$

It is seen that for constant diffusion coefficient κ, we recover the five-point scheme (4.2.11) for Laplace equation.

4.2.2 Mixed Derivatives

Mixed derivatives of any order can be discretized in much the same way by using for $\partial/\partial x$ and $\partial/\partial y$ the various formulas and their possible combinations described above.

The simplest, second order central formula for the mixed derivative is obtained from applying central differences in both directions x and y:

$$(u_{xy})_{ij} = \frac{\partial}{\partial x}\left(\frac{u_{i,\, j+1} - u_{i,\, j-1}}{2\Delta y} + O(\Delta y^2)\right)$$

$$= \frac{1}{4\Delta x\Delta y}(u_{i+1,\, j+1} - u_{i-1,\, j+1} - u_{i+1,\, j-1} + u_{i-1,\, j-1}) + O(\Delta x^2, \Delta y^2)$$

$$(4.2.18)$$

which is illustrated by the molecule of Figure 4.2.6.

Other combinations are possible, for instance by combining a central difference in the x-direction and a first order forward difference in the y-direction:

$$(u_{xy})_{ij} = \frac{1}{2\Delta x\Delta y}(u_{i+1,\, j+1} - u_{i-1,\, j+1} - u_{i+1,\, j} + u_{i-1,\, j}) + O(\Delta x^2, \Delta y) \quad (4.2.19)$$

which is first order in Δy and second order in Δx. This formula is represented in Figure 4.2.7.

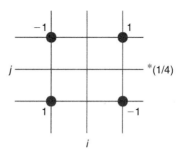

Figure 4.2.6 *Computational molecule for the second order accurate, mixed derivative formula (4.2.18).*

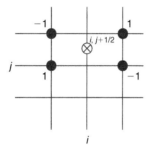

Figure 4.2.7 *Mixed derivative formula (4.2.19). All coefficients to be multiplied by 1/2.*

Similar formulas can be obtained by applying a backward difference in y, instead, or by changing the roles of x and y, leading to a formula which is second order in Δy and first order in Δx (see Problem P.4.8).

As mentioned previously a same formula can be of higher order if considered as an approximation of the derivative at an appropriate mid-point. The difference formula of equation (4.2.19) is indeed a second order approximation to $(u_{xy})_{i,j+1/2}$, which is centrally located, within the four involved mesh point.

This can be seen by applying a Taylor expansion of all the mesh point values of this equation around $u_{i,j+1/2}$. For instance, for the points $(i \pm 1, j+1)$, we have

$$u_{i\pm1,j+1} = u_{i,j+1/2} + \left(\pm\Delta x\frac{\partial}{\partial x} + \frac{\Delta y}{2}\frac{\partial}{\partial y} \right) u_{i,j+1/2}$$

$$+ \frac{1}{2}\left(\pm\Delta x\frac{\partial}{\partial x} + \frac{\Delta y}{2}\frac{\partial}{\partial y} \right)^2 u_{i,j+1/2} + O(\Delta x^3, \Delta y^3)$$

$$= u_{i,j+1/2} \pm \Delta x(u_x)_{i,j+1/2} + \frac{\Delta y}{2}(u_y)_{i,j+1/2} + \frac{1}{2}\Delta x^2(u_{xx})_{i,j+1/2}$$

$$+ \frac{1}{8}\Delta y^2(u_{yy})_{i,j+1/2} \pm \frac{\Delta x\Delta y}{2}(u_{xy})_{i,j+1/2} + O(\Delta x^3, \Delta y^3) \quad (4.2.20)$$

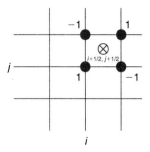

Figure 4.2.8   *Mixed derivative formulas (4.2.23) and (4.2.24).*

Similarly for the points $(i \pm 1, j)$, we have

$$u_{i\pm1,j} = u_{i,j+1/2} + \left(\pm\Delta x \frac{\partial}{\partial x} - \frac{\Delta y}{2}\frac{\partial}{\partial y}\right) u_{i,j+1/2}$$

$$+ \frac{1}{2}\left(\pm\Delta x \frac{\partial}{\partial x} - \frac{\Delta y}{2}\frac{\partial}{\partial y}\right)^2 u_{i,j+1/2} + O(\Delta x^3, \Delta y^3)$$

$$= u_{i,j+1/2} \pm \Delta x (u_x)_{i,j+1/2} - \frac{\Delta y}{2}(u_y)_{i,j+1/2} + \frac{1}{2}\Delta x^2 (u_{xx})_{i,j+1/2}$$

$$+ \frac{1}{8}\Delta y^2 (u_{yy})_{i,j+1/2} \mp \frac{\Delta x \Delta y}{2}(u_{xy})_{i,j+1/2} + O(\Delta x^3, \Delta y^3) \qquad (4.2.21)$$

By forming the differences of equation (4.2.19), we obtain

$$(u_{xy})_{i,j+1/2} = \frac{1}{2\Delta x \Delta y}(u_{i+1,j+1} - u_{i-1,j+1} - u_{i+1,j} + u_{i-1,j}) + O(\Delta x^2, \Delta y^2) \qquad (4.2.22)$$

demonstrating the second order accuracy of this formula.

A first order formula in both x and y is obtained from first order forward differences in both directions, leading to

$$(u_{xy})_{ij} = \frac{1}{\Delta x \Delta y}(u_{i+1,j+1} - u_{i,j+1} - u_{i+1,j} + u_{i,j}) + O(\Delta x, \Delta y) \qquad (4.2.23)$$

Observe that the same formula (4.2.23) will give a second order accurate estimation of the mixed derivative taken at the point $(i + 1/2, j + 1/2)$ (see Problem P.4.9) that is

$$(u_{xy})_{i+1/2,j+1/2} = \frac{1}{\Delta x \Delta y}(u_{i+1,j+1} - u_{i,j+1} - u_{i+1,j} + u_{i,j}) + O(\Delta x^2, \Delta y^2) \qquad (4.2.24)$$

These formulas are represented on Figure 4.2.8.

Applying backward differences in both directions, leads to

$$(u_{xy})_{ij} = \frac{1}{\Delta x \Delta y}(u_{i-1,j-1} - u_{i,j-1} - u_{i-1,j} + u_{i,j}) + O(\Delta x, \Delta y) \qquad (4.2.25)$$

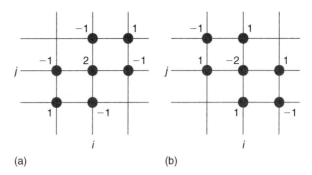

Figure 4.2.9 *Second order mixed derivative approximations (a) equation (4.2.26) and (b) equation (4.2.27).*

Since the truncation errors of equations (4.2.25) and (4.2.23) are equal but of opposite signs (see Problem P.4.10) the sum of the two expressions will lead to a second order accurate formula for the mixed derivative:

$$(u_{xy})_{ij} = \frac{1}{2\Delta x \Delta y}(u_{i+1, j+1} - u_{i, j+1} - u_{i+1, j} + u_{i-1, j-1} - u_{i, j-1} - u_{i-1, j}$$
$$+ 2u_{i,j}) + O(\Delta x^2, \Delta y^2) \tag{4.2.26}$$

This formula is represented in Figure 4.2.9a and, compared to the central approximation (4.2.18) shown in Figure 4.2.6, has a non-zero coefficient for u_{ij}. This might be advantageous in certain cases by enhancing the weight of the u_{ij} coefficients in the matrix equations obtained after discretization, that is enhancing the diagonal dominance, see for instance O'Carroll (1976).

An alternative to the last formulation is obtained by a different combination of forward and backward differences, leading to the second order approximation for $(u_{xy})_{ij}$, shown in Figure 4.2.9b:

$$(u_{xy})_{ij} = \frac{1}{2\Delta x \Delta y}(u_{i+1, j} - u_{i+1, j-1} + u_{i, j+1} + u_{i, j-1} - u_{i-1, j+1} + u_{i-1, j}$$
$$- 2u_{i, j}) + O(\Delta x^2, \Delta y^2) \tag{4.2.27}$$

It can also be seen, by adding up the two last expressions, that we recover the fully central second order approximation (4.2.18). Many other formulas can be found in the literature, for instance in Mitchell and Griffiths (1980).

4.3 FINITE DIFFERENCE FORMULAS ON NON-UNIFORM GRIDS

Up to now, we have introduced finite difference formula on uniform grids where the distance between adjacent mesh points is constant over the whole grid. In practice however, this will seldom be the case, as more often the grid has to adapt to the geometrical boundaries or to the flow physics, making uniform grids unpractical.

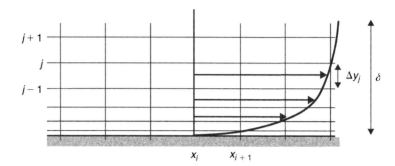

Figure 4.3.1 *Representative grid in a boundary layer region above a solid wall. The velocity profile is superposed on the grid, which is selected to be uniform in the x-direction and non-uniform in the y-direction.*

We refer you to the grid examples shown in the general introduction to this book, where you can observe a selected cross-section of realistic grids.

A very representative example is provided by the grids in boundary layer regions around solid surfaces (see Figure 4.3.1). In a laminar boundary layer, it is known from Prandtl's analysis that the boundary layer thickness δ over a flat plate scales with x like the inverse of the square root of the Reynolds number. We refer you to your basic course of Fluid Mechanics for the background behind the boundary layer properties. That is

$$\delta \sim \sqrt{\frac{\nu x}{U_\infty}} = \frac{x}{\sqrt{Re_x}} \quad \text{with} \quad Re_x = \frac{U_\infty x}{\nu}$$

where ν is the kinematic viscosity (in m²/s) and U_∞ is the velocity outside the boundary layer. It is also known that the ratio of the velocity gradients in the normal and streamwise directions is of the order of the square root of this same Reynolds number:

$$\frac{\partial u}{\partial y} \bigg/ \frac{\partial u}{\partial x} \sim \sqrt{Re_x}$$

After having generated a grid, if we wish to ensure that the velocity variations in the x- and y-directions are of the same order over the mesh distances Δx and Δy, respectively, we should generate a grid with an aspect ratio $\Delta x/\Delta y$ of the order of

$$\frac{\Delta y}{\Delta x} \sim \frac{1}{\sqrt{Re_x}}$$

For a realistic Reynolds number of say, 1 million, this ratio is of the order of 1000!

Hence, we would need to generate cells where Δy is thousand times smaller than Δx. In practical terms, for a plate of unit length and 101 mesh points in the x-direction, that is $\Delta x = 0.01$, Δy should be of the order of 10^{-5}!

This simple analysis shows that a uniform mesh would require unrealistic small cell sizes, to be equal to the smallest cell size of 10^{-5}, leading to millions of mesh points.

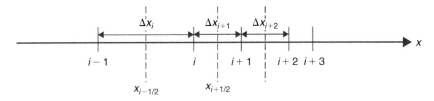

Figure 4.3.2   *Arbitrary mesh point distribution in one-dimensional space x.*

Another lesson of this analysis is that we should not generate uniform grids in the y-direction, but have Δy increase progressively when moving away from the wall as the intensity of the gradients progressively reduces to values of the same order as in the x-direction, as shown in Figure 4.3.1.

An approach often used in practice is to define a clustering factor r, such that

$$y_{j+1} = y_1 r^j \qquad (4.3.1)$$

This creates mesh cells growing progressively in the y-direction as shown in Figure 4.3.1. Typical values of the clustering factor are $r \sim 1.1$–1.5.

Due to the strong gradients in the normal directions, it is essential for reasons of accuracy to position at least 10–20 points in the boundary layer regions.

The derivation of accurate difference formulas for non-uniform grids is therefore very important in CFD, since this is the situation most often found in practice. Hence, great care should be given to the choice of appropriate discretization formulas on these grids.

It is very essential that you give a serious attention to the related problem of loss of accuracy on non-uniform grids, if either you write your own code, or if you use existing, commercial or other, CFD codes to study practical fluid problems, when you evaluate the results of the simulations.

We will go along the following steps:

- Firstly, derive formulas valid on non-uniform grids.
- Secondly, evaluate the loss of accuracy related to the FD formula on non-uniform grids.
- Thirdly, and more importantly, provide guidelines on the grid properties and on grid quality in order to minimize the errors related to the non-uniform grids.

For one-dimensional non-uniform grids, we can define different types of configurations, depending on the way we position the mid-point values and the points where the function values are evaluated.

Figure 4.3.2 represents a mesh arrangement, where the points $(i \pm 1/2)$ are the mid-points of the intervals $(i, i \pm 1)$ and is defined as a *cell vertex* configuration in the finite volume context. If we consider the cell $(i - 1/2, i + 1/2)$ mesh point i is not at its center due to the grid non-uniformity. An alternative option, known as *cell centered*, is shown in Figure 4.3.3.

In order to derive FD formula for non-uniform mesh sizes in a one-dimensional space defined by the grid of Figure 4.3.2, we refer again to the Taylor series expansions, written for the mesh points $(i \pm 1)$, $(i \pm 2)$:

$$u_{i-1} = u_i - \Delta x_i (u_x)_i + \frac{\Delta x_i^2}{2} (u_{xx})_i - \frac{\Delta x_i^3}{6} (u_{xxx})_i + \cdots$$

$$u_{i+1} = u_i + \Delta x_{i+1} (u_x)_i + \frac{\Delta x_{i+1}^2}{2} (u_{xx})_i + \frac{\Delta x_{i+1}^3}{6} (u_{xxx})_i + \cdots$$

$$u_{i-2} = u_i - (\Delta x_i + \Delta x_{i-1})(u_x)_i + \frac{(\Delta x_i + \Delta x_{i-1})^2}{2} (u_{xx})_i$$

$$- \frac{(\Delta x_i + \Delta x_{i-1})^3}{6} (u_{xxx})_i + \cdots$$

$$u_{i+2} = u_i + (\Delta x_{i+1} + \Delta x_{i+2})(u_x)_i + \frac{(\Delta x_{i+1} + \Delta x_{i+2})^2}{2} (u_{xx})_i$$

$$+ \frac{(\Delta x_{i+1} + \Delta x_{i+2})^3}{6} (u_{xxx})_i + \cdots \qquad (4.3.2)$$

where the notation

$$\Delta x_i = x_i - x_{i-1} \qquad (4.3.3)$$

is introduced.

In presence of a non-uniform grid, many options are open for difference formulas, generalizing the formulas derived in Section 4.3.1 for uniform grids.

The examples derived in the following sections are representative of the additional variety that arises when the grid has non-constant cell sizes.

4.3.1 Difference Formulas for First Derivatives

A one-sided, first order, formulas for the first derivative can be defined as follows:

Forward difference

$$(u_x)_i = \frac{u_{i+1} - u_i}{\Delta x_{i+1}} - \frac{\Delta x_{i+1}}{2} (u_{xx})_i \qquad (4.3.4)$$

Backward difference

$$(u_x)_i = \frac{u_i - u_{i-1}}{\Delta x_i} + \frac{\Delta x_i}{2} u_{xx} \qquad (4.3.5)$$

Central differences
If we take the simple average of the two formulas above, which is often done in finite difference computer programs, as well as in finite volume or finite element methods, we obtain a form of central difference,

$$(u_x)_i = \frac{1}{2} \left[\frac{u_{i+1} - u_i}{\Delta x_{i+1}} + \frac{u_i - u_{i-1}}{\Delta x_i} \right] - \frac{\Delta x_{i+1} - \Delta x_i}{4} (u_{xx})_i$$

$$- \frac{\Delta x_i^2 + \Delta x_{i+1}^2}{12} (u_{xxx})_i \qquad (4.3.6)$$

This formula reduces to equation (4.1.7) on a uniform grid, but the important observation is the appearance of the third term, which is of first order if the difference in adjacent cell sizes $(\Delta x_{i+1} - \Delta x_i)$ remains finite. If the mesh size varies abruptly, for instance if $\Delta x_{i+1} \sim 2\Delta x_i$ the above formula will be *only first order accurate*.

This is a general property of finite difference approximations on non-uniform meshes. If the mesh size does not vary smoothly, a loss of accuracy is unavoidable.

We will come back to this very important point at the end of this section.

A more elaborate formula can be defined, by combining formulas (4.3.4) and (4.3.5) in order to eliminate the first order truncation error, referring to the Taylor expansions of equation (4.3.2). This leads to the following formula (see also Problems P.4.10 and P.4.13):

$$
\begin{aligned}
(u_x)_i ={}& \frac{1}{\Delta x_{i+1} + \Delta x_i} \left[\frac{\Delta x_i}{\Delta x_{i+1}} (u_{i+1} - u_i) + \frac{\Delta x_{i+1}}{\Delta x_i} (u_i - u_{i-1}) \right] \\
& - \frac{\Delta x_i \Delta x_{i+1}}{6} (u_{xx})_i
\end{aligned}
\tag{4.3.7}
$$

which is *second order for any grid size distribution*. The price to pay for this property is a formula which is more complicated as it is formed by a weighted average of the one-sided formulas, based on the sizes of the *adjacent cells*.

It has to be mentioned here that these weighted averages are very difficult to generalize in two or three dimensions, while the simple average of formula (4.3.6) remains straightforward in multidimensional structured or unstructured grids. For practical applications, we have to restrict ourselves to simple expressions, easily extendable to arbitrary dimensions.

This explains why this simple average is most widely applied, with the risk of reduction of the order of accuracy. Therefore, guidelines in order to minimize the unfavorable effect of this approach are required and will be given in the following.

4.3.1.1 *Conservative FD formulas*

Referring to the grid distribution of Figure 4.3.2, an alternative expression for the first derivative can be written under the following form, instead of (4.3.6), based on the function values at the mid-points $(i \pm 1/2)$:

$$
(u_x)_i = \frac{u_{i+1/2} - u_{i-1/2}}{x_{i+1/2} - x_{i-1/2}}
$$

$$
x_{i+1/2} - x_{i-1/2} = \frac{1}{2}(\Delta x_i + \Delta x_{i+1}) = \frac{1}{2}(x_{i+1} - x_{i-1})
\tag{4.3.8}
$$

This formula is also said to be *conservative*.

We have mentioned in Chapter 1, the importance of the conservative form of the conservation laws of Fluid Mechanics. This important property has also to be satisfied at the discrete level, that is after the discretization of the equations and the *conservative discretizations* will be extensively discussed in general terms in the next Chapter 5, in relation with the finite volume method. However, we wish to provide you already here with a first glimpse at these essential properties.

*Stated here in relation with structured grids, a difference formula is said to be in **conservative form**, if it is written as the difference of two quantities defined on opposite cell-faces, where in addition the cell-face quantities are not dependent on the cell in which the face is considered.*

Observe that equation (4.3.7), although having second order accuracy does not satisfy this condition of **conservative discretization.**

Formula (4.3.8) is at best of first order accuracy on a non-uniform grid, since point i is not at the center of the interval $(i - 1/2, i + 1/2)$ (see Problem P.4.11). This is readily seen from the Taylor expansions (4.3.2), leading to the formula:

$$(u_x)_i = \frac{u_{i+1/2} - u_{i-1/2}}{x_{i+1/2} - x_{i-1/2}} - \frac{\Delta x_{i+1} - \Delta x_i}{4}(u_{xx})_i$$

$$- \frac{\Delta x_{i+1}^2 - \Delta x_{i+1}\Delta x_i + \Delta x_i^2}{24}(u_{xxx})_i \tag{4.3.9}$$

Applying the approximation $u_{i+1/2} = (u_i + u_{i+1})/2$ to equation (4.3.8), leads to the following central difference formula:

$$(u_x)_i = \frac{u_{i+1} - u_{i-1}}{x_{i+1} - x_{i-1}} \tag{4.3.10}$$

Performing a Taylor expansion, the truncation error becomes

$$(u_x)_i = \frac{u_{i+1} - u_{i-1}}{x_{i+1} - x_{i-1}} - \frac{\Delta x_{i+1} - \Delta x_i}{2}(u_{xx})_i$$

$$- \frac{\Delta x_{i+1}^2 - \Delta x_{i+1}\Delta x_i + \Delta x_i^2}{6}(u_{xxx})_i \tag{4.3.11}$$

The same comments as stated in relation with formula (4.3.6) are valid here, namely that this formula is only first order accurate on a general non-uniform grid, unless the variation of the grid size is very smooth, namely $(\Delta x_{i+1} - \Delta x_i) \sim O(\Delta x^2)$.

Observe also that the first term of the truncation error of formula (4.3.9) is lower than the corresponding term of formula (4.3.11) showing that on an irregular grid this formula will be slightly more accurate.

4.3.2 A General Formulation

The general expression (4.3.8) can be applied to generate a whole family of FD formulas by defining a general interpolation rule for the cell-face values. The function values at the cell 'faces' $(i \pm 1/2)$ are defined by a linear interpolation from the mesh point values, following:

$$u_{i+1/2} = u_i + \alpha_i(u_{i+1} - u_i) + \beta_i(u_i - u_{i-1}) \tag{4.3.12}$$

General conditions can be written for this interpolation formula to be at least second order accurate (see Problem P.4.17). On a uniform mesh, the condition reduces to $\alpha_i + \beta_i = 1/2$.

On a uniform mesh, $\alpha_i = 1/2$, $\beta_i = 0$, corresponding to the 'central' choice $u_{i+1/2} = (u_i + u_{i+1})/2$, reproduces the second order central difference (4.1.7); the choice $\alpha_i = \beta_i = 0$ reproduces the first order backward difference (4.1.6), while selecting $\alpha_i = 0$, $\beta_i = 1/2$ leads to the second order backward difference (4.1.16).

On the non-uniform mesh, the backward difference obtained with $\alpha_i = \beta_i = 0$ gives the following expression, as an alternative to equation (4.3.5):

$$(u_x)_i = \frac{u_i - u_{i-1}}{(x_{i+1} - x_{i-1})/2} \qquad (4.3.13)$$

with the truncation error

$$\frac{u_i - u_{i-1}}{(x_{i+1} - x_{i-1})/2} = \frac{\Delta x_{i+1} + \Delta x_i}{2\Delta x_i}(u_x)_i - \frac{\Delta x_i}{2}(u_{xx})_i + \frac{\Delta x_i^2}{6}(u_{xxx})_i + O(\Delta x^3) \qquad (4.3.14)$$

The first term on the right-hand side has a coefficient different from one on a non-uniform grid, showing that this backward difference formula reduces to **zero order of accuracy** on a general non-uniform grid and **is therefore not acceptable**.

Similarly, the option $\alpha_i = 0$, $\beta_i = 1/2$, which gives the second order backward difference (4.1.16) on a uniform mesh, leads to the formula:

$$(u_x)_i = \frac{3u_i - 4u_{i-1} + u_{i-2}}{(x_{i+1} - x_{i-1})/2} \qquad (4.3.15)$$

with the truncation error

$$\frac{3u_i - 4u_{i-1} + u_{i-2}}{(x_{i+1} - x_{i-1})/2} = +\frac{3\Delta x_i - \Delta x_{i-1}}{\Delta x_i + \Delta x_{i+1}}(u_x)_i$$
$$-\frac{(\Delta x_i + \Delta x_{i-1})^2 - 4\Delta x_i^2}{2(\Delta x_i + \Delta x_{i+1})}(u_{xx})_i + O(\Delta x^2) \qquad (4.3.16)$$

Hence, this formula is also of **zero order of accuracy** on an arbitrary grid, although it is second order on a uniform mesh. Again, this formula should *not* be used.

The main observation is that difference formulas generally can loose at least one order of accuracy, and sometimes two, on general non-uniform grids.

In order to achieve second order accuracy on arbitrary grids, one has to consider difference formulas that are formally of higher order on uniform grids, involving additional mesh points.

Let us therefore consider a family of finite difference formulas for derivatives at point i on a four point support, whereby point $(i-2)$ is added to the basic set $(i-1, i, i+1)$. Hence, we look for an expression of the form:

$$(u_x)_i = \frac{au_{i-2} + bu_{i-1} + cu_i + du_{i+1}}{(\Delta x_i + \Delta x_{i+1})/2} \qquad (4.3.17)$$

Applying the procedure of Section 4.1.2 based on the Taylor expansions (4.3.2), we obtain the following conditions for a second order accurate formula:

$$a + b + c + d = 0$$

$$d\Delta x_{i+1} - b\Delta x_i - a(\Delta x_i + \Delta x_{i-1}) = \frac{1}{2}(\Delta x_{i+1} + \Delta x_i) \qquad (4.3.18)$$

$$d\Delta x_{i+1}^2 + b\Delta x_i^2 + a(\Delta x_i + \Delta x_{i-1})^2 = 0$$

or, defining the mesh ratio $r_i = \Delta x_{i+1}/\Delta x_i$

$$a + b + c + d = 0$$

$$dr_i - b - a\left(1 + \frac{1}{r_{i-1}}\right) = \frac{1}{2}(1 + r_i) \qquad (4.3.19)$$

$$dr_i^2 + b + a\left(1 + \frac{1}{r_{i-1}}\right)^2 = 0$$

These relations define a one-parameter difference formula of second order accuracy for the first derivative (see Problem P.4.13).

On a uniform mesh, the formulas are defined by the coefficients $(a, b, c, d) = (a, -3a - 1/2, 3a, -a + 1/2)$ and the value $a = 1/6$ leads to the unique third order formula on a uniform grid.

Observe that the three point formula with $a = 0$ leads to the unique second order scheme (4.3.7), which is not conservative, since it cannot be written under the form (4.3.8).

The combination of second order and conservativity therefore requires an additional degree of freedom, provided by the coefficient a. Assuming the interpolation relation (4.3.12), formula (4.3.8) can be written under the form (4.3.17) and identifying the coefficients leads to

$$
\begin{aligned}
a &= \beta_{i-1} \\
b &= -(1 + \beta_i - \alpha_{i-1} + \beta_{i-1}) \\
c &= 1 + \beta_i - \alpha_{i-1} - \alpha_i \\
d &= \alpha_i
\end{aligned}
\qquad (4.3.20)
$$

This has to be seen as a set of equations for the unknown functions $\alpha(r)$ and $\beta(r)$, with $\alpha_i = \alpha(r_i)$ and $\beta_i = \beta(r_i)$, since these interpolation coefficients are non-dimensional functions of Δx_i and Δx_{i+1}. Similarly, $\alpha_{i-1} = \alpha(r_{i-1})$ and $\beta_{i-1} = \beta(r_{i-1})$. Eliminating b from the two last equations (4.3.19) leads to the following relation between $\alpha_i = \alpha(r_i)$ and $\beta_{i-1} = \beta(r_{i-1})$:

$$d = \alpha(r_i) = \frac{1}{2r_i} - \frac{\beta(r_{i-1})(1 + r_{i-1})}{r_{i-1}^2} \frac{1}{r_i(1 + r_i)} \qquad (4.3.21)$$

The terms depending on r_{i-1} have to be equal to a constant K, since r_i and r_{i-1} are independent variables. Hence, we have

$$\beta(r) = K\frac{r^2}{(1 + r)} \qquad \alpha(r) = \frac{1 + r - 2K}{2r(1 + r)} \qquad (4.3.22)$$

The constant K is easily determined, for instance by combining the second equation (4.3.19) with the second equation (4.3.20), leading to the unique value $K = 1/2$. This choice of interpolation coefficients, namely

$$\beta_i \equiv \beta(r_i) = \frac{r_i^2}{2(1 + r_i)} \quad \alpha_i \equiv \alpha(r_i) = \frac{1}{2(1 + r_i)} \tag{4.3.23}$$

provides a second order, conservative formula for the first derivative on an arbitrary mesh (See Problem P.4.15).

It has been applied to the numerical simulation of two-dimensional nonlinear wave solutions of the Euler and Navier–Stokes equations by Cain and Bush (1994), showing excellent accuracy on stretched grids.

On a uniform mesh, this formula reduces to

$$(u_x)_i = \frac{u_{i+1} + 3u_i - 5u_{i-1} + u_{i-2}}{4\Delta x} + \frac{\Delta x^2}{12}(u_{xxx})_i - \frac{\Delta x^3}{8}(u_{xxxx})_i \tag{4.3.24}$$

and is only second order accurate.

Note that on this fourth-point mesh, a formula with third order accuracy can be obtained by adding this condition to the system (4.3.19) (see Problem P.4.16). This formula is not conservative on a general grid and reduces to the following formula on a uniform grid:

$$(u_x)_i = \frac{2u_{i+1} + 3u_i - 6u_{i-1} + u_{i-2}}{6\Delta x} - \frac{\Delta x^3}{12}(u_{xxxx})_i \tag{4.3.25}$$

Two-dimensional extensions on non-uniform Cartesian grids are straightforward, applying the ratios (4.3.23) in each direction separately.

Second derivatives

A three-point, central difference formula for the second derivative is obtained in the simplest way by subtracting formulas (4.3.4) and (4.3.5), leading to

$$(u_{xx})_i = \frac{2}{\Delta x_{i+1} + \Delta x_i} \left[\frac{(u_{i+1} - u_i)}{\Delta x_{i+1}} - \frac{(u_i - u_{i-1})}{\Delta x_i} \right]$$
$$+ \frac{\Delta x_{i+1} - \Delta x_i}{3}(u_{xxx})_i - \frac{\Delta x_{i+1}^3 + \Delta x_i^3}{12(\Delta x_{i+1} + \Delta x_i)} \left(\frac{\partial^4 u}{\partial x^4} \right)_i \tag{4.3.26}$$

On a uniform grid, this formula reduces to the second order accurate finite difference (4.1.18). However, as with equation (4.3.6), the presence of a truncation error proportional to the difference of two consecutive mesh lengths reduces the accuracy to first order, under similar conditions as mentioned above.

4.3.3   Cell-Centered Grids

Another popular choice for a finite volume discretization consists in selecting the mesh points as the cell 'faces', which are then labeled at half-integer index values

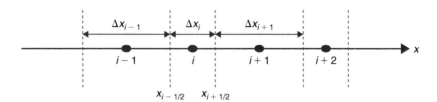

Figure 4.3.3 *Finite volume subdivision of a non-uniform, one-dimensional mesh point distribution. Cell-centered approach.*

$(i \pm 1/2)$, and the function values are defined at the centers of the cells. This *cell-centered* approach is represented in Figure 4.3.3 and is typical of a finite volume approach, to be introduced more extensively in Chapter 5.

It is characterized by the property that the nodes i at which the function values are defined are at the centers of the cells $(i - 1/2, i + 1/2)$, implying that the cell 'faces' $i \pm 1/2$ are not at the mid-points of the intervals $(i, i \pm 1)$.

Compare this mesh layout to the cell vertex of Figure 4.3.2 and observe the differences in the definition of the cell sizes.

Hence, the approximation $u_{i+1/2} = (u_i + u_{i+1})/2$ is only first order accurate on a non-uniform mesh, since combining the Taylor expansions of u_i and u_{i+1} around point i, we have

$$
u_{i+1/2} = \frac{1}{2}(u_i + u_{i+1}) - \frac{1}{4}(\Delta x_{i+1} - \Delta x_i)u_x \Big|_i
$$

$$
- \frac{1}{16}(\Delta x_{i+1}^2 + 2\Delta x_{i+1}\Delta x_i - \Delta x_i^2)u_{xx}\Big|_i + O(\Delta x^3) \tag{4.3.27}
$$

The following approximation for $u_{i+1/2}$ is second order accurate on an arbitrary mesh,

$$
u_{i+1/2} = \frac{\Delta x_{i+1}u_i + \Delta x_i u_{i-1}}{\Delta x_{i+1} + \Delta x_i} - \frac{1}{8}\Delta x_i \Delta x_{i+1} u_{xx}|_i + O(\Delta x^3) \tag{4.3.28}
$$

but again requires a weighted average, as in equation (4.3.7), which is difficult to generalize to multidimensional grids.

As we will see more in detail in Chapter 5, in cell-centered finite volume methods, all quantities and gradients are evaluated at cell faces, based on cell-center quantities. On the other hand, on a mesh such as depicted in Figure 4.3.3, the derivative u_{xi} is calculated from formulas based on cell-face values. For instance, a straightforward formula would be

$$
(u_x)_i = \frac{u_{i+1/2} - u_{i-1/2}}{\Delta x_i} \tag{4.3.29}
$$

leading to

$$
(u_x)_i = \frac{u_{i+1/2} - u_{i-1/2}}{\Delta x_i} - \frac{1}{24}\Delta x_i^2 (u_{xxx})_i \tag{4.3.30}
$$

based on the Taylor expansion of $u_{i\pm1/2}$ around point i. However, in practice, these are not the values at hand during a computation. If the values of formula (4.3.27) are used, a more straightforward formula would be

$$(u_x)_i = \frac{u_{i+1} - u_{i-1}}{\Delta x_i + (\Delta x_{i+1} + \Delta x_{i-1})/2} \tag{4.3.31}$$

leading to

$$(u_x)_i = \frac{u_{i+1} - u_{i-1}}{\Delta x_i + (\Delta x_{i+1} + \Delta x_{i-1})/2} - \frac{1}{4}(\Delta x_{i+1} - \Delta x_{i-1})(u_{xx})_i$$

$$- \frac{1}{24}(\Delta x_{i+1}^2 - \Delta x_{i-1}\Delta x_{i+1} + \Delta x_i^2 + \Delta x_{i-1}\Delta x_i + \Delta x_{i-1}^2$$

$$+ \Delta x_i \Delta x_{i+1})(u_{xxx})_i \tag{4.3.32}$$

which is first order on the non-uniform mesh.

4.3.4 Guidelines for Non-uniform Grids

The errors due to the grid non-uniformity will be minimized for **smoothly varying grids**, defined in such a way that the size variation between consecutive cells is of second order in the grid size. That is, if

$$\Delta x_{i+1} - \Delta x_i = O(\Delta x_i^2) \tag{4.3.33}$$

then formulas such as (4.3.6), (4.3.11) will be of second order accuracy, as on a uniform grid.

However, a grid variation, such as defined by equation (4.3.1), will not satisfy this condition, since this grid variation leads to, translated in the x-direction:

$$\Delta x_{i+1} - \Delta x_i = \Delta x_i(r - 1) = O(\Delta x_i) \tag{4.3.34}$$

Hence, this will reduce the formulas, such as (4.3.6), (4.3.11) to first order accuracy. However, as the coefficient is multiplied by $(r - 1)$, which is generally a small number, the error could remain small and acceptable. In addition, keep in mind that these errors are proportional to some derivative, second or higher, of the function u. Therefore, the impact of the errors due to the mesh non-uniformity will depend on the local flow properties. In regions where the flow variation is smooth, this impact will be reduced and could eventually be neglected. On the other hand, in regions with strong variations, where the gradients are important these additional errors can become critical and have to be severely controlled.

Additional guidelines can be stated here:

- Avoid discontinuities in grid size of adjacent cells.
- Always use laws for the grid size variation, defined by analytical, continuous functions of the associated coordinate, to minimize the numerical error. For

instance, clustering laws such as $\Delta x_{i+1} = r\Delta x_i$ where r is a constant factor, typically 1.1–2 are widely used.
- Pay a particular attention to the grid smoothness and grid density in regions of strong flow variations, for instance around a leading edge of an airfoil profile, or around stagnation points.

A4.4 GENERAL METHOD FOR FINITE DIFFERENCE FORMULAS

The following two sections are of interest for more advanced applications of finite difference methods and cover a methodology for the generation of arbitrary FD formulas with prescribed order (Section A4.4) and the generation of high order compact, implicit finite difference formulas (Section A4.5).

They are marked by the letter A, for 'Advanced' and can be included or not in the basic introductory course.

General procedures developed in order to generate finite difference formulas to any order of accuracy and a general theory can be found in Hildebrand (1956). This approach is based on the definition of the following difference operators:

Displacement operator E

$$Eu_i = u_{i+1} \tag{4.4.1a}$$

Forward difference operator δ^+

$$\delta^+ u_i = u_{i+1} - u_i \tag{4.4.1b}$$

Backward difference operator δ^-

$$\delta^- u_i = u_i - u_{i-1} \tag{4.4.1c}$$

Central difference operator δ

$$\delta u_i = u_{i+1/2} - u_{i-1/2} \tag{4.4.1d}$$

Central difference operator $\bar{\delta}$

$$\bar{\delta} u_i = \frac{1}{2}(u_{i+1} - u_{i-1}) \tag{4.4.1e}$$

Averaging operator μ

$$\mu u_i = \frac{1}{2}(u_{i+1/2} + u_{i-1/2}) \tag{4.4.1f}$$

Differential operator D

$$Du = u_x \equiv \frac{\partial u}{\partial x} \qquad (4.4.1g)$$

From these definitions some obvious relations can be defined between these operators, such as

$$\delta^+ = E - 1 \qquad (4.4.2)$$

$$\delta^- = 1 - E^{-1} \qquad (4.4.3)$$

where the inverse displacement operator E is introduced, defined by

$$E^{-1}u_i = u_{i-1} \qquad (4.4.4)$$

This leads to the following relations

$$\delta^- = E^{-1}\delta^+ \qquad (4.4.5)$$

and

$$\delta^+\delta^- = \delta^-\delta^+ = \delta^+ - \delta^- = \delta^2 \qquad (4.4.6)$$

With the general definition, n being positive or negative:

$$E^n u_i = u_{i+n} \qquad (4.4.7)$$

we also have

$$\delta = E^{1/2} - E^{-1/2} \qquad (4.4.8)$$

$$\bar{\delta} = \frac{1}{2}(E - E^{-1}) \qquad (4.4.9)$$

and

$$\mu = \frac{1}{2}(E^{1/2} + E^{-1/2}) \qquad (4.4.10)$$

Any of the above difference operators taken to a given power n, is interpreted as n repeated actions of this operator. For instance:

$$\delta^{+2} = \delta^+\delta^+ = E^2 - 2E + 1 \qquad (4.4.11)$$

$$\delta^{+3} = (E - 1)^3 = E^3 - 3E^2 + 3E - 1 \qquad (4.4.12)$$

A4.4.1 Generation of Difference Formulas for First Derivatives

The key to the operator technique for generating finite difference formulas lies in the relation between the derivative operator D and the finite displacement operator E.

This relation is obtained from the Taylor expansion:

$$u(x + \Delta x) = u(x) + \Delta x\, u_x(x) + \frac{\Delta x^2}{2!} u_{xx}(x) + \frac{\Delta x^3}{3!} u_{xxx}(x) + \cdots \tag{4.4.13}$$

or in operator form:

$$Eu(x) = \left(1 + \Delta x D + \frac{(\Delta x D)^2}{2!} + \frac{(\Delta x D)^3}{3!} + \cdots \right) u(x) \tag{4.4.14}$$

This last relation can be written formally as

$$Eu(x) = e^{\Delta x D} u(x) \tag{4.4.15}$$

and therefore one has symbolically

$$E = e^{\Delta x D} \tag{4.4.16}$$

This relation has to be interpreted as giving identical results when acting on the exponential function e^{ax} and on any polynomial of degree n. In this latter case, the expansion on the right-hand side has only n terms and therefore *all the expressions to be defined in the following are exact up to n terms for polynomials of degree n.*
   The basic operation is then to use equation (4.4.16) in the inverse way, leading to

$$\Delta x D = \ln E \tag{4.4.17}$$

A4.4.1.1   *Forward differences*

Formulas for forward differences are obtained by introducing the relation (4.4.3) between E and the forward operator δ^+. We obtain after a formal development of the ln function:

$$\Delta x D = \ln E = \ln(1 + \delta^+)$$
$$= \delta^+ - \frac{\delta^{+2}}{2} + \frac{\delta^{+3}}{3} - \frac{\delta^{+4}}{4} + \cdots \tag{4.4.18}$$

The order of accuracy of the approximation increases with the number of terms kept in the right-hand side. The first neglected term gives the **truncation error**. For instance, keeping the first term only, leads to the first order formula (4.1.5) and a truncation error equal to $(\Delta x\, u_{xx}/2)$. If the first two terms are considered, we obtain the second order formula (4.1.17) with the truncation error $(\Delta x^2 u_{xxx}/3)$:

$$(u_x)_i = \frac{-3u_i + 4u_{i+1} - u_{i+2}}{2\Delta x} + \frac{\Delta x^2}{3} u_{xxx} \tag{4.4.19}$$

Hence, this relation leads to the definition of various forward finite difference formulas for the first derivative with increasing order of accuracy. As the forward difference operator can be written as $\delta^+ = \Delta x\, u_x + O(\Delta x^2)$, the first neglected operator δ^{+n} is of order n showing that the associated truncation error is $O(\Delta x^{n-1})$.

A4.4.1.2 *Backward differences*

Similarly, backward difference formulas can be obtained with increasing order of accuracy, by application of the relation (4.4.3):

$$\Delta x D = \ln E = -\ln(1 - \delta^-)$$

$$= \delta^- + \frac{\delta^{-2}}{2} + \frac{\delta^{-3}}{3} + \frac{\delta^{-4}}{4} + \cdots \tag{4.4.20}$$

To second order accuracy we have

$$(u_x)_i = Du_i = \frac{3u_i - 4u_{i-1} + u_{i-2}}{2\Delta x} + \frac{\Delta x^2}{3} u_{xxx} \tag{4.4.21}$$

A4.4.1.3 *Central differences*

Central difference formulas are obtained from equation (4.4.8)

$$\delta u_i = u_{i+1/2} - u_{i-1/2} = (E^{1/2} - E^{-1/2})u_i$$

and therefore

$$\delta = e^{\Delta x D/2} - e^{-\Delta x D/2} = 2 \sin h\left(\frac{\Delta x D}{2}\right) \tag{4.4.22}$$

which, through inversion, leads to

$$\Delta x D = 2 \sin h^{-1}\delta/2 = 2\left[\frac{\delta}{2} - \frac{1}{2\cdot 3}\left(\frac{\delta}{2}\right)^3 + \frac{1\cdot 3}{2\cdot 4\cdot 5}\left(\frac{\delta}{2}\right)^5\right.$$

$$\left. - \frac{1\cdot 3\cdot 5}{2\cdot 4\cdot 6\cdot 7}\left(\frac{\delta}{2}\right)^7 + \cdots\right]$$

$$= \delta - \frac{\delta^3}{24} + \frac{3\delta^5}{640} - \frac{5\delta^7}{7168} + \cdots \tag{4.4.23}$$

This formula generates a family of central difference approximations to the first order derivative $(u_x)_i$ based on the values of the function u at half-integer mesh point locations. By keeping only the first term, we obtain, with second order accuracy

$$(u_x)_i = \frac{u_{i+1/2} - u_{i-1/2}}{\Delta x} - \frac{\Delta x^2}{24} u_{xxx} \tag{4.4.24}$$

Keeping the first two terms, we obtain a fourth order accurate approximation

$$(u_x)_i = \frac{-u_{i+3/2} + 27u_{i+1/2} - 27u_{i-1/2} + u_{i-3/2}}{24\Delta x} + \frac{3}{640}\Delta x^4 \frac{\partial^5 u}{\partial x^5} \tag{4.4.25}$$

To derive central differences involving only integer mesh points, we could apply the above procedure to the operator $\bar{\delta}$. From equation (4.4.9) we have

$$\bar{\delta} = \frac{1}{2}(E - E^{-1}) = \frac{1}{2}(e^{\Delta xD} - e^{-\Delta xD}) = \sin h(\Delta xD) \tag{4.4.26}$$

and therefore, in function of $\bar{\delta}$,

$$\Delta xD = \sin h^{-1}\bar{\delta}$$
$$= \bar{\delta} - \frac{\bar{\delta}^3}{6} + \frac{3}{2 \cdot 4 \cdot 5}\bar{\delta}^5 + \cdots \tag{4.4.27}$$

This formula can be used to replace equation (4.4.23) for the central difference at point i. However, although the first term is the second order central difference approximation (4.1.7), the next term leads to a fourth order formula for $(u_x)_i$ involving the four points $i - 3$, $i - 1$, $i + 1$, $i + 3$. This is of no interest for numerical computations since we would expect a fourth order formula for $(u_x)_i$ to involve the points $i - 2$, $i - 1$, $i + 1$, $i + 2$. This can be obtained from the identity:

$$\mu^2 = 1 + \frac{\delta^2}{4} \tag{4.4.28}$$

After multiplication of equation (4.4.23) by

$$1 = \mu\left(1 + \frac{\delta^2}{4}\right)^{-1/2} = \mu\left(1 - \frac{\delta^2}{8} + \frac{3\delta^4}{128} - \frac{5\delta^6}{1024} + \cdots\right) \tag{4.4.29}$$

we obtain the relation

$$\Delta xD = \mu\left(\delta - \frac{1}{3!}\delta^3 + \frac{1^2 2^2}{5!}\delta^5 - \cdots\right)$$
$$= \bar{\delta}\left(1 - \frac{\delta^2}{3!} + \frac{2^2}{5!}\delta^4 - \frac{2^2 \cdot 3^2}{7!}\delta^6 + \cdots\right) \tag{4.4.30}$$

Hence, we obtain the following second and fourth order accurate central difference approximations to the derivative $(u_x)_i$, with integer mesh point values:

$$(u_x)_i = \frac{u_{i+1} - u_{i-1}}{2\Delta x} - \frac{\Delta x^2}{6}u_{xxx} \tag{4.4.31}$$

and

$$(u_x)_i = \frac{-u_{i+2} + 8u_{i+1} - 8u_{i-1} + u_{i-2}}{2\Delta x} + \frac{\Delta x^4}{30}\frac{\partial^5 u}{\partial x^5} \tag{4.4.32}$$

A4.4.2 Higher Order Derivatives

Applying the operator technique, an unlimited number of finite difference formulas can be applied to obtain second and higher order derivatives. From equation

(4.4.18) we have the one-sided, forward difference formula, see for instance Ames (1977):

$$\left(\frac{\partial^n u}{\partial x^n}\right)_i = D^n u_i = \frac{1}{\Delta x^n}[\ln(1 + \delta^+)]^n u_i$$

$$= \frac{\delta^{+n}}{\Delta x^n}\left[1 - \frac{n}{2}\delta^+ + \frac{n(3n+5)}{24}\delta^{+2} - \frac{n(n+2)(n+3)}{48}\delta^{+3} + \cdots\right]$$

(4.4.33)

In terms of the backward difference operator δ^-, we have

$$\left(\frac{\partial^n u}{\partial x^n}\right)_i = -\frac{1}{\Delta x^n}[\ln(1 - \delta^-)]^n u_i$$

$$= \frac{1}{\Delta x^n}\left(\delta^- + \frac{\delta^{-2}}{2} + \frac{\delta^{-3}}{3} + \frac{\delta^{-4}}{4} + \cdots\right)^n u_i$$

$$= \frac{\delta^{-n}}{\Delta x^n}\left[1 + \frac{n}{2}\delta^- + \frac{n(3n+5)}{24}\delta^{-2} + \frac{n(n+2)(n+3)}{48}\delta^{-3} + \cdots\right]u_i$$

(4.4.34)

Central difference formulas for higher order derivatives can also be obtained, through

$$D^n u_i = \left(\frac{2}{\Delta x}\sin \mathrm{h}^{-1}\delta/2\right)^n u_i$$

$$= \frac{1}{\Delta x^n}\left[\delta - \frac{\delta^3}{24} + \frac{3}{640}\delta^5 - \frac{5}{7168}\delta^7 + \cdots\right]^n u_i$$

$$= \frac{1}{\Delta x^n}\delta^n\left[1 - \frac{n}{24}\delta^2 + \frac{n}{5760}(22 + 5n)\delta^4 \right.$$

$$\left. - \frac{n}{45}\left(\frac{5}{7} + \frac{n-1}{5} + \frac{(n-1)(n-2)}{81}\right)\delta^6 + \cdots\right]u_i \quad (4.4.35)$$

For n even, this equation generates difference formulas with the function values at the integer mesh points. For n uneven, the difference formulas involve points at half-integer mesh points. To involve only points at integer values of i for n uneven we define, using equation (4.4.28):

$$D^n u_i = \frac{\mu}{[1 + \delta^2/4]^{1/2}}\left(\frac{2}{\Delta x}\sin \mathrm{h}^{-1}\delta/2\right)^n u_i$$

$$= \frac{\mu}{\Delta x^n}\delta^n\left[1 - \frac{n+3}{24}\delta^2 + \frac{(5n+27)(n+5)}{5760}\delta^4 + \cdots\right]u_i \quad (4.4.36)$$

A4.4.2.1 *Second order derivative*

For instance, second order derivative formulas are

$$(u_{xx})_i = \frac{1}{\Delta x^2} \left(\delta^{-2} + \delta^{-3} + \frac{11}{12}\delta^{-4} + \frac{5}{6}\delta^{-5} + \cdots \right) u_i \tag{4.4.37}$$

$$(u_{xx})_i = \frac{1}{\Delta x^2} \left(\delta^{+2} - \delta^{+3} + \frac{11}{12}\delta^{+4} - \frac{5}{6}\delta^{+5} + \cdots \right) u_i \tag{4.4.38}$$

$$(u_{xx})_i = \frac{1}{\Delta x^2} \left(\delta^2 - \frac{\delta^4}{12} + \frac{\delta^6}{90} - \frac{\delta^8}{560} + \cdots \right) u_i \tag{4.4.39}$$

$$(u_{xx})_i = \frac{\mu}{\Delta x^2} \left(\delta^2 - \frac{5\delta^4}{24} + \frac{259}{5760}\delta^6 + \cdots \right) u_i \tag{4.4.40}$$

These equations define four families of difference operators for the second derivative, to various orders of accuracy. By maintaining only the first term, we obtain the following difference formulas:

Forward difference – first order accurate

$$(u_{xx})_i = \frac{1}{\Delta x^2}(u_{i+2} - 2u_{i+1} + u_i) - \Delta x u_{xxx} \tag{4.4.41}$$

Backward difference – first order accurate

$$(u_{xx})_i = \frac{1}{\Delta x^2}(u_{i-2} - 2u_{i-1} + u_i) + \Delta x u_{xxx} \tag{4.4.42}$$

Central difference – integer points – second order accurate

$$(u_{xx})_i = \frac{1}{\Delta x^2}(u_{i+1} - 2u_i + u_{i-1}) - \frac{\Delta x^2}{12}\left(\frac{\partial^4 u}{\partial x^4}\right) \tag{4.4.43}$$

Central difference – half integer mesh points – second order accurate

$$(u_{xx})_i = \frac{1}{2\Delta x^2}(u_{i+3/2} - u_{i+1/2} - u_{i-1/2} + u_{i-3/2}) - \frac{5}{24}\Delta x^2\left(\frac{\partial^4 u}{\partial x^4}\right) \tag{4.4.44}$$

With the exception of the last one, these difference approximations for the second derivative involve three mesh points like the first derivatives. The one-sided difference formulas are only first order accurate, while the central differences always lead to a higher order of accuracy.

By keeping the two first terms of the above formulas, we obtain difference formulas with a higher order of accuracy:

Forward difference – second order accurate

$$(u_{xx})_i = \frac{1}{\Delta x^2}(2u_i - 5u_{i+1} + 4u_{i+2} - u_{i+3}) + \frac{11}{12}\Delta x^2\left(\frac{\partial^4 u}{\partial x^4}\right) \tag{4.4.45}$$

Backward difference – second order accurate

$$(u_{xx})_i = \frac{1}{\Delta x^2}(2u_i - 5u_{i-1} + 4u_{i-2} - u_{i-3}) - \frac{11}{12}\Delta x^2 \left(\frac{\partial^4 u}{\partial x^4}\right) \tag{4.4.46}$$

Central difference – integer points – fourth order accurate

$$(u_{xx})_i = \frac{1}{12\Delta x^2}(-u_{i+2} + 16u_{i+1} - 30u_i + 16u_{i-1} - u_{i-2}) + \frac{\Delta x^4}{90}\left(\frac{\partial^6 u}{\partial x^6}\right) \tag{4.4.47}$$

Central difference – half-integer mesh points – fourth order accurate

$$(u_{xx})_i = \frac{1}{48\Delta x^2}(-5u_{i+5/2} + 39u_{i+3/2} - 34u_{i+1/2} - 34u_{i-1/2}$$

$$+ 39u_{i-3/2} - 5u_{i-5/2}) + \frac{259}{5760}\Delta x^4 \left(\frac{\partial^6 u}{\partial x^6}\right) \tag{4.4.48}$$

This last formula is of little practical use since it requires six mesh points to obtain a fourth order accurate approximation to the second derivative at point i, while the previous formula (4.4.47) requires only four mesh points.

A more complex operator, often occurring in second order differential problems is $\partial_x[k(x)\partial_x u]$. A central difference formula of second order accuracy with three mesh points is given by

$$\frac{\partial}{\partial x}\left[k(x)\frac{\partial}{\partial x}\right]u_i = \frac{1}{\Delta x^2}\delta^+(k_{1-1/2}\delta^-)u_i + O(\Delta x^2) \tag{4.4.49}$$

which takes the explicit form

$$\frac{\partial}{\partial x}\left[k(x)\frac{\partial}{\partial x}\right]u_i = \frac{1}{\Delta x^2}[k_{i+1/2}(u_{i+1} - u_i) - k_{i-1/2}(u_i - u_{i-1})] + O(\Delta x^2) \tag{4.4.50}$$

An equivalent formula is obtained by inverting the forward and backward

$$\frac{\partial}{\partial x}\left[k(x)\frac{\partial}{\partial x}\right]u_i = \frac{1}{\Delta x^2}\delta^-(k_{1+1/2}\delta^+)u_i + O(\Delta x^2) \tag{4.4.51}$$

leading to the same expression (4.4.50).

A4.4.2.2 *Third order derivatives*

Approximations for third derivatives are obtained from the above general expressions. To the lowest orders of accuracy one has the following difference formulas:

Forward difference

$$\left(\frac{\partial^3 u}{\partial x^3}\right)_i \equiv (u_{xxx})_i$$

$$= \frac{1}{\Delta x^3}(u_{i+3} - 3u_{i+2} + 3u_{i+1} - u_i) - \frac{\Delta x}{2}\left(\frac{\partial^4 u}{\partial x^4}\right) \tag{4.4.52}$$

or with the second order accuracy

$$(u_{xxx})_i = \frac{1}{2\Delta x^3}(-3u_{i+4} + 14u_{i+3} - 24u_{i+2} + 18u_{i+1} - 5u_i) + \frac{21}{12}\Delta x^2 \left(\frac{\partial^5 u}{\partial x^5}\right)$$

$$(4.4.53)$$

Backward difference

$$(u_{xxx})_i = \frac{1}{\Delta x^3}(u_i - 3u_{i-1} + 3u_{i-2} - u_{i-3}) + \frac{\Delta x}{2}\left(\frac{\partial^4 u}{\partial x^4}\right)$$

$$(4.4.54)$$

or with second order accuracy

$$(u_{xxx})_i = \frac{1}{2\Delta x^3}(5u_i - 18u_{i-1} + 24u_{i-2} - 14u_{i-3} + 3u_{i-4}) - \frac{21}{12}\Delta x^2 \left(\frac{\partial^5 u}{\partial x^5}\right)$$

$$(4.4.55)$$

Central difference – half-integer points

$$(u_{xxx})_i = \frac{1}{\Delta x^3}(u_{i+3/2} - 3u_{i+1/2} + 3u_{i-1/2} - u_{i-3/2}) - \frac{\Delta x^2}{8}\left(\frac{\partial^5 u}{\partial x^5}\right) \quad (4.4.56)$$

This is a second order accurate approximation to the third derivative, and a fourth order accuracy is obtained from the following formula:

$$(u_{xxx})_i = \frac{1}{8\Delta x^3}(-u_{i+5/2} + 13u_{i+3/2} - 34u_{i+1/2} + 34u_{i-1/2}$$

$$- 13u_{i-3/2} + u_{i-5/2}) + \frac{37}{1920}\Delta x^4 \left(\frac{\partial^7 u}{\partial x^7}\right)$$

$$(4.4.57)$$

Central difference – integer mesh point

$$(u_{xxx})_i = \frac{1}{2\Delta x^3}(u_{i+2} - 2u_{i+1} + 2u_{i-1} - u_{i-2}) - \frac{\Delta x^2}{4}\left(\frac{\partial^5 u}{\partial x^5}\right)$$

$$(4.4.58)$$

or with fourth order accuracy

$$(u_{xxx})_i = \frac{1}{8\Delta x^3}(-u_{i+3} + 8u_{i+2} - 13u_{i+1} + 13u_{i-1} - 8u_{i-2} + u_{i-3})$$

$$+ \frac{7}{120}\Delta x^4 \left(\frac{\partial^7 u}{\partial x^7}\right)$$

$$(4.4.59)$$

A4.4.2.3 *Fourth order derivatives*

To the lowest order of accuracy, we have the following approximations:

Forward difference – first order accurate

$$\left(\frac{\partial^4 u}{\partial x^4}\right)_i \equiv (u_{xxxx})_i$$

$$= \frac{1}{\Delta x^4}(u_{i+4} - 4u_{i+3} + 6u_{i+2} - 4u_{i+1} + u_i) - 2\Delta x \left(\frac{\partial^5 u}{\partial x^5}\right) \quad (4.4.60)$$

Backward difference – first order accurate

$$\left(\frac{\partial^4 u}{\partial x^4}\right)_i = \frac{1}{\Delta x^4}(u_i - 4u_{i-1} + 6u_{i-2} - 4u_{i-3} + u_{i-4}) + 2\Delta x \left(\frac{\partial^5 u}{\partial x^5}\right) \quad (4.4.61)$$

Central difference – second order accurate

$$(u_{xxxx})_i = \frac{1}{\Delta x^4}(u_{i+2} - 4u_{i+1} + 6u_i - 4u_{i-1} + u_{i-2}) + \frac{\Delta x^2}{6}\left(\frac{\partial^6 u}{\partial x^6}\right) \quad (4.4.62)$$

Obtaining these formulas is left as an exercise to the reader (see Problems P.4.18–P.4.21).

A4.5 IMPLICIT FINITE DIFFERENCE FORMULAS

Implicit formulas are defined as expressions where derivatives at different mesh points appear simultaneously. Their essential advantage comes from the high order of accuracy that is generated when derivatives at different mesh points are related to each other. The price to be paid is that we generate an algebraic system for the approximated derivatives, which cannot be written in an explicit way. The above expressions can be used to generate these high order implicit formulas for the derivative operators in the following way.

A4.5.1 General Approach

For instance, equation (4.4.30) gives, with a fourth order accuracy

$$\Delta x D = \mu \delta \left(1 - \frac{\delta^2}{6}\right) + O(\Delta x^5) \quad (4.5.1)$$

or by a formal operation, to the same order of accuracy

$$\Delta x D = \frac{\mu \delta}{1 + \delta^2/6} + O(\Delta x^5) \quad (4.5.2)$$

This formula is a rational fraction or PADE differencing approximation, Kopal (1961).

The interpretation of these two last formulas are quite distinct from each other. Equation (4.5.1), applied to u_i leads to the fourth order formula (4.4.32), while equation (4.5.2) is to be interpreted after multiplication of both sides by the operator $(1 + \delta^2/6)$:

$$(1+\delta^2/6)Du_i = \frac{1}{6}[(u_x)_{i+1}+4(u_x)_i+(u_x)_{i-1}] = \frac{u_{i+1} - u_{i-1}}{2\Delta x} + O(\Delta x^4) \quad (4.5.3)$$

The left-hand side has an implicit structure and this formula has the important property of involving only three spatial points while being of the same fourth order as equation (4.4.32) which requires five mesh points. These schemes are called sometimes Hermitian schemes and can also be obtained from a finite element formulation (see Chapter 5).

Similar procedures can be applied to generate other implicit formulas; for instance equation (4.4.18) leads to

$$\Delta x D = \delta^+ - \frac{1}{2}\delta^{+2} + O(\Delta x^3) = \delta^+\left(1 - \frac{1}{2}\delta^+\right) + O(\Delta x^3)$$

$$= \frac{\delta^+}{1 + \frac{1}{2}\delta^+} + O(\Delta x^3) \quad (4.5.4)$$

After multiplication by $(1 + \delta^+)/2$, we obtain the two point implicit relation, of second order accuracy

$$\frac{1}{2}[(u_x)_i + (u_x)_{i+1}] = \frac{u_{i+1} - u_i}{\Delta x} + O(\Delta x^2) \quad (4.5.5)$$

Formulas such as (4.5.3) or (4.5.5) do not allow the explicit determination of the numerical approximations to the derivatives $(u_x)_i$. Instead these formulas have to be written for all the mesh points and solved simultaneously as an algebraic system of equations for the unknowns $(u_x)_i, i = 1, \ldots, N$.

For instance, the fourth order implicit approximation (4.5.3) for $(u_x)_i$ will be obtained from the solution of the *tridiagonal* system:

$$
\begin{vmatrix}
\cdot & & & \\
\cdot & & & \\
1 & 4 & 1 & \\
& 1 & 4 & 1 \\
& & 1 & 4 & 1 \\
& & & \cdot
\end{vmatrix}
\cdot
\begin{vmatrix}
\cdot \\
(u_x)_{i-2} \\
(u_x)_{i-1} \\
(u_x)_i \\
(u_x)_{i+1} \\
(u_x)_{i+2} \\
\cdot
\end{vmatrix}
= \frac{3}{\Delta x}
\begin{vmatrix}
\cdot \\
u_i - u_{i-2} \\
u_{i+1} - u_{i-1} \\
u_{i+2} - u_i \\
\cdot
\end{vmatrix}
\quad (4.5.6)
$$

while equation (4.5.5) leads to a *bidiagonal* system:

$$
\begin{vmatrix}
\cdot & & \\
\cdot & & \\
1 & 1 & \\
& 1 & 1 \\
& & 1 & 1 \\
& & & \cdot
\end{vmatrix}
\cdot
\begin{vmatrix}
\cdot \\
(u_x)_{i-2} \\
(u_x)_{i-1} \\
(u_x)_i \\
(u_x)_{i+1} \\
(u_x)_{i+2} \\
\cdot
\end{vmatrix}
= \frac{2}{\Delta x}
\begin{vmatrix}
\cdot \\
u_i - u_{i-1} \\
u_{i+1} - u_i \\
u_{i+2} - u_{i+1} \\
\cdot
\end{vmatrix}
\quad (4.5.7)
$$

As a consequence, the numerical value of $(u_x)_i$ obtained as solution of the above systems, is influenced by all the mesh point values u_j.

This explains why these formulas are of higher order of accuracy than the corresponding explicit formulas involving the same number of mesh points. When applied to practical flow problems, the function values and the derivatives are considered as unknowns. They are obtained as solutions of an algebraic system formed by adding the basic equations to be solved to the above implicit relations.

Along the same lines, we obtain implicit formulas for second order derivatives with a higher order of accuracy and a number of mesh point values limited to two or three. From equation (4.4.39), we have, to fourth order accuracy

$$
(u_{xx})_i = \frac{1}{\Delta x^2} \delta^2 \left(1 - \frac{\delta^2}{12} \right) u_i + O(\Delta x^4)
$$

$$
= \frac{1}{\Delta x^2} \frac{\delta^2 u_i}{(1 + \delta^2/12)} + O(\Delta x^4) \tag{4.5.8}
$$

Multiplying formally by $(1 + \delta^2/12)$ we obtain the implicit, compact expression for the second order derivative

$$
(1 + \delta^2/12)(u_{xx})_i = \frac{1}{\Delta x^2} \delta^2 u_i + O(\Delta x^4) \tag{4.5.9a}
$$

or

$$
\frac{1}{12}[(u_{xx})_{i+1} + 10(u_{xx})_i + (u_{xx})_{i-1}] = \frac{1}{\Delta x^2}(u_{i+1} - 2u_i + u_{i-1}) + O(\Delta x^4) \tag{4.5.9b}
$$

Here again a tridiagonal system is to be solved in order to calculate $(u_x)_i$ from the mesh point values u_j.

There is no way of obtaining an implicit relation for the second derivatives with only values at the two mesh points i and $i + 1$, without involving also first derivative values (Hirsh, 1975, 1983).

A4.5.2   General Derivation of Implicit Finite Difference Formula's for First and Second Derivatives

Implicit finite difference relations for first and second derivatives have been derived by various methods and given a variety of names. Many formulas can be found in Collatz (1966), under the name of *Mehrstellen* method or *Hermitian* method by analogy with Hermitian finite elements. We have already mentioned the name of Pade approximations and recently a large number of applications to the solution of fluid-mechanical equations have been developed by Krause (1971), Hirsh (1975), Lele (1992) under the name of *compact* methods; Rubin and Graves (1975), Rubin and Khosla (1977) under the name of *spline* methods; Adam (1975, 1977), Ciment and Leventhal (1975), Leventhal (1980) as *(operator) compact implicit* (OCI) methods. However, following Peyret (1978) – see also Peyret and Taylor (1982) – all the implicit formulas can be derived in a systematic way from a Taylor series expansion.

With a limitation to three-point expressions, the general form of an implicit finite difference relation between a function and its first two derivatives would be

$$a_+ u_{i+1} + a_0 u_i + a_- u_{i-1} + b_+ (u_x)_{i+1} + b_0 (u_x)_i + b_- (u_x)_{i-1}$$
$$+ c_+ (u_{xx})_{i+1} + c_0 (u_{xx})_i + c_- (u_{xx})_{i-1} = 0 \tag{4.5.10}$$

Developing all the variables in a Taylor series about point i, we have the following expansion, for equal mesh spacing

$$u_{i\pm1} = u_i \pm \Delta x (u_x)_i + \frac{\Delta x^2}{2} (u_{xx})_i \pm \frac{\Delta x^3}{6} (u_{xxx})_i + \frac{\Delta x^4}{24} \left(\frac{\partial^4 u}{\partial x^4} \right)_i \pm \frac{\Delta x^5}{5!} \left(\frac{\partial^5 u}{\partial x^5} \right)_i$$
$$+ \frac{\Delta x^6}{6!} \left(\frac{\partial^6 u}{\partial x^6} \right)_i \pm \frac{\Delta x^7}{7!} \left(\frac{\partial^7 u}{\partial x^7} \right)_i + \frac{\Delta x^8}{8!} \left(\frac{\partial^8 u}{\partial x^8} \right)_i + \cdots \tag{4.5.11}$$

$$(u_x)_{i\pm1} = (u_x)_i \pm \Delta x (u_{xx})_i + \frac{\Delta x^2}{2} (u_{xxx})_i \pm \frac{\Delta x^3}{6} \left(\frac{\partial^4 u}{\partial x^4} \right)_i + \frac{\Delta x^4}{24} \left(\frac{\partial^5 u}{\partial x^5} \right)_i$$
$$\pm \frac{\Delta x^5}{5!} \left(\frac{\partial^6 u}{\partial x^6} \right)_i + \frac{\Delta x^6}{6!} \left(\frac{\partial^7 u}{\partial x^7} \right)_i \pm \frac{\Delta x^7}{7!} \left(\frac{\partial^8 u}{\partial x^8} \right)_i + \cdots \tag{4.5.12}$$

$$(u_{xx})_{i\pm1} = (u_{xx})_i \pm \Delta x (u_{xxx})_i + \frac{\Delta x^2}{2} \left(\frac{\partial^4 u}{\partial x^4} \right)_i \pm \frac{\Delta x^3}{6} \left(\frac{\partial^5 u}{\partial x^5} \right)_i + \frac{\Delta x^4}{24} \left(\frac{\partial^6 u}{\partial x^6} \right)_i$$
$$\pm \frac{\Delta x^5}{5!} \left(\frac{\partial^7 u}{\partial x^7} \right)_i + \frac{\Delta x^6}{6!} \left(\frac{\partial^8 u}{\partial x^8} \right)_i + \cdots \tag{4.5.13}$$

When introduced in the implicit relation (4.5.10), one can request the coefficients up to the third order derivative of the truncation error to vanish, in order to obtain at least second order accuracy, for the second derivatives. This leads to the conditions

$$a_+ + a_0 + a_- = 0$$

$$\Delta x (a_+ - a_-) + b_+ + b_0 + b_- = 0$$

$$\frac{\Delta x^2}{2} (a_+ + a_-) + \Delta x (b_+ - b_-) + c_+ + c_0 + c_- = 0$$

$$\frac{\Delta x^3}{6} (a_+ - a_-) + \frac{\Delta x^2}{2} (b_+ + b_-) + \Delta x (c_+ - c_-) = 0 \tag{4.5.14}$$

from which one can choose to eliminate a_+, a_0, a_-, and b_0, for instance, (other choices are obviously possible, see Problem P.4.22),

$$a_+ = \frac{1}{2\Delta x}\left[-5b_+ - b_- + \frac{2}{\Delta x}(2c_- - 4c_+ - c_0)\right]$$

$$a_0 = \frac{2}{\Delta x}\left[b_+ - b_- + \frac{1}{\Delta x}(c_+ + c_0 + c_-)\right]$$

$$a_- = \frac{1}{2\Delta x}\left[b_+ + 5b_- + \frac{2}{\Delta x}(2c_+ - 4c_- - c_0)\right] \tag{4.5.15}$$

$$b_0 = 2(b_+ + b_-) + \frac{6}{\Delta x}(c_+ - c_-) = 0$$

and the truncation error R reduces to

$$
\begin{aligned}
R = {}& \frac{\Delta x^3}{4!}\left[2(b_+ - b_-) + \frac{10}{\Delta x}(c_+ + c_-) - \frac{2}{\Delta x}c_0\right]\frac{\partial^4 u}{\partial x^4} \\
&+ \frac{\Delta x^4}{5!}\left[2(b_+ + b_-) + \frac{14}{\Delta x}(c_+ - c_-)\right]\frac{\partial^5 u}{\partial x^5} \\
&+ \frac{\Delta x^5}{6!}\left[4(b_+ - b_-) + \frac{28}{\Delta x}(c_+ + c_-) - \frac{2}{\Delta x}c_0\right]\frac{\partial^6 u}{\partial x^6} \\
&+ \frac{\Delta x^6}{7!}\left[4(b_+ + b_-) + \frac{36}{\Delta x}(c_+ - c_-)\right]\frac{\partial^7 u}{\partial x^7} \\
&+ \frac{\Delta x^7}{8!}\left[6(b_+ - b_-) + \frac{54}{\Delta x}(c_+ + c_-) - \frac{2}{\Delta x}c_0\right]\frac{\partial^8 u}{\partial x^8} \tag{4.5.16}
\end{aligned}
$$

Hence, one has a four-parameter family of implicit relations (one parameter may always be set arbitrarily to one since equation (4.5.10) is homogeneous). These parameters can be selected on the basis of various conditions, according to the number of derivatives and mesh points one wishes to maintain in the implicit relation or by imposing a minimum order of accuracy. For instance, the second order relation (4.5.7) is obtained with $b_+ = b_- = b_0 = 0$ and by selecting $c_+ = c_- = 1$, $c_0 = 10$.

As can be seen from the expression of the truncation error, the highest order of accuracy that can be achieved is six. This is obtained by imposing the coefficients of the three first terms in R to vanish. This gives the relations:

$$b_+ = -\frac{1}{\Delta x}(8c_+ + c_-)$$

$$b_- = \frac{1}{\Delta x}(c_+ + 8c_-) \tag{4.5.17}$$

$$c_0 = -4(c_+ + c_-)$$

Inserted into the above formulas, a one-parameter family of implicit relations is obtained between the function u and its first two derivatives, with $\alpha = c_+/c_-$,

$$\frac{3}{2\Delta x^2}(13 + 3\alpha)u_{i+1} - \frac{24}{\Delta x^2}(1 + \alpha)u_i + \frac{3}{2\Delta x^2}(3 + 13\alpha)u_{i-1}$$
$$- \frac{1}{\Delta x}(8 + \alpha)(u_x)_{i+1} - \frac{1}{\Delta x}(1 - \alpha)(u_x)_i + \frac{1}{\Delta x}(1 + 8\alpha)(u_x)_{i-1}$$
$$+ (u_{xx})_{i+1} - 4(1 + \alpha)(u_{xx})_i + \alpha(u_{xx})_{i-1} = 0 \qquad (4.5.18)$$

with the truncation error

$$R = \frac{8\Delta x^5}{7!}(1 - \alpha)\frac{\partial^7 u}{\partial x^7} + \frac{\Delta x^6}{7!}(1 + \alpha)\frac{\partial^8 u}{\partial x^8} \qquad (4.5.19)$$

The unique, implicit relation of order six, is obtained from $\alpha = 1$

$$\frac{24}{\Delta x^2}(u_{i+1} - 2u_i + u_{i-1}) - \frac{9}{\Delta x}[(u_x)_{i+1} - (u_x)_{i-1}]$$
$$+ (u_{xx})_{i+1} - 8(u_{xx})_i + (u_{xx})_{i-1} = 0 \qquad (4.5.20)$$

with a truncation error

$$R = \frac{2}{8!}\Delta x^6\frac{\partial^8 u}{\partial x^8} \qquad (4.5.21)$$

Implicit relations with first derivatives only, are obtained from $c_+ = c_0 = c_- = 0$, and can therefore be at most fourth order accurate. From equations (4.5.15) and (4.5.16) we obtain the one parameter family, with $\beta = b_-/b_+$

$$\frac{1}{2\Delta x}(-5 - \beta)u_{i+1} + \frac{2}{2\Delta x}(1 - \beta)u_i + \frac{1}{2\Delta x}(1 + 5\beta)u_{i-1}$$
$$+ (u_x)_{i+1} + 2(1 - \beta)(u_x)_i + \beta(u_x)_{i-1} = 0 \qquad (4.5.22)$$

with a truncation error

$$R = \frac{\Delta x^3}{12}(1 - \beta)\frac{\partial^4 u}{\partial x^4} + \frac{\Delta x^4}{60}(1 + \beta)\frac{\partial^5 u}{\partial x^5} \qquad (4.5.23)$$

For $\beta = 1$, one obtains the unique fourth order relation (4.5.3). For other choices of β, the formula is only third order accurate.

Two-point implicit difference formulas

The most general two-point relation, with at least second order accuracy for the second derivatives, is obtained from $a_- = b_- = c_- = 0$. We obtain the one-parameter family of relations, from (4.514) with $\gamma = b_+/(\Delta x a_+)$

$$\frac{1}{\Delta x^2}(u_i - u_{i+1}) + \frac{1 + \gamma}{\Delta x}(u_x)_i - \frac{\gamma}{\Delta x}(u_x)_{i+1}$$
$$+ \frac{1}{6}[(1 + 3\gamma)(u_{xx})_{i+1} + (2 + 3\gamma)(u_{xx})_i] = 0 \qquad (4.5.24)$$

with the truncation error

$$R = \frac{\Delta x^2}{12} \left(\gamma + \frac{1}{2} \right) \frac{\partial^4 u}{\partial x^4} + \frac{\Delta x^4}{24} \left(\gamma + \frac{7}{15} \right) \frac{\partial^5 u}{\partial x^5} \qquad (4.5.25)$$

For $\gamma = -1/2$, we have the unique third order accurate relation,

$$\frac{1}{\Delta x^2}(u_{i+1} - u_i) - \frac{1}{2\Delta x}[(u_x)_{i+1} + (u_x)_{i+1}] + \frac{1}{12}[(u_{xx})_{i+1} - (u_{xx})_i] = 0$$

$$(4.5.26)$$

with the truncation error

$$R = -\frac{\Delta x^3}{720} \frac{\partial^5 u}{\partial x^5} \qquad (4.5.27)$$

Many other formulas can be derived, according to the points or (and) the derivatives we wish to isolate.

CONCLUSIONS AND MAIN TOPICS TO REMEMBER

This chapter four has introduced you to the basis of numerical discretization, namely the finite difference method. Although it is only applicable to structured grids, it remains the reference to all numerical analysis steps.

The main topics to remember are the following:

- The Taylor expansion for continuous functions is the key to the evaluation of the order of accuracy of FD formulas.
- For any derivative, we always have an infinite number of possible FD formulas, depending on the number of mesh points we decide to involve in the formula and on the expected order of accuracy, which can be arbitrarily high.
- Depending on the position of the points involved in the FD formula for point i, we distinguish between backward, forward, central or mixed FD formulas.
- Although the order of accuracy of an FD formula is uniquely defined by the related Taylor expansion, it is very important to remember that the effective order of the same formula can be different if interpreted as an approximation in the mesh point or at the mid-cell point. For instance, a first order backward formula for a first derivative at point i, will provide a second order approximation of the same first derivative at the mid-point $(i - 1/2)$.
- The extension to two-dimensional partial derivatives, as described in Section 4.2, should require your careful attention, as the number of possible discretizations increases with the number of space dimensions. An interesting example is given by the FD formula for the Laplace equation, which appears currently in CFD. Observe the simplicity of these formulas.
- Section 4.3 on FD formulas for non-uniform grids is of utmost importance, as most of the CFD applications are performed on non-uniform grids. You learn in this section, how and why a non-negligible loss of accuracy of any FD formula

can appear with 'brutal' changes of cell sizes. This loss of accuracy is often limited to one order, but could also be more severe for strong discontinuous grid changes or for inappropriate formulas.

- The important recommendation for avoiding a significant loss of accuracy on non-uniform grids is to avoid discontinuous variations of cell size and to ensure that the grid variations are defined by smooth analytical functions, such as a power law with a fixed cell size ratio.
- Take notice of the methodologies developed in Sections A4.4 and A4.5, as it provides a general framework for the derivation of FD formulas of any order of accuracy, for any derivative of order n, on a uniform grid.

REFERENCES

Adam, Y. (1975). A Hermitian finite difference method for the solution of parabolic equations. *Comp. Math. Appl.*, 1, 393–406.

Adam, Y. (1977). Highly accurate compact implicit methods and boundary conditions. *J. Comp. Phy.*, 24, 10–22.

Ames, W.F. (1977). *Numerical Methods for Partial Differential Equations*, 2nd edn. Academic Press, New York.

Cain A.B. and Bush R.H. (1994). Numerical wave propagation analysis for stretched grids. *32nd AIAA Aerospace Sciences Meeting*, Paper AIAA-94-0172.

Ciment, M. and Leventhal, S.H. (1975). Higher order compact implicit schemes for the wave equation. *Math. Comp.*, 29, 885–944.

Collatz, L. (1966). *The Numerical Treatment of Differential Equations*, 3rd edn. Springer Verlag, Berlin.

Dahlquist, G. and Bjorck, A. (1974). *Numerical Methods*. Prentice Hall, New Jersey.

Hildebrand, F.B. (1956). *Introduction to Numerical Analysis*. Mc Graw Hill, New York.

Hirsh, R.S. (1975). Higher order accurate difference solutions of fluid mechanics problems by a compact differencing scheme. *J. Comp. Phys.*, 19, 20–109.

Hirsh, R.S. (1983). Higher order approximations in fluid mechanics-compact to spectral, Von Karman Institute Lecture Series 1983-06, Brussels, Belgium.

Kopal, Z. (1961). *Numerical Analysis*. J. Wiley & Sons, New York.

Lele, S.K. (1992). Compact finite differences with spectral-like resolution. *J. Comput. Phys.*, 103(1), 16–42.

Krause, E. (1971). Mehrstellen Verfahren zur Integration der Grenzschicht gleichungen. *DLR Mitteilungen*, 71(13), 109–140.

Leventhal, S.H. (1980). The operator compact implicit method for reservoir simulation. *J. Soc. Pet. Eng.*, 20, 120–128.

O'Carroll, M.J. (1976). Diagonal dominance and SOR performance with skew nets. *Int. J. Numer. Meth. Eng.*, 10, 225–240.

Peyret, R. (1978). A Hermitian finite difference method for the solution of the Navier–Stokes equations. *Proceedings First of the Conference on Numerical Methods in Laminar and Turbulent Flows*, Pentech Press, Plymouth (UK) pp 43–54.

Peyret, R. and Taylor, T.D. (1982). *Computational Methods for Fluid Flow*. Springer Verlag, New York.

Rubin, S.G. and Graves, R.A. (1975). Viscous flow solutions with a cubic spline approximation. *Comput. Fluid.*, 3, 1–36.

Rubin, S.G. and Khosla, P.K. (1977). Polynomial interpolation method for viscous flow calculations. *J. Comp. Phys.*, 24, 217–246.

PROBLEMS

P.4.1 Evaluate numerically, the first and second derivatives of $\cos(x)$, $\sin(x)$, $\exp(x)$ at $x=0$, with forward, backward and central differences of first and second order each, by applying the formula of Section 4.1. Compare the error with the estimated truncation error. Take $\Delta x = 0.1$.

P.4.2 Show the second order accuracy of the formulas (4.1.9) and (4.1.10) by applying a Taylor expansion of u_i and $u_{i\pm1}$ around $u_{i\pm1/2}$.

P.4.3 Apply a Taylor series expansion to a mixed backward formula for the first derivative:

$$(u_x)_i = \frac{1}{\Delta x}(au_{i-2} + bu_{i-1} + cu_i + du_{i+1})$$

Derive the family of second order accurate formulas and the corresponding truncation error in function of the coefficient d. Obtain the unique third order accurate upwind-biased scheme and determine the corresponding truncation error.

Hint: Show that $(d = 1/2 - a; b = -3a - 1/2; c = 3a)$ and that the second order truncation error is equal to $\Delta x^2 u_{xxx} (1/6 - a)$. The unique third order scheme is obtained for $a = 1/6$:

$$(u_x)_i = \frac{1}{6\Delta x}(u_{i-2} - 6u_{i-1} + 3u_i + 2u_{i+1}) + \frac{\Delta x^3}{12}\left(\frac{\partial^4 u}{\partial x^4}\right)_i$$

P.4.4 Show, by applying the Taylor expansions such as (4.2.1), that formula (4.2.10) is indeed of second order accuracy.

P.4.5 Derive the truncation errors of the 'rotated' Laplacian given in equation (4.2.14), by applying Taylor expansions of all the points involved around point (i, j).

P.4.6 Show that the computational molecule for the Laplace operator $\Delta^{(3)}$ on a uniform mesh, satisfies formula (4.2.16) for an arbitrary b.

P.4.7 Find the truncation errors of the mixed derivative formulas (4.2.18) and proof its second order accuracy by applying Taylor expansions.

P.4.8 Obtain formulas similar to equation (4.2.19), by applying a backward difference in y, instead of a forward difference or by changing the roles of x and y leading to a difference formula of first order in x and second order in y.

P.4.9 Obtain the formula of equation (4.2.24), by applying Taylor series expansions of the points involved around the point $(i + 1/2, j + 1/2)$, following the development of equations (4.2.20) and (4.2.21).

P.4.10 Obtain the formula (4.3.7) and show, by a Taylor expansion, that it is second order accurate.

P.4.11 Consider the variables $u_{i\pm1/2}$ as the basic unknowns on the mesh of Figure 4.3.2 and apply equation (4.3.8) to estimate the first derivative u_x at point i. Show, by expanding the 'face values' $u_{i\pm1/2}$ around point i, that this formula is only first order accurate and obtain equation (4.3.11).

P.4.12 Apply the Taylor expansion for the non-uniform mesh of Figure 4.3.2 to the backward difference formulas (4.3.13) and (4.3.15) and obtain the expansions

(4.3.14) and (4.3.16). Explain why these formulas are of zero order of accuracy.

P.4.13 Obtain the system (4.3.19) and solve in function of a, obtaining the one-parameter family of difference formulas for the first derivative u_x. derive the truncation errors up to fourth order. Show that on a uniform grid the formulas are defined by the coefficients $(a, b, c, d) = (a, -3a-1/2, 3a, -a+1/2)$ and show that the value $a = 1/6$ leads to the unique third order formula on a uniform grid. Derive also the general difference formulas.

Show also that equation (4.3.7) is the unique second order formula on the support $(i - 1, i, i + 1)$, referring to the mesh of Figure 4.3.2.

We suggest using mathematical tools, such as MAPLE, Mathematica or MATLAB to solve this problem.

Hint: Obtain the following coefficients

$$b = -\frac{1}{2}r_i - \frac{(1 + r_{i-1})(1 + r_i r_{i-1} + r_{i-1})}{(1 + r_i)r_{i-1}^2}a$$

$$c = \frac{(1 + r_i r_{i-1} + r_{i-1})}{r_i r_{i-1}^2}a - \frac{1}{2}\frac{(1 - r_i^2)}{r_i}$$

$$d = \frac{1}{2}r_i - \frac{(1 + r_{i-1})}{r_i(1 + r_i)r_{i-1}^2}a$$

The truncation error is given by

$$TE = \left[-\frac{(r_{i-1} + 1)(r_{i-1} + r_i r_{i-1} + 1)}{3(1 + r_i)r_{i-1}^3}a + \frac{r_i}{6} \right] \Delta x_i^2 u_{xxx}$$

$$+ \left[\frac{(r_{i-1} + 1)(r_{i-1} + r_i r_{i-1} + 1)(1 + 2r_{i-1} - r_i r_{i-1})}{12(1 + r_i)r_{i-1}^4}a \right.$$

$$\left. - \frac{r_i(1 - r_i)}{24} \right] \Delta x_i^3 u_{xxxx}$$

On a uniform grid, the difference formula becomes

$$u_x = \frac{1}{\Delta x} \left[au_{i-2} - \left(3a + \frac{1}{2} \right) u_{i-1} + 3au_i + \left(\frac{1}{2} - a \right) u_{i+1} \right]$$

$$+ \left(a - \frac{1}{6} \right) \Delta x^2 u_{xxx} - \frac{a}{2} \Delta x^3 u_{xxxx}$$

P.4.14 Obtain one-sided forward or backward second order formulas, involving three mesh points by applying Taylor expansions, for any cell size distribution.

For a forward formula, obtain

$$(u_x)_i = \left[\frac{\Delta x_{i+1} + \Delta x_{i+2}}{\Delta x_{i+2}\Delta x_{i+1}}(u_{i+1} - u_i) - \frac{\Delta x_{i+1}}{\Delta x_{i+2}(\Delta x_{i+1} + \Delta x_{i+2})}(u_{i+2} - u_i)\right]$$
$$+ \frac{\Delta x_{i+1}(\Delta x_{i+1} + \Delta x_{i+2})}{6}(u_{xxx})_i + \cdots$$

Explain why this formula is not in conservative form.

P.4.15 Apply the relations (4.3.23) combined to the relations (4.3.20) to obtain the unique second order conservative difference formula for the first derivative on a non-uniform grid. Compare with the formula (4.3.7).

  Hint: Show that the numerator of formula (4.3.17) can be written as

$$a(u_{i-2} - u_i) + b(u_{i-1} - u_i) + d(u_{i+1} - u_i)$$

Apply the findings to the interpolation relation (4.3.12).

P.4.16 Consider the general four-point formula (4.3.17) for the first derivative on a non-uniform mesh and add the condition for third order accuracy to the system (4.3.18), by applying the consistency relations. Obtain the general formula in function of the mesh ratios r_i and r_{i-1}, by solving the system for the four coefficients (a,b,c,d). Show that on a uniform grid, the formula reduces to equation (4.3.25).

  We suggest using mathematical tools, such as MAPLE, Mathematica or MATLAB to solve this problem.

  Hint: The condition for third order accuracy to be added to the system is

$$-a\left(1 + \frac{1}{r_{i-1}}\right)^3 + dr_i^3 - b = 0$$

The solution is

$$a = \frac{r_i(1 + r_i)r_{i-1}^3}{2(r_{i-1} + 1)(r_{i-1} + r_i r_{i-1} + 1)} \qquad b = -\frac{1}{2}(1 + r_{i-1})r_i$$

$$c = \frac{1}{2}\frac{(1 + r_i)(2r_i r_{i-1} + r_i - 1 - r_{i-1})}{(1 + r_{i-1})r_i} \qquad d = \frac{1}{2}\frac{(1 + r_{i-1})}{r_i(1 + r_{i-1} + r_i r_{i-1})}$$

P.4.17 Consider the general interpolation formula (4.3.12) and derive the relation between the coefficients α and β, for this formula to be at least second order accurate. Derive these conditions separately for the cell-vertex grid of Figure 4.3.2 and the cell-centered grid of Figure 4.3.3. Derive also the conditions for the formula to be third order accurate and show that on a uniform grid we have $\alpha = 1/8$; $\beta = 3/8$.

  Hint: Expand the left and right-hand side terms in a Taylor expansion around point i. Obtain the following conditions:

Cell-vertex case

$$\alpha_i \Delta x_{i+1} + \beta_i \Delta x_i = \frac{1}{2}\Delta x_{i+1} \quad \text{or} \quad \alpha_i r_i + \beta_i = \frac{r_i}{2} \quad \text{for second order accuracy}$$

$$\alpha_i = \frac{(r_i + 2)}{4(r_i + 1)} \quad \beta_i = \frac{r_i^2}{4(r_i + 1)} \quad \text{for third order accuracy}$$

Cell-centered case

$$\alpha_i(\Delta x_i + \Delta x_{i+1}) + \beta_i(\Delta x_i + \Delta x_{i-1}) = \Delta x_i \quad \text{or}$$

$$\alpha_i(1 + r_i) + \beta_i(1 + r_{i-1}) = 1 \quad \text{for second order accuracy}$$

$$\alpha_i = \frac{(2r_{i-1} + 1)}{(1 + r_i)(1 + r_i + 2r_{i-1}r_i)} \quad \beta_i = \frac{r_{i-1}^2 r_i}{(1 + r_{i-1})(1 + r_i + 2r_{i-1}r_i)}$$

for third order accuracy.

P.4.18 Apply a Taylor series expansion to the general form:

$$(u_x)_i = au_{i+2} + bu_{i+1} + cu_i + du_{i-1} + eu_{i-2}$$

and obtain the central fourth order accurate finite difference approximation to the first derivative $(u_x)_i$, at mesh point i. Repeat the same procedure, to obtain an approximation to the second derivative for $(u_{xx})_i$, with the same mesh points. Show that the formula is also fourth order accurate. Calculate the truncation error for both cases.

Hint: Show that we have

$$(u_x)_i = \frac{1}{12\Delta x}(-u_{i+2} + 8u_{i+1} - 8u_{i-1} + u_{i-2}) + \frac{\Delta x^4}{30}\left(\frac{\partial^5 u}{\partial x^5}\right)$$

$$(u_{xx})_i = \frac{1}{12\Delta x^2}(-u_{i+2} + 16u_{i+1} - 30u_i + 16u_{i-1} - u_{i-2}) + \frac{\Delta x^4}{90}\left(\frac{\partial^6 u}{\partial x^6}\right)$$

P.4.19 Repeat Problem P.4.18, for the third and fourth derivatives and obtain

$$(u_{xxx})_i = \frac{1}{2\Delta x^3}(u_{i+2} - 2u_{i+1} + 2u_{i-1} - u_{i-2}) - \frac{\Delta x^2}{4}\left(\frac{\partial^5 u}{\partial x^5}\right)$$

$$(u_{xxxx})_i = \frac{1}{\Delta x^4}(u_{i+2} - 4u_{i+1} + 6u_i - 4u_{i-1} + u_{i-2}) - \frac{\Delta x^2}{6}\left(\frac{\partial^6 u}{\partial x^6}\right)$$

P.4.20 Obtain formulas (4.4.52) to (4.4.59).

P.4.21 Obtain formulas (4.4.60) to (4.4.62).

P.4.22 Derive a family of compact implicit finite difference formulas by eliminating the coefficients a_+, b_+, c_+ and a_0 from the system (4.5.14). Derive the truncation error and obtain the formulas with the highest order of accuracy.

P.4.23 Find the highest order implicit difference formula, involving second derivatives at only one point. Write this expression as an explicit relation for $(u_{xx})_i$ and derive the truncation error.

Hint: Select $c_+ = c_- = 0$, $c_0 = 1$, obtain

$$(u_{xx})_i = -\frac{1}{2\Delta x}[(u_x)_{i+1} - (u_x)_{i-1}] + \frac{2}{\Delta x^2}(u_{i+1} - 2u_i + u_{i-1})$$

The truncation error is found to be

$$R = \frac{\Delta x^4}{360}\left(\frac{\partial^6 u}{\partial x^6}\right)$$

and the formula is fourth order accurate.

P.4.24 Derive a family of implicit difference formulas involving no second derivatives.
Hint: Select $c_+ = c_- = c_0 = 0$ and set $\alpha = b_+/b_-$. Obtain the scheme

$$\alpha(u_x)_{i+1} + 2(1+\alpha)(u_x)_i + (u_x)_{i-1}$$

$$= \frac{1}{2\Delta x}[(1+5\alpha)u_{i+1} + 4(1-\alpha)u_i - (5+\alpha)u_{i-1}]$$

with the truncation error

$$R = \frac{\Delta x^3}{12}(\alpha - 1)\left(\frac{\partial^4 u}{\partial x^4}\right) + \frac{\Delta x^4}{12}(\alpha - 1)\left(\frac{\partial^4 u}{\partial x^4}\right)$$

Chapter 5

Finite Volume Method and Conservative Discretization with an Introduction to Finite Element Method

OBJECTIVES AND GUIDELINES

The Finite Volume Method (FVM) is the name given to the technique by which the integral formulation of the conservation laws is discretized directly in the physical space.

It is the most widely applied method today in CFD, and this is likely to remain so in the foreseeable future. The reason behind the appeal to the FVM lies in its generality, its conceptual simplicity and its ease of implementation for arbitrary grids, structured as well as unstructured.

We consider it therefore as important, even for a first course in CFD, that you become knowledgeable with the basics of the FVM, as it has also many implications for the understanding of the nature and properties of the results obtained from a CFD simulation. As we will see in the following, the FVM is based on *cell-averaged values*, which appear as a most fundamental quantity in CFD. This distinguishes the FVM from the finite difference and finite element methods, where the main numerical quantities are the *local* function values at the mesh points.

Once a grid has been generated, the FVM consists in associating a local *finite volume*, also called *control volume*, to each mesh point and applying the integral conservation law to this local volume. This is a first major distinction from the finite difference approach, where the discretized space is considered as a set of points, while in the FVM the discretized space is formed by a set of small cells, one cell being associated to one mesh point.

An essential advantage of the FVM is connected to the very important concept of *conservative discretization*. We have already introduced the notion of conservation form of the flow equations in Chapter 1, and we refer you to Section 1.1 for a quick reminder of these properties. It is of very great importance to maintain the global conservation of the basic flow quantities, mass, momentum and energy, at the discrete level and this puts conditions on the way the discretization process of the equations is performed. *The FVM has the great advantage that the conservative discretization is automatically satisfied*, through the direct discretization of the integral form of the conservation laws. This most fundamental property for numerical schemes, and its precise meaning will be discussed in Section 5.1.

The finite volume method takes its full advantage on an arbitrary mesh, where a large number of options are open for the definition of the control volumes on which the conservation laws are expressed. Modifying the shape and location of the control volumes associated to a given mesh point, as well as varying the rules and accuracy for the evaluation of the fluxes through the control surfaces, gives a considerable flexibility to the finite volume method. This explains the generality of the FVM, and of course a certain number of rules have to be satisfied during these operations. It will be the objective of Section 5.2 to introduce you to these rules.

Here we draw your attention on Section 5.2.3, which is the most fundamental contribution of this chapter. By an exact and rigorous, although simple and straight-forward, derivation, the most general form of a numerical scheme is presented. It provides the significance of the input as well as the output of any numerical scheme. *Section 5.2.3 is to be studied very carefully and to be remembered whenever you deal with CFD, as a developer or as a user.*

Section 5.3 introduces some practical aspects of the implementation of the FVM, in particular the various choices available for the flux calculations at the cell interfaces, while numerical formulas that can be applied for the estimation of surfaces and volumes on an arbitrary grid will be handled in Chapter 6.

As a more advanced topic, we introduce the *Finite Element Method* (FEM) and its application to fluid dynamic conservation laws in Section 5.4. The FEM is widely used in structural mechanics and many books can be found providing an excellent introduction to its basics. See for instance the books by Zienkiewicz and Taylor (2000), in three volumes, and you can also consult the web site http://ohio.ikp.liu.se/fe for an extensive database on the FEM past and most recent literature.

Hence, we restrict ourselves here to a short summary of the main properties. For those of you who would want to extend their knowledge of the FEM for fluid dynamic applications, we recommend particularly Sections 5.4.3–5.4.5, where the *weighted residual method* is introduced with its *Galerkin method* variant, as applied to the general form of a conservation law. Finally, Section 5.4.6 indicates the link between the weighted residual method and the finite volume method, seen as a particular case of this more general method.

Figure 5.0.1 shows the suggested roadmap to this chapter. For an introductory course, the Sections 5.1–5.3 are very important and strongly recommended, while we will leave Section 5.4 on the FEM as optional for a more advanced level. This section will therefore be marked with A, based on our notation conventions.

5.1 THE CONSERVATIVE DISCRETIZATION

From the general presentation of Chapter 1, we know that the flow equations are the expression of a conservation law. Their general form for a scalar quantity U, with volume sources Q, is given by equation (1.1.1), incorporating eventual surface sources into the flux term:

$$\frac{\partial}{\partial t}\int_\Omega U \, d\Omega + \oint_S \vec{F} \cdot d\vec{S} = \int_\Omega Q \, d\Omega \qquad (5.1.1)$$

The essential property of this formulation is the presence of the surface integral and the fact that the time variation of U inside the volume, *only depends on the surface values of the fluxes*.

Hence for an arbitrary subdivision of the volume into say, three subvolumes, we can write the conservation law for each subvolume and recover the global conservation law

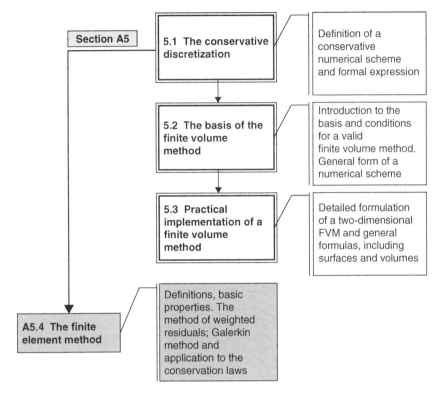

Figure 5.0.1   *Content and guide to this chapter.*

by adding up the three subvolume conservation laws. Indeed, referring to Figure 5.1.1, the above equation written for the subvolumes Ω_1, Ω_2, Ω_3 becomes

$$\frac{\partial}{\partial t} \int_{\Omega_1} U \, d\Omega + \oint_{ABCA} \vec{F} \cdot d\vec{S} = \int_{\Omega_1} Q \, d\Omega$$

$$\frac{\partial}{\partial t} \int_{\Omega_2} U \, d\Omega + \oint_{DEBD} \vec{F} \cdot d\vec{S} = \int_{\Omega_2} Q \, d\Omega \qquad (5.1.2)$$

$$\frac{\partial}{\partial t} \int_{\Omega_3} U \, d\Omega + \oint_{AEDA} \vec{F} \cdot d\vec{S} = \int_{\Omega_3} Q \, d\Omega$$

When summing the surface integrals, the contributions of the internal lines ADB and DE always appear twice, but with opposite signs and will cancel in the addition of the three subvolume conservation laws. Indeed, for volume Ω_2 for instance we have a contribution of the fluxes:

$$\int_{DE} \vec{F} \cdot d\vec{S}$$

while for Ω_3 we have a similar term:

$$\int_{ED} \vec{F} \cdot d\vec{S} = -\int_{DE} \vec{F} \cdot d\vec{S}$$

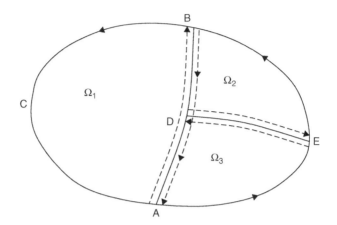

Figure 5.1.1 *Conservation laws for subvolumes of volume $\Omega_1, \Omega_2, \Omega_3$.*

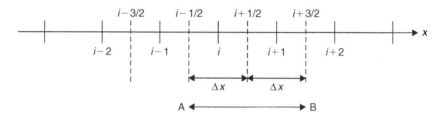

Figure 5.1.2 *Subdivision of a one-dimensional space into mesh cells.*

This essential property has to be satisfied by the numerical discretization of the flux contributions in order for a scheme to be conservative. When this is not the case, that is if after summation of the discretized equations over a certain number of adjacent mesh cells, the resulting equation still contains flux contributions from inside the global cell, the discretization is said to be ***non-conservative***, and the internal flux contributions appear as ***numerical internal volume sources***.

Let us illustrate this on a one-dimensional form of the conservation law, written here as follows, where f is the x-component of the flux vector:

$$\frac{\partial u}{\partial t} + \frac{\partial f}{\partial x} = q \tag{5.1.3}$$

Referring to Figure 5.1.2, we define here a finite volume mesh by associating a finite cell to each mesh point, which we define here by taking the 'cell faces' as the mid-points. Hence for cell (i), the 'cell faces' are the mid-points $(i - 1/2)$ and $(i + 1/2)$.

With a central difference applied to this finite volume mesh of Figure 5.1.2, the following discretized equation is obtained at point i, assuming constant cell sizes $\Delta x_i = \Delta x_{i+1} = \Delta x_{i-1} = \Delta x$

$$\frac{\partial u_i}{\partial t} + \frac{f_{i+1/2} - f_{i-1/2}}{\Delta x} = q_i \tag{5.1.4}$$

The same discretization applied to the point $(i + 1)$, will give

$$\frac{\partial u_{i+1}}{\partial t} + \frac{f_{i+3/2} - f_{i+1/2}}{\Delta x} = q_{i+1} \tag{5.1.5}$$

and for $(i - 1)$

$$\frac{\partial u_{i-1}}{\partial t} + \frac{f_{i-1/2} - f_{i-3/2}}{\Delta x} = q_{i-1} \tag{5.1.6}$$

The sum of these three equations is a consistent discretization of the conservation law for the cell AB $(i - 3/2, i + 3/2)$:

$$\frac{\partial}{\partial t} \left(\frac{u_i + u_{i+1} + u_{i-1}}{3} \right) + \frac{f_{i+3/2} - f_{i-3/2}}{3\Delta x} = \frac{q_i + q_{i+1} + q_{i-1}}{3} \tag{5.1.7a}$$

since the flux contributions at internal points have canceled out. This is sometimes called the 'telescoping property' for the flux terms (Roache, 1972).

As an exercise, let us write a possible scheme directly for the cell AB, if we consider the cell AB associated to point i. This implies that we do not consider anymore on this coarsest grid, the intermediate points $i - 1$ and $i + 1$. Hence we would write:

$$\frac{\partial u_i}{\partial t} + \frac{f_{i+3/2} - f_{i-3/2}}{3\Delta x} = q_i \tag{5.1.7b}$$

We see that the flux balance is identical to the result of equation (5.1.7a), obtained by adding up the contributions form the 'internal' of AB. This is exactly the conservative property we wish to emphasize here.

If we consider now the **non-conservative** form of equation (5.1.3), by expressing the flux derivative as

$$\frac{\partial f}{\partial x} = \left(\frac{\partial f}{\partial u} \right) \frac{\partial u}{\partial x} \triangleq a(u) \frac{\partial u}{\partial x}$$

defining hereby the **Jacobian function** $a(u)$ as the derivative of the flux function with respect to the variable u: $a(u) = \partial f / \partial u$. For instance, if $f = u^2/2$, as with the Burgers equation identified in Section 3.1, then $a(u) = u$.

The non-conservative form is then written as

$$\frac{\partial u}{\partial t} + a(u) \frac{\partial u}{\partial x} = q \tag{5.1.8}$$

Both formulations (5.1.3) and (5.1.8) are mathematically equivalent for arbitrary, nonlinear fluxes, *but their numerical implementation is not*.

Applying, for instance a second order central difference at mesh point i, on the finite volume mesh of Figure 5.1.2, we obtain

$$\frac{\partial u_i}{\partial t} + a_i \frac{u_{i+1/2} - u_{i-1/2}}{\Delta x} = q_i \tag{5.1.9a}$$

where a_i is an estimate of the value of $a(u_i)$.

If similar equations are written for $(i + 1)$ and $(i - 1)$

$$\frac{\partial u_{i+1}}{\partial t} + a_{i+1} \frac{u_{i+3/2} - u_{i+1/2}}{\Delta x} = q_{i+1}$$

$$\frac{\partial u_{i-1}}{\partial t} + a_{i-1} \frac{u_{i-1/2} - u_{i-3/2}}{\Delta x} = q_{i-1}$$

(5.1.9b)

and summed, a discretized equation for the cell AB of Figure 5.1.2 is obtained

$$\frac{\partial}{\partial t} \left(\frac{u_i + u_{i+1} + u_{i-1}}{3} \right) + a_i \frac{u_{i+3/2} - u_{i-3/2}}{6\Delta x} - \frac{q_i + q_{i+1} + q_{i-1}}{3}$$

$$= -(a_{i+1} - a_i) \frac{u_{i+3/2} - u_{i+1/2}}{6\Delta x} + (a_i - a_{i-1}) \frac{u_{i-1/2} - u_{i-3/2}}{6\Delta x}$$

(5.1.10a)

A direct discretization of equation (5.1.8) on the cell AB would have given, referring to equation (5.1.7b):

$$\frac{\partial u_i}{\partial t} + a_i \frac{u_{i+3/2} - u_{i-3/2}}{6\Delta x} = q_i$$

(5.1.10b)

Hence, the right-hand side of equation (5.1.10a), which results from the fact that the flux contributions at the internal points of the cell AB do not cancel, appears as an additional source term, which somehow cannot be distinguished by the computer program from the physical sources q_i.

We see that the discretization of the non-conservative form of the equation gives rise to internal sources. These terms can be considered (by performing a Taylor expansion) as a discretization to second order of a term proportional to $\Delta x^2 (a_x u_x)_x$ at mesh point i. For continuous flows, these numerical source terms are of the same order as the truncation error and hence could be neglected. However, numerical experiments and comparisons consistently show that non-conservative formulations are generally less accurate than conservative ones, particularly in presence of strong gradients. For discontinuous flows, such as transonic flows with shock waves, these numerical source terms can become important across the discontinuity and give rise to large errors. This is indeed the case and the discretization of the non-conservative form will not lead to the correct shock intensities. Therefore, in order to obtain, in the numerical computation, the correct discontinuities, such as the Rankine–Hugoniot relations for the Euler equations, it has been shown by Lax (1954) that it is necessary to discretize the conservative form of the flow equations.

5.1.1 Formal Expression of a Conservative Discretization

The conservativity requirement on equation (5.1.3) will be satisfied if the scheme can be written under the form:

$$\frac{\partial u_i}{\partial t} + \frac{f^*_{i+1/2} - f^*_{i-1/2}}{\Delta x} = q_i$$

(5.1.11)

where f^* is called the **numerical flux**, and is a function of the values of u at $(2k)$ neighboring points:

$$f_{i+1/2}^* = f^*(u_{i+k}, \ldots, u_{i-k+1}) \tag{5.1.12}$$

In addition, the consistency of equation (5.1.11) with the original equation, requires that, when all the u_{i+k} are equal, we should have

$$f^*(u, \ldots, u) = f(u) \tag{5.1.13}$$

The generalization to multidimensions is straightforward, the above conditions must hold separately for all the components of the flux vector. The importance of this formalization of the conservativity condition is expressed by the following fundamental theorem of Lax and Wendroff (1960).

Theorem: If the solution u_i of the discretized equation (5.1.11) converges boundedly almost everywhere to some function $u(x,t)$ when Δx, Δt tend to zero, then $u(x,t)$ is a **weak solution** of equation (5.1.3).

This theorem guarantees that when the numerical solution converges, it will converge to a solution of the basic equations, with the correct satisfaction of the Rankine–Hugoniot relations in presence of discontinuities. Indeed, by comparing the derivation of the Rankine–Hugoniot relations in Section 2.7.1 with the weak formulation of the basic flow equations, with $W = 1$ in equation (5.4.36), it is obvious that these relations are satisfied by the weak solutions, since the starting point of the derivation is the integral form of the conservation law.

5.2 THE BASIS OF THE FINITE VOLUME METHOD

Historically, the finite volume method has apparently been introduced in the field of numerical fluid dynamics independently by Mc Donald (1971) and MacCormack and Paullay (1972) for the solution of two-dimensional, time-dependent Euler equations and extended by Rizzi and Inouye (1973) to three-dimensional flows.

The strength of the FVM is its direct connection to the physical flow properties. Indeed, the basis of the method relies on the **direct discretization of the integral form of the conservation law**. This distinguishes the FVM significantly from the finite difference method, described in the previous chapter, since the latter discretizes the *differential form* of the conservation laws. As indicated in Chapter 1, the integral form is the most general expression of a conservation law, as it does not require the fluxes to be continuous (property which is not satisfied for instance along shock waves or along free surfaces). This is why we can state that the FVM is close to the physics of the flow system.

The FVM requires setting up the following steps:

- Subdivide the mesh, obtained from the space discretization, into finite (small) volumes, one control volume being associated to each mesh point.
- Apply the integral conservation law to each of these finite volumes.

This approach is fully justified by the generality of the integral form, as seen in Chapter 1, where we have pointed out that the integral form of the conservation law is valid for any volume Ω. It can therefore be applied to each individual finite volume. However, we have to ensure that when adding the contributions of a certain number of neighboring finite volumes, we do not loose the consistency required for a valid discretization, as well as the conservativity property introduced in the previous section. These conditions are now explained in what follows.

5.2.1   Conditions on Finite Volume Selections

Due to its generality the finite volume method can handle any type of mesh, structured as well as unstructured. Moreover, an additional degree of freedom appears through the way we relate the control volumes to the grid. This has then an impact on the position of the points at which the function values will be defined.

Considering Figure 5.2.1, we can define, *for the same mesh*, either

- *A cell-centered approach*, where the unknowns are at the **centers** of the mesh cells and the grid lines define the finite volumes and surfaces. An obvious choice

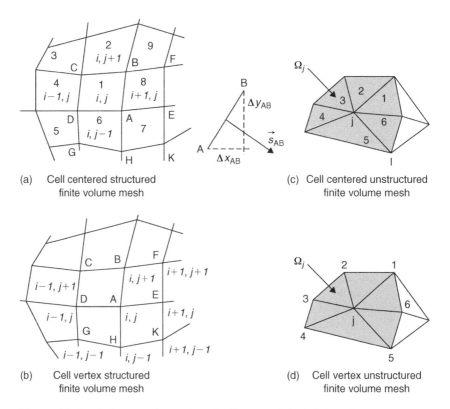

(a)   Cell centered structured finite volume mesh

(c)   Cell centered unstructured finite volume mesh

(b)   Cell vertex structured finite volume mesh

(d)   Cell vertex unstructured finite volume mesh

Figure 5.2.1   *Cell-vertex (bottom) and cell-centered (top) finite volume configurations for structured and unstructured grids.*

for the control volumes is indeed to make them coincide with the mesh cells. Here the variables are associated with a cell, as on Figure 5.2.1a and c. The flow variables are averaged values over the cell and can be considered as representative of some point inside the cell, for instance the central point of the cell.

- *A cell-vertex approach*, where the unknowns are defined at the ***corners*** of the mesh. Here the variables are attached to the mesh points, i.e. to the cell vertices, as shown on Figure 5.2.1b and d. A larger flexibility exists for the definition of the control volumes. Referring to Figure 5.2.1b, an obvious choice would be to consider the four cells having mesh point (i,j) in common, as the control volume GHKEFBCDG, associated to point (i,j). Many other choices are however possible and two of them are shown on Figure 5.2.2. Figure 5.2.2a, from Mc Donald (1971), selects an hexagonal control volume, while Denton (1975) used a trapezoidal control surface covering two half mesh cells (Figure 5.2.2b).

The following constraints on the choice of the Ω_J volumes for a consistent finite volume method have to be satisfied:

(i) Their sum should cover the entire domain Ω_J .

(ii) The subdomains Ω_J are allowed to overlap with the conditions that each part of the surface Γ_J appears as part of an even number of different subdomains such that the overall integral conservation law holds for any combination of adjacent subdomains.

(iii) Fluxes along a cell surface have to be computed by formulas independent of the cell in which they are considered.

Requirement (iii) ensures that the conservative property is satisfied, since the flux contributions of internal boundaries will cancel when the contributions of the associated finite volumes are added.

Referring to Figure 5.2.3, the cells 1,2,3,4 have no common sides and their sum does not cover the whole volume. In addition, the sides are not common to two volumes. The cells 5,6,7 overlap, but have no common surfaces. Hence the conservative property will not be satisfied.

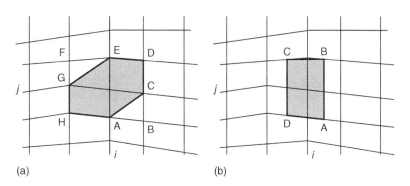

(a)   (b)

Figure 5.2.2   *Examples of two-dimensional control surfaces with cell-vertex finite volume method: (a) Mc Donald (1971) (b) Denton (1975).*

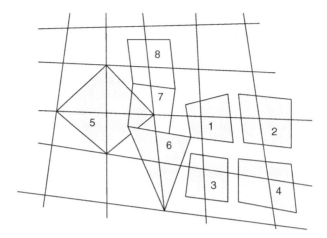

Figure 5.2.3 *Incorrect finite volume decomposition.*

5.2.2 Definition of the Finite Volume Discretization

The integral conservation law (5.1.1) is applied to each control volume Ω_J, associated to mesh point J, defining hereby the discretized equation for the unknowns U_J attached to that same vertex or cell:

$$\frac{\partial}{\partial t}\int_{\Omega_J} U \, d\Omega + \oint_{S_J} \vec{F} \cdot d\vec{S} = \int_{\Omega_J} Q \, d\Omega \qquad (5.2.1)$$

The advantage of this method, especially in absence of sources terms, is that the fluxes are calculated only on two-dimensional surfaces instead of in the three-dimensional space.

Equation (5.2.1) is replaced by its discrete form, where the volume integrals are expressed as the averaged values over the cell and where the surface integral is replaced by a sum over all the bounding faces of the considered volume Ω_J:

$$\frac{\partial}{\partial t}(U_J \Omega_J) + \sum_{\text{faces}} \vec{F} \cdot \Delta \vec{S} = Q_J \Omega_J \qquad (5.2.2)$$

Referring to Figure 5.2.1a and to cell 1(i,j), we would identify U_J with $U_{i,j}$, Ω_J with the area of ABCD and the flux terms are summed over the four sides AB,BC,CD,DA. On the mesh of Figure 5.2.1d, Ω_J is the doted area of the triangles having node J in common and the flux summation extends over the six sides 12,23,34,45,56,61. This is the general formulation of the finite volume method and the user has to define, for a selected Ω_J, how to estimate the volume and cell face areas of the control volume Ω_J and how to approximate the fluxes at the faces. We will discuss some of the most current options, in two and three dimensions.

Equation (5.2.2) shows several interesting features which distinguish the interpretation of finite volume methods from the finite difference and finite element approaches:

1. The coordinates of point J, that is the precise location of the variable U_J inside the control volume Ω_J, do not appear explicitly. Consequently, U_J is

not necessarily attached to a fixed point inside the control volume and can be considered as an average value of the flow variable U over the control cell. This is the interpretation taken in Figure 5.2.1a. The first term of equation (5.2.2) represents therefore the time rate of change of the averaged flow variable over the selected finite volume. This will be further specified in the next section.

2. The mesh coordinates appear only in the determination of the cell volume and side areas. Hence, referring to Figure 5.2.1a, and considering for instance the control cell ABCD around point 1, only the coordinates of A,B,C,D will be needed.

3. In absence of source terms, the finite volume formulation expresses that the variation of the average value U over a time interval Δt is equal to the sum of the fluxes exchanged between neighboring cells. For stationary flows, the numerical solution is obtained as a result of the balance of all the fluxes entering or leaving the control volume. That is,

$$\sum_{\text{faces}} (\vec{F} \cdot \Delta \vec{S}) = 0 \tag{5.2.3}$$

When adjacent cells are considered, for instance cells ABCD and AEFB on Figure 5.2.1a, the flux through face AB contributes to the two cells but with opposite signs. It is therefore convenient to program the method by sweeping through the cell faces and, when calculating the flux through side AB, to add this contribution to the flux balance of cell 1 and subtract it from the flux balance of cell 8. This guarantees automatically global conservation.

4. The finite volume method also allows a natural introduction of boundary conditions, for instance at solid walls where certain normal components are zero. For instance, for the mass conservation equation, $\vec{F} = \rho \vec{v}$ and at a solid boundary $\vec{F} \cdot d\vec{S} = 0$. Hence, the corresponding contribution to equation (5.2.2) or (5.2.3) would vanish.

5.2.3 General Formulation of a Numerical Scheme

A general and important interpretation of any numerical, conservative scheme, generalizing equation (5.1.11), is obtained directly from the integral conservation laws. The formulation that follows is valid for all possible cases, with structured grids or unstructured grids, either cell-centered or cell-vertex defined variables, as summarized in Figures 5.2.4.

If the integral form of the conservation law (5.2.1) is integrated from $t = n\Delta t$ to $(n+1)\Delta t$ for a control volume Ω_J associated to a node or cell J, we obtain

$$\int_{\Omega_J} U \, d\Omega \Big|^{n+1} = \int_{\Omega_J} U \, d\Omega \Big|^{n} - \sum_{\text{faces}} \int_{n}^{n+1} (\vec{F} \cdot \Delta \vec{S})_f \, dt + \int_{n}^{n+1} dt \int_{\Omega_J} Q \, d\Omega$$

$$\tag{5.2.4}$$

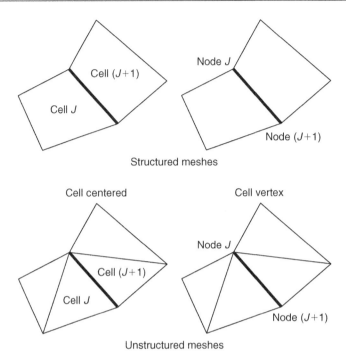

Figure 5.2.4   *Cell-centered and cell-vertex cells for structured and unstructured grids.*

Introducing the **cell-averaged** conservative variable $\overline{U}_J^n$ and source $\overline{Q}_J^n$ at time $n\Delta t$, the *cell- and time-averaged* sources $\overline{Q}_J^*$, together with the **numerical flux** $\vec{F}^*$ over each side, defined by

$$\overline{U}_J^n \triangleq \frac{1}{\Omega_J} \int_{\Omega_J} U \, d\Omega_J \bigg|^n \qquad \overline{Q}_J \triangleq \frac{1}{\Omega_J} \int_{\Omega_J} Q \, d\Omega_J \tag{5.2.5a}$$

$$\vec{F}^* \cdot \Delta\vec{S} \triangleq \frac{1}{\Delta t} \int_n^{n+1} \vec{F} \cdot \Delta\vec{S} \, dt \quad \overline{Q}_J^* \triangleq \frac{1}{\Delta t} \int_n^{n+1} \overline{Q}_J \, dt \tag{5.2.5b}$$

the conservative discretization takes the form:

$$\left[\overline{U}_J \Omega_J \right]^{n+1} = \left[\overline{U}_J \Omega_J \right]^n - \Delta t \sum_{\text{faces}} \vec{F}^* \cdot \Delta\vec{S} + \Delta t \overline{Q}_J^* \Omega_J \tag{5.2.6}$$

This is an **exact relation** for the time evolution of the space averaged conservative variables $\overline{U}_J^n$ over cell J. It is also important to observe that there is no mesh point associated to $\overline{U}_J^n$ which is attached only to cell J. The numerical flux $\vec{F}^*$ identifies completely a scheme by the way it approximates the time-averaged physical flux along each cell face.

As already mentioned above, in order to fulfill conservation at the discretized level, the estimation of the numerical flux at a given cell face must be independent of the cell to which it belongs.

The absence of a time index on the balance of fluxes and on the source term is meant to indicate that one can choose between n for an ***explicit scheme*** and $(n+1)$ for an ***implicit scheme***.

The above formulation of a numerical scheme can be generalized if one considers that the space discretization is completely defined by its numerical flux, leaving open the choice of the time integration. A general numerical scheme can then be defined as a system of ordinary differential equations in time by

$$\frac{\mathrm{d}}{\mathrm{d}t}\left[\overline{U}_J \Omega_J\right] = -\sum_{\text{faces}} \vec{F}^* \cdot \Delta\vec{S} + \overline{Q}_J^* \Omega_J \equiv -R_J \tag{5.2.7}$$

The right-hand side defines the ***residual*** R_J as the balance of fluxes over a cell J plus the source term contribution.

The scheme (5.2.6) is obtained by applying a forward difference in time to the left-hand side of (5.2.7) and typical examples of other time integration schemes, such as Runge–Kutta methods or linear multistep methods will be discussed in Chapter 9.

In practice, when the results of a CFD simulation has to be analyzed and post-processed, we need to assign the cell-averaged values to a mesh point, for instance the center of the cell. ***This introduces an error, which is generally of second order***, which becomes then part of the discretization error.

Alternative formulation of the conservative condition

Extending the subdivisions of equation (5.1.2) to an arbitrary number of cells, $J = 1$–N, and summing over all the cells it is seen, after cancellation of the contributions from all the internal cell faces, that the sum

$$\sum_{J=1}^{N} \int_{\Omega_J} \frac{\partial U}{\partial t} \mathrm{d}\Omega_J = -\oint_S \vec{F} \cdot \mathrm{d}\vec{s} + \int_\Omega Q \, \mathrm{d}\Omega \tag{5.2.8}$$

will contain only contributions from the fluxes along the parts of the cells belonging to the boundaries of the domain and from the sources. Therefore, the conservative condition can be expressed as a requirement on the transient time evolution of the scheme. Note that for stationary sources and boundary fluxes, the right-hand side of this equation vanishes at convergence.

Defining $\Delta\overline{U}_J/\Delta t$ as the average value of $\partial U/\partial t$ over the cell Ω_J, conservation of the scheme requires that, at each time step, the following condition is to be satisfied:

$$\sum_{J=1}^{N} \frac{\Delta\overline{U}_J}{\Delta t} \Omega_J = \text{boundary and source terms} \tag{5.2.9}$$

Hence, the sum of $\Delta\overline{U}_J\Omega_J/\Delta t$ over all the cells may not contain contributions from inside the domain, with the exception of the source terms (see also Problem P.5.13).

5.3 PRACTICAL IMPLEMENTATION OF FINITE VOLUME METHOD

In this section we wish to develop some practical formulas, that can be used for the effective calculation of surfaces and volumes, in two or three dimensions. When the general finite volume method is applied on a Cartesian grid, several finite difference formulas are recovered.

We recommend to study these links very carefully, as comparing the two methods can provide you with a better understanding of the main properties of each of the two approaches. The following will also illustrate the generality of the FVM.

5.3.1 Two-Dimensional Finite Volume Method

Equation (5.2.1), considered for control cell ABCD of Figure 5.2.1a, can be written as

$$\frac{\partial}{\partial t}\int_{\Omega_{ij}} U\, d\Omega + \oint_{ABCD} (f\, dy - g\, dx) = \int_{\Omega_{ij}} Q\, d\Omega \tag{5.3.1}$$

where f and g are the Cartesian components of the flux vector F. Equation (5.3.1) is the most appropriate for a direct discretization. The surface vector for a side AB can be defined as

$$\vec{S}_{AB} = \Delta y_{AB}\vec{1}_x - \Delta x_{AB}\vec{1}_y = (y_B - y_A)\vec{1}_x - (x_B - x_A)\vec{1}_y \tag{5.3.2}$$

and we obtain the finite volume equation for cell Ω_{ij}

$$\frac{\partial}{\partial t}(U_{ij}\Omega_{ij}) + \sum_{ABCD} [f_{AB}(y_B - y_A) - g_{AB}(x_B - x_A)] = Q_{ij}\Omega_{ij} \tag{5.3.3}$$

The sum over ABCD extends over the four sides of the quadrilateral ABCD. For a general quadrilateral ABCD, the area Ω can be evaluated from the vector products of the diagonals. As seen from Figure 5.3.1 the parallelogram 1234 built on the diagonals

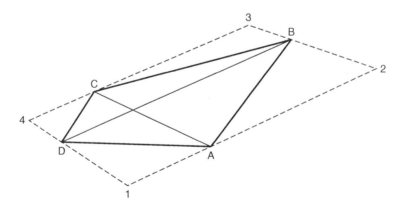

Figure 5.3.1   *Area of an arbitrary plane quadrilateral.*

is twice the area of the quadrilateral ABCD. Hence with $\vec{x}_{AB} = \vec{x}_B - \vec{x}_A$ where $\vec{x}_A$ is the position vector of point A:

$$\Omega_{ABCD} = \frac{1}{2} |\vec{x}_{AC} \times \vec{x}_{BD}|$$

$$= \frac{1}{2}[(x_C - x_A)(y_D - y_B) - (x_D - x_B)(y_C - y_A)] \tag{5.3.4}$$

$$= \frac{1}{2}(\Delta x_{AC} \Delta y_{BD} - \Delta x_{BD} \Delta y_{AC})$$

The right-hand side of equation (5.3.4) should be positive for a cell ABCD where A,B,C,D are located counterclockwise.

Evaluation of fluxes through cell faces

The evaluation of flux components along the sides, such as f_{AB}, g_{AB}, depends on the selected scheme, as well as on the location of the flow variables with respect to the mesh.

As will be seen more in detail in the following chapters and in Volume II for the systems of Euler and Navier–Stokes equations, one can distinguish essentially between central and upwind discretization schemes. Central schemes are based on local flux estimations, while upwind schemes determine the cell face fluxes according to the propagation direction of the convection velocity:

(a) For *central schemes and cell-centered finite volume methods*, following alternatives can be considered:
1. Average of fluxes:

$$f_{AB} = \frac{1}{2}(f_{ij} + f_{i+1,j}) \tag{5.3.5}$$

$$f_{ij} = f(U_{ij}) \tag{5.3.6}$$

2. Since the flux components are generally nonlinear functions of U, the following choice is not identical to equation (5.3.5):

$$f_{AB} = f\left(\frac{U_{ij} + U_{i+1,j}}{2}\right) \tag{5.3.7}$$

3. Take f_{AB} as the average of the fluxes in A and B,

$$f_{AB} = \frac{1}{2}(f_A + f_B) \tag{5.3.8}$$

where either the variables are evaluated in A and B, for instance

$$U_A = \frac{1}{4}(U_{ij} + U_{i+1,j} + U_{i+1,j-1} + U_{i,j-1}) \tag{5.3.9}$$

and

$$f_A = f(U_A) \tag{5.3.10}$$

or the fluxes are averaged, as

$$f_A = \frac{1}{4}(f_{ij} + f_{i+1,j} + f_{i+1,j-1} + f_{i,j-1}) \tag{5.3.11}$$

Observe that formulas (5.3.7) and (5.3.10) will generally lead to schemes requiring a lower number of flux evaluations compared to the application of formulas (5.3.5) and (5.3.11).

(b) *For central schemes and cell-vertex finite volume methods*, formulas (5.3.7) and (5.3.8) are straightforward approximations to the flux f_{AB}. The choice (5.3.8) corresponds to the application of a trapezium formula for the integral:

$$\int_{AB} f \, dy = \frac{1}{2}(f_A + f_B)(y_B - y_A)$$

By summing the contributions of these integrals over the four sides of cell ABCD of Figure 5.2.1b, we obtain

$$\oint_{ABCD} \vec{F} \cdot d\vec{S} = \frac{1}{2}[(f_A - f_C)\Delta y_{DB} + (f_B - f_D)\Delta y_{AC}$$
$$- (g_A - g_C)\Delta x_{DB} + (g_B - g_D)\Delta x_{AC}] \tag{5.3.12}$$

Example E.5.3.1: Central scheme on Cartesian mesh

Over a Cartesian, uniform mesh the above finite volume formulation is identical to a finite difference formula. Indeed, with

$$\Delta y_{AB} = y_{i+1/2,j+1/2} - y_{i+1/2,j-1/2} = \Delta y$$
$$\Delta x_{AB} = 0 \quad \Delta x_{BC} = -\Delta x \tag{E5.3.1}$$
$$\Omega_{ij} = \Delta x \, \Delta y \quad \Delta y_{BC} = 0$$

we obtain, writing $f_{AB} = f_{i+1/2,j}$ and similarly for the other components:

$$\frac{\partial}{\partial t} U_{ij} \Delta x \, \Delta y + (f_{i+1/2,j} - f_{i-1/2,j})\Delta y + (g_{i,j+1/2} - g_{i,j-1/2})\Delta x$$
$$= Q_{ij} \Delta x \, \Delta y \tag{E5.3.2}$$

After division by $\Delta x \, \Delta y$ this reduces to the central difference form:

$$\frac{\partial U_{ij}}{\partial t} + \frac{(f_{i+1/2,j} - f_{i-1/2,j})}{\Delta x} + \frac{(g_{i,j+1/2} - g_{i,j-1/2})}{\Delta y} = Q_{ij} \tag{E5.3.3}$$

We have still to define how to calculate the flux components at the side centers $f_{i\pm1/2,j}, g_{i,j\pm1/2}$. With the choice (5.3.5) applied to Figure 5.2.1a, equation (E.5.3.3) becomes

$$\frac{\partial U_{ij}}{\partial t} + \frac{(f_{i+1,j} - f_{i-1,j})}{2\Delta x} + \frac{(g_{i,j+1} - g_{i,j-1})}{2\Delta y} = Q_{ij} \tag{E5.3.4}$$

while (5.3.8) with (5.3.11) leads to the formula:

$$\frac{\partial U_{ij}}{\partial t} + \frac{1}{4}\left[2\frac{(f_{i+1,j} - f_{i-1,j})}{2\Delta x} + \frac{(f_{i+1,j+1} - f_{i-1,j+1})}{2\Delta x} + \frac{(f_{i+1,j-1} - f_{i-1,j-1})}{2\Delta x}\right]$$

$$+ \frac{1}{4}\left[2\frac{(g_{i,j+1} - g_{i,j-1})}{2\Delta y} + \frac{(g_{i+1,j+1} - g_{i+1,j-1})}{2\Delta y} + \frac{(g_{i-1,j+1} - f_{i-1,j+1})}{2\Delta y}\right]$$

$$= Q_{ij} \tag{E5.3.5}$$

The central finite volume method therefore leads to second order accurate space discretizations on Cartesian meshes.

Observe that f_{ij}, g_{ij} do not appear in equation (E.5.3.4), and if $(i + j)$ is even this equation contains only nodes with $(i + j)$ odd. Hence even and odd numbered nodes are decoupled, and this could lead to oscillations in the solution. This decoupling is not present with formula (E.5.3.5). For applications to cell-vertex meshes, the reader is referred to Problems P.5.1–P.5.4.

(c) *For upwind schemes and cell-centered finite volume methods*, a convective flux is evaluated in function of the propagation direction of the associated convection speed. The latter is determined by the flux Jacobean

$$\vec{A}(U) = \frac{\partial \vec{F}}{\partial U} = a(U)\vec{1}_x + b(U)\vec{1}_y \tag{5.3.13}$$

with $a(U) = \partial f/\partial U$ and $b(U) = \partial g/\partial U$.

The simplest upwind scheme takes the cell side flux equal to the flux generated in the upstream cell. This expresses that the cell side flux is fully determined by contributions transported in the direction of the convection velocity.

Considering Figure 5.2.1a, we could define

$$(\vec{F} \cdot \vec{S})_{AB} = (\vec{F} \cdot \vec{S})_{ij} \qquad \text{if } (\vec{A} \cdot \vec{S})_{AB} > 0$$

$$(\vec{F} \cdot \vec{S})_{AB} = (\vec{F} \cdot \vec{S})_{i+1,j} \quad \text{if } (\vec{A} \cdot \vec{S})_{AB} < 0 \tag{5.3.14}$$

(d) For upwind schemes and cell-vertex finite volume methods, figure 5.2.1b, we could define

$$(\vec{F} \cdot \vec{S})_{AB} = (\vec{F} \cdot \vec{S})_{CD} \quad \text{if } (\vec{A} \cdot \vec{S})_{AB} > 0$$

$$(\vec{F} \cdot \vec{S})_{AB} = (\vec{F} \cdot \vec{S})_{EF} \quad \text{if } (\vec{A} \cdot \vec{S})_{AB} < 0 \tag{5.3.15}$$

When applied to the control volume GHKEFBCD of figure 5.2.1b, we obtain contributions from points such as $(i - 2, j)$ and $(i, j - 2)$, for positive convection speeds. This leads to schemes with an unnecessary large support for the same accuracy. Therefore, this option is not applied in practice (see Problem P.5.12).

Example E.5.3.2: Upwind scheme on Cartesian mesh

We consider the discretization of the two-dimensional linear convection equation

$$\frac{\partial U}{\partial t} + a\frac{\partial U}{\partial x} + b\frac{\partial U}{\partial y} = 0 \quad \text{with} \quad a > 0 \text{ and } b > 0 \tag{E5.3.6}$$

by a finite volume formulation on the cell ABCD of figure 5.2.1a, defined as a Cartesian cell following example E.5.3.1.

The fluxes are defined by $f = aU$ and $g = bU$ and with the choice of equation (5.3.14), we have for AB and CD taken as vertical sides:

$$(\vec{F} \cdot \vec{S})_{AB} = f_{ij}\Delta y = aU_{ij}\Delta y$$
$$(\vec{F} \cdot \vec{S})_{CD} = -f_{i-1,j}\Delta y = -aU_{i-1,j}\Delta y \tag{E5.3.7}$$

and similarly for the two horizontal sides BC and DA.

$$(\vec{F} \cdot \vec{S})_{BC} = g_{ij}\Delta x = bU_{ij}\Delta x$$
$$(\vec{F} \cdot \vec{S})_{DA} = -g_{i,j-1}\Delta x = -bU_{i,j-1}\Delta x \tag{E5.3.8}$$

The resulting scheme, obtained after division by the cell area $\Delta x \Delta y$, is only first order accurate and is a straightforward generalization of the first order upwind scheme:

$$\frac{\partial U_{ij}}{\partial t} + \frac{(f_{ij} - f_{i-1,j})}{\Delta x} + \frac{(g_{ij} - g_{i,j-1})}{\Delta y} = 0 \tag{E5.3.9}$$

or

$$\frac{\partial U_{ij}}{\partial t} + \frac{a}{\Delta x}(U_{ij} - U_{i-1,j}) + \frac{b}{\Delta y}(U_{ij} - U_{i,j-1}) = 0 \tag{E5.3.10}$$

Non-uniform mesh

Although the finite volume formulation applies to arbitrary grids, the above formulas for the determination of the fluxes nevertheless imply some regularity of the mesh. Referring for instance to formula (5.3.5) or (5.3.7) as applied to cell-centered finite volume methods, and interpreting the cell averaged values U_{ij} in Figure 5.2.1a as mid-cell values, it is seen that these formulas perform an arithmetic average of the fluxes (or the variables) on both sides of the cell face AB. This leads to a second order approximation on a Cartesian mesh (see Example E.5.3.1), if AB is at mid-distance from the cell centers 1 and 8.

However, this will seldom be the case on non-uniform meshes, as shown on Figure 5.3.2a and a loss of accuracy will result from the application of these formulas. Similar considerations apply to equations (5.3.9) and (5.3.11), based on the assumption that point A is in the center of the cell 1678.

An analysis of the truncation errors for certain finite volume discretizations on non-uniform meshes can be found in Arts (1984) and more general analysis can be found in Turkel (1985), Turkel et al. (1985) and Roe (1987).

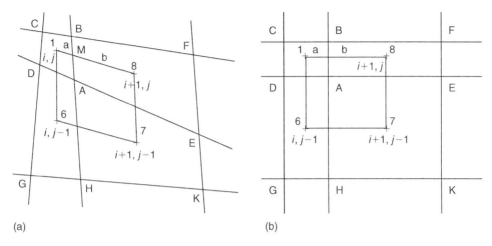

(a)   (b)

Figure 5.3.2 *Non-uniform finite volume meshes: (a) non-uniform finite volume mesh (b) orthogonal non-uniform finite volume mesh.*

We will come back to this very important issue in the Chapter 6, when dealing with grid properties.

5.3.2 Finite Volume Estimation of Gradients

It is often necessary, in finite volume discretizations, to define numerically cell averages of derivatives of mesh variables. In particular with the Navier–Stokes equations, the viscous flux components are functions of the velocity gradients and we have to estimate appropriate values of these gradients on the cell faces.

A general procedure, valid for an arbitrary control volume in two and three dimensions, can be derived by application of the Gauss divergence theorem. This theorem can be considered as defining the average of the gradient of a scalar U in function of its values at the boundaries of the considered volume.

Since for an arbitrary volume Ω

$$\int_\Omega (\vec{\nabla} U)d\Omega = \oint_S U \, d\vec{S} \tag{5.3.16}$$

where S is the closed boundary surface, we can define the averaged gradients as

$$\left(\overline{\frac{\partial U}{\partial x}}\right)_\Omega \equiv \frac{1}{\Omega}\int_\Omega \frac{\partial U}{\partial x} d\Omega = \frac{1}{\Omega}\oint_S U \vec{1}_x \cdot d\vec{S} \tag{5.3.17}$$

and

$$\left(\overline{\frac{\partial U}{\partial y}}\right)_\Omega \equiv \frac{1}{\Omega}\int_\Omega \frac{\partial U}{\partial y} d\Omega = \frac{1}{\Omega}\oint_S U \vec{1}_y \cdot d\vec{S} \tag{5.3.18}$$

For two-dimensional control cells Ω we obtain

$$\left(\frac{\overline{\partial U}}{\partial x}\right)_\Omega = \frac{1}{\Omega}\oint_S U\,dy = -\frac{1}{\Omega}\oint_S y\,dU \tag{5.3.19}$$

after partial integration. Similarly, the averaged y-derivatives are obtained from

$$\left(\frac{\overline{\partial U}}{\partial y}\right)_\Omega = -\frac{1}{\Omega}\oint_S U\,dx = \frac{1}{\Omega}\oint_S x\,dU \tag{5.3.20}$$

Considering as an example the control cell of Figure 5.2.1d and applying trapezoidal integration formulas along each side, the following formulas are obtained

$$\left(\frac{\overline{\partial U}}{\partial x}\right)_\Omega = \frac{1}{\Omega}\oint_S U\,dy$$

$$= -\frac{1}{\Omega}\oint_S y\,dU = \frac{1}{2\Omega}\sum_I (U_I + U_{I+1})(y_{I+1} - y_I)$$

$$= -\frac{1}{2\Omega}\sum_I (y_{I+1} + y_I)(U_{I+1} - U_I)$$

$$= \frac{1}{2\Omega}\sum_I U_I(y_{I+1} - y_{I-1}) = -\frac{1}{2\Omega}\sum_I y_I(U_{I+1} - U_{I-1}) \tag{5.3.21}$$

where the summation extends over all the vertices, from 1 to 6 with $U_0 = U_6$ and $U_7 = U_1$. The two last relations are obtained by rearranging the sums, as can be seen from the relation:

$$(U_1 + U_2)(y_2 - y_1) + (U_2 + U_3)(y_3 - y_2) + \cdots = U_2(y_3 - y_1) + \cdots$$

$$= -y_2(U_3 - U_2) + \cdots \tag{5.3.22}$$

The corresponding relations for the y-derivatives are derived after replacing x by y and changing the signs of the various expressions:

$$\left(\frac{\overline{\partial U}}{\partial y}\right)_\Omega = -\frac{1}{\Omega}\oint_S U\,dx$$

$$= \frac{1}{\Omega}\oint_S x\,dU = -\frac{1}{2\Omega}\sum_I (U_I + U_{I+1})(x_{I+1} - x_I)$$

$$= \frac{1}{2\Omega}\sum_I (x_{I+1} + x_I)(U_{I+1} - U_I)$$

$$= -\frac{1}{2\Omega}\sum_I U_I(x_{I+1} - x_{I-1}) = \frac{1}{2\Omega}\sum_I x_I(U_{I+1} - U_{I-1}) \tag{5.3.23}$$

The area of the cells can be obtained by formulas similar to the above, by noting that for $U = x$ the left-hand side of equation (5.3.21) is equal to 1. Hence, the following expressions can be used for the estimation of the area of an arbitrary cell:

$$\Omega = \oint_S x \, dy = -\oint_S y \, dx = \frac{1}{2} \sum_I (x_I + x_{I+1})(y_{I+1} - y_I)$$

$$= -\frac{1}{2} \sum_I (y_{I+1} + y_I)(x_{I+1} - x_I)$$

$$= \frac{1}{2} \sum_I x_I (y_{I+1} - y_{I-1})$$

$$= -\frac{1}{2} \sum_I y_I (x_{I+1} - x_{I-1}) \tag{5.3.24}$$

For an arbitrary quadrilateral ABCD, as shown on Figure 5.3.2, an interesting formula is obtained by applying the third of the above relations, noticing that the differences Δy can be grouped for opposite nodes, leading to

$$\oint_{ABCD} U \, dy = \frac{1}{2}[(U_A - U_C)(y_B - y_A) - (U_B - U_D)(y_A - y_C)] \tag{5.3.25}$$

and

$$\left(\overline{\frac{\partial U}{\partial x}}\right)_{ABCD} = \frac{(U_A - U_C)(y_B - y_D) - (U_B - U_D)(y_A - y_C)}{(x_A - x_C)(y_B - y_D) - (x_B - x_D)(y_A - y_C)} \tag{5.3.26}$$

with a similar relation for the y-derivative:

$$\left(\overline{\frac{\partial U}{\partial y}}\right)_{ABCD} = \frac{(U_B - U_D)(x_A - x_C) - (U_A - U_C)(x_B - x_D)}{(x_A - x_C)(y_B - y_D) - (x_B - x_D)(y_A - y_C)} \tag{5.3.27}$$

The vector version of the divergence relation, written for an arbitrary vector $\vec{a}$ is also of interest

$$\int_\Omega (\vec{\nabla} \cdot \vec{a}) d\Omega = \oint_S \vec{a} \cdot d\vec{S} \tag{5.3.28}$$

since it can be applied, in particular for the derivation of formulas for cell face areas and volumes. For a two-dimensional cell, taking $\vec{a} = \vec{x}$ with $\vec{\nabla} \cdot \vec{x} = 2$ leads to the formula

$$2\Omega = \oint_S \vec{x} \cdot d\vec{S} = \oint_S (x \, dy - y \, dx) \tag{5.3.29}$$

which reproduces the above relations when a trapezium formula is applied. Applications of this relation to three-dimensional volumes are discussed in Chapter 6.

Example E.5.3.3: Two-dimensional diffusion equation

We consider the two-dimensional diffusion equation:

$$\frac{\partial U}{\partial t} + \frac{\partial}{\partial x}\left(k\frac{\partial U}{\partial x}\right) + \frac{\partial}{\partial y}\left(k\frac{\partial U}{\partial y}\right) = 0 \tag{E5.3.11}$$

with diffusive flux components $f = k\,\partial U/\partial x$ and $g = k\,\partial U/\partial y$, where k is a constant.

We would like to construct a finite volume discretization on the mesh of Figure 5.2.1a, considered as Cartesian, by expressing the balance of fluxes around the cell ABCD with the choice

$$f_{AB} = \frac{1}{2}(f_A + f_B) \tag{E5.3.12}$$

and an evaluation of the derivatives $\partial U/\partial x$ and $g = \partial U/\partial y$ in the cell corners A,B. Equation (5.3.3) for cell (i,j) is written here as

$$\left(\frac{\partial U}{\partial t}\right)_{ij}\Delta x\,\Delta y + (f_{AB} - f_{CD})\Delta y + (g_{BC} - g_{DA})\Delta x = 0 \tag{E5.3.13}$$

For point A, the derivatives of U are taken as the average value over the cell 1678 and with equation (5.3.26):

$$f_A = k\left(\frac{\partial U}{\partial x}\right)_A = \frac{k}{2\Delta x}(U_{i+1,j} + U_{i+1,j-1} - U_{ij} - U_{i,j-1}) \tag{E5.3.14}$$

A similar relation is obtained for point B

$$f_B = k\left(\frac{\partial U}{\partial x}\right)_B = \frac{k}{2\Delta x}(U_{i+1,j} + U_{i+1,j+1} - U_{ij} - U_{i,j+1}) \tag{E5.3.15}$$

and the flux contribution through the side AB is given by the sum of the two equations (E5.3.13) and (E5.3.14) multiplied by Δy.

The contributions of the other sides are obtained in a similar way, for instance the flux through BC is given by the sum:

$$g_{BC}\Delta x = \frac{1}{2}(g_B + g_C)\Delta x \tag{E5.3.16}$$

with

$$g_B = k\left(\frac{\partial U}{\partial y}\right)_B = \frac{k}{2\Delta y}(U_{i+1,j+1} + U_{i,j+1} - U_{ij} - U_{i+1,j}) \tag{E5.3.17}$$

A similar relation is obtained for point C:

$$g_c = k\left(\frac{\partial U}{\partial y}\right)_C = \frac{k}{2\Delta y}(U_{i,j+1} + U_{i-1,j+1} - U_{ij} - U_{i-1,j}) \tag{E5.3.18}$$

Finally, equation (E5.3.13) becomes with $\Delta x = \Delta y$

$$\frac{\partial U_{ij}}{\partial t} + k\frac{U_{i+1,j+1} + U_{i+1,j-1} + U_{i-1,j+1} + U_{i-1,j-1} - 4U_{ij}}{4\Delta x^2} = 0 \qquad \text{(E5.3.19)}$$

This scheme corresponds to the discretization of Figure 4.2.3 for the Laplace operator. Note that the alternative, simpler choice

$$f_{AB} = k\left(\frac{\partial U}{\partial x}\right)_{AB} = \frac{k}{\Delta x}(U_{i+1,j} - U_{ij}) \qquad \text{(E5.3.20)}$$

leads to the standard finite difference discretization of the diffusion equation, corresponding to Figure 4.2.2:

$$\frac{\partial U_{ij}}{\partial t} + k\frac{U_{i+1,j} + U_{i,j-1} + U_{i-1,j} + U_{i,j+1} - 4U_{ij}}{4\Delta x^2} = 0 \qquad \text{(E5.3.21)}$$

Extensions to three-dimensional volumes, typical of current unstructured or structured grids, such as tetrahedra, pyramids, prisms or hexahedra are shifted to Chapter 6.

A.5.4 THE FINITE ELEMENT METHOD

The finite element method originated from the field of structural analysis as an outcome of many years of research mainly between 1940 and 1960. The concept of 'elements' can be traced back to the techniques used in stress calculations whereby a structure is subdivided in small substructures of various shapes and reassembled after each 'element' had been analyzed. The development of this technique and its formal elaboration led to the introduction of what is now called the finite element method by Turner et al. (1956) in a paper dealing with the properties of a triangular element in plane stress problems. The expression finite elements itself was introduced by Clough (1960).

After having been applied with great success to a variety of problems in linear and nonlinear structural mechanics, it appeared very soon that the method could be used also to solve continuous field problems (Zienkiewicz and Cheung, 1965). From then on, the finite element method came out as a general approximation method for the numerical solution of physical problems described by field equations in continuous media, containing actually many of the finite difference schemes as special cases. Today, after the initial developments in an engineering framework, mathematicians have put the finite element method in a very elegant, rigorous, formal framework, with precise mathematical conditions for existence and convergence criteria and exactly derived error bounds. Due to the particular character of finite element discretizations the appropriate mathematical background is functional analysis and an excellent introduction to the mathematical formulation of the method can be found in Strang and Fix (1973), Oden and Reddy (1976) and more advanced treatments in Oden (1972) and Ciarlet (1978).

With regard to fluid flow problems a historical general introduction is to be found in Chung (1978), Baker (1983) and more advanced developments are analyzed in Temam (1977), Girault and Raviart (1979) and Thomasset (1981). A recent addition to the literature on the applications of finite elements to fluid flow problems is to be found in the books by Zienkiewicz and Taylor (2000), in Volume 3.

You can also consult the web site http://ohio.ikp.liu.se/fe for an extensive database on the FEM past and most recent literature.

The finite element approximation has two major common points with the finite volume method, namely that

- The space discretization is considered as a set of volumes or cells, called *elements* in the finite element tradition, as opposed to a set of points as with finite difference methods.
- It requires an integral formulation as a starting point which, as will be seen in section, can be considered as a generalization of the FVM.

The finite element method is based on the following steps:

1. Discretize the space in contiguous elements of arbitrary shapes, that is typical of unstructured grids. These *elements* are actually the control cells defined in the previous sections related to the FVM. Therefore, the FEM is 'naturally' appropriate to unstructured grids, more particularly to cell-vertex grids, as seen in next section. We will often refer in this section to the elements of the mesh, instead of the cells, to follow the FEM terminology. However, keep in mind that these two names refer to the same basic components of a finite volume mesh.
2. Define in each element a parametric representation of the unknown variables, based on families of *interpolating or shape functions*, associated to each element or cell.
3. Define an integral formulation of the equations to be solved to each element (cell) of the discretized space.

We will now develop these three steps more explicitly.

A.5.4.1 Finite Element Definition of Interpolation Functions

The FEM is based on the definition of function values attached to the nodes of the mesh, where the numerical value of the unknown functions, and eventually their derivatives, will have to be determined.

Therefore, the FEM requires a cell-vertex definition. The total number of unknowns at the nodes, function values and eventually their derivatives are called the *degrees of freedom* of the numerical problem or *nodal values*. The field variables are approximated by linear combinations of known *basis functions* also called *shape functions*, *interpolations functions* or *trial functions*.

If $\tilde{u}$ is an approximate numerical solution of the unknown function $u(\vec{x})$ we define its parametric representation as a linear superposition of basis functions N_I:

$$\tilde{u}(\vec{x}) = \sum_I u_I N_I(\vec{x}) \tag{5.4.1}$$

where the summation extends over all the nodes I. To each node I, we associate one basis function N_I. The functions $N_I(\vec{x})$ can be quite general with varying degrees of continuity at the inter-element boundaries. Methods based on defining the interpolation functions on the whole domain, for instance as trigonometric functions leading to Fourier series, are used in **collocation methods** and in **spectral methods** where the $N_I(\vec{x})$ can be defined as orthogonal polynomials of Legendre, Chebyshev or similar types. Other possible choices are spline functions, leading to **spline interpolation methods**. In these cases, the coefficients u_I are obtained from the expansions in series of the basis functions.

The properties of finite element methods are based on a very specific choice with the following properties:

(i) The u_I coefficients are *the numerical values of the unknowns at node I*, i.e. we require that:

$$\tilde{u}(\vec{x}_I) = u_I \tag{5.4.2}$$

Consequently, we have for any point $\vec{x}_J$:

$$N_I(\vec{x}_J) = \delta_{IJ} \tag{5.4.3}$$

implying also that the interpolation functions N_I must be equal to one at the point I to which they are associated.

(ii) In standard finite element methods, the interpolation functions are chosen to be *locally defined polynomials within each element, being zero outside the considered element*. As a consequence the local interpolation functions satisfy the following additional conditions on each element (e), with I being a node of (e):

$$N_I^{(e)}(\vec{x}) = 0 \quad \text{if } \vec{x} \text{ not in element } (e) \tag{5.4.4}$$

since u_I are the values of the unknowns at node number I.

(iii) An additional condition is provided by the requirement to represent exactly a constant function $u(x) = $ constant. Hence, this requires

$$\sum_I N_I^{(e)}(\vec{x}) = 1 \quad \text{for all } \vec{x} \in (e) \tag{5.4.5}$$

(iv) The global function N_I is obtained by assembling the contributions $N_I^{(e)}$ of all the elements to which node I belongs. This condition connects the various basis functions within an element and the allowed polynomials will be strongly dependent on the number of nodes within each element.

The above conditions (5.4.2)–(5.4.5) define the properties of the local polynomial interpolation functions used in finite element approximations. Two families of elements are generally considered according to their degree of inter-element continuity and to the associated nodal values.

If the nodal values are defined by the values of the unknown functions, then C° continuity at the inter-element boundary is sufficient for systems described by partial differential equations not higher than two. These elements and their associated shape functions are then called *Lagrangian elements*.

If first order partial derivatives of the unknown functions are to be considered as additional unknowns (or degrees of freedom), the inter-element continuity up to the highest order of these derivatives will generally be imposed and the elements satisfying these conditions are *Hermitian elements*. When the required continuity conditions are satisfied along every point of the inter-element boundary, the element is called *conforming*. This condition is sometimes relaxed and elements whereby this continuity condition is imposed only at a limited number of points of the boundary are said to be *non-conforming*.

A.5.4.1.1 *One-dimensional linear elements*

The simplest element has a *piecewise linear interpolation* function and contains two nodes. Referring to Figure 5.4.1, the element between nodes i and $i - 1$ is denoted as element 1 and the adjacent element between i and $i + 1$ as element 2. They have node i in common and have respective lengths Δx_i and Δx_{i+1}.

Considering element 1, the relation (5.4.3) gives two conditions and we obtain for the basis functions at the nodes i and $i - 1$ of element 1, the linear form:

$$N_i^{(1)}(x) = \frac{x - x_i}{\Delta x_i} \quad N_{i-1}^{(1)}(x) = \frac{x_i - x}{\Delta x_i} \tag{5.4.6}$$

For element 2 we have the following linear shape functions:

$$N_i^{(2)}(x) = \frac{x_{i+1} - x}{\Delta x_{i+1}} \quad N_{i-1}^{(1)}(x) = \frac{x - x_i}{\Delta x_{i+1}} \tag{5.4.7}$$

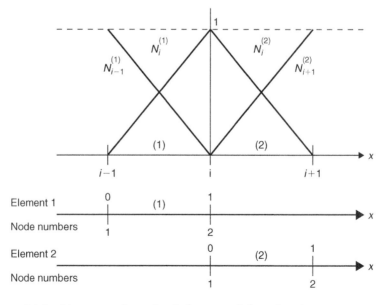

Figure 5.4.1 *Linear one-dimensional element and shape functions.*

The global shape function N_i, associated with node i, is obtained by assembling $N_i^{(1)}$ and $N_i^{(2)}$, and is illustrated in Figure 5.4.1. It is zero for $x \geq x_{i+1}$ and $x \leq x_{i-1}$. If we define a local coordinate ξ within each element through the mapping:

$$\xi = \frac{x - x_{i-1}}{\Delta x_i} \tag{5.4.8}$$

the interpolation functions take the universal form:

$$N_1(\xi) = 1 - \xi \quad N_2(\xi) = \xi \tag{5.4.9}$$

where node 1 corresponds to mesh point $i - 1$ with $\xi = 0$ and node 2 to mesh point i with $\xi = 1$. Similarly for element 2, the mapping between the subspace (x_i, x_{i+1}) and the subspace $(0,1)$ of the ξ-space.

$$\xi = \frac{x - x_i}{\Delta x_{i+1}} \tag{5.4.10}$$

leads to

$$N_i^{(2)}(\xi) = 1 - \xi = N_1(\xi) \quad N_{i+1}^{(2)}(\xi) = \xi = N_2(\xi) \tag{5.4.11}$$

Hence this mapping allows the base functions to be defined through the universal forms given by equation (5.4.9), independently of the physical coordinates. They are determined by the nature of the element and the number of nodal points. The explicit form of the shape function in physical space is reconstructed from the knowledge of the mapping functions $\xi(x)$ between the considered element and the normalized reference element $(0, 1)$:

$$N_i^{(1)}(x) = N_i^{(1)}(\xi(x)) \tag{5.4.12}$$

where the relation $\xi(x)$ is defined by equation (5.4.8) or (5.4.10).

The function $u(x)$ is approximated on element 1 by the linear representation

$$\tilde{u}(x) = u_{i-1} N_{i-1}^{(1)}(x) + u_i N_i^{(1)}(x) \tag{5.4.13}$$

or

$$\tilde{u}(x) = u_{i-1} + \frac{x - x_{i-1}}{\Delta x_i}(u_i - u_{i-1}) \quad x_{i-1} \leq x \leq x_i \tag{5.4.14}$$

On element 2, $u(x)$ is approximated by

$$\tilde{u}(x) = u_i N_i^{(2)}(x) + u_{i+1} N_{i+1}^{(2)}(x)$$

$$= u_i + \frac{x - x_i}{\Delta x_{i+1}}(u_{i+1} - u_i) \quad x_i \leq x \leq x_{i+1} \tag{5.4.15}$$

The derivative of $\tilde{u}$ is approximated in this finite element representation by

$$\frac{\partial \tilde{u}}{\partial x} = \sum_i u_i \frac{\partial N_i}{\partial x} \tag{5.4.16}$$

and in particular at node i, we have within element 1:

$$(\tilde{u}_x)_i = \left(\frac{\partial \tilde{u}}{\partial x}\right)_i^{(1)} = \frac{u_i - u_{i-1}}{\Delta x_i} \tag{5.4.17}$$

which corresponds to the first order accurate, backward difference formula (4.1.6). Considered in element 2, we have the approximation

$$(\tilde{u}_x)_i = \left(\frac{\partial \tilde{u}}{\partial x}\right)_i^{(2)} = \frac{u_{i+1} - u_i}{\Delta x_{i+1}} \tag{5.4.18}$$

which is identical to the forward difference formula (4.1.5). It should be observed that the derivatives of $\tilde{u}$ are not continuous at the boundary between two elements, since by definition, the interpolation functions have only C^0-continuity, as can be seen from Figure 5.4.1.

It is therefore customary, in finite element approximations, to define a 'local' approximation to the derivative $(\tilde{u}_x)_i$ by an average of the two 'element' approximations (5.4.17) and (5.4.18). If a simple arithmetic average is taken, the resulting formula is

$$(\tilde{u}_x)_i = \frac{1}{2}\left[\left(\frac{\partial \tilde{u}}{\partial x}\right)_i^{(1)} + \left(\frac{\partial \tilde{u}}{\partial x}\right)_i^{(2)}\right] = \frac{1}{2}\left[\frac{u_{i+1} - u_i}{\Delta x_{i+1}} + \frac{u_i - u_{i-1}}{\Delta x_i}\right] \tag{5.4.19}$$

and has a dominant truncation error equal to

$$\frac{\Delta x_i - \Delta x_{i+1}}{4}(u_{xx})_i$$

as seen from equation (4.3.6). This approximation is only strictly second order accurate on a uniform mesh.

However, if each 'element' approximation is weighted by the relative length of the other element, that is if we define

$$(\tilde{u}_x)_i = \frac{1}{\Delta x_{i+1} + \Delta x_i}\left[\Delta x_{i+1}\left(\frac{\partial \tilde{u}}{\partial x}\right)_i^{(1)} + \Delta x_i\left(\frac{\partial \tilde{u}}{\partial x}\right)_i^{(2)}\right] \tag{5.4.20}$$

we obtain formula (4.3.7), which is second order accurate on an arbitrary mesh.

Since any linear function can be represented exactly on the element, we can also express the linear mapping $\xi(x)$ or $x(\xi)$ in function of the linear base functions (5.4.9). It is easily verified that equation (5.4.8) or (5.4.10) can be written as

$$x = \sum_i x_i N_i(\xi) \tag{5.4.21}$$

where the sum extends over the two nodes of the element.

This particular mapping is called an ***isoparametric mapping*** and illustrates a general procedure in finite element methods which applies also to two- and three-dimensional elements.

Polynomials of higher order, in particular quadratic, or cubic, are also currently defined, but they require of course more points, as more conditions (5.4.3) have to be specified.

The most currently used elements are widely described in the finite element literature and their derivation and properties can be found in most textbooks on the subject (see for instance Huebner, 1975; Zienkiewicz and Taylor, 2000).

A.5.4.1.2 *Two-dimensional linear elements*

The simplest element in a two-dimensional space is a triangle, to which we can associate linear interpolation functions.

Having three points available, the nodes of the triangle, we have three conditions to satisfy the equations (5.4.2) to (5.4.5). Referring to Figure 5.4.2 of an arbitrary triangle 123, the linear shape function associated to, say node 1, is defined as

$$N_1(x,y) = \frac{y_2 - y_3}{2A}(x - x_2) + \frac{x_3 - x_2}{2A}(y - y_2)$$

$$2A = (x_1 - x_2)(y_2 - y_3) + (y_1 - y_2)(x_3 - x_2) \qquad (5.4.22)$$

where A is the area of the triangle.

We leave it as an exercise to verify that all the conditions (5.4.2)–(5.4.5) are indeed satisfied (See Problem P.5.22).

With these definitions, we have the following derivative values:

$$\frac{\partial N_1}{\partial x} = (y_2 - y_3)/2A \qquad \frac{\partial N_1}{\partial y} = (x_3 - x_2)/2A \qquad (5.4.23)$$

Note that these derivatives are constants, as the functions are linear over each triangle.

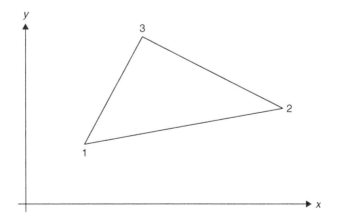

Figure 5.4.2   *Linear triangular element.*

A.5.4.2 Finite Element Definition of the Equation Discretization: Integral Formulation

This is the most essential and specific step of the finite element approximation since it requires the definition of an *integral* formulation of the physical problem equivalent to the field equations to be solved.

Two possibilities are open for that purpose, either a *variational principle* can be found expressing the physical problem as the extremum of a functional, either an integral formulation is obtained from the differential system through a **weak formulation**, also called the **method of weighted residuals**. Although many physical models can be expressed through a variational equation, for instance, the potential flow model, it is well known that it is not always possible to find a straightforward variational principle for all physical problems for instance for the Navier–Stokes equations. Therefore the weak formulation, or method of weighted residuals, is the most general technique which allows to define in all cases an equivalent integral formulation. Actually, in situations where discontinuous solutions are possible such as shock waves in transonic flows, the integral formulation is the only one which is physically meaningful since the derivatives of the discontinuous flow variables are not defined.

Since the weak formulations involve the definition of functionals, where restricted continuity properties are allowed on the functions involved, it is important to define in a clear and precise way the functional spaces as well as their appropriate norms, in order to define the correct convergence properties and error bounds. It is not our intention here to enter into these mathematical aspects and the interested reader will find these detailed developments in the references mentioned.

A.5.4.3 The Method of Weighted Residuals or Weak Formulation

In order to illustrate the principles of this approach, we will consider first the classical example of the two-dimensional quasi-harmonic equation:

$$\frac{\partial}{\partial x}\left(k\frac{\partial u}{\partial x}\right) + \frac{\partial}{\partial y}\left(k\frac{\partial u}{\partial y}\right) = q \tag{5.4.24}$$

written as $L(u) = q$, where L represents the differential operator.

If $\tilde{u}(x, y)$ is an approximation to the solution u, the quantity R, called the **residual**:

$$R(\tilde{u}) \overset{\Delta}{=} L(\tilde{u}) - q$$
$$= \vec{\nabla} \cdot (k\vec{\nabla}\tilde{u}) - q \tag{5.4.25}$$

is different from zero, otherwise $\tilde{u}(x, y)$ would be the analytical solution. Any resolution algorithm will converge if it drives the residual R toward zero, although this value will never be reached in a finite number of operations. Hence, the residual appears as a measure of the accuracy or of the error of the approximation $\tilde{u}(x, y)$. Since this error cannot be made to vanish simultaneously in all the points of the discretized domain, a 'best' solution can be extracted by requiring that some weighted average of the residuals over the domain should be identically zero.

If $W(x)$ is some weight function, with appropriate smoothness properties, the *method of weighted residuals* or *weak formulation*, requires

$$\int_\Omega W \cdot R(\tilde{u})\, d\Omega = 0 \qquad\qquad (5.4.26)$$

Applied to equation (5.4.25), this condition becomes

$$\int_\Omega W \vec{\nabla} \cdot (k\vec{\nabla}\tilde{u}) d\Omega = \int_\Omega qW\, d\Omega \qquad\qquad (5.4.27)$$

An essential step in this approach is the *integration by part of the second order derivative* terms, according to Green's theorem:

$$-\int_\Omega [k\vec{\nabla}\tilde{u} \cdot \vec{\nabla}W] d\Omega + \oint_\Gamma k\frac{\partial \tilde{u}}{\partial n} W\, d\Gamma = \int_\Omega qW\, d\Omega \qquad\qquad (5.4.28)$$

where the normal derivative along the boundary Γ of the domain appears in the right-hand side. Equation (5.4.27) becomes

$$\int_\Omega W \vec{\nabla} \cdot (k\vec{\nabla}\tilde{u}) d\Omega = -\int_\Omega \left[k\vec{\nabla}\tilde{u} \cdot \vec{\nabla}W \right] d\Omega + \oint_\Gamma k\frac{\partial \tilde{u}}{\partial n} W\, d\Gamma \qquad\qquad (5.4.29a)$$

or in condensed notation

$$-(k\vec{\nabla}\tilde{u}, \vec{\nabla}W) + \left(k\frac{\partial \tilde{u}}{\partial n}, W \right)_\Gamma - (q, W) = 0 \qquad\qquad (5.4.29b)$$

where the inner functional product (f, g) is defined by

$$(f, g) \overset{\Delta}{=} \int_\Omega fg\, d\Omega \qquad\qquad (5.4.30)$$

Equation (5.4.29) is the mathematical formulation of the weighted residual method, and is also called the weak formulation of the problem.

According to the choice of the weighting functions W, also called **test functions**, different methods are obtained. From numerical point of view, equation (5.4.29) is an algebraic equation and therefore, in any method, as many weighting functions as unknown variables (degrees of freedom) will have to be defined within the chosen subspace of test functions. That is, a unique correspondence will have to be established between each nodal value and a corresponding weighting function, in such a way that one equation of the type (5.4.29) is defined for each nodal value. Note also that if Ω_W is the subspace of the test functions, then the weighted residual equation (5.4.26) expresses the condition that the projection of the residual in the subspace of the test functions is zero, i.e. the residual is orthogonal to the subspace Ω_W.

A.5.4.4 The Galerkin Method

The most widely applied method is the **Galerkin** method in which the weighting functions are taken equal to the interpolation functions $N_I(x)$. This is also called the Bubnow–Galerkin method, to be distinguished from the Petrov–Galerkin method in which the test functions are different from the interpolation functions N_I.

For each of the M degrees of freedom, with the finite element representation

$$\tilde{u}(\vec{x}, t) = \sum_I u_I(t) N_I(\vec{x}) \quad (I = 1, ..., M) \tag{5.4.31}$$

and the choice $W = N_J(x)$ in order to obtain the discretized equation for node J, we obtain from equation (5.4.29)

$$-\sum_I u_I \int_{\Omega_J} \left[k\vec{\nabla}N_I \cdot \vec{\nabla}N_J \right] d\Omega + \oint_\Gamma k\frac{\partial \tilde{u}}{\partial n} N_J \, d\Gamma = \int_{\Omega_J} q \, N_J \, d\Omega \tag{5.4.32}$$

where Ω_J is the subdomain of all elements containing node J and the summation over I covers all the nodes of Ω_J, (see Figure 5.1.1). The matrix

$$K_{IJ} = \int_{\Omega_J} \left[k\vec{\nabla}N_I \cdot \vec{\nabla}N_J \right] d\Omega \stackrel{\Delta}{=} (k\vec{\nabla}N_I, \vec{\nabla}N_J) \tag{5.4.33}$$

is called the **stiffness matrix.** For linear problems whereby k is independent of u, it will depend only on the geometry of the mesh and the chosen elements.

Equation (5.4.32) can also be obtained from the Rayleigh–Ritz method for homogeneous boundary conditions. This is a general property, namely, the Rayleigh–Ritz method applied to a variational formulation leads to the same system of numerical equations as the Galerkin-weighted residual method.

Example E.5.4.1: One-dimensional equation

Consider the one-dimensional form of equation (5.4.24)

$$\frac{\partial}{\partial x}\left(k\frac{\partial u}{\partial x} \right) = q \tag{E5.4.1}$$

and a Galerkin weak formulation with linear elements. Applying equation (5.4.32), we have explicitly with the linear shape functions (5.4.9), and with $\Delta x_{i+1} = \Delta x_i = \Delta x$:

$$-\sum_{j=i-1}^{i+1} u_j \int_{i-1}^{i+1} k\frac{\partial N_j}{\partial x}\frac{\partial N_i}{\partial x} dx - \int_{i-1}^{i+1} q \, N_i dx = 0 \tag{E5.4.2}$$

Performing the integrations, with the shape function derivatives equal to $\pm 1/\Delta x$, we obtain

$$k_{i+1/2}\frac{u_{i+1} - u_i}{\Delta x} - k_{i-1/2}\frac{u_i - u_{i-1}}{\Delta x} = \frac{1}{6}(q_{i-1} + 4q_i + q_{i+1}) \tag{E5.4.3}$$

where

$$k_{i+1/2} = \int_{i}^{i+1} k \, dx \tag{E5.4.4}$$

and a similar expression for $k_{i-1/2}$. If a linear variation within each element is assumed for k, than

$$k_{i+1/2} = \frac{1}{2}(k_i + k_{i+1}) \tag{E5.4.5}$$

It is interesting to note that the left hand side of equation (E.5.4.3) is identical to the central second order finite difference formula (4.2.17). In this latter case the right-hand side would be equal to q_i while in the finite element Galerkin approach we obtain an average weighted over the three nodal points $i-1, i, i+1$. This is a typical property of the weighted residual Galerkin method.

Observe also that linear elements lead to second order accurate discretizations. *It is a general rule, on uniform meshes, that elements of order p lead to discretizations of order of accuracy p + 1.*

Example E.5.4.2 : Laplace equation on a triangular uniform mesh

A triangulation of a uniform Cartesian mesh ($\Delta x = \Delta y$) can be defined as on Figure 5.4.3 Node J is associated to the mesh coordinates (i, j) and the Laplace equation is considered with Dirichlet boundary conditions:

$$\Delta u = q$$
$$u = u_0 \quad \text{on } \Gamma \tag{E5.4.6}$$

The Galerkin equation (5.4.32) becomes

$$-\sum_I u_I \int_{\Omega_J} \left[\frac{\partial N_I}{\partial x} \cdot \frac{\partial N_J}{\partial x} + \frac{\partial N_I}{\partial y} \cdot \frac{\partial N_J}{\partial y} \right] dx \, dy = \int_{\Omega_J} q \, N_J \, dx \, dy \tag{E5.4.7}$$

There is no boundary integral, since the weight functions are taken to vanish on the boundaries. The integration domain covers all the triangles containing node $J(i, j)$, i.e. triangles 1–6. The summation extends over all the nodes of these triangles.

With the linear shape functions defined by (5.4.22) we have for $J = (i, j)$ in triangle 1 of Figure 5.4.3:

$$N_{ij}^1 \equiv N_J^1 = 1 - \frac{x - x_{ij}}{\Delta x}$$

$$N_{i+1,j}^1 = 1 + \frac{x - x_{i+1,j}}{\Delta x} - \frac{y - y_{i+1,j}}{\Delta y}$$

$$N_{i+1,j+1}^1 = 1 + \frac{y - y_{i+1,j+1}}{\Delta y} \tag{E5.4.8}$$

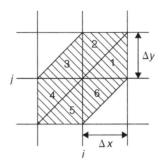

Figure 5.4.3 *Finite element domain formed by six linear triangular elements.*

Similarly in triangles 2 and 3:

$$N_{ij}^2 = 1 - \frac{y - y_{ij}}{\Delta y}$$

$$N_{ij}^3 = 1 + \frac{x - x_{ij}}{\Delta x} - \frac{y - y_{ij}}{\Delta y} \tag{E5.4.9}$$

The contributions from triangle 1 to the stiffness matrix K_{IJ} are obtained as follows, with the notation $K_{IJ} = K_{ij}^{ij}$ and $\Delta x = \Delta y$:

$$K_{ij}^{i+1,j(1)} = \int_1 \left[\frac{\partial N_{ij}^1}{\partial x} \cdot \frac{\partial N_{i+1,j}^1}{\partial x} + \frac{\partial N_{ij}^1}{\partial y} \cdot \frac{\partial N_{i+1,j}^1}{\partial y} \right] dx \, dy$$

$$= \int \frac{-1}{\Delta x} \cdot \frac{1}{\Delta x} dx \, dy = -\frac{1}{2} \tag{E5.4.10}$$

$$K_{ij}^{i+1,j+1(1)} = 0$$

$$K_{ij}^{ij(1)} = \frac{1}{2} \tag{E5.4.11}$$

Adding the contributions from all the triangles, we obtain

$$K_{ij}^{ij} = 4 \quad K_{ij}^{i+1,j} = K_{ij}^{i-1,j} = K_{ij}^{i,j-1} = K_{i,j+1}^{ij} = -1$$

$$K_{ij}^{i+1,j+1} = K_{i-1,j-1}^{ij} = 0 \tag{E5.4.12}$$

and equation (E.5.4.7) becomes for a linear variation of the source term q within each triangle

$$-4u_{ij} + u_{i+1,j} + u_{i-1,j} + u_{i,j+1} + u_{i,j-1}$$

$$= \frac{\Delta x^2}{12}(6q_{ij} + q_{i+1,j} + q_{i-1,j} + q_{i,j+1} + q_{i,j-1} + q_{i+1,j+1} + q_{i-1,j-1})$$

$$\tag{E5.4.13}$$

Compared to the finite difference discretization of the Laplace operator, it is seen that the left-hand side is identical to the five-point molecule of Figure 4.2.2. In a finite difference method, the left-hand side would be equal to q_{ij}, while the finite element method generates an average of the source term values at the points surrounding the node (i, j).

A.5.4.5 Finite Element Galerkin Method for a Conservation Law

Consider the conservation law of the form:

$$\frac{\partial U}{\partial t} + \vec{\nabla} \cdot \vec{F} = Q \tag{5.4.34}$$

where F is the flux vector, containing only convective contributions with the following initial, boundary conditions on the domain Ω with boundary $\Gamma = \Gamma_0 \cup \Gamma_1$

$$\begin{aligned}
U(\vec{x}, 0) &= U_0(\vec{x}) & \text{for } t = 0 \; \vec{x} \in \Omega \\
U(\vec{x}, t) &= U_1(\vec{x}) & \text{for } t \geq 0 \; \vec{x} \in \Gamma_0 \\
\vec{F} \cdot \vec{1}_n \equiv F_n &= g & \text{for } t \geq 0 \; \vec{x} \in \Gamma_1
\end{aligned} \tag{5.4.35}$$

Defining a weak formulation, with $W = 0$ on Γ_0,

$$\int_{\Omega} \frac{\partial U}{\partial t} W \, d\Omega + \int_{\Omega} (\vec{\nabla} \cdot \vec{F}) W \, d\Omega = \int_{\Omega} Q W \, d\Omega \tag{5.4.36}$$

followed by an integration by parts on the flux term, leads to

$$\int_{\Omega} \frac{\partial U}{\partial t} W \, d\Omega - \int_{\Omega} (\vec{F} \cdot \vec{\nabla}) W \, d\Omega + \int_{\Gamma} W \vec{F} \cdot d\vec{S} = \int_{\Omega} Q W \, d\Omega \tag{5.4.37}$$

The finite element representation is defined by

$$U = \sum_I U_I(t) N_I(\vec{x}) \tag{5.4.38}$$

and since the flux term F is generally a nonlinear function of U, it is preferable to define also a separate representation for the fluxes F as

$$\vec{F} = \sum_I \vec{F}_I N_I(\vec{x}) \tag{5.4.39}$$

The discretized equation for node J is obtained via the Galerkin method, $W = N_J$, leading to

$$\sum_I \frac{dU_I}{dt} \int_{\Omega_J} N_I N_J \, d\Omega - \sum_I \vec{F}_I \int_{\Omega_J} (N_I \cdot \vec{\nabla}) N_J \, d\Omega + \int_{\Gamma_1} g N_J \, d\Gamma = \int_{\Omega_J} Q N_J \, d\Omega \tag{5.4.40}$$

where Ω_J is the subdomain of all elements containing node J and the summation over I covers all the nodes of Ω_J, Figure 5.4.4. The matrix of the time-dependent term is called the **mass matrix** M_{IJ}

$$M_{IJ} = \int_{\Omega_J} N_I N_J \, d\Omega \tag{5.4.41}$$

and the stiffness matrix

$$\vec{K}_{IJ} = \int_{\Omega_J} N_I \vec{\nabla} N_J \, d\Omega \tag{5.4.42}$$

is not symmetric anymore. Hence equation (5.4.40) becomes

$$\sum_I M_{IJ} \frac{dU_I}{dt} - \sum_I \vec{F}_I \cdot \vec{K}_{IJ} = \int_{\Omega_J} Q N_J \, d\Omega - \int_{\Gamma_1} g N_J \, d\Gamma \tag{5.4.43}$$

If the flux F contains in addition a diffusive term of the form $(-\kappa \vec{\nabla} u)$ then the term $(\vec{\nabla} \cdot \kappa \vec{\nabla} u)$ will be treated following equation (5.4.29). In finite difference discretizations, the time-dependent term will generally reduce to dU_J/dt, corresponding to a diagonal mass matrix, while the present formulation leads to an average over the various nodes in Ω_J. The presence of this mass matrix complicates the resolution of the system of ordinary differential equations in time (5.4.43).

A rigorous way of diagonalizing the mass matrix is to introduce 'orthogonal' interpolation functions and to apply a Petrov–Galerkin method with these new functions, following Hirsch and Warzee (1978). A more currently applied approximation, called **mass lumping**, consists in replacing M_{IJ} by the sum over I, of its elements at fixed J, this sum along the elements of a line being concentrated on the main diagonal. That is,

$$M_{IJ}^{(\text{lump})} = \left[\sum_i M_{iJ} \right] \delta_{iJ} \tag{5.4.44}$$

The modified equation (5.4.43), obtained in this way, is close to a finite volume formulation.

A.5.4.6 Subdomain Collocation: Finite Volume Method

The **collocation methods**, domain and point collocation, both use the residual equation (5.4.36) *without* partial integration on the weighting function W. If a subdomain Ω_J is attached to each nodal point J, with the corresponding weighting function defined by (see Figure 5.4.4)

$$\begin{aligned} W_J(\vec{x}) &= 0 \quad \vec{x} \notin \Omega_J \\ W_J(\vec{x}) &= 1 \quad \vec{x} \in \Omega_J \end{aligned} \tag{5.4.45}$$

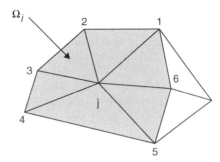

Figure 5.4.4 *General triangular elements*

then the residual equation (5.4.36) becomes

$$\int_{\Omega_J} \frac{\partial U}{\partial t} d\Omega + \int_{\Omega_J} \vec{\nabla} \cdot \vec{F} d\Omega = \int_{\Omega} Q d\Omega \qquad (5.4.46)$$

A more interesting form of this equation is obtained after application of Gauss theorem on the flux term, leading to the conservation equation in integral form written for each subdomain Ω_J limited by the closed surface Γ_J.

$$\int_{\Omega_J} \frac{\partial U}{\partial t} d\Omega + \oint_{\Gamma_J} \vec{F} \cdot d\vec{S} = \int_{\Omega_J} Q d\Omega \qquad (5.4.47)$$

This equation, which can be obtained directly from equation (5.4.37), is the basic equation for the finite volume method as we have seen in the first part of this chapter.

Example E.5.4.3: Conservation law on linear triangles

We apply the Galerkin equation (5.4.40) with linear triangles for a node J not adjacent to the boundary Γ_J of the domain and in absence of source terms. Referring to Figure 5.4.4, the domain Ω_J contains six triangles and the nodes numbered 1–6 around node J. We will also lump the mass matrix, leading to the approximation (see Problem P.5.17):

$$M_{IJ} = \frac{\Omega_J}{3} \delta_{IJ} \qquad (E5.4.14)$$

The flux terms can be written as

$$\int_{\Omega_J} (\vec{F} \cdot \vec{\nabla}) N_J d\Omega = \sum_I \vec{F}_I \int_{\Omega_J} N_I \cdot \vec{\nabla} N_J d\Omega$$

$$= \sum_I f_I \int_{\Omega_J} N_I \cdot \frac{\partial N_J}{\partial x} d\Omega + \sum_I g_I \int_{\Omega_J} N_I \cdot \frac{\partial N_J}{\partial y} d\Omega$$

$$(E5.4.15)$$

where f and g are the Cartesian components of F.

With the definitions (5.4.23) applied to the triangle J_{12} of figure 5.4.4, we have

$$\frac{\partial N_J}{\partial x} = (y_1 - y_2)/2A_{12} \qquad \frac{\partial N_J}{\partial y} = (x_2 - x_1)/2A_{12} \qquad (\text{E5.4.16})$$

where A_{12} is the area of triangle J_{12}. Since these derivatives are constants, the integrals in equation (E.5.4.16) reduce to the integrals over N_I, which are equal to $A_{12}/3$.

Hence, the contribution to the flux term of the Galerkin formulation from triangle J_{12}, becomes

$$\int_{J_{12}} \vec{F} \cdot \vec{\nabla} N_J \, d\Omega = \frac{y_1 - y_2}{6}(f_1 + f_2 + f_J) - \frac{x_1 - x_2}{6}(g_1 + g_2 + g_J) \qquad (\text{E5.4.17})$$

Summing over all the triangles, we obtain

$$-\int_{\Omega_J} \vec{F} \cdot \vec{\nabla} N_J \, d\Omega = \frac{1}{3} \sum_{\text{sides}} (f_{12}\Delta y_{12} - g_{12}\Delta x_{12}) \qquad (\text{E5.4.18})$$

with

$$\begin{aligned} f_{12} = \tfrac{1}{2}(f_1 + f_2) \quad g_{12} = \tfrac{1}{2}(g_1 + g_2) \\ \Delta y_{12} = y_2 - y_1 \quad \Delta x_{12} = x_2 - x_1 \end{aligned} \qquad (\text{E5.4.19})$$

The contributions from f_J, g_J cancel out of the summation on all the triangles, since

$$\sum_{\text{sides}} \Delta y_{12} = \sum_{\text{sides}} \Delta x_{12} = 0 \qquad (\text{E5.4.20})$$

because Γ_J is a closed contour.

Introducing this expression in the Galerkin equation (5.4.40) leads to the following discretized scheme:

$$\Omega_J \frac{dU_J}{dt} + \sum_{\text{sides}} (f_{12}\Delta y_{12} - g_{12}\Delta x_{12}) = 0 \qquad (\text{E5.4.21})$$

which is nothing else than a finite volume discretization on the hexagonal contour Γ_J as seen in Section 5.3.

Alternative formulations

If the flux terms of equation (E.5.4.21) are recombined, for instance by assembling terms such as

$$(f_1 + f_2)(y_2 - y_1) + (f_2 + f_3)(y_3 - y_2) + \cdots = f_2(y_3 - y_1) + \cdots$$
$$= -y_2(f_3 - f_2) + \cdots$$

we obtain the alternative formulation

$$\Omega_J \frac{dU_J}{dt} + \frac{1}{2} \sum_l [f_l(y_{l+1} - y_{l-1}) - g_l(x_{l+1} - x_{l-1})] = 0 \qquad \text{(E5.4.22)}$$

where the summation extends over all the nodes, with $y_0 = y_6$ and $y_7 = y_1$ (and similarly for the x coordinates). We can also write equation (E.5.4.22) as

$$\Omega_J \frac{dU_J}{dt} + \frac{1}{2} \sum_l [(f_l - f_J)(y_{l+1} - y_{l-1}) - (g_l - g_J)(x_{l+1} - x_{l-1})] = 0$$

$$\text{(E5.4.23)}$$

since all the contributions to f_J and g_J vanish, because of

$$\sum_l f_J(y_{l+1} - y_{l-1}) = f_J \sum_l (y_{l+1} - y_{l-1}) = 0 \qquad \text{(E5.4.24)}$$

The other formulation becomes

$$\Omega_J \frac{dU_J}{dt} - \frac{1}{2} \sum_l [y_l(f_{l+1} - f_{l-1}) - x_l(g_{l+1} - g_{l-1})] = 0 \qquad \text{(E5.4.25)}$$

or alternatively

$$\Omega_J \frac{dU_J}{dt} + \frac{1}{2} \sum_l [(y_J - y_l)(f_{l+1} - f_{l-1}) - (x_J - x_l)(g_{l+1} - g_{l-1})] = 0$$

$$\text{(E5.4.26)}$$

with the conventions $f_0 = f_6$ and $f_7 = f_1$ (and similarly for g). The above relations can also be derived from the results of Problem P.5.18. Note that these relations are independent of the number of triangles inside Ω_J and apply therefore to any polygonal contour.

CONCLUSIONS AND MAIN TOPICS TO REMEMBER

This chapter has been dealing with the most widely applied methodology in CFD, namely the finite volume method. For more advanced readers, we have also introduced the basics of the finite element method and in particular its application to the conservation laws.

The finite volume method (FVM) refers to the very important property of conservation at the discrete level and leads also to a fundamental understanding of any numerical scheme.

The main topics to remember are the following:

- The integral form of the conservation law is the basis of the FVM and satisfies automatically the discrete conservation property.

- Look carefully at the finite volume treatment of the non-conservative form of the one-dimensional convection equation, as described in Section 5.1. It is of great importance that you understand and recognize the numerical conservativity properties, and its general form, as shown by equation (5.1.11).
- When defining an FVM, the space discretization divides the space in a number of small volumes or cells, instead of arrays of points. A certain number of conditions have to be satisfied by this finite volume space discretization. A key property for an FVM to satisfy automatically conservativity is that the flux calculation on each face has to be independent of the volumes shared by this face.
- Section 5.2.3 is of most essential significance and we recommend you to study very deeply all the consequences of this general formulation of a numerical scheme. Retain that any numerical scheme delivers as output cell-averaged approximations and that assigning these values to a mesh point, as required for post-processing, introduces a numerical error, generally of second order.
- The concept of numerical flux is also critical for finite volume methods. All possible schemes can always be written under the form of a numerical flux.
- Another key element in the FVM is the evaluation of the cell face flux approximations, as defined by the numerical flux. A large number of choices are possible, some of them are listed in Section 5.3. Important to remember here is that the choices presented are only the tip of the iceberg, and we will see in the following chapters many more options.
- As the FVM is perfectly adapted to arbitrary grids, the formulas for the evaluation of surfaces and volumes establish a general framework for discretization formulas applicable to non-uniform grids.
- On a more advanced level, pay attention to the links between the finite element and finite volume methods.

REFERENCES

Arts, T. (1984). On the consistency of four different control surfaces used for finite area blade-to blade calculations. *In. J. Numer. Meth. Fluids*, 4, 1083–1096.

Baker, A.J. (1983). *Finite Element Computational Fluid Mechanics*. Hemisphere Publ. Co McGraw-Hill, New York.

Chung, T.J. (1978). *Finite Element Analysis in Fluid Dynamics*. McGraw-Hill, New York.

Ciarlet, P.G. (1978). *The Finite Element Method for Elliptic Problems. Studies in Mathematics and Application*, Vol.4. North-Holland Publ. Co., Amsterdam.

Clough, R.W. (1960). The finite element method in plane stress analysis. *Proceedings of the 2nd ASCE Conference on Electronic Computation. J. Struct. Div. ASCE*, 345–378.

Denton, J. (1975). *A Time Marching Method for Two and Three Dimensional Blade to Blade flows*. Aeronautical Research Council, R&M 3775.

Girault, V. and Raviart, P.A. (1979). *Finite Element Approximation of the Navier–Stokes Equation. Lecture Notes in Mathematics*, Vol. 749. Springer Verlag, New York.

Glowinski, R. (1983). *Numerical Methods for Non-Linear Variational Problems*, 2nd edn. Springer Verlag, New York.

Hirsch, Ch. and Warzee, G. (1978). An orthogonal finite element method for transonic flow calculations. *Proceedings of the 6th International Conference on Numerical Methods in Fluid Dynamics. Lecture Notes in Physics*, Vol.90. Springer Verlag, New York.

Huebner, K.H. (1975). *The Finite Element Method for Engineers.* J. Wiley & Sons, New York.

Lax, P.D. (1954). Weak solutions of non-linear hyperbolic equations and their numerical computation. *Comm. Pure Appl. Math.* 7, 159–193.

Lax, P.D. and Wendroff, B. (1960). Systems of conservation laws. *Comm. Pure Appl. Math.* 13, 217–237.

MacCormack, R.W. and Paullay, A.J. (1972). Computational efficiency achieved by time splitting of finite difference operators. AIAA Paper 72–154.

McDonald, P.W. (1971). The computation of transonic flow through Two-dimensional gas turbine cascades. ASME Paper 71-GT-89.

Oden, J.T. (1972). *Finite Elements of Non Linear Continua.* McGraw-Hill, New York.

Oden, J.T. and Reddy, J.N. (1976). *An Introduction to the Mathematical Theory of Finite Elements.* J. Wiley & Sons, New York.

Roache, P.J. (1972). *Computational Fluid Dynamics.* Hermosa Publ., Albuquerque, New Mexico.

Rizzi, A.W. and Inouye, M. (1973). Time split finite volume method for three-dimensional blunt-body flows. *AIAA J.* 11, 1478–1485.

Roe, P.L. (1987). Error estimates for cell-vertex solutions of the compressible Euler equations. ICASE Report No. 87-6. NASA Langley Research Center.

Strang, G. and Fix, G. (1973). *An Analysis of the Finite Element Method.* Prentice Hall, New Jersey.

Temam, R. (1977). *Navier–Stokes Equations.* North-Holland Publ. Co., Amsterdam.

Thomasset, F. (1981). *Implementation of Finite Element Methods for Navier–Stokes Equations. Springer Series in Computational Physics*, Springer Verlag, New York.

Turkel, E. (1985). Accuracy of schemes with non-uniform meshes for compressible fluid flows. ICASE Report N^1/4 85-43. NASA Langley Research Center.

Turkel, E., Yaniv, S. and Landau, U. (1985). Accuracy of schemes for the Euler equations with non-uniform meshes. ICASE Report N^1/4 85-59. NASA Langley Research Center; also in *SIAM J. Sci. Statist. Computi.*

Turner, M.J., Clough, R.W., Martin, H.C. and Topp, L.P. (1956). Stiffness and deflection analysis of complex structures. *J. Aeron. Soc.* 23, 805.

Zienkiewicz, O.C. and Cheung, Y.K. (1965). Finite elements in the solution of field problems. *Engineer*, 507–10.

Zienkiewicz, O.C. and Taylor, R.L. (2000) *Finite Element Method – Vol. 1: The Basis; Vol. 2: Solid Mechanics; Vol. 3: Fluid Dynamics.* Elsevier, London.

PROBLEMS

P.5.1 Apply the finite volume formula (5.2.2) to the contour ACDEGH of Figure 5.2.1a and derive the discretization for node (i, j). Compare with the formula (E.5.3.5) when the variables are defined at the nodes of the control volume, with the side fluxes defined by the average of the corner points value, i.e.:

$$f_{AC} = \frac{1}{2}(f_A + f_C)$$

P.5.2 Apply the finite volume method to the contour of Figure 5.2.1b, and compare the different assumptions for the evaluation of fluxes at the mid-side. Derive four schemes by combining the two options for the vertical sides with the two options for the horizontal sides:

$$f_{AB} = \frac{1}{2}(f_{i+1,j} + f_{ij})$$

or

$$f_{AB} = \frac{1}{6}(f_{i+1,j} + f_{i+1,j+1} + f_{i+1,j-1} + f_{ij} + f_{i,j+1} + f_{i,j-1})$$

with

$$f_{DA} = f_{i,j-1}$$

or

$$f_{DA} = \frac{1}{4}(2f_{i,j-1} + f_{i+1,j-1} + f_{i-1,j-1})$$

with similar expressions for g and for the two other sides.

P.5.3 Determine the different formulas obtained in Problem P.5.2 when the mesh is Cartesian and compare with the results of Problem P.5.1.

P.5.4 Develop a finite volume discretization for mesh point A(i,j), with the control volume BCDGHKEF of Figure 5.2.1b. Compare the results from the evaluation of the side fluxes by the following three options, written for instance for side K(E)F:

$$f_{KF}\Delta y_{KF} = \frac{1}{2}(f_{i+1,j+1} + f_{i+1,j-1})(y_{i+1,j+1} - y_{i+1,j-1}) \tag{a}$$

or

$$f_{KF}\Delta y_{KF} = f_{KE}\Delta y_{KE} + f_{EF}\Delta y_{EF}$$
$$= \frac{1}{2}(f_{i+1,j} + f_{i+1,j-1})(y_{i+1,j} - y_{i+1,j-1})$$
$$+ \frac{1}{2}(f_{i+1,j+1} + f_{i+1,j})(y_{i+1,j+1} - y_{i+1,j}) \tag{b}$$

or

$$f_{KF}\Delta y_{KF} = f_E\Delta y_{KF} = f_{i+1,j}(y_{i+1,j+1} - y_{i+1,j-1}) \tag{c}$$

Compare the three results for a Cartesian mesh and refer also to Example E.5.3.1.

P.5.5 Apply the results (5.3.21) in order to derive average values of the first derivatives $\partial f/\partial x$ and $\partial f/\partial y$ over the triangle J_{12} of Figure 5.2.1d. Compare with the expressions obtained in equations (E.5.4.21), (E.5.4.22) and with the results of Problem P.5.11. Note that the results are identical and comment on the reason behind the validity of the derivation by the finite element method with linear triangles.

P.5.6 Consider the two-dimensional diffusion equation treated in Example E.5.3.3 with diffusive flux components $f = k\partial u/\partial x$ and $g = k\partial u/\partial y$, where k is a function of the coordinates. Construct the discrete equation by the finite volume approach on the mesh of Figure 5.2.1a, considered as Cartesian, by generalizing the development of Example E.5.3.3. Consider the quadrilateral control surface ABCD for the mesh point $1(i,j)$ and consider the values of k defined at the corners of the cell, that is in A,B,C,D. If necessary define

$$k_{AB} = \frac{1}{2}(k_A + k_B)$$

P.5.7 Consider the diffusion equation of the previous problem and apply it to the cell BCDGHKEF of Figure 5.2.1b, considered as Cartesian, with constant k. Define the derivatives on the cell sides by one sided formulas, from inside the control cell. Apply successively the three options of Problem P.5.4 and compare with the results of Example E.5.3.3. Show in particular that the options a, b, c reproduce the schemes derived in this example.

    Hint: For a point F, define the derivatives as

$$\left(\frac{\partial u}{\partial x}\right)_F = \frac{u_{i+1,j+1} - u_{i,j+1}}{\Delta x}$$

$$\left(\frac{\partial u}{\partial y}\right)_F = \frac{u_{i+1,j+1} - u_{i+1,j}}{\Delta x}$$

and similar relations for the other points.

P.5.8 Apply equation (5.3.29) to the quadrilateral ABCD of Figure 5.3.1 and take point A as origin of the position vector $\vec{x}$. Show that the contour integral reduces to the contributions along BC and CD with $\vec{x} = \vec{x}_{AC}$ and that

$$2\Omega = \vec{x}_{AC} \cdot \oint_{BCD} d\vec{S}$$

By working out the integral, obtain the relation (5.3.4)

$$\Omega_{ABCD} = \frac{1}{2}(\Delta x_{AC}\Delta y_{BD} - \Delta x_{BD}\Delta y_{AC})$$

Hint: Observe that with A as origin the position vector is aligned with the sides AB and AD and hence normal to the vector dS. Therefore, there are no contributions from these two sides.

P.5.9 Repeat Problem P.5.8 for triangle ABC of Figure 5.3.1 and show that we can write

$$2\Omega_{ABC} = \vec{x}_{AB} \cdot \oint_{BC} d\vec{S}$$

obtaining

$$\Omega_{ABC} = \frac{1}{2}(\Delta x_{AB}\Delta y_{BC} - \Delta x_{BC}\Delta y_{AB})$$

P.5.10 Consider the quadrilateral BDHE on Figure 5.3.2b and apply the relations (6.2.29) in order to define the average value of the x-derivative of a function U. Consider $y_E = y_D$ and obtain

$$\left(\frac{\partial U}{\partial x}\right)_{BDHE} = \frac{U_E - U_D}{x_E - x_D}$$

Comment on the accuracy of this formula when applied to point A.

P.5.11 Repeat Problem P.5.10 for the contour BCDGHKEF of Figure 5.2.1b by applying formulas (5.3.26) and (5.3.27). Obtain the following approximation:

$$\left(\frac{\partial U}{\partial x}\right) = \frac{U_E - U_D}{x_E - x_D} + \frac{\Delta y_{BA}}{2\Delta y_{BH}}\frac{U_F - U_C}{x_F - x_C} + \frac{\Delta y_{AH}}{2\Delta y_{BH}}\frac{U_K - U_G}{x_K - x_G}$$

Derive also the corresponding expression for a Cartesian mesh.

P.5.12. Apply the upwind flux evaluation (5.3.15) to derive a finite volume scheme for the cell GHKEFBCD of Figure 5.2.1b, considered as Cartesian. Compare the obtained discretization with the results of Example E.5.3.2.

P.5.13 Show that the integral conservation law over the one-dimensional domain $a \leq x \leq b$, applied to the one-dimensional conservation law

$$\frac{\partial u}{\partial t} + \frac{\partial f}{\partial x} = 0$$

with the condition $f(a) = f(b)$, reduces to the condition

$$\int_a^b u \, dx = \text{const.}$$

with time.

Apply this condition to the to a discretized x-space, with an arbitrary mesh point distribution, as in Figure 4.3.2, and show that this condition reduces to

$$\frac{1}{2}\sum_i \Delta u_i (x_{i+1} - x_{i-1}) = 0$$

where

$$\Delta u_i \equiv u_i^{n+1} - u_i^n = \frac{\partial u_i}{\partial t}\Delta t$$

Hint: apply a trapezoidal rule to evaluate the integral

$$\frac{\partial}{\partial t}\int_a^b u \, dx = 0$$

and rearrange the sum to isolate the u_i-terms.

P.5.14 Apply the Galerkin method, with linear elements to the first order equation

$$a\frac{\partial u}{\partial x} = q$$

Show that on a uniform mesh, $\Delta x_i = \Delta x_{i+1} = \Delta x$, we obtain the same discretization as with central differences.

P.5.15 Work out all the calculations of Example E.5.4.2.

P.5.16 Show by an explicit calculation, that the average value of a quantity U over an element is approximated for linear triangles by

$$\frac{1}{\Omega}\int_\Omega U \, d\Omega = \frac{1}{3}\sum_{I=1}^{3} U_I$$

P.5.17 Calculate the mass matrix elements attached to node J of Figure 5.4.4, with linear triangles. Show that one obtains equation (E.5.4.14) for the lumped mass approximation.

Hint: Obtain the following matrix, for a triangle of area A

$$M_{IJ} = \frac{A}{12}\begin{vmatrix} 2 & 1 & 1 \\ 1 & 2 & 1 \\ 1 & 1 & 2 \end{vmatrix}$$

P.5.18 Referring to Figure 5.4.4, show that the average of $\partial f/\partial x$ and $\partial g/\partial y$ over the domain Ω_J covered by the six linear triangles, can be defined as

$$\left(\overline{\frac{\partial f}{\partial x}}\right)_{\Omega_J} = \frac{1}{\Omega_J}\int_{\Omega_J}\frac{\partial f}{\partial x} d\Omega$$

$$= \frac{1}{2\Omega_J}\sum_I f_I(y_{I+1} - y_{I-1})$$

where the summation extends over all the nodes of the contour Γ_J, and

$$\left(\overline{\frac{\partial g}{\partial y}}\right) = \frac{1}{\Omega_J}\int_{\Omega_J}\frac{\partial g}{\partial y} d\Omega$$

$$= -\frac{1}{2\Omega_J}\sum_I g_I(x_{I+1} - x_{I-1})$$

Hint: Calculate the average values for each triangle J_{12} by taking $f = \sum f_I N_I$. Show that for each triangle we have

$$\int_{J_{12}}\frac{\partial f}{\partial x} d\Omega = \frac{1}{2}[f_1(y_2 - y_J) + f_2(y_J - y_1) + f_J(y_1 - y_2)]$$

and sum these contributions over all the triangles.

P.5.19 Apply the results of Example E.5.4.3 to a quadrilateral domain such as $J234$ on Figure 5.4.1b, discarding node J, to obtain the following discretization of the flux integral on this quadrilateral.

$$\oint_{J234} \vec{F}.d\vec{S} = \frac{1}{2}[(f_2 - f_4)(y_3 - y_J) + (f_3 - f_J)(y_4 - y_2)$$

$$-(g_2 - g_4)(x_3 - x_J) - (g_3 - g_J)(x_4 - x_2)]$$

P.5.20 Proof equation (5.4.21) for linear one-dimensional elements.

P.5.21 Apply the Galerkin method with linear elements to the conservation equation

$$\frac{\partial u}{\partial t} + \frac{\partial f}{\partial x} = 0$$

following Section A54.4. Obtain the implicit formulation:

$$\frac{1}{6}\left[\frac{du_{i-1}}{dt} + 4\frac{du_i}{dt} + \frac{du_{i+1}}{dt}\right] + \frac{1}{2\Delta x}(f_{i+1} - f_{i-1}) = 0$$

P.5.22 Show that the shape functions (5.4.22) satisfy the conditions (5.4.2)–(5.4.5) for two-dimensional linear triangular elements.

Chapter 6

Structured and Unstructured Grid Properties

OBJECTIVES AND GUIDELINES

Grid generation is a major component in setting up a CFD simulation and if you consider Figure I.2.1 of the general introduction, you will find the explanation for its crucial role in the overall CFD process. It is the first necessary step, as no simulation can be started without having defined an appropriate mesh point distribution. Methods have therefore been developed to help the CFD user in generating grids in the best possible way. The importance of grid properties cannot be emphasized enough. We have already stressed in Chapter 4 the effects of non-uniform grids and the potential loss of accuracy of the most current finite difference formulas associated to grid non-uniformities. Although this analysis was restricted to one-dimensional grids, you can easily imagine that these effects will be amplified on irregular 2D and 3D grids, of the type shown in the Introduction.

This emphasizes the essential role played by grid properties in the overall accuracy of a CFD simulation.

The software methods in support of the grid generation process are complicated to develop, particularly for general geometries, as they require sophisticated programming and mathematical methodologies. The relevant software tools call upon algebraic geometry theories, mathematical surface definitions, normals and curvature estimations, coordinate transformations, topological properties, etc. The interested reader will find detailed descriptions of grid generation methods in the following references: Thompson (1984), Thompson et al. (1985), (1999), Dale and King (1993) for structured grids and George and Borouchaki (1998), Frey and George (1999) for unstructured grids.

We consider these topics as being outside the scope of this introduction to CFD, although you will need to generate grids for the applications to be covered in the last part of this text, in Chapters 11 and 12. We have therefore chosen not to take you through the different mathematical techniques applied to generate grids. Instead, our objective is to present you with an overview of the possible grid configurations as applied in practice.

You can refer to the following web sites for a widely documented overview of most of the available grid generators: http://www-users.informatik.rwth-aachen.de/~roberts/meshgeneration.html; http://www.andrew.cmu.edu/user/sowen/mesh.html.

Next to the commercial offers, a large variety of freely available grid generators can be found and downloaded for particular applications.[1]

---

[1] We suggest exercising great care when downloading free software tools. Check carefully the last update of the visited web site and verify that the software you are interested in is still being maintained by the author(s). If not, we recommend avoiding its use, as you will face great risks of encountering software bugs and problems, for which no help would be available.

Despite the numerous available software codes to support the task of grid generation, it still can be a very time-consuming exercise, particularly when dealing with complex geometries. Consequently, automatic grid generation, with an adequate control of grid quality, has become one of the major objectives of modern CFD, both for structured as well as unstructured grids.

As mentioned already in the Introduction and in Chapter 5, we distinguish between **structured** and **unstructured** grids. The former is composed of families of intersecting lines, one for each space dimension (two families of lines in 2D and three in 3D), where each mesh point is located at the intersection of one line, and only one line, of each family.

Unstructured grids, on the contrary, refer to arbitrary distributions of mesh points, where the points are connected by triangles, quadrilaterals or polygons in 2D, or by various polyhedrals in 3D (tetrahedra, prisms, pyramids, hexahedra or arbitrary polyhedrals).

The objective of this chapter is to introduce you to the various types of grids, both structured and unstructured, that you can select or encounter in practice. For a given geometrical configuration, many different grid topologies can be defined and there are always several alternative possibilities for a given application and flow model. The choice of a mesh topology is often a matter of personal choice and it is therefore of importance to acquire knowledge of the various possibilities offered by grid generation tools. The first choice is between a structured or an unstructured grid, based on the properties of the flow solver and on the level of geometrical complexity. The second choice is, within either of these two families, to select an adapted grid topology or element type. We will cover these various options in this chapter, which is subdivided as follows:

Section 6.1 describes and illustrates various options for structured grids, while Section 6.2 focuses on unstructured grid configurations.

Once a grid is established, the flow solvers have to evaluate surfaces and volumes, based on the mesh point coordinates and element shapes, as required by the application of finite volume methods to cells of arbitrary shapes. Guidelines and a few easy to implement formulas for the numerical evaluation of surfaces and volumes are provided in Section 6.3, as an extension to the formulas already given in Chapter 5.

As already mentioned, grid quality has a direct impact on the overall accuracy of the CFD results. An evaluation of grid related numerical errors for arbitrary 2D grids is presented to serve as a basis for the establishment of a few general guidelines toward minimizing grid-related error sources. This forms the subject of Section 6.4, which offers recommendations concerning grid properties, to be followed as best as possible in order to minimize grid-related losses of accuracy.

Figure 6.0.1 summarizes the roadmap to this chapter.

6.1 STRUCTURED GRIDS

Structured grids can be considered as most 'natural' for flow problems as the flow is generally aligned with the solid bodies and we can imagine the grid lines to follow in some sense the streamlines, at least conceptually, when not possible realistically.

It has to be emphasized that structured grids will, compared to unstructured grids, often be more efficient from CFD point of view, in terms of accuracy, CPU time and

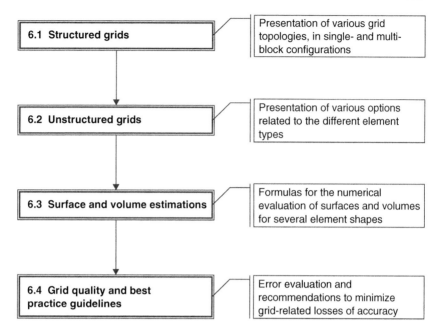

| 6.1 **Structured grids** | Presentation of various grid topologies, in single- and multi-block configurations |

| 6.2 **Unstructured grids** | Presentation of various options related to the different element types |

| 6.3 **Surface and volume estimations** | Formulas for the numerical evaluation of surfaces and volumes for several element shapes |

| 6.4 **Grid quality and best practice guidelines** | Error evaluation and recommendations to minimize grid-related losses of accuracy |

Figure 6.0.1   *Roadmap to this Chapter.*

memory requirement. The reason behind the development of unstructured CFD codes is essentially connected to the time required to generate good quality block-structured grids on complex geometries. This task, with the best available software tools, can easily take weeks or months of engineering time and the associated engineering costs are considered as prohibitive industrially. Hence, the requirement for automatic grid generation tools has become essential for the further development of industrial CFD. This explains largely the preference given nowadays to unstructured CFD solvers, due to the availability of general-purpose automatic grid generation methods.

However, it remains also possible to generate automatic block-structured grids, when restricted to well-defined families of topologies. Examples are shown in the following.

The ideal mesh is a Cartesian distribution, where all the points are equidistant and where all the cells are perfect cubes, with $\Delta x = \Delta y = \Delta z$. This grid will be associated with the highest possible accuracy of the discretized formulas, where the finite volume method leads to the same formulas as finite differences. Hence, all evaluations of grid qualities will be done by comparing a selected cell to the ideal cubic cell.

When curved solid surfaces are present, they cannot be part of the Cartesian mesh lines and we have two options: either we keep the Cartesian structure of the grids or we move away from the ideal and introduce curvilinear grids in order to fit the grid lines to the solid surfaces. We call these types of grids '***body fitted***'. In the former case, we have to define a particular treatment to the cells cutting the solid surface. In the latter case, we have to generate grids that follow the solid surfaces, for instance by defining curvilinear coordinates (ξ, η, ζ) that would be constant along the lines of mesh points in the physical space and Cartesian in the mathematical space formed by these variables. Various topologies of the grid lines can be defined, and will be presented in this section.

The drawback of structured grids is a form of stiffness connected to the fact that adding a point locally implies adding lines of each family through that point, which will therefore affect the whole domain. In complex geometries, this can be very detrimental and render the grid generation process quite cumbersome. One way to ease these constraints is to define multi-block grids, each block covering a subset of the computational domain with its own structured grid. This can be further generalized when the connectivity of the points at the block interfaces is relaxed by allowing 'non-matching' lines at the inter-block boundaries. This provides maximum flexibility to block-structured grids.

Another way is to allow for overlapping grids, each grid being attached to a solid body, when multiple moving bodies are present, or to separate blocks. Both ways imply sophisticated treatment for the interpolation of the numerical flow variables between two independent grids, with the requirement to satisfy constraints of conservation and accuracy.

6.1.1 Cartesian Grids

As mentioned above, uniform Cartesian grids are the ideal solution from the point of view of accuracy and they should be applied whenever possible. It is a valid option when the solid walls are parallel to the Cartesian axes, or in absence of solid walls in free space.

Cartesian grids are often applied in aero-acoustic computations, where high order schemes are required for an accurate simulation of the propagation of acoustic pressure waves (see for instance the review paper by Tam (2004)).

6.1.2 Non-uniform Cartesian Grids

Variable mesh sizes
A first variant on the ideal Cartesian uniform grid is to allow for variable values of the mesh spacing, for instance in boundary layers where a strong clustering near solid walls is required. Figure 6.1.1 shows an example of a Cartesian grid applied to the flow simulation over and in a rectangular cavity.

Quadtree-Octree grid
A second variant consists of allowing for local refinements with '*hanging nodes*', also called non-conformal grids, obtained by subdividing an initial Cartesian grid in sub-cells, either uniformly or non-isotropically, as shown in Figure 6.1.2. This leads to a *quadtree* structure in 2D and an *octree* structure in 3D.

In presence of curved boundaries, Cartesian grids still remain an option, with the advantage that the grid generation process is trivial, while minimizing numerical errors. However, the treatment of the curved solid boundaries requires special attention.
Several options can be considered:

- *Method 1*: The Cartesian type grid on both sides of the surface is maintained and a numerical procedure is defined in the flow solver to handle the physical boundary conditions (Figure 6.1.3). This is called the **immersed boundary method** (see for instance Mittal and Iaccarino (2005) for a review).

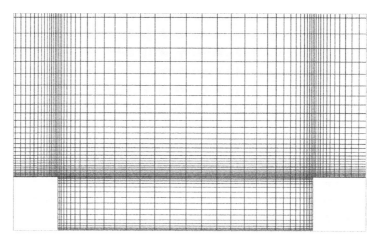

Figure 6.1.1 *Cartesian grid with non-uniform cell sizes for a cavity.*

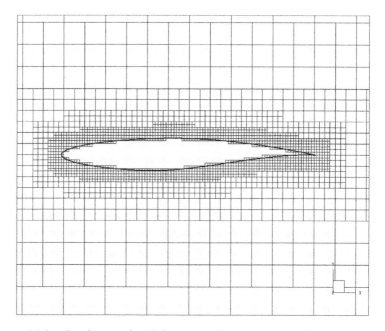

Figure 6.1.2 *Quadtree grid, with hanging nodes, around an airfoil, with staircase boundary approximation.*

- *Method 2*: The Cartesian cells outside the computational domain are removed, replacing hereby the solid boundaries by a ***staircase shape***; this is the case with Figure 6.1.2.
- *Method 3*: The intersection of the solid surface with the Cartesian cells is defined, leading to boundary cells of arbitrary shapes, called ***cut-cells***; see Figure 6.1.4, from Aftosmis et al. (2000). This requires the application of a finite volume discretization on the cut-cell faces.

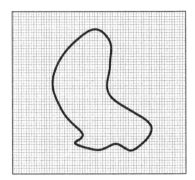

Figure 6.1.3 *Cartesian mesh around a solid boundary with Immersed Boundary Method.*

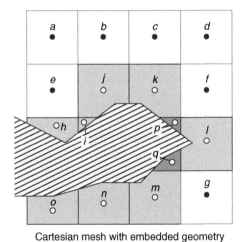

Cartesian mesh with embedded geometry

Figure 6.1.4 *Cut-cell configuration; from Aftosmis et al. (2000). AIAA copyright.*

6.1.3 Body-Fitted Structured Grids

In this approach, the grid is made curvilinear to adapt as far as possible to the geometries. It calls upon more sophisticated methods to generate the grids in order to satisfy requirements on smoothness and continuity of cell sizes. Depending on the orientation of the grid lines, various configurations can be selected, indicated by the letter to which they resemble the most. We refer in this context to grids of H-type, C-type, O-type, I-type and their various combinations.

H-mesh
The grid lines are curvilinear, approaching a set of horizontal and vertical lines in a pseudo-orthogonal configuration, with a topology that can be associated to the letter H (see Figure 6.1.5 for a representative example).

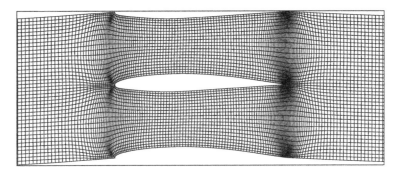

Figure 6.1.5 *Structured curvilinear body-fitted grid of the H-type.*

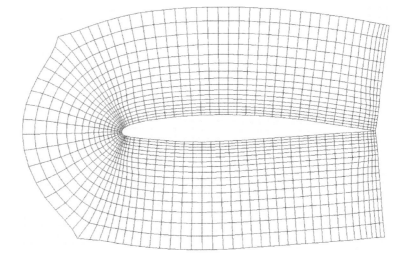

Figure 6.1.6 *Structured curvilinear body-fitted grid of the C-type.*

C-mesh

The grid lines are curvilinear, surrounding the geometry, with a topology that can be associated to the letter C, on one side (for instance around the leading edge of the airfoil), but remaining open at the other end of the computational domain. This can be adapted to concentrate grid lines in the wake region of an airfoil or wing (see Figure 6.1.6 for a representative example).

O-mesh

The grid lines are curvilinear, surrounding completely the geometry, with a topology that can be associated to the letter O. This option allows an accurate mesh point distribution around both leading and trailing edges of external aerodynamic configurations, such as wings and airfoil sections (see Figure 6.1.7 for a representative example).

I-mesh

In the particular case of highly staggered turbomachinery blade sections, the quality requirement of nearly orthogonal cells is better fulfilled with grid lines

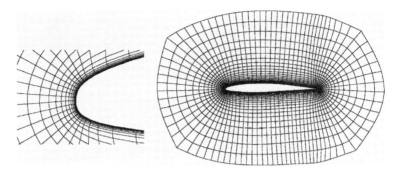

Figure 6.1.7 *Structured curvilinear body-fitted grid of the O-type.*

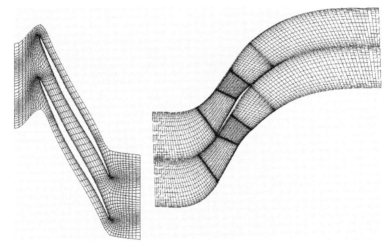

Figure 6.1.8 *Structured curvilinear body-fitted grid of the I-type, for turbomachinery blades. Courtesy Numeca Int.*

nearly orthogonal to the blade sections, leading to a I-type topology, as shown on Figure 6.1.8.

6.1.4 Multi-block Grids

In order to increase the flexibility, the range of application and the easiness of the meshing process of structured grids, combinations of basic topologies can offer significant advantages, in terms of achieving higher grid quality or adaptation to more complex topologies. In this strategy, different mesh topologies are applied in different regions of the computational domain, leading to ***multi-block*** configurations.

Matching and non-matching boundaries between blocks
Normally, we would attempt to satisfy the condition of full matching mesh lines between the blocks, whereby the mesh lines cross the block boundaries in a continuous

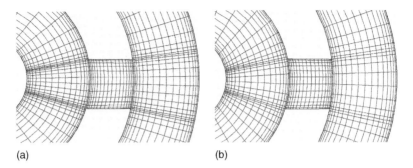

(a)   (b)

Figure 6.1.9 *Representation of (a) matching and (b) non-matching block boundary interfaces of a multi-block-structured grid, with a channel connecting two circular ducts.*

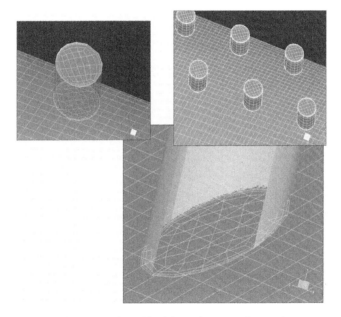

Figure 6.1.10 *Full non-matching block boundary interfaces of a multi-block-structured grid. Courtesy Numeca Int.*

way. However, in order to exploit maximally the potential of block-structured grids, the additional flexibility of allowing for ***non-matching block interfaces*** offers significant advantages. The price of this enhanced flexibility is the necessity for the flow solver to handle with sufficient accuracy the transfer of information through the non-matching interface, requiring sophisticated interpolation routines between two totally independent surface grids.

This is illustrated in Figure 6.1.9, where the choice between the two options is still available. This is not always the case and Figure 6.1.10 shows an example where

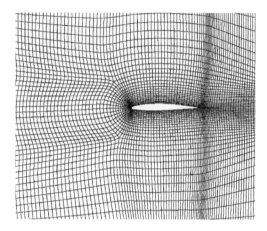

Figure 6.1.11   *Structured curvilinear body-fitted grid of the C–H type.*

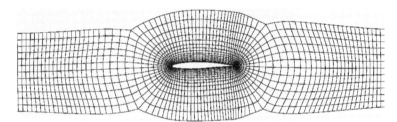

Figure 6.1.12   *Structured multi-block body-fitted grid of the H–O–H type.*

smaller exhaust pipes are connected on a larger duct. In this case, both components can be meshed optimally and freely connected in a non-matching mode.

C–H mesh
This combines a C-mesh around the body and an H-mesh in the upstream region as shown in Figure 6.1.11.

H–O–H mesh
In this configuration, an O-mesh is kept around the body, while H-topologies are defined in the upstream and downstream regions (see Figure 6.1.12).

'Butterfly' grids for internal flows
High quality structured grids with internal flow configurations, such as complex ducts, are difficult to ensure and a high level of flexibility is required. One of the options is obtained by the so-called '*butterfly*' topology shown in Figure 6.1.13, for a simple duct section.

It can also be applied to bulbs of a rotating axis, in order to avoid a singular mesh line on the axis of rotation, at zero radius. Figure 6.1.14 shows combinations of block-structured grids, obtained with the automatic grid generator Autogrid™ from Numeca Int. (http://www.numeca.be), applied to the pump inducer of the liquid hydrogen pump of the VULCAIN engine of the European ARIANE 5 rocket launcher.

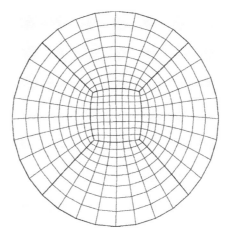

Figure 6.1.13   *Structured multi-block body-fitted grid of the 'butterfly' type.*

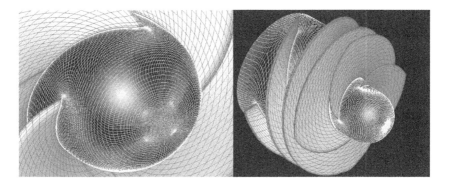

Figure 6.1.14   *Multi-block grid for the inducer of the hydrogen pump of the European ARIANE 5 rocket launcher. The figure on the left is a zoom on the 'butterfly' mesh on the bulb. Courtesy SNECMA and Numeca Int.*

O–H grids with matching and non-matching periodic boundaries

For internal turbomachinery flow simulations, a high degree of mesh flexibility is required and various combinations can be considered to enhance the quality of the grids. For instance, a combination O–H, associated with either matching (a) or non-matching (b) periodic boundaries for a turbine blade row are shown on Figure 6.1.15.

Figures 6.1.16 and 6.1.17 show two industrial examples, respectively, of an industrial heat exchanger combining matching and non-matching multi-block interfaces.

Overset grids

Another alternative to flexible block-structured grid generation is the technique of **overset grids**, also called '***chimera***' technique, where independent generated grids around a fixed or moving body are made to overlap with a background fixed grid. This technique is largely applied with several bodies in relative motion where a mesh is

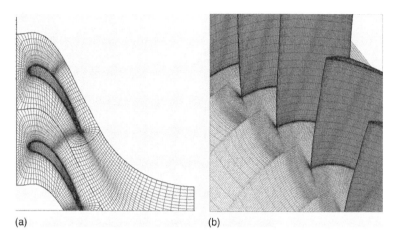

(a)   (b)

Figure 6.1.15 *Structured curvilinear body-fitted grid of the O–H type: (a) matching periodic boundaries and (b) non-matching periodic boundaries. Courtesy Numeca Int.*

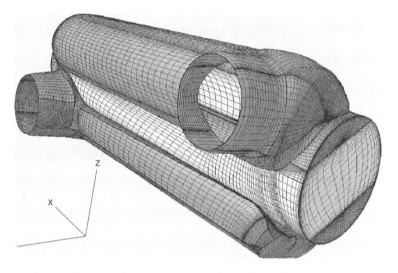

Figure 6.1.16 *Structured multi-block grid of an industrial heat exchanger combining matching and non-matching multi-block interfaces. Courtesy Atlas Copco and Numeca Int.*

attached to each body. The drawback is related to the necessity for an accurate interpolation between three-dimensional overlapping grids. This is extremely challenging, particularly if conservative interpolations are required.

Note that the overset principle can equally be applied with unstructured grids, although it was developed initially for structured grids (Steger et al., 1983; Benek et al., 1985).

Figure 6.1.18 shows overlapping grids around moving parts of a flying structure, with a zoom on a section of the overlapping region.

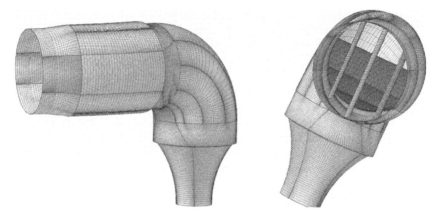

Figure 6.1.17 *Structured multi-block grid of an industrial inlet ducting with guide vanes, combining matching and non-matching multi-block interfaces. Courtesy Atlas Copco and Numeca Int.*

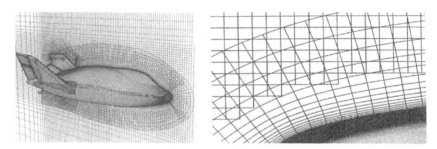

Figure 6.1.18 *Overlapping grids around moving parts of a flying structure, with a zoom on a section of the overlapping region.*

6.2 UNSTRUCTURED GRIDS

Unstructured grids have progressively become the dominating approach to industrial CFD, due to the impossibility to generate automatically block-structured grids on arbitrary geometries. It is indeed nearly impossible, for topology-connected reasons, to envisage an automatic block-structured grid generator without an a priori knowledge of the involved topologies. However, this is possible with unstructured grids and therefore unstructured flow solvers for the Navier–Stokes equations have gained wide acceptance.

Although on a same regular distribution of points, an unstructured grid, formed for instance by triangles in 2D, will tend to have a lower accuracy than the corresponding structured grid, as will be shown in Section 6.4, this trend has arisen because of the industrial requirements for automatic grid generation tools.

One of the advantages of unstructured grids is the possibility to perform local refinements in a certain region, without affecting the grid point distribution outside that region. This opens the way for flexible ***grid adaptation*** by local refinement or local coarsening, based on some criteria associated either to some flow gradients or to some error estimation. Grid adaptation is based on the addition or removal of mesh points in order to increase the accuracy in regions of strong flow variations and by removing points in regions where the solutions has already reached an acceptable accuracy. This process has as objective to optimize the number of grid points for a certain level of accuracy.

The space domain can be discretized by subdivision of the continuum into elements of arbitrary shape and size. Since any polygonal structure with rectilinear or curved sides can finally be reduced to triangular and quadrilateral elements, they form the basis for the most current space subdivision in 2D space. Cells with an arbitrary number of faces can also be considered, resulting from a dual grid construction, or from an agglomeration process of groups of cells into coarser cells, as required by multigrid methods. The only restriction is that the elements may not overlap and have to cover the complete computational domain.

Most of the unstructured grid generators applied in practice are focused on the generation of basic cell shapes formed by:

- triangle/tetrahedra elements;
- hybrid elements involving combinations of tetrahedra, pyramids and prisms, the latter being concentrated near the solid surfaces;
- quadrilaterals and hexahedra.

6.2.1 Triangle/Tetrahedra Cells

Various methods are available to generate triangular/tetrahedral grids around arbitrary bodies. Most of them require an initial surface triangulation, which has to be generated first, before launching the generation of the volume mesh. See for instance the books by George and Borouchaki (1998) and Frey and George (1999) for an overview.

The following examples are obtained with the system DELANDO developed by Jens-Dominik Müller (see http://www.cerfacs.fr/~muller/delaundo.html).

Figure 6.2.1 shows a two-dimensional unstructured grid with triangular cells, around an airfoil with flaps.

An example of a complex tetrahedral grid is shown on Figure 6.2.2, generated for the simulation of the electrochemical plating of a system of decorative cronium wheels.

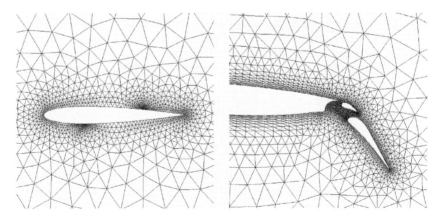

Figure 6.2.1 *Example of an unstructured triangular grid.*

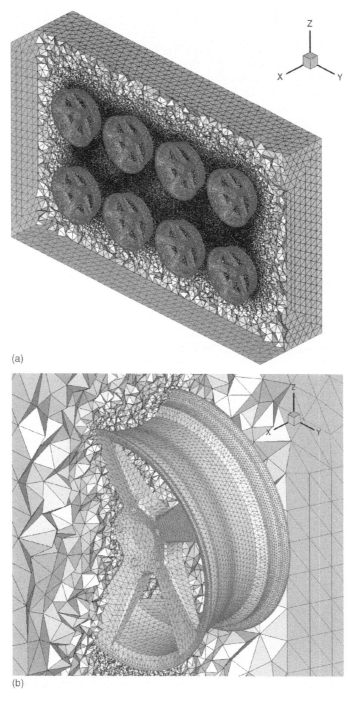

(a)

(b)

Figure 6.2.2 *Example of an unstructured tetrahedral grid for the simulation of electrochemical plating of a system of Decorative Cronium wheels. (a) Full view (b) Zoom on one of the wheels. Courtesy Von Karman Institute and Vrije Universiteit Brussel, Computational Electrochemistry Group.*

6.2.2 Hybrid Grids

The main difficulty with triangular/tetrahedral grids is connected to the boundary layer requirements of high Reynolds number flows, where the grid density in the normal direction has to be adapted to the boundary layer velocity profiles. As seen in Section 4.3, the ratio of mesh sizes should optimally be of the other $\Delta x/\Delta y \sim \sqrt{(Re)}$, where Δx and Δy are the representative mesh sizes in the streamwise and normal directions, respectively. This implies mesh aspect ratios $\Delta x/\Delta y$ of the order of 1000, for typical industrial flows, which would lead to very poorly configured triangles with height to base ratios of that order and, consequently, a significant loss of accuracy.

To avoid this problem, hybrid grids have been developed, whereby layers of quadrilaterals or prisms are generated in the near-wall region, by a form of extrusion process out of the triangulated surface grid. This is shown on Figure 6.2.3, for a 2D case, from the same reference as the previous figure http://www.cerfacs.fr/~muller/hip.html.

A three-dimensional example of a hybrid grid for a gas-turbine stator with rows of film cooling holes in the leading edge region is shown on Figure 6.2.4. The grids are obtained with the CENTAUR™ grid generator from Centaursoft (http://www.centaursoft.com), showing different views of the hybrid grid. The top figure shows the 3D view and the other figures show a 2D section with a close-up view of the leading edge region.

6.2.3 Quadrilateral/Hexahedra Cells

It is known from numerous simulations (see also Section 6.4) that hexahedra offer significant advantages compared to tetrahedral cells, in terms of memory requirements and accuracy. For tetrahedral grids, the ratio of the number of cells to the number of vertices is close to 6, not taking into account the boundaries. (for a two-dimensional triangulation this ratio is of the order of 3); while this ratio remains close to one for hexahedral cells.

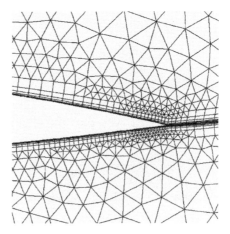

Figure 6.2.3 *Example of an unstructured hybrid grid showing the regular quadrilateral type structure near the solid walls.*

Automatic generation of unstructured hexahedra is quite challenging and some examples from the HEXPRESS™ generator from Numeca Int. (http://www. numeca.be) are shown in the following figures. In this approach, the surface mesh is obtained from the volume mesh, through a projection step of a Cartesian/octree volume mesh, whereby the initial generation of a surface mesh is not required. Figure 6.2.5 shows a 2D quadrilateral grid around an airfoil, while Figure 6.2.6 shows a 3D unstructured hexahedral grid of a valve system, with cuts displaying the internal cells.

Similarly, Figure 6.2.7 shows the hexahedral mesh for a complex dusting system, with appropriate cuts to visualize the internal volume cells.

6.2.4   Arbitrary Shaped Elements

The most general unstructured grid configuration is obtained with cells having an arbitrary number of faces. They can be defined either by considering the ***dual mesh*** of a base grid formed by simple shapes, or by an ***agglomeration process*** of cells.

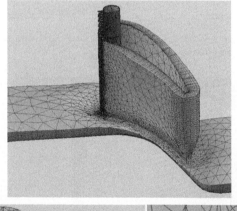

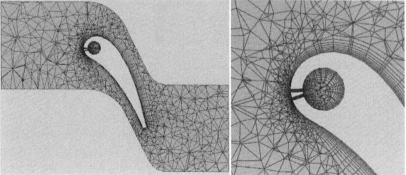

Figure 6.2.4   *Three-dimensional hybrid grid of a turbine blade with film cooling configuration, with a 2D section and a close-up view of the leading edge region. From* http://www.centaursoft.com.

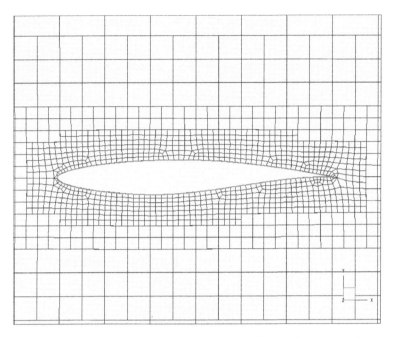

Figure 6.2.5 *Unstructured quadrilateral grid around an airfoil, obtained with the HEXPRESS™ generator from Numeca Int.*

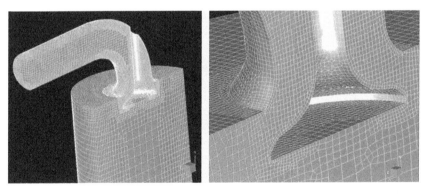

Figure 6.2.6 *3D unstructured hexahedral grid of a valve system, with cuts displaying the internal cells, obtained with the HEXPRESS™ generator from Numeca Int.*

Two-dimensional dual grids are obtained by joining the midpoints of cell edges to the center of the cells, as illustrated in Figure 6.2.8. Alternatively, they can be defined by joining the midpoint normals to the cell edges to form a polygonal cell.

The agglomeration process is generally applied for multigrid convergence acceleration methods (see Chapter 10) on unstructured grids, in grouping cells into coarser cells, typically with a ratio close to 8. This can be applied to tetrahedral or hexahedral grids, as seen on the examples of Figures 6.2.9 and 6.2.10.

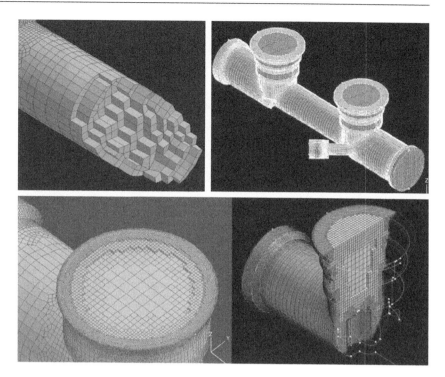

Figure 6.2.7 *Unstructured hexahedral mesh for a complex dusting system, with cuts to showing the internal volume cells, obtained with the HEXPRESS™ generator from Numeca Int.*

6.3 SURFACE AND VOLUME ESTIMATIONS

Non-Cartesian grids are predominantly applied with finite volume methods, and as seen in Chapter 5, this requires the numerical evaluation of cell volumes and face areas, including the direction of the normals to the faces.

In this section, we intend to provide you with a few additional formulas for the evaluation of these geometrical quantities, up to second order accuracy.

In 2D cases, we will consider that we have either triangles or quadrilaterals and the area of the cells is the only quantity we need, as the length of the edges is trivially defined by the distance between its end-points. In case of cells of arbitrary shape, it will be considered that they are subdivided in triangles and or quadrilaterals.

In three dimensions, we will assume that the geometrical space is subdivided in tetrahedra, pyramids, prisms or hexahedra. When a face is formed by three nodes, i.e. by a triangle, as is the case with an arbitrary triangulation, then the calculation of the area is straightforward as three points lie always in the same plane. However, for a quadrilateral face, the four points forming the face are not necessarily coplanar, which can be a source of error, as there is more than one way to evaluate the cell face area. Care has to be exercised in the evaluation of the cell volume and face areas, in order to ensure that the sum of the computed volumes of adjacent cells is indeed equal to the total volume of the combined cells.

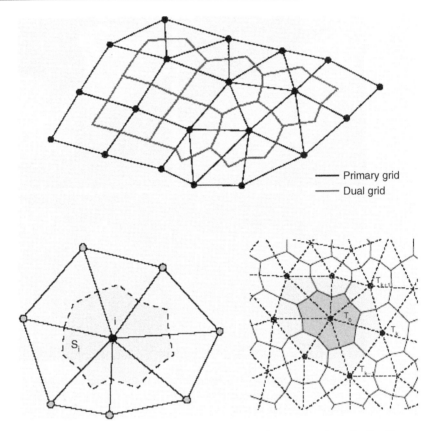

Figure 6.2.8 *Different forms of dual grids with arbitrary number of cells. From Barth (2003).*

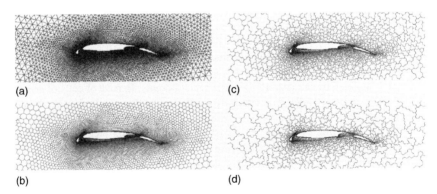

Figure 6.2.9 *Triangular source grid (a) around a three-component airfoil showing the dual grid (b), agglomerated grids (c) and second agglomerated grid (d). From Lassaline and Zingg (2003). Copyright © 2003 by J.V. Lassaline and D.W. Zingg.*

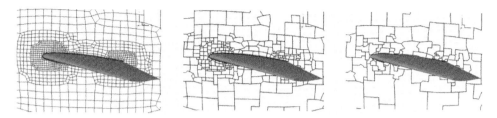

Figure 6.2.10   *Two levels of agglomerated grids from an initial unstructured hexahedral grid. Courtesy Numeca Int.*

6.3.1   Evaluation of Cell Face Areas

Some formulas have already been derived in Sections 5.3 and A5.4 when we introduced the finite volume and finite element methods. Hence, the formulas presented here can be considered as a generalization, providing additional options.

An important property of the area vector S attached to a cell face is derived from the divergence theorem. Equation (5.3.16) with $U = 1$ becomes, where the integral loops over all the faces of the cell:

$$\oint_S d\vec{S} = 0 \quad \text{or} \quad \sum_{\text{faces}} \vec{S}_{\text{faces}} = 0 \tag{6.3.1}$$

This relation indicates that the sum over all the face surface normals of any closed cell must be zero. In addition, it shows that the outward surface vector of a given face belonging to the closed surface S:

$$\vec{S}_{\text{face}} = \int_{\text{face}} d\vec{S} \tag{6.3.2}$$

is only dependent on the boundaries of the face.

Hence, for face ABCD of Figure 6.3.1, we could apply equation (5.3.4), reproduced here:

$$\vec{S}_{\text{ABCD}} = \frac{1}{2}(\vec{x}_{\text{AC}} \times \vec{x}_{\text{BD}}) \tag{6.3.3}$$

also when the points A, B, C, D are not coplanar, i.e. when AC and BD do not intersect. Other alternative formulas are obtained by considering the area of the quadrilateral as the sum of the areas of the two triangles ABC and CDA. Since the surface of the triangle CDA is defined by half of the area of the parallelogram constructed on two of its sides, we have

$$\begin{aligned}
\vec{S}_{\text{CDA}} &= \frac{1}{2}(\vec{x}_{\text{AC}} \times \vec{x}_{\text{CD}}) \\
&= \frac{1}{2}(\vec{x}_{\text{CD}} \times \vec{x}_{\text{DA}}) \\
&= \frac{1}{2}(\vec{x}_{\text{DA}} \times \vec{x}_{\text{AC}})
\end{aligned} \tag{6.3.4}$$

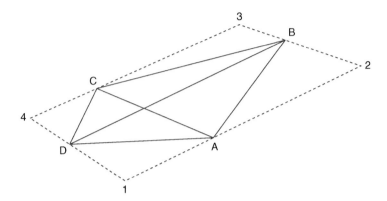

Figure 6.3.1 *Arbitrary quadrilateral face.*

and similarly for the other triangles, in an anti-clockwise rotation of the triangle points. Hence, the sum of the two triangles ABC and CDA is equal to

$$\vec{S}_{ABCD} = \frac{1}{2}[(\vec{x}_{AB} \times \vec{x}_{BC}) + (\vec{x}_{CD} \times \vec{x}_{DA})] \tag{6.3.5}$$

This last formula expresses the surface vector $\vec{S}_{ABCD}$ as the average of the surface vectors of the two parallelograms constructed on the adjacent sides (AB, BC) and (CD, DA). In the general case, the two normals will not be in the same direction since (ABC) and (CDA) are not in the same plane. Hence, equation (6.3.5) takes the vector $\vec{S}_{ABCD}$ as the average vector of these two surface vectors, while equation (6.3.3) expresses $\vec{S}_{ABCD}$ as the vector product of the two diagonals.

Note that the two formulas lead to identical results, even for non-coplanar cell faces, which can be seen by considering vector relations such as $\vec{x}_{AC} = \vec{x}_{AB} + \vec{x}_{BC}$ and the properties of the vector products.

Similar to equation (6.3.5), we have, considering the sum of the triangles BCD and DAB:

$$\vec{S}_{ABCD} = \frac{1}{2}[(\vec{x}_{BC} \times \vec{x}_{CD}) + (\vec{x}_{DA} \times \vec{x}_{AB})] \tag{6.3.6}$$

Another formula is obtained by averaging equations (6.3.5) and (6.3.6):

$$\vec{S}_{ABCD} = \frac{1}{4}[(\vec{x}_{AB} + \vec{x}_{CD}) \times (\vec{x}_{BC} + \vec{x}_{AD})] \tag{6.3.7}$$

All these formulas give identical results and are applied in practical computations. The first formula, equation (6.3.3) being the less expensive in number of arithmetic operations.

6.3.2 Evaluation of Control Cell Volumes

We indicate here a few simple formulas for the volume estimation of basic cells, such as tetrahedra, pyramids or hexahedra. Referring to Figure 6.3.2a, we see that a

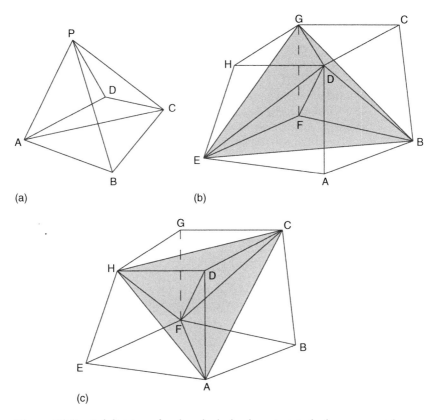

(a)

(b)

(c)

Figure 6.3.2   *Subdivision of an hexahedral volume in tetrahedra or pyramids.*

pyramid can be subdivided in two tetrahedra, while hexahedra can be subdivided in three pyramids, or in tetrahedra.

As the tetrahedron forms the basic component, we first define its volume as follows.

Volume of a tetrahedron

The volume of a tetrahedron Ω_{PABC} is obtained by applying the Gauss formula for the divergence of an arbitrary vector $\vec{a}$:

$$\int_{\Omega} \vec{\nabla} \cdot \vec{a} \, d\Omega = \oint_{S} \vec{a} \cdot d\vec{S} \tag{6.3.8}$$

Taking the vector $\vec{a}$ equal to the position vector $\vec{x}$, we obtain the following general formula for cell volumes, since $\vec{\nabla} \cdot \vec{x} = 3$,

$$\Omega = \frac{1}{3} \oint_{S} \vec{x} \cdot d\vec{S} \tag{6.3.9}$$

Applying this formula to the tetrahedron PABC, leads to

$$\Omega_{PABC} = \frac{1}{3} \int_{PABC} \vec{x} \cdot d\vec{S} = \frac{1}{3} \sum_{faces} \vec{x} \cdot \vec{S}_{faces} \tag{6.3.10}$$

In this formula, the vector $\vec{x}$ has its end-point in the corresponding face, and can be approximated by the center of gravity of the face. However, formula (6.3.10) simplifies considerably if we select the origin of the position vector as one of the vertices of the tetrahedron, for instance in point P. We can write then, where $\vec{x}_{(P)}$ represents the vector originating in P:

$$\Omega_{PABC} = \frac{1}{3}\vec{x}_{(P)} \cdot \vec{S}_{ABC} \tag{6.3.11}$$

This relation results from the fact that $\vec{x}_{(P)}$ lies in the three faces containing P and is therefore orthogonal to the associated surface vector, which makes the scalar product vanish. The only remaining contribution comes from the face ABC, which is opposite to P. Since any vector from P to an arbitrary end-point Q in ABC can be decomposed in the vector PA plus a vector AQ in ABC, this contribution will vanish as the vector AQ is orthogonal to the surface vector of triangle ABC. Hence, with $\vec{x}_{(P)} = \vec{x}_{PA}$, we have

$$\Omega_{PABC} = \frac{1}{6}\vec{x}_{PA} \cdot (\vec{x}_{AB} \times \vec{x}_{BC}) = \frac{1}{6}\vec{x}_{PA} \cdot (\vec{x}_{BC} \times \vec{x}_{CA}) \tag{6.3.12}$$

Equation (6.3.12) can also be expressed as a determinant:

$$\Omega_{PABC} = \frac{1}{6} \begin{vmatrix} x_P & y_P & z_P & 1 \\ x_A & y_A & z_A & 1 \\ x_B & y_B & z_B & 1 \\ x_C & y_C & z_C & 1 \end{vmatrix} \tag{6.3.13}$$

Volume of a pyramid
In a similar way, for a pyramid, PABCD, we have

$$\Omega_{PABCD} = \frac{1}{3} \int_{PABCD} \vec{x} \cdot d\vec{S} = \frac{1}{3}\vec{x}_{(P)} \cdot \vec{S}_{ABCD} \tag{6.3.14}$$

Since ABCD is not necessarily coplanar, $\vec{x}_{(P)}$ has to be estimated by an appropriate approximation. For instance

$$\vec{x}_{(P)} = \frac{1}{4}(\vec{x}_{PA} + \vec{x}_{PB} + \vec{x}_{PC} + \vec{x}_{PD}) \tag{6.3.15}$$

and with the expression (6.3.3) for $\vec{S}_{ABCD}$, we obtain

$$\begin{aligned}\Omega_{PABCD} &= \frac{1}{24}(\vec{x}_{PA} + \vec{x}_{PB} + \vec{x}_{PC} + \vec{x}_{PD}) \cdot (\vec{x}_{AC} \times \vec{x}_{BD}) \\ &= \frac{1}{12}(\vec{x}_{PA} + \vec{x}_{PB}) \cdot (\vec{x}_{AC} \times \vec{x}_{BD})\end{aligned} \tag{6.3.16}$$

If the face ABCD is coplanar, then equation (6.2.44) reduces to

$$\Omega_{PABCD} = \frac{1}{6}\vec{x}_{PA} \cdot (\vec{x}_{AC} \times \vec{x}_{BD}) \tag{6.3.17}$$

The volume formulas for the pyramids are actually expressed as the sum of the two tetrahedra.

Volume of a hexahedron

Different formulas can be applied to obtain the volume of a hexahedral cell, the most current approach consisting in a subdivision in tetrahedra or in pyramids. Referring to Figure 6.3.2, the hexahedron can be divided into three pyramids, for instance with point D as summit,

$$\Omega_{HEX} = \Omega_{DABFE} + \Omega_{DBCGF} + \Omega_{DEFGH} \tag{6.3.18}$$

Dividing each pyramid into two tetrahedra, leads to a decomposition of the hexahedron into six tetrahedra, originating for instance in D, as

$$\Omega_{HEX} = \Omega_{DABE} + \Omega_{DBFE} + \Omega_{DBCG} + \Omega_{DBGF} + \Omega_{DEFG} + \Omega_{DEGH} \tag{6.3.19}$$

Extreme care has to be exercised in the evaluation of the tetrahedra volumes since the sign of the volumes Ω_{PABC} in equations (6.3.11) and (6.3.12) depends on the orientation of the triangular decomposition. In addition, when the cell surfaces are not coplanar the same diagonal has to be used in the evaluations of the tetrahedra in the two cells that share this surface. Otherwise gaps or overlaps would occur in the summation of volumes. A useful guideline, in order to avoid sign errors, consists in applying a right-hand rotation (screwdriver) rule from the base toward the summit of each tetrahedron.

Another alternative is to decompose the volume of the hexahedron into five tetrahedra originating in D for instance, referring to Figure 6.3.2b, as

$$\Omega_{HEX} = \Omega_{DABE} + \Omega_{DBCG} + \Omega_{DEGH} + \Omega_{DBGE} + \Omega_{FBEG} \tag{6.3.20}$$

In this decomposition four tetrahedra have D as summit and one tetrahedron originates in point F, opposite to D. Considering the same two points D and F as references, there is a unique, second decomposition into five tetrahedra shown on Figure 6.3.2c:

$$\Omega_{HEX} = \Omega_{FACB} + \Omega_{FAEH} + \Omega_{FCHG} + \Omega_{FAHC} + \Omega_{DACH} \tag{6.3.21}$$

For a general hexahedral volume, where points of a same cell face are not coplanar, the two formulas (6.3.20) and (6.3.21) will not give identical volume values. It is therefore recommended to take an average of both.

In this context, it is interesting to observe that volumes of hexahedral cells can also be evaluated from a finite element isoparametric trilinear transformation, applying a $2 \times 2 \times 2$ Gauss point integration rule. Although very tedious to prove analytically, numerical experiments consistently show that this finite element procedure leads to volume values equal to the average of the two formulas (6.3.20) and (6.3.21).

An investigation of more elaborate decompositions of hexahedral volumes in pyramids can be found in Davies and Salmond (1985), while some of the above-mentioned decompositions are also discussed in Rizzi and Ericksson (1981), Kordulla and Vinokur (1983).

6.4 GRID QUALITY AND BEST PRACTICE GUIDELINES

The most critical issue of CFD simulations is the inevitable loss of accuracy due to grid non-uniformities, particularly with currently used schemes, which are generally of second order accuracy on uniform grids.

This topic has already been introduced in Section 4.3, in the case of one-dimensional grids. We recommend you to turn back to this section, if needed, in order to refresh your perception of the analysis and the impact of non-uniform grids on accuracy. With two- or three-dimensional arbitrary grids, the situation is clearly more complicated. We can nevertheless draw some very useful guidelines from an elegant analysis of the errors generated with second order finite volume methods on an arbitrary 2D grid, due to Roe (1987).

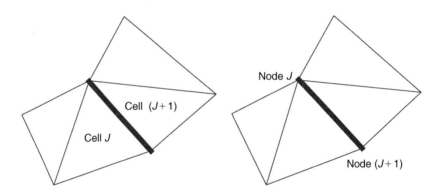

Figure 6.4.1 *Cell-centered and cell-vertex configurations of a 2D grid.*

6.4.1 Error Analysis of 2D Finite Volume Schemes

We consider an arbitrary 2D grid, either cell centered or cell vertex, as illustrated in Figure 6.4.1 and we look at the estimation of the flux through the face $(J, J + 1)$.

We apply a finite volume method, following equation (5.2.7), which reduces to equation (5.3.3) in 2D. In absence of source terms, the residual reduces to the sum of the fluxes over all the sides (faces). For each side, the following quantity has to be estimated, as a contribution to the residual:

$$\int_{J,J+1} \vec{F} \cdot d\vec{S} = \int_{J,J+1} f \, dy - g \, dx \qquad (6.4.1)$$

Hence, the residual becomes, when applying a trapezium formula for the integration:

$$R_J = \sum_{\text{faces}} \frac{1}{2}[(f_J + f_{J+1})(y_{J+1} - y_J) - (g_J + g_{J+1})(x_{J+1} - x_J)] \qquad (6.4.2)$$

which is second order accurate on a uniform grid, as shown in Section 5.3.

On a non-uniform grid, Roe's analysis provides an expression for the truncation error ε_J, associated to the residual R_J, by applying Taylor expansions. The following expression is obtained

$$\Omega_J \varepsilon_J = -E_2 f_{xx} + 2E_3 f_{xy} - E_3 f_{yy} - E_1 g_{xx} + 2E_2 g_{xy} - E_3 g_{yy} \qquad (6.4.3)$$

where the subscripts on the flux components f and g, indicate second order derivatives and Ω_J is the volume (area) of the cell J.

The E-coefficients are defined as follows:

$$E_1 = \frac{1}{12} \sum_{\text{sides}} (\Delta x)^3 \quad E_2 = \frac{1}{12} \sum_{\text{sides}} (\Delta x)^2 \Delta y$$

$$E_4 = \frac{1}{12} \sum_{\text{sides}} (\Delta y)^3 \quad E_3 = \frac{1}{12} \sum_{\text{sides}} (\Delta y)^2 \Delta x \qquad (6.4.4)$$

where Δx and Δy are the respective coordinate differences between the end-points of the cell side.

The following properties can be derived from these expressions, writing h for either Δx or Δy:

- The errors are of the order $O(h^3)/h^2$ on an arbitrary mesh. *Hence, the scheme reduces to first order*.
- The general condition for maintaining second order accuracy on an arbitrary grid, is expressed by the conditions:

$$\sum_{\substack{\text{opposite}\\\text{faces}}} (\Delta x) = O(h^2) \qquad \sum_{\substack{\text{opposite}\\\text{faces}}} (\Delta y) = O(h^2) \qquad (6.4.5)$$

- *On a parallelogram, i.e. for a quadrilateral cell with parallel sides, the E-coefficients vanish and the scheme remains second order. However, if the same points are connected by triangles, i.e. on a triangular grid, the coefficients cannot be made to vanish.*
- The consequence of this important property is that quadrilateral/hexahedral grids will tend to lead to a higher accuracy than triangular/tetrahedral grids, for the same level of mesh non-uniformity.
- Conditions (6.4.5) are satisfied on grids with a high degree of continuity in the cell size. This has been called '*analytical grids*' by Turkel (1985). In non-mathematical terms, this very important condition requires that *the cell size h is a continuous function with smooth variations*.

6.4.2 Recommendations and Best Practice Advice on Grid Quality

In addition to the grid size continuity, other typical multidimensional quantities influence the global accuracy of the CFD simulation, related to the distortion of a cell from the ideal Cartesian shape. Typical quantities are aspect ratio $\Delta x/\Delta y$; skewness factor measured by the angle between adjacent sides/faces and their various 3D generalizations.

Although it is not easy to quantify the effect of these cell shape quantities on the local accuracy, it is safe to keep in mind that highly distorted cells will always influence negatively the accuracy of the solution, and consequently it will also deteriorate the convergence rate of the numerical schemes.

Based on the above considerations, we can offer the following recommendations regarding grid quality:

- Avoid *absolutely* discontinuities in grid cell size. Any sudden jump in grid size could reduce the local accuracy to order zero.
- Ensure that the grid sizes vary in a continuous way in all directions. This is not always easy to achieve, as you can observe from the various examples shown in the introduction to this book and in this chapter.
- Minimize grid distortion, avoiding concave cells or cells with angles between adjacent edges that are too far away from orthogonality. If these angles are reduced to a few degrees, poor accuracy is guaranteed.
- Avoid cells with one or more very short edges, except in boundary layers where high aspect ratios are acceptable, provided the cells are sufficiently close to orthogonality to the solid surface.

The importance of these recommendations is particularly critical in flow regions with high gradients, i.e. in regions where the flow variables undergo rapid variations. On the other hand, in regions where the flow is quasi-uniform, these recommendations can be relaxed.

CONCLUSIONS AND MAIN TOPICS TO REMEMBER

This chapter has focused on a qualitative presentation of the various grid types that can be encountered or applied in practice, and to formulas for the estimation of surfaces and volumes.

If you develop a grid generator or apply an existing one, you have to decide between structured and unstructured grids, with their respective advantages and drawbacks.

The main topics to remember are:

- Structured grids if easily obtained, will offer the highest guarantee for optimal accuracy.
- Unstructured grids will give you higher flexibility for automatic generation of grids around complex geometries.
- Give the highest care to the quality of the grids, avoiding grid discontinuities and stretched distorted grids, particularly in regions where the flow is expected to vary significantly.

REFERENCES

Aftosmis, M.J., Berger, M.J. and Adomavicius, G. (2000). A parallel multilevel method for adaptively refined Cartesian grids with embedded boundaries. *38th Aerospace Sciences Meeting*, AIAA Paper 2000-0808.

Barth, T.J. (2003). Numerical Methods and Error Estimation for Conservation Laws on Structured and Unstructured Meshes. Lecture Series 2003-4, Von Karman Institute for Fluid Dynamics, Brussels, Belgium.

Benek, J., Buning, P. and Steger, J. (1985). A 3-D chimera grid embedding technique, AIAA Paper 85-1523.

Davies, D.E., Salmond, D.J. (1985). "Calculation of the volume of a general hexahedron for flow predictions". AIAA Journal, Vol. 23, pp. 954–956.

Frey, P.J. and George, P.L. (1999). *Mesh Generation*. Hermes Science Publishing, Oxford, UK.

George, P.-L. and Borouchaki, H. (1998). *Delaunay Triangulation and Meshing*. Editions Hermes Science Publishing, Oxford, UK.

Kordulla, W., Vinokur, M. (1983). "Efficient Computation of Volume in Flow Predictions". AIAA Journal, Vol. 21, pp. 917–918.

Lassaline, J. and Zingg, D. (2003). An investigation of directional-coarsening and line implicit smoothing applied to agglomeration multigrid. *16th AIAA Computational Fluid Dynamics Conference*, AIAA Paper 2003-3435.

Mittal, R. and Iaccarino, G. (2005). Immersed boundary methods. *Annual Review Fluid Mechanics*, Vol. 37, pp. 239–261.

Rassineux, A. (1998). Generation and optimization of unstructured tetrahedral meshes by advancing front technique. *Int. J. Numer. Meth. Eng.*, 41, 651–674.

Rizzi, A.W., Eriksson, L.E. (1981). "Transfinite Mesh Generation and Damped Euler Equation Algorithm for Transonic Flow around Wing-Body Configurations". Proc. AIAA 5th Computational Fluid Dynamics Conference, AIAA Paper 81-0999, pp. 43–68.

Roe, P.L. (1987). Error estimates for cell-vertex solutions of the compressible Euler equations. ICASE Report No. 87-6, NASA Langley Research Center.

Steger, J.L., Dougherty, F.C. and Benek, J.A. (1983). A chimera grid scheme. *Advances in Grid Generation*, K. Ghia and U. Ghia (Eds.). ASME-FED-Vol. 5, The American Society of Mechanical Engineers, pp. 59–69.

Tam, C. (2004). Computational aeroacoustics: an overview of computational challenges and applications. *Int. J. Comput. Fluid Dynam.*, 18(6), 547–567.

Thompson, J.F. (1984). A survey of grid generation techniques computational fluid dynamics. *AIAA J.*, 22, 1505–1523.

Thompson, J.F., Warzi, Z.U. and Mastin, C. (1985). *Numerical, Grid Generation: Foundations and Applications*. North-Holland.

Thompson, J.F., Soni, B. and Weatherill, N. (1999). *Handbook of Grid Generation*. CRC Press.

Turkel, E. (1985). Accuracy of schemes with non-uniform meshes for compressible fluid flows. ICASE Report N1/4 85-43, NASA Langley Research Center.

Part III

The Analysis of Numerical Schemes

Let us go back again to Figures I.2.1 and I.3.1 of the general introduction, where you can follow the succession of the steps and components required to set up a CFD model. We have now reached the third part, after following the steps of the previous chapters toward the development of a numerical simulation of our flow problem. Hence we are, at this stage, faced with a set of discretized equations. This set of equations, defining the ***numerical scheme***, has now to be analyzed for its properties, i.e. we have to investigate the validity of the selected discretization and its accuracy by attempting to quantify the associated numerical errors.

To achieve these objectives, we will define a certain number of concepts, such as ***consistency, stability and convergence***, which form the basis for the quantitative assessment of validity and accuracy of a numerical scheme.

These fundamental concepts are all that is required and they have to provide us with the assurance that the results of the computer simulation indeed represent a valid approximation of our 'reality'.

What do we mean with this statement?

Consider one of the examples from the general introduction or from Chapter 2. If you want, for instance, to investigate by CFD the flow around a car, or an aircraft, or a building, with the full turbulent Navier–Stokes equations and tens of millions of mesh points, you will wind up with a considerable amount of numbers as output. ***The main question we have to ask ourselves is the following: how can we be assured that the hundreds of millions of numbers generated are a valid approximation of the 'reality' we are trying to describe by our simulation, and not just an arbitrary set of numbers***.

This is indeed a fundamental question, as we have to make sure that our numerical simulation will approach the real flow, when we refine the mesh or increase the accuracy of our simulation. This is very much the same as with experiments, where we have also to make sure that our experimental equipment and our data processing tools, provide a valid representation of the flow configuration we are measuring, while the level of accuracy will depend on the quality and sensitivity of the measurement equipment. This fundamental question, remains unchanged for any mathematical model we decide to apply, simple potential flow, Euler equations for inviscid fluids, etc. ***For any mathematical model, we must have the total assurance that the numerical data we obtain indeed are a valid approximation of the 'exact' solutions, which generally we will not even know!***

This requirement is called ***convergence***. If not fulfilled, we could never claim any confidence in the obtained simulation results.

In addition to this fundamental requirement on the relation between the numerical results and the 'reality' we are simulating represented by the exact solutions of our mathematical model, a numerical scheme has to satisfy conditions of ***consistency*** between the mathematical model and its discretization. This means that we are not allowed to perform arbitrary discretization choices without verifying that when the space and time steps $\Delta \vec{x}$, Δt tend to zero, we indeed recover the original mathematical model equations. However, for finite values of the space and time steps, as is the case for any simulation, the difference between the mathematical equations and the discretized equations represents the ***truncation error*** of the scheme. The study of this truncation error is an important source of information on the expected accuracy of a scheme, and will be translated into the important concept of the ***equivalent (or modified) differential equation*** of a numerical scheme.

The third condition, ***stability***, has to ensure that the numerical scheme does not allow errors to grow indefinitely. We have to realize indeed that a computer is a machine with 'finite arithmetic', i.e. operates with a finite number of digits (e.g. 7 digits on a 32-bit machine in single precision, or 14 in double precision). This means very practically that a simple fraction such as 1/3, will never be represented exactly but will be affected by ***round-off errors***, and these errors may not grow during the progress of the computation. This condition is the stability condition of the scheme. As will be seen, the analysis of stability, will, as a by-product, lead to a methodology to quantify the nature and the level of the discretization errors.

A most fundamental property, which makes our lives much easier with regard to the basic requirement of convergence, is provided by ***Peter Lax's equivalence theorem. This fundamental theorem states, in short, that a scheme that is proven to be consistent and stable will automatically satisfy the convergence requirement***. In other words, once we ensure consistency and stability of an algorithm, we have the assurance we are looking for, namely we are certain that the numbers we obtain out of the computer are indeed a valid representation of the 'reality' described by the selected mathematical model.

In practical computations, additional conditions may have to be required on the behavior of the numerical solution. For instance, many physical quantities, such as density, concentrations, turbulent kinetic energy, ..., may never become physically negative, but there is nothing in the above-mentioned properties of consistency and stability, that ensures that the numerical values of these quantities will always satisfy the related physical requirements, and this even for a stable and convergent solution! Therefore, additional conditions may have to be imposed on a numerical scheme, called ***monotonicity conditions***, whose role is to ensure that these properties are also satisfied at the discrete level. This will require an in-depth analysis, as it will be shown that linear schemes of order of accuracy higher than one, that is second order or higher, always will generate numerical solutions that *do not* satisfy the conditions for a monotone behavior. This is known as the Godunov theorem. The cure to this situation will require the introduction of new concepts, guided by nonlinear properties of ***limiters***, to get around the 'curse' of this Godunov theorem.

This part is a most fundamental step in the development and the understanding of an algorithm, and its associated errors.

It forms a crucial step for the developer of a CFD method, but also for the user of a CFD code. When you apply an existing code, commercial or not, it is essential that you develop an understanding of the possible errors generated by the applied algorithm, in order to allow you to evaluate and to assess the accuracy of the obtained numerical results.

This part is organized in two chapters:

1. Chapter 7 introduces the basic definitions and the most widely used methods for the analysis of consistency, stability and accuracy of a numerical scheme. Various methods for the analysis of stability are available, the most popular and useful one being the Von Neumann method based on a Fourier analysis in space of the errors of the numerical solution. Next to the assessment of stability, this approach allows also a profound investigation of the accuracy and the error structure of a numerical scheme.
2. Chapter 8 will organize the various conditions to be imposed on a scheme. From the analysis of Chapter 7, we will derive conditions for the establishment of families of schemes having a pre-determined support and order of accuracy. In addition, a particular attention will be given to the conditions and methods that can be developed and applied to achieve the very important property of monotonic schemes. This will lead us to the definition of *high-resolution schemes*, combining monotonicity and higher order of accuracy through the introduction of the nonlinear *limiters*. The presented methodology will allow you to either select a new scheme with pre-selected properties, or to evaluate an existing scheme.

Chapter 7

Consistency, Stability and Error Analysis of Numerical Schemes

OBJECTIVES AND GUIDELINES

If you go back to the first examples of numerical schemes in Section 4.1, which we hope you have programmed for the suggested test cases, you will have noticed the complexity of the world of numerical discretization, as some schemes happen to be completely unstable and hence useless, while others are acceptable under certain conditions. You have also noticed that the obtained numerical solutions can contain large errors, which are not always acceptable neither. This diversity raises some basic questions with regard to the properties of discretized equations:

1. What conditions do we have to impose on a numerical scheme to obtain an acceptable approximation to the differential problem?
2. Why do the various schemes have widely different behaviors and how can we predict their stability limits?
3. For a stable calculation, such as shown on Figure 4.1.5, how can we obtain quantitative information on the *accuracy* of the numerical simulation?

To provide answers to these questions, it is necessary to define, more precisely, the requirements to be imposed on a numerical scheme.

These requirements are defined as *consistency, stability and convergence*.

They cover different aspects of the relations between the analytical and the discretized equations, between the numerical solution and the exact, analytical solution of the differential equations representing the mathematical model.

Since all fluid flow equations can be classified as elliptic, parabolic or hyperbolic, typical examples of each of them will cover the whole range of possible systems. To analyze the properties of numerical schemes and sustain the methodology to be developed, we will rely on the simplified mathematical flow models, already introduced in Section 3.1.

Section 7.1 will introduce rigorous definitions of the key concepts of consistency, stability and convergence and focus more specifically on the analysis of the consistency conditions. Although the verification of this condition is very straightforward, it leads to the definition of two very essential concepts, namely the *truncation error* of a scheme and the associated *equivalent differential equation*. The latter is of great importance as it contains all the information of the numerical errors generated by a given scheme, even if it is not always possible to explicitly derive them.

Section 7.2 is focusing on the most widely applied method for the analysis of the stability of a numerical scheme, namely the Von Neumann method, based on

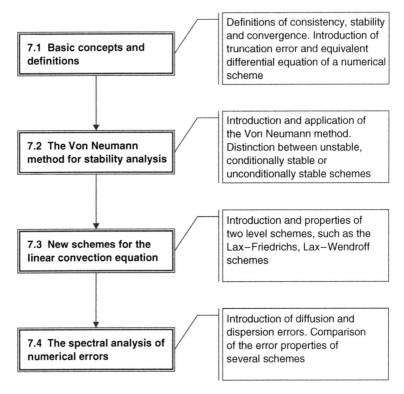

Figure 7.0.1 *Roadmap to this chapter.*

a Fourier decomposition of the errors or of the numerical solution itself. It will provide an easy way for distinguishing between **unstable, conditionally stable or unconditionally stable** schemes. The Von Neumann method will be applied to various schemes, including cases where three time levels appear simultaneously, on several model equations from pure convection to diffusion.

The analysis of some of the most straightforward discretizations already introduced in Chapter 4, will lead us to generate new schemes, with advanced properties, in Section 7.3. It will provide us with an extended range of possibilities, with the very important family of the Lax–Wendroff schemes, which opened the way to the modern approach to CFD. The stability analysis methods introduced in Section 7.2 will be applied to these new schemes.

Section 7.4 will introduce a most important component of numerical analysis, namely the **quantitative** evaluation and prediction of the numerical errors attached to the computed solutions, through their decomposition into **dispersion** and **diffusion** errors. This will permit an in-depth understanding and prediction of the behavior of the numerical solutions for a variety of test cases, such as propagating discontinuities or propagating waves. It will also lead to guidelines and recommendations for ensuring a given accuracy with a selected scheme.

Figure 7.0.1 summarizes the roadmap to this chapter.

7.1   BASIC CONCEPTS AND DEFINITIONS

As stated in the introduction, three main criteria have to be defined and satisfied, representing different aspects of the interrelations between the analytical mathematical model, the numerical scheme and their solutions. These three criteria are consistency, stability and convergence:

1. **Consistency** is a condition on the *numerical scheme*, namely that *the numerical scheme must tend to the differential equation, when time and space steps tend to zero.*
2. **Stability** is a condition on the *numerical solution*, namely that *all errors, such as round-off errors (due to the finite arithmetic of the computer) must remain bounded when the iteration process advances. That is, for finite values of Δt and Δx, the error (defined as the difference between the numerical solution and the exact solution of the numerical scheme) has to remain bounded, when the number of time steps n tends to infinity.*

If we consider the error $\bar{\varepsilon}_i^n$ as the **difference between the computed solution u_i^n and the exact solution of the discretized equation $\bar{u}_i^n$** :

$$\bar{\varepsilon}_i^n = u_i^n - \bar{u}_i^n \tag{7.1.1}$$

the stability condition can be formulated by the requirement that any error $\bar{\varepsilon}_i^n$ between u and $\bar{u}$ should remain uniformly bounded for $n \to \infty$ at fixed Δt.

Hence, the stability condition can be written as.

$$\lim_{n \to \infty} |\bar{\varepsilon}_i^n| \leq K \quad \text{at fixed } \Delta t \tag{7.1.2}$$

with K independent of n.

This stability condition is a requirement solely on the numerical scheme and does not involve any condition on the differential equation.

Actually, the stability condition (7.1.2) has to be valid for any kind of error.

The rigorous definition of stability involves a number of subtleties and more advanced mathematical concepts, which are out of the scope at this introductory stage. One point worth mentioning however, in relation with the above condition, is that it does not ensure that the error will not become unacceptably large at fixed intermediate times $t^n = n\Delta t$, while still remaining bounded in the general sense as not tending to infinity. Practical examples showing that the condition (7.1.2) is not always sufficient to control all error sources, will be discussed in Section 8.2.2, in relation with the convection–diffusion equation.

A more general definition of stability, introduced by Lax and Richtmyer (1956) and developed in Richtmyer and Morton (1967), is based on the time behavior of the solution itself instead of the error's behavior. It will be applied in the next section, in relation with the Von Neumann method for stability analysis, and also in Chapter 9, Section 9.2 when dealing with the general formulation of time integration methods. This stability criterion states that **any component of the initial solution should not be amplified without bound, at fixed values of $t^n = n\Delta t$, in particular for $n \to \infty$, Δt with $n\Delta t$ fixed.**

3. *Convergence* is a condition on the numerical solution: *we have to be sure that the output of the simulation is a correct representation of the model we solve, i.e. the numerical solution must tend to the exact solution of the mathematical model, when time and space steps tend to zero.*

The mathematical formulation of the convergence condition states that the numerical solution u_i^n should approach the *exact solution* $\tilde{u}(x,t)$ of the differential equation, at any point $x_i = i \cdot \Delta x$ and time $t^n = n \cdot \Delta t$ when Δx and Δt tend to zero, i.e. when the mesh is refined, x_i and t^n being fixed. This condition implies that i and n tend to infinity while Δx and Δt tend to zero, such that the products $i\Delta x$ and $n\Delta t$ remain constant.

Here we define the error $\tilde{\varepsilon}_i^n$ as the **difference between the computed solution and the exact solution of the analytical equation representing the selected mathematical model**:

$$\tilde{\varepsilon}_i^n = u_i^n - \tilde{u}(i\Delta x, n\Delta t) \tag{7.1.3}$$

This error has to satisfy the following convergence condition

$$\lim_{\substack{\Delta x \to 0 \\ \Delta t \to 0}} \left| \tilde{\varepsilon}_i^n \right| = 0 \quad \text{at fixed values of } x_i = i\Delta x \text{ and } t^n = n\Delta t \tag{7.1.4}$$

Note here that the stability and convergence conditions do not refer to the same errors.

Clearly the conditions of consistency, stability and convergence are related to each other and the precise relation is contained in the fundamental **Equivalence Theorem of Lax**, a proof of which can be found in the now classical book of Richtmyer and Morton (1967).

Equivalence Theorem of Lax: For a well-posed initial value problem and a consistent discretization scheme, stability is the necessary and sufficient condition for convergence.

This fundamental theorem shows that in order to analyze a time dependent or initial value problem, two tasks have to be performed:

1. Analyze the consistency condition; this leads to the determination of the order of accuracy of the scheme and its truncation error.
2. Analyze the stability properties.

From these two steps, convergence can be established without additional analysis.

This is a crucial property, as it ensures that it suffices to test for the stability of a consistent scheme, to ensure that the numerical solution will provide a valid representation of the 'reality' we wish to simulate numerically.

These interrelations are summarized in Figure 7.1.1 which expresses, in short, that the consistency condition defines a relation between the differential equation and its discrete formulation; that the stability condition establishes a relation between the computed solution and the exact solution of the discretized equations; while the convergence condition connects the computed solution to the exact solution of the differential equation.

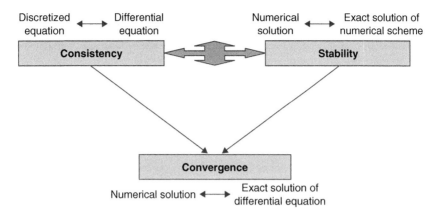

Figure 7.1.1   *Relations between consistency, stability and convergence.*

We will introduce and describe the associated methodology on the simplified time dependent one-dimensional models introduced in Chapter 3, taking as first reference some of the schemes introduced in Section 4.1.

7.1.1   Consistency Condition, Truncation Error and Equivalent Differential Equation of a Numerical Scheme

Consistency expresses that the discretized equations should tend to the differential equations to which they are related, when Δt and Δx tend to zero. This rather simple statement will nevertheless lead us to the introduction of two essential properties of numerical discretizations, namely the ***truncation error*** of a numerical scheme and the associated ***equivalent differential equation (EDE)***. They express a new vision of the relations between the numerical and the exact solutions, the numerical scheme and the differential equation, through the truncation error. The equivalent differential equation will then appear as the strongest expression of these links.

One of the major conclusions of the analysis to follow is that the numerical solution, because of the discretization errors, does *not* satisfy the mathematical model equations, but is instead a solution of the equivalent differential equation. Hence, any information on the properties of the solution of this EDE, will provide us with an information on the behavior and associated errors of the numerical solution.

In order to guide you through these steps, we apply the following methodology:

7.1.1.1   *Methodology*

In order to check for consistency, we consider the discretized equation for the unknown u_i^n. This equation contains values at other points $(i+j)$ and other time levels $(n+k)$:

- The function values u_{i+j}^{n+k} in the numerical scheme are developed in a Taylor series around the value u_i^n and the high order terms are maintained in substituting these developments back in the numerical equation.

- An equation is hereby obtained expressing the numerical scheme as the mathematical model equation plus additional terms, resulting from the Taylor series. These additional terms are called the **truncation error**, and noted ε_T.
- The truncation error will have the form:

$$\varepsilon_T = O(\Delta t^q, \Delta x^p) \tag{7.1.5}$$

where p and q are the lowest values occurring in the development of the truncation error. This defines the **order of accuracy of the scheme**. Equation (7.1.5) tells us indeed that the considered scheme is of order q in time and p in space.

The condition for consistency can be stated as follows: *A scheme is consistent if the truncation error tends to zero for Δt, Δx tending to zero.* This can also be stated as the requirement that the orders of accuracy in time and in space, should be positive for any combinations of Δt and Δx when they both tend to zero.

As will be seen next, we can select u_i^n as representing either the exact solution of the mathematical model, or as the exact solution of the numerical scheme. This will lead to different, but complementary interpretations.

To illustrate the methodology and the steps leading to the **equivalent differential equation** of the numerical scheme, we consider first the linear convection model

$$u_t + au_x = 0 \tag{7.1.6}$$

and we select, as a first example, the second order central discretization in space and a first order explicit forward difference in time. This is the scheme (4.1.28), repeated here for convenience:

$$\frac{u_i^{n+1} - u_i^n}{\Delta t} + \frac{a}{2\Delta x}(u_{i+1}^n - u_{i-1}^n) = 0 \tag{7.1.7}$$

Let us apply the above methodology to this first example.
Step 1: We write the following Taylor expansions

$$u_i^{n+1} = u_i^n + \Delta t(u_t)_i^n + \frac{\Delta t^2}{2}(u_{tt})_i^n + \frac{\Delta t^3}{6}(u_{ttt})_i^n + \cdots \tag{7.1.8}$$

$$u_{i+1}^n = u_i^n + \Delta x(u_x)_i^n + \frac{\Delta x^2}{2}(u_{xx})_i^n + \frac{\Delta x^3}{6}(u_{xxx})_i^n + \cdots \tag{7.1.9}$$

$$u_{i-1}^n = u_i^n - \Delta x(u_x)_i^n + \frac{\Delta x^2}{2}(u_{xx})_i^n - \frac{\Delta x^3}{6}(u_{xxx})_i^n + \cdots \tag{7.1.10}$$

where the x and t subscripts indicate partial derivatives.
Step 2: We substitute these developments in equation (7.1.7), obtaining

$$\frac{u_i^{n+1} - u_i^n}{\Delta t} + a\frac{u_{i+1}^n - u_{i-1}^n}{2\Delta x} - (u_t + au_x)_i^n = \frac{\Delta t}{2}(u_{tt})_i^n + \frac{\Delta x^2}{6}a(u_{xxx})_i^n$$

$$+ O(\Delta t^2, \Delta x^4) \tag{7.1.11}$$

The right-hand side of this consistency relation represents the **truncation error** ε_T, equal to

$$\varepsilon_T = \frac{\Delta t}{2}(u_{tt})_i^n + a\frac{\Delta x^2}{6}(u_{xxx})_i^n + O(\Delta t^2, \Delta x^4) \tag{7.1.12}$$

The truncation error (TE) is therefore defined as the difference between the numerical scheme and the differential equation.

It is seen from the above equation that the right-hand side vanishes when Δt and Δx tend to zero and therefore scheme (7.1.7) is consistent. As expected, its accuracy is first order in time and second order in space since the right-hand side goes to zero as the first power of Δt and the second power of Δx.

Note however that if a relation is established between Δt and Δx, when they both tend to zero then the *overall* accuracy of the scheme might be different. If $\Delta t/\Delta x$ is kept constant, then the scheme has a global first order accuracy, while it would be second order, if $\Delta t/\Delta x^2$ would be kept constant.

The consistency equation (7.1.11) can be interpreted in two equivalent ways.

First interpretation of the consistency condition

The Taylor expansion is performed around the exact solution of the differential equation, i.e. around $\tilde{u}(i\Delta x, n\Delta t) \equiv \tilde{u}_i^n$, where $\tilde{u}(x,t)$ is the analytical solution. Equation (7.1.11) then reduces to

$$\frac{\tilde{u}_i^{n+1} - \tilde{u}_i^n}{\Delta t} + a\frac{\tilde{u}_{i+1}^n - \tilde{u}_{i-1}^n}{2\Delta x} = \tilde{\varepsilon}_T \tag{7.1.13}$$

This relation shows that the exact solution $\tilde{u}_i^n$ does *not* satisfy the difference equation exactly, but is solution of a modified scheme, with the truncation error in the right-hand side.

*We can also view equation (7.1.13) as a definition of the truncation error: the truncation error is equal to the **residual** of the discretized equation for values of $\tilde{u}_i^n$ equal to the exact, analytical solution.*

Second interpretation of the consistency condition

The Taylor expansion is performed around the *exact solution of the discretized equation $\bar{u}_i^n$*. In this case equation (7.1.11) reduces to

$$(\bar{u}_t + a\bar{u}_x)_i^n = -\frac{\Delta t}{2}(\bar{u}_{tt})_i^n - a\frac{\Delta x^2}{6}(\bar{u}_{xxx})_i^n - O(\Delta t^2, \Delta x^4) \equiv -\bar{\varepsilon}_T \tag{7.1.14}$$

This relation shows that the exact solution of the discretized equation does not satisfy exactly the differential equation at finite values of Δt and Δx (which is always the case in practical computations).

However, the solution of the numerical scheme satisfies an **equivalent differential equation (EDE)**, also sometimes called **modified differential equation**, which differs from the original (differential) equation by a truncation error represented by the terms on the right-hand side.

Actually, the *equivalent differential equation* is not provided exactly by equation (7.1.14), as the right-hand side contains higher order derivatives in time and in space. In order to gain a better insight in this equation and learn something of its properties, the rule is to eliminate the lowest order time derivatives in the truncation error, up to higher order correction terms, by applying the equivalent differential equation itself to replace them by equivalent space derivatives.

Applying this rule to the above equation (7.1.14), we obtain

$$(\bar{u}_t)_i^n = -a(\bar{u}_x)_i^n + \mathrm{O}(\Delta t, \Delta x^2) \tag{7.1.15}$$

where all the remaining terms are proportional to Δt and Δx^2, to the lowest order. By taking the time derivative of this relation, we have

$$(\bar{u}_{tt})_i^n = -a(\bar{u}_{xt})_i^n + \mathrm{O}(\Delta t, \Delta x^2) = -a((\bar{u}_t)_x)_i^n + \mathrm{O}(\Delta t, \Delta x^2) \tag{7.1.16}$$

To eliminate the time derivative in the u_{xt} term of the right-hand side, we apply once more equation (7.1.15), leading to

$$(\bar{u}_{tt})_i^n = -a((\bar{u}_t)_x)_i^n + \mathrm{O}(\Delta t, \Delta x^2) = +a^2(\bar{u}_{xx})_i^n + \mathrm{O}(\Delta t, \Delta x^2) \tag{7.1.17}$$

Hence, the truncation error can be written as

$$\bar{\varepsilon}_T = a^2 \frac{\Delta t}{2}(\bar{u}_{xx})_i^n + a\frac{\Delta x^2}{6}(\bar{u}_{xxx})_i^n + \mathrm{O}(\Delta t^2, \Delta x^4) \tag{7.1.18}$$

Up to the lowest order, the *equivalent differential equation* (7.1.14) becomes

$$\bar{u}_t + a\bar{u}_x = -\frac{\Delta t}{2}a^2\bar{u}_{xx} + \mathrm{O}(\Delta t^2, \Delta x^2) \tag{7.1.19}$$

*What we see here is that the equation satisfied by the exact numerical solution $\bar{u}$ is not the original convection equation, but is instead a convection–diffusion equation, with a **numerical diffusion** (also called **numerical viscosity**) coefficient equal here to $(-a^2\Delta t/2)$.*

This shows why the corresponding scheme is **unstable**. Indeed, the right-hand side represents a diffusion term with a *negative* viscosity coefficient equal to $(-a^2\Delta t/2)$. A positive viscosity is known to damp oscillations and strong gradients; a negative viscosity on the other hand, will amplify exponentially any disturbance, describing explosion phenomena. Since the exact numerical solution satisfies the above equation, it means that its behavior is unstable.

Hence, the determination of the equivalent differential equation and, in particular the truncation error, provides essential information on the behavior of the numerical solution.

General rules for obtaining the equivalent differential equation

We can now summarize the rules for the derivation of the equivalent differential equation of a numerical scheme.

Denoting by $D(U) = 0$ the mathematical model we wish to solve numerically and by $N(U_i^n) = 0$ the numerical scheme, we proceed as follows:

- Perform the consistency analysis and obtain the truncation error ε_T, by generalizing equation (7.1.11) as

$$N(U_i^n) - D(U_i^n) = \varepsilon_T \tag{7.1.20}$$

- Consider the exact solution of the numerical scheme $\overline{U}_i^n$ defined by

$$N(\overline{U}_i^n) \equiv 0 \tag{7.1.21}$$

leading to the differential equation satisfied by the numerical exact solution

$$D(\overline{U}_i^n) = -\overline{\varepsilon}_T \tag{7.1.22}$$

- Replace the lowest time derivatives in the truncation error by space derivatives, obtained by applying equation (7.1.22) to perform this replacement.
- The *equivalent differential equation* is defined as the equation obtained after that replacement step, restricted to the lowest order terms of the modified truncation error, which contains now only space derivatives.
- If we could solve this equivalent differential equation, we would know the complete behavior and error properties of the numerical solution of our scheme.

Let us illustrate this with a second example, also treated in Section 4.1, namely the first order upwind scheme (FOU) for the linear convection equation, equation (4.1.32), repeated here for convenience:

$$\frac{u_i^{n+1} - u_i^n}{\Delta t} + \frac{a}{\Delta x}(u_i^n - u_{i-1}^n) = 0 \tag{7.1.23}$$

Introducing the Taylor expansions (7.1.8) and (7.1.10) in this scheme, we obtain, following the above steps (see Problem P.7.1), the equivalent differential equation:

$$\frac{\partial \overline{u}}{\partial x} + a\frac{\partial \overline{u}}{\partial x} = \frac{a\Delta x}{2}\left(1 - \frac{a\Delta t}{\Delta x}\right)\frac{\partial^2 \overline{u}}{\partial x^2} \equiv \nu_{num}\frac{\partial^2 \overline{u}}{\partial x^2} \tag{7.1.24}$$

Here again we observe that the numerical solution obeys a convection–diffusion equation, instead of the expected convection equation. The discretization has indeed introduced a **numerical diffusion or numerical viscosity** equal to

$$\nu_{num} \equiv \frac{a\Delta x}{2}\left(1 - \frac{a\Delta t}{\Delta x}\right) \tag{7.1.25}$$

This numerical viscosity has to be positive for the solution of this equation to be damped in time. Otherwise, as in the previous example, where the numerical viscosity is negative, the numerical solution will grow indefinitely with time and the scheme will be unstable. Hence for stability of the first order upwind scheme we should have

$$0 \leq \frac{a\Delta t}{\Delta x} \leq 1 \tag{7.1.26}$$

This stability condition requires $a > 0$ and the **Courant or CFL (Courant Friedrichs–Lewy) number** $\sigma = a\Delta t/\Delta x$ to be lower (or equal) than one. This condition

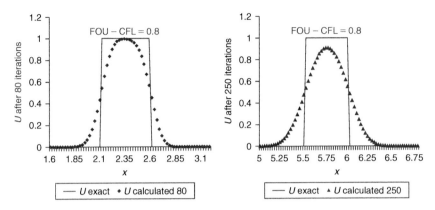

Figure 7.1.2 *Numerical solution of a traveling block obtained with the FOU scheme after 80 and 250 time steps, with CFL = 0.8.*

has a very deep physical significance, which we will analyze in the next section. Note that the scheme is unstable for negative velocities $a < 0$.

It is seen that for constant Courant number, the first order upwind scheme (FOU) has a numerical viscosity of the order of $O(\Delta x)$, which is generally excessive, as it is only first order in time and space. Therefore the FOU scheme has a poor accuracy, being too diffusive. This can also be seen on the example of Figure 7.1.2 showing the numerical solution of a traveling block after 80 and 250 time steps, for a CFL number $\sigma = 0.8$. The progressive diffusion is clearly seen, because of the numerical viscosity of this scheme, and this diffusion will of course continue when the number of time step iterations increases.

HANDS-ON TASK 1

We suggest that you work this example further out by applying the program you developed in Section 4.1, running the FOU scheme for 500 and 1000 time steps.

More general developments and properties of the EDE are presented in Chapter 8, Section 8.1.

7.2 THE VON NEUMANN METHOD FOR STABILITY ANALYSIS

Once consistency has been verified, the next step is to establish the stability behavior of the numerical scheme. With the Lax equivalence theorem, we will then have the assurance that the whole simulation will satisfy convergence, in the stability domain of the scheme.

Many methods have been developed for the analysis of stability, nearly all of them restricted to linear problems. But even with this restriction, the investigation of stability for initial, boundary value problems can be extremely complicated, particularly in the presence of boundary conditions and their numerical representation.

To separate the influence of the boundary conditions from the main stability analysis, we can consider a slightly different problem with periodic boundary conditions.

This is the basis of the Von Neumann stability analysis that has emerged as the most widely applied and the most relevant of the stability methods. One could envisage a second step consisting in analyzing the stability effects of the real boundary conditions; although it can be sometimes be done, it turns out that it is generally not necessary. Some guidelines towards the analysis of the influence of boundary conditions on stability are given in Chapter 9, section 9.1.

The method developed by John Von Neumann has an interesting history. It started in Los Alamos during World War II as he was part of the team of high-level scientists developing the first nuclear device. Consequently, the method was classified as 'secret', until its brief description in Crank and Nicholson (1947) and in a publication by Charney et al. (1950).

Biographical note

*John Von Neumann was born in Hungary in 1903 and studied mathematics in Budapest and then at the ETH in Zurich (where Albert Einstein also had graduated in 1900). He moved to Germany where he became Professor at the University of Berlin, orienting his work progressively toward applied mathematics. After the Nazi regime came to power in 1933, J. Von Neumann was forced to leave Germany, as many other Jewish scientists and joined Princeton University (where Albert Einstein had also found a proper scientific environment after his forced departure from Nazi Germany). When called upon to join the Los Alamos team, he was charged with evaluation of the aerodynamic effects of an explosion and the estimation of the generated shock waves. He then developed what can be considered as the first modern approach to numerical solutions of the Euler equations. In addition to the now famous stability analysis method that bears his name, J. Von Neumann also introduced the concept of **artificial viscosity** to **capture** shock discontinuities.*

In addition J. Von Neumann can also be considered as the inventor of the modern computer. During 1944–1945 he conceives the structure of an electronic computing engine that would be 'programmed' to execute instructions and store them in an electronic 'memory'. The first computer, based on these principles, called EDSAC (Electronic Data Storage Automatic Calculator), was built at the University of Cambridge (UK) in 1949. J. Von Neumann died in 1957 in the USA.

Today, J. Von Neumann is rightfully considered as the 'father' of modern CFD.

See the biography of S. Ulam (1958) and the book by W. Aspray (1990) for more details.

The key innovation of the Von Neumann stability analysis is the expansion of the error, or the numerical solution, in a finite Fourier series in the spatial frequency domain. To achieve this property, the computational domain on the x-axis of length L, is repeated periodically for instance by mapping it to the domain $(-L,0)$ and therefore all quantities, the solution, as well as the errors, can be developed in a finite Fourier series over the domain $(-L, +L)$.

As will be seen in Section 7.4, this Fourier expansion will lead furthermore to a quantitative estimation of the errors in function of the frequency content of the initial data and of the solution.

7.2.1   Fourier Decomposition of the Solution

If $\bar{u}_i^n$ is the exact solution of the difference equation and u_i^n the actual computed solution, the difference might be due to round-off errors and to errors in the initial data.

Hence,

$$u_i^n = \bar{u}_i^n + \bar{\varepsilon}_i^n \tag{7.2.1}$$

where $\bar{\varepsilon}_i^n$ indicates the error at time level n in mesh point i. By definition, the exact solution $\bar{u}_i^n$ satisfies exactly the numerical scheme equation and therefore, the errors $\bar{\varepsilon}_i^n$ are also solutions of the same discretized equations.

This can easily be seen; if $N(u_i^n) = 0$ represents the *linear* numerical scheme, the exact solution of the numerical scheme $\bar{u}_i^n$ satisfies $N(\bar{u}_i^n) \equiv 0$ and introducing equation (7.2.1), we obtain

$$N(u_i^n) = N(\bar{u}_i^n + \bar{\varepsilon}_i^n) = N(\bar{u}_i^n) + N(\bar{\varepsilon}_i^n) = N(\bar{\varepsilon}_i^n) = 0 \tag{7.2.2}$$

Hence, the errors $\bar{\varepsilon}_i^n$ do satisfy the same equation as the numerical solution u_i^n.

Therefore, assuming that the exact numerical solution is uniformly bounded, any unbounded behavior of the error will reflect itself upon the numerical solution.

Hence, we can indifferently analyze the stability by studying the behavior of the errors, or the numerical solution itself.

We will select here the second option.

If the boundary conditions are considered as periodic, we can expand the solution u_i^n into a Fourier series in space, at each time level n. Since the space domain is of finite length, we will need a *discrete* Fourier representation, with a finite sum over all the wave numbers that can be represented on the discretized space, that is on the set of mesh points.

As the wave number represents the number of wavelengths within a range of 2π, we have to ask ourselves what are the wavelengths of a discrete function that can be 'seen' or represented, on the existing mesh?

In a one-dimensional domain of length L, we reflect the region $(0,L)$ onto the negative part $(-L,0)$ as a way to create a periodicity $2L$.

What is the shortest wavelength we can represent on the uniform mesh with spacing Δx?

Referring to Figure 7.2.1, we see that the shortest resolvable wavelength is equal to $\lambda_{min} = 2\Delta x$. Indeed, if you imagine a shorter wavelength between two mesh points, say $\lambda = \Delta x$, it will require a point inside this interval located midway between the mesh points, and hence it cannot be recognized on the finite grid, as there is no such a mesh point.

The associated wave number $k = 2\pi/\lambda$ reaches its maximum wave number $k_{max} = \pi/\Delta x$. The largest wavelength on the other hand corresponds to $\lambda_{max} = 2L$, covering the whole space domain. The associated wave number attains its minimum value $k_{min} = \pi/L$. All the other harmonics have to be multiples of this smallest value and we have as many harmonics as mesh points.

With the mesh index i, ranging from 0 to N, with $x_i = i \cdot \Delta x$ and

$$\Delta x = L/N \tag{7.2.3}$$

all the N harmonics represented on a finite mesh are given by

$$k_j = jk_{min} = j\frac{\pi}{L} = j\frac{\pi}{N\Delta x} \quad j = 0, \ldots, N \tag{7.2.4}$$

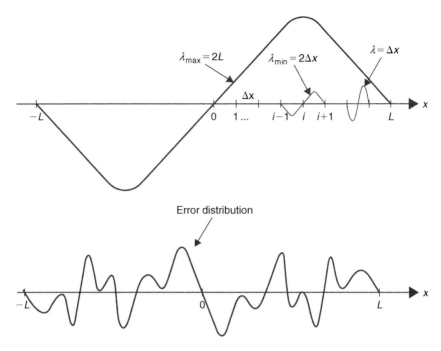

Figure 7.2.1 *Frequency range of finite spectrum on interval (−L,L).*

with j ranging from 1 to N. We add the value $j = 0$ to take into account a possible constant value of the solution.

The product $k_j \Delta x$ is represented as a phase angle

$$\phi \equiv k_j \Delta x = \frac{j\pi}{N} \qquad (7.2.5)$$

and covers the domain $(-\pi, \pi)$, in steps of π/N, when including the reflected region $(-L, 0)$.

Figure 7.2.1 shows a representative behavior of the error, which contains the whole spectrum of frequencies. Depending on the particular scheme, the frequency content might vary, ranging form domination of low frequencies up to high frequencies. For instance, if the error varies strongly between two mesh points, it will be seen as dominated by short wavelengths of the size $2\Delta x$ and hence having a dominating high-frequency content.

The region around $\phi = 0$ corresponds to the low frequencies while the region close to $\phi = \pi$ is associated with the high-frequency range of the spectrum. In particular the value $\phi = \pi$ corresponds to the highest-frequency resolvable on the mesh, namely the frequency associated to the shortest wavelength $2\Delta x$.

Any function on the finite mesh, such as the full solution u_i^n, will be decomposed in a finite Fourier series as

$$u_i^n = \sum_{j=-N}^{N} V_j^n e^{Ik_j \cdot x_i} = \sum_{j=-N}^{N} V_j^n e^{Ik_j \cdot i\Delta x} = \sum_{j=-N}^{N} V_j^n e^{Iij\pi/N} \qquad (7.2.6)$$

where $I = \sqrt{-1}$ and V_j^n is the amplitude of the jth harmonic. The harmonic associated to $j = 0$, represents a constant function in space.

Note that this decomposition separates the time and space dependence of the solution. The time behavior is represented by the amplitudes V^n, which contain the full time dependence, while the Fourier modes contain the full space dependence.

The Von Neumann method relies on the observation that for linear schemes, the discretized equation (7.2.2), which is satisfied by the solution and by the error, must also be satisfied by each individual harmonic.

Hence, introducing an arbitrary harmonic in the numerical scheme, the stability condition of Von Neumann will be expressed by the following condition: *the amplitude of any harmonic may not grow indefinitely in time, i.e. when n tends to infinity.*

7.2.2 Amplification Factor

As the Von Neumann stability condition requires that amplitudes V^n do not grow indefinitely, for any value j, we define an *amplification factor*

$$G \triangleq \frac{V^{n+1}}{V^n} \tag{7.2.7}$$

that has to satisfy the following *Von Neumann stability condition*:

$$|G| \leq 1 \quad \text{for all values of } \phi_j = k_j \Delta x = j\frac{\pi}{N} \quad j = -N, \ldots, +N \tag{7.2.8}$$

Note that this amplification factor is function of the scheme parameters and of the phase angle ϕ and is independent of n.

The methodology for the application of the Von Neumann stability condition can be summarized as follows:

7.2.2.1 *Methodology*

- Replace in the numerical scheme all the terms of the form u_{i+m}^{n+k} by

$$u_{i+m}^{n+k} \Rightarrow V^{n+k} e^{I(i+m)\varphi} \tag{7.2.9}$$

For a two time level scheme, involving the time steps n and $(n + 1)$, k will be limited by $k = 1$. For a three time level scheme, k will take the values $k = -1$, $k = 0$ and $k = 1$.
- As all the terms contain the factor $e^{Ii\varphi}$, the next step is to simplify all the terms by this factor.
- From the obtained relation, derive the explicit form of the amplification factor G.
- The stability conditions, which are conditions on the scheme parameters, are to be obtained from the condition (7.2.8). As we have removed the index j from the phase angle, it can be considered from now on as taking any value between

$(-\pi, +\pi)$. Hence we reformulate the Von Neumann stability condition as follows:

$$|G| \leq 1 \quad \text{for all values of } \phi \text{ in the range } (-\pi, +\pi) \qquad (7.2.10)$$

Let us now apply this methodology to some of the schemes derived in Section 4.1 for the linear convection equation $u_t + au_x = 0$.

First example: Second order central discretization in space with explicit first order difference in time

This scheme is given by equation (7.1.7), which we write here in an explicit form:

$$u_i^{n+1} = u_i^n - \frac{\sigma}{2}(u_{i+1}^n - u_{i-1}^n) \qquad (7.2.11)$$

where the parameter σ is the *unique parameter of the scheme*, and has already been defined as the Courant or CFL number:

$$\sigma = \frac{a\Delta t}{\Delta x} \qquad (7.2.12)$$

Observe here that the numerical solution will not depend on Δx or on Δt separately, but only on the non-dimensional CFL number. Therefore in practice, all computations will be performed at constant values of σ, that is for Δt values proportional to the mesh spacing Δx.

Let us now apply the general methodology step by step:

The first step consists in applying the replacement (7.2.9), leading to

$$V^{n+1}e^{Ii\phi} = V^n e^{Ii\phi} - \frac{\sigma}{2}[V^n e^{I(i+1)\phi} - V^n e^{I(i-1)\phi}] \qquad (7.2.13)$$

The second step is the simplification by $e^{Ii\varphi}$, leading to

$$V^{n+1} = V^n - V^n \frac{\sigma}{2}[e^{I\phi} - e^{-I\phi}] \qquad (7.2.14)$$

The third step, the derivation of the amplification factor, is here straightforward,

$$G \equiv \frac{V^{n+1}}{V^n} = 1 - \frac{\sigma}{2} \cdot 2I \sin\phi = 1 - I\sigma \sin\phi \qquad (7.2.15)$$

The stability condition (7.2.10) requires the modulus of G to be lower or equal to one. For the present example

$$|G|^2 = G \cdot G^* = 1 + \sigma^2 \sin^2\phi \geq 1 \qquad (7.2.16)$$

which is always equal or larger than one and the stability condition is clearly never satisfied. Hence, the centered scheme (7.2.11) for the convection equation with forward difference in time is ***unconditionally unstable***.

Second example: Second order central discretization in space with implicit first order difference in time

This scheme is defined by the following equation:

$$u_i^{n+1} = u_i^n - \frac{\sigma}{2}(u_{i+1}^{n+1} - u_{i-1}^{n+1}) \tag{7.2.17}$$

Combining now the first and the second steps, we obtain

$$V^{n+1} = V^n - V^{n+1}\frac{\sigma}{2}[e^{I\phi} - e^{-I\phi}] \tag{7.2.18}$$

The third step leads us to the following expression for the amplification factor:

$$G = 1 - \frac{\sigma}{2}G(e^{I\phi} - e^{-I\phi}) \quad \text{or} \quad G = \frac{1}{1 + I\sigma \sin \phi} \tag{7.2.19}$$

The modulus of G is always lower than one, for all values of σ, since

$$|G|^2 = G \cdot G^* = \frac{1}{1 + \sigma^2 \sin^2 \phi} \leq 1 \quad \text{for all } \phi \tag{7.2.20}$$

and therefore the implicit scheme (7.2.17) is **unconditionally stable**.

Third example: FOU scheme – first order backward discretization in space with explicit first order difference in time

This is the first order upwind scheme (FOU) already discussed in Section 7.1.1, given by equation (7.1.23), which we rewrite here as

$$u_i^{n+1} = u_i^n - \sigma(u_i^n - u_{i-1}^n) \tag{7.2.21}$$

Combining here the first, second and third steps, we obtain

$$G = 1 - \sigma(1 - e^{-I\phi}) = 1 - \sigma + \sigma \cos \phi - I\sigma \sin \phi$$
$$= 1 - 2\sigma \sin^2 \phi/2 - I\sigma \sin \phi \tag{7.2.22}$$

In order to analyze the stability of the scheme (7.2.8), i.e. the regions where the modulus of the amplification factor G is lower than one, a representation of G in the complex plane is a convenient approach. Writing ξ and η, respectively for the real and imaginary parts of G, we have

$$\xi \equiv Re\,G = 1 - 2\sigma \sin^2 \phi/2 = (1 - \sigma) + \sigma \cos \phi$$
$$\eta \equiv Im\,G = -\sigma \sin \phi \tag{7.2.23}$$

which can be considered as parametric equations for G with ϕ as parameter. We recognize the parametric equations of a circle centered on the real axis ξ at $(1 - \sigma)$ with radius σ. In the complex plane of G, the stability condition (7.2.10) states

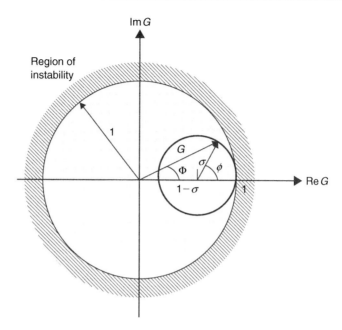

Figure 7.2.2 *Complex G-plane representation of upwind scheme (7.2.8), with unit circle defining the stability region.*

that the curve representing G for all values of $\phi = k\Delta x$, should remain within the unit circle (see Figure 7.2.2). It is clearly seen from this figure that the scheme is stable for

$$0 \leq \sigma \leq 1 \tag{7.2.24}$$

Hence, the FOU scheme (7.2.21) is **conditionally stable** and we recover the Courant, Friedrichs, Lewy condition, or in short the 'CFL-condition' already obtained from the analysis of the equivalent differential equation.

This condition for stability was introduced for the first time in 1928 in a article by Courant et al. (1928). This article can be considered as laying the foundations of the concepts of convergence and stability for finite difference schemes, although the authors were using finite difference concepts as a mathematical tool for proving existence theorems of continuous problems.

Observe that the upwind scheme (7.2.21) is unstable for $a < 0$ (see also Problem P.7.4).

Fourth example: Implicit FOU scheme – first order backward discretization in space with implicit first order difference in time

This scheme is defined by the following equation:

$$u_i^{n+1} = u_i^n - \sigma(u_i^{n+1} - u_{i-1}^{n+1}) \tag{7.2.25}$$

Applying the now familiar steps, we obtain

$$G = \frac{1}{1 + \sigma(1 - e^{-I\phi})} \tag{7.2.26}$$

and the stability condition leads to

$$G \cdot G^* = \frac{1}{(1 - \sigma + \sigma \cos \phi)^2 + \sigma^2 \sin^2 \phi} \leq 1 \quad \text{for all values of } \phi \tag{7.2.27}$$

Hence, this implicit FOU scheme is ***unconditionally stable***.

7.2.3 Comments on the CFL Condition

This fundamental stability condition of most explicit schemes for wave and convection equations expresses that the distance at covered during the time interval Δt, by the disturbances propagating with speed a, should be lower than the minimum distance between two mesh points. Referring to Figure 7.2.3 and to Chapter 3, we recognize the line PQ as the characteristic $dx/dt = a$ of the hyperbolic linear convection equation through P. By adding the propagation case for negative values of a, we define the domain of dependence of the differential equation in P, for convection velocities of either signs by the domain PQCQP. On the other hand, we can consider that the numerical scheme also defines a *numerical domain of dependence* of P which is the domain between PAC, since the solution at time level $(n + 1)$ depends on the points i and $(i - 1)$.

The CFL stability condition $\sigma < 1$ expresses that the mesh ratio $\Delta t / \Delta x$ has to be chosen in such a way that *the domain of dependence of the differential equation should be entirely contained in the numerical domain of dependence of the discretized equations*.

In other words, the numerical scheme defining the approximation u_i^{n+1} in mesh point i must be able to include all the physical information which influences the behavior of the system in this point. If this is not the case, as seen on Figure 7.2.3b then a change in physical conditions in the regions AQ and BQ would not be seen by the numerical scheme and therefore the difference between the exact solution and the numerical solution could be made arbitrary large. *Hence, the scheme will not be able to converge. This interpretation is generally applied for two- and three-dimensional problems when it appears difficult to express analytically the stability conditions from the derived amplification factor or matrix. It will then lead at least to a necessary condition.*

We also understand hereby why the first order upwind scheme (FOU) (7.2.21) is unstable when a is negative. Indeed, in this case the variable u is convected from right to left and the solution in point i can only depend on the upstream points such as $(i + 1)$, and the physical domain of dependence is on the right side of point P. Hence, the scheme (7.2.21) which involves only the point $(i - 1)$ is outside the domain of dependence. In other words, this scheme is contrary to the physics it aims to describe, when $a < 0$.

In this case the space derivative has to be discretized with a forward difference.

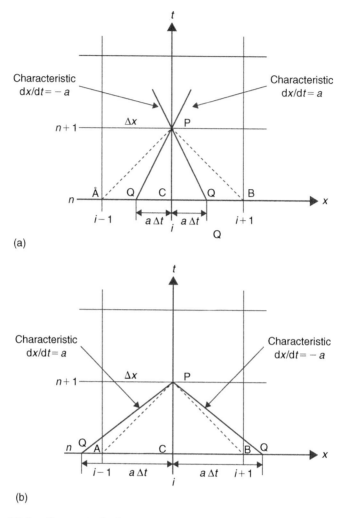

Figure 7.2.3 *Geometrical, characteristic interpretation of the CFL condition (a) σ < 1 (stable condition) and (b) σ > 1 (unstable condition).*

Fifth example: Time-dependent diffusion equation with explicit first order difference in time and central discretization in space

Let us consider as next example the diffusion equation

$$\frac{\partial u}{\partial t} = \alpha \frac{\partial^2 u}{\partial x^2} \quad \text{or} \quad u_t = \alpha u_{xx} \tag{7.2.28}$$

We select an explicit first order forward difference in time and a central, second order discretization of the second space derivative, leading to

$$u_i^{n+1} = u_i^n + \frac{\alpha \Delta t}{\Delta x^2}(u_{i+1}^n - 2u_i^n + u_{i-1}^n) \tag{7.2.29}$$

The amplification factor is obtained, following the steps outlined in the methodology, as

$$G = 1 - 4\beta \sin^2 \phi/2 \qquad (7.2.30)$$

with

$$\beta = \frac{\alpha \Delta t}{\Delta x^2} \qquad (7.2.31)$$

The stability condition is

$$|1 - 4\beta \sin^2 \phi/2| \leq 1$$

which is satisfied for

$$-1 \leq 1 - 4\beta \sin^2 \phi/2 \leq 1$$

that is

$$0 \leq \beta \leq \frac{1}{2} \qquad (7.2.32)$$

Hence, the above scheme is stable for

$$\alpha \geq 0 \quad \text{and} \quad \beta = \frac{\alpha \Delta t}{\Delta x^2} \leq \frac{1}{2} \qquad (7.2.33)$$

The first condition expresses the stability of the physical problem, since for $\alpha < 0$, i.e. for a negative diffusion coefficient, the analytical solution is exponentially increasing with time. The second condition provides the *conditional stability* of this explicit scheme.

Sixth example: Time-dependent diffusion equation with implicit first order difference in time and central discretization in space

If we select an implicit scheme, i.e., taking the space difference at time level $(n + 1)$ instead of n, as in equation (7.2.29), we obtain the scheme

$$u_i^{n+1} = u_i^n + \frac{\alpha \Delta t}{\Delta x^2}(u_{i+1}^{n+1} - 2u_i^{n+1} + u_{i-1}^{n+1}) \qquad (7.2.34)$$

leading to the amplification factor:

$$G = \frac{1}{1 + 4\beta \sin^2 \phi/2} \qquad (7.2.35)$$

Hence, this implicit scheme is *unconditionally stable* for $\alpha > 0$.

General comments

In summary, it is seen that schemes can have *conditional stability, unconditional stability or unconditional instability.*

You will have noticed that explicit schemes lead at best to conditional stability while implicit schemes are generally unconditionally stable.

Conditional stability puts a limit on the time step, expressing that one cannot progress too rapidly in time in order to maintain stability. This is generally a severe requirement, in particular for convection dominated equations, as the convection speed a can be very large. For compressible flows it is of the order of the speed of sound and the allowable time step becomes very small. For instance with 101 mesh points, $\Delta x = 0.01$ and for $a = 350$ m/s we obtain a time step limit of $\Delta t < 3 \times 10^{-5}$ s.

Implicit methods have on the other hand no time step restrictions, and allow to progress in time more rapidly, but the cost per iteration is much higher, as algebraic systems have to be solved at each iteration.

The choice between explicit or implicit schemes is still a source of intense debate today and you will find in the literature, and in the practice of CFD codes, advocates of both options. Actually, as implicit schemes have higher CPU costs per time step and require also more memory, the balance will actually depend on the optimal value of the product [(cost/iteration) times (number of iterations)], to be moderated by the memory requirements.

For stationary flow problems, we will see in Chapter 10 with the multigrid method and in Volume II, that techniques are available to accelerate the numerical transients while iterating toward the steady state solution, which make explicit schemes more competitive.

For diffusion equations, the explicit time step restriction $\Delta t < \Delta x^2 / 2\alpha$ is generally not so severe, depending of course on the values of the diffusion coefficient. For a value of $\alpha = 10^{-3}$ m^2/s, $\Delta x = 0.01$, we would have $\Delta t < 5 \times 10^{-2}$ s.

The Von Neumann method offers an easy and simple way of assessing the stability properties of linear schemes with constant coefficients, when the boundary conditions are assumed periodic.

The problem of stability for a linear problem with constant coefficients is now well understood when the influence of boundaries can be neglected or removed. This is the case either for an infinite domain or for periodic conditions.

However, as soon as we have to deal with non-constant coefficients and (or) nonlinear terms in the basic equations, the information on stability becomes limited. Hence, we have to resort to a local stability analysis, with frozen values of the nonlinear and non-constant coefficients, such as to make the formulation linear. In any case, linear stability is a necessary condition for nonlinear problems, but is certainly not sufficient.

7.3   NEW SCHEMES FOR THE LINEAR CONVECTION EQUATION

The explicit schemes considered up to this point for the linear convection equation, do not appear to be very satisfactory. Indeed the central scheme with forward time difference, equation (7.2.11), is *useless* as it is an unstable scheme and the conditionally stable first order upwind scheme (FOU), equations (7.1.23) or (7.2.21), is of poor accuracy, as seen from Figure 7.1.2, since it has too much numerical dissipation resulting from its limited accuracy of first order.

Hence, we still have to look for better schemes to be able to handle convection phenomena with higher accuracy. Since the first order scheme is not adequate, we have to look for second order accuracy, both in space and time.

7.3.1 The Leapfrog Scheme for the Convection Equation

A first option has already been briefly mentioned in Section 4.1, equation (4.1.40), where both time and space derivatives are discretized by second order central difference formulas. This leads to the so-called *leapfrog scheme*, defined by

$$\frac{u_i^{n+1} - u_i^{n-1}}{2\Delta t} + \frac{a}{2\Delta x}(u_{i+1}^n - u_{i-1}^n) = 0 \tag{7.3.1}$$

or

$$u_i^{n+1} = u_i^{n-1} + \sigma(u_{i+1}^n - u_{i-1}^n) \tag{7.3.2}$$

This scheme is of second order accuracy, with a truncation error $O(\Delta t^2, \Delta x^2)$ (see Problem P.7.2).

This scheme is called leapfrog, because of the particular structure of its computational molecule (Figure 7.3.1) where the new solution u_i^{n+1} is obtained by starting from the value at point i and time level $(n-1)$, then taking into account the values at time level n to the right and to the left, with u_i^n not contributing to the computation u_i^{n+1}.

This scheme has the drawback of having to treat three time levels simultaneously and in order to start the calculation, two time levels $n = 0$ and $n = 1$ have to be known. This requires using another two level scheme to start the calculations.

Applying the methodology described in the previous section for the Von Neumann stability analysis, we obtain

$$V^{n+1} = V^{n-1} + \sigma V^n(e^{I\phi} - e^{-I\phi}) \tag{7.3.3}$$

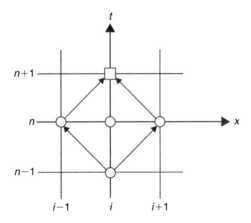

Figure 7.3.1 *Computational molecule for the leapfrog scheme.*

whereby we write for the amplitudes V^n of a single harmonic

$$G = \frac{V^{n+1}}{V^n} = \frac{V^n}{V^{n-1}} \tag{7.3.4}$$

since G is independent of n. When this is introduced in the three level scheme, a quadratic equation for G is obtained

$$G - 1/G = -\sigma(e^{I\phi} - e^{-I\phi}) \tag{7.3.5}$$

with the two solutions

$$G = -I\sigma \sin \phi \pm \sqrt{1 - \sigma^2 \sin^2 \phi} \tag{7.3.6}$$

If $\sigma > 1$, the scheme is unstable, since the term under the square root can become negative, making G purely imaginary and in magnitude larger than one. This is best seen for the particular value $\phi = \pi/2$.

For $|\sigma| \leq 1$, the term under the square root is always real and therefore

$$G \cdot G^* = |G|^2 = Re(G)^2 + Im(G)^2 = (1 - \sigma^2 \sin^2 \phi) + \sigma^2 \sin^2 \phi = 1 \tag{7.3.7}$$

Hence, the leapfrog scheme is **neutrally stable**, since

$$|G| = 1 \quad \text{for} \quad |\sigma| \leq 1 \tag{7.3.8}$$

7.3.2 Lax–Friedrichs Scheme for the Convection Equation

This scheme was introduced by Lax (1954), as a way of stabilizing the unstable forward in time, central scheme (7.2.11). It consists in replacing u_i in the right-hand side by the average value $(u_{i-1} + u_{i+1})/2$. With this substitution an error of the order of Δx is introduced, reducing this scheme to first order in space.

For the single convection equation we obtain the following scheme, schematically represented in Figure 7.3.2.

$$u_i^{n+1} = \frac{1}{2}(u_{i+1}^n + u_{i-1}^n) - \frac{\sigma}{2}(u_{i+1}^n - u_{i-1}^n) \tag{7.3.9}$$

A consistency analysis to obtain the truncation error, confirms that the Lax–Friedrichs (LF) scheme is indeed reduced to first order accuracy in space and time (see Problem P.7.3).

The amplification factor is obtained by inserting the single harmonic $V^n e^{Ii\phi}$, leading to

$$G = \cos \phi - I\sigma \sin \phi \tag{7.3.10}$$

In the complex G-plane, the amplification factor G is represented by an ellipse centered at the origin, with a horizontal semi-axis equal to 1 and a vertical semi-axis equal to σ, (Figure 7.3.3).

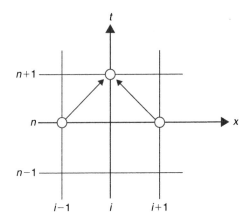

Figure 7.3.2 *Lax–Friedrichs scheme for convection equations.*

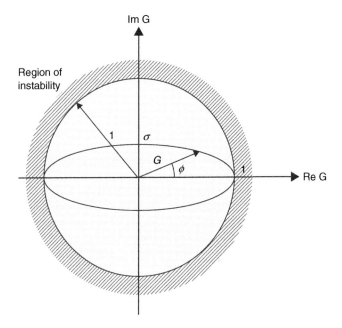

Figure 7.3.3 *Stability region for the Lax–Friedrichs scheme.*

This leads again to the CFL stability condition $|\sigma| \leq 1$, as the G-ellipse will remain inside the stability circle of radius 1, as long as $|\sigma| \leq 1$.

7.3.3 The Lax–Wendroff Scheme for the Convection Equation

The scheme of the previous example is of first order accuracy, which is insufficient for practical purposes.

The first scheme to be introduced historically with second order accuracy in space and time, with two time levels, is due to Lax and Wendroff (1960).

The introduction of this scheme (LW) was a landmark in the history of CFD, as it can be considered as the advent of many modern concepts of CFD. It has been the most widely applied scheme for aeronautical applications, up to the end of the 1980s under various forms, some of them being presented in more details in Volume II. Actually the first introduction of the Finite Volume method in CFD, was done on the basis of a two-step variant of the LW schemes, introduced by W. MacCormack in 1969; see the introduction of Section 5.2 for the relevant references.

The original derivation of Lax and Wendroff was based on a Taylor expansion in time up to the second order, such as to achieve second order accuracy in time.

$$u_i^{n+1} = u_i^n + \Delta t(u_t)_i + \frac{\Delta t^2}{2}(u_{tt})_i + O(\Delta t^3)$$ (7.3.11)

The basis of the LW method is to keep the second time derivative in the discretization, and replacing all time derivatives by equivalent spatial derivatives. This is straightforward for the first time derivative, while the second derivative is obtained by taking the time derivative of the convection equation $u_t + au_x = 0$, resulting in

$$u_{tt} = -a(u_x)_t = -a(u_t)_x = +a^2 u_{xx}$$ (7.3.12)

Introduced in equation (7.3.11), we obtain

$$u_i^{n+1} = u_i^n - a\Delta t(u_x)_i + \frac{a^2 \Delta t^2}{2}(u_{xx})_i + O(\Delta t^3)$$ (7.3.13)

If we discretize all the space derivative with second order central formulas in mesh point i, we obtain

$$u_i^{n+1} = u_i^n - \frac{\sigma}{2}(u_{i+1}^n - u_{i-1}^n) + \frac{\sigma^2}{2}(u_{i+1}^n - 2u_i^n + u_{i-1}^n)$$ (7.3.14)

Observe that the third term, which stabilizes the instability generated by the first two terms, is the discretization of an additional dissipative term of the form $(a^2 \Delta t/2)u_{xx}$. If you look back at the equivalent differential equation of this unstable scheme, equation (7.1.19), we can interpret the Lax–Wendroff scheme as the discretization of a modified convection equation obtained by adding the lowest order truncation error term (which is responsible for the instability) to the left-hand side. In other terms, the Lax–Wendroff scheme is obtained by discretizing the following corrected differential model:

$$u_t + au_x + \frac{\Delta t}{2}a^2 u_{xx} = 0$$ (7.3.15)

Taking into account the truncation error (7.1.18), we also see that the dominating term of the truncation error of the Lax–Wendroff scheme is proportional to the third space derivative, leading to the equivalent differential equation of the Lax–Wendroff scheme

$$\bar{u}_t + a\bar{u}_x + \frac{\Delta t}{2}a^2 \bar{u}_{xx} = a\frac{\Delta x^2}{6}(\bar{u}_{xxx}) + O(\Delta t^2, \Delta x^4)$$ (7.3.16)

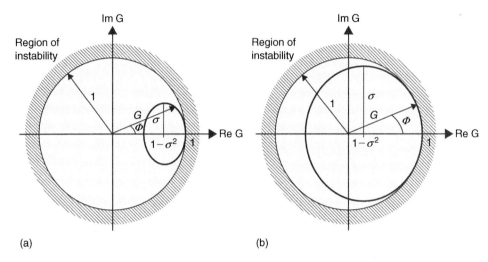

(a)   (b)

Figure 7.3.4 *Polar representation of the amplification factor for Lax–Wendroff scheme: (a): $\sigma < 1$ and $\sigma^2 < 1/2$ and (b): $\sigma < 1$ and $\sigma^2 > 1/2$.*

The amplification matrix from the Von Neumann method is now readily obtained, as

$$G = 1 - \frac{\sigma}{2}(e^{I\phi} - e^{-I\phi}) + \frac{\sigma^2}{2}(e^{I\phi} - 2 + e^{-I\phi})$$

$$= 1 - I\sigma \sin\phi - \sigma^2(1 - \cos\phi) \tag{7.3.17}$$

Writing ξ and η, respectively, for the real and imaginary parts of G, we have

$$\xi \equiv \mathrm{Re}\, G = (1 - \sigma^2) + \sigma^2 \cos\phi$$

$$\eta \equiv \mathrm{Im}\, G = -\sigma \sin\phi \tag{7.3.18}$$

In the complex G-plane, this represents an ellipse centered on the real axis at the abscissa $(1 - \sigma^2)$ and having semi-axis length of σ^2 along the real axis and σ along the vertical imaginary axis. Hence, this ellipse will always be contained in the unit circle if the CFL condition is satisfied (Figure 7.3.4). For $\sigma = 1$, the ellipse becomes identical to the unit circle.

The stability condition is therefore

$$|\sigma| \leq 1 \tag{7.3.19}$$

What have we achieved at this point?

We have now three new explicit schemes for the linear convection equation, in addition to the conditionally stable FOU scheme (7.2.21), namely the first order Lax–Friedrichs (LF) scheme and the two second order schemes, leapfrog and Lax–Wendroff (LW). All three of them are also conditionally stable under the CFL condition (7.3.19).

Let us now compare the results obtained with these four schemes on some representative, but also challenging test cases.

With any flow model, one of the most critical situations occurs in regions with very strong local variations. For instance, the inviscid flow around a wing section undergoes a strong local velocity gradient in the leading edge region, where over a very short distance of a few percent of the chord downstream of the stagnation point, the velocity increases from zero to values larger than the incoming velocity (see Figure 7.3.5[1]).

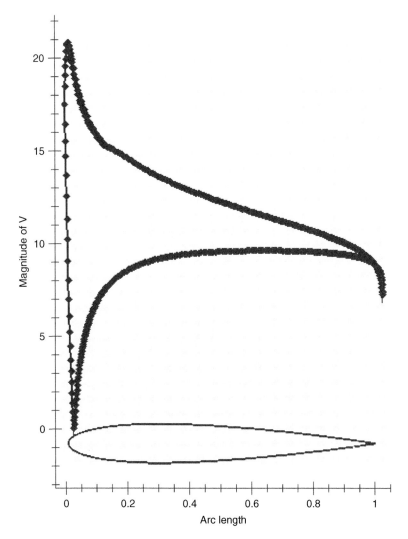

Figure 7.3.5   *Inviscid flow around the leading edge of a NACA0012 airfoil section, at 8 degree incidence.*

---

[1] The author thanks Benoit Leonard, from NUMECA Int, for providing this figure.

Still more challenging is the occurrence of shock waves where the flow variables are submitted to a ***discontinuous jump*** of velocity and pressure.

Therefore the numerical representation of a jump in the flow variables is one of the most difficult tasks facing the numerical simulation. This will be our first test case, where the initial distribution of the variable u is formed by a Heaviside function from 1 to 0.

Another challenging situation is represented by a moving wave of a certain frequency. As can easily be expected, higher frequencies lead to higher difficulties for their numerical simulation. Hence, we will consider a traveling sinusoidal wave packet as our second test case, and we will look at two different frequencies. The sinusoidal wave packet will be chosen to correspond to a selected value of the wave number and hence to a fixed value of the phase angle ϕ for a given mesh size Δx.

HANDS-ON TASK 2

Generalize the program you have written to obtain the results of Figure 4.1, to include the three new schemes LF, LW and Leapfrog, and apply them to the test cases defined here. Obtain the results of Figures 7.3.6–7.3.8.

Figure 7.3.6 compares the computed results for the propagating discontinuity at a Courant number of 0.8 after 80 time steps on a mesh size $\Delta x = 0.05$. The strong dissipation of the first order upwind and Lax–Friedrichs schemes is clearly seen from the way the discontinuity is smoothed out. Observe also the 'double' solution obtained with the Lax–Friedrichs scheme, illustrating the odd–even decoupling already discussed in Section 4.2 (Figure 4.2.3). Looking at Figure 7.3.2, it can be seen that u_i^{n+1} does not depend on u_i^n but on the neighbouring points u_{i+1}^n and u_{i-1}^n. These points also influence the solutions $u_{i+2}^{n+1}, u_{i+4}^{n+1}, \ldots$, while u_i^n will influence independently the points $u_{i+1}^{n+1}, u_{i+3}^{n+1}, \ldots$ The solutions obtained at the even and odd numbered points can therefore differ by a small amount without preventing convergence and such a difference appears on the LF solution shown on Figure 7.3.6.

The second order Lax–Wendroff and leapfrog schemes show a better representation of the discontinuity, with a significant reduction of the numerical diffusion. But another problem appears, namely the generation of numerical oscillations in front of the discontinuity. We also observe that the leapfrog scheme generates stronger oscillations compared to the Lax–Wendroff scheme.

The results of the moving wave packet simulations are shown in Figures 7.3.7 and 7.3.8 where the four schemes are compared for a CFL number of 0.8 after 80 time steps on a mesh of 101 points, over a domain length $L = 2$ between $x = 0$ and $x = 2$; i.e. for $\Delta x = 0.02$. The first example (Figure 7.3.7) corresponds to two sinusoidal periods over a distance of 1, hence a wavelength $\lambda = 0.5$. The corresponding wave number $k = 2\pi/\lambda = 4\pi$ and the associated phase angle $\phi = k\Delta x$ is equal to $\phi = \pi/12.5$. The second case (Figure 7.3.8) of higher frequency has four periods over the distance of 1, i.e. $\lambda = 0.25$ and a phase angle $\phi = \pi/6.25$.

You immediately can see that results with the two first order schemes, FOU and LF, are catastrophic and become worse as the wave frequency increases!

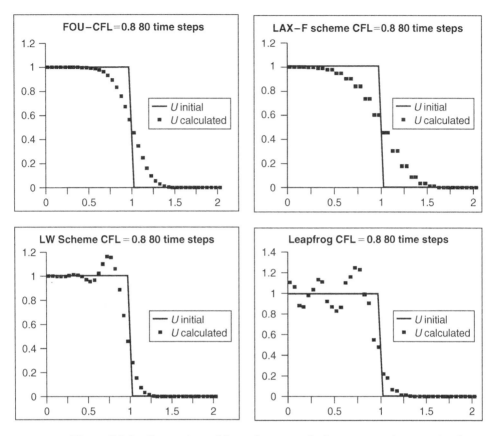

Figure 7.3.6 *Comparison of four schemes on the linear convection equation for a propagating discontinuity.*

The lesson to be learned form these results is that *due to the excessive numerical dissipation of first order schemes, they become totally useless for time-dependent propagation problems of this kind and should be avoided*.

The second order schemes, leapfrog and LW, look much better, particularly at the lowest frequency, where the numerical solutions are excellent. A small problem however is the appearance of oscillations in the numerical solution at the front of the propagating wave. This is due to the discontinuity in the slope of the initial solution at the points $x = 0.25$ and $x = 1.5$. This behavior is similar to the case of the propagating discontinuity of Figure 7.3.6.

The same computations performed at a higher frequency, corresponding to a phase angle of $\phi = \pi/6.25$, shown in Figure 7.3.8 indicate larger deviations compared to the exact solution. The Lax–Wendroff scheme has an error in the amplitude as well as a phase shift of the numerical solution. The leapfrog scheme has a better behavior with regard to the amplitude of the wave, which is correctly reproduced by the scheme, but here also we observe an increased level of numerical oscillations generated at the initial slope discontinuity. A phase error is also to be observed with the leapfrog scheme.

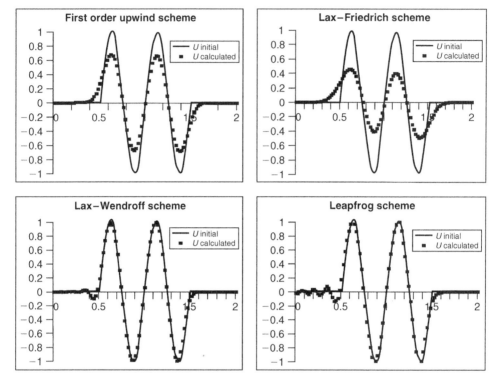

Figure 7.3.7 *Comparison of four schemes on the linear convection equation for a propagating wave packet for* $\phi = \pi/12.5$.

So, summarizing our observations, we note that:

- The first order schemes are of very poor accuracy.
- The second order schemes provide a significant better accuracy, particularly at the lower frequencies.
- The second order schemes generate numerical oscillations as soon as function or slope discontinuities in the solution are present. These oscillations being stronger with the leapfrog scheme, compared to LW.
- The generated numerical errors are very sensitive to the frequency, i.e. to the phase angle.
- These results, although obtained on a simple one-dimensional linear convection equation, are very representative of real flow simulations. You will encounter these same effects also with the full 3D Navier–Stokes simulations.

The main questions arising from these representative examples are:

- How can we *explain* and *predict* the differences between these four schemes and their dependence with frequency?
- Why do numerical oscillations appear with the second order schemes and not with the first order schemes?

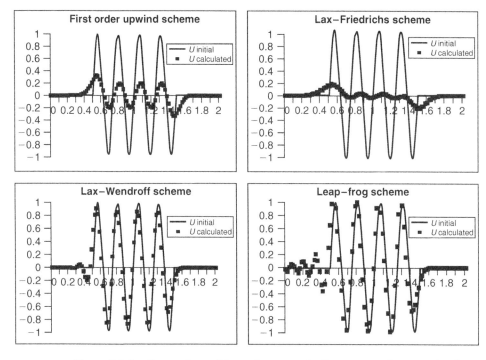

Figure 7.3.8 *Comparison of four schemes on the linear convection equation for a propagating wave packet for $\phi = \pi/6.25$.*

- Although discontinuities in the solution and slopes are present at both ends of the solution, why do we see numerical oscillations only on the upstream side?

The response to these questions is of fundamental importance, not only for the understanding of numerical schemes, but even more importantly, for our capacity to analyze a numerical solution, to evaluate its accuracy and hence, its degree of reliability. It will also determine the level of confidence we can attach to a numerical simulation on a given mesh.

In order to provide an answer to these questions, we have to develop a methodology for error analysis of a numerical solution. This is the content of the next section.

7.4 THE SPECTRAL ANALYSIS OF NUMERICAL ERRORS

When we perform a Von Neumann stability analysis, we obtain a specific expression of the amplification factor G as a function of the phase angle ϕ and of the parameters of the scheme. *Hence we can consider that $G(\phi$, scheme parameters) is a uniquely defined 'blueprint' of a numerical scheme and should contain the complete information concerning its behavior, and in particular the information on the generated numerical errors.*

What errors are we referring to in this context?

We have seen that the truncation error defines the global discretization error of the scheme, but it does *not* provide a direct evaluation of the errors on the solutions. Indeed, the comparisons shown in the previous section refer to the difference between the numerical solution and the exact solution of the mathematical model. That is, we looked at the differences

$$\tilde{\varepsilon} \overset{\Delta}{=} u_i^n - \tilde{u}_i^n \tag{7.4.1}$$

where $\tilde{u}_i^n = \tilde{u}(x_i, t^n)$ is the exact analytical solution at point x_i and time t^n.

The numerical representation of the time evolution of the solution is fully contained in the amplitudes V^n for each harmonic corresponding to the wave number k. However, up to now we had no need to be more specific, as the definition of the amplification factor (7.2.7), did not require it.

Since we wish now to go a step further and gain additional information on the errors, we have to be more specific about the time dependence of V^n both for the numerical and the analytical solutions.

Considering first the analytical solution, we always can represent it by a Fourier decomposition in space, similar to the approach followed for the Von Neumann analysis. It is known form the physics of plane waves that to each harmonic k of a spatial Fourier decomposition, a pulsation function $\tilde{\omega} = \tilde{\omega}(k)$ can be attached, called the **dispersion relation**; the explicit dependence $\tilde{\omega} = \tilde{\omega}(k)$ depending on the equations to be solved. We refer you to Section 3.4 for the introduction and discussion of this representation.

For instance, for the linear convection equation $u_t + au_x = 0$, $\tilde{\omega} = \tilde{\omega}(k) = ak$, and a representative solution is the plane wave:

$$\tilde{u}_i^n = \hat{V} e^{Ikx_i} e^{-I\tilde{\omega} t^n} \tag{7.4.2}$$

For a more general solution, the harmonic k, has the same form, at time level $t^n = n\Delta t$, as

$$(\tilde{u}_i^n)_k = \hat{V}(k) e^{-I\tilde{\omega} n\Delta t} e^{Ik(i\Delta x)} \tag{7.4.3}$$

where we designate from now on the exact dispersion relation as $\tilde{\omega} = \tilde{\omega}(k)$, obtained from the differential system as solution of the eigenvalue equation (3.4.17).

The function $\hat{V}(k)$ is obtained from the Fourier decomposition of the initial solution at $t = 0$. If we define $u(x, 0) = f(x)$ at $t = 0$ we have

$$\hat{V}(k) = \frac{1}{2L} \int_{-L}^{L} f(x) e^{-Ikx} dx \tag{7.4.4}$$

This form of the exact solution is very similar to the representation of the numerical solution applied for the Von Neumann analysis, equation (7.2.6), where a single harmonic is described as

$$(u_i^n)_k = V^n e^{Ik(i\Delta x)} \tag{7.4.5}$$

Assuming that the initial solution is represented exactly in the numerical scheme, with the exception of round-off errors, we can assume that $\hat{V}(k)$ is exactly seen on the discretized mesh. Hence, we can represent the numerical amplitude V^n, in a very similar way to equation (7.4.3), as

$$V^n = \hat{V}(k)e^{-I\omega n\Delta t} = \hat{V}(k)(e^{-I\omega\Delta t})^n \tag{7.4.6}$$

where the approximate relation between ω and k, $\omega = \omega(k)$ is obtained from the amplification matrix G. It will be called the **numerical dispersion relation** of the scheme.

Indeed, from the definition of the amplification matrix, we can write, **since G is independent of n**:

$$V^n = GV^{n-1} = (G)^2 \cdot V^{n-2} = \ldots = (G)^n \cdot V^0 = (G)^n \cdot \hat{V}(k) \tag{7.4.7}$$

where we put G between brackets to indicate that $(G)^n$ is G to the power n and *not* G with n as superscript.

Comparing with equation (7.4.6), we can identify $(G)^n$ with the exponential in this equation, leading to the expression, **which will be considered as the definition of $\omega(k)$**:

$$G \overset{\Delta}{=} e^{-I\omega\Delta t} \tag{7.4.8}$$

Referring to equation (7.4.3), we can also rewrite the analytical solution in a similar way as

$$\tilde{V}^n = (e^{-I\bar{\omega}\Delta t})^n \hat{V}(k) \overset{\Delta}{=} (\tilde{G})^n \hat{V}(k) \tag{7.4.9}$$

defining hereby an **exact amplification function**:

$$\tilde{G} = e^{-I\bar{\omega}\Delta t} \tag{7.4.10}$$

This allows us now to investigate the nature and frequency spectrum of the numerical errors. Since ω is a complex function, the amplification matrix can be separated in an amplitude $|G|$ and a phase Φ:

$$G = |G|e^{-I\Phi} \tag{7.4.11}$$

A similar decomposition, performed for the exact solution (7.4.10), leads to

$$\tilde{G} = |\tilde{G}|e^{-I\tilde{\Phi}} \tag{7.4.12}$$

Under this form, we see that the modulus of G will influence the amplitude of the solution, while the phase of G, will influence the phase of the solution, since

$$V^n = GV^{n-1} = |G|e^{-I\Phi}V^{n-1} \tag{7.4.13}$$

The error in amplitude, called the ***diffusion or dissipation error***, is defined by the ratio of the computed amplitude to the exact amplitude.

$$\varepsilon_D = \frac{|G|}{|\tilde{G}|} \tag{7.4.14}$$

The error on the phase of the solution, the ***dispersion error***, can be defined as the difference

$$\varepsilon_\phi = \Phi - \tilde{\Phi} \tag{7.4.15}$$

suitable for pure parabolic problems, where $\tilde{\Phi} = 0$, in absence of convective terms. For convection dominated problems, the definition

$$\varepsilon_\phi = \Phi/\tilde{\Phi} \tag{7.4.16}$$

is better adapted.

In particular for hyperbolic problems, such as the scalar convection equation, the exact solution is a single wave propagating with the velocity a. Hence

$$\tilde{\Phi} = ka\Delta t \tag{7.4.17}$$

7.4.1 Error Analysis for Hyperbolic Problems

A hyperbolic problem such as the convection equation $u_t + au_x = 0$ represents a wave traveling at constant speed a without damping, i.e. with constant amplitude. The exact solution for a wave of the form (7.4.3) is given by

$$\tilde{\omega} = ak \tag{7.4.18}$$

and

$$\tilde{u} = \hat{V} e^{Ikx} e^{-Iakt} \tag{7.4.19}$$

The exact amplification factor reduces here to

$$|\tilde{G}| = 1 \quad \text{and} \quad \tilde{\Phi} = ak\Delta t = \frac{a\Delta t}{\Delta x} k\Delta x = \sigma\phi \tag{7.4.20}$$

and

$$\tilde{G} = e^{-I\sigma\phi} \tag{7.4.21}$$

This confirms that the exact solution of the linear convection equation is an initial shape propagating at the constant velocity a, without reduction of the amplitude.

Since in the numerical solution an initial wave amplitude is damped by a factor $|G|$ per time step, as can be seen from equation (7.4.7), the error in amplitude, ***the diffusion error***, will be given by the modulus of the amplification factor

$$\varepsilon_D = |G| \tag{7.4.22}$$

The phase of the numerical solution is determined by Φ, which can be considered as defining a numerical convection speed $a_{num} = \Phi/k\Delta t$, by direct comparison with the relations (7.4.20)

$$a_{num} = \frac{\Phi}{k\Delta t} = \frac{a\Phi}{\sigma\phi} \qquad (7.4.23)$$

Therefore the error in phase, i.e. *the dispersion error*:

$$\varepsilon_\phi = \frac{\Phi}{ak\Delta t} = \frac{\Phi}{\sigma\phi} \equiv \frac{a_{num}}{a} \qquad (7.4.24)$$

can be interpreted as defining the ratio between the numerical and physical convection speeds, indicating that the physical convection (or propagation) speed will be modified by the dispersion error ε_ϕ.

When the dispersion error is larger than one, $\varepsilon_\phi > 1$, the phase error is a leading error and the numerical convection velocity, a_{num}, is larger than the exact one. This means that the computed solution appears to move faster than the physical one. On the other hand, when $\varepsilon_\phi < 1$, the phase error is said to be a lagging error and the computed solution travels at a lower velocity than the physical one.

Let us now analyze the error properties of the four schemes we have been comparing in the previous section, in order to explain their observed behavior. The procedure is as follows:

- Obtain the expression of the amplification factor G. For stability we decompose G in its real and imaginary parts.
- For the error analysis, we have to write G in 'polar' form, i.e. defining its module and phase. The relation between the two representations is straightforward:

$$G = (\mathrm{Re}\,G) + I(\mathrm{Im}\,G) = |G|e^{-I\Phi}$$

$$|G| = \sqrt{(\mathrm{Re}\,G)^2 + (\mathrm{Im}\,G)^2} \qquad \tan\Phi = -\mathrm{Im}\,G/\mathrm{Re}\,G \qquad (7.4.25)$$

- Define the diffusion and dispersion errors by their definitions (7.4.22), (7.4.24) and analyze their behavior in function of frequency or phase angle ϕ.

A general comment can be made here, concerning the conflicting requirements between stability and accuracy. *Accuracy requires the modulus of G to be as close to one as possible, but stability requires |G| to be lower than one, i.e. have a larger diffusion error*.

7.4.1.1   *Error analysis of the explicit First Order Upwind scheme (FOU)*

The amplification factor for this scheme is defined by equation (7.2.22). Its modulus is given by

$$|G| = [(1-\sigma+\sigma\cos\phi)^2 + \sigma^2\sin^2\phi]^{1/2} = [1-4\sigma(1-\sigma)\sin^2\phi/2]^{1/2} \quad (7.4.26)$$

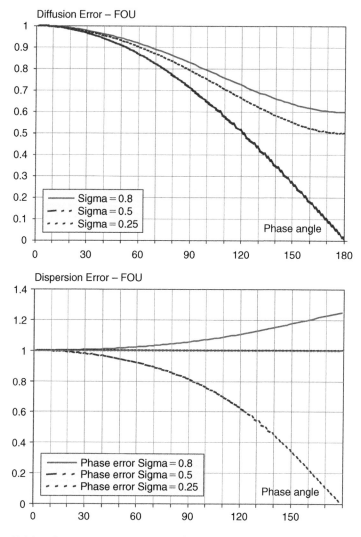

Figure 7.4.1 *Cartesian representation of amplitude and phase errors in function of phase angle, in degrees, for the first order upwind scheme.*

and the phase error is

$$\varepsilon_\phi = \frac{\tan^{-1}\left[(\sigma \sin \phi)/(1 - \sigma + \sigma \cos \phi)\right]}{\sigma \phi} \tag{7.4.27}$$

These errors are represented in Figure 7.4.1, in a Cartesian diagram in function of the phase angle ϕ, or equivalently frequency, for constant values of the CFL number within the stability range.

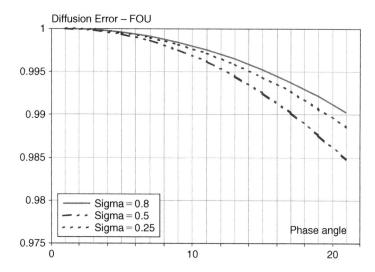

Figure 7.4.2 *Zoom of amplitude errors in the low frequency region, in function of phase angle, in degrees, for the first order upwind scheme.*

For $\sigma = 0.5$, the phase error $\varepsilon_\phi = 1$, but for $\sigma < 0.5$, $\varepsilon_\phi < 1$ indicating a lagging error, while the numerical speed of propagation becomes larger than the physical speed when $\varepsilon_\phi > 1$ for Courant numbers $\sigma > 0.5$ (see also Problem P.7.7).

We observe that the diffusion error, which is always equal to 1 for $\phi = 0$, for consistency reasons, drops rapidly when the frequency increases. This explains the dramatic behavior of the FOU scheme in the examples shown in the previous section.

Remember here that the highest frequency that can be represented on the mesh corresponds to the shortest wavelength $\lambda = 2\Delta x$, i.e. to $\phi = \pi$. Hence we designate the *regions around $\phi = 0$ and $\phi = \pi$ as the low and high frequency regions*, respectively.

At the lowest phase angle $\phi = \pi/12.5 = 14.4$ deg, treated in Figure 7.3.7, the diffusion error is equal to 0.995, as can be seen from the zoom on the low frequency region shown in Figure 7.4.2. Hence, we can estimate quantitatively the decrease in amplitude after the 80 iterations at $\sigma = 0.8$ observed in Figure 7.3.7.

In general, after n iterations, the amplitude is reduced by a factor $|G|^n$, as shown by equation (7.4.7). Hence in this case the amplitude will be decreased by $(0.995)^{80} = 0.67$, which is the value you can find on Figure 7.3.7 for this first order upwind scheme. At the highest frequency of Figure 7.3.8, $\phi = \pi/6.25 = 28.8°$, the diffusion error is equal to 0.98 and the amplitude is decreased by $(0.98)^{80} = 0.2$, which corresponds to what you observe on Figure 7.3.8 for this FOU scheme.

This very rapid decease of the modulus of G, is typical for first order schemes and explains the observed behavior.

Observe also the very strong sensitivity of the numerical amplitude variation to the values of the diffusion error. An error of only 1%, i.e. a value of $|G| = 0.99$, leads after 100 time steps to a reduction in amplitude of 36.6% of the initial value. Even for an error of 0.5%, i.e. $|G| = 0.995$, the amplitude reduction is of 60%.

Reversing the question, if we would require an amplitude reduction less than 1% after 100 time steps, we should have $|G| > 0.9999$!

7.4.1.2 Error analysis of the Lax–Friedrichs scheme for the convection equation

The accuracy of the scheme is obtained from the modulus and phase of the amplification factor (7.3.10)

$$|G| = [\cos^2 \phi + \sigma^2 \sin^2 \phi]^{1/2}$$
$$\Phi = \tan^{-1} (\sigma \tan \phi) \tag{7.4.28}$$

This defines the dissipation error

$$\varepsilon_D = |G| = [\cos^2 \phi + \sigma^2 \sin^2 \phi]^{1/2} \tag{7.4.29}$$

and the dispersion error

$$\varepsilon_\phi = \frac{\Phi}{\sigma\phi} = \frac{\tan^{-1} (\sigma \tan \phi)}{\sigma\phi} \tag{7.4.30}$$

As can be seen the choice $\sigma = 1$ gives the exact solution, but lower values of σ will generate amplitude and phase errors.

Cartesian representations of $|G|$ and ε_ϕ in function of the phase angle $\phi = k\Delta x$, are shown in Figure 7.4.3. For small values of σ, the amplitudes are strongly damped, indicating that this scheme is generating a strong numerical dissipation, even larger than the FOU scheme. The phase error is everywhere larger or equal one, showing a leading phase error. Observe that for $\phi = \pi$ we have $\varepsilon_\phi = 1/\sigma$, as seen from equation (7.4.30) (See also Problem P.7.6).

An interesting difference with the FOU scheme is the behavior of the diffusion error for the highest frequency $\phi = \pi$. It is seen here that the LF scheme has no diffusion error for $\phi = \pi$, while this error takes its maximum value for the FOU scheme. This explains the double solution of Figure 7.3.6, which corresponds to an odd–even decoupling, i.e. an error oscillation of wavelength $2\Delta x$, or a phase angle $\phi = \pi$. Hence, in order to avoid these high frequency oscillations, we need to have $G(\pi) < 1$. When this is satisfied the scheme is said to be **dissipative in the sense of Kreiss** (Kreiss, 1964).

7.4.1.3 Error analysis of the Lax–Wendroff scheme for the convection equation

We refer here to the amplification factor (7.3.17), from which we derive the dissipation error as:

$$|G| = [1 - 4\sigma^2(1 - \sigma^2)\sin^4 \phi/2]^{1/2} \tag{7.4.31}$$

and the phase error as

$$\varepsilon_\phi = \frac{\tan^{-1} [(\sigma \sin \phi)/(1 - 2\sigma^2 \sin^2 \phi/2)]}{\sigma\phi} \tag{7.4.32}$$

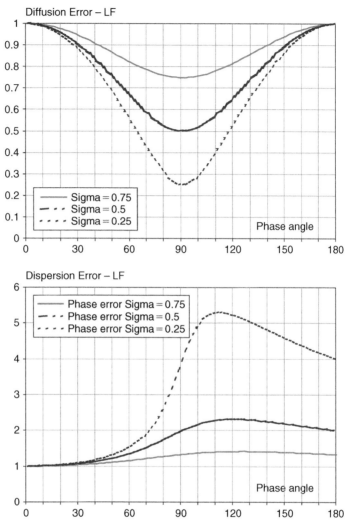

Figure 7.4.3 *Amplitude and dispersion errors for the Lax–Friedrichs scheme, in function of phase angle in degrees, applied to the linear convection equation.*

The diffusion and dispersion errors are represented in Figure 7.4.4, in function of phase angle (in degrees) for constant values of the CFL number σ.

To the lowest order, we can expand the above expressions around $\phi = 0$, leading to

$$|G| \approx 1 - \frac{\sigma^2}{8}(1 - \sigma^2)\phi^4 + O(\phi^6)$$

$$\varepsilon_\phi \approx 1 - \frac{1}{6}(1 - \sigma^2)\phi^2 + O(\phi^4) \qquad (7.4.33)$$

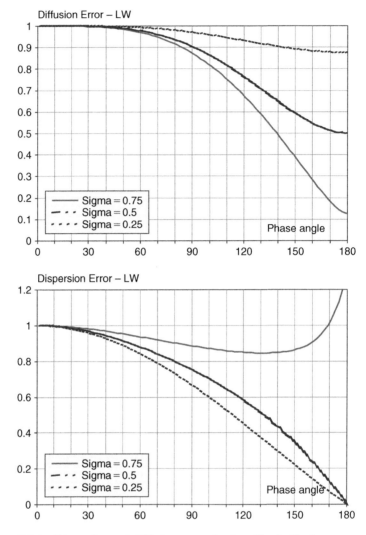

Figure 7.4.4 *Dispersion and diffusion errors for Lax–Wendroff scheme.*

The relative phase error is mostly lower than one, indicating a dominating lagging phase error and showing that the numerical convection velocity is lower than the physical one.

This explains why the numerical solution of the LW scheme lags behind the exact solution, as seen in Figure 7.3.8.

In terms of accuracy, we also clearly see the reasons behind the frequency dependence of the numerical solutions, seen by comparing Figures 7.3.7 and 7.3.8. From the zoom on the diffusion error, shown in Figure 7.4.5, we can again estimate quantitatively the amplitude errors.

Observe first that the region where the diffusion error is very close to one, which we can consider as the '***accuracy region***' of the scheme, is much larger than with the

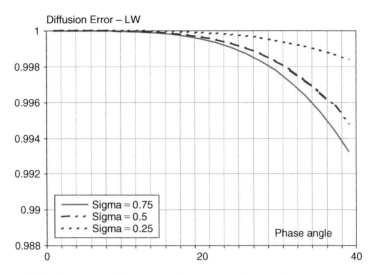

Figure 7.4.5   *Zoom on diffusion error for the Lax–Wendroff scheme.*

FOU scheme, as seen by comparing with Figure 7.4.2. This is typical of second order schemes.

Actually, the higher the order of the scheme, the larger this accuracy region.

At the higher frequency of Figure 7.3.8, where $\phi = \pi/6.25 = 28.8°$, the diffusion error, for $\sigma = 0.8$, is $|G| \sim 0.9985$. This leads to the amplitude reduction, after 80 time steps of 0.89, which corresponds to what you can observe on Figure 7.3.8.

7.4.1.4   *Error analysis of the leapfrog scheme for the convection equation*

The leapfrog scheme is the three time level scheme defined by equation (7.3.2), with second order accuracy in space and time. It has the interesting property of being conditional stable with $|G| = 1$, i.e. ***this scheme has no diffusion errors.***

Hence, this explains the behavior seen in Figure 7.3.8, where you can observe that the numerical amplitude remains equal to 1.

For this reason, the leapfrog scheme is very useful for long-term simulations, and is applied in some weather forecasting codes, where the accuracy of the predicted weather for a longer term of 3 to 4 days is essential for the reliability of the predictions.

The dispersion error is given by

$$\varepsilon_\phi = \pm \frac{\tan^{-1}\left[(\sigma \sin \phi)/\sqrt{1 - \sigma^2 \sin^2 \phi}\right]}{\sigma\phi} = \pm \frac{\sin^{-1}(\sigma \sin \phi)}{\sigma\phi} \tag{7.4.34}$$

and its dispersion error is shown on Figure 7.4.6.

We observe that this error is below 1 and again explains why the numerical solution is lagging behind the exact solution on Figures 7.3.7 and 7.3.8.

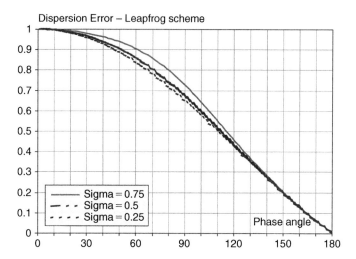

Figure 7.4.6 *Dispersion error for the leapfrog scheme in function of the phase angle in degrees, at constant CFL numbers.*

The leapfrog scheme should give accurate results when the function u has a smooth variation, since the amplitudes are correctly modeled, so more that for low frequencies the phase error is close to one.

However, the fact that this scheme is neutrally stable, $|G| = 1$ for all $\sigma < 1$ has also unfavorable consequences, as high frequency errors will not be damped, this scheme being non-dissipative in the sense of Kreiss. This explains why the oscillations on Figures 7.3.6–7.3.8 are larger with the leapfrog scheme, compared to the LW scheme.

When applied to the inviscid Burgers equation $u_t + uu_x = 0$, the computations become in certain circumstances unstable, as can be seen from Figure 7.4.7. This figure shows the computed solutions of Burgers equation for a stationary shock, after 10, 20 and 30 time steps at a Courant number of 0.8 and a mesh size of $\Delta x = 0.05$. The continuous line indicates the exact solution.

The amplitude of the errors increases continuously and the solution is completely destroyed after 50 time steps. The instability is entirely due to the nonlinearity of the equation, since the same scheme applied to the linear convection equation does not diverge, although strong oscillations are generated. In the present case, the high frequency errors are generated by the fact that the shock is located on a mesh point. This point has zero velocity and with an initial solution passing through this point, a computed shock structure is enforced with this internal point fixed, creating high frequency errors at the two adjacent points. This is clearly seen on Figure 7.4.7, looking at the evolution of the computed solutions.

For these reasons, the leapfrog scheme is not recommended for simulations of high-speed flows where high gradients and shocks can occur.

7.4.2 The Issue of Numerical Oscillations

The origin of the oscillations observed with the second order LW and leapfrog schemes, *cannot* be explained by the error properties discussed up to here.

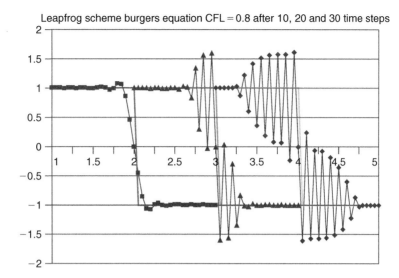

Figure 7.4.7 *Solutions of Burgers' equation with the leapfrog scheme after 10, 20, 30 time steps, for a stationary shock located on a mesh point.*

Their explanation will require the introduction of an additional property, called monotonicity. This will be discussed in details in Chapter 8.

But what we *can* explain here is why the oscillations appear only at the upstream end of the discontinuities with the LW and leapfrog schemes. The numerical oscillations are clearly of high frequency as they are dominated by point-to-point variations of wavelength $2\Delta x$, that is associated to the high frequency $\phi = \pi$.

Since the phase error is predominantly below one and largest in the high frequency region for both schemes, the associated errors will have a convection velocity significantly lower than the physical one and hence travel at a much lower velocity. They will therefore tend to accumulate on the upstream side of the moving discontinuity. This is what we observe in Figures 7.3.6–7.3.8 with the two second order schemes.

With the leapfrog scheme, the high frequency errors tend to remain stationary since $\varepsilon_\phi \to 0$ *for* $\phi \to \pi$ and since they are not damped, they accumulate on the upstream side and destroy the accuracy of the numerical solution. The leapfrog scheme therefore generates stronger high frequency oscillations compared to the Lax–Wendroff scheme, whose amplification factor is lower than one at the phase angle $\phi = \pi$, where $G(\pi) = 1 - 2\sigma^2$.

If the phase error is dominantly higher than one in the high frequency region, then the numerical oscillations will accumulate on the downstream side of the discontinuities. This will be the case with the second order upwind (SOU) scheme of Warming and Beam (1976), obtained by repeating the procedure leading to the Lax–Wendroff scheme, but introducing second order backward differences in equation (7.3.13), instead of central differences. This leads to the scheme

$$u_i^{n+1} = u_i^n - \frac{\sigma}{2}(3u_i^n - 4u_{i-1}^n + u_{i-2}^n) + \frac{\sigma^2}{2}(u_i^n - 2u_{i-1}^n + u_{i-2}^n) \qquad (7.4.35)$$

The stability analysis gives the amplification function (see Problem P.7.12):

$$G = 1 - I\sigma[1 + 2(1-\sigma)\sin^2 \phi/2]\sin\phi - 2\sigma[1 - (1-\sigma)\cos\phi]\sin^2 \phi/2 \quad (7.4.36)$$

It is easily shown that this scheme is conditionally stable under the CFL condition:

$$0 \leq \sigma \leq 2 \quad (7.4.37)$$

Observe that the scheme is unstable for negative convection velocities, and has an upper CFL limit of 2.

The diffusion and dispersion errors are given by

$$\varepsilon_D = |G| = \sqrt{1 - \sigma(1-\sigma)^2(2-\sigma)(1-\cos\phi)^2} \quad (7.4.38)$$

and

$$\varepsilon_\phi = \frac{1}{\sigma\phi}\tan^{-1}\frac{\sigma[1 + 2(1-\sigma)\sin^2 \phi/2]\sin\phi}{1 - 2\sigma[1 - (1-\sigma)\cos\phi]\sin^2 \phi/2} \quad (7.4.39)$$

and are shown in Figure 7.4.8 for CFL numbers 0.25, 0.5, 0.75 and 1.5.

The interesting observation here is that the dispersion error is predominantly above one for $\sigma < 1$ and below one for $\sigma > 1$.

Hence, for CFL numbers below one, for instance $\sigma = 0.5$, the numerical solution will move faster than the physical one and the numerical oscillations will be concentrated on the downstream side of the propagating wave form. This is seen on Figures 7.4.9 and 7.4.10. As you can further see from these figures, this second order scheme also generates numerical oscillations, similarly to the LW and leapfrog schemes.

Figure 7.4.9 shows the convection of a square wave with the second order upwind scheme of Warming and Beam scheme for $\sigma = 0.5$ and $\sigma = 1.5$. You can clearly see the phenomena discussed here.

Figure 7.4.10 is obtained with the Warming and Beam scheme for $\sigma = 0.5$ after 80 time steps with $\Delta x = 0.02$, and should be compared with Figure 7.3.8.

7.4.3 The Numerical Group Velocity

The group velocity of a wave packet, containing more than one frequency, has been defined in Chapter 3, equation (3.4.29), and is also the velocity at which the energy of the wave is traveling. For a one-dimensional wave, we have

$$\tilde{v}_G(k) = \frac{d\tilde{\omega}}{dk} \quad (7.4.40)$$

defining the exact group velocity as the derivative of the time frequency with respect to the wave number k. For a linear wave, the group velocity is equal to the phase speed a, as $\tilde{\omega} = ak$.

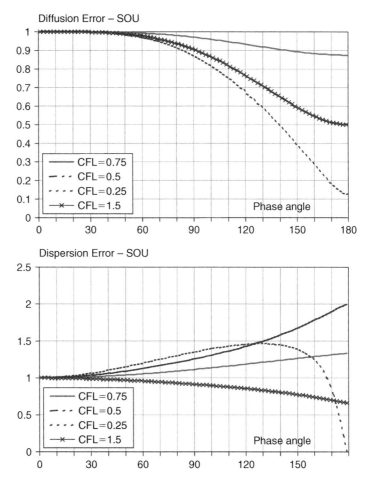

Figure 7.4.8 *Diffusion and dispersion errors for the Warming and Beam SOU scheme.*

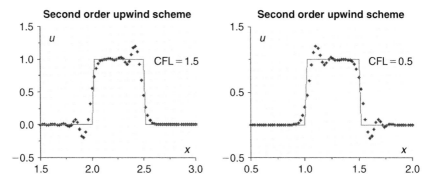

Figure 7.4.9 *Convection of a square wave with the second order upwind scheme of Warming and Beam scheme for σ = 0.5 and σ = 1.5.*

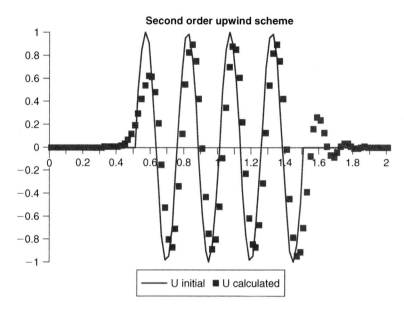

Figure 7.4.10 *Numerical solution obtained with the Warming and Beam scheme for $\sigma = 0.5$ after 80 time steps.*

By writing the numerical dispersion relation $\omega = \omega(k)$, we can define the **numerical group velocity** as

$$v_G(k) = \left(\frac{d\omega}{dk}\right) \qquad (7.4.41)$$

which represents the traveling speed of numerical wave packets centered around the wave number k.

Since the errors generated by a numerical scheme generally contain a variety of frequencies it is more likely that they will travel at the numerical group velocity instead of the numerical phase speed a_{num}, defined by (7.4.23).

Let us apply this to the leapfrog scheme. We introduce equation (7.4.8) into (7.3.6), to relate ω to the other parameters, leading to

$$G = e^{-I\omega\Delta t} \equiv \cos\omega\Delta t - I\sin\omega\Delta t = -I\sigma\sin\phi \pm \sqrt{1 - \sigma^2\sin^2\phi} \qquad (7.4.42)$$

Identifying the imaginary parts, leads to the numerical dispersion relation

$$\sin\omega\Delta t = \sigma\sin\phi \qquad (7.4.43)$$

from which we derive

$$v_G(k) = a\frac{\cos\phi}{\cos\omega\Delta t} = a\frac{\cos\phi}{\sqrt{1 - \sigma^2\sin^2\phi}} \qquad (7.4.44)$$

For low frequencies, the group velocity is close to the phase speed a, but for the high frequencies $\phi \approx \pi$, the group velocity is close to $-a$ indicating that the high

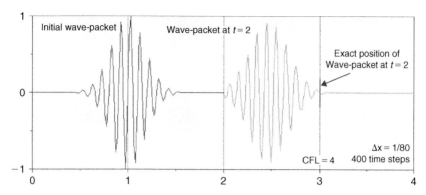

Figure 7.4.11   *Solution of the linear propagation of an exponential wave packet by the leapfrog scheme, after 400 time steps, for $\phi = k\Delta x = \pi/4$, $\Delta x = 1/80$.*

wave number packets will travel in the opposite direction to the wave phase speed a. An instructive example is provided by the exponential wave packet

$$u(x, t = 0) = \exp(-\alpha x^2)\sin 2\pi kx \tag{7.4.45}$$

shown on Figure 7.4.11 for a phase angle $\phi = k\Delta x = \pi/4$ corresponding to a wavelength of $\lambda = 8\Delta x$. The solution of the linear wave equation $u_t + au_x = 0$ with the leapfrog scheme is shown on the same figure after 400 time steps, for a Courant number of 0.4 and $\Delta x = 1/80$ for $a = 1$. If the initial solution is centered at $x = 1$, the exact solution should be located around $x = 3$ at time $t = 400\Delta t = 2$. However the numerical solution is seen to have traveled only to the point $x = 2.475$, which indicates a propagating speed of 0.7375 instead of the phase speed $a = 1$. This corresponds exactly to the computed group velocity from equation (7.4.44) which gives a value of $v_G = 0.7372$ at $\phi = \pi/4$.

These properties of the group velocity should be kept in mind when analyzing numerical data associated to high-frequency solutions.

More details on the applications of the concept of group velocity to the analysis of numerical schemes can be found in the interesting monograph of Vichnevetsky and Bowles (1982), Trefethen (1982), Cathers and O'Connor (1985). This last reference presents a detailed, comparative analysis of the group velocity properties of various finite element and finite difference schemes, applied to the one-dimensional, linear convection equation. Trefethen (1983, 1984) has derived some important relations between the group velocity and the stability of numerical boundary conditions of hyperbolic problems. Trefethen's results can be expressed by the condition that the numerical boundary treatment should not allow group velocities at these boundaries to transport energy into the computational domain. We refer you to the original references for more details and derivations.

HANDS-ON TASK 3

Extend the program you have written under the Hands-On Task 2, to apply it to the Warming and Beam scheme and to the Burgers equation $u_t + uu_x = 0$ with the test cases defined up to now. Obtain also the results of Figures 7.4.7–7.4.11.

7.4.4 Error Analysis for Parabolic Problems

Let us consider now the error analysis for the explicit, central discretization of the heat diffusion equation $u_t = \alpha u_{xx}$, given by equation (7.2.29) with the amplification factor obtained from equation (7.2.30).

The exact solution corresponding to a wave number k is obtained by searching a solution of the type

$$\tilde{u} = \hat{V} e^{-I\tilde{\omega}t} e^{Ikx} \tag{7.4.46}$$

Inserting into the model equation, we have

$$\tilde{\omega}(k) = -I\alpha k = -I\beta\phi^2/\Delta t \tag{7.4.47}$$

The exact solution of this parabolic problem is associated with a purely imaginary eigenvalue $\tilde{\omega}$, i.e. with an exponential decay in time of the initial amplitude, if $\alpha > 0$.

$$\tilde{u} = \hat{V} e^{Ikx} e^{-\alpha k^2 t} \tag{7.4.48}$$

Since the amplification matrix G is real, there is no dispersion error for this scheme. The error in the amplitude is measured by the diffusion error

$$\varepsilon_D = \frac{1 - 4\beta \sin^2 \phi/2}{e^{-\beta\phi^2}} \tag{7.4.49}$$

Expanding in powers of ϕ, we obtain

$$\begin{aligned}
\varepsilon_D &= \frac{1 - \beta\phi^2 + \beta\phi^4/12 + \cdots}{1 - \beta\phi^2 + (\beta\phi^2)^2/2 + \ldots} = 1 - \frac{\beta^2\phi^4}{2} + \frac{\beta\phi^4}{12} + \ldots \\
&= 1 - \frac{\alpha^2 k^4 \Delta t^2}{2} + \frac{\alpha k^4 \Delta t \Delta x^2}{12} + \ldots
\end{aligned} \tag{7.4.50}$$

For the low frequencies, $\phi \sim 0$, the error in amplitude remains small; while at high frequencies, $\phi \approx \pi$, the error could become unacceptably high, in particular for the larger values of $\beta \leq 1/2$. However, for $\beta = 1/6$, the two first terms of the expansion cancel, and the error is minimized, becoming of higher order, namely of the order $O(\Delta t^2, \Delta x^4)$ for constant values of $\beta = \alpha\Delta t/\Delta x^2$ and proportional to k^6.

It is seen that the error is proportional to the fourth and sixth power of the wave number, indicating that the high frequencies are computed with large errors. However, the amplitudes of these high frequencies are strongly damped since they are equal to $e^{-\alpha k^2 t}$. Therefore, this will generally not greatly effect the overall accuracy, with the exception of situations where the initial solution $u(x, 0)$ contains a large number of high-frequency components (see also Problem P.7.5).

7.4.5 Lessons Learned and Recommendations

The error analysis of the different schemes provided important lessons and guidelines, as well as recommendations as to how ensure a required accuracy. Although based on the simplest one-dimensional convection equation, the recommendations hereafter are of general validity.

The main observations can be summarized as follows:

- First order schemes generate high levels of errors, which make them unsuitable for unsteady problems and are too diffusive for steady problems.
- Second order schemes offer the possibility for acceptable accuracies and are presently the best compromise between number of mesh points and accuracy level.
- However, at higher frequencies, even second order schemes generate high error levels. In addition they generate unwanted oscillations around function and slope discontinuities.
- The following guidelines should be followed for the control of accuracy when using second order schemes. Taking Figure 7.4.5 as a representative reference for all second order schemes, we can establish a phase angle limit ϕ_{lim}, defined by the boundary of the region where the diffusion error remains very close to one. For a second order scheme, we can safely select a value of $\phi_{lim} \sim 10°$ or $\phi_{lim} \sim \pi/18$.
- The key quantity defining the accuracy of time-dependent flows is the *number of mesh points per wavelength* λ, $N_\lambda = \lambda/\Delta x$, and we require that the associated phase angle remains below the imposed limit. Hence we recommend:

$$\varphi = k\,\Delta x = \frac{2\pi}{\lambda}\Delta x \leq \varphi_{lim} \tag{7.4.51}$$

or

$$N_\lambda = \frac{\lambda}{\Delta x} \geq \frac{2\pi}{\varphi_{lim}} \tag{7.4.52}$$

For a value of $\phi_{lim} = \pi/18 \, (=10°)$, we need at least 36 points points/wavelength, while a limit value of $\phi_{lim} = \pi/12 \, (=15°)$ leads to 24 points/wavelength. This is satisfied for the example with the 2 wavelengths of Figure 7.3.7, where a slight error in phase can still be seen. The case with 4 waves (Figure 7.3.8) has only 12 points/wavelength, corresponding to a phase angle of 30°, which is clearly inadequate for a satisfactory accuracy. Note that in practice, this can be a severe requirement on the minimum grid size.

- *For current applications with second order schemes, we recommend a minimum of 25 points per wavelength.*
- For unsteady problems, the above recommendations have to be followed strictly, by avoiding first order schemes and by controlling the number of points/wavelength with second order schemes.
- For steady problems, the errors we are monitoring are associated to the transients, and therefore we wish them to be damped as rapidly as possible, and $|G|$ values much smaller than one are beneficial, provided the truncation error is acceptable. This explains the good convergence properties of first order schemes for steady state problems, although their accuracy will generally be poor.

CONCLUSIONS AND MAIN TOPICS TO REMEMBER

This chapter has introduced you to the basic methodology for understanding 'what is behind' numerical schemes, in particular the consistency and stability analysis and the subsequent estimation of the associated errors.

The main topics to remember are the following:

- The fundamental equivalence theorem of Lax ensures that consistency and stability guarantees that the numerical results will converge and hence be a valid approximation of the 'reality' we wish to simulate.
- The consistency condition leads to the observation that the numerical solution does not satisfy the mathematical model equation we are attempting to solve. Instead the numerical solution satisfies an equivalent (or modified) differential equation (EDE) which incorporates the dominating terms of the truncation error.
- The analysis of this EDE leads to crucial information on the behavior of the numerical solution and can provide guidelines to the stability conditions.
- The Von Neumann method for stability analysis is the most reliable and easily applied method.
- Numerical schemes can be conditionally stable, implying a restriction on the time step, or unconditionally stable or just unstable.
- The CFL stability condition expresses that the physical domain of dependence of a hyperbolic problem, must always be totally included in the numerical domain of dependence.
- For hyperbolic, convection dominated problems, we distinguish diffusion errors, which are errors on amplitude from dispersion errors, which are errors on the numerical convection or propagation velocity. Both are function of frequency and scheme parameters.
- The critical parameter for ensuring a certain level of accuracy for time-dependent convection problems is the number of mesh points per wavelength, with a minimum of 25 for second order schemes.
- First order schemes should be avoided for time-dependent problems, as their results are generally enticed with large errors.
- Parabolic, diffusion dominated problems, will generally have mostly diffusion errors.

REFERENCES

Aspray, W. (1990). *John Von Neumann and the Origins of Modern Computing*, MIT Press, Cambridge, MA.

Cathers, B., O'Connor, B.A. (1985). The group velocity of some numerical schemes. *Int. J. Numer. Method. Fluid.*, 5, 201–224.

Charney, J.G., Fjortoft, R., Von Neumann, J. (1950). Numerical integration of the barotropic vorticity equation. *Tellus*, 2, 237–254.

Courant, R., Friedrichs, K.O., Lewy, H. (1928). Uber die Partiellen Differenzgleichungen der Mathematischen Physik. *Mathematische Annalen*, 100, 32–74. *Englisch Translation in IBM Journal*, 1967, 215–234.

Crank, J., Nicholson, P. (1947). A practical method for numerical evaluation of solutions of partial differential equations of the heat conduction type. *Proceedings of the Cambridge Philosophical Society*, Vol. 43, pp. 50–67.

Kreiss, H.O. (1964). On difference approximations of the dissipative type for hyperbolic differential equations. *Comm. Pure Appl. Math.*, 17, 335–353.

Lax, P.D. (1954). Weak solutions of nonlinear hyperbolic equations and their numerical computation. *Comm. Pure Appl. Math.*, 7, 159–193.

Lax, P.D., Richtmyer, R.D. (1956). Survey of the stability of linear finite difference equations. *Comm. Pure Appl. Math.*, 17, 267–293.

Lax, P.D., Wendroff, B. (1960). Systems of conservation laws. *Comm. Pure Appl. Math.*, 13, 217–237.

MacCormack, R.W. (1969). The effect of viscosity in hypervelocity impact cratering. *AIAA Paper*, 69–354.

Richtmyer, R.D., Morton, K.W. (1967). *Difference Methods for Initial Value Problems*, 2nd edn. Interscience Publication J. Wiley & Sons, London.

Trefethen, L.N. (1982). Group velocity in finite difference schemes. *SIAM Rev.*, 24, 113–136.

Trefethen, L.N. (1983). Group velocity interpretation of the stability theory of Gustafsson, Kreiss and Sundstrom. *J. Comput. Phys.*, 49, 199–217.

Trefethen, L.N. (1984). Instability of difference models for hyperbolic initial boundary value problems. *Comm. Pure Appl. Math.*, 37, 329–367.

Ulam, S. [John Von Neumann (1903–1957)] (1958). Bull. Amer. Math. Soc., 64, 1–49.

Vichnevetsky, R., Bowles, J.B. (1982). *Fourier Analysis of Numerical Approximations of Hyperbolic Equations*. SIAM Publications, Philadelphia.

Warming, R.F., Beam, R.W. (1976). "Upwind Second Order Difference Schemes and Applications in Aerodynamic Flows". AIAA Journal, Vol. 24, pp. 1241–1249.

PROBLEMS

P.7.1 Determine the equivalent differential equation and the truncation error for the first order upwind scheme. Follow all the steps leading to the equivalent differential equation of this scheme (7.1.24), as described in Section 7.1.1.

Hint: Obtain

$$u_t + au_x = \frac{a\Delta x}{2}(1 - \sigma)u_{xx} + \frac{a\Delta x^2}{6}(1 - \sigma)(2\sigma - 1)u_{xxx} + O(\Delta x^4)$$

P.7.2 Derive the truncation error for the leapfrog scheme (7.3.1), applying the methodology of Section 7.1.1. Show that the scheme is indeed of second order accuracy in space and time.

Hint: Obtain

$$u_t + au_x = \frac{a\Delta x^2}{6}(\sigma^2 - 1)\frac{\partial^3 u}{\partial x^3} - \frac{a\Delta x^4}{120}(9\sigma^2 - 1)(\sigma - 1)\frac{\partial^5 u}{\partial x^5} + O(\Delta x^6)$$

P.7.3 Derive the truncation error for the Lax–Friedrichs scheme (7.3.9) and show that this scheme is also only first order accurate in space and time. Derive the truncation error dominating term and show that it corresponds to a numerical viscosity equal to $v_{num} = (a\Delta x/2\sigma)(1 - \sigma^2)$ and compare with the FOU scheme.

Hint: Obtain

$$u_t + au_x = \frac{a\Delta x}{2\sigma}(1 - \sigma^2)u_{xx} + \frac{a\Delta x^2}{3}(1 - \sigma^2)u_{xxx} + O(\Delta x^3)$$

P.7.4 Apply a forward space differencing, with a forward time difference (Euler method) to the convective equation $u_t + au_x = 0$. Analyze the stability with Von Neumann's method and show that the scheme is unconditionally unstable for $a > 0$, and conditionally stable for $a < 0$. Derive also the equivalent differential equation and show why this scheme is unstable when $a > 0$.

P.7.5 Solve the one-dimensional diffusion equation $u_t = \alpha u_{xx}$ for the following conditions, with k an integer:

$$u(x, 0) = \sin k\pi x \quad 0 < x < 1$$

$$u(0, t) = 0$$

$$u(1, t) = 0$$

applying the explicit central scheme (7.2.29).

Compare with the exact solution for different values of β, in particular $\beta = 1/3$ and $\beta = 1/6$ (which is the optimal value). Consider initial functions with different wave numbers k, namely $k = 1, 5, 10$.

The exact solution is $u = \exp(-\alpha k^2 \pi^2 t) \sin(k\pi x)$.

Choose $x_i = i\Delta x$ with i ranging from 0 to 30.

Make plots of the computed and of exact solutions in function of x.

Perform the calculations for 5 and 10 time steps and control the error by comparing with equation (7.4.50) for the diffusion error for $\beta = 1/3$.

Calculate the higher order terms in ε_D, for $\beta = 1/6$, by taking more terms in the expansion.

P.7.6 Calculate the amplitude and phase errors for Lax–Friedrichs scheme (7.3.9) after 10 time steps, for an initial wave of the form

$$u(x, 0) = \sin k\pi x \quad 0 < x < 1$$

for $k = 1, 10$.

Consider $\Delta x = 0.02$ and a velocity $a = 1$.

Perform the calculations for $\sigma = 0.25$ and $\sigma = 0.75$. Plot the computed and exact solutions for these various cases; compare and comment the results.

Hint: The exact solution is $u = \sin \pi k(x - t)$. The exact numerical solution is $\bar{u}_i^n = |G|^n \sin \pi k(x_i - a_{num} n\Delta t)$ where $a_{num} = a\varepsilon_\phi$ is the numerical speed of propagation (see equation (7.4.30)).

Show that we can write $\bar{u}_i^n = |G|^n \sin[\pi k(x_i - an\Delta t) + n(\tilde{\Phi} - \Phi)]$, where $\tilde{\Phi}$ is the exact phase, given by equation (7.4.20).

P.7.7 Apply the same problem as P.7.6 to the first order upwind scheme (7.2.21).

P.7.8 Apply the same problem as P.7.6 to the leapfrog scheme (7.3.2).

P.7.9 Apply the central difference in time (leapfrog scheme) to the diffusion equation $u_t = \alpha u_{xx}$ with the space differences of second order accuracy

$$u_i^{n+1} - u_i^{n-1} = 2\frac{\alpha \Delta t}{\Delta x^2}(u_{i+1}^n - 2u_i^n + u_{i-1}^n)$$

Calculate the amplification matrix and show that the scheme is *unconditionally unstable*.

P.7.10 Apply the leapfrog scheme with the upwind space discretization of the convection equation $u_t + au_x = 0$. This is the scheme

$$u_i^{n+1} - u_i^{n-1} = 2\sigma(u_i^n - u_{i-1}^n)$$

Calculate the amplification matrix, and show that the scheme is *unconditionally unstable*.

P.7.11 Find the numerical group velocity for the first order upwind (FOU), Lax–Friedrichs (LF) and Lax–Wendroff (LW) schemes for the linear convection equation, applying the relation (7.4.41) to the dispersion relation of the scheme.

Plot the ratios v_G/a, in function of ϕ and observe the deviations from the exact value of 1.

Hint: Obtain

$$\text{FOU: } v_G = \frac{a[(1-\sigma)\cos\phi + \sigma]}{[1+\sigma(\cos\phi - 1)]^2 + \sigma^2 \sin^2\phi}$$

$$\text{LF: } v_G = a/[\cos^2\phi + \sigma^2 \sin^2\phi]$$

$$\text{LW: } v_G = \frac{a[(1-2\sigma^2 \sin^2\phi/2)\cos\phi + \sigma^2 \sin^2\phi]}{(1-2\sigma^2 \sin^2\phi/2)^2 + \sigma^2 \sin^2\phi}$$

P.7.12 Consider the second order upwind scheme of Warming and Beam, equation (7.4.35). Apply a stability Von Neumann analysis and obtain the amplification factor (7.4.36). show that the scheme is conditionally stable under the CFL condition $0 \le \sigma \le 2$.

Chapter 8

General Properties and High-Resolution Numerical Schemes

OBJECTIVES AND GUIDELINES

In Chapter 7, we learned about the stability and error properties of numerical schemes and we have provided a methodology for the quantitative estimation of the associated errors. We have also derived some guidelines on mesh sizes to achieve a preset level of accuracy, particularly with the very demanding simulation of time-dependent problems.

In addition, we studied a certain number of second order schemes for the convection equations, in particular the leapfrog and the celebrated Lax–Wendroff schemes, both based on central difference formulas, and we have briefly mentioned the second order upwind scheme of Warming and Beam.

At this stage you might have asked yourself about the eventual existence of other schemes? This would be totally justified, as we have noticed already in Chapter 4, that an unlimited number of finite difference formulas can be defined and that for every mathematical model, an unlimited number of schemes could indeed be written down. However, they will not be equally acceptable in practice, as we have learned from Chapter 7. Stability limits, error properties can vary significantly between various schemes and it would be very useful to rely on guidelines for the evaluation of the best-adapted schemes for a given application.

In response to this objective, we will introduce in this chapter a general approach to derive conditions on families of schemes having a predetermined support and order of accuracy. The presented methodology will allow you to either select a new scheme with preset properties, or to evaluate an existing scheme.

However, we are still faced with some unanswered questions and remaining issues, such as the unwanted appearance of wiggles in the numerical solutions with the second order schemes. This is a major problem, which has marked the history of CFD for more than 30 years and required, for its resolution, a deep mathematical understanding of the properties of nonlinear conservation equations and of nonlinear numerical schemes. It has finally led to the introduction of nonlinear components, called *limiters*, even for the simplest linear convection equation, in order to develop *high resolution, monotone* schemes without numerical wiggles.

We will also answer, in this chapter, the following questions:

- *Why do numerical oscillations (wiggles) appear in second (and higher) order schemes for the convection equation and why do they not occur with first order schemes?*
- *Why are wiggles absent with second, or higher, order schemes when applied to the diffusion equation?*

- *How could we remove these unwanted effects, which can severely spoil the quality and reliability of the numerical solution?*

We will focus essentially on hyperbolic systems, that is convection type problems. You might wonder why do we focus so strongly on the convection dominated models?

This is a direct consequence, as seen in Chapter 1, of the high Reynolds number behavior of the overwhelming majority of flow systems, in nature and in technology.

You will remember that the Reynolds number is a measure of the ratio between convection and diffusion effects and therefore, a high Reynolds number means that convection is the dominating feature of the flow systems. It is also the most challenging part, since the convection terms of the Navier–Stokes equations are associated to propagation phenomena and moreover are the main sources of nonlinearities. Diffusion terms on the other hand are mostly linear, at least for constant flow properties and due to their intrinsic nature as elliptic, Laplace type operators, they should always be discretized with central second order schemes. In addition, these diffusion terms do not generate numerical wiggles, as they tend to diffuse any strong variation. You can actually observe the typical behavior of diffusion phenomena on the results of the first order upwind scheme, as shown on Figures 7.1.2, although in this case the diffusion is of numerical origin.

This chapter is organized as follows: Section 8.1 will introduce a general formulation of two-level schemes, with arbitrary support and we will derive general properties on the coefficients of the scheme to achieve a prescribed order of accuracy. We will also establish an explicit relation between the stability analysis, the error evaluation and the equivalent differential equation of the considered scheme. In addition an important property, known as the *accuracy barrier*, will give an indication to the maximum accuracy that can be achieved by a stable scheme on the considered support. Two more advanced subsections provide additional relations and properties on the last two topics.

Section 8.2 is focusing on the analysis of two one-parameter families of schemes, respectively of first order on a 3-point support and of second order on a 4-point support. Applications will be given for the convection and convection-diffusion equations.

Section 8.3 is focused on the critical issue of the numerical oscillations appearing in second (and higher) order schemes. We will explain why these over- and undershoots appear, based on the celebrated Godunov theorem, and will derive the methodology towards their cure. This will require a deep understanding of the fundamental concept of *monotonicity* properties and related terminology. The outcome of the analysis will be the introduction of *high-resolution schemes*, containing nonlinear terms, based on *limiter functions*, which eliminate the unwanted numerical oscillations of the computed solution, while remaining of order higher than one.

Section 8.4 will introduce the finite volume form of the investigated schemes, as a synthesis and a framework towards the generalization to nonlinear and multi-dimensions, as required for practical applications, identifying them by a uniquely associated *numerical flux*. It will also introduce an alternative vision and description of limiters, through the concept of *normalized variables*.

Figure 8.0.1 summarizes the roadmap to this chapter.

You will notice that this chapter has a somewhat more 'theoretical' flavor than the previous chapters. We need this approach to be able to address properties such as limits on order of accuracy in function of number of points, the critical issue of the

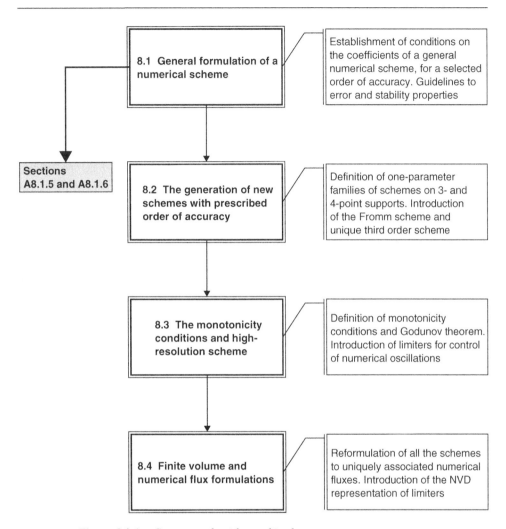

Figure 8.0.1 *Content and guides to this chapter.*

unwanted numerical oscillations of higher order schemes. It should also give you a 'feeling' for the flexibility of numerical schemes and to guide you towards criteria for their selection, for instance when dealing with applications requiring higher orders of accuracy, such as Computational Aero-Acoustics (CAA), where currently schemes with 7 to 15 points are applied for very high orders of accuracy on Cartesian grids.

8.1 GENERAL FORMULATION OF NUMERICAL SCHEMES

We have seen up to now several schemes, both for the convection and diffusion models, of first and second order accuracy. We know however from Chapter 4 that the number of possible schemes is unlimited, depending on the number of points involved in the space and time discretizations. So, we can ask ourselves what are the possible alternatives to the schemes already described and what conditions can we impose on a scheme, in terms of order of accuracy, stability and other conditions, such as monotonicity, for a better control of the unavoidable numerical errors.

This is the objective of this section. We will limit ourselves to explicit, two-time-level schemes, as schemes with more time levels or implicit schemes lead to more complicated mathematics. Aspects of implicit schemes will be addressed in Chapter 9.

We will first define a general form of an explicit two-time-level scheme, with arbitrary coefficients and derive the conditions to reach an order of accuracy p, in time and space. Then, we will derive conditions on these coefficients for stability, based on the Von Neumann method and subsequently obtain expressions for the diffusion and dispersion errors, particularly for the linear convection model equation.

8.1.1 Two-Level Explicit Schemes

A general two-level explicit scheme can be written as a linear combination of mesh point values at level n

$$u_i^{n+1} = \sum_j b_j u_{i+j}^n .$$ (8.1.1)

where the sum over j involves all the mesh points defining the numerical scheme; (see Figure 8.1.1). The range of j, covering ju points upwind and jd points downwind on the x-axis, is called the **support of the scheme**. The total number of support points, including point i, is $M = ju + jd + 1$.

Note that this general formulation is valid for any one-dimensional model, pure convection, diffusion or convection–diffusion.

This equation represents the way the new function value of u at point i, at level $n + 1$ is obtained from the known function values at time level n. The coefficient b_j

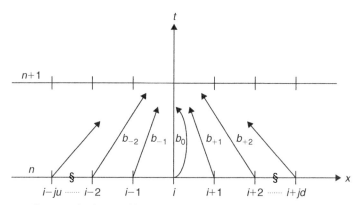

Support of scheme with ju upstream and jd downstream points

Figure 8.1.1 *Weight coefficients of contributions of function values at level n to solution at level $n + 1$.*

is the weight of the contribution of point $(i + j)$ to the new value at point i and time level $n + 1$.

If we consider the linear convection equation $u_t + au_x = 0$, the schemes already discussed correspond to the following coefficients.

First order upwind scheme (7.2.25)

The support is determined by the values of $j(-1, 0)$, with the coefficients

$$b_{-1} = \sigma \quad b_0 = 1 - \sigma \tag{8.1.2}$$

and the scheme is written as

$$u_i^{n+1} = \sigma u_{i-1}^n + (1 - \sigma)u_i^n \tag{8.1.3}$$

Lax–Friedrichs scheme (7.3.9)

The support is determined by the values of $j(-1, 0, 1)$, with the coefficients

$$b_{-1} = \frac{1}{2}(1 + \sigma) \quad b_0 = 0 \quad b_1 = \frac{1}{2}(1 - \sigma) \tag{8.1.4}$$

and the scheme is written as

$$u_i^{n+1} = \frac{1}{2}(1 + \sigma)u_{i-1}^n + \frac{1}{2}(1 - \sigma)u_{i+1}^n \tag{8.1.5}$$

Lax–Wendroff scheme (7.3.14)

The support is determined by the values of $j(-1, 0, 1)$, with the coefficients

$$b_{-1} = \frac{\sigma}{2}(1 + \sigma) \quad b_0 = 1 - \sigma^2 \quad b_1 = -\frac{\sigma}{2}(1 - \sigma) \tag{8.1.6}$$

and the scheme can be written as

$$u_i^{n+1} = \frac{\sigma}{2}(1 + \sigma)u_{i-1}^n + (1 - \sigma^2)u_i^n - \frac{\sigma}{2}(1 - \sigma)u_{i+1}^n \tag{8.1.7}$$

This is illustrated on Figure 8.1.2 for the upwind, Lax–Friedrichs and Lax–Wendroff schemes.

The leap-frog scheme cannot be put into this form since it is a three-level scheme and the general development to follow will not be valid in this case. For an implicit scheme, we could also write an expression similar to equation (8.1.1) with the terms on the right-hand side taken at time level $(n + 1)$.

For the diffusion equation $u_t = au_{xx}$ and the explicit scheme (7.2.29)

The support is again $j(-1, 0, 1)$ and the coefficients are:

$$b_{-1} = \beta = \alpha \Delta t / \Delta x^2 \quad b_0 = 1 - 2\beta \quad b_1 = \beta \tag{8.1.8}$$

The b_j coefficients are obviously not arbitrary and have to satisfy a certain number of consistency conditions, depending on the order of accuracy of the scheme. *If p is*

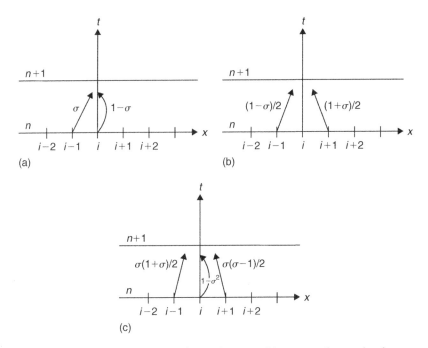

Figure 8.1.2   *Weight coefficients of contributions of function values at level n to solution at level n + 1, for: (a) first order upwind scheme, (b) Lax–Friedrichs scheme and (c) Lax–Wendroff scheme.*

the order of the scheme, there are clearly (p + 1) relations to be satisfied. A first condition is obtained from the requirement that a constant should be a solution of the numerical scheme. This leads to the first consistency condition:

$$\sum_{j} b_j = 1 \tag{8.1.9}$$

The other conditions can be derived from a consistency analysis, following the guidelines of Chapter 7, obtaining the truncation error and the equivalent differential equation for scheme (8.1.1).

We introduce the Taylor series expansion in space,

$$u^n_{i+j} = u^n_i + \sum_{m=1}^{\infty} \frac{(j \cdot \Delta x)^m}{m!} \left(\frac{\partial^m u}{\partial x^m} \right) \tag{8.1.10}$$

and in time written as

$$u^{n+1}_i = u^n_i + \sum_{m=1}^{\infty} \frac{(\Delta t)^m}{m!} \left(\frac{\partial^m u}{\partial t^m} \right) \tag{8.1.11}$$

Inserting these developments into equation (8.1.1), leads to

$$\Delta t \frac{\partial u}{\partial t} + \sum_{m=2}^{\infty} \frac{(\Delta t)^m}{m!} \left(\frac{\partial^m u}{\partial t^m} \right) = \sum_{j} (b_j \cdot j \Delta x) \frac{\partial u}{\partial x} + \sum_{j} b_j \sum_{m=2}^{\infty} \frac{(j \cdot \Delta x)^m}{m!} \left(\frac{\partial^m u}{\partial x^m} \right)$$

$$(8.1.12)$$

To go further, we need to specify the mathematical model we are simulating.

If we consider the *diffusion equation* $u_t = \alpha u_{xx}$, then there is no first space derivative and the second consistency relation requires that

$$\sum_{j} (b_j \cdot j) = 0 \qquad (8.1.13)$$

The first term of the m-summation is a second derivative, which has to be equal to the diffusion coefficient for consistency, leading to the relation, with $\beta = \alpha \, \Delta t / \Delta x^2$

$$\sum_{j} (b_j \cdot j^2) = 2\beta \qquad (8.1.14)$$

If we consider a support $j(-1, 0, 1)$, then the relations (8.1.9), (8.1.13) and (8.1.14) lead to the coefficients (8.1.8). The remaining terms form the truncation error.

8.1.2 Two-Level Schemes for the Linear Convection Equation

Our main interest lies however with the linear convection equation, for the reasons already explained.

In this case, the coefficient of the first space derivative in equation (8.1.12) has to be equal to $(-a \, \Delta t)$ for consistency. Hence, a second consistency condition is obtained as

$$\sum_{j} (b_j \cdot j) = -\sigma \qquad (8.1.15)$$

With this condition, the development (8.1.12) becomes

$$\frac{\partial u}{\partial t} + a \frac{\partial u}{\partial x} = - \sum_{m=2}^{\infty} \frac{(\Delta t)^{m-1}}{m!} \left(\frac{\partial^m u}{\partial t^m} \right) + \frac{1}{\Delta t} \sum_{j} b_j \sum_{m=2}^{\infty} \frac{(j \cdot \Delta x)^m}{m!} \left(\frac{\partial^m u}{\partial x^m} \right) \quad (8.1.16)$$

The full equivalent differential equation is obtained by replacing the time derivatives in equation (8.1.16) by space derivatives derived from the equivalent differential equation. Rewriting equation (8.1.16) by replacing the summation index m by n,

$$\frac{\partial u}{\partial t} = -a \frac{\partial u}{\partial x} + \frac{1}{\Delta t} \sum_{j} b_j \sum_{n=2}^{\infty} \frac{(j \cdot \Delta x)^n}{n!} \left(\frac{\partial^n u}{\partial x^n} \right) - \sum_{n=2}^{\infty} \frac{(\Delta t)^{n-1}}{n!} \left(\frac{\partial^n u}{\partial t^n} \right) \quad (8.1.17)$$

we can evaluate the time derivatives by expanding the first terms to the power m. To the lowest order, we have by applying the Taylor development of $(x + y + z)^m$ as (HOT stands for Higher Order Terms)

$$(x + y + z)^m = x^m + mx^{m-1}y + mx^{m-1}z + \text{HOT}$$

In the third line, we have replaced the time derivative by the first term of this line

$$\frac{\partial^m u}{\partial t^m} = \left[-a \frac{\partial}{\partial x} + \frac{1}{\Delta t} \sum_j b_j \sum_{n=2}^{\infty} \frac{(j \cdot \Delta x)^n}{n!} \left(\frac{\partial^n}{\partial x^n} \right) - \sum_{n=2}^{\infty} \frac{\Delta t^{n-1}}{n!} \left(\frac{\partial^n}{\partial t^n} \right) \right]^m u$$

$$= (-a)^m \frac{\partial^m u}{\partial x^m} + m(-a)^{m-1} \frac{1}{\Delta t} \sum_j b_j \sum_{n=2}^{\infty} \frac{(j \cdot \Delta x)^n}{n!} \left(\frac{\partial^{n+m-1}}{\partial x^{n+m-1}} \right) u$$

$$- m(-a)^{m-1} \frac{\partial^{m-1}}{\partial x^{m-1}} \sum_{n=2}^{\infty} \frac{\Delta t^{n-1}}{n!} \left(\frac{\partial^n}{\partial t^n} \right) u + \text{HOT}$$

$$= (-a)^m \frac{\partial^m u}{\partial x^m} + m(-a)^{m-1} \frac{\Delta x}{\Delta t} \sum_j \sum_{n=2}^{\infty} b_j j^n \frac{\Delta x^{n-1}}{n!} \left(\frac{\partial^{n+m-1}}{\partial x^{n+m-1}} \right) u$$

$$- m(-a)^{m-1} \sum_{n=2}^{\infty} \frac{\Delta t^{n-1}}{n!} (-a)^n \left(\frac{\partial^n}{\partial x^n} \right) u + \text{HOT}$$

$$= (-a)^m \frac{\partial^m u}{\partial x^m} + m(-a)^{m-1} \frac{\Delta x}{\Delta t} \sum_{n=2}^{\infty} \frac{\Delta x^{n-1}}{n!} \left\{ \sum_j b_j j^n - (-\sigma)^n \right\}$$

$$\times \left(\frac{\partial^{n+m-1}}{\partial x^{n+m-1}} \right) u + \text{HOT} \tag{8.1.18}$$

Introducing this relation in the development (8.1.16), we obtain the following equivalent differential equation for the scheme (8.1.1):

$$\frac{\partial u}{\partial t} + a \frac{\partial u}{\partial x} = \frac{1}{\Delta t} \sum_j b_j \sum_{m=2}^{\infty} \frac{(j \cdot \Delta x)^m}{m!} \left(\frac{\partial^m u}{\partial x^m} \right) - \sum_{m=2}^{\infty} (-a)^m \frac{(\Delta t)^{m-1}}{m!} \left(\frac{\partial^m u}{\partial x^m} \right)$$

$$- \sum_{m=2}^{\infty} \frac{(\Delta t)^{m-1}}{m!} m(-a)^{m-1} \frac{\Delta x}{\Delta t} \sum_{n=2}^{\infty} \frac{\Delta x^{n-1}}{n!} \left\{ \sum_j b_j j^n - \left(-\frac{a\Delta t}{\Delta x} \right)^n \right\}$$

$$\times \left(\frac{\partial^{n+m-1}}{\partial x^{n+m-1}} \right) u + \text{HOT}$$

$$= \frac{\Delta x}{\Delta t} \sum_{m=2}^{\infty} \frac{(\Delta x)^{m-1}}{m!} \left[\sum_{j} b_j j^m - \left(-\frac{a \Delta t}{\Delta x} \right)^m \right] \left(\frac{\partial^m u}{\partial x^m} \right)$$

$$- \frac{\Delta x}{\Delta t} \sum_{m=2}^{\infty} \frac{1}{(m-1)!} (-\sigma)^{m-1} \sum_{n=2}^{\infty} \frac{\Delta x^{m+n-2}}{n!}$$

$$\times \left\{ \sum_{j} b_j j^n - \left(-\frac{a \Delta t}{\Delta x} \right)^n \right\} \left(\frac{\partial^{n+m-1}}{\partial x^{n+m-1}} \right) u + \text{HOT} \qquad (8.1.19)$$

The right-hand side of this equation is the truncation error of the scheme, where the first term is the dominating contribution.

Additional conditions can now be derived, by expressing that the scheme will be of order of accuracy p. This will be satisfied if all the dominating terms with order lower than p vanish in the truncation error. Hence this leads to the following consistency conditions on the b_j coefficients, which have to satisfy the following $(p+1)$ relations, with $\sigma = a \, \Delta t / \Delta x$ being the CFL number

$$\sum_{j} b_j j^m = (-\sigma)^m \quad \text{for } m = 0, 1, \dots, p \qquad (8.1.20)$$

The relations for $m = 0$ and 1 reproduce the conditions (8.1.9) and (8.1.15).

These $p + 1$ conditions will define $(p + 1)$ b_j coefficients for a given support and a given order of accuracy p.

The remaining terms in the first sum of equation (8.1.19) therefore starts at $m = p + 1$, leading to the first non-zero term of the truncation error:

$$\text{TE} = \frac{\Delta x}{\Delta t} \left[\sum_{j} b_j j^{p+1} - (-\sigma)^{p+1} \right] \frac{(\Delta x)^p}{(p+1)!} \left(\frac{\partial^{p+1} u}{\partial x^{p+1}} \right) + \text{HOT} \qquad (8.1.21)$$

If the remaining terms in equation (8.1.18) are taken into account, the higher order terms of the truncation error can be derived. Considering the conditions (8.1.20), the summation on n of the higher order terms of TE starts at $n = p + 1$. Hence, the first contribution starts at $m = 2$, leading to a term proportional to Δx^{p+1}.

Writing the general form of the equivalent differential equation for a scheme of order p as

$$u_t + a u_x = \sum_{m=p}^{\infty} a_{m+1} \Delta x^m \left(\frac{\partial^{m+1} u}{\partial x^{m+1}} \right) \qquad (8.1.22)$$

the second highest term of order Δx^{p+1} can be written as follows, introducing the quantity α_{p+1} defined by

$$\alpha_{p+1} = \frac{\Delta x}{\Delta t} \left[\sum_{j} b_j j^{p+1} - (-\sigma)^{p+1} \right] \frac{1}{(p+1)!} \qquad (8.1.23)$$

leading to

$$a_{p+1} = \alpha_{p+1} \tag{8.1.24}$$

and, as seen from equation (8.1.19):

$$a_{p+2} = \alpha_{p+2} + \sigma\alpha_{p+1} \tag{8.1.25}$$

For a second order scheme, $(p=2)$, we find for the next term associated to $m=3$

$$a_{p+3} = \alpha_{p+3} + \sigma a_{p+2} + \frac{\sigma^2}{2}a_{p+1} \tag{8.1.26}$$

Higher order terms can be obtained by further inspection, if necessary (see also Problem P.8.1).

8.1.3 Amplification Factor, Error Estimation and Equivalent Differential Equation

The stability of the scheme can be partly analyzed by an investigation of the properties of the truncation error and the equivalent differential equation, as shown initially by Hirt (1968).

Examples have been given in Section 7.1, where the explicit, central scheme for the convection equation, was shown to correspond to a negative numerical viscosity coefficient and hence could only be unstable and where the conditional stability of the first order upwind scheme was also derived.

Generally, this method will lead to necessary conditions for stability, although sufficient conditions can in some cases also be derived, in particular for hyperbolic equations.

Warming and Hyett (1974) have shown that for a wide range of schemes, necessary as well as sufficient stability conditions can be derived from the coefficients of the equivalent differential equation. In these cases, the method of the equivalent differential equation can be considered on equal foot with the Von Neumann method, with regard to the stability analysis.

This analysis can be extended to nonlinear conservation laws, with attention to the nonlinear contributions from the physical fluxes. It appears that they can generate additional dissipation, but also anti-dissipation which can reduce under certain conditions the oscillatory behavior of higher order schemes, without being sufficient to remove it totally.[1]

---

[1] More detailed investigations, relating the structure of the truncation error to the stability of the scheme have been developed by Warming and Hyett (1974) and in a very systematic way by Yanenko and Shokin (1969). The extensive application of the equivalent differential equation developed by the Russian authors, called the method of Differential Approximation, can be found in a book by Shokin (1983). A most important application of this method is the analysis of the nature and properties of the truncation error. In particular the errors generated from non-linear terms can be investigated by this approach and schemes can be defined in order to minimize the non-linear error sources (Lerat and Peyret, 1974, 1975; Lerat, 1979; Shokin, 1983).

In order to investigate the stability and accuracy of the general scheme we rewrite the equivalent differential equation (8.1.22), by separating the even and odd order derivatives. The motivation of this step is connected to the properties of the errors, since we will show next that *the even order derivatives contribute to the diffusion error, while the odd order derivatives define the dispersion errors.* Hence, we write, where the a_{2l} and a_{2l+1} coefficients are uniquely defined by the b_j coefficients:

$$u_t + au_x = \sum_{l=1}^{\infty} \left[a_{2l} \left(\frac{\partial^{2l} u}{\partial x^{2l}} \right) \Delta x^{2l-1} + a_{2l+1} \left(\frac{\partial^{2l+1} u}{\partial x^{2l+1}} \right) \Delta x^{2l} \right] \qquad (8.1.27)$$

If the scheme is of order p in space, the first non-zero coefficient is proportional to Δx^p. Therefore,

$$a_{2l} = a_{2l-1} = 0 \quad \text{for } 2l, \ 2l - 1 < p \qquad (8.1.28)$$

The connection to the stability analysis is based on the fundamental property of the equivalent differential equation, as representing the behavior of the numerical solution, as seen in Section 7.1. *Therefore the analytical amplification factor of the above equation for a harmonic k, obtained by inserting a solution of the form $e^{-I\omega t}e^{Ikx}$ represents the Von Neumann amplification factor of the numerical scheme.*

Hence, we have, following the definition (7.4.8):

$$G = e^{-I\omega \Delta t} \qquad (8.1.29)$$

where

$$I\omega = Iak - \frac{1}{\Delta x} \sum_{l} [a_{2l}(-)^l \phi^{2l} + I(-)^l a_{2l+1} \phi^{2l+1}] \qquad (8.1.30)$$

Observe that the even derivatives lead to a real component of ω, and the uneven coefficients contribute to the imaginary part.

If you refer to equations (7.4.11)–(7.4.13), you will notice that the diffusion error is defined by the real part and the dispersion error by the imaginary part of ω. Hence we see the very important association between even order derivatives and dissipation of the scheme, through the diffusion error; while the uneven derivative terms contribute to the dispersion error.

From the definition of the numerical errors, following Section 7.4, we write the amplification factor (8.1.29) as

$$G = |G|e^{-I\Phi} \qquad (8.1.31)$$

where the modulus of G is defined by the real part of ω, and the phase of G, is fully defined by the imaginary part.

This defines directly the diffusion error as

$$\varepsilon_D = |G| = \exp\left[\sum_{l} (-)^l a_{2l} \phi^{2l} \frac{\Delta t}{\Delta x} \right] \qquad (8.1.32)$$

and the dispersion error as

$$\varepsilon_\phi = \frac{\Phi}{ak\,\Delta t} = 1 - \frac{1}{a}\sum_l (-)^l \phi^{2l} a_{2l+1} \tag{8.1.33}$$

Observe again that the dispersion error contains only odd order coefficients, while the diffusion error is totally defined by the even order coefficients.

A Taylor expansion as a function of ϕ, around $\phi = 0$, gives from equation (8.1.32) the amplitude of the diffusion error, for small values of ϕ

$$\varepsilon_D = |G| = 1 + \sum_l (-)^l a_{2l}\phi^{2l}\frac{\Delta t}{\Delta x} + \cdots \tag{8.1.34}$$

A necessary condition for stability is directly obtained form the diffusion error, since it has to be lower than one for stability. Hence, to the lowest order, the necessary condition for stability, is

$$(-)^r a_{2r} < 0 \quad \text{with } 2r = (2l)_{\min} \tag{8.1.35}$$

if $2r$ is the lowest even derivative of the expansion.

For a first order scheme, $p = 1$, the lowest value of $2l$ is 2, hence $r = 1$. In this case the coefficient a_2 will be different from zero, and correspond to a term $a_2 u_{xx}$. This term is interpreted as a **numerical viscosity** generated by the discretization error of the scheme, and **has to be positive for stability**.

For second or third order schemes, $p = 2$ or 3, the lowest value of $2l$ is equal to 4 and $r = 2$ and the first non-zero term in the expansion of the truncation error is a third order derivative $a_3 u_{xxx}$, associated to a dispersion error. Since the phase error can be of either sign, no stability condition can be deduced from the dispersion term.

However, the fourth order derivative term $a_4(\partial^4 u/\partial x^4)$ contributes a dissipation error

$$\varepsilon_D = \exp\left[a_4\phi^4\frac{\Delta t}{\Delta x}\right] \tag{8.1.36}$$

and the coefficient a_4 has to be negative, according to equation (8.1.35).

Dissipative schemes

An important criterion for stability when the coefficients are not constant or when the problem is not linear is related to the dissipation property in the sense of Kreiss, as mentioned in Section 7.4. In particular, this property requires that the amplification factor, and hence the diffusion error, be different from one for the high frequency waves associated to the $2\Delta x$ waves, or $\phi \approx \pi$. From equation (8.1.44), we have

$$G(\pi) = \sum_j b_j \cos j\pi = \sum_{j\,\text{even}} b_j - \sum_{j\,\text{odd}} b_j \tag{8.1.37}$$

The condition

$$|G(\pi)| \le 1$$

is satisfied if

$$-1 \leq \left(\sum_{j\text{even}} b_j - \sum_{j\text{odd}} b_j \right) \leq 1 \tag{8.1.38}$$

or, taking into account the consistency condition, $\Sigma_j b_j = 1$, we obtain (Roe, 1981)

$$0 \leq \sum_{j\text{even}} b_j \leq 1 \tag{8.1.39}$$

The equality signs on these limits are valid for stability, but have to be excluded for the Kreiss' dissipative property. For Lax–Friedrichs scheme $b_0 = 0$ and the condition (8.1.39) is not satisfied, indicating that this scheme is not dissipative in the sense of Kreiss. For the upwind scheme, $b_0 = 1 - \sigma$, and the scheme is stable and dissipative for $0 < \sigma < 1$. For Lax–Wendroff scheme, $b_0 = 1 - \sigma^2$, and the scheme is stable and dissipative for $0 < |\sigma| < 1$.

Some more advanced additions are to be found in Section A8.1.5, at the end of this section.

8.1.4 Accuracy Barrier for Stable Scalar Convection Schemes

We can ask ourselves, when we select the number of support points, what is the maximum possible order of accuracy.

Referring to the general explicit scheme, we can be more specific by indicating separately the number of upstream and downstream points, denoted by ju and jd. The summation over j extends from $-ju$ to $+jd$ including $j = 0$; hence the support of the scheme is $(i - ju, i - ju + 1, \ldots, i - 1, i, i + 1, \ldots, i + jd - 1, i + jd)$ containing $M = (ju + jd + 1)$ points. The general form of the scheme is rewritten as

$$u_i^{n+1} = \sum_{j=-ju}^{+jd} b_j u_{i+j}^n \tag{8.1.40}$$

As stated above, if p is the order of accuracy of the scheme, there are $(p + 1)$ consistency conditions and hence $(M - p - 1) = (ju + jd - p)$ free parameters to select. Therefore p has to satisfy

$$p \leq ju + jd \tag{8.1.41}$$

indicating that the maximum possible order of accuracy of the scheme (8.1.40) is $p_{\text{max}} = ju + jd$.

That is, the number of points in the support, excluding point i, defines the maximum possible order of accuracy.

However, additional conditions are to be considered if we add the requirement of stability and we could wonder as to what is the maximum order of accuracy for a *stable* scheme with ju upstream and jd downstream points.

Several investigations have been pursued in order to provide a response to this question. The first result was obtained by Iserles (1982), showing that stable schemes have to satisfy the following condition, when $a > 0$, referred to as the **accuracy barrier**,

$$p \leq 2 \min (ju, jd + 1) \tag{8.1.42}$$

Table 8.1.1 *Maximum order of accuracy of an explicit stable scheme for the linear convection equation, using ju upstream and jd downstream points.*

ju \ jd	0	1	2	3	4
0	0	0	0	0	0
1	1	2	2	2	2
2	2	3	4	4	4
3	2	4	5	6	6
4	2	4	6	7	8

Additional analysis along this line can be found in Iserles and Strang (1983), Jeltsch (1985), Jeltsch and Smit (1987).

The maximum order of accuracy for a stable scheme is obtained with the equal signs and is satisfied for $ju = jd$; $ju = jd + 1$ or $ju = jd + 2$, as seen from the following Table 8.1.1.

A remarkable property emerges from this result for a ***pure upwind explicit scheme***, that is for $jd = 0$ when the convection speed $a > 0$. In this case, *the maximum order of accuracy cannot exceed two*, whatever the number of upwind points, although the spatial accuracy of the space discretization for $ju \geq 3$ can be higher than 2.

On the other hand, an explicit scheme with two upstream points and one downstream point, $ju = 2$, $jd = 1$, can achieve third order accuracy, indicating that the addition of the point $(i + 1)$, although outside the physical domain of dependence, increases the maximum order of accuracy. Hence, restricting the numerical schemes to follow strictly some of the physical propagation properties, such as the domain of dependence, is not necessarily the most effective choice. A ***pure downwind scheme*** with $ju = 0$ for $a > 0$ is unstable reflecting the CFL condition which forbids schemes going *against* the domain of dependence rule expressed in Section 7.2.

On the other hand, the ***central schemes*** with $ju = jd$, reach *the highest accuracy with the lowest number of points*, for a fixed value of ju, as can be seen from looking at the diagonal of Table 8.1.1. With $ju = jd = s$, j covers the interval $(-s, +s)$ and the number of support mesh points is $M = 2s + 1$.

The consistency relations (8.1.20) represent $(p + 1)$ conditions for the $M = 2s + 1$ coefficients b_j to obtain a scheme of order of accuracy p. Therefore we can achieve a ***maximum accuracy of*** $p_{max} = 2s$.

In the case of a symmetric scheme, these consistency relations form a linear non-homogeneous system of the Vandermonde type, which can be solved exactly, using Cramer's rule, see Karni (1994), defining completely the central schemes of order $2s$. The general form of the b_j coefficients is defined by

$$b_j = \prod_{\substack{k=-s \\ k \neq j}}^{k=s} \frac{\sigma + k}{k - j} \tag{8.1.43}$$

For $s = 1$, we recover the Lax–Wendroff scheme, while a fourth order accurate scheme is obtained for $s = 2$.

The restrictions expressed by equation (8.1.42) are connected to the choice of an *explicit scheme involving two time levels*. Therefore, other choices for the time integration, to be discussed in Chapter 9, will *not* necessarily suffer from the same restriction.

Some more advanced considerations for the accuracy barrier of multi-time-level schemes are presented in Section A8.1.6, to be found at the end of this section.

A8.1.5 An Addition to the Stability Analysis

An exact expression for the amplification matrix of the scheme (8.1.1) is obtained by applying the Von Neumann method, following the guidelines seen in the previous chapter.

For a Fourier mode $u_{i+j}^n \Rightarrow V^n e^{I(i+j)\phi}$ ($\phi = k\Delta x$) and the definition of the amplification factor $G(\phi) = V^{n+1}/V^n$, we obtain

$$G(\phi) = \sum_j b_j e^{Ij\phi} = \sum_j b_j \cos j\phi + I \sum_j b_j \sin j\phi \qquad (8.1.44)$$

The stability condition requires that the modulus of the amplification factor should be lower or equal to one. This modulus can be written directly as follows, since all the b_j coefficients are real, as

$$|G(\phi)|^2 = \left| \sum_j b_j e^{Ij\phi} \right|^2 = \left| \sum_j b_j \cos j\phi \right|^2 + \left| \sum_j b_j \sin j\phi \right|^2$$

$$= \sum_j \sum_k b_j b_k \cos(j-k)\phi \qquad (8.1.45)$$

Note that $(j - k)$ ranges from $(-2ju, 2jd)$, generating $(ju + jd)^2 = (M - 1)^2$ terms. For $ju = jd = 1$, we have at most $m = 4$ terms.

With the trigonometric relation

$$\cos m\phi = 1 + \sum_{l=1}^{m} (-)^l \frac{2^{2l}}{(2l)!} \prod_{k=0}^{l-1} (m^2 - k^2) \sin^{2l} \phi/2 \qquad (8.1.46)$$

equation (8.1.45) can be written as a polynomial in the variable

$$z = \sin^2 \phi/2 \qquad (8.1.47)$$

under the form

$$|G(\phi)|^2 = 1 - \sum_{l=1}^{m} \beta_l z^l \qquad (8.1.48)$$

The choice for formula (8.1.46) is due to the presence of the unity term in the right-hand side, which allows to explicit the term '1', since for $\phi = 0$, we have $z = 0$ and $|G(\phi)| = 1$ as expected, as a consequence of the consistency condition (8.1.9), expressing that the sum of the b_j coefficients is equal to one.

The polynomial in z is at the most of degree m, where m is the total number of points included in the scheme, located at the left and at the right of mesh point i. That is, $m = M - 1 = ju + jd$ is equal to the total number of points involved at level n, excluding point i. The coefficients β_l contain the sum of products of (two) b_j coefficients.

It is seen from equation (8.1.48) that a term z^r can always be factored out from the sum over m, leading to an expression of the form

$$|G(\phi)|^2 = 1 - z^r S(z) \tag{8.1.49}$$

where $S(z)$ is a polynomial at most of order $s = m - r$.

The polynomial $S(z)$ determines the Von Neumann stability of the scheme, since the stability condition $|G(\phi)| \leq 1$ will be satisfied for the following necessary and sufficient conditions

$$0 \leq S(z) \leq 1 \quad \text{for } 0 < z \leq 1 \tag{8.1.50}$$

Warming and Hyett (1974) showed that this condition is equivalent to the requirement

$$S(0) \geq 0 \quad \text{and} \quad S(z) \geq 0 \quad \text{for } 0 < z \leq 1 \tag{8.1.51}$$

As an example, for all schemes based on three points including point i, $m = 2$ and therefore $r = 1$ or 2 and $s = 1$ or 0.

For first order schemes, $r = 1$ and for second order schemes $r = 2$. In particular, the Lax–Wendroff scheme corresponds to $s = 0$ and the polynomial $S(z)$ reduces to a constant $S(0) = -16b_1 b_{-1} = 4\sigma^2(1 - \sigma^2)$, whereby we recover the CFL condition $|\sigma| \leq 1$.

The second condition (8.1.51) might not be easy to achieve for high order schemes, when $S(z)$ is a high order polynomial, but can readily be worked out if $S(z)$ is at most of degree one. In this case, the conditions $S(0) > 0$ and $S(1) > 0$ are both necessary and sufficient for stability. For instance, for $r = 1$ we obtain for stability (see Problem P.8.4).

$$S(1) = 2\frac{\Delta t}{\Delta x}\left[\frac{1}{3}a_2 - \frac{\Delta t}{\Delta x}a_2^2 - a_4\right] \geq 0 \tag{8.1.52}$$

An extension of this analysis for two-level implicit schemes can be found in Warming and Hyett (1974).

A8.1.6 An Advanced Addition to the Accuracy Barrier

Jeltsch and his coworkers have been investigating extensions of Iserles' theorem (8.1.41) to multi-time-level schemes, with K time levels of the form

$$\sum_{k=0}^{K} \sum_{j=-ju_k}^{+jd_k} b_j^k u_{i+j}^{n+k} = 0 \tag{8.1.53}$$

with the consistency condition

$$\sum_{k=0}^{K}\sum_{j=-ju_k}^{+jd_k} b_j^k = 0 \qquad (8.1.54)$$

where the b_j^k coefficients depend only on the Courant number $\sigma = a\,\Delta t/\Delta x$.

Note that the explicit two-time-level scheme (8.1.40) is obtained for $K = 1$, $ju_1 = jd_1 = 0$ with $b_0^1 = -1$.

For other values of the coefficients of the highest time index K, b^K, when $j \neq 0$, equation (8.1.53) represents an **_implicit scheme_** (see Figure 8.1.3).

The maximum order of accuracy of the considered schemes, excluding the stability restrictions, is equal to the total number of variables minus the two consistency conditions given by equation (8.1.54) and by the generalization of condition (8.1.15), written as

$$\sum_{k=0}^{K}\sum_{j=-ju_k}^{+jd_k} jb_j^k = \sigma \sum_{k=0}^{K}\sum_{j=-ju_k}^{+jd_k} kb_j^k \qquad (8.1.55)$$

Hence, we obtain

$$p_{max} = JU + JD - 2 \qquad (8.1.56)$$

where JU and JD are respectively the sum of the number of upstream and downstream points at all time levels with respect to the characteristic issued from the point i at time level $n + K$. As seen from Figure 8.1.3 we have

$$JU = \sum_{k=0}^{K} ju_k \quad JD = \sum_{k=0}^{K} jd_k + K \qquad (8.1.57)$$

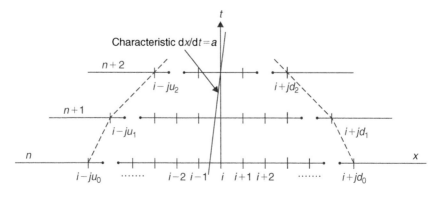

Figure 8.1.3 *Support of multi-level scheme with ju_k upstream and ju_k downstream points at time level k.*

provided the characteristic intersects the mesh at time level n between i and $i-1$. If this is not the case, the counts for JU and JD can easily be adapted accordingly.

This extension appears to be extremely difficult to prove in general and at the time of writing, the following accuracy barrier for stable multi-time-level schemes (8.1.53) has only been proven in special cases and is conjectured for general cases to be given by

$$p \leq 2\min(JU,JD) \tag{8.1.58}$$

See Strang and Iserles (1983), Jeltsch and Smit (1987), Jeltsch (1988), Jeltsch, Kiani and Raczek (1991).

More results can be proven for three-level-schemes, $K=2$, following Jeltsch, Renaut and Smit (1997), where the relation (8.1.58) is demonstrated based on the validity of a conjecture related to the multiplicity of order stars. The concept of order stars, introduced by Wanner, Hairer and Nørsett (1978) and further developed by Hairer and Wanner (1991), Iserles and Nørsett (1991), allows an elegant geometrical interpretation of the relationship between accuracy and stability in the space of complex analytical functions. We refer the interested reader to the original references for more details about the mathematical definition and properties of order stars.

The three-level schemes are implicit if $ju_2 > 0$ and $jd_2 > 0$ and explicit only for $ju_2 = jd_2 = 0$. Jeltsch, Renaut and Smit (1997) restrict their analysis to *convex* schemes with an increasing stencil for which the number of points at a given time level is increasing with decreasing time index k, as displayed on Figure 8.1.3, i.e.

$$ju_2 \leq ju_1 \leq ju_0 \quad \text{with } ju_0 - ju_1 \leq ju_1 - ju_2$$
$$jd_2 \leq jd_1 \leq jd_0 \quad \text{with } jd_0 - jd_1 \leq jd_1 - jd_2 \tag{8.1.59}$$

Hence, this excludes for instance the implicit Euler scheme (7.2.25), which is unconditionally stable, or the explicit leapfrog scheme for which $ju_0 = jd_0 = 0, ju_1 = jd_1 = 1$ together with $ju_2 = jd_2 = 0$.

Stronger results are obtained for the convex schemes with maximal accuracy p_{max}, in particular for $a > 0$, the authors show that for stability

$$p_{max} \leq 2JU \quad \text{when } a > 0 \tag{8.1.60}$$

for which the above mentioned conjecture is shown to be true. For other cases, the relation (9.3.28) is proven, assuming the conjectured proposition.

8.2 THE GENERATION OF NEW SCHEMES WITH PRESCRIBED ORDER OF ACCURACY

An interesting outcome of the analysis derived in the previous section is the possibility to generate new families of schemes, having a given support and a given order of accuracy. The following derivation is based on an elegant analysis by Phil Roe (1981).

The consistency conditions (8.1.20), for a scheme of order of accuracy p, define $(p+1)$ relations for the b_j coefficients. If the support of the scheme covers M points there are M values b_j, and $(M-p-1)$ coefficients b_j can be chosen arbitrarily. If $M = p+1$, there is a unique solution and therefore there is always a unique scheme on a support of M points having the maximum possible order of accuracy of $p = M - 1$. However, as seen above, not all of these schemes are valid, as some can be unstable, if they do not satisfy the accuracy barrier of Iserles.

When $M > p + 1$, families of schemes with $(M - p - 1)$ parameters can be defined by analyzing all the possible solutions of the system of equations (8.1.20).

8.2.1 One-Parameter Family of Schemes on the Support $(i - 1, i, i+1)$

Let us consider the 3-point support $j = -1, 0, 1$, $(M = 3)$, on which we can define a *one-parameter family of first order accurate schemes* by satisfying the two first conditions (8.1.20):

$$b_{-1} + b_0 + b_1 = 1$$
$$b_{-1} - b_1 = \sigma \qquad (8.2.1)$$

If, following Roe (1981), the schemes are identified by the set $S(b_{-1}, b_0, b_1)$, the Lax–Friedrichs scheme corresponds to $S_{LF}((1+\sigma)/2, 0, (1-\sigma)/2)$ and the first order upwind scheme to $S_{FOU}(\sigma, (1-\sigma), 0)$. The unstable, central scheme (7.1.7) is represented by $S_C(\sigma/2, 1, -\sigma/2)$.

If we consider a particular solution of this system, a general solution can be generated by adding an arbitrary multiple of the homogeneous solution of the system (8.2.1) to this particular solution. The homogenous system is obtained by putting to zero all the right-hand side terms and is defined by

$$b_{-1} + b_0 + b_1 = 0$$
$$b_{-1} - b_1 = 0 \qquad (8.2.2)$$

It has the solution, $b_{-1} = b_1 = -b_0/2$.

Because of the homogeneity, this solution is always defined up to a constant, which we call here γ. If we write the homogenous solution as $H(b_{-1}, b_0, b_1)$, we have

$$H(b_{-1}, b_0, b_1) = \gamma H(1, -2, 1) \qquad (8.2.3)$$

It is of interest to consider the different solutions as corrections to the unstable central scheme, as already commented for the Lax–Wendroff scheme. Hence we select the scheme S_C, as the particular solution, writing the general expression as

$$S(b_{-1}, b_0, b_1) = S_C(\sigma/2, 1, -\sigma/2) + \gamma H(1, -2, 1) \qquad (8.2.4)$$

This equation describes a family of first order schemes. For instance, with $\gamma = \sigma/2$ we recover the upwind scheme and with $\gamma = 1/2$ the Lax–Friedrichs scheme.

Note that because of the form (8.2.4) the choice of the particular solution is not constrained, as another choice would only have as consequence a redefinition of the parameter γ.

Hence, all the possible schemes on the 3-point support $(i - 1, i, i+1)$ with first order accuracy are defined by the parameter γ and the b_j values $b_{-1} = \sigma/2 + \gamma$, $b_0 = 1 - 2\gamma$, $b_1 = \gamma - \sigma/2$.

They can be written as follows:

$$u_i^{n+1} = u_i^n - \frac{\sigma}{2}(u_{i+1}^n - u_{i-1}^n) + \gamma(u_{i+1}^n - 2u_i^n + u_{i-1}^n) \tag{8.2.5}$$

where γ appears as a numerical viscosity coefficient. The third term, which stabilizes the instability generated by the first two terms, is the discretization of a **numerical viscosity** term of the form $\gamma \Delta x^2 u_{xx}$.

The stability condition can be obtained, through a direct analysis, as developed in Chapter 7. Alternatively we can apply the results obtained in Section 8.1, in particular from equations (8.1.39) and (8.1.35).

From the first of these equations, we obtain

$$0 \leq b_0 = 1 - 2\gamma \leq 1 \tag{8.2.6}$$

leading to the condition

$$0 \leq \gamma \leq \frac{1}{2} \tag{8.2.7}$$

We notice, what we know already from the earlier chapters, that the coefficient γ of the numerical viscosity must be positive in order to represent a diffusion or dissipation effect, as opposed to an 'explosion' behavior for negative diffusion coefficient.

In addition, an upper limit is set on the γ-coefficient for stability, namely $\gamma = 1/2$, which corresponds to the Lax–Friedrich scheme.

Since the condition (8.1.39) is only necessary, it cannot represent the whole of the stability requirements. Considering the more general condition (8.1.35) with $r = 1$ and

$$\alpha_2 = \frac{\Delta x}{\Delta t}\left[\sum_j b_j j^2 - (-\sigma)^2\right]\frac{1}{2} = \frac{\Delta x}{\Delta t}\left[(b_{-1} + b_1) - \sigma^2\right]\frac{1}{2}$$

$$= \frac{\Delta x}{2\Delta t}\left[2\gamma - \sigma^2\right] \geq 0 \tag{8.2.8}$$

leads to the condition

$$2\gamma - \sigma^2 \geq 0 \tag{8.2.9}$$

Summarizing the two results, we find the stability condition as

$$\frac{\sigma^2}{2} \leq \gamma \leq \frac{1}{2} \tag{8.2.10}$$

Note that this condition implies the CFL condition $|\sigma| \leq 1$ for stability.

From the above considerations, adding the third consistency condition (8.1.20), which reads here for the support $j(-1, 0, 1)$,

$$b_{-1} + b_1 = \sigma^2 \tag{8.2.11}$$

gives three conditions for the three coefficients, leading to the unique explicit scheme on the domain $j(-1, 0, 1)$, with second order accuracy and two-time-levels, that is *centrally defined with respect to the mesh point i*. This is the Lax–Wendroff scheme corresponding to $\gamma = \sigma^2/2$. It represents the lowest limit of the stability region of γ, as seen from equation (8.2.10).

8.2.2 The Convection–Diffusion Equation

We can apply the above results to analyze and gain an understanding of a critical issue related to the central discretization of the convection–diffusion equation:

$$u_t + au_x = \alpha u_{xx} \tag{8.2.12}$$

The simplest scheme combines first order in time and second order in space to ensure that the steady state solutions will be second order accurate. Considering a central difference for the convection term and the standard central difference for the diffusion term, we obtain the scheme

$$u_i^{n+1} = u_i^n - \frac{\sigma}{2}(u_{i+1}^n - u_{i-1}^n) + \beta(u_{i+1}^n - 2u_i^n + u_{i-1}^n) \tag{8.2.13}$$

with $\beta = \alpha \, \Delta t/\Delta x^2$.

This scheme is of the form (8.2.5), where here the parameter γ has a physical interpretation as representing a physical instead of a purely numerical diffusion.

The stability condition (8.2.10) becomes here

$$\sigma^2 \leq 2\beta \leq 1 \tag{8.2.14}$$

leading to the stability condition on the time step

$$\Delta t \leq \min\left(\frac{2\alpha}{a^2}, \frac{\Delta x^2}{2\alpha}\right) \tag{8.2.15}$$

An important quantity in simulations of viscous flows is the **cell Reynolds number** (also called **cell Peclet number** when the diffusivity coefficient is not the kinematic viscosity), based on the cell size Δx, defined by

$$\mathrm{Re}_{\Delta x} \triangleq \frac{a \Delta x}{\alpha} \tag{8.2.16}$$

The stability condition can be rewritten, taking into account the relation

$$\frac{\sigma}{\beta} = \frac{a \, \Delta x}{\alpha} = \mathrm{Re}_{\Delta x} \tag{8.2.17}$$

as

$$\sigma \leq \frac{\mathrm{Re}_{\Delta x}}{2} \leq \frac{1}{\sigma} \tag{8.2.18}$$

This stability region is shown in the diagram $(\sigma, \text{Re}_{\Delta x})$ in Figure 8.2.1, showing that there is a stability domain for all values of the cell Reynolds number.

This is of interest in view of the history behind this scheme and its stability conditions. It is a remarkable fact that historically, a first Von Neumann stability condition was incorrectly derived by Fromm (1964), and quoted in the book of Roache (1972), as well as in some later textbooks.

The correct results have been obtained initially by Hirt (1968) applying a different approach, but remained largely unnoticed, and a regain of concern has generated a variety of publications for one- and multidimensional stability analyses of the discretized convection–diffusion equation, see for instance the papers of Rigal (1979), Leonard (1980), Chan (1984), Hindmarsh, Gresho and Griffiths (1984).

This erroneous condition was stated as

$$\sigma \le 2\beta \le 1 \tag{8.2.19}$$

to be compared with the correct condition (8.2.14). This condition had a remarkable consequence, considering the relation (8.2.17) leading to a condition

$$\frac{\sigma}{\beta} = \text{Re}_{\Delta x} \le 2 \tag{8.2.20}$$

which implies a restriction on the mesh size, instead of the expected stability restriction on the time step.

This incorrect concept of a mesh size limitation for stability, has generated an intense discussion between experts and considerable confusion, see for instance Thompson et al. (1985) for an additional clarification.

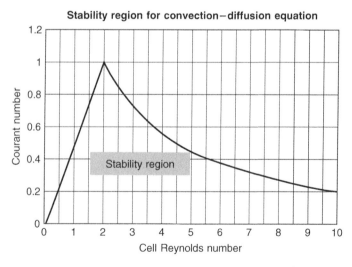

Figure 8.2.1 *Stability region for the central scheme for the convection–diffusion equation*

Although the stability condition is now correctly known, the reason behind the numerous discussions behind the $Re_{\Delta x}$ restriction is due to the fact that when this scheme is applied to obtain a steady solution with $Re_{\Delta x} > 2$, numerical oscillations appear, that can become very large.

Computations performed by Griffiths, Christie and Mitchell (1980) as well as by Hindmarsh, Gresho and Griffiths (1984) showed indeed growing oscillations for $Re_{\Delta x} > 2$, although the calculations were performed inside the stability region of the scheme. This has been interpreted at occasions as a lack of stability, re-enforcing the belief in a stability condition on the cell Reynolds number. As a consequence a re-evaluation of the definition of stability was required and a theoretical discussion of the impact on the definition of stability can be found in Morton (1980).

We mention here these historical facts, because they reflect some of the fundamental questions that have marked the evolution of CFD and because the convection–diffusion equation is of fundamental importance to the whole of CFD, as it represents the structure of the Navier–Stokes equations and of any conservation law, as we have seen in Chapter 1.

So let us try to understand what is happening with this scheme, in particular when $Re_{\Delta x} > 2$, when we wish to apply the scheme (8.2.13) to obtain a steady solution.

At steady state, the exact numerical solution satisfies the following steady numerical scheme

$$a\frac{u_{i+1} - u_{i-1}}{2\Delta x} = \frac{\alpha}{\Delta x^2}(u_{i+1} - 2u_i + u_{i-1}) \tag{8.2.21}$$

which can also be written as

$$\left(1 - \frac{Re_{\Delta x}}{2}\right)u_{i+1} - 2u_i + \left(1 + \frac{Re_{\Delta x}}{2}\right)u_{i-1} = 0 \tag{8.2.22}$$

This equation can be solved exactly, by a method called the **normal mode** method introduced by Godunov and Ryabenkii (1964) to generate exact solutions of finite different equations of stationary problems. The method consists in looking for solutions of the form:

$$u_i = \kappa^i \tag{8.2.23}$$

By introducing this solution in equation (8.2.22), we obtain the quadratic equation for κ, as

$$\left(1 - \frac{Re_{\Delta x}}{2}\right)\kappa^2 - 2\kappa + \left(1 + \frac{Re_{\Delta x}}{2}\right) = 0 \tag{8.2.24}$$

which has the two solutions

$$\kappa_1 = 1 \quad \kappa_2 = \frac{2 + Re_{\Delta x}}{2 - Re_{\Delta x}} \tag{8.2.25}$$

When $Re_{\Delta x} > 2$, the denominator of the second solution is negative.

Hence, the general solution of the numerical scheme (8.2.22) is given by

$$\bar{u}_i = A + B\left(\frac{2 + Re_{\Delta x}}{2 - Re_{\Delta x}}\right)^i \tag{8.2.26}$$

When $Re_{\Delta x} > 2$, the second term will be positive for even values of i and negative on uneven points, while its absolute values is increasing with the mesh point number i. Hence, the numerical solution will alternate between increasingly positive and negative values and the amplitude of the oscillations will grow as the mesh point number increases. This can indeed become very large, for instance for a value of $Re_{\Delta x} = 3$, the amplitude will grow exponentially like 5^i.

The analytical solution of the differential equation $u_t = \alpha u_{xx}$, with the boundary conditions $u(0) = 0$, $u(1) = 1$, is

$$\tilde{u}(x) = \frac{e^{xRe} - 1}{e^{Re} - 1} \tag{8.2.27}$$

where Re is the global Reynolds number $Re = aL/\alpha$. The exact solution is shown on the left side of Figure 8.2.2, for Re-numbers from 1 to 10, showing the growing boundary layer behavior of the solution.

The exact and numerical solutions are shown on Figure 8.2.2, for $Re = 50$, with 8 mesh points, for which $Re_{\Delta x} = 50/8 = 6.25$ showing the oscillatory behavior of the numerical solution (points marked CDS).

In order to eliminate this problem, one option is to reduce mesh size in order to always satisfy $Re_{\Delta x} < 2$, but this requires adaptive grids or very fine grids. Another approach is to change the scheme to a '*hybrid scheme*', by which the convective term is discretized by a first order upwind formula when $Re_{\Delta x} > 2$ and centrally when $Re_{\Delta x} < 2$. This is obtained by the scheme (8.3.17) and the result is shown on Figure 8.2.2 as HOC. But this leads generally to large regions where the scheme is reduced to first order.

Another alternative, which we recommend, is to introduce nonlinear limiters to enforce monotonicity, as we will see in Section 8.3.

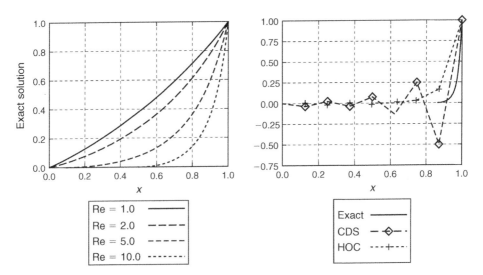

Figure 8.2.2 *Left diagram: Exact solution for Re-numbers from 1 to 10, Right diagram: Numerical solution at Re = 50, with 8 mesh points, for which $Re_{\Delta x} = 50/8 = 6.25$, with the central scheme, marked CDS. The solution with the hybrid scheme is marked HOC.*

8.2.3 One-Parameter Family of Schemes on the Support $(i-2, i-1, i, i+1)$

By repeating the same methodology on the support $(i-2, i-1, i, i+1)$, we can define a one-parameter family of schemes with *second order accuracy*, identified as $S(b_{-2}, b_{-1}, b_0, b_1)$.

Second order upwind scheme on the support $(i-2, i-1, i)$
If we make a first attempt to a scheme with $b_1 = 0$, for $a > 0$, of the form,

$$u_i^{n+1} = b_0 u_i^n + b_{-1} u_{i-1}^n + b_{-2} u_{i-2}^n \tag{8.2.28}$$

we have the three conditions for the three coefficients:

$$\begin{aligned} b_0 + b_{-1} + b_{-2} &= 1 \\ -2b_{-2} - b_{-1} &= -\sigma \\ 4b_{-2} + b_{-1} &= \sigma^2 \end{aligned} \tag{8.2.29}$$

There is only one solution, namely

$$b_{-2} = \frac{\sigma}{2}(\sigma - 1) \quad b_{-1} = \sigma(2 - \sigma) \quad b_0 = \frac{1}{2}(1 - \sigma)(2 - \sigma) \tag{8.2.30}$$

leading to the second order accurate upwind scheme of Warming and Beam, already introduced in Chapter 7.

The truncation error is obtained from the above relations, leading to the equivalent differential equation:

$$u_t + a u_x = \frac{a}{6} \Delta x^2 (1 - \sigma)(2 - \sigma) u_{xxx} - \frac{a}{8} \Delta x^3 \sigma (1 - \sigma)^2 (2 - \sigma) u_{xxxx} + O(\Delta x^4) \tag{8.2.31}$$

The stability condition (8.1.35) reduces to the condition:

$$\sigma(2 - \sigma) \geq 0 \quad \text{or} \quad 0 \leq \sigma \leq 2 \tag{8.2.32}$$

Both the Lax–Wendroff and the Warming and Beam schemes are the *unique* schemes of second order accuracy on the supports $(i-1, i, i+1)$ and $(i-2, i-1, i)$, respectively.

Family of second order schemes on the support $(i-2, i-1, i, i+1)$
Since we dispose of four coefficients we can satisfy the first three consistency conditions (8.1.20), ensuring second order accuracy. These conditions define the following system

$$\begin{aligned} b_1 + b_0 + b_{-1} + b_{-2} &= 1 \\ b_1 - b_{-1} - 2b_{-2} &= -\sigma \\ b_1 + b_{-1} + 4b_{-2} &= \sigma^2 \end{aligned} \tag{8.2.33}$$

which can be solved, with for instance b_{-2} as parameter, leading to the solution

$$b_0 = 3b_{-2} + 1 - \sigma^2$$
$$b_1 = -b_{-2} + \frac{\sigma}{2}(\sigma - 1) \qquad\qquad (8.2.34)$$
$$b_{-1} = -3b_{-2} + \frac{\sigma}{2}(\sigma + 1)$$

Writing again the general scheme $S(b_{-2}, b_{-1}, b_0, b_1)$ as a particular solution plus the general homogenous solution of system (8.2.33), we have first to find this homogenous solution $H(b_{-2}, b_{-1}, b_0, b_1)$. It is easily seen that this solution is, up to a factor $H(-1, 3, -3, 1)$.

We know that the case $b_{-2} = 0$ is the Lax–Wendroff scheme, as it is the unique second order scheme on the support $(i - 1, i, i + 1)$, identified here as

$$S_{LW}(0, \sigma(1 + \sigma)/2,\ (1 - \sigma^2),\ -\sigma(1 - \sigma)/2) \qquad\qquad (8.2.35)$$

Hence, if we take the Lax–Wendroff scheme S_{LW} as the particular solution we can write the general one-parameter family of second order schemes on the support $(i - 2, i - 1, i, i + 1)$, as

$$S(b_{-2}, b_{-1}, b_0, b_1) = S_{LW} + \gamma H(-1, 3, -3, 1) \qquad\qquad (8.2.36)$$

By comparing with the solution (8.2.34), we see that the parameter γ is equal to the coefficient $(-b_{-2})$.

The one-parameter family of second order schemes is written as

$$u_i^{n+1} = u_i^n - \frac{\sigma}{2}(u_{i+1}^n - u_{i-1}^n) + \frac{\sigma^2}{2}(u_{i+1}^n - 2u_i^n + u_{i-1}^n)$$
$$+ \gamma(-u_{i-2}^n + 3u_{i-1}^n - 3u_i^n + u_{i+1}^n) \qquad\qquad (8.2.37)$$

The last term can be interpreted as the discretization of an additional dispersion term of the form $(\gamma \Delta x^3 u_{xxx})$.

In order to derive conditions on the parameter γ and on stability, we apply the methodology of Section 8.1.

We leave it to you as an exercise to derive the amplification factor, the dispersion and diffusion errors and to look for the series expansion of these errors in function of the phase angle ϕ, as a way to obtain the first two terms of the equivalent differential equation (8.1.27), based on the expansions (8.1.33) and (8.1.34). (see Problem P.8.8).

The diffusion error is obtained as

$$\varepsilon_D = 1 - \frac{1}{2}\left[\gamma(1 - 2\sigma) + \frac{1}{4}\sigma^2(1 - \sigma^2)\right]\phi^4$$
$$+ \left[\frac{\gamma^2}{2} + \frac{\gamma}{12}(1 - 5\sigma + 3\sigma^2) + \frac{\sigma^2}{48}(1 - \sigma^2)\right]\phi^6 + O(\phi^8) \qquad (8.2.38)$$

The necessary condition for stability (8.1.35) requires the first coefficient to be positive, i.e.

$$\gamma(1 - 2\sigma) + \frac{1}{4}\sigma^2(1 - \sigma^2) \geq 0 \qquad\qquad (8.2.39)$$

As γ is a function of the CFL number σ, this condition is not so useful, as it stands. However, for selected values of γ this condition will lead to necessary stability conditions on σ. We can simplify it by introducing the parameter Γ, defined by

$$\Gamma = \frac{2\gamma}{\sigma(1-\sigma)}$$

leading to

$$\Gamma(1 - 2\sigma) + \frac{1}{2}\sigma(1 + \sigma) \geq 0$$

The dispersion error is given by

$$\varepsilon_\phi = 1 + \frac{1}{6}[6\gamma - \sigma(1 - \sigma^2)]\phi^2$$

$$-\left[\frac{\gamma}{4\sigma}(1 - 2\sigma + 2\sigma^2) - \frac{1}{120}(1 - \sigma^2)(6\sigma^2 + 1)\right]\phi^4 + O(\phi^6) \quad (8.2.40)$$

Collecting the first terms, we obtain the equivalent differential equation of the scheme (8.2.38):

$$u_t + au_x = \frac{a}{6}[6\gamma - \sigma(1 - \sigma^2)]\Delta x^2 u_{xxx}$$

$$-\frac{a}{8}\left[\gamma(1 - 2\sigma) + \frac{1}{4}\sigma^2(1 - \sigma^2)\right]\Delta x^3 u_{xxxx} + O(\Delta x^4) \quad (8.2.41)$$

From equation (8.2.30) we see that the Warming and Beam scheme is defined by the value of γ equal to the corresponding $(-b_{-2})$ value, $\gamma = \sigma(1 - \sigma)/2$. That is we have

$$S_{WB}(b_{-2}, b_{-1}, b_0, b_1) = S_{LW} + \frac{\sigma(1-\sigma)}{2}H(-1, 3, -3, 1) \quad (8.2.42)$$

Fromm's schemes

We have seen in Chapter 7 that the Lax–Wendroff and Warming–Beam schemes have phase errors of opposite signs in the range $0 < \sigma < 1$. Therefore an attempt to derive a scheme with reduced dispersion error could be obtained by taking the arithmetic average of the Lax–Wendroff S_{LW} and the Warming–Beam second order upwind schemes, S_{WB}.

This is the scheme of Fromm (1968), which has indeed a significant reduced dispersion error. It is defined by $\gamma = \sigma(1 - \sigma)/4$, leading to

$$S_F\left(\frac{\sigma(\sigma - 1)}{4}, \frac{\sigma(5 - \sigma)}{4}, \frac{(1 - \sigma)(4 + \sigma)}{4}, \frac{\sigma(\sigma - 1)}{4}\right) \quad (8.2.43)$$

Under the form (8.2.42), Fromm's scheme is written as a correction to the Lax–Wendroff scheme

$$u_i^{n+1} = u_i^n - \frac{\sigma}{2}(u_{i+1}^n - u_{i-1}^n) + \frac{\sigma^2}{2}(u_{i+1}^n - 2u_i^n + u_{i-1}^n)$$

$$+ \frac{\sigma(1-\sigma)}{4}(-u_{i-2}^n + 3u_{i-1}^n - 3u_i^n + u_{i+1}^n) \quad (8.2.44)$$

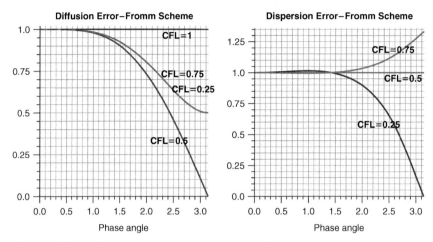

Figure 8.2.3 *Diffusion and dispersion errors for Fromm's scheme, for CFL = 0.25; 0.5; 0.75 and 1.*

or under the form

$$u_i^{n+1} = u_i^n - \frac{\sigma}{4}(u_{i+1}^n + 3u_i^n - 5u_{i-1}^n + u_{i-2}^n) + \frac{\sigma^2}{4}(u_{i+1}^n - u_i^n - u_{i-1}^n + u_{i-2}^n)$$

$$(8.2.45)$$

This scheme is stable for $0 \leq \sigma \leq 1$.

The diffusion and dispersion errors are shown on Figure 8.2.3. You can clearly see, by comparing with Figures 7.4.4 and 7.4.8, that the dispersion error is significantly reduced, as it remains close to 1 up to a phase angle over $(\pi/2)$ (see Problem P.8.5).

The unique third order scheme

As seen from Table 8.1.1, the maximum accuracy for schemes with the support $(i-2, i-1, i, i+1)$ is of order 3. This unique third order accurate scheme, Warming, Kutler, Lomax (1973), is obtained by adding the fourth consistency condition to the system (8.2.33) (see Problem P.8.9).

This fourth equation is given by

$$-8b_{-2} - b_{-1} + b_1 = -\sigma^3 \tag{8.2.46}$$

and when introduced in the solution (8.2.34) leads to

$$b_{-2} = -\gamma = \frac{\sigma}{6}(\sigma^2 - 1) \tag{8.2.47}$$

This gives the scheme

$$S_3 \left(\frac{\sigma(\sigma^2 - 1)}{6}, \frac{\sigma(2 - \sigma)(\sigma + 1)}{2}, \frac{(2 - \sigma)(1 - \sigma^2)}{2}, \frac{\sigma(\sigma - 1)(2 - \sigma)}{6} \right)$$

$$(8.2.48)$$

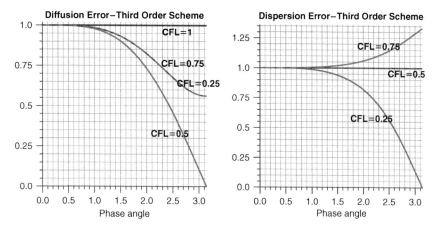

Figure 8.2.4 *Diffusion and dispersion errors for the third order scheme, for CFL = 0.25; 0.5; 0.75 and 1.*

or, written as

$$u_i^{n+1} = u_i^n - \frac{\sigma}{2}(u_{i+1}^n - u_{i-1}^n) + \frac{\sigma^2}{2}(u_{i+1}^n - 2u_i^n + u_{i-1}^n)$$

$$+ \frac{\sigma(1 - \sigma^2)}{6}(-u_{i-2}^n + 3u_{i-1}^n - 3u_i^n + u_{i+1}^n) \qquad (8.2.49)$$

This can also be seen from equation (8.2.41), where the third order term disappears, leading to the equivalent differential equation

$$u_t + au_x = -\frac{a}{24}[\sigma(1 - \sigma^2)(2 - \sigma)]\Delta x^3 u_{xxxx} + O(\Delta x^4) \qquad (8.2.50)$$

confirming that the scheme is indeed of third order accuracy.

This scheme is also stable for the CFL condition $0 \leq \sigma \leq 1$.

Its error properties are quite close to Fromm's scheme, although the latter is only second order accurate. This can be seen by comparing with Figure 8.2.4, showing the diffusion and dispersion errors for this third order scheme.

8.3 MONOTONICITY OF NUMERICAL SCHEMES

We are now ready to address the major problem that we have observed with numerical schemes of second or higher orders of accuracy, namely the appearance of an oscillatory behavior, as seen on the solutions presented in Chapter 7, see Figures 7.3.6–7.3.8. This creates regions of high errors, compared to the exact solution, which are not only un-esthetical but can lead to non-physical values. This can occur with any quantity that is physically bounded, such as density or turbulent kinetic energy, which have to remain positive, or values such as 'volume of fluid' or 'combustion mixture fraction' which have to remain between zero and one in order to be physically meaningful. With these types of variables, the occurrence of numerical wiggles might lead to unphysical values that could arm considerably the reliability of the numerical simulation.

Hence, accurate numerical schemes should not allow such an oscillatory behavior to occur.

A systematic analysis of the conditions required by a scheme to satisfy this requirement has been developed, initiated by Godunov (1959) who introduced the important property of *monotonicity*. There is an extensive literature on this very important topic of CFD and many different definitions and criteria can be found. For nonlinear equations the criterion of bounded total variation, under the name of *Total Variation Diminishing (TVD)*, of the solution has been introduced by Harten (1983), (1984) as a general concept to ensure that unwanted oscillations are not generated by a numerical scheme. Spekreijse (1987) expressed monotonicity as a *positivity condition*, while a *Convection Boundedness Criterion (CBC)* was explicitly formulated by Gaskell and Lau (1988) based on the concept of *Normalized Variables* introduced by Leonard (1979), (1988). A more recent general analysis, has been developed by Jameson (1993), (1995a, b), based on a definition of *Local Extremum Diminishing* (LED) schemes.

We will not enter here in the subtleties of the theoretical differences between these various concepts, as they basically all lead to the same conditions, at least for linear schemes. A discussion of the differences between these various definitions, when applied to nonlinear scalar, or systems of, conservation laws can be found in the books of Laney (1998) and Leveque (2002).

We will therefore refer in the following essentially to the notion of *monotonicity*.

The monotonicity requirement for a numerical scheme can be stated as the requirement that no new extrema be created by the numerical scheme, other than those eventually present in the initial solution. *In other words, the numerical solution should have a monotone behavior, whereby the new solution value u_i^{n+1} at time index (n+1) should not reach values outside the range covered by the solution values $(u_{i+j})^n$ at time step n.*

Clearly, the observed oscillatory behavior does not satisfy this property, while the solutions of the first order upwind and Lax–Friedrichs solutions do show the required monotone behavior.

8.3.1 Monotonicity Conditions

We will now derive the explicit conditions on a numerical scheme to satisfy this requirement.

Consider the general explicit scheme with two time levels, which we repeat here for convenience:

$$u_i^{n+1} = \sum_j b_j u_{i+j}^n \tag{8.3.1}$$

We know that the b_j coefficients have to satisfy the consistency condition (8.1.9), also repeated here

$$\sum_j b_j = 1 \tag{8.3.2}$$

Theorem: the monotonicity condition is satisfied if all the b_j coefficients are non-negative.
Indeed, if

$$b_j \geq 0 \quad \text{for all } j \tag{8.3.3}$$

the condition (8.3.2) implies that all $b_j \leq 1$. Hence, the new solution u_i^{n+1} is the convex sum, i.e. a weighted average with positive coefficients lower than one, of the old solution. If u_{max}^n and u_{min}^n are the boundaries of the variation range of u_{i+j}^n then for

$$u_{min}^n \leq u_{i+j}^n \leq u_{max}^n \tag{8.3.4}$$

we obtain, after multiplying by the positive b_j coefficient and summing over all values of j,

$$\sum_j b_j u_{min}^n \leq u_{i+j}^{n+1} = \sum_j b_j u_{i+j}^n \leq \sum_j b_j u_{max}^n \tag{8.3.5}$$

or, taking equation (8.3.2) into account

$$u_{min}^n \leq u_i^{n+1} \leq u_{max}^n \tag{8.3.6}$$

This relation shows that the new solution at time step $(n+1)$ is also contained within the same range. This excludes therefore any oscillatory behavior whereby new values would exceed the initial variation range of the solution at an earlier time step.

An alternative, but equivalent, way of looking at the monotonicity condition (8.3.3) is obtained by rewriting the general scheme (8.3.1) with the condition (8.3.2) as follows:

$$u_i^{n+1} = u_i^n + \sum_j b_j(u_{i+j}^n - u_i^n) \tag{8.3.7}$$

If u_i^n is a local minimum all $(u_{i+j}^n - u_i^n) \geq 0$, and if the monotonicity condition is satisfied, that is all the b_j are positive, then we have $u_i^{n+1} \geq u_i^n$, indicating that the local minimum will not decrease. Similarly, if u_i^n is a local maximum, then $(u_{i+j}^n - u_i^n) \leq 0$; consequently $u_i^{n+1} \leq u_i^n$ and the local maximum will not increase.

Let us look at this monotonicity property for certain of the schemes we have encountered, for the diffusion and the convection equations.

Monotonicity condition for the diffusion equation
If we consider the diffusion equation (7.2.28) and the explicit scheme (7.2.29)

$$u_i^{n+1} = u_i^n + \beta(u_{i+1}^n - 2u_i^n + u_{i-1}^n) \quad \beta = \frac{\alpha \, \Delta t}{\Delta x^2} \tag{8.3.8}$$

The b_j coefficients are

$$b_{-1} = \beta \quad b_0 = 1 - 2\beta \quad b_1 = \beta \tag{8.3.9}$$

and we observe that all the b_j coefficients are positive in the stability region $\beta \leq 1/2$. Hence, this *scheme is monotone*.

Monotonicity condition for the convection equation: first order schemes
The general form for first order schemes on the 3-point support $(i-1, i, i+1)$ is defined by equation (8.2.5), with the following coefficients:

$$b_{-1} = \gamma + \frac{\sigma}{2} \quad b_0 = 1 - 2\gamma \quad b_1 = \gamma - \frac{\sigma}{2} \tag{8.3.10}$$

with the stability condition (8.2.10), repeated here

$$\frac{\sigma^2}{2} \leq \gamma \leq \frac{1}{2} \tag{8.3.11}$$

The condition for monotonicity is satisfied when all coefficients are non-negative, that is for the parameter γ satisfying the following relation

$$\frac{\sigma}{2} \leq \gamma \leq \frac{1}{2} \tag{8.3.12}$$

The limits of the monotonicity range of this artificial viscosity coefficient are precisely the upwind ($\gamma = \sigma/2$) and Lax–Friedrichs ($\gamma = 1/2$) schemes. Both these schemes are monotone.

This example shows that first order schemes are not always monotone, in particular for values of γ between $\sigma^2/2$ and $\sigma/2$, this scheme is stable but not monotone.

Monotonicity condition for the convection equation: second order schemes
The second order Lax–Wendroff scheme is defined by

$$u_i^{n+1} = u_i^n - \frac{\sigma}{2}(u_{i+1}^n - u_{i-1}^n) + \frac{\sigma^2}{2}(u_{i+1}^n - 2u_i^n + u_{i-1}^n) \tag{8.3.13}$$

with the coefficients

$$b_{-1} = \frac{\sigma}{2}(1 + \sigma) \quad b_0 = 1 - \sigma^2 \quad b_1 = -\frac{\sigma}{2}(1 - \sigma) \tag{8.3.14}$$

Within the stability region $|\sigma| \leq 1$, the coefficient b_1 is always negative when $\sigma > 0$ and the scheme is ***non-monotone***.

Similarly, the Warming and Beam scheme defined by the coefficients (8.2.30) is ***non-monotone***, since the coefficient b_{-2} is negative when $\sigma < 1$, while b_0 is negative when $1 < \sigma < 2$.

More generally, the one-parameter family of second order schemes (8.2.37), has the coefficients

$$b_{-2} = -\gamma \quad b_{-1} = \frac{\sigma}{2}(1 + \sigma) + 3\gamma \quad b_0 = 1 - \sigma^2 - 3\gamma$$

$$b_1 = -\frac{\sigma}{2}(1 - \sigma) + \gamma \tag{8.3.15}$$

It is easily seen that these schemes can never be monotone, since the positivity condition on the coefficients requires $\gamma < 0$ and on the other hand from b_1, $\gamma > \sigma(1 - \sigma)/2$. This can never be satisfied within the CFL stability condition $|\sigma| < 1$, which implies $\gamma > 0$. Hence these schemes are ***non-monotone***.

Monotonicity condition for the convection–diffusion equation

The discussion on the oscillations associated with the cell Reynolds number of the previous section has a straightforward explanation, looking at the coefficients of the scheme (8.2.13):

$$b_{-1} = \sigma/2 + \beta = \beta\left(\frac{\mathrm{Re}_{\Delta x}}{2} + 1\right) \quad b_0 = 1 - 2\beta,$$

$$b_1 = \beta - \sigma/2 = \beta\left(1 - \frac{\mathrm{Re}_{\Delta x}}{2}\right) \tag{8.3.16}$$

We see that the scheme is ***non-monotone*** for $\mathrm{Re}_{\Delta x} > 2$, since the coefficient b_1 is negative in this case, which explains the appearance of oscillations. On the other hand, when $\mathrm{Re}_{\Delta x} < 2$, the scheme is monotone in its stability domain $2\beta < 1$ and no oscillations will appear.

If we look at a scheme where the convection term is discretized by a first order upwind scheme, we obtain the following scheme, for $a > 0$

$$u_i^{n+1} = u_i^n - \sigma(u_i^n - u_{i-1}^n) + \beta(u_{i+1}^n - 2u_i^n + u_{i-1}^n) \tag{8.3.17}$$

The coefficients of this scheme are

$$b_{-1} = \sigma + \beta = \beta(\mathrm{Re}_{\Delta x} + 1) \quad b_0 = 1 - \sigma - 2\beta = 1 - \beta(\mathrm{Re}_{\Delta x} + 2)$$

$$b_1 = \beta \tag{8.3.18}$$

with the stability condition (see Problem P.8.12):

$$0 \leq \sigma + 2\beta \leq 1 \tag{8.3.19}$$

Hence, all the b-coefficients are positive and the scheme is monotone.

This explains its use in the hybrid approach, to eliminate the oscillations generated with the central scheme when $Re_{\Delta x} > 2$, as seen by the solution HOC in figure 8.2.2.

The introduction of monotonicity is a significant achievement, as we have finally found the explanation for the behavior of the results shown by the Figures 7.3.6–7.3.8, or for the $Re_{\Delta x}$ problem occurring with the central explicit scheme of the convection–diffusion equation discussed in the previous section.

The answer to the questions raised has become clear: the numerical oscillations are the consequence of the non-monotone behavior of the second order schemes considered.

We also understand why the first order schemes do not generate these wiggles, and also why the diffusion terms, centrally discretized, are also free of unwanted oscillations.

We have now to find a cure to these non-monotone effects. This will be the subject of the following Section 8.3.4, but before addressing this problem, we have to look at these issues in more general terms. The crucial information is based on the Godunov theorem, which states that all linear monotone schemes for the convection equation are necessarily first order accurate. Hence how could we make second or higher order schemes to have a monotone behavior?

The other issue we have to deal with is the generalization of the monotonicity concepts for more general time discretizations than the two-level explicit schemes discussed up to now.

8.3.2 Semi-Discretized Schemes or Method of Lines

The above analysis is based on the explicit form of the schemes (8.3.1) and we could question the generality of the derived monotonicity properties.

For instance we could ask ourselves: could we possibly reduce or eliminate numerical oscillations by considering implicit instead of explicit schemes; or by considering other formulas for the time derivative than the forward difference leading to the explicit scheme formulations?

Referring to equation (8.3.7) we can view this equation as an explicit time discretization of the *semi-discretized form* of the scheme (also called *method of lines*), *whereby the space discretization is separated from the time integration*, as follows:

$$\frac{du_i}{dt} = \sum_j \beta_j u_{i+j} \tag{8.3.20}$$

where the right-hand side represents only the discretized space derivatives.

Equation (8.3.20) is a system of ordinary differential equations in time, which can be integrated in various ways.

This important general approach to the discretization of conservation laws will be discussed in more details in Chapter 9.

The consistency condition for the semi-discretized system, replacing equation (8.3.2), is obviously given by

$$\sum_j \beta_j = 0 \tag{8.3.21}$$

expressing that the scheme has to remain valid for a constant valued function $u = 1$. Hence, the scheme can be also written as

$$\frac{du_i}{dt} = \sum_{j, j \neq 0} \beta_j (u_{i+j} - u_i) \qquad (8.3.22)$$

obtained from

$$\beta_0 = -\sum_{j, j \neq 0} \beta_j \qquad (8.3.23)$$

The monotonicity, positivity, or Local Extremum Diminishing (LED) condition requires that

$$\beta_j \geq 0 \quad \text{for all } j \neq 0 \qquad (8.3.24)$$

The above demonstration applies here without change, i.e. if u_i is a local minimum and (8.3.24) is verified, then $du_i/dt \geq 0$ and the minimum cannot decrease, and similarly a local maximum cannot increase for a monotone scheme.

It is important to note here that this condition depends *only* on the space discretization and does not involve the coefficient β_0 associated to $j = 0$ (point i). **Therefore condition (8.3.24) does not guarantee that the time integrated scheme is monotone, since an additional condition has to be added, related to β_0.**

For instance, if an explicit, forward time difference is applied to equation (8.3.22) then, comparing to the formulation (8.3.1), we have, observing that $\beta_0 < 0$ for a monotone scheme

$$\begin{aligned} b_j &= \Delta t \beta_j \geq 0 \qquad \text{for } j \neq 0 \\ b_0 &= 1 - \Delta t |\beta_0| \end{aligned} \qquad (8.3.25)$$

and the monotonicity condition (8.3.3) requires that in addition the scheme also satisfies $\Delta t |\beta_0| \leq 1$, which can be considered as a stability condition. For the linear convection equation ($\Delta t |\beta_0|$) is the CFL number (see also Problem P.8.13).

Let us consider as an example the case of the convection–diffusion equation, centrally discretized, which we write as

$$\frac{du_i}{dt} = -a \frac{u_{i+1} - u_{i-1}}{2 \, \Delta x} + \frac{\alpha}{\Delta x^2} (u_{i+1} - 2u_i + u_{i-1}) \qquad (8.3.26)$$

Writing this scheme into the general form (8.3.22), we obtain

$$\beta_{-1} = \frac{\alpha}{\Delta x^2} \left(1 + \frac{\text{Re}_{\Delta x}}{2} \right) \qquad \beta_1 = \frac{\alpha}{\Delta x^2} \left(1 - \frac{\text{Re}_{\Delta x}}{2} \right) \qquad (8.3.27)$$

This confirms again that the scheme is non-monotone when $\text{Re}_{\Delta x} > 2$, which leads to an oscillatory behavior for *any* stable time integration scheme.

This is an important observation as it indicates that in order to remove the unwanted oscillations, we have to act on the space discretization and not on the time integration.

If we consider the first order upwind difference for the convection term, we obtain the scheme, for $a > 0$,

$$\frac{du_i}{dt} = -a\frac{u_i - u_{i-1}}{\Delta x} + \frac{\alpha}{\Delta x^2}(u_{i+1} - 2u_i + u_{i-1}) \tag{8.3.28}$$

the coefficients

$$\beta_{-1} = \frac{\alpha}{\Delta x^2}(1 + \text{Re}_{\Delta x}) \quad \beta_1 = \frac{\alpha}{\Delta x^2} \tag{8.3.29}$$

indicating that this scheme is monotone.

Summarizing what we have learned from this section, we realize that the dominating problem for the monotonicity of a scheme comes from the convection term, as the diffusion, also by its intrinsic physical properties, will not give rise to numerical oscillations. Hence in order to find cures, we have to focus on the convection terms.

This explains why the following fundamental theorem concerning monotonicity, particularly for the convection equation, is of major significance.

8.3.3 Godunov's Theorem

Theorem: All linear monotone schemes for the convection equation are necessarily first order accurate.

This important theorem, due to Godunov (1959), can be demonstrated in various ways and we will focus here on the explicit schemes of the form (8.3.1).

The condition for second order accuracy of this scheme is given by the second consistency condition (8.1.20) for $m = 2$,

$$\sum_j b_j j^2 = \sigma^2 \tag{8.3.30}$$

We will show that this condition **cannot** be satisfied for a monotone scheme, where all b_j coefficients are non-negative. Instead, for a monotone scheme, we would have

$$\sum_j b_j j^2 \geq \sigma^2 \tag{8.3.31}$$

indicating that the first term of the right-hand side of the equivalent differential equation (8.1.21) for $p = 1$, is positive and hence that the scheme is only first order accurate. Indeed, if we take the square of the consistency relation for $m = 1$

$$\sum_j b_j j = -\sigma \tag{8.3.32}$$

we obtain successively

$$\sigma^2 = \left[\sum_j b_j j\right]^2 = \left[\sum_j \left(j\sqrt{b_j}\right)\sqrt{b_j}\right]^2 \leq \left(\sum_j j^2 b_j\right)\left(\sum_j b_j\right) = \sum_j j^2 b_j \tag{8.3.33}$$

The inequality results from applying the Schwartz inequality, expressing that the square of a sum of products is lower or equal to the product of the sums of the squares (or in other words that the scalar product of two vectors is lower or equal to the product of their lengths $\vec{a} \cdot \vec{b} \leq |\vec{a}||\vec{b}|$). The last equal sign results from the consistency condition (8.3.2).

If we separate the upwind from the downwind points, following the form (8.1.40), the consistency condition (8.3.32) can be rewritten as follows:

$$\sum_j b_j j = -\sigma = \sum_{j=1}^{jd} b_j j - \sum_{j=1}^{ju} b_j j < 0 \tag{8.3.34}$$

If the scheme is monotone, all the b_j coefficients are non-negative and each of the two summations is positive, which implies that the upwind points have a greater 'weight' than the downwind points, i.e. *the scheme has to be upwind biased to be monotone*, since this equation indicates that, for $a > 0$

$$0 \leq \sum_{j=1}^{jd} b_j j \leq \sum_{j=1}^{ju} b_j j \tag{8.3.35}$$

The theorem of Godunov can also be demonstrated for more general discretizations of the nonlinear conservation laws, see for instance Harten, Hyman and Lax (1976).

The concepts of monotonicity and Godunov's theorem are of considerable importance, with the awareness that *all linear* second, or higher, order schemes for the convection equation will necessarily generate undesirable oscillations in presence of discontinuities or very high gradients of the solution or its derivatives. On the other hand, practical applications require *high-resolution schemes*, that is schemes of at least second order accuracy without numerical oscillations and therefore a way around Godunov's theorem has to be found.

There are not many options available and the only way to circumvent this theorem is to move away from the linearity property of the schemes.

As a consequence, only *nonlinear* schemes can be made to be monotone while being at the same time essentially of higher order accuracy.

This will be made possible by the introduction of nonlinear components, known as *limiters*.

8.3.4   High-Resolution Schemes and the Concept of Limiters

The basic idea behind the concept of limiters is to control the process of generation of over- and undershoots by preventing gradients to exceed certain limits, or to change sign between adjacent points.

In this way, the non-monotone schemes will be 'controlled', at each time step and within each cell, such as to keep the gradients within the proper bounds. This approach towards high-resolution schemes consists actually in preventing the generation of oscillations by acting somehow on their generation mechanism.

This concept was initially introduced by Van Leer (1973, 1974) and Boris and Book (1973, 1976). Various approaches can be followed and a general framework has been set by Van Leer (1977a, b, 1979) in a series of papers leading to second order upwind schemes without numerical oscillations and where many of the ideas at the basis of modern high-resolution schemes have been developed.

To illustrate the main idea behind the limiters, let us consider a second order difference for the convection equation $u_t + au_x = 0$ and the formulation, $a > 0$

$$\frac{du_i}{dt} = -\frac{a}{2\Delta x}(u_{i+1} - u_{i-1}) = -\frac{a}{2\Delta x}[-(u_{i-1} - u_i) + (u_{i+1} - u_i)] \qquad (8.3.36)$$

The last term is in the form of equation (8.3.22) and we can see that one of the coefficients is negative, indicating that this scheme is non-monotone.

Since the concept of limiters is based on monitoring the ratio of successive gradients, we write he above scheme under the form:

$$\frac{du_i}{dt} = -\frac{a}{2\Delta x}\left[1 - \frac{u_{i+1} - u_i}{u_i - u_{i-1}}\right](u_i - u_{i-1}) \qquad (8.3.37)$$

We observe that this scheme could become monotone if the term between the square brackets would remain positive, that is if the ratio between successive gradients remains lower than 1.

$$r_i \overset{\Delta}{=} \frac{u_{i+1} - u_i}{u_i - u_{i-1}} \leq 1 \qquad (8.3.38)$$

This is illustrated in Figure 8.3.1, where the ratio $r = a/b$, with $a = u_{i+1} - u_i$ and $b = u_i - u_{i-1}$. The figure on the left satisfies the condition for monotonicity (8.3.38) as $a < b$, while the figure on the right does not, since $a > b$. On this figure at point $i + 1$, the ratio of gradients becomes negative, showing a pattern of over- and undershoots, with $r_{i+1} < 0$.

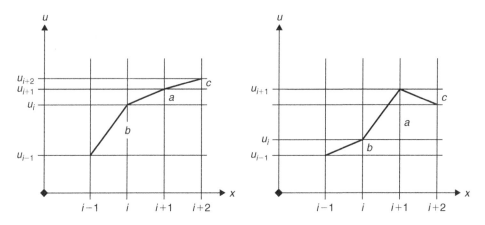

Figure 8.3.1 *Illustration of the relation between successive gradients.*

This guides the methodology behind the definition of limiter functions Ψ, as functions of ratio of successive gradients $\Psi = \Psi(r_i)$, defined to ensure that conditions such as (8.3.38) are satisfied.

This methodology can be defined as follows:

1. *Select a first order monotone space discretization, generally the first order upwind discretization, as reference and write the higher order scheme as the monotone scheme plus additional terms.*
2. *Multiply the additional terms by a 'limiting function' Ψ, expressed as a function of a ratio of successive gradients.*
3. *Express the monotonicity conditions to derive the conditions on the 'limiters'.*

We have seen in the conclusion of Section 8.3.2 that the non-monotone conditions are predominantly conditioned by the space discretization and not by the time integration. Therefore, we will focus on the spatial gradients to develop the methodology for high-resolution schemes and for the definition of limiters.

To illustrate the main idea behind the limiters, let us consider a second order backward difference for the convection equation $u_t + au_x = 0$ and the formulation, for $a > 0$

$$\frac{du_i}{dt} = -\frac{a}{2\Delta x}(3u_i - 4u_{i-1} + u_{i-2}) = -\frac{a}{2\Delta x}[3(u_i - u_{i-1}) - (u_{i-1} - u_{i-2})]$$

$$= -\frac{a}{2\Delta x}[-4(u_{i-1} - u_i) + (u_{i-2} - u_i)] \tag{8.3.39}$$

The last equality is in the form of equation (8.3.22) and we can see that one of the coefficients is negative, indicating that this scheme is non-monotone.

Let us apply the proposed methodology, step by step.

Step 1: Rewrite the scheme as a correction to the monotone first order upwind difference, writing the correction terms in the form of successive gradients

$$\frac{du_i}{dt} = \underbrace{-\frac{a}{\Delta x}(u_i - u_{i-1})}_{\text{First order monotone upwind scheme}} - \frac{a}{\Delta x}\left[+\frac{1}{2}(u_i - u_{i-1}) - \frac{1}{2}(u_{i-1} - u_{i-2})\right] \tag{8.3.40}$$

Step 2: Multiply the two non-monotone terms by functions $\Psi(r_i)$ and $\Psi(r_{i-1})$, where

$$r_{i-1} = \frac{u_i - u_{i-1}}{u_{i-1} - u_{i-2}} \qquad r_i = \frac{u_{i+1} - u_i}{u_i - u_{i-1}} \tag{8.3.41}$$

leading to

$$\frac{du_i}{dt} = -\frac{a}{\Delta x}\left[(u_i - u_{i-1}) + \frac{1}{2}\Psi(r_i)(u_i - u_{i-1}) - \frac{1}{2}\Psi(r_{i-1})(u_{i-1} - u_{i-2})\right]$$

$$= -\frac{a}{\Delta x}\left[1 + \frac{1}{2}\Psi(r_i) - \frac{1}{2}\frac{\Psi(r_{i-1})}{r_{i-1}}\right](u_i - u_{i-1}) \tag{8.3.42}$$

Step 3: Derive the conditions on the limiters

This scheme will be monotone if the term in the brackets is positive, that is the limiters should satisfy the condition

$$\frac{\Psi(r_{i-1})}{r_{i-1}} - \Psi(r_i) \leq 2 \tag{8.3.43}$$

A detailed analysis of the properties of the Ψ limiters has been given by Roe (1981, 1984), Sweby (1984) and also, from a different standpoint, by Roe (1985).

The above functional relations can be satisfied by a large variety of Ψ functions. However, a certain number of constraints can be identified or imposed. First of all, we restrict Ψ to be a positive function, i.e.

$$\Psi(r) \geq 0 \quad \text{for } r \geq 0 \tag{8.3.44}$$

In addition, when $r < 0$, that is when an extremum is encountered in the variation of the solution u, it seems logical to set $\Psi = 0$ corresponding to a zero slope in the interval considered. This avoids non-monotone behaviors with changes of slope directions, at the expense of a certain loss of accuracy.

Note the important property that when $\Psi = 0$ the scheme reduces locally to first order accuracy.

Hence, we set

$$\Psi(r) = 0 \quad \text{for } r \leq 0 \tag{8.3.45}$$

With these assumptions, we have the sufficient condition

$$0 \leq \Psi(r) \leq 2r \tag{8.3.46}$$

A logical requirement is the **symmetry property**, which expresses that forward and backward gradients are treated in the same way:

$$\frac{\Psi(r)}{r} = \Psi\left(\frac{1}{r}\right) \tag{8.3.47}$$

which implies, from equation (8.3.43), the sufficient condition:

$$\Psi(r) \leq 2 \tag{8.3.48}$$

Therefore, the second order upwind scheme will be monotone if the limiting function Ψ lies within the shaded area of Figure 8.3.2, which summarizes the above relations as

$$0 \leq \Psi(r) \leq \min(2r, 2) \tag{8.3.49}$$

Figure 8.3.2, displaying the limiter in function of the gradient ratio r, is called the **Sweby diagram**.

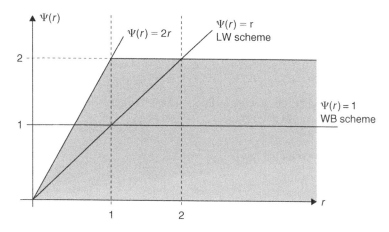

Figure 8.3.2   *Monotonicity region for limiter functions, based on relation (8.3.49).*

Limiters for the second order upwind (SOU) scheme of Warming and Beam

Let us re-apply the methodology to the unique second order upwind (SOU) scheme of Warming and Beam on the support $(i - 2, i - 1, i)$. We will subsequently apply it to the centrally discretized Lax–Wendroff scheme on the support $(i - 1, i, i + 1)$, with an amazing consequence, namely that, after introduction of the limiters, the high-resolution versions of these two schemes will be viewed as leading to the same results.

We consider the Warming and Beam scheme with the coefficients defined by equation (8.2.30):

$$u_i^{n+1} = u_i^n + \sigma(2 - \sigma)(u_{i-1}^n - u_i^n) + \frac{\sigma}{2}(\sigma - 1)(u_{i-2}^n - u_i^n) \tag{8.3.50}$$

Step 1: rewrite the scheme as a correction to the monotone first order upwind scheme, with the additional terms under the form of differences between adjacent points

$$u_i^{n+1} = \underbrace{u_i^n - \sigma\,(u_i^n - u_{i-1}^n)}_{\text{Monotone scheme}} \underbrace{- \frac{\sigma}{2}(1 - \sigma)(u_i^n - u_{i-1}^n) + \frac{\sigma}{2}(1 - \sigma)(u_{i-1}^n - u_{i-2}^n)}_{\text{Non-monotone terms}}$$

$$\tag{8.3.51}$$

Step 2: multiply the two non-monotone terms by functions $\Psi(r_i)$ and $\Psi(r_{i-1})$, leading to

$$u_i^{n+1} = u_i^n - \sigma\,(u_i^n - u_{i-1}^n) - \frac{\sigma}{2}(1 - \sigma)\Psi(r_i)\,(u_i^n - u_{i-1}^n)$$

$$+ \frac{\sigma}{2}(1 - \sigma)\,\Psi(r_{i-1})(u_{i-1}^n - u_{i-2}^n)$$

$$= u_i^n - \sigma\left\{1 + \frac{1}{2}(1 - \sigma)\left[\Psi(r_i) - \frac{\Psi(r_{i-1})}{r_{i-1}}\right]\right\}(u_i^n - u_{i-1}^n) \tag{8.3.52}$$

Step 3: this 'limited' scheme will be monotone if the term between the curled brackets is positive, leading to the condition

$$\frac{\Psi(r_{i-1})}{r_{i-1}} - \Psi(r_i) \leq \frac{2}{1-\sigma} \tag{8.3.53}$$

This relation is satisfied by the sufficient relation, Roe (1981), Roe and Baines (1981), Sweby (1984):

$$0 \leq \Psi(r) \leq \min\left(\frac{2r}{\sigma}, \frac{2}{1-\sigma}\right) \tag{8.3.54}$$

which generalizes the sufficient condition (8.3.49) on the Ψ function.

This condition is more adequate for time-dependent problems, although in practice the simpler condition (8.3.49) is more often applied, being also more suitable for stationary problems.

Observe that the 'unlimited', non-monotone second order upwind scheme of Warming and Beam corresponds to $\Psi = 1$.

Remark: There is a large amount of flexibility in the choice of the gradient ratios, as defined by equation (8.3.41). In this equation the ratios are based on comparing upwind gradients, but one could also define downstream ratios, such as

$$R_i = \frac{u_i - u_{i-1}}{u_{i+1} - u_i} = \frac{1}{r_i} \quad R_{i+1} = \frac{u_{i+1} - u_i}{u_{i+2} - u_{i+1}} = \frac{1}{r_{i+1}} \tag{8.3.55}$$

Limiters for the Lax–Wendroff scheme

Applying the methodology to the Lax–Wendroff scheme, the three steps develop as follows:

Step 1: rewrite the LW scheme as a correction to the monotone first order upwind scheme,

$$u_i^{n+1} = \underbrace{u_i^n - \sigma(u_i^n - u_{i-1}^n)}_{\text{Monotone scheme}} \underbrace{- \frac{\sigma}{2}(1-\sigma)(u_{i+1}^n - u_i^n) + \frac{\sigma}{2}(1-\sigma)(u_i^n - u_{i-1}^n)}_{\text{Non-monotone terms}} \tag{8.3.56}$$

Step 2: multiply the two non-monotone terms by functions $\Psi(R_i)$ and $\Psi(R_{i-1})$, leading to

$$u_i^{n+1} = u_i^n - \sigma(u_i^n - u_{i-1}^n) - \frac{\sigma}{2}(1-\sigma)\Psi(R_i)(u_{i+1}^n - u_i^n)$$

$$+ \frac{\sigma}{2}(1-\sigma)\Psi(R_{i-1})(u_i^n - u_{i-1}^n)$$

$$u_i^{n+1} = u_i^n - \sigma(u_i^n - u_{i-1}^n)\left\{1 + \frac{1}{2}(1-\sigma)\left[\frac{\Psi(R_i)}{R_i} - \Psi(R_{i-1})\right]\right\}(u_i^n - u_{i-1}^n) \tag{8.3.57}$$

Step 3: this 'limited' scheme will be monotone if the term between the curled brackets is positive, leading to the condition:

$$\Psi(R_{i-1}) - \frac{\Psi(R_i)}{R_i} \le \frac{2}{(1-\sigma)} \tag{8.3.58}$$

The interesting observation is that, with the symmetry property (8.3.47) and the relation (8.3.55), this condition is identical to (8.3.53).

If we take the WB scheme as reference, the 'limited' version (8.3.42) reproduces the LW scheme for $\Psi(r) = r$, as can be readily verified and recovers the WB scheme for $\Psi(r) = 1$. The same conditions can be expressed in function of the $R_i = 1/r_i$ ratios, taking into account the symmetry property. Indeed with the relation

$$\Psi(r_i) = \Psi\left(\frac{1}{R_i}\right) = \frac{\Psi(R_i)}{R_i} \tag{8.3.59}$$

the LW scheme is recovered for $\Psi(R) = 1$, while the WB scheme is obtained for $\Psi(R) = R$.

In order to narrow down further the choices for limiters, we should consider the case of a linear solution, for which $r = 1$. In this case, the second order schemes should be satisfied exactly, which implies that the limiter functions should be equal to 1, in order to recover the 'non-limited' version.

Hence, we require all limiters to satisfy the additional condition

$$\Psi(1) = 1 \tag{8.3.60}$$

which is a necessary requirement for second order accuracy on smooth solutions. In addition, Leonard (1988) has shown that if the slope of the (Ψ, r) curve in this point $(1, 1)$ is equal to 3/4, than the scheme is locally third order accurate.

As seen in Section 8.2 any linear second order explicit scheme on the support $(i-2, i-1, i, i+1)$ can be viewed as a linear combination of the Warming and Beam and the Lax–Wendroff schemes. Consequently, any second order 'limited' scheme could be based on a limiter function Ψ, which lies between the lines $\Psi = r$ and $\Psi = 1$. This restricts the region of validity of the limiters to the domain shown on Figure 8.3.3, for second order explicit monotone schemes.

As reported by Sweby (1984) the regions of Figure 8.3.2 outside the domain between $\Psi = r$ and $\Psi = 1$ are theoretically acceptable, but lead to schemes which are '*over-compressive*', that is turning sine waves into square wave forms.

Actually, the analysis of the monotonicity of the Lax–Wendroff schemes was at the basis of the generalization of the concept of limiters, Davis (1984), Roe (1984), Sweby (1984). It may seem surprising that this led to the same constraints on the limiter function Ψ as the second order upwind schemes, but demonstrates, on the other hand, the generality of the hereby defined limiters.

In addition, as observed by Sweby (1984), the fact that the second order (in time) explicit upwind scheme of Warming and Beam and the second order explicit Lax–Wendroff schemes can both be made monotone by the same set of limiters, justifies the subset of the monotonicity region of Figure 8.3.3, where the 'limited' explicit schemes remain globally second order accurate in time and space.

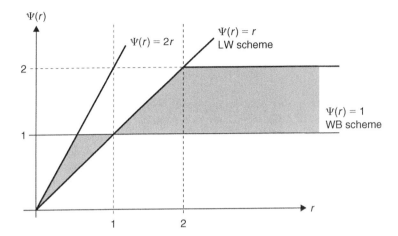

Figure 8.3.3 *Region of monotonicity for second order schemes.*

The drawback of the limiters is that they reduce the scheme locally to first order at extrema, when $r = 0$. This is justified when it serves to suppress oscillations, since first order schemes are generally monotone, but if the exact solution has extrema, like a sine wave, then the solution will be locally deformed by the action of the limiters.

This has to be considered as the 'price' to pay for the achievement of high-resolution schemes.

It has to be said that the development of high-resolution schemes is one of the most remarkable achievements of the history of CFD.

Various limiter functions have been defined in the literature and are currently applied.

Van Leer (1974) proposed initially the formula

$$\Psi(r) = \frac{r + |r|}{1 + r} \tag{8.3.61}$$

shown in Figure 8.3.4a.

A similar limiter, with a smoother behavior, has been applied by Van Albada et al. (1982),

$$\Psi(r) = \frac{r^2 + r}{1 + r^2} \tag{8.3.62}$$

It has the property of tending to 1 for large values of r, while the Van Leer limiter tends to 2 asymptotically.

The lowest boundary of the considered TVD domain is an often applied limiter. It is shown on Figure 8.3.4b and can be represented by

$$\Psi(r) = \begin{cases} \min(r, 1) & \text{if } r \geq 0 \\ 0 & \text{if } r \leq 0 \end{cases} \tag{8.3.63}$$

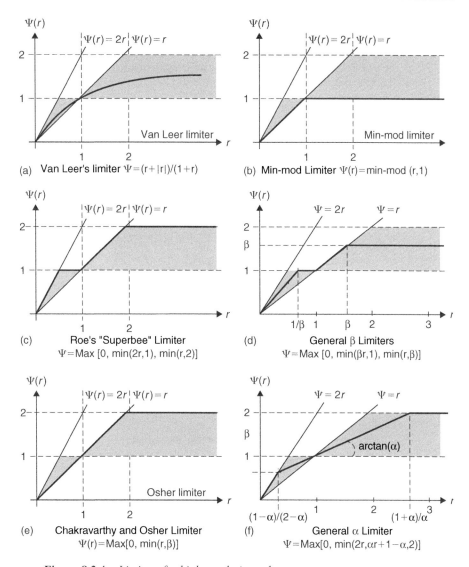

Figure 8.3.4 *Limiters for high-resolution schemes.*

and is a particular case of the ***min-mod*** function, defined as the function which selects the number with the smallest modulus from a series of numbers when they all have the same sign, and zero otherwise. For two arguments

$$\min \bmod(x, y) = \begin{cases} x & \text{if } |x| < |y| \quad \text{and} \quad x.y > 0 \\ y & \text{if } |x| < |y| \quad \text{and} \quad x.y > 0 \\ 0 & \text{if } x.y < 0 \end{cases} \qquad (8.3.64)$$

or in compact form

$$\min \bmod(x, y) = \operatorname{sgn}(x). \max\left[0, \min\left(|x|, \operatorname{sgn}(x).y\right)\right] \qquad (8.3.65)$$

Equation (8.3.63) can therefore be written as $\Psi(r) = \min \bmod (1, r)$.

The upper limit of the second order monotonicity domain has been considered by Roe, Roe and Baines (1981), Roe (1985) under the nickname of **superbee** and shown to have excellent resolution properties for jump discontinuities. It is shown on Figure 8.3.4c and defined by

$$\Psi(r) = \max \left[0, \min (2r, 1), \min (r, 2) \right] \tag{8.3.66}$$

This limiter actually amplifies certain contributions, when $\Psi > 1$, while remaining within the monotonicity bounds. This explains the property of this 'superbee' limiter in counteracting the excessive spreading of jump discontinuities. It has a remarkable property, shown by Roe and Baines (1983), *namely that it allows linear discontinuities to propagate indefinitely without numerical diffusion*. This is illustrated on Figures 8.3.6 and 8.3.7.

The limits of the monotonicity region are members of a family of limiters, based on a single parameter β, in the range $1 \leq \beta \leq 2$, Sweby (1984),

$$\Psi(r) = \max \left[0, \min (\beta r, 1), \min (r, \beta) \right] \quad 1 \leq \beta \leq 2 \tag{8.3.67}$$

These β-limiters are shown on Figure 8.3.4d. It is seen by inspection that it reduces to the min-mod limiter for $\beta = 1$, and to the Superbee limiter for $\beta = 2$.

All of these limiters share the symmetry property (8.3.47).

Let us also mention the limiter used by Chakravarthy and Osher (1983):

$$\Psi(r) = \max \left[0, \min (r, \beta) \right] \quad 1 \leq \beta \leq 2 \tag{8.3.68}$$

shown on Figure 8.3.4e. Note that this limiter does not satisfy the symmetry condition (8.3.47), excepted for $\beta = 1$, where it reduces to the min-mod limiter.

More limiters can be found in the literature; see for instance Waterson and Deconinck (1995, 2007) for a general review.

In particular, the SMART limiter of Gaskell and Lau (1988), controls the slope of the (Ψ, r) curve in point $(1, 1)$, putting it equal to the optimum value of 0.75, for local third order accuracy. This SMART limiter is defined by

$$\Psi(r) = \begin{cases} 0 & \text{for } r \leq 0 \\ 2r & \text{for } 0 \leq r \leq \dfrac{1}{5} \\ \dfrac{3}{4}r + \dfrac{1}{4} & \text{for } \dfrac{1}{5} \leq r \leq 5 \\ 4 & \text{for } 5 \leq r \end{cases} \tag{8.3.69}$$

or in a compact form as $\Psi(r) = \max \left[0, \min (2r, 0.75r + 0.25, 4) \right]$.

Note that this limiter does not respect the sufficient condition (8.3.48) ($\Psi \leq 2$), but it neither satisfies the symmetry property.

If we impose the condition (8.3.49), but add a control of the slope (α) of the limiter in point $(1, 1)$, we can define the following ALFA family, in the range $0 \leq \alpha \leq 1$, as follows:

$$\Psi(r) = \max \left[0, \min (2r, \alpha r + 1 - \alpha, 2) \right] \tag{8.3.70}$$

which is shown on Figure 8.3.4f. Note that α indicates the slope of the line through $(1, 1)$, but is not the angle of that line with the horizontal direction. Actually this angle is equal to $\arctan \alpha$.

This family of limiters has been considered initially by Roe and Baines (1981), and later taken up by Jeng and Payne (1995) and by Arora and Roe (1997), showing it has excellent properties.

For $\alpha = 0$, it is close to the superbee limiter but restricted to a maximum value $\Psi = 1$, while $\alpha = 1$ reproduces the Chakravarthy and Osher limiter at $\beta = 2$. As it follows the $\Psi = 2r$ limit for small values of r, like the superbee limiter, it will behave in a 'compressive way' in this region, while having a smooth behavior around point $(1, 1)$.

The value $\alpha = 0.5$ satisfies the symmetry condition, and actually corresponds to the Fromm scheme, for which $\Psi(r) = (1 + r)/2$. It is mentioned by Van Leer (1979), in association with the development of the **Monotone Upstream-centered Scheme for Conservation Laws (MUSCL)** and referred to in the literature as the MUSCL limiter. For $\alpha = 3/4$, the SMART limiter is followed, up to a maximum value of $\Psi = 2$. The value $\alpha = 2/3$ has been considered by Koren (1993).

A fully symmetric ALFA family can be considered, satisfying (8.3.47), leading to

$$\Psi(r) = \max\left[0, \min\left(2r, \alpha r + 1 - \alpha, (1 - \alpha)r + \alpha, 2\right)\right] \tag{8.3.71}$$

The value $\alpha = 3/4$ has been introduced by Lien and Leschziner (1994) and considered by Waterson and Deconinck (1995, 2007) under a general form similar to (8.3.71). This family has the advantage of symmetry, but suffers from the slope discontinuity at the point $r = 1$, $\Psi = 1$.

Much of the recent and current research is oriented at the non-trivial extension of the concept of limiters to arbitrary unstructured grids. This extension, important in practice, will be dealt within Volume II. The impatient reader will find additional information for instance in Barth (1993), Venkatakrishnan and Barth (1993), Jasak et al. (1999), Darwish and Moukalled (2000).

It is important to observe at this point that all these high-resolution schemes are strictly nonlinear due to the dependence on the Δu ratios, even when applied to the linear convection equation.

More insight into the action of the limiters is obtained by considering the specific contribution of the second order terms to the new solution at point i at time level $n + 1$.

The term $(u_i - u_{i-1})$ is modified by a nonlinear correction $\Psi(u_i - u_{i-1})$. With the 'min-mod' limiter, we actually set the following restrictions:

- If the gradient $(u_{i+1} - u_i)/\Delta x < (u_i - u_{i-1})/\Delta x$, that is if $r < 1$, then $\Psi(r) = r$ and the contribution $(u_i - u_{i-1})$ to the solution at time step $(n + 1)$ is replaced by the smaller quantity $(u_{i+1} - u_i)$.
- If $r > 1$, the contribution $(u_i - u_{i-1})$ remains unchanged.
- If the slopes of consecutive intervals change sign, the updated point i receives no contribution from the upstream interval.

With the superbee limiter, on the other hand, some contributions are enhanced instead of reduced, while remaining within the monotonicity region. If $r < 1/2$, $\Psi = 2r$ and the contribution $(u_i - u_{i-1})$ to the solution at time step $(n + 1)$ is replaced by the smaller quantity $2(u_{i+1} - u_i)$, while for $1/2 < r < 1$, the larger quantity is kept. For

$1 < r < 2$, $\Psi = r$ and again the larger quantity is transferred as contribution to the updated solution, Finally, for $r > 2$ the smaller quantity $2(u_i - u_{i-1})$ is transferred. The specific effect of the limiters on smooth flows can be seen from a comparison of the Figures 8.3.5, which display the results of the convection of a low frequency sinusoidal wave. The linear convection equation is solved with the second order limited upwind scheme (8.3.52), applying the min-mod and the superbee limiters. Figure 8.3.5a is obtained with the first order upwind scheme and the excessive dissipation inherent to all first order schemes is apparent, when compared to the exact solution. Figure 8.3.5b shows the improvement obtained with the standard second order upwind scheme (8.3.50), at the expense of oscillations appearing at the slope discontinuities, typical of all second order schemes. The introduction of the limiters in the second order upwind scheme removes completely the oscillations, producing monotone profiles. However, the min-mod limiter reduces locally the accuracy of the solution around the extrema, as seen on Figure 8.3.5c, bringing it close to first order, because of the condition (8.3.45). Finally, Figure 8.3.5d shows the behavior of the superbee limiter where its over-compressive property is clearly seen.

The maxima are flattened and the gradients are made steeper. This is well adapted for sharp discontinuities but not too adequate for smooth profiles.

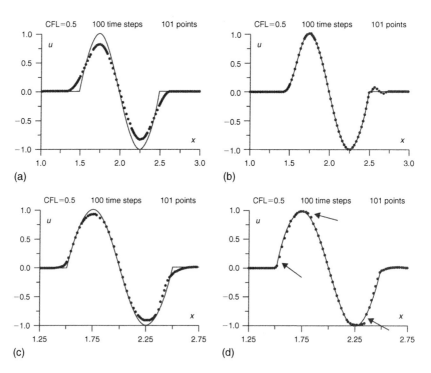

Figure 8.3.5   *Effects of limiters on the linear convection of a sinusoidal wave (a) first order upwind scheme (b) second order upwind scheme (c) second order upwind scheme with min-mod limiter (d) second order upwind scheme with superbee limiter.*

The effects of the limiters on discontinuities can also be seen from the convection of a square wave. Figures 8.3.6 and 8.3.7 compare the linear convection of a square wave, after 120 and 400 time steps at a Courant number of 0.5. Figure 8.3.6a is obtained with the first order upwind scheme, showing its excessive diffusion; Figure 8.3.6b is obtained with the second order upwind scheme, showing the strong oscillations around the discontinuities. Figures 8.3.6c–e are computed with the min-mod, Van Leer and superbee limiters, respectively, and generate monotone profiles.

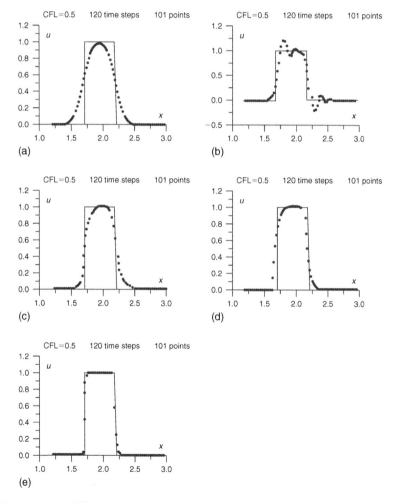

Figure 8.3.6 *Effects of limiters on the linear convection of a square wave after 120 time steps: (a) first order upwind scheme, (b) second order upwind scheme, (c) second order upwind scheme with min-mod limiter, (d) second order upwind scheme with Van Leer limiter and (e) second order upwind scheme with superbee limiter.*

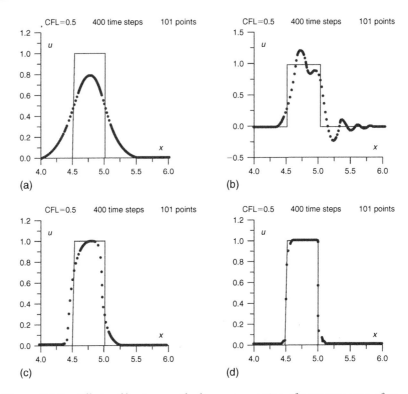

Figure 8.3.7   *Effects of limiters on the linear convection of a square wave after 400 time steps: (a) first order upwind scheme, (b) second order upwind scheme, (c) second order upwind scheme with Van Leer limiter and (d) second order upwind scheme with superbee limiter.*

The min-mod limiter however is still too diffusive, while the superbee limiter produces excellent results, with extremely sharp discontinuities. The Van Leer limiter has properties between the previous two. Superbee maintains the sharpness of the profile indefinitely, as can be seen by comparing with a similar calculation after 400 time steps, shown on Figure 8.3.7c. The points in the transition region are practically unchanged from time step 120 to step 400, while it is seen that the Van Leer limiter still continues to generate a small, but continuous, diffusion of the transition profiles.

Figure 8.3.8 shows results obtained with the high-resolution LW scheme (8.3.57) on the same test case as the previous figures for the linear convection equation. Here again the superbee limiter leads to very sharp, non-diffusive transition profiles, while the other limiters, the min-mod and the Van Leer limiters, still have some diffusive components. Note in particular the symmetrical shape of the profiles, compared to the similar profiles obtained with the upwind method, which show traces of the upwind discretizations.

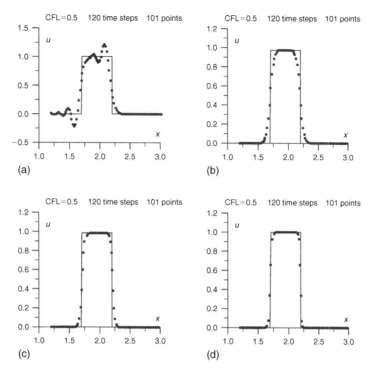

Figure 8.3.8   *Effects of limiters on the linear convection of a square wave after 120 time steps: (a) standard LW scheme, (b) second order high-resolution LW scheme with min-mod limiter, (c) second order high-resolution LW scheme with Van Leer limiter and (d) second order high-resolution LW scheme with superbee limiter.*

HANDS-ON TASK 4

Extend the program you have written under the Hands-On Task 2 (Chapter 7), introducing the limiters to obtain high-resolution schemes. Obtain the various results displayed in Figures 8.3.5–8.3.9.

Introduce also other limiters and test various cases. In particular, apply the high-resolution schemes to the test cases of the propagating waves of 7.3.7 and 7.3.8. Observe the positive effect of the limiters on the reduction of the dispersion errors, in particular with the LW and WB schemes.

Limiters for time-dependent problems

Time-dependent problems, in particular moving discontinuities such as moving shocks, or free surface problems between two fluids such as surface waves in ship hydrodynamics or sloshing problems in tanks, put very severe constraints on the numerical schemes, in order to be able to follow the motion of the discontinuity over longer times. The main challenge is to avoid the effects of the numerical diffusion of the interface.

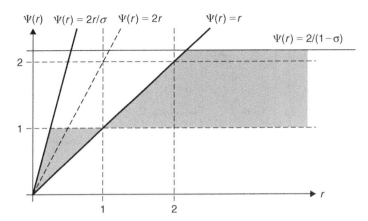

Figure 8.3.9 *Region of monotonicity for CFL-dependent limiters and second order schemes.*

In these cases it is recommended to consider the CFL-dependent conditions (8.3.54) and introduce an explicit CFL effect in the limiters. If we add the condition for second order accuracy on the limiters (8.3.60), we obtain an extended validity domain for the limiters, where the upper limit is given by $\Psi = 2/(1 - \sigma)$ instead of $\Psi = 2$ and the left side limit is now defined by $\Psi = 2r/\sigma$ instead of $\Psi = 2r$. This leads to the validity domain shown in Figure 8.3.9, to be compared with Figure 8.3.3.

The upper limit of this region is the generalization of the superbee limiter and has been nicknamed as the ***ultrabee*** limiter by Roe and Baines (1981). It is defined by

$$\Psi(r) = \max \left[0, \min\left(\frac{2r}{\sigma}, 1 \right), \min\left(r, \frac{2}{1 - \sigma} \right) \right] \tag{8.3.72}$$

This limiter is also considered by Leonard (1991), as part of the ***ultimate*** strategy for limiters.

The generalization of the ALFA family (8.3.70) to CFL dependent conditions has been initially considered by Roe and Baines (1981), and applied by Jeng and Paine (1995), Arora and Roe (1997), under the form

$$\Psi(r) = \max \left[0, \min\left(\frac{2r}{\sigma}, \alpha(r - 1) + 1, \frac{2}{1 - \sigma} \right) \right] \tag{8.3.73}$$

where the parameter α can be made CFL dependent.

The value $\alpha = (2 - \sigma)/3$ is advocated by Arora and Roe (1997) as it corresponds to the third order scheme (8.2.49) (see Problem P.8.17). These authors also recommend to restrict the limiter region to the stationary limits $\Psi = 2$ and $\Psi = 2r$ for nonlinear waves. This leads to the alternative

$$\Psi(r) = \max \left[0, \min\left(2r, \alpha(r - 1) + 1, 2 \right) \right] \quad \text{with } \alpha = \alpha(\sigma) \tag{8.3.74}$$

These limiters are represented in Figure 8.3.10, that you can compare with Figure 8.3.4.

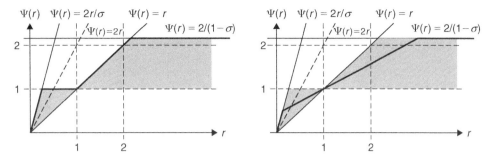

Figure 8.3.10 *Ultrabee and CFL-dependent ALFA family of limiters.*

As you will have noticed at this stage, a very large variety of limiters can be defined and applied and some guidelines can be put forward based on experiences with many test cases:

- Limiters with continuous slopes generally favor convergence.
- For sharp discontinuities, compressive limiters such as superbee or ultrabee show an excellent behavior in removing numerical diffusion.
- For smooth discontinuities, continuous limiters, such as Van Leer or Van Albada are a good choice.

Another approach to the definition and interpretation of limiters has been developed by Gaskell and Lau (1988) and Leonard (1988), (1991), based on a different estimation of gradient ratios, as *Normalized Variables* and a graphical representation in a *Normalized Variable Diagram (NVD)*. This approach will be presented in the next section, where the one-to-one relation to the limiters defined in this section and the Sweby diagram will be shown.

8.4 FINITE VOLUME FORMULATION OF SCHEMES AND LIMITERS

The objective of this section is to set the framework towards the generalization of the various methods and schemes discussed up to now for the simplified linear models, to:

- More general time integration methods, which will form the subject of Chapter 9;
- Nonlinear conservation laws;
- The multidimensional configurations.

We refer you at this stage to Chapter 5 and the Finite Volume method, which offers the appropriate framework and where two essential concepts have been introduced: the definition of numerical fluxes and their face value definition, based on cell-averaged quantities. The finite volume formulation of a general conservative numerical scheme, is defined by two components:

1. The numerical flux which identifies completely the scheme, formulated either in the explicit form (5.2.6) or when the space and time discretizations are separated, defined by equation (5.2.7) which represents only the space discretization.

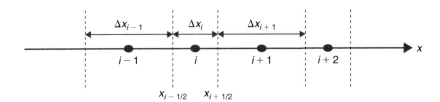

Figure 4.3.3 *Finite volume subdivision of a non-uniform, one-dimensional mesh point distribution. Cell-centered approach.*

2. The definition of cell-face values, as these are the sole contributions to the numerical scheme, in function of the cell-averaged quantities, which are the only available variables.

Let us specify this for the one-dimensional case, as illustrated by the cell-centered mesh of Figure 4.3.3, which we reproduce here for convenience.

8.4.1 Numerical flux

Considering the one-dimensional conservation law

$$\frac{\partial u}{\partial t} + \frac{\partial f}{\partial x} = 0 \qquad (8.4.1)$$

we define the numerical flux at cell face $(i + 1/2)$, as a function of the surrounding function values

$$f^*_{i+1/2} = f^*_{i+1/2}(u_i, u_{i+1}, u_{i-1}, \ldots.) \qquad (8.4.2)$$

with the consistency condition, that for equal values of u, the numerical flux should reduce to the physical flux $f(u)$. That is,

$$f^*_{i+1/2}(u, u, u, \ldots.) = f(u) \qquad (8.4.3)$$

The explicit finite volume scheme (5.2.6) reduces to the following form:

$$u_i^{n+1} = u_i^n - \frac{\Delta t}{\Delta x_i}(f^*_{i+1/2} - f^*_{i-1/2})^n \qquad (8.4.4)$$

If we keep the time integration separate, then equation (5.2.7) reduces to the system of ordinary differential equations in time

$$\frac{du_i}{dt} = -\frac{1}{\Delta x_i}(f^*_{i+1/2} - f^*_{i-1/2}) \qquad (8.4.5)$$

For linear convection $f = au$; for the Burgers equation $f = u^2/2$, while for the linear convection–diffusion model, $f = au - a\partial u/\partial x$.

The important observation is that every scheme is uniquely identified by its numerical flux, that is by the way the cell-face value $u_{i+1/2}$ is interpolated between the surrounding points.

If we refer to the explicit form (8.4.4), we can write all the schemes defined up to now in the above conservative forms, summarized in the following Table 8.4.1. We define, for the linear case $f = au$.

$$f^*_{i+1/2} \stackrel{\Delta}{=} au_{i+1/2} \tag{8.4.6}$$

and the rules for defining the cell-face values of the variable u, will be representative of the rules to be applied in more general cases on the numerical flux.

Note that the dependence on the CFL number is a direct result of the combined space and time discretization.

We leave it as an exercise to you to put all other schemes, not listed in the table, under this form.

See also problem P.8.17.

The second form (8.4.5) is actually more general, as it is based on a two-step discretization approach, namely:

1. First step define the space discretization in a conservative form, for instance by the finite volume approach, where the choice of the numerical flux, represents completely the space discretization.
2. The second step is the choice of the time integration. This will be discussed in Chapter 9, but is not considered here.

This approach has already been introduced occasionally in Section (8.3.2), equation (8.3.20) or (8.3.22), and also in Section 8.3.4, equation (8.3.39).

We wish now to generalize this approach by considering, a general scheme on the support $(i - 2, i - 1, i, i + 1)$, along the lines of Section 8.2, concentrating only on the space discretization.

Hence if we write the scheme under the form (8.4.5), then the cell-face value $u_{i+1/2}$ should be defined as an interpolation between the function values at the points $(i - 1, i, i + 1)$; while $u_{i-1/2}$ will be defined as an interpolation between the function values at the points $(i - 2, i - 1, i)$.

As a guideline, we can look at the finite difference formulas of second and third order accuracy for the first derivatives, as derived in Chapter 4. In particular, we consider the analysis of Section 4.3.2, where equation (4.3.12) is the general formula we are looking for.

We rewrite this equation as

$$u_{i+1/2} = u_i + \alpha_i(u_{i+1} - u_i) + \beta_i(u_i - u_{i-1}) \tag{8.4.7}$$

with an additional condition between the two parameters, to ensure second order accuracy (see Problem P.4.17). On uniform grids, the condition becomes

$$\alpha + \beta = \frac{1}{2} \tag{8.4.8}$$

leading to the equivalent form

$$u_{i+1/2} = \frac{1}{2}(u_i + u_{i+1}) - \beta(u_{i+1} - 2u_i + u_{i-1}) \tag{8.4.9}$$

Table 8.4.1 Definition of numerical fluxes for explicit schemes.

Scheme	Definition	Numerical flux $f^*_{i+1/2} = a u_{i+1/2}$
FOU	$u_i^{n+1} = u_i^n - \sigma(u_i^n - u_{i-1}^n)$	$u_{i+1/2} = u_i$
Warming–Beam (SOU)	$u_i^{n+1} = u_i^n - \sigma(u_i^n - u_{i-1}^n) - \dfrac{\sigma}{2}(1-\sigma)(u_i^n - u_{i-1}^n)$ $+ \dfrac{\sigma}{2}(1-\sigma)(u_{i-1}^n - u_{i-2}^n)$	$u_{i+1/2} = u_i + \dfrac{1}{2}(1-\sigma)(u_i - u_{i-1})$
Lax–Wendroff (LW)	$u_i^{n+1} = u_i^n - \sigma(u_i^n - u_{i-1}^n) - \dfrac{\sigma}{2}(1-\sigma)(u_{i+1}^n - u_i^n)$ $+ \dfrac{\sigma}{2}(1-\sigma)(u_i^n - u_{i-1}^n)$	$u_{i+1/2} = u_i + \dfrac{1}{2}(1-\sigma)(u_{i+1} - u_i)$
Family of first order schemes (8.2.5)	$u_i^{n+1} = u_i^n - \sigma(u_i^n - u_{i-1}^n)$ $- \left(\dfrac{\sigma}{2} - \gamma\right)[(u_{i+1}^n - u_i^n) - (u_i^n - u_{i-1}^n)]$	$u_{i+1/2} = u_i + \dfrac{1}{2}\left(1 - \dfrac{2\gamma}{\sigma}\right)(u_{i+1} - u_i)$
Family of second order schemes (8.2.37)	$u_i^{n+1} = u_i^n - \sigma(u_i^n - u_{i-1}^n) - \dfrac{\sigma}{2}(1-\sigma)(u_{i+1}^n - u_i^n)$ $+ \dfrac{\sigma}{2}(1-\sigma)(u_i^n - u_{i-1}^n)$ $+ \gamma[(u_{i+1} - 2u_i + u_{i-1}) - (u_{i-2}^n - 2u_{i-1}^n + u_i^n)]$	$u_{i+1/2} = u_i + \dfrac{1}{2}(1-\sigma)(u_{i+1} - u_i)$ $- \dfrac{\gamma}{\sigma}[(u_{i+1} - u_i) - (u_i - u_{i+1})]$

This formula expresses the cell-face value as a central discretization (first term) plus a correction proportional to the second order difference, controlled by the coefficient β.

In the literature, it is customary to write the cell-face values under the form known as the κ-scheme, corresponding to $\beta = (1-\kappa)/4$, for uniform grids

$$u_{i+1/2} = u_i + \frac{\varepsilon}{4}[(1 - \kappa)(u_i - u_{i-1}) + (1 + \kappa)(u_{i+1} - u_i)] \qquad (8.4.10)$$

This general form covers all the possible schemes on the considered 4-point support. The value $\varepsilon = 0$ gives the first order upwind discretization, as identified in the first line of Table 8.4.1.

When $\varepsilon = 1$, we have second order discretizations, with some well-known particular cases: $\kappa = -1$ reproduces the second order upwind space discretization, already seen in equation (8.3.40); while $\kappa = 1$ is a central discretization and $\kappa = 0$ corresponds to Fromm's scheme. The value $\kappa = 1/2$ or $\beta = 1/8$ leads to a third order accurate interpolation of the cell-face value, which is known as the Quick scheme, Leonard (1979), while the third order accurate scheme (8.2.49) corresponds to $\kappa = 1/3$. There various schemes are summarized in Table 8.4.2.

The generalization to an arbitrary flux is straightforward, and is obtained by replacing the u-variables by the flux function.

Hence, for the first order upwind scheme, we would write $f^*_{i+1/2} = f_i$; while for the κ-scheme we would define $f^*_{i+1/2} = f_i + \frac{1}{4}[(1 - \kappa)(f_i - f_{i-1}) + (1 + \kappa)(f_{i+1} - f_i)]$.

For scalar nonlinear fluxes $f = f(u)$, such as the Burgers equation with $f = u^2/2$, we define the 'convection' velocity $a(u)$ by $a(u) = \partial f/\partial u$, with $\partial f/\partial x = a(u)\partial u/\partial x$; leading to the quasi-linear form of equation (8.4.1), as $\partial u/\partial t + a(u)\partial u/\partial x = 0$.

Note the various forms in the right column. All the expressions correspond to different approximations for the first derivative, and when writing them as a correction to the central difference, the additional term represents a discretization of the third derivative u_{xxx}, with different coefficients. This is in accordance with the derivations of Section 8.2.3 and equation (8.2.37), in particular.

Comment on the Quick scheme

There is some controversy in the literature, as to the claim of the third order accuracy of the Quick scheme. As seen from Problem P.4.17, the formula in the middle column of the above table is indeed a third order approximation for the mid-cell value $u_{i+1/2}$, as is shown by comparing with the Taylor expansion of $u_{i+1/2}$ around u_i. However, when considered as a formula for the first derivative based solely on the mesh point values, and applying Taylor expansions of the points $u_{i-2}, u_{i-1}, u_{i+1}$, around u_i, the formula for the first order derivative shown in the last column of table 8.4.2 is only second order accurate. Referring to Problem P.4.13 in Chapter 4, the finite difference formula for the first derivative is of third order for the parameter $a = 1/6$; while the Quick scheme corresponds to $a = 1/8$, leading to a dominating truncation error equal to $-1/8\Delta x^2 \cdot u_{xxx}$.

On the other hand, if we would work with cell-face values $u_{i+1/2}$ and $u_{i-1/2}$ as basic variables, then the Quick approximation would indeed lead to third order accuracy. However, this is rarely the case in practice, where in many codes the mesh point variables or the cell-averaged values are the reference quantities.

The *introduction of the limiters* on the cell-face values, or on the numerical fluxes, follows the steps described in the previous section.

Table 8.4.2 Definition of numerical fluxes for space discretized fluxes.

Scheme	Numerical flux $f^*_{i+1/2} = au_{i+1/2}$	Scheme
	$a > 0$	$\dfrac{du_i}{dt} = -\dfrac{1}{\Delta x}(f^*_{i+1/2} - f^*_{i-1/2})$
FOU	$u_{i+1/2} = u_i$	$\dfrac{du_i}{dt} = -\dfrac{a}{\Delta x}(u_i - u_{i-1})$
κ – scheme	$u_{i+1/2} = u_i + \dfrac{1}{4}[(1-\kappa)(u_i - u_{i-1})$ $+ (1+\kappa)(u_{i+1} - u_i)]$	$\dfrac{du_i}{dt} = -\dfrac{a}{\Delta x}(u_i - u_{i-1}) - \dfrac{a}{4\Delta x}[(1-\kappa)(u_i - 2u_{i-1})$ $+ (1+\kappa)(u_{i+1} - 2u_i + u_{i-1})]$
Central scheme $\kappa = 1$	$u_{i+1/2} = u_i + \dfrac{1}{2}(u_{i+1} - u_i) = \dfrac{1}{2}(u_{i+1} + u_i)$	$\dfrac{du_i}{dt} = -\dfrac{a}{\Delta x}(u_i - u_{i-1}) - \dfrac{a}{2\Delta x}(u_{i+1} - 2u_i + u_{i-1})$ $= -\dfrac{a}{2\Delta x}(u_{i+1} - u_{i-1})$
Upwind scheme $\kappa = -1$	$u_{i+1/2} = u_i + \dfrac{1}{2}(u_i - u_{i-1})$	$\dfrac{du_i}{dt} = -\dfrac{a}{\Delta x}(u_i - u_{i-1}) - \dfrac{a}{2\Delta x}(u_i - 2u_{i-1} + u_{i-2})$ $= -\dfrac{a}{2\Delta x}(3u_i - 4u_{i-1} + u_{i-2})$ $= -\dfrac{a}{2\Delta x}(u_{i+1} - u_{i-1})$ $\quad - \dfrac{a}{2\Delta x}(-u_{i+1} + 3u_i - 3u_{i-1} + u_{i-2})$

Fromm scheme

$\kappa = 0$

$$u_{i+1/2} = u_i + \frac{1}{4}(u_i - u_{i-1}) + \frac{1}{4}(u_{i+1} - u_i)$$

$$\frac{du_i}{dt} = -\frac{a}{\Delta x}(u_i - u_{i-1}) - \frac{a}{4\Delta x}[u_{i+1} - u_i - u_{i-1} + u_{i-2}]$$

$$= -\frac{a}{4\Delta x}(u_{i+1} + 3u_i - 5u_{i-1} + u_{i-2})$$

$$= -\frac{a}{2\Delta x}(u_{i+1} - u_{i-1}) - \frac{a}{4\Delta x}(-u_{i+1} + 3u_i - 3u_{i-1} + u_{i-2})$$

Quick scheme

$\kappa = 1/2$

$$u_{i+1/2} = u_i + \frac{1}{8}(u_i - u_{i-1}) + \frac{3}{8}(u_{i+1} - u_i)$$

$$\frac{du_i}{dt} = -\frac{a}{\Delta x}(u_i - u_{i-1}) - \frac{a}{8\Delta x}(3u_{i+1} - 5u_i + u_{i-1} + u_{i-2})$$

$$= -\frac{a}{8\Delta x}(3u_{i+1} + 3u_i - 7u_{i-1} + u_{i-2})$$

$$= -\frac{a}{2\Delta x}(u_{i+1} - u_{i-1}) - \frac{a}{8\Delta x}(-u_{i+1} + 3u_i - 3u_{i-1} + u_{i-2})$$

Third order scheme

$\kappa = 1/3$

$$u_{i+1/2} = u_i + \frac{1}{6}(u_i - u_{i-1}) + \frac{1}{3}(u_{i+1} - u_i)$$

$$\frac{du_i}{dt} = -\frac{a}{\Delta x}(u_i - u_{i-1}) - \frac{a}{6\Delta x}(2u_{i+1} - 3u_i + u_{i-2})$$

$$= -\frac{a}{6\Delta x}(2u_{i+1} + 3u_i - 6u_{i-1} + u_{i-2})$$

$$= -\frac{a}{2\Delta x}(u_{i+1} - u_{i-1}) - \frac{a}{6\Delta x}(-u_{i+1} + 3u_i - 3u_{i-1} + u_{i-2})$$

As the cell-face values are written as the first order terms plus the additional non-monotone terms, we can define directly the second step as follows:

$$u_{i+1/2} = u_i + \frac{1}{4}[(1 - \kappa)\Psi(r_i) + (1 + \kappa)r_i\Psi(r_i)](u_i - u_{i-1}) \tag{8.4.11}$$

leading to

$$u_{i+1/2} = u_i + \frac{1}{4}[(1 - \kappa) + (1 + \kappa)r_i]\Psi(r_i)(u_i - u_{i-1}) \tag{8.4.12}$$

In terms of the more general formulation (8.4.7), we would write

$$u_{i+1/2} = u_i + [\beta_i + \alpha_i r_i]\Psi(r_i)(u_i - u_{i-1}) \tag{8.4.13}$$

This form is more suitable for applications on non-uniform grids. See again Problem P.4.17 for an example of the dependence of these coefficients on non-uniform grid cell sizes.

The above choice for the limited high-resolution form is by far not unique. An alternative often applied is based on defining the second limiter on the inverse ratio $1/r_i$, leading to

$$u_{i+1/2} = u_i + \frac{1}{4}\left[(1 - \kappa)\Psi(r_i) + (1 + \kappa)r_i\Psi\left(\frac{1}{r_i}\right)\right](u_i - u_{i-1}) \tag{8.4.14}$$

For the more general interpolation, we write

$$u_{i+1/2} = u_i + \left[\beta_i\Psi(r_i) + \alpha_i r_i\Psi\left(\frac{1}{r_i}\right)\right](u_i - u_{i-1}) \tag{8.4.15}$$

If the limiters Ψ satisfy the symmetry condition (8.3.47), equation (8.4.14) reduces to, for all values of κ:

$$u_{i+1/2} = u_i + \frac{1}{2}\Psi(r_i)(u_i - u_{i-1}) \tag{8.4.16}$$

For expression (8.4.15) the expression reduces to

$$u_{i+1/2} = u_i + (\beta_i + \alpha_i)\Psi(r_i)(u_i - u_{i-1}) \tag{8.4.17}$$

which is identical to (8.4.16) on a uniform grid, for which $(\alpha + \beta) = 1/2$.

We can regroup the various options into a single form, defining a composed limiter Φ, by

$$u_{i+1/2} = u_i + \frac{1}{2}\Phi(r_i)(u_i - u_{i-1}) \tag{8.4.18}$$

Depending on the selected option, we would have

$$\Phi(r_i) \triangleq \frac{1}{2}[(1 - \kappa) + (1 + \kappa)r_i]\Psi(r_i) \tag{8.4.19}$$

or

$$\Phi(r_i) \overset{\Delta}{=} \frac{1}{2}\left[(1-\kappa)\Psi(r_i) + (1+\kappa)r_i\Psi\left(\frac{1}{r_i}\right)\right]$$

(8.4.20)

or

$$\Phi(r_i) \overset{\Delta}{=} 2[\beta_i + \alpha_i r_i]\Psi(r_i)$$

(8.4.21)

or

$$\Phi(r_i) \overset{\Delta}{=} 2\left[\beta_i\Psi(r_i) + \alpha_i r_i\Psi\left(\frac{1}{r_i}\right)\right]$$

(8.4.22)

If the limiters Ψ satisfy the symmetry condition (8.3.47), and we consider uniform grids, then $\Phi = \Psi$ for the options (8.4.20) and (8.4.22). Hence, we will consider this case in the following.

All the conditions on the limiters derived in the previous Section 8.3 remain valid.

8.4.2 The Normalized Variable Representation

An alternative representation of the nonlinear limiters has been introduced by Gaskell & Lau (1988) and Leonard (1988). It is based on the definition of *Normalized Variables (NV)*, which lead to a reduction of the number of explicit variables in the relation between the cell-face values and the surrounding points. This reduction is of particular interest for non-uniform grids.

The reduced variable ϕ, is defined as follows:

$$\phi \overset{\Delta}{=} \frac{u - u_{i-1}}{u_{i+1} - u_{i-1}}$$

(8.4.23)

Denoting by a subscript f the cell-face value, and by C the reference point i, the corresponding normalized variables are

$$\phi_f = \frac{u_{i+1/2} - u_{i-1}}{u_{i+1} - u_{i-1}} \qquad \phi_C = \frac{u_i - u_{i-1}}{u_{i+1} - u_{i-1}}$$

(8.4.24)

The relation between the gradient ratio r_i and ϕ_C, is readily obtained from the definitions, as

$$\frac{1}{\phi_C} = \frac{u_{i+1} - u_{i-1}}{u_i - u_{i-1}} = 1 + r_i$$

(8.4.25)

Note that the normalized variable is a ratio of differences and not a ratio of gradients, particularly if the grid is non-uniform. Referring to Figure 8.3.1, in point i, we have $\phi_C = b/(a+b)$ and in point $(i+1)$, we have $(\phi_C)_{i+1} = a/(a+c)$. Hence, when an extremum appears, as in point $(i+1)$ on the figure on the right, then $\phi_C > 1$.

If we consider the functional dependence of the cell-face value $u_{i+1/2}$, it is translated in a relation of the form $\phi_f = F(\phi_C)$ **which is only function of** ϕ_C, at least

Table 8.4.3 *Correspondence between the Sweby diagram and the NVD for various reference schemes on uniform grids.*

LIMITER relation	NVD relation	Comment
$\Psi(r) = 0$	$\phi_f = \phi_C$	First order monotone upwind scheme
$\Psi(r) = 2r$	$\phi_f = 1$	Upper bound of monotonicity region
$\Psi = 2$	$\phi_f = 2\phi_C$	Bound for limiter domain
$\Psi = 1$	$\phi_f = 3\phi_C/2$	Second order non-monotone upwind (SOU) discretization ($\kappa = -1$)
$\Psi(r) = r$	$\phi_f = (1 + \phi_C)/2$	Central (Lax–Wendroff) scheme ($\kappa = 1$)
$\Psi(r) = (1+r)/2$	$\phi_f = \phi_C + 1/4$	Fromm scheme ($\kappa = 0$) defined by the average of SOU and LW
$\Psi(r) = (1+3r)/4$	$\phi_f = 3/4(\phi_C + 1/2)$	Quick scheme ($\kappa = 1/2$)
$\Psi(r) = (1+2r)/3$	$\phi_f = (\phi_C + 5/2)/3$	Third order scheme ($\kappa = 1/3$)

for uniform grids. On non-uniform grids, an additional dependence on the ratio of consecutive cell sizes has to be added.

For the unlimited case, corresponding to $\Psi = 1$, we have a linear dependence between the normalized variables, as summarized in Table 8.4.3.

Referring to the form (8.4.18), the general relation between the normalized variables and the limiters Ψ is given by

$$\phi_f = \left[1 + \frac{1}{2}\Phi(r_i) \right]\phi_C \tag{8.4.26}$$

which reduces, for the case $\Phi = \Psi$ to

$$\phi_f = \left[1 + \frac{1}{2}\Psi(r_i) \right]\phi_C \tag{8.4.27}$$

With this definition, we can translate the Sweby diagram into a ***Normalized Variable Diagram (NVD)***, where $\phi_f = F(\phi_C)$ is displayed in function of ϕ_C.

We restrict ourselves in the following to the case of a uniform grid for simplicity. The extension to the more general cases will be straightforward and left as an exercise.

The condition for second order accuracy, namely that all limiters should contain the point ($r = 1$, $\Psi = 1$), is translated into the condition on the normalized variables that it should contain the point ($\phi_C = 1/2$, $\phi_f = 3/4$).

We can now express the monotonicity conditions and the various limiters in the NVD diagram, applying the above correspondence relations. In particular, the monotonicity conditions (8.3.49) and (8.3.60), defining the region of monotonicity for second order schemes, as shown on Figure 8.3.3, translate into the shaded region of Figure 8.4.1, with the correspondences derived form Table 8.4.3. To perform this translation, we have to notice that the region around $r = 0$ corresponds to very large values of ϕ_C, as a consequence of relation (8.4.25). Indeed, when r covers the range of positive values, or $0 < r < \infty$, the variable ϕ_C covers the range $1 > \phi_C > 0$. Accordingly, the region of large values of r is concentrated around the origin of the NVD diagram.

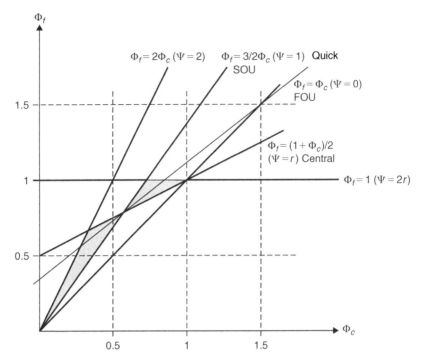

Figure 8.4.1 *Region of monotonicity for second order schemes in the NVD representation.*

Hence, the monotonicity region $r > 1$ to the right of point $(1,1)$ in Figure 8.3.3, is found in the region $0 < \phi_C < 0.5$; while the region $0 < r < 1$ is translated into the region $0.5 < \phi_C < \infty$.

In Figure 8.4.1, you can see the lines corresponding to the schemes listed in Table 8.4.3 and the shaded region corresponds to the shaded monotonicity region of Figure 8.3.3. Notice that this region represents a small part of the NVD diagram, as opposed to the Sweby diagram where the different regions and limiters are much more visible.

Some of the limiters shown on Figure 8.3.4 are displayed in Figure 8.4.2, based on the correspondence between the limiters listed in Table 8.4.4.

Note that all the limiters satisfy the condition corresponding to $\Psi = 0$ for $r < 0$, which translates into $\phi_f = \phi_C$ for $\phi_C > 1$.

Observe that the SMART limiter is outside the shaded area of Figure 8.4.1, for $0 < \phi_C < 0.3$, corresponding to the upper limit of $\Psi = 4$ as seen from equation (8.3.69).

Compared to the (Ψ, r) representation of Figure 8.3.4, the different limiters in the above representation are more difficult to distinguish visually, since the monotonicity region is concentrated into a small subregion of the NV diagram.

Therefore we would prefer the Sweby diagram, such as shown on Figures 8.3.4, to distinguish between different limiters as this representation offers indeed more visibility. But in any case, the two representations are fully equivalent, where the

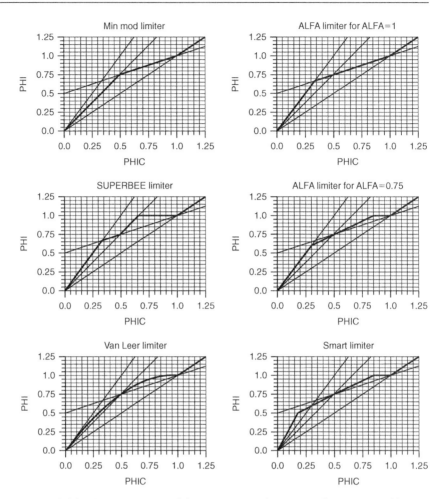

Figure 8.4.2   *Representation of the most current limiters in the NVD variables.*

NVD option has a more direct physical meaning, as the variables ϕ_f and ϕ_C are direct translations of the interpolation rules for the cell-face values.

CONCLUSIONS AND MAIN TOPICS TO REMEMBER

This Chapter contains many topics of general significance for numerical schemes and is of importance for the understanding of some key properties, such as monotonicity. This central concept has led us to understand the origin of the numerical oscillations occurring with second or higher order schemes. Based on this understanding, the cures to these unwanted effects have been developed, leading to high-resolution schemes, based on the introduction of nonlinear limiters.

We have analyzed a general form of two level schemes on an arbitrary support, which led us to a synthetic view, under the form of general conditions on the coefficients of the scheme. In addition, the stability, accuracy, diffusion and dispersion

Table 8.4.4 *Definition of limiters in the NVD diagram*

LIMITER function	NVD definition	Comment
$\Psi(r) = \max[0, \min(r, 1)]$	$\phi_f = \max[0, \min(3\phi_C/2, (\phi_C + 1)/2, 1)]$	Min-mod limiter
$\Psi(r) = \max[0, \min(r, 2), \min(2r, 1)]$	$\phi_f = \max[0, \min(1, 3\phi_C/2), \min(2\phi_C, (\phi_C + 1)/2)]$	Superbee limiter
$\Psi(r) = (r + \lvert r \rvert)/(1 + r)$	$\phi_f = \phi_C[(1 - \phi_C) + \lvert(1 - \phi_C)\rvert]$	Van Leer limiter
$\Psi(r) = (r + r^2)/(1 + r^2)$	$\phi_f = (\phi_C - 1)/(1 - 2\phi_C + 2\phi_C^2)$	Van Albada limiter
$\Psi(r) = \max[0, \min(2r, (3r + 1)/4, 4)]$	$\phi_f = \max[0, \min(3(\phi_C + 1/2)/4, 3\phi_C, 1, \phi_C)]$	SMART limiter
$\Psi(r) = \max[0, \min(r, \beta), \min(\beta r, 1)]$ $1 \leq \beta \leq 2$	$\phi_f = \max[0, \min(\beta/2 + (1 - \beta/2)\phi_C, 3\phi_C/2),$ $\min((1 + \beta/2)\phi_C, (\phi_C + 1)/2), \phi_C]$	β-family
$\Psi(r) = \max[0, \min(2r, \alpha r + 1 - \alpha, 2)]$	$\phi_f = \max[0, \min((3/2 - \alpha)\phi_C + \alpha/2, 2\phi_C, 1)]$	α-family

errors are all fully defined by these coefficients. Some of the important points are the following:

- A very close relation exists between the equivalent differential equation of a scheme and its diffusion and dispersion errors. In particular, remember that even order derivatives in the equivalent differential equation contribute to the real parts of the amplification factor, while the uneven coefficients contribute to the imaginary part.
- Consequently, the even order coefficients totally define the diffusion error and the odd order coefficients determine the dispersion error.
- Inversely, after having derived the dispersion and diffusion errors of a numerical scheme, you can easily recover the dominating terms of its equivalent differential equation, by expanding these errors in a power series around $\phi = 0$. By identifying the terms in these developments with equations (8.1.33) and (8.1.34), you immediately obtain the equivalent differential equation of the scheme under consideration.

We have also learned about the accuracy barrier, that is the maximum order of accuracy that can be achieved by a stable explicit scheme on a given support. In general, the more points the higher the possible order of accuracy. Note in particular that central schemes reach the highest possible accuracy, and that combining upwind and downwind points is the next good choice for higher accuracy, on a given support.

The weakest choice is the pure upwind case, where only one-sided points are considered. Here we have found that an explicit scheme with two-time-levels cannot be of order higher than 2, whatever the number of points involved in the scheme.

Section 8.2 focuses on the analysis of the most current schemes applied in practice, defined either on 3- or 4-point supports. Study carefully the properties of the one-parameter families of schemes of first order (on 3 points) and second order (on 4 points). The presented analysis offers a unified approach, which should help you understand the common points and the differences between selected schemes.

A most significant contribution of this chapter is the concept and conditions for monotonicity introduced in Section 8.3. This is of utmost importance as it identifies the source of the numerical wiggles that we have encountered with second, or higher, order schemes. Study carefully the celebrated Godunov theorem, which shows that monotone linear schemes can be at best of first order accuracy. We have also learned in this section how to easily identify the non-monotone contributions to a higher order scheme.

Based on this knowledge, we have introduced one of the most significant advances made in the history of CFD, namely the development of high-resolution schemes (HRS), which avoid the limitations of Godunov's theorem, through the introduction of nonlinear limiters in the numerical schemes, whose role is to control the gradients of the solution by avoiding unphysical growth.

Study carefully these sections, and work out the Hands-on task, as for any of the selected limiters, you will be able to discover that the practical implementation is quite easy and straightforward. The essential points to remember are:

- HRS combine second order accuracy with monotonicity, avoiding hereby the appearance of oscillations.

- As the limiters depend on ratios of solution gradients, the schemes have become nonlinear.
- The key property of the HRS is that they reduce, through the nonlinear action of the limiters, locally to first order accuracy, in the vicinity of extrema. They can therefore loose some accuracy near extrema present in the solutions.
- HRS are a major achievement of CFD and should be applied when necessary, in order to have more flexibility with regard to grid requirements.
- An additional observation is the unification of the second order LW and WB schemes after application of the limiters.

Finally the last section opens the way towards the introduction of general time integration methods, to nonlinear conservation laws of fluid mechanics and their multidimensional components, through the numerical flux concept. This is tied to the general finite volume approach, based on estimations of cell-face values. With these cell-face values, an alternative view and representation of the limiters, through the definition of normalized variables, is introduced. This approach has the advantage of providing a more direct interpretation of the 'limited' interpolation of the cell-face values, although the visibility of the limiters in the NVD representation is more difficult to identify.

REFERENCES

Arora, M., Roe, P.L. (1997). A well-behaved TVD limiter for high-resolution calculations of unsteady flow. J. Comput. Phys., 132, 3–11.

Barth, T.J. (1993). Recent Development in High Order k-Exact Reconstruction on Unstructured Meshes. AIAA paper 93-0668.

Boris, J.P., Book D.L. (1973). Flux corrected transport: I. SHASTA, a fluid transport algorithm that works. J. Comput. Phys., 11, 38–69.

Boris, J.P., Book D.L. (1976). Solution of the continuity equation by the method of flux corrected transport. J. Comput. Phys., 16, 85–129.

Chakravarthy, S.R., Osher S. (1983). High Resolution Applications of the Osher Upwind Scheme for the Euler Equations. *Proceedings of AIAA 6th Computational Fluid Dynamics Conference*, AIAA Paper 83–1943, pp.363–373.

Chan, T.F. (1984). Stability analysis of finite difference schemes for the advection–diffusion equation. SIAM J. Num. Anal., 21, 272–283.

Darwish, M.S. (1993). A new high-resolution scheme based on the normalized variable formulation. *Num. Heat. Trans. B*, 24, 353–371.

Darwish, M.S. and Moukalled, F. (2000). B-EXPRESS: A new Bounded EXtrema PREServing Strategy for Convective Schemes, *Num. Heat. Trans. B*, 37 (2), 227–246.

Fromm, J. (1964): The time dependent flow of an incompressible viscous fluid. In *Methods in Computational Physics*, Vol. 3. Academic Press, New York, pp. 345–382.

Fromm, J.E. (1968). A method for reducing dispersion in convective difference schemes. *J. Comput. Phys.*, 3, 176–189.

Gaskell, P.H., Lau, A.K.C. (1988). Curvature compensated convective transport: SMART, a new boundedness preserving transport algorithm. *Int. J. Num. Meth. Fluids*, 8, 617–641.

Godunov, S.K. (1959). A difference scheme for numerical computation of discontinuous solution of hydrodynamic Equations. *Math. Sbornik*, 47, 271–306 (in Russian). Translated *US Joint Publ. Res. Service*, JPRS 7226 (1969).

Godunov, S.K., Ryabenkii, V.S. (1964). *The Theory of Difference Schemes*. North Holland Publ., Amsterdam.

Griffiths, D.F., Christie, I. and Mitchell, A.R. (1980). Analysis of error growth for explicit difference schemes in conduction–convection problems. *Int. J. Num. Meth. Eng.,* 15, 1075–1081.

Hairer E. and Wanner G. (1991), *Solving ordinary differential equations II*. Springer-Verlag, Berlin.

Harten, A., Hyman J.M. and Lax P.D. (1976). On finite difference approximations and entropy conditions for shocks. *Commun. Pure Appli. Math.*, 29, 297–322.

Harten, A. (1983). High resolution schemes for hyperbolic conservation laws. *J. Comput. Phys.*, 49, 357–393.

Harten, A. (1984). On a class of high resolution total variation stable finite difference schemes. *SIAM J. Num. Anal.*, 21, 1–23.

Hindmarsh, A.C., Gresho, P.M., Griffiths, D.F. (1984). The stability of explicit euler time integration for certain finite difference approximations of the multidimensional advection diffusion equation. *Int. J. Num. Meth. Fluid.*, 4, 853–897.

Hirt, C.W. (1968). Heuristic stability theory for finite difference equations. *J. Comput. Phys.*, 2, 339–355.

Iserles, A. (1982). Order stars and a saturation theorem for first order hyperbolics, *IMA J. Num. Anal.*, 2, 49–61.

Iserles, A., Strang, G. (1983). The optimal accuracy of difference schemes, *Trans. Amer. Math. Soc.*, 277, 779–803.

Iserles, A., Nørsett, S.P. (1991). *Order stars*. Chapman & Hall, Boca Raton, Florida, USA

Jasak, H., Weller, H.G. and Gosman, A.D. (1999). High resolution NVD differencing scheme for arbitrarily unstructured meshes. *Int. J. Numer. Meth. Fluid.*, 31, 431–449.

Jameson, A. (1993). Artificial diffusion, upwind biasing, limiters and their effect on accuracy and multigrid convergence in transonic and hypersonic flows. *AIAA 11th Computational Fluid Dynamics Conference*, AIAA Paper 93-3359.

Jameson, A. (1995a). Analysis and design of numerical schemes for gas dynamics 1: artificial diffusion, upwind biasing, limiters and their effect on multigrid convergence. *Int. J. Comp. Fluid Dynamic.*, 4, 171–218.

Jameson, A. (1995b). Analysis and design of numerical schemes for gas dynamics 2: Artificial diffusion and discrete shock structure. *Int. J. Comp. Fluid Dynamic.*, 5, 1–38.

Jeltsch, R. (1985). Stability and accuracy of difference schemes for hyperbolic problems. *J. Comput. Appl. Math.*, 12–13, 91–108.

Jeltsch, R., Smit, J.H. (1987). Accuracy barriers of difference schemes for hyperbolic equations. *SIAM J. Num. Anal.*, Vol.24, pp.1–11.

Jeltsch, R. (1988). Order barriers for difference schemes for linear and nonlinear hyperbolic problems, *Numerical Analysis, Proceedings 1987 Dundee Conference*, Griffiths et al. (Eds.), Pitman.

Jeltsch, R., Kiani P. and Raczek K. (1991). Stability of a family of multi-time-level difference schemes for the advection equation, *Numer. Math.*, 60, 77–95.

Jeltsch, R., Renaut, R.A. and Smit J.H. (1997). An accuracy barrier for stable three-time-level difference schemes for hyperbolic equations, *Research Report* 97–10, Seminar für Angewandte Mathematik, Swiss Federal Institute of Technology ETH-Zurich.

Jeng, Y.N., Payne U.J., (1995). An adaptive TVD limiter. *J. Comput. Phys.*, 118, 229–241.

Karni, S. (1994). On the group velocity of symmetric and upwind numerical schemes, *Int. J. Num. Meth. Fluid.*, 18, 1073–1081.

Koren, B. (1998). A robust upwind discretization method for advection, diffusion and source terms, In *Numerical Methods for Advection–Diffusion Problems*, C.B. Vreugdenhil and B. Koren (Eds.). Vieweg, Braunschweig, p. 117.

Laney, C. (1998). *Computational Gasdynamics*. Cambridge University press, Cambridge, UK.

Lax, P.D. (1973). *Hyperbolic Systems of Conservation Laws and the Mathematical Theory of Shock Waves*. SIAM Publications, Philadelphia.

Leonard, B.P. (1979). A stable and accurate convective modeling procedure based on quadratic upstream interpolation. *Comp. Meth. Appl. mech. engr.*, 19, 58–98.

Leonard, B.P. (1980). Note on the Von Neumann stability of the explicit FTCS convection diffusion equation. *Appl. Math. Model.*, 4, 401–402.

Leonard, B.P. (1988). Simple High-Accuracy Resolution Program (SHARP) for convective modeling of discontinuities. *Int. J. Num. Meth. Eng.*, 8, 1291–1318.

Leonard, B.P., (1991). The ULTIMATE conservative difference scheme applied to unsteady one-dimensional advection. *Comp. Meth. Appl. Mech. Eng.*, 88, 17–74.

Lerat, A. (1979). *Numerical Shock Structure and Non Linear Corrections for Difference Schemes in Conservation Form*. Lecture Notes in Physics, Vol 20. Springer Verlag, New York, pp. 345–351.

Lerat, A. and Peyret, R. (1974). Non centered schemes and shock propagation problems. *Comput. Fluid.*, 2, 35–52.

Lerat, A. and Peyret, R. (1975). *The Problem of Spurious Oscillations in the Numerical Solution of the Equations of Gas Dynamics*. Lecture Notes in Physics, Vol. 35. Springer Verlag, New York, pp. 251–256.

Leveque, R.J. (2002). *Finite Volume Methods for Hyperbolic Problems.* Cambridge University Press, Cambridge, UK.

Lien, F.S. and Leschziner, M.A. (1994). Upstream monotonic interpolation for scalar transport with application to complex turbulent flows. *Int. J. Num. Meth. Eng.*, 19, 527.

Morton, K.W. (1980). Stability of finite difference approximations to a diffusion–convection equation. *Int. J. Num. Meth. Eng.*, 15, 677–683.

Rigal, A. (1978). Stability analysis of explicit finite difference schemes for the Navier–Stokes equations. *Int. J. Num. Meth. Eng.*, 14, 617–620.

Richtmyer, R.D. and Morton, K.W. (1967). *Difference Methods for Initial Value Problems*. J. Wiley & Sons, New York.

Roache, P.J. (1972). *Computational Fluid Dynamics.* Hermosa Publication. Albuquerque, USA.

Roe, P.L. (1981). *Numerical Algorithms for the Linear Wave Equation*. Royal Aircraft Establishment, TR 81047, April 1981.

Roe, P.L. (1984). *Generalized Formulation of TVD Lax–Wendroff Schemes, ICASE Report* 84-53, NASA CR-172478, NASA Langley Research Center.

Roe, P.L. (1985). Some Contributions to the Modeling of Discontinuous Flows, *Proceedings of 1985 AMS-SIAM Summer Seminar on Large Scale Computing in Fluid Mechanics, Lectures in Applied Mathematics*, Vol.22, Springer Verlag, New York, pp. 163–193.

Roe, P. L. and Baines, M.J. (1981). Algorithms for advection and shock problems, *Proceedings of the Fourth GAMM Conference on Numerical Methods in Fluid Mechanics*, Notes on Numerical Fluid Mechanics, Vol. 5, H. Viviand (Ed.). Vieweg, p. 281.

Roe, P.L. and Baines, M.J. (1983). Asymptotic behavior of some non-linear schemes for linear advection. *Proceedings of the Fifth GAMM Conference on Numerical Methods in Fluid Mechanics*, Notes on Numerical Fluid Mechanics, Vol. 7, M. Pandolfi and R. Piva (Eds.). Vieweg, p. 283.

Shokin, Yu.I. (1983). *The Method of Differential Approximation.* Springer Verlag, New York.

Spekreijse, S.P. (1987). Multigrid solution of monotone second order discretizations of hyperbolic conservation laws. *Math. Comp.*, 49, 135–155.

Strang, G. and Iserles, A. (1983). Barriers to stability. *SIAM J. Num. Anal.*, 20, 1251–1257.

Sweby, P.K. (1984). High resolution schemes using flux limiters for hyperbolic conservation laws. *SIAM J. Num. Anal.*, 21, 995–1011.

Thompson, H.D., Webb, B.W. and Hoffmann, J.D. (1985). The Cell Reynolds Number Myth. *Int. J. Num. Meth. Fluid.*, 5, 305–310.

Van Albada, G.D., Van Leer, B. and Roberts, W.W. (1982). A comparative study of computational methods in cosmic gas dynamics. *Astron. Astrophy.*, 108, 76–84.

Van Leer, B.(1973). *Towards the Ultimate Conservative Difference Scheme. I: The Quest of Monotinicity*, Lecture Notes in Physics, Vol.18, pp.163–168.

Van Leer, B.(1974). Towards the ultimate conservative difference scheme. II: Monotinicity and Conservation Combined in a Second Order Scheme *J. Comput. Phys.*, 14, 361–370.

Van Leer, B.(1977a). Towards the ultimate conservative difference scheme. III: Upstream-centered finite difference schemes for ideal compressible flow. *J. Comput. Phys.*, 23, 263–275.

Van Leer, B. (1977b). Towards the Ultimate Conservative Difference Scheme. IV. A New Approach to Numerical Convection, *J. Comput. Phys.*, 23, 276–299.

Van Leer, B.(1979). Towards the ultimate conservative difference scheme. V: A second order sequel to Godunov's method, *J. Comput. Phys.*, 32, 101–136.

Venkatakrishnan, V. and Barth, T.J. (1993). Application of Direct Solvers to Unstructured Meshes, *J. Comput. Phys.*, 105, 83–91.

Wanner G., Hairer E. and Nørsett S.P. (1978). *Order Stars and Stability Theorems, BIT*, Vol. 18, pp. 475–489.

Warming, R.F., Kutler, P. and Lomax, H. (1973). Second and third order non centered difference schemes for non linear hyperbolic equations. *AIAA J.*, 11, 189–195.

Warming, R.F., Hyett, B.J. (1974). The modified equation approach to the stability and accuracy of finite difference methods. *J. Comput. Phys.*, 14, 159–179.

Warming, R.F., and Beam, R.W. (1976). Upwind second order difference schemes and applications in aerodynamic flows. *AIAA J.*, 24, 1241–1249.

Waterson, N.P. and Deconinck, H. (1995). A unified approach to the design and application of bounded higher-order convection schemes. In *Numerical Methods in Laminar Turbulent Flow*, Volume IX, pp. 203–214. C. Taylor and P. Durbetaki (Eds). Pineridge Press, Swansea, U.K.

Waterson, N.P. and Deconinck, H. (2007). Design principles for bounded higher-order convection scheme – a unified approach. *J. Comput. Phys.* (In Press).

PROBLEMS

P.8.1 Apply the methodology of Section 8.1 to obtain the first three terms of the equivalent differential equation for the upwind and Lax–Friedrichs schemes. Compare with the power expansions of $|G|$ and ε_ϕ.

Hint: Obtain for the
FOU scheme

$$u_t + au_x = \frac{a\Delta x}{2}(1 - \sigma)u_{xx} + \frac{a\Delta x^2}{6}(2\sigma - 1)(1 - \sigma)u_{xxx}$$

$$+ \frac{a\Delta x^3}{24}(1 + 6\sigma^2 - 6\sigma)(1 - \sigma)u_{xxxx} + O(\Delta x^4)$$

Lax–Friedrichs scheme   .

$$u_t + au_x = \frac{\Delta x^2}{2\Delta t}(1 - \sigma^2)u_{xx} + \frac{a\Delta x^2}{3}(1 - \sigma^2)u_{xxx} + O(\Delta x^3)$$

P.8.2 Develop the general relations of Section 8.1, for a three-level scheme of the form

$$u_i^{n+1} = u_i^{n-1} + \sum_j b_j u_{i+j}^n$$

by applying equation (8.1.11) at $t = (n+1)\Delta t$ and $t = (n-1)\Delta t$, to the convective linear equation $u_t + au_x = 0$. Obtain the consistency relations on the b_j coefficients.

P.8.3 Apply the form (8.1.49) of the modulus squared of the Von Neumann amplification factor, to the first order upwind, Lax–Friedrichs and Lax–Wendroff schemes, for the linear convection equation. Determine for each case the polynomial $S(z)$.

P.8.4 Obtain the relation (8.1.52) for the linear convection equation.

P.8.5 Obtain the amplification factor for the second order upwind schemes of Fromm and Warming and Beam. Generate a polar plot of the amplitude and phase errors and analyze the differences, in particular for the dispersion and diffusion errors. Derive, by identification with the power expansion of the dispersion and diffusion error around $\phi = 0$, the dominating terms of the equivalent differential equation (see also Problem P.8.8).

P.8.6 Solve the convective equation $u_t + au_x = 0$ for a sinus wave packet with four cycles, on meshes with 10, 15, 20 points per wavelength with $a = 1$. The wave packet is defined by

$$
\begin{aligned}
u(t = 0) &= \sin(2\pi kx) \quad && 0 < x < 1 \text{ and } k = 4 \\
&= 0 \quad && x < 0 \quad\quad \text{ and } x > 1
\end{aligned}
$$

Compute the numerical transported wave after 50, 100 and 200 time steps and plot the exact solutions with the numerical solution. Take $\sigma = 0.1, 0.5, 0.9$ and compare the results with the second order upwind scheme and Fromm's schemes with the results you have obtained in Chapter 7.

Hint: The exact solution after n time steps is

$$
\begin{aligned}
u^n &= \sin 2\pi k(x - n \cdot \Delta t) \quad && n\Delta t < x < (n+1)\Delta t \\
&= 0 \quad && x < n\Delta t \text{ and } x > (n+1)\Delta t
\end{aligned}
$$

P.8.7 Repeat the same problem for a moving discontinuity

$$u(t = 0) = 1 \quad x < 0$$
$$= 0 \quad x > 0$$

after 10, 50, 100 time steps, for $\sigma = 0.1, 0.5, 0.9$.

P.8.8 Apply the analysis of Section 8.1 to obtain the equivalent differential equations, the lowest order terms of the dissipation and dispersion errors for the general scheme (8.2.37). Derive the amplification factor, the dispersion and diffusion errors and the series expansion of these errors in function of the phase angle ϕ, as a way to obtain the first two terms of the equivalent differential equation (8.1.27), based on the expansions (8.1.33) and (8.1.34). Apply this to the Fromm scheme and compare with the two other schemes, Lax–Wendroff and Warming–Beam, S_{LW}, S_{WB}.

P.8.9 Derive the third order scheme (8.2.49) and apply the analysis of Section 8.1 to obtain the fist two terms of the equivalent differential equation, as well as the dispersion and diffusion first terms in function of phase angle.

P.8.10 Solve Problem P.8.6 with the third order accurate scheme (8.2.49). Compare and analyze the results.

P.8.11 Solve Problem P.8.7 with the third order accurate scheme. Compare and analyze the results.

P.8.12 Obtain the stability condition (8.3.19) for the upwind-discretized convection–diffusion equation (8.3.17).

P.8.13 Consider the linear convection equation $u_t + au_x = 0$ and the semi-discretized schemes obtained from applying successively the following space discretizations to the space term $(-au_x)$:
1. first order upwind difference,
2. second order backward difference,
3. second order central difference.

Writing the schemes in the form (8.3.22) obtain the β_j coefficients and the conditions for monotonicity.

Apply an explicit, forward difference for the time discretization and obtain the form (8.3.1) and all the b_j coefficients. Define the additional stability and monotonicity condition for b_0. Show that only scheme (i) is monotone and that when applied to the second order schemes (ii) and (iii), the schemes are unstable.

Derive the amplification factors for these schemes.

P.8.14 Repeat the Problems P.8.6 and P.8.7 after introduction of limiters. Compare the results with min-mod, and superbee limiters.

P.8.15 Repeat the Problem P.8.14 with the α-family of limiters, in particular compare $\alpha = 1/3$ and $\alpha = 3/8$. Repeat with the SMART limiter to evaluate the effects of the upper bounds on this limiter.

P.8.16 Derive the correspondence Table 8.4.4 between the Sweby and NVD diagrams.

P.8.17 Consider the general family of second order schemes on the support $(i - 2, i - 1, i, i + 1)$, as expressed in function of the γ-parameter in Table 8.4.1:
1. Express the cell-face value $u_{i+1/2}$ under the form (8.4.7) and show that for this CFL-dependent extrapolation, second order accuracy is satisfied by the condition $\alpha + \beta = (1 - \sigma)/2$.

2. Introduce the limiters under the form (8.4.17) and define the function $\Psi(r)$ associated to the linear schemes of Lax–Wendroff, Warming–Beam, Fromm and for the third order scheme. Express also the limiters in function of the inverse gradient ratio $R_i = 1/r_i$

Hint: Referring to Table 8.4.1, write the schemes under the form

$$u_{i+1/2} = u_i + \frac{1}{2}(1 - \sigma)[K + (1 - K)r_i](u_i - u_{i-1}) \quad K = 2\gamma/(1 - \sigma)$$

generalized to

$$u_{i+1/2} = u_i + \frac{1}{2}(1 - \sigma)\Psi(r_i)(u_i - u_{i-1}) = u_i + \frac{1}{2}(1 - \sigma)\frac{\Psi(r_i)}{r_i}(u_{i+1} - u_i)$$

$$\stackrel{\Delta}{=} u_i + \frac{1}{2}(1 - \sigma)\Psi(R_i)(u_{i+1} - u_i) \quad \text{where } R_i = \frac{1}{r_i}$$

Obtain the following correspondence:

LW scheme	$K = 0$	$\Psi(r) = 1$	$\Psi(R) = R$
WB scheme	$K = 1$	$\Psi(r) = r$	$\Psi(R) = 1$
Fromm scheme	$K = 1/2$	$\Psi(r) = (1 + r)/2$	$\Psi(R) = (1 + R)/2$
Third order scheme	$K = (1 + \sigma)/3$	$\Psi(r) = (1 + \sigma)/3$ $+ r(2 - \sigma)/3$	$\Psi(R) = R(1 + \sigma)/3$ $+ (2 - \sigma)/3$

P.8.18 Convert the Symmetric ALFA limiters (8.3.71) and the CFL-dependent limiters (8.3.72) and (8.3.73) to the NVD representation.

P.8.19 Determine the exact solutions of the scheme obtained by discretizing the convection term in the convection–diffusion equation $au_x = \alpha u_{xx}$, with a first order backward formula (for $a > 0$). Show that the solution will always remain oscillation free for $a > 0$.

Analyze the exact numerical solutions for the boundary conditions $u(0) = u_0$, $u(L) = u_L$, for $N \cdot \Delta x = Ly$ applying the normal mode method and compare with equation (8.2.26). Calculate the error $\varepsilon_i = u - u_i$ and show that the scheme is only first order accurate.

Hint: The scheme is

$$a(u_i - u_{i-1}) = \frac{\alpha}{\Delta x}(u_{i+1} - 2u_i + u_{i-1})$$

and obtain $\kappa_1 = 1, \kappa_2 = (1 + \text{Re}_{\Delta x})$.

Part IV

The Resolution of Numerical Schemes

We have now reached the fourth part in the succession of steps and components required to set up a CFD model, and we are, at this stage, faced with a set of space-discretized, or semi-discretized, equations.

As seen in Part III, Section 8.4, we have to make an essential decision as to the time dependence of our numerical formulation. If the physical problem is time dependent, there is obviously no choice; the mathematical initial value problem has to be discretized in time and the numerical solution has to be time accurate. On the other hand, for physical stationary problems, we can decide either to discretize the equations in space and deal with a time-independent numerical scheme or maintain the time dependency and discretize the equations in space and time, but aim only at the time asymptotic, steady, numerical solution. *Remember that we have advocated that last option, as a standard discretization approach for convection–diffusion equations, such as the conservation laws.*

When we select to discretize the time-dependent form, we can either develop a combined space and time discretization, such as the Lax–Wendroff schemes, as seen in the previous chapters, or perform first a separate space discretization, based on the definition of the numerical flux, leading to a *semi-discretized set of ordinary differential equations (ODEs) in time*.

In the next two chapters we will investigate some of the most currently applied techniques for either the resolution of semi-discretized equations that is, systems of ordinary differential equations in time (Chapter 9) or algebraic systems of equations to which they reduce for stationary or implicit time integration problems (Chapter 10).

A large number of techniques have been developed to define and analyze numerical methods for systems of ODEs in time, or more precisely, initial value problems. The most widely applied methods, as well as the techniques for the analysis of their stability properties and the related error properties, will be presented in Chapter 9.

By combining an arbitrary time integration method with an arbitrary space discretization, we are faced with all kinds of possible schemes. We have therefore to generalize the analysis methods derived in Chapters 7 and 8, to these general situations. This will be the main subject of this Chapter, which will lead to a generalization of the Von Neumann method for stability and error analysis. The methodology to be presented has many common features with the Von Neumann method, as it will rely on the same spectral Fourier decomposition of the solutions and their errors.

With a stationary formulation in space or with implicit time integration, we will deal with an algebraic system of equations. This algebraic system has to be solved in the lowest possible number of numerical operations. Various iterative techniques for achieving this goal will be introduced in Chapter 10.

A bridge between stationary and non-stationary formulations can be established, which enlarges the family of methods to be applied to obtain solutions to physical stationary problems. We can indeed select pseudo-time operators in order to accelerate the convergence of the scheme and an illustration is provided by the preconditioning methods. This bridge has also as important consequence that an iterative method can be analyzed by the techniques used for the stability analysis of time-dependent schemes, for instance by their Fourier symbol.

From the spectral analysis of the errors of an iterative method, appropriate operations can be defined in order to damp selectively certain frequency domains of the error spectrum. This philosophy is at the basis of the most powerful of the presently available convergence acceleration techniques, namely the ***multigrid method*** to be presented in Chapter 10.

Chapter 9

Time Integration Methods for Space-discretized Equations

OBJECTIVES AND GUIDELINES

If we look back at some of the schemes introduced and analyzed in the previous chapters, for instance for the convection equation, we notice that various combinations of space and time discretizations show widely different behaviors. Referring to Chapter 7 and to the central second order space discretization for the convection term $a\partial u/\partial x$, we have seen that associated to a forward time difference in an explicit scheme, the resulting scheme is unstable; while associated to a backward time difference (leading to an implicit scheme) it is unconditionally stable, and when selecting a central difference in time (the leapfrog scheme) it is conditionally stable under the CFL condition. Similar properties have been seen with other space discretizations.

This raises several questions:

- How can we understand, and predict, these properties?
- What criteria are to be satisfied by a time integration method, applied to a given space discretization, in order to lead to a stable and accurate scheme.
- Inversely, if we pre-select a time integration method, what are the properties of space discretizations required in order to generate a stable scheme, with the lowest possible diffusion and dispersion errors.

The methodologies allowing to answer these questions and to guide you through this selection process form the essential objectives of this chapter.

We consider here the general formulation of a numerical scheme, as already introduced in Section 8.4, where we separate the time from the space discretizations. Hereby, we first perform a space discretization, defined for instance by a numerical flux in a general finite volume formulation, leading to a system of first order ordinary differential equations (ODEs) in time. In the second step we have to integrate this system in time, by selecting an appropriate method.

A general methodology for the analysis of the association of a given space discretization with a time integration method, can be defined by the following steps:

1. Express the space discretization into a matrix form, including the boundary conditions.
2. Perform a spectral analysis of its representative matrix, obtaining hereby its *eigenvalue spectrum*.

3. This leads us to a first set of **stability conditions on the space discretization** by expressing that the **exact time integrated solution** of the considered space discretized system of ODEs should not grow exponentially with time. We will show that the ODE system is stable when the eigenvalues Ω of the space discretization matrix have non-positive real parts, that is, are located in the left half, imaginary axis included, of the complex Ω-plane.

These three steps will form the subject of Section 9.1.
Next, we proceed as follows:

- Select a time integration method for the semi-discretized system of ODEs and define the methodology for its stability analysis, as a function of the eigenvalues of the space discretization.
- The stability condition of the time integration scheme has to be compatible with the range of the eigenvalue spectrum, namely the stability region of the time discretization method must include the whole spectrum of eigenvalues.

This will be developed in Section 9.2.

- Finally we apply the methodology to some of the most current methods of interest in CFD, such as the family of implicit multistep methods, the family of explicit predictor–corrector methods or the multistage Runge–Kutta methods.

This will be considered in section 9.3 and illustrated by various examples.

As a glimpse of typical issues related to multidimensional methods, an advanced Section A9.4 is devoted to some specific properties of implicit schemes for multidimensional problems, in particular when applying **alternating direction implicit** methods (ADI), also called **approximate factorization** methods.

The roadmap to this chapter is summarized in Figure 9.0.1.

9.1 ANALYSIS OF THE SPACE-DISCRETIZED SYSTEMS

The general form of a conservation law, can be considered mathematically as an initial, boundary value differential problem, over the domain Ω, with boundary Γ

$$\frac{\partial u}{\partial t} + \vec{\nabla} \cdot \vec{F} = 0$$

Initial condition: $u(\vec{x}, 0) = u^0(\vec{x})$ for $t = 0$ and $\vec{x} \in \Omega$

Boundary condition: $u(\vec{x}, t) = g(\vec{x}, t)$ for $t > 0$ and $\vec{x} \in \Gamma$

$$(9.1.1)$$

Referring to Sections 8.3.2 and 8.4.1, for a one-dimensional case, we can discretize the flux differential operator via the general definition of a numerical flux and obtain the system of ordinary differential equations in time, written as equation (8.4.5), which we repeat here for convenience

$$\frac{du_i}{dt} = -\frac{1}{\Delta x_i}(f^*_{i+1/2} - f^*_{i-1/2})$$

$$(9.1.2)$$

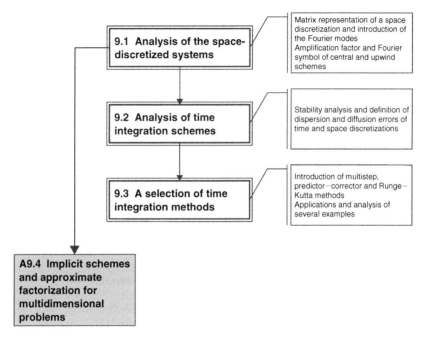

Figure 9.0.1 *Roadmap to this chapter.*

If we group all the mesh point values u_i in a column vector U, the right-hand side can be written in the matrix form

$$\frac{\mathrm{d}U}{\mathrm{d}t} = S \cdot U + Q \tag{9.1.3}$$

where S is the matrix representing the space discretization and Q contains contributions from the boundary conditions. The matrix S therefore depends on the space discretization *and* on the boundary conditions. Within the finite volume framework, see equation (5.2.7) in Chapter 5, the right-hand side is the **residual**, defined as the balance of fluxes though all the cell faces plus sources and boundary condition terms.

Equation (9.1.3) is a system of ordinary differential equations (ODE) in time and forms a **semi-discretized representation**, also called **method of lines**.

Important note
The semi-discretized approach has to be compared with the two-step explicit schemes derived in Sections 8.1 and 8.2, with the Lax–Wendroff scheme as typical reference. These schemes are characterized by a combined space and time discretization, which is translated by their specific behavior when applied to obtain a steady state solution. Take for instance the Lax–Wendroff scheme

$$u_i^{n+1} = u_i^n - \frac{\sigma}{2}(u_{i+1}^n - u_{i-1}^n) + \frac{\sigma^2}{2}(u_{i+1}^n - 2u_i^n + u_{i-1}^n)$$

where σ is the CFL number, $\sigma = a\Delta t/\Delta x$. When the steady state is reached, u_i is solution of the system

$$-(u_{i+1}^n - u_{i-1}^n) + \sigma(u_{i+1}^n - 2u_i^n + u_{i-1}^n) = 0$$

which contains σ as parameter, that is Δt, although the time step has no physical significance for a time-independent problem. This is mathematically correct, but on the conceptual level it is somehow disturbing to generate a numerical solution depending on a non-physical and non-relevant parameter.

This explains the success of the semi-discretized approach in contemporary CFD, whereby the space and time discretizations are kept totally separated. Here, the steady state solution is independent of Δt, since it is solution of the system $S \cdot U + Q = 0$, which does not contain any non-relevant, non-physical parameters, such as Δt.

Note that the following developments are valid for an arbitrary number of space variables, when U is defined accordingly.

When the problem is linear, S will be independent of U, otherwise $(S \cdot U)$ is a nonlinear function of U. This option and the issue of the necessary linearization will be discussed in Section 9.3.

Let us illustrate the derivation of the semi-discretized form (9.1.3) and its S-matrix by some examples of the representative linear diffusion and convection operators, associated to specific boundary conditions.

9.1.1 The Matrix Representation of the Diffusion Space Operator

For the linear diffusion equation $u_t = au_{xx}$, a central difference of the space second derivative leads to the semi-discretized scheme of second order accuracy:

$$\frac{du_i}{dt} = \frac{\alpha}{\Delta x^2}(u_{i+1} - 2u_i + u_{i-1}) \tag{9.1.4}$$

In order to derive the complete matrix S, we have to include the boundary conditions and their numerical implementation:

(a) *Dirichlet boundary conditions*
    For a domain on the x-axis between $(0, +L)$, with $(N + 1)$ mesh points, such that $x_i = i\Delta x$ and $\Delta x = L/N$, a Dirichlet type of condition will be defined by

$$u_0 = u(0, t) = a \quad u_N = u(+L, t) = b \tag{9.1.5}$$

For the first mesh point, $i = 1$, the discretized equation will be

$$\frac{du_1}{dt} = \frac{\alpha}{\Delta x^2}(u_2 - 2u_1 + a) \tag{9.1.6}$$

where the mesh point value u_0 has been replaced by its imposed boundary value. Similarly the equation for the point $i = N - 1$ can be written as

$$\frac{du_{N-1}}{dt} = \frac{\alpha}{\Delta x^2}(b - 2u_{N-1} + u_{N-2}) \tag{9.1.7}$$

This completes the determination of the matrix S representing the space discretization and the system of ordinary differential equations can be written as

$$\frac{dU}{dt} = \frac{\alpha}{\Delta x^2}
\begin{vmatrix}
-2 & 1 & \cdot & \cdot & & & \cdot \\
1 & -2 & 1 & & & & \cdot \\
& 1 & -2 & 1 & & & \\
& & \cdot & \cdot & \cdot & & \\
& & & \cdot & & & \\
& & & 1 & -2 & 1 & \\
& \cdot & & & 1 & -2
\end{vmatrix}
\begin{vmatrix}
u_1 \\
\cdot \\
u_{i-1} \\
u_i \\
u_{i+1} \\
\cdot \\
u_{N-1}
\end{vmatrix}
+
\begin{vmatrix}
a\alpha/\Delta x^2 \\
\cdot \\
\cdot \\
\cdot \\
\cdot \\
\cdot \\
b\alpha/\Delta x^2
\end{vmatrix}$$

$$\equiv S \cdot U + Q \tag{9.1.8}$$

Observe that the introduction of the boundary conditions has led to the definition of the non-homogeneous term Q containing the boundary contributions.

(b) *Neumann boundary conditions*

When Neumann boundary conditions are imposed at an end-point, for instance at $x = 0$, with a Dirichlet condition at $x = L$, we have

$$\frac{\partial u}{\partial x} = a \quad \text{at} \quad x = 0 \quad \text{and} \quad u_N = u(+L, t) = b \tag{9.1.9}$$

This Neumann condition has of course to be discretized, for instance with a one-sided difference, so as to include only points inside the computational domain.

The first equation, for mesh point $i = 0$, would then be written with a forward difference of the derivative condition as

$$\frac{1}{\Delta x}(u_1 - u_0) = a \tag{9.1.10}$$

This equation provides a relation for u_0, to be applied in the discretized equation for mesh point $i = 1$. We obtain in this way the following equation, instead of equation (9.1.6)

$$\frac{du_1}{dt} = \frac{\alpha}{\Delta x^2}(u_2 - u_i) - \frac{a\alpha}{\Delta x} \tag{9.1.11}$$

The matrix equation (9.1.8) takes here the following form,

$$\frac{dU}{dt} = \frac{\alpha}{\Delta x^2}
\begin{vmatrix}
-1 & 1 & \cdot & \cdot & & & \cdot \\
1 & -2 & 1 & & & & \cdot \\
& 1 & -2 & 1 & & & \\
& & \cdot & \cdot & \cdot & & \\
& & & \cdot & & & \\
& & & 1 & -2 & 1 & \\
& \cdot & & & 1 & -2
\end{vmatrix}
\begin{vmatrix}
u_1 \\
\cdot \\
u_{i-1} \\
u_i \\
u_{i+1} \\
\cdot \\
u_{N-1}
\end{vmatrix}
+
\begin{vmatrix}
-a\alpha/\Delta x \\
\cdot \\
\cdot \\
\cdot \\
\cdot \\
\cdot \\
b\alpha/\Delta x^2
\end{vmatrix}$$

$$\tag{9.1.12}$$

Notice that the form of the matrix S depends on the selected implementation of the boundary condition. If instead of the first order forward difference (9.1.10), we had selected a second order forward difference, we would obtain a different matrix S, with different properties.

This is a first indication that the stability conditions of a numerical scheme can be influenced by the boundary conditions and their numerical implementation.

This is a major issue in CFD and should retain your careful attention, either if writing a code, or when using an existing one. The detailed analysis of the impact of boundary conditions on a numerical scheme is very complicated, since it is in general not possible to derive analytical expressions for the eigenvalues. Therefore, the knowledge of the adequacy of certain boundary conditions, or certain implementations, can mostly be gained by practical experience. *Problems occurring with an implementation of boundary conditions would be an indication that the mathematical problem is not well-posed either numerically, or analytically.*

(c) *Periodic boundary conditions*

A third type of boundary condition, with great practical and theoretical importance, is the imposition of periodic conditions. This implies that

$$u_0 = u_N \quad \text{and} \quad u_{-1} = u_{N-1} \tag{9.1.13}$$

The first discretized equation is written here for $i = 0$ (instead of $i = 1$), as

$$\frac{du_0}{dt} = \frac{\alpha}{\Delta x^2}(u_1 - 2u_0 + u_{N-1}) \tag{9.1.14}$$

and the last equation, written for $i = N - 1$, becomes

$$\frac{du_{N-1}}{dt} = \frac{\alpha}{\Delta x^2}(u_0 - 2u_{N-1} + u_{N-2}) \tag{9.1.15}$$

The resulting matrix structure is similar to the matrix obtained from Dirichlet conditions, but with the addition of a coefficient 1 in the upper right and lower left corners. This is typical for *'periodic' matrices*:

$$\frac{dU}{dt} = \frac{\alpha}{\Delta x^2}
\begin{vmatrix}
-2 & 1 & \cdot & \cdot & & 1 \\
1 & -2 & 1 & & & \cdot \\
 & 1 & -2 & 1 & & \\
 & & \cdot & \cdot & \cdot & \\
 & & & \cdot & & \\
 & & & 1 & -2 & 1 \\
1 & & & & 1 & -2
\end{vmatrix}
\begin{vmatrix}
u_0 \\
\cdot \\
u_{i-1} \\
u_i \\
u_{i+1} \\
\cdot \\
u_{N-1}
\end{vmatrix} \tag{9.1.16}$$

Notice here that there is no Q matrix generated with periodic boundary conditions. This is an important property, as it will make it easier to derive analytical eigenvalues, as we will see next.

9.1.2 The Matrix Representation of the Convection Space Operator

For convection equations, the problem of the boundary conditions is much more severe and its impact on the schemes is often dominating. The reason for this strong influence is to be found in the physical nature of the convection or propagation phenomenon, since they describe 'directional' phenomena in space. For instance, in one-dimension, the propagation or convection of the quantity u in the positive x-direction, when $a > 0$, implies that the value of u at the downstream end point of the computational region $0 < x < L$, is determined by the upstream behavior of u and cannot be imposed arbitrarily. However, from numerical point of view, depending on the space discretization, we might need information on $u_N = u(x = L)$ in order to close the algebraic system of equations. In this case, the condition to be imposed on u_N cannot be taken from physical sources and has to be defined numerically. This is called a **numerical boundary condition**. The choice of this numerical boundary condition is critical for the whole scheme. Intuitively we suspect that a good choice should be compatible with the physics of the problem but the stability analysis is the ultimate criterion. For $a > 0$, we have obviously to impose, on physical grounds, boundary values at the left end of the domain; while when a is negative, the propagation occurs in the negative x-direction and the physical boundary condition has to be defined at the right end of the domain.

This is a difficult task, since even for very simple cases, we cannot find analytically the eigenvalues of the matrix S. For a hyperbolic system of equations, with the simultaneous presence of positive and negative propagation speeds, as is the case for the Euler equations in subsonic flows, the problem becomes even more complex. More detailed information, in the framework of the Euler equations, will be presented in Chapter 11 and in Volume II.

The reader interested in more mathematical-oriented work will refer to the work of Kreiss (1968, 1970), Gustafsson et al. (1972), Gustafsson and Kreiss (1979) and to an interesting review of Helen Yee (1981) where additional references can be found.

Upwind space discretization

Let us consider a first order upwind space discretization for the linear convection equation $u_t + au_x = 0$, with $a > 0$. We obtain the following system of ODEs in time

$$\frac{du_i}{dt} = -\frac{a}{\Delta x}(u_i - u_{i-1}) \tag{9.1.17}$$

for the interior points, also called the **interior scheme**. A boundary value

$$u(x = 0, t) = g(t) \tag{9.1.18}$$

will be imposed at the left end boundary, when $a > 0$.

For the first mesh point $i = 1$, we would write see Figure 9.1.1.

$$\frac{du_1}{dt} = -\frac{a}{\Delta x}(u_1 - u_0) = -\frac{a}{\Delta x}u_1 + \frac{a}{\Delta x}g(t) \tag{9.1.19}$$

At, the downstream end, the equation for $i = N$ is

$$\frac{du_N}{dt} = -\frac{a}{\Delta x}(u_N - u_{N-1}) \tag{9.1.20}$$

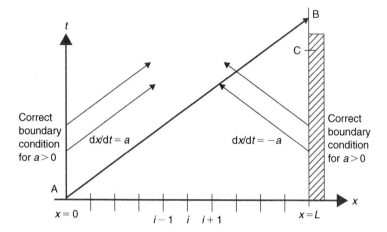

Figure 9.1.1 *Boundary conditions for hyperbolic problems.*

and no additional, numerical condition is necessary. The semi-discretized system of ODEs in time becomes in this case

$$\frac{dU}{dt} = \frac{-a}{\Delta x} \begin{vmatrix} 1 & 0 & \cdot & \cdot & & & \cdot \\ -1 & 1 & 0 & & & & \\ & -1 & 1 & \cdot & & & \\ & & \cdot & \cdot & \cdot & & \\ & & & \cdot & & & \\ & & & -1 & 1 & 0 & \\ \cdot & & & & -1 & 1 \end{vmatrix} \begin{vmatrix} u_1 \\ \cdot \\ u_{i-1} \\ u_i \\ u_{i+1} \\ \cdot \\ u_N \end{vmatrix} + \begin{vmatrix} -ag(t)/\Delta x \\ \cdot \\ \cdot \\ \cdot \\ \cdot \\ \cdot \\ \cdot \end{vmatrix} \qquad (9.1.21)$$

In order to illustrate the dominating influence of boundary conditions, let us look at what happens if we impose the physical condition at the downstream end. In this case, there is no information on u_0, to start equation (9.1.19) and the last equation for u_N would be overruled by the boundary condition $u_N = g(t)$. However, the last equation

$$\frac{du_{N-1}}{dt} = -\frac{a}{\Delta x}(u_{N-1} - u_{N-2}) \qquad (9.1.22)$$

is decoupled from the physical information at u_N. Hence, there is no way of ensuring that a numerical boundary condition at u_0, will lead to a downstream value satisfying the boundary condition $u_N = g(t)$ and the problem is not well-posed. This is illustrated in Figure 9.1.1, where the characteristic $dx/dt = a$ is shown along which u is constant. If A represents the value of u_0, the corresponding value of u_N is B and imposing $u_N = C$ for instance, is not compatible with $u_0 = A$.

When $a < 0$, in order to stabilize the scheme we have to impose a boundary condition at $i = N$, $u_N = g(t)$ and apply a forward space discretization.

Central space discretization with periodic boundary condition

If we assume periodic boundary conditions, in the form of equation (9.1.13), we obtain for the end-points $i = 0$ and $i = N - 1$,

$$\frac{du_0}{dt} = -\frac{a}{2\Delta x}(u_1 - u_{-1}) = -\frac{a}{2\Delta x}(u_1 - u_{N-1})$$

$$\frac{du_{N-1}}{dt} = -\frac{a}{2\Delta x}(u_N - u_{N-2}) = -\frac{a}{2\Delta x}(u_0 - u_{N-2})$$

(9.1.23)

leading to the following matrix form

$$\frac{dU}{dt} = \frac{-a}{2\Delta x}
\begin{vmatrix}
0 & 1 & \cdot & \cdot & & & -1 \\
-1 & 0 & 1 & & & & \cdot \\
& -1 & 0 & 1 & & & \\
& & \cdot & \cdot & \cdot & & \\
& & & \cdot & & & \\
& & & & -1 & 0 & 1 \\
1 & & & & & -1 & 0
\end{vmatrix}
\begin{vmatrix}
u_0 \\
\cdot \\
u_{i-1} \\
u_i \\
u_{i+1} \\
\cdot \\
u_{N-1}
\end{vmatrix}$$

(9.1.24)

Here again we observe that periodic boundary conditions do not generate a source term Q, as seen from the above equation, as well as from equation (9.1.16).

9.1.3 The Eigenvalue Spectrum of Space-discretized Systems

It is well known in matrix theory that the properties of a matrix are fully contained in its eigenvalue and eigenvector spectrum.

The analysis of a matrix, or an operator, through the eigenvalues and eigenvectors, represents therefore a most profound investigation of their properties. The complete 'identity' of a matrix is 'X-rayed' through the eigenvalue analysis and its spectrum can be viewed as a unique form of identification.

The stability analysis of the semi-discretized system (9.1.3) will consequently be based on the eigenvalue structure of the matrix S, since we will show here that its exact solution is directly determined by the eigenvalues and eigenvectors of S.

This is a very fundamental property and we recommend you to study carefully the development hereafter.

Let $\Omega_j, j = 1, N$, be the N eigenvalues of the $(N \times N)$ matrix S solution of the eigenvalue equation

$$\det |S - \Omega I| = 0$$

(9.1.25)

and $V^{(j)}$ the associated eigenvectors, a solution of

$$S \cdot V^{(j)}(\vec{x}) = \Omega_j V^{(j)}(\vec{x}) \quad \text{(no summation on } j)$$

(9.1.26)

N is the total number of mesh points and there are as many eigenvalues and associated eigenvectors as mesh points.

Observe that all these quantities depend only on the space coordinates.

The matrix S is supposed to be of rank N and hence, have N linearly independent eigenvectors. Each eigenvector consists of a set of mesh point values $v_i^{(j)}$, that is

$$V^{(j)} = \begin{vmatrix} \cdot \\ v_{i-1}^{(j)} \\ v_i^{(j)} \\ v_{i+1}^{(j)} \\ \cdot \end{vmatrix} \qquad (9.1.27)$$

The $(N \times N)$ matrix T formed by the N columns $V^{(j)}$, diagonalizes the matrix S, since all the equations (9.1.26) for the N eigenvalues can be grouped as

$$S \cdot T = T \cdot \Omega \qquad (9.1.28)$$

where Ω is the diagonal matrix of the eigenvalues

$$\Omega = \begin{vmatrix} \Omega_1 & & & & \\ & \Omega_2 & & & \\ & & \cdot & & \\ & & & \cdot & \\ & & & & \Omega_N \end{vmatrix} \qquad (9.1.29)$$

Hence, we have

$$\Omega = T^{-1} \cdot S \cdot T \qquad (9.1.30)$$

Since, the eigenvectors $V^{(j)}$ form a complete set of basis vectors in the considered space of the mesh-point functions, we can always write the **exact solution** $\overline{U}$ of equation (9.1.3) as a linear combination of the $V^{(j)}$ eigenvectors,

$$\overline{U}(t, \vec{x}) = \sum_{j=1}^{N} \overline{U}_j(t) \cdot V^{(j)}(\vec{x}) \qquad (9.1.31)$$

where the $\overline{U}_j$ coefficients depend only on time, while the full space dependence is contained exclusively in the eigenvectors $V^{(j)}(\vec{x})$. This is called a **modal decomposition**.

Similarly the non-homogenous term Q, assumed independent of time, can be decomposed in the same modal space of the $V^{(j)}$ eigenvectors

$$Q = \sum_{j=1}^{N} Q_j \cdot V^{(j)} \qquad (9.1.32)$$

The time dependent coefficients $\overline{U}_j(t)$ are obtained from the differential system (9.1.3), by inserting (9.1.31), leading to the *modal equation*

$$\frac{d\overline{U}_j}{dt} = \Omega_j \overline{U}_j + Q_j \quad \text{(no summation on } j\text{)} \tag{9.1.33}$$

The exact solution of this modal equation is easily obtained as the sum of the solution of the *homogenous modal equation*

$$\frac{d\overline{U}_{jT}}{dt} = \Omega_j \overline{U}_{jT} \quad \text{(no summation on } j\text{)} \tag{9.1.34}$$

$$\overline{U}_{jT}(t) = c_{0j} e^{\Omega_j t} \tag{9.1.35}$$

and a particular solution, for instance a solution of the 'steady state' equation

$$\Omega_j \overline{U}_{jS} + Q_j = 0 \quad \text{(no summation on } j\text{)} \tag{9.1.36}$$

$$\overline{U}_{jS} = -\frac{Q_j}{\Omega_j} \tag{9.1.37}$$

We obtain hereby the *modal solution*

$$\overline{U}_j(t) = c_{0j} e^{\Omega_j t} - \frac{Q_j}{\Omega_j} \tag{9.1.38}$$

The coefficient c_{0j} is related to the coefficients of the expansion of the initial solution $u^0(\vec{x})$ at $t = 0$, and its matrix representation

$$U^0 = \begin{vmatrix} \cdot \\ u^0_{i-1} \\ u^0_i \\ u^0_{i+1} \\ \cdot \end{vmatrix} \tag{9.1.39}$$

in series of the basis vectors $V^{(j)}$

$$U^0 = \sum_{j=1}^{N} U^0_j \cdot V^{(j)} \tag{9.1.40}$$

Considering the solution (9.1.38) at $t = 0$, we have

$$c_{0j} = U^0_j + Q_j / \Omega_j \tag{9.1.41}$$

and the solution (9.1.38) can be written as

$$\overline{U}_j(t) = U^0_j e^{\Omega_j t} + \frac{Q_j}{\Omega_j}(e^{\Omega_j t} - 1) \tag{9.1.42}$$

Notice here the essential property of the eigenvalues of the space discretization matrix S, which determine completely the time behavior of the semi-discretized system (9.1.3) of ODEs in time.

The exact solution of the system of ordinary differential equations (9.1.3) is obtained now by inserting the modal components (9.1.42) in the summation (9.1.31).

Stability condition

The ODE system will be **well-posed, or stable**, if its exact solution (9.1.31) remains bounded. This implies that all the modal components are also bounded, since if any of them would grow exponentially with time, the full solution would also have this unwanted behavior.

Hence, we have to require that the exponential terms in the modal components (9.1.42) do not grow exponentially with time, that is we have to require that the real part of the eigenvalues be negative or zero.

The stability condition is therefore

$$\text{Re}(\Omega_j) \leq 0 \quad \text{for all } j$$ (9.1.43)

In addition, if an eigenvalue is zero, it has to be a simple eigenvalue.

In the complex Ω-plane, the stability conditions states that the eigenvalue spectrum has to be restricted to the left half plane, including the imaginary axis.

If the space discretization, including the boundary conditions, leads to eigenvalues with non-positive real parts, the system of ODEs will be stable at the level of the semi-discretized formulation and a stable time integration method will always be possible.

Each mode j contributes a time behavior of the form $\exp(\Omega_j t)$ to the time dependent part of the solution, called the *transient solution*. The remaining part of the general solution is the modal decomposition of the *steady state solution*, defined as the solution remaining for large values of time when all the eigenvalues have real negative components. In this case the transient will damp out in time, and asymptotically we would have from (9.1.42) and (9.1.43)

$$\lim_{t \to \infty} \overline{U}(t) = -\sum_{j=1}^{N} \frac{Q_j}{\Omega_j} \cdot V^{(j)} = \overline{U}_S$$ (9.1.44)

which is a solution of the stationary problem,

$$S \cdot \overline{U}_S + Q = 0$$ (9.1.45)

For stationary problems, solved by a time-dependent formulation, we are interested in obtaining the steady state solution in the shortest possible time, that is, with the lowest possible number of time steps. This will be the case, if the eigenvalues Ω_j have large negative real parts, since $\exp(-|\text{Re } \Omega_j|t)$ will rapidly decrease in time. On the other hand, if $(\text{Re } \Omega_j)$ is negative but close to zero the corresponding mode will decrease very slowly and a large number of time steps will be necessary to reach the stationary conditions.

Unfortunately, in practical problems, the spectrum of Ω_j is very wide, including very large and very small magnitudes simultaneously. In this case when the ratio

$|\Omega_{max}|/|\Omega_{min}|$, called the **condition number** of the matrix, is very much larger than one, the convergence to the steady state is dominated by the eigenvalues close to the minimum Ω_{min}, and this would lead to very slow convergence. One method to accelerate convergence, consists in a **preconditioning technique**, whereby equation (9.1.45) is multiplied by a matrix M, such that the preconditioned matrix MS has a spectrum of eigenvalues with a more favorable condition number and larger negative values of Ω_{min}. This is an important technique for accelerating the convergence of numerical algorithms to steady state solutions and several examples will be discussed in the next chapter.

9.1.4 Matrix Method and Fourier Modes

If you go back to the explicit forms of the matrices derived in Sections 9.1.1 and 9.1.2, you can observe that they strongly depend on the boundary conditions and their specific implementation. Consequently, the same will be true for their eigenvalue and eigenvector structure and therefore the stability conditions can be significantly affected by the nature and implementation of these boundary conditions.

Hence, this provides a general method for the analysis of the influence of boundary conditions on stability of numerical schemes, also known as the **matrix method for stability analysis**. In addition, through the numerical estimation of the eigenvalues, we could also evaluate effects such as non-uniform meshes or non-constant coefficients. It is basically a more general approach than the Von Neumann method based on the assumption of periodic boundary conditions leading to a Fourier modal decomposition.

The price of this generality is its complication, as it is quite difficult, and mostly impossible, to derive analytical expressions for the eigenvalues for general boundary conditions. However, it is a current practice to derive the eigenvalues numerically, by applying for instance, existing numerical or symbolic algebra codes, such as MAPLE, Mathematica or MATLAB.

Therefore we will assume from now on that we apply periodic boundary conditions, with the immediate consequence that the eigenvectors are the Fourier modes, as defined in Chapter 7.

Hence, we can obtain the expressions of the eigenvalues in a straightforward way, following the methodology developed in Chapter 7 for the Von Neumann analysis.

The eigenvectors are defined by the Fourier modes, following Section 7.2.1, equation (7.2.6), for a one-dimensional space variable x

$$V^{(j)}(x) = e^{Ik_j \cdot x} \tag{9.1.46}$$

At point x_i, this eigenvector becomes

$$v_i^{(j)} = e^{Ik_j \cdot i\Delta x} = e^{Iij\pi/N} = e^{Ii\phi_j} \tag{9.1.47}$$

with

$$\phi_j \equiv k_j \Delta x = \frac{j\pi}{N} \quad j = -N, \dots, N \tag{9.1.48}$$

When applied to the space discretization operator S, the resulting eigenvalues result from

$$S \cdot e^{Ik_j \cdot i \Delta x} = \Omega(\phi_j) e^{Ik_j \cdot i \Delta x} \tag{9.1.49}$$

The eigenvalue $\Omega(\phi_j)$ is also called the **Fourier symbol** of the space discretization matrix. Its derivation is identical to the derivation of the Von Neumann amplification factor defined in Chapter 7.

Let us apply this to some of the well-known schemes for the diffusion and convection operators.

Example E.9.1.1: Diffusion operator (9.1.4)

The central discretized diffusion operator is defined by

$$\frac{du_i}{dt} = \frac{\alpha}{\Delta x^2}(u_{i+1} - 2u_i + u_{i-1}) \overset{\Delta}{=} S \cdot u_i$$

With the eigenfunctions (9.1.46), we obtain for this system of ODEs

$$
\begin{aligned}
S \cdot e^{Ik_j \cdot i \Delta x} &= \frac{\alpha}{\Delta x^2}(e^{Ik_j \cdot (i+1)\Delta x} - 2e^{Ik_j \cdot i \Delta x} + e^{Ik_j \cdot (i-1)\Delta x}) \\
&= \frac{\alpha}{\Delta x^2}(e^{Ik_j \cdot \Delta x} - 2 + e^{-Ik_j \cdot \Delta x})e^{Ik_j \cdot i \Delta x} \\
&= \frac{2\alpha}{\Delta x^2}(\cos \phi_j - 1)e^{Ik_j \cdot i \Delta x} \equiv \Omega(\phi_j)e^{Ik_j \cdot i \Delta x}
\end{aligned}
\tag{E.9.1.1}
$$

The eigenvalues are real and negative, covering the range $(-4\alpha/\Delta x^2, 0)$,

$$\Omega(\phi_j) = \frac{2\alpha}{\Delta x^2}(\cos \phi_j - 1) \tag{E.9.1.2}$$

Example E.9.1.2: First order upwind space discretization for the convection equation (9.1.17)

With

$$\left(a \frac{\partial u}{\partial x}\right)_i = -\frac{a}{\Delta x}(u_i - u_{i-1}) = S \cdot u_i$$

we obtain

$$
\begin{aligned}
S \cdot e^{Ik_j \cdot i \Delta x} &= -\frac{a}{\Delta x}(e^{Ik_j \cdot i \Delta x} - e^{Ik_j \cdot (i-1)\Delta x}) = -\frac{a}{\Delta x}(1 - e^{-Ik_j \cdot \Delta x})e^{Ik_j \cdot i \Delta x} \\
&= -\frac{a}{\Delta x}(1 - \cos \phi_j + I \sin \phi_j)e^{Ik_j \cdot i \Delta x} \equiv \Omega(\phi_j)e^{Ik_j \cdot i \Delta x}
\end{aligned}
\tag{E.9.1.3}
$$

$$\Omega(\phi_j) = -\frac{a}{\Delta x}(1 - \cos \phi_j + I \sin \phi_j) \tag{E.9.1.4}$$

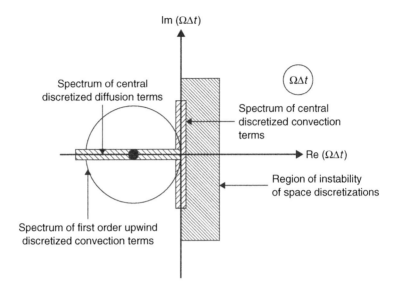

Figure 9.1.2   *Representation of stability regions in the complex* Ω *(space discretization) plane.*

The eigenvalues are complex, with a negative real part, covering the range $[(-2a/\Delta x, 0) + I(-a/\Delta x, +a/\Delta x)]$. In the complex Ω-plane, the eigenvalues are on a circle of radius $a/\Delta x$ centered at $(-a/\Delta x, 0)$, when ϕ_j covers the range $(-\pi, +\pi)$ (see Figure 9.1.2).

Example E.9.1.3: Central space discretization for the convection equation

With

$$\left(a\frac{\partial u}{\partial x}\right)_i = -\frac{a}{2\Delta x}(u_{i+1} - u_{i-1}) \stackrel{\Delta}{=} S \cdot u_i$$

we obtain

$$S \cdot e^{Ik_j \cdot i\Delta x} = -\frac{a}{2\Delta x}(e^{Ik_j \cdot (i+1)\Delta x} - e^{Ik_j \cdot (i-1)\Delta x})$$

$$= -I\frac{a}{\Delta x}\sin\phi_j e^{Ik_j \cdot i\Delta x} \equiv \Omega(\phi_j)e^{Ik_j \cdot i\Delta x} \qquad (E.9.1.5)$$

$$\Omega(\phi_j) = -I\frac{a}{\Delta x}\sin\phi_j \qquad (E.9.1.6)$$

The eigenvalues are purely imaginary, covering the range $I[(-a/\Delta x, +a/\Delta x)]$.

Clearly, these three cases satisfy the stability condition and are represented in Figure 9.1.2.

In Figure 9.1.2, the vertical imaginary axis of the Ω-plane contains the spectrum of the central discretized convection operator, while the negative real axis contains the spectrum of the central discretized diffusion terms. Note also that the spectrum of the first order upwind scheme for the convection equation covers the same part of the imaginary axis as central differencing, but with the inclusion of a negative real part.

It is important here to observe that a negative real part will generate a contribution $e^{Re(\Omega)t}$ in the solution (9.1.38). This creates a damping of the solution of the space discretized equations, there where the analytical solution of the convection equation $u_t + au_x = 0$, is a pure wave with no damping.

Hence, the appearance of negative real parts in the eigenvalues of the space discretization of convective, hyperbolic equations indicates that this discretization has generated a numerical dissipation.

This is to be put in relation with the numerical viscosity appearing in the equivalent differential equation, as well as to the presence of even order derivatives in the development of the truncation error, as seen in Chapter 7.

9.1.5 Amplification Factor of the Semi-discretized System

As we have seen from equations (9.1.34) and (9.1.35), the stability of the semi-discretized system (9.1.3) is completely determined by the time behavior of the transient and consequently, it is sufficient for this purpose to investigate the time behavior of the homogeneous part of equation (9.1.3).

Assuming therefore $Q = 0$, the exact transient solution of the homogeneous semi-discretized equation at time level $t = n\Delta t$ can be written as, following equation (9.1.42):

$$\overline{U}_T(n\Delta t) = \sum_{j=1}^{N} \overline{U}_{Tj}(n\Delta t) \cdot V^{(j)} = \sum_{j=1}^{N} U_j^0 e^{\Omega_j n\Delta t} \cdot V^{(j)} \qquad (9.1.50)$$

Hence, for an arbitrary mode Ω, we can define an *amplification factor* $\overline{G}(\Omega)$ *of the exact solution of the semi-discretized system of ODEs in time* by

$$\overline{U}_{Tj}(n\Delta t) \overset{\Delta}{=} \overline{G}(\Omega_j) \cdot \overline{U}_{Tj}((n-1)\Delta t) = U_j^0 e^{\Omega_j n\Delta t}$$

$$= e^{\Omega_j \Delta t} [U_j^0 e^{\Omega_j(n-1)\Delta t}] \qquad (9.1.51)$$

leading to

$$\overline{G}(\Omega) = e^{\Omega \Delta t} \qquad (9.1.52)$$

The stability condition (9.1.43) ensures that the transient will not be amplified and that the modulus of the amplification function is lower than or equal to one:

$$|\overline{G}| \le 1 \quad \text{for all } \phi \in [-\pi, +\pi] \qquad (9.1.53)$$

A major consequence of this analysis is that we can investigate the full system of semi-discretized equations by isolating a single mode. Indeed, since the exact solution

(9.1.31) with (9.1.42) is expressed as a contribution from all the modes of the initial solution, which have propagated or (and) diffused with the eigenvalue Ω_j, and a contribution from the source terms Q, all the properties of the time integration schemes, and most essentially their stability properties, can be analyzed separately for each mode with the scalar modal equation (9.1.33), dropping the subscript j, written here as

$$\frac{dw}{dt} = \Omega w \tag{9.1.54}$$

We introduce the notation w to designate an arbitrary component of the modal equations, as its *canonical form*. The space operator S is replaced by an eigenvalue Ω and the 'modal' equation (9.1.54) will serve as the basic equation for the analysis of the stability of a time integration method.

9.1.6   Spectrum of Second Order Upwind Discretizations of the Convection Operator

We have seen in Section 8.4.1, several options for second order upwind biased discretizations of the convection term, summarized in Table 8.4.2.

Following the methodology of Section 9.1.4, we can easily find the Fourier symbols, or eigenvalues, of the second order upwind schemes listed in this table. They are summarized in Table 9.1.1, and we leave it to you to work out the derivation.

We compare the eigenvalue spectra in the complex $\Omega\Delta t$-plane, for several values of the CFL number in Figure 9.1.3 for the five listed schemes at CFL $= 0.5$ and CFL $= 1$. Observe that the range on the negative real axis indicates the level of numerical dissipation generated by the upwind biased discretization. The largest dissipation is generated by the second order upwind scheme ($\kappa = -1$), while the lowest dissipation is provided, as can be expected, by the third order scheme ($\kappa = -1/3$). The same properties can be observed at CFL $= 1$. You also can see from these figures the higher range of the eigenvalues in the vertical direction, for some of the schemes, compared to the first and second order upwind schemes. This will lead to a reduction of the maximum CFL value when associated with conditionally stable time integration methods.

9.2   ANALYSIS OF TIME INTEGRATION SCHEMES

In the previous section, we have developed guidelines to the analysis of the properties of a selected space discretization. We have shown that its eigenvalue spectrum completely defines its behavior, and we have derived the conditions required on the space discretization, to obtain a stable, or **well-posed**, semi-discretized system of ODEs in time. As a consequence, we can select an arbitrary mode, represented by the canonical modal equation (9.1.54), to move to next step, namely the analysis of the stability conditions of time integration numerical methods, associated to an eigenvalue Ω.

Table 9.1.1 Fourier eigenvalues of several higher order upwind biased discretizations of the convection term.

Scheme	Space discretization scheme for $a(\partial u/\partial x)$	Fourier symbol (eigenvalue) $\Omega\Delta t$ $\sigma = a(\Delta t/\Delta x)$
First order upwind (FOU)	$\dfrac{du_i}{dt} = -\dfrac{a}{\Delta x}(u_i - u_{i-1})$	$\Omega\Delta t = -\sigma[1 - \cos\phi + I\sin\phi]$
Second order upwind (SOU) $\kappa = -1$	$\dfrac{du_i}{dt} = -\dfrac{a}{2\Delta x}(3u_i - 4u_{i-1} + u_{i-2})$	$\Omega\Delta t = -\sigma[1 - \cos\phi]^2 + I(2 - \cos\phi)\sin\phi]$
Fromm scheme $\kappa = 0$	$\dfrac{du_i}{dt} = -\dfrac{a}{4\Delta x}(u_{i+1} + 3u_i - 5u_{i-1} + u_{i-2})$	$\Omega\Delta t = -\dfrac{\sigma}{2}[(1 - \cos\phi)^2 + I(3 - \cos\phi)\sin\phi]$
QUICK scheme $\kappa = 1/2$	$\dfrac{du_i}{dt} = -\dfrac{a}{8\Delta x}(3u_{i+1} + 3u_i - 7u_{i-1} + u_{i-2})$	$\Omega\Delta t = -\dfrac{\sigma}{4}[2(1 - \cos\phi)^2 + I(5 - \cos\phi)\sin\phi]$
Third order scheme $\kappa = 1/3$	$\dfrac{du_i}{dt} = -\dfrac{a}{6\Delta x}(2u_{i+1} + 3u_i - 6u_{i-1} + u_{i-2})$	$\Omega\Delta t = -\dfrac{\sigma}{3}[(1 - \cos\phi)^2 + I(4 - \cos\phi)\sin\phi]$

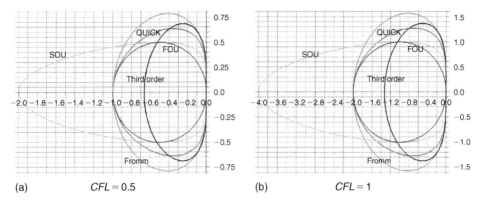

Figure 9.1.3   *Fourier symbol (eigenvalues) for the five upwind biased discretizations of the convection term, listed in Table 9.1.1, at CFL = 0.5 and CFL = 1.*

9.2.1   Stability Regions in the Complex Ω Plane and Fourier Modes

In order to define a general integration method, we introduce the time shift operator $\overline{E}$ ($\overline{E}^k$ is $\overline{E}$ to the power k)

$$\overline{E}w^n = w^{n+1}$$

$$\overline{E}^k w^n = w^{n+k} \tag{9.2.1}$$

We write a general time integration method of the canonical equation (9.1.54) under the form

$$w^{n+1} = P(\overline{E}, \Omega\Delta t) \cdot w^n \tag{9.2.2}$$

where the operator P depends on the single parameter $\Omega\Delta t$. We will see next that the operator P can always be written as a rational fraction of $\overline{E}$.

The operator P is in fact the **numerical amplification factor** and will be compared with the exact amplification factor (9.1.52) to assess the accuracy.

Indeed, repeating the operator P successively, we obtain

$$w^n = [P(\overline{E}, \Omega\Delta t)] \cdot w^{n-1} = \cdots = [P(\overline{E}, \Omega\Delta t)]^n \cdot w^0 \tag{9.2.3}$$

where w^0 is the initial solution at $t = 0$.

The stability condition requires that the solution w^n remains bounded, which will be satisfied if the operator P^n (P to the power n) remains uniformly bounded for all n and Δt, in particular for $n \to \infty$, $\Delta t \to 0$ with $n\Delta t$ fixed. That is, we should have, in a selected norm, for finite T

$$||P^n|| < K \quad \text{for } 0 < n\Delta t < T \tag{9.2.4}$$

with K independent of n and Δt.

The norm of P^n is often very difficult to analyze and instead a **necessary, but not always sufficient** condition can be obtained from a local mode analysis on the eigenvalues of P.

If we designate the eigenvalues of P by z_P, solution of

$$z_P = P(z_P, \Omega \Delta t) \tag{9.2.5}$$

the necessary stability condition is that all the eigenvalues z_P should be of modulus lower than, or equal to, one.

$$\boxed{|z_P| \leq 1 \quad \text{for all eigenvalues } \Omega} \tag{9.2.6}$$

When some values lie on the unit circle, $|z_P| = 1$, they have to be simple, otherwise the solution would increase with time as t^m where $(m+1)$ is the multiplicity of the eigenvalue.

This condition establishes a relation between the selected time discretization and the space discretization as characterized by the eigenvalues Ω of the space discretization scheme, since

$$z_P = z_P(\Omega \Delta t) \tag{9.2.7}$$

The condition (9.1.43) on the space discretization has to be always satisfied for the system of ordinary differential equations (9.1.3) to be stable (or well-posed).

For all the space discretizations, which satisfy this condition, the associated numerical discretization in time will be stable if condition (9.2.6) is satisfied.

The space discretization generates a representative set of eigenvalues $\Omega \Delta t$ which cover a certain region of the complex $\Omega \Delta t$ plane – to be situated on the left side of the plane including the imaginary axis for stability. If we consider the trace of every root $z_P(\Omega \Delta t)$, as $\Omega \Delta t$ covers the whole spectrum, in a complex z-plane, this trace will be represented by some line which has to remain inside a circle of radius one. If some roots come outside the stability circle when $\Omega \Delta t$ covers the range of its spectrum, the scheme is unstable (Figure 9.2.1).

A method with two or more time levels steps, will generate more than one eigenvalue; in particular a two-step method, which involves three time levels, will generate two solutions, a three-step method, involving four time levels, will generate three solutions and so on, and this for the same value of $\Omega \Delta t$.

When more than one value is present, the consistency of the scheme requires that one of the eigenvalues should represent an approximation to the physical behavior. This solution of the eigenvalue equation, called the **principal solution**, is to be recognized by the fact that it tends to one when $\Omega \Delta t$ goes to zero. Denoting by z_{P1} the principal ('physical') solution we have

$$\lim_{\Omega \Delta t \to 0} z_{P1}(\Omega) = 1 \tag{9.2.8}$$

The other solution, called the **spurious solution**, represents a 'non-physical' time behavior of the numerical solution introduced by the scheme, and could destroy its stability. For instance, we will see in the following section that the leapfrog scheme has two roots since it is a three-level scheme, and its first root is accurate up to second

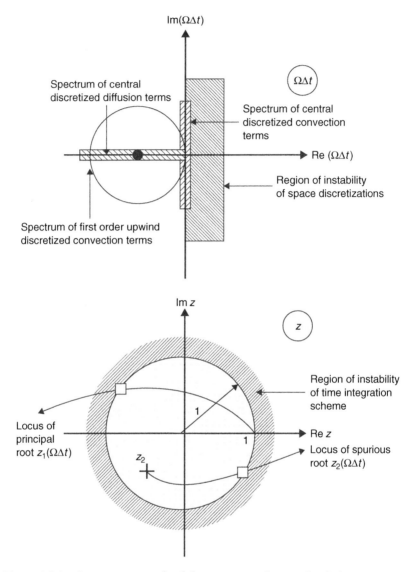

Figure 9.2.1   *Representation of stability regions in the complex Ω (space discretization) and z (time discretization) planes.*

order, but its spurious root starts at -1, and is the one that generates the instability when $(\Omega \Delta t) < 0$, that is when dissipation is present in the scheme, either physically for diffusion equations or numerically for certain discretizations of the convection equation.

An alternative way to (9.2.2) is to write the time integration method under the form

$$P_1(\overline{E})w^n = 0 \tag{9.2.9}$$

which by comparison establishes the relation,

$$P = \overline{E} + P_1 \tag{9.2.10}$$

Hence, the amplification factor z_P is solution of the relation $z_P = z_P + P_1(z_P)$ and therefore the eigenvalues z_P are solutions of the **characteristic polynomial**

$$P_1(z_P) = 0 \tag{9.2.11}$$

9.2.2 Error Analysis of Space and Time Discretized Systems

In order to define the errors of the combined space and time discretizations, we should compare the numerical amplification factor z_p with the exact amplification factor. Let us consider equation (9.2.3), which becomes

$$w^n = z_P^n(\Omega \Delta t) \cdot w^0 \tag{9.2.12}$$

where z_P^n is z_P to the power n. This amplification factor of the scheme, associated to the mode Ω, is an approximation to the exact amplification factor $\overline{G}$, defined by equation (9.1.52) and represents the errors introduced by the selected time integration. Hence, $z_P(\Omega \Delta t)$ is the numerical approximation to $e^{\Omega \Delta t}$, which we can write as

$$z_P(\Omega \Delta t) \approx e^{\Omega \Delta t} = 1 + \Omega \Delta t + \frac{(\Omega \Delta t)^2}{2!} + \frac{(\Omega \Delta t)^3}{3!} + \frac{(\Omega \Delta t)^4}{4!} + \cdots \tag{9.2.13}$$

Therefore, if we write the Taylor expansion of $z_P(\Omega \Delta t)$, the first term in the expansion of $z_P(\Omega \Delta t)$ which differs from the Taylor development of the exponential defines the order of the method.

For instance, for the explicit Euler method, $z_P(\Omega \Delta t) = 1 + \Omega \Delta t$, which is clearly of first order accuracy. The backward Euler method is defined by

$$z_P = \frac{1}{1 - \Omega \Delta t} \approx 1 + \Omega \Delta t + (\Omega \Delta t)^2 + \cdots \tag{9.2.14}$$

and since the quadratic term is not equal to the quadratic term in equation (9.2.13), the method is only first order accurate.

9.2.2.1 *Diffusion and dispersion errors of the time integration*

Following the approach of Section 7.4, we define here the **diffusion and dispersion errors of the time integration method**, after writing the numerical factor $z_P = P$, as

$$z_P = |z_P| e^{I \Phi_P} \tag{9.2.15}$$

The global error is defined, after we decompose the eigenvalue $\Omega = \text{Re}(\Omega) + I\,\text{Im}(\Omega)$ in its real and imaginary parts, by

$$\varepsilon^{\mathrm{T}} = \frac{z_P}{\overline{G}} = \frac{|z_P|}{e^{\text{Re}(\Omega \Delta t)}} \frac{e^{I \Phi_P}}{e^{I\,\text{Im}(\Omega \Delta t)}} \tag{9.2.16}$$

The error in amplitude, the ***diffusion or dissipation error***, is defined by the ratio of the modulus of $|z_P|$ and the amplitude of the exact solution of the semi-discretized system (9.1.52)

$$\varepsilon_D^T = \frac{|z_P|}{|\overline{G}|} = \frac{|z_P|}{e^{\mathrm{Re}(\Omega \Delta t)}} \tag{9.2.17}$$

and the phase error, or ***dispersion error***, will be defined by the ratio of the phase Φ_P of z_P, to the phase of this exact solution, that is $\mathrm{Im}(\Omega \Delta t)$

$$\varepsilon_\phi^T = \frac{\Phi_P}{\mathrm{Im}(\Omega \Delta t)} \tag{9.2.18}$$

The superscript T indicates that the respective errors are related to the selected time discretization.

Since all quantities in these equations are function of $(\Omega \Delta t)$, the error (9.2.16) of the time integration method can also be represented in the complex $(\Omega \Delta t)$-plane.

9.2.2.2 *Diffusion and dispersion errors of space and time discretization*

The errors just derived take as reference the exact solution of the semi-discretized system, for a given space discretization, uniquely characterized by its eigenvalue spectrum. However, the space discretization itself generates dissipation and dispersion errors, which can be evaluated by comparing with the exact solution of the analytical model. This depends of course on the considered model.

For the reference linear convection model $u_t + au_x = 0$, we have seen in Section 7.4 that the analytical amplification factor is given by

$$\tilde{G} = e^{-I\tilde{\omega}\Delta t} = e^{-Iak\Delta t} \tag{9.2.19}$$

where k is the wave number of the space component of the wave solution e^{Ikx}.

The space discretization error is defined, with the superscript S referring to the space discretization errors, by

$$\varepsilon^S = \frac{\overline{G}}{e^{-Iak\Delta t}} = \frac{e^{\mathrm{Re}(\Omega\Delta t)}}{1} \frac{e^{I\,\mathrm{Im}(\Omega\Delta t)}}{e^{-Iak\Delta t}}$$

$$\stackrel{\Delta}{=} \varepsilon_D^S \frac{e^{I\,\mathrm{Im}(\Omega\Delta t)}}{e^{-Iak\Delta t}}$$

$$= \varepsilon_D^S e^{-Iak\Delta t(\varepsilon_\phi^S - 1)} \tag{9.2.20}$$

the spatial diffusion and dispersion errors being defined as

$$\varepsilon_D^S \stackrel{\Delta}{=} \frac{e^{\mathrm{Re}(\Omega\Delta t)}}{1} = e^{\mathrm{Re}(\Omega\Delta t)} \qquad \varepsilon_\phi^S \stackrel{\Delta}{=} -\frac{\mathrm{Im}(\Omega\Delta t)}{ak\Delta t} \tag{9.2.21}$$

The **total errors of the combined space and time discretizations** are obtained by comparing the amplification factor z_P with the analytical factor (9.2.19), leading to

$$\varepsilon \stackrel{\Delta}{=} \frac{z_P}{e^{-Iak\Delta t}} = \frac{z_P}{\overline{G}} \frac{\overline{G}}{e^{-Iak\Delta t}} = \varepsilon^T \cdot \varepsilon^S \tag{9.2.22}$$

The global dissipation and dispersion errors are therefore defined by

$$\varepsilon_D = \varepsilon_D^T \cdot \varepsilon_D^S = |z_P| \quad \text{and} \quad \varepsilon_\phi = \varepsilon_\phi^T \cdot \varepsilon_\phi^S = -\frac{\Phi_P}{ak\Delta t} = -\frac{\Phi_P}{\sigma\phi} \tag{9.2.23}$$

This generalizes the definitions derived from the Von Neumann analysis and allows to separately evaluate the error sources from the space and from the time discretizations.

We can see here the very important relation between the eigenvalues Ω and the errors. *The real part of Ω will contribute to the dissipation error, while its imaginary part determines the dispersion error.* This imaginary part, divided by the wave number k, is a measure of the **numerical wave speed** and *the dispersion error is defined as the ratio between the numerical and physical wave speeds.*

9.2.2.3 *Relation with the equivalent differential equation*

The development of Section 8.1.3, defining the general form of the equivalent differential equation of a numerical scheme for the linear convection equation, equation (8.1.27), and its relation with the dispersion and dissipation errors, remains totally valid, when extended to an arbitrary combination of time and space discretizations.

Therefore, we have hereby a simple way of obtaining the equivalent differential equation of the scheme resulting from the time integration of the system of space discretized ODEs. *To achieve this objective, it suffices to develop the expressions of the dispersion and diffusion errors in power series of the phase angle around $\phi = 0$, and identify the terms, following equations (8.1.33) and (8.1.34).*

Note that the diffusion error expansion will provide the coefficients of the even order derivatives of the equivalent differential equation, while the expansion of the dispersion error will provide the odd order derivative coefficients.

Each time discretization scheme is represented by a unique relation between z_P and Ω, determined by the dependence $z_P = z_P(\Omega)$. When the stability limit for z_P, equation (9.2.6), is introduced in this equation, under the form $z_P = e^{\mathrm{I}\theta}$, $0 < \theta < 2\pi$, representing the stability circle of radius 1, a corresponding stability region is defined in the $(\Omega\Delta t)$-plane, through the mapping function $z_P = z_P(\Omega)$.

For a two-level scheme, where P is independent of $\overline{E}$ the stability curve is given by the condition $|P(\Omega)| = 1$.

Let us illustrate these relations with the examples of the previous section based on the Fourier modes, associated to periodic boundary conditions, and some of the time discretization methods applied in the previous chapters, in particular the forward, backward and central differences in time.

9.2.3 Forward Euler Method

The forward Euler method corresponds to a forward difference in time in an explicit mode. Applied to the canonical modal equation (9.1.54), it is defined by

$$w^{n+1} - w^n = \Omega\Delta t\, w^n \tag{9.2.24}$$

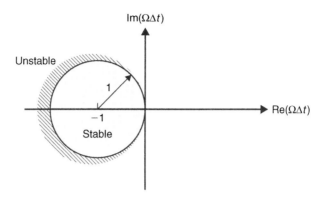

Figure 9.2.2 *Stability region for the Euler explicit method in the complex ($\Omega \Delta t$) plane.*

leading to

$$z_P = P = 1 + \Omega \Delta t \qquad (9.2.25)$$

The scheme is therefore stable for all space discretizations associated to an eigenvalue spectrum such that

$$|(1 + \Omega \Delta t)| \leq 1 \qquad (9.2.26)$$

or

$$[1 + \text{Re}(\Omega \Delta t)]^2 + [\text{Im}(\Omega \Delta t)]^2 \leq 1 \qquad (9.2.27)$$

In the ($\Omega \Delta t$) plane, this stability region is a circle of radius one centered around $\Omega \Delta t = -1$ (Figure 9.2.2).

Diffusion operator (9.1.4)
The eigenvalues are given by equation (E.9.1.2) and since all Ω-values are real and negative, the stability condition is

$$-2 \leq -|\text{Re}(\Omega \Delta t)| \leq 0 \qquad (9.2.28)$$

or

$$0 \leq \frac{\alpha \Delta t}{\Delta x^2} \leq \frac{1}{2} \qquad (9.2.29)$$

as derived by the Von Neumann method in Chapter 7.

Central space discretization for the convection equation
With equation (E.9.1.6), $\Omega \Delta t$ becomes

$$\Omega \Delta t = -I \frac{a \Delta t}{\Delta x} \sin \phi = -I \sigma \sin \phi \qquad (9.2.30)$$

with the CFL number $\sigma = a \Delta t / \Delta x$

All eigenvalues Ω are purely imaginary, and are outside the stability circle of the forward Euler method. The scheme is therefore unstable.

Upwind space discretization for the convection equation (9.1.17)
The eigenvalues are given by equation (E.9.1.4),

$$\Omega\Delta t = -\sigma(1 - \cos\phi + I\sin\phi) \tag{9.2.31}$$

In the complex $\Omega\Delta t$-plane all the eigenvalues are located along a circle centered at $-\sigma$, with radius σ. This circle will be inside the stability circle of radius 1, if σ lies between 0 and 1. We recover the CFL condition of the upwind explicit scheme as seen in Chapter 7.

$$0 \leq \sigma \leq 1 \tag{9.2.32}$$

9.2.4 Central Time Differencing or Leapfrog Method

The leapfrog method corresponds to a central difference in time, with an explicit formulation, leading to a three level, two-step method:

$$w^{n+1} - w^{n-1} = 2\Omega\Delta t\, w^n \tag{9.2.33}$$

Hence

$$P(\overline{E}, \Omega\Delta t) = \overline{E}^{-1} + 2\Omega\Delta t \tag{9.2.34}$$

The eigenvalues z_P are obtained from

$$z_P = 2\Omega\Delta t + 1/z_P \tag{9.2.35}$$

This quadratic equation results from the three levels in time

$$z_P^2 - 2\Omega\Delta t\, z_P - 1 = 0$$

leading to the two solutions

$$z_P = \Omega\Delta t \pm \sqrt{(\Omega\Delta t)^2 + 1} \tag{9.2.36}$$

which behave as follows, when $\Delta t \to 0$

$$z_1 = \Omega\Delta t + \sqrt{(\Omega\Delta t)^2 + 1} = 1 + \Omega\Delta t + \frac{(\Omega\Delta t)^2}{2} - \frac{(\Omega\Delta t)^4}{8} + \cdots \tag{9.2.37}$$

$$z_2 = \Omega\Delta t - \sqrt{(\Omega\Delta t)^2 + 1} = -1 + \Omega\Delta t - \frac{(\Omega\Delta t)^2}{2} + \frac{(\Omega\Delta t)^4}{8} + \cdots \tag{9.2.38}$$

The first solution is the physical one, while the second one starts at -1 and is the *spurious value*. Observe, by comparing (9.2.37) with the expansion (9.2.13), that the

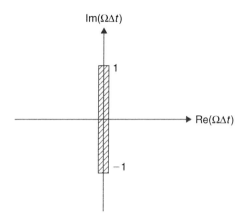

Figure 9.2.3   *Stability region for the leapfrog method.*

first three terms are exact and therefore the scheme is second order accurate in time, as we would expect indeed from the central difference.

From

$$2(\Omega \Delta t) = z_p - \frac{1}{z_p} \tag{9.2.39}$$

with the stability limit $z_p = e^{I\theta}$, we obtain

$$\Omega \Delta t = I \sin \theta \tag{9.2.40}$$

The stability region of the leapfrog method is therefore a strip of amplitude ± 1 along the imaginary axis, Figure 9.2.3.

It is seen immediately that the diffusion operator with its real negative eigenvalues or the upwind convection operators, which also generate a negative real component, lead to unstable schemes when solved by the leapfrog method. Since negative values of $(\Omega \Delta t)$ correspond to the presence of a dissipative mechanism (or numerical viscosity), for hyperbolic equations, it results that the leapfrog scheme is totally unadapted to the presence of dissipative terms.

On the other hand, the leapfrog scheme applied to the central differenced convective equation, which does not generate dissipation is stable under the CFL condition.

9.2.5   Backward Euler Method

The backward Euler method corresponds to a backward difference in time in an implicit mode. Applied to the canonical modal equation (9.1.54), it is defined by

$$w^{n+1} - w^n = \Omega \Delta t \, w^{n+1} \quad \text{or} \quad (1 + \Omega \Delta t) w^{n+1} = w^n \tag{9.2.41}$$

leading to

$$P(\overline{E}, \Omega \Delta t) = 1 + \Omega \Delta t \cdot \overline{E} \tag{9.2.42}$$

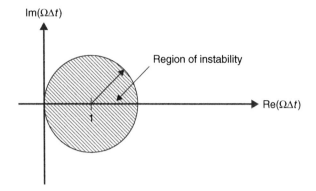

Figure 9.2.4 *Stability region for implicit Euler scheme.*

The eigenvalue z_P is obtained from

$$z_P = 1 + \Omega\Delta t \cdot z_P \quad \text{or} \quad z_P = \frac{1}{1 - \Omega\Delta t} \tag{9.2.43}$$

For $z_P = e^{I\theta}$, the limit of the stability region in the $\Omega\Delta t$-plane is defined by

$$(\Omega\Delta t) = 1 - e^{-I\theta} \tag{9.2.44}$$

and represents a circle centered on $\Omega\Delta t = 1$ of radius one, Figure 9.2.4.

Since for $|z_P| < 1$, $|1 - \Omega\Delta t| > 1$, the stability region is *outside* the circle and is seen to cover even regions in the Ω-plane where the space-discretized equations are unstable. ***Hence, all the schemes seen up to now will be stable with the implicit Euler time integration.***

The approach developed here allows also us to analyze *separately* the properties of the space discretization and of the time integration. For a given discretization in space, we can select appropriate time integrations with stability regions containing the spatial eigenvalue spectrum. Hence, a space discretization cannot be said to be unstable by itself, at least if the condition $\mathrm{Re}\,\Omega \leq 0$ is satisfied. It is only when it is coupled to a time integration that we can decide upon stability.

For instance, the explicit Euler scheme has a stability region that does not contain the imaginary axis of the $\Omega\Delta t$-plane, and will not be stable for a $\Omega\Delta t$-spectrum on the imaginary axis like the one generated by the central discretization of the convection equation. However, it will be stable, conditionally, for the diffusion equation if all the real eigenvalues are contained within the range $-2 < (\Omega\Delta t) < 0$, or for the upwind discretization of the convection term, under a CFL condition. Similarly, the leapfrog scheme is unstable in this latter case since its stability region is limited to a segment of the imaginary axis. Hence, as already noticed, the leapfrog scheme is totally unadapted for dissipative problems, while the Euler explicit scheme is not suitable for non-dissipative space discretizations.

HANDS-ON TASK 5

Apply the methodology described above to analyze the combinations of the three time integration methods of Sections 9.2.3–9.2.5, with various discretizations of the convection and diffusion model equations. For the linear convection equation, consider the central, as well as the first and second order upwind discretizations for the space term. For the diffusion equation, consider a central discretization of the second derivative term.

Develop the following steps:

- *Convection terms*: Find the trajectories of the eigenvalues $\Omega\Delta t$ in the complex $\Omega\Delta t$-plane, by applying a Fourier analysis. Consider the range of phase angles $\phi = k\Delta x$ in the interval $(-\pi, +\pi)$ for the three discretizations options (central, first and second order upwind). Plot the results for values of the CFL number, CFL $= 0.25, 0.5, 0.75, 1, 1.5$.
- Define the stability conditions for all the combinations by comparing with the stability domain of the time integration method.
- Determine the dissipation and dispersion errors for the same combinations and display them in function of the phase angle for the various values of the CFL number.
- Comment and analyze these results.
- *Diffusion terms*: Repeat the same development for the central discretization of the second derivative term.
- Plot the eigenvalue trajectories in the $\Omega\Delta t$-plane for several values of the scheme parameter $\beta = \alpha\Delta t/\Delta x^2$, in the range $(0, 1)$, in function of the phase angle $\phi = k\Delta x$ in the interval $(-\pi, +\pi)$.
- Define the stability conditions for the three considered time integration methods.
- Determine the dissipation and dispersion errors for the same combinations and display them in function of the phase angle for the various values of the β-parameter number.
- Comment and analyze these results.

Apply symbolic mathematical software tools, such as MAPLE, Mathematica or MATLAB.

9.3   A SELECTION OF TIME INTEGRATION METHODS

A large number of methods are available for the solution of the system of ordinary differential equations obtained after the space discretization.

If you have to choose a time integration method, the first question to ask yourself is: Do I favor an explicit or an implicit method? The debate between implicit and explicit methods is still open and no definitive recommendation can be given, as both options are valid. We can state the pros and cons as follows:

Explicit methods
- Are at best conditionally stable and have a limitation on the maximum allowable time step $(\Delta t)_{stab}$. This maximum time step is generally quite small, particularly

for convection dominated compressible flows, with $(\Delta t)_{\text{stab}} < (\text{CFL}) \, \Delta x/a$, where a is of the order of the speed of sound.

- For $\text{CFL} = 1$, $\Delta x \approx 10^{-2}$ m and $a \approx 330$ m/s, we would have $(\Delta t)_{\text{stab}} < 3.10^{-5}$ s.
- The CPU cost per iteration is low, since no matrices have to be inverted.
- But the low stability limit will require a large number of time steps.

Implicit methods

- Are generally unconditionally stable and large time steps can be applied, theoretically even tending to infinity.
- In practice, due to nonlinearities of the flow equations, time step restrictions will nevertheless appear. Moreover, for unsteady flow problems accuracy requirements will also tend to restrict the maximum time step.
- The resulting time step will nearly always be significantly higher than the explicit CFL-based time limit $(\Delta t)_{\text{stab}}$.
- The CPU cost per iteration of implicit methods is significantly higher compared to explicit methods, since matrices have to be inverted at each time step. These matrices also lead to requirements for additional memory, depending on the selected resolution method. Some of the available methods are described in Chapter 10.
- One of the guidelines toward the selection of the time step is related to the lowest physical time scale $(\Delta t)_{\text{phys}}$ that the simulation intends to capture:
  - If the physical time step $(\Delta t)_{\text{phys}}$ is of the same order as the stability limit $(\Delta t)_{\text{stab}}$, then explicit methods are clearly the most appropriate choice.
  - If the physical time step $(\Delta t)_{\text{phys}}$ is significantly higher than the stability limit $(\Delta t)_{\text{stab}}$, then implicit methods are a valid option. This will be the case for steady state problems.
- For nonlinear problems, linearization techniques have to be applied with implicit methods, which might restrict the linear stability conditions.

Finally, the choice between explicit and implicit methods will be guided by the product (cost per time step)*(number of time steps), as implicit methods require a lower number of time steps, but are more costly for each time step.

Therefore, we have to balance the higher allowable time step against the higher number of operations necessary to resolve the implicit algebraic system of equations.

We can classify the most relevant time integration methods for steady or unsteady flow simulations into three groups:

1. **Multistep methods**, which allow *implicit time integration* options.
2. **Predictor–corrector methods**, restricted in their simplest form to second order, are explicit, and easy to program:
   - They have a widespread range of applications, the most popular being the explicit McCormack scheme.
   - They can be considered, for nonlinear problems, as an approximation to the implicit linearization techniques.

3. ***Runge–Kutta multistage methods***, which are *explicit* and allow arbitrary high orders in time:
   – They are in particular well adapted to convection problems with central discretizations.
   – They apply without additional modifications to nonlinear and multidimensional problems.

In terms of stability terminology, the time integration schemes are said to be:

- ***Unconditionally unstable***: If the stability domain is found in the right-hand side of the complex $\Omega \Delta t$-plane.
- ***Conditionally stable***: If the stability domain covers part of the left-hand side of the complex $\Omega \Delta t$-plane.
- ***A-stable or unconditionally stable***: If the domain of stability covers the whole of the left-hand side of the complex $\Omega \Delta t$-plane.

In practical applications, the vector U contains a large number of variables, namely $N = m \cdot M$; M is the total number of mesh points where the m independent variables have to be determined. Generally, these variables have to be stored for each time level and increasing the number of time levels could rapidly put severe restrictions on the allowable space variables and mesh points. It is therefore very exceptional to consider applications with more than three time levels (two-step methods) to fluid mechanical problems, the overwhelming majority of schemes being limited to one-step methods with two time levels.

In the 'advanced' Section A9.4, we introduce you to some multidimensional problems, solved by implicit multistep methods. In these cases, the matrices are too large for direct inversions and ***factorization methods***, reducing the problem to a succession of one-dimensional implicit operators, can be defined. These methods are also known as ***Alternating Direction Implicit*** (ADI) methods. The factorization process is by no means trivial, since it can destabilize the numerical scheme, as will be shown for the three-dimensional convection equation with central differenced space derivatives.

We refer you to the literature for more details and general presentations. In particular, Gear (1971), Lambert (1973–1991), Dahlquist and Bjorck (1974), Beam and Warming (1982). The following books also present very general and complete presentations of methods for ODEs, Hairer et al. (1993), Hairer and Wanner (1996), Hundsdorfer and Verwer (2003).

9.3.1   Nonlinear System of ODEs and their Linearization

For a general system of ordinary differential equations (9.1.3), the right-hand side may, and often will be, nonlinear. If we write the system in the more general form:

$$\frac{dU}{dt} = H(U) \tag{9.3.1}$$

where H is a nonlinear function or matrix operator, acting on the vector U as defined previously, the vector $H(U, t)$ generalizes the space-discretized operators $(SU + Q)$

of equation (9.1.3), and is represents the residual, divided by the local cell volume (see equation (5.2.7)).

The link with the linear form, where the S-matrix is independent of U, can be obtained through an adequate linearization procedure.

Performing a series development around a reference value U_e, we can write

$$H(U) = H(U_e) + \left(\frac{\partial H}{\partial U}\right)_e \cdot (U - U_e) + O(U - U_e)^2 \tag{9.3.2}$$

where $(\partial H/\partial U)$ is the **Jacobian** of the operator H with respect to the dependent variable U. We introduce the notation

$$J(U) = \frac{\partial H}{\partial U} \tag{9.3.3}$$

and if we select the reference solution U_e as the steady state solution of $H(U_e) = 0$, we can rewrite the system (9.3.1) as

$$\frac{dU}{dt} = J(U_e)(U - U_e) \tag{9.3.4}$$

Comparing with equation (9.1.3), we see that the matrix S is identical to the Jacobian matrix J(U$_e$) and the eigenvalue spectrum is determined by the Jacobian eigenvalues. The Jacobian plays therefore a central role in the stability properties of the selected schemes. *Note here that J(U$_e$) is the Jacobian of the space-discretized flux balance of the conservation law, that is the Jacobian of the residual, and NOT the analytical Jacobian of the flux terms.*

Hence, we will apply the general methodology, as defined in the previous sections, for the stability analysis, by the following steps:

- Find the Fourier mode eigenvalues Ω (the Fourier symbol) of the Jacobian matrix of the space-discretized operator $H(U)$.
- Introduce the canonical modal equation:

$$\frac{dw}{dt} = \Omega \cdot w \tag{9.3.5}$$

Define the selected time integration method by its operator P, by rewriting the selected method under the form of equations (9.2.2) or (9.2.9), repeated here

$$w^{n+1} = P(\overline{E}, \Omega\Delta t) \cdot w^n \quad \text{or} \quad P_1(\overline{E}, \Omega\Delta t) \cdot w^n = 0$$

- Find the eigenvalues of the operator P, following (9.2.5), or solve the characteristic equation (9.2.11), and apply the stability condition $|z| = |P| \leq 1$ for the subsequent stability analysis of the combined time and space-discretized scheme.

We will now analyze several important methods.

9.3.2   General Multistep Method

Linear multistep methods have been analyzed systematically by Dahlquist (1963) and with special attention to fluid mechanical simulations by Beam and Warming (1976–1980), who developed them further into operational computer codes for the Euler and Navier–Stokes equations. See for instance Pulliam (1984) for a description of the structure and properties of these codes.

Following Beam and Warming (1980), we introduce the most general consistent *two-step method (three time levels)* under the form of a three-parameter family:

$$(1 + \xi)U^{n+1} - (1 + 2\xi)U^n + \xi U^{n-1} = \Delta t[\theta H^{n+1} + (1 - \theta + \phi)H^n - \phi H^{n-1}]$$

$$(9.3.6)$$

The three parameters have clear distinct roles:

1. ξ controls the order of the finite difference formula for dU/dt.
2. θ controls the implicitness of the method: $\theta = 0$ generates explicit methods.
3. ϕ controls the number of time levels of the space-discretized terms: $\phi = 0$ reduces to a two time level family when $\xi = 0$.

The characteristic polynomial becomes

$$P_1(\overline{E}) = (1 + \xi)\overline{E}^2 - (1 + 2\xi)\overline{E} + \xi - \Omega\Delta t[\theta\overline{E}^2 + (1 - \theta + \phi)\overline{E} - \phi]   (9.3.7)$$

It can be shown that for second order accuracy in time, we should have the relation

$$\xi = \theta + \phi - 1/2 \tag{9.3.8}$$

and if, in addition

$$\xi = 2\phi - 1/6 \tag{9.3.9}$$

the method is third order accurate. These conditions for third order accuracy reduce to $\theta = \phi + 1/3$ and $\xi = 2\phi - 1/6$.

Finally a unique fourth order accurate method is obtained for

$$\theta = -\phi = -\xi/3 = 1/6 \tag{9.3.10}$$

This is easily shown from a Taylor series expansion of the principal root of the characteristic polynomial (9.3.7) (see Problem P.9.7).

A certain number of properties have been proven by Dahlquist (1963) with regard to the order of accuracy of A-stable methods:

- The two-step scheme (9.3.6) is A-stable if and only if

$$\theta \geq \phi + 1/2$$
$$\xi \geq -1/2 \tag{9.3.11}$$
$$\xi \leq \theta + \phi - 1/2$$

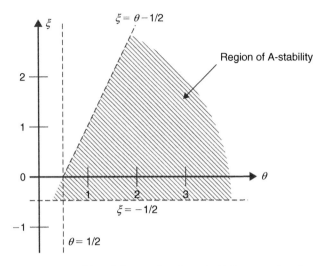

Figure 9.3.1 *Domain of A-stability in the plane (ξ, θ) for second order, two-step methods with $\phi = 0$.*

This is shown on Figure 9.3.1, for the family of second order schemes, satisfying (9.3.8):

- An A-stable linear multistep method cannot have an order of accuracy higher than two.
- The trapezoidal scheme $(\xi = \phi = 0, \theta = 1/2)$ has the smallest truncation error of all second order A-stable methods.

A classification of several well-known methods is listed in Table 9.3.1. Note that the schemes with $\theta = 0$ are explicit. *It is also seen that not all the implicit methods $(\theta \neq 0)$ are A-stable, showing that implicitness does not ensure always that the schemes will be unconditionally stable.*

Figure 9.3.2 shows the stability domain of some of the methods listed in the table.

The Adams–Moulton method is implicit but conditionally stable, within the domain limited on the real axis to $B = -6$, while the backward difference method is unconditionally stable, as its stability domain is outside the curve on the right side of the $\Omega \Delta t$-plane. The same is true for the A-stable Adams-type method. The Adams–Bashworth method on the other hand, is conditionally stable, with an upper bound on the real negative axis of $B = -1$, and will not be suitable for centrally discretized convection terms, whose spectrum is on the imaginary axis. We need to add dissipation terms in order to push the eigenvalues of the space discretization in the left side of the $\Omega \Delta t$-plane.

In unsteady CFD applications, some of the most widely used methods of the three-parameter family (9.3.6) are the methods of Adams–Bashworth (explicit), and the implicit Crank–Nicholson and Gear or Backward Differencing methods.

They are often written in incremental form, for the unknowns $\Delta U^n = U^{n+1} - U^n$, as

$$(1 + \xi)\Delta U^n - \xi \Delta U^{n-1} = \Delta t[\theta H^{n+1} + (1 - \theta + \phi)H^n - \phi H^{n-1}] \quad (9.3.12)$$

Table 9.3.1 *Partial list of one- and two-step methods, from Beam and Warming (1982).*

θ	ξ	ϕ	Method	Order	Comment/Method
0	0	0	Euler explicit	1	Explicit $\quad U^{n+1} - U^n = \Delta t H^n$
1	0	0	Backward Euler	1	A-stable $\quad U^{n+1} - U^n = \Delta t H^{n+1}$
1/2	0	0	Trapezoidal or Crank–Nicholson method	2	A-stable $\quad U^{n+1} - U^n = \Delta t(H^{n+1} + H^n)/2$
1	1/2	0	Backward differencing or Gear method	2	A-stable $\quad 3U^{n+1} - 4U^n + U^{n-1} = 2\Delta t H^{n+1}$
3/4	0	−1/4	Adams type	2	A-stable $\quad U^{n+1} - U^n = \Delta t(3H^{n+1} + H^{n-1})/4$
1/3	−1/2	−1/3	Lees type	2	A-stable $\quad U^{n+1} - U^{n-1} = 2\Delta t(H^{n+1} + H^n + H^{n-1})/3$
1/2	−1/2	−1/2	Two-step trapezoidal	2	A-stable $\quad U^{n+1} - U^{n-1} = \Delta t(H^{n+1} + H^{n-1})$
5/9	−1/6	−2/9	A-contractive	2	A-stable $\quad 5U^{n+1} - 4U^n - U^{n-1} = 2\Delta t(5H^{n+1} + 2H^n + 2H^{n-1})/3$
0	−1/2	0	Leapfrog	2	Explicit $\quad U^{n+1} - U^n = 2\Delta t H^n$
0	0	1/2	Adams–Bashworth	2	Explicit $\quad U^{n+1} - U^n = \Delta t(3H^n - H^{n-1})/2$
0	−5/6	−1/3	Most accurate explicit	3	$U^{n+1} + 4U^n - 5U^{n-1} = 2\Delta t(2H^n + H^{n-1})$
1/3	−1/6	0	Third order implicit	3	$5U^{n+1} - 4U^n - U^{n-1} = 2\Delta t(H^{n+1} + 2H^n)$
5/12	0	1/12	Adams–Moulton	3	Implicit $\quad U^{n+1} - U^n = \Delta t(5H^{n+1} + 8H^n - H^{n-1})/12$
1/6	−1/2	−1/6	Milne	4	Implicit $\quad U^{n+1} - U^{n-1} = 2\Delta t(H^{n+1} + 4H^n + H^{n-1})/6$

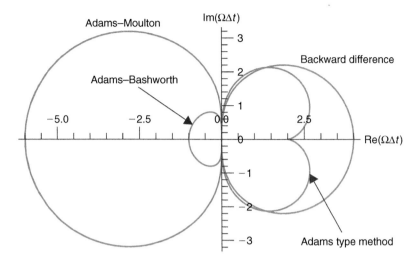

Figure 9.3.2   *Stability regions in the $\Omega\Delta t$-plane for some of the schemes of Table 9.3.1.*

The subset $\phi = 0$

A particular family of schemes, extensively applied, are the schemes with $\phi = 0$.

$$(1 + \xi)U^{n+1} - (1 + 2\xi)U^n = \Delta t[\theta H^{n+1} + (1 - \theta)H^n] - \xi U^{n-1} \qquad (9.3.13)$$

They are second order accurate in time for $\xi = \theta - 1/2$.

When applied to the Euler and Navier–Stokes equations with central space differencing, these schemes have become known as the **Beam and Warming schemes**, when a Jacobian linearization is introduced. The characteristic polynomial reduces to

$$P_1(z) = (1 + \xi)z^2 - (1 + 2\xi)z + \xi - (\Omega\Delta t)z(\theta z + 1 - \theta) = 0 \qquad (9.3.14)$$

For fixed values of θ, the stability domain in the $\Omega\Delta t$-plane is obtained from the condition $|z| < 1$.

For $\xi = 0$, we obtain a two-level, one-step, scheme, known as the **generalized trapezoidal method**,

$$U^{n+1} - U^n = \Delta t[\theta H^{n+1} + (1 - \theta)H^n] \qquad (9.3.15)$$

For $\theta = 1/2$, the method is second order in time and is known as the **Trapezium method** or the **Crank–Nicholson method** when applied to the diffusion equation:

$$U^{n+1} - U^n = \frac{\Delta t}{2}(H^{n+1} + H^n) \qquad (9.3.16)$$

The root of the characteristic equation (9.3.14), after removal of the trivial root $z = 0$, is

$$z = \frac{1 + (1 - \theta)\Omega\Delta t}{1 - \theta(\Omega\Delta t)} \qquad (9.3.17)$$

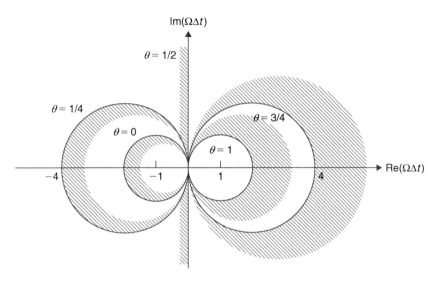

Figure 9.3.3 *Stability regions for the two-step schemes (9.3.13) with $\xi = \phi = 0$.*

The stability curves are displayed in Figure 9.3.3. For $\theta = 0$, we recover the explicit Euler scheme and for $\theta = 1$ the implicit Euler scheme. The stability regions are *inside* the curves at the left ($\theta < 1/2$) and *outside* the curves at the right ($\theta > 1/2$). This confirms that the scheme (9.3.13) is A-stable for $\theta > 1/2$. As a consequence, the generalized trapezoidal scheme with $\theta > 1/2$, applied to the centrally discretized convection equation, will be stable, since its eigenvalues are on the imaginary axis (see also Problems P.9.10 and P.9.11).

Observe that for $0 \leq \theta < 1/2$ the generalized trapezoidal schemes (9.3.15) are only conditionally stable, although they are implicit.

You can see from Figure 9.3.3 that in this range $0 \leq \theta < 1/2$ the stability domain does not contain any part of the imaginary axis. *Therefore these schemes are unstable for the convection equation with central space differencing, but are suitable for diffusion equations and upwind discretized space terms.*

For $\theta = 1/2$ the line coincides with the imaginary axis and the stability region is the left half plane. The characteristic root is obtained from equation (9.3.17) as

$$z = \frac{1 + (\Omega \Delta t)/2}{1 - (\Omega \Delta t)/2} \quad \text{or} \quad \Omega \Delta t = 2\frac{z-1}{z+1} \tag{9.3.18}$$

9.3.2.1 *Beam and Warming schemes for the convection equation*

The Beam and Warming (1976, 1980) scheme for CFD applications applies the generalized trapezoidal method (9.3.15) to the central discretization of the convection flux, leading to

$$u_i^{n+1} - u_i^n = \Delta t[\theta H^{n+1} + (1 - \theta)H^n]$$

$$H = -\frac{a}{2\Delta x}(u_{i+1} - u_{i-1}) \tag{9.3.19}$$

With $\Omega\Delta t = -I\sigma \sin \phi$, we obtain

$$z = 1 - \frac{I\sigma \sin \phi}{1 + I\sigma\theta \sin \phi} \tag{9.3.20}$$

The diffusion error is

$$\varepsilon_D = |z| = \sqrt{\frac{1 + \sigma^2(\theta - 1)^2 \sin^2 \phi}{1 + \sigma^2\theta^2 \sin^2 \phi}} \tag{9.3.21}$$

and the dispersion error is given by

$$\varepsilon_\phi = \frac{1}{\sigma\phi} \tan^{-1}\left(\frac{\sigma \sin \phi}{1 + \sigma^2\theta(\theta - 1) \sin^2 \phi}\right) \tag{9.3.22}$$

The Crank–Nicholson or trapezium method, $\theta = 1/2$, leads to $|z| = 1$ for all ϕ and consequently has no dissipation error with the centrally discretized convection operator. This is a very favorable property. However, the lack of dissipation makes the scheme marginally stable and quite sensitive to nonlinearities and high frequency error sources. Therefore, the scheme will require some additional *artificial dissipation* to be added to the convective flux terms (see Section 9.3.6).

Note that all these schemes have a large lagging phase error, as can be seen from Figure 9.3.4. Figure 9.3.4 compares the behavior of the Euler implicit and the backward Gear implicit methods, at four different CFL values, namely CFL $= 0.5$, 2, 10 and 100, showing the dissipation and the dispersion errors in function of the phase angle from 0 to π. Both schemes are unconditionally stable. Observe that all the schemes have $z = 1$ at $\phi = \pi$, indicating a lack of dissipation of the high frequency errors.

Of particular significance are the increasingly large errors when the CFL number increases. Remember that the objective of implicit methods is to allow high values of the time steps to gain faster convergence.

Hence, we have to ask ourselves if this is acceptable? The answer depends on the type of problem we are solving:

- If we solve a steady state problem, then we wish the transients, which are of numerical significance only, to be damped as fast as possible. This will be the case at the high CFL numbers and fast convergence will be obtained with these schemes.
- For unsteady flow problems, these errors are unacceptable and the high CFL values are not acceptable. Hence, these schemes loose much of their interest, since reducing the CFL number implies a reduction of the time step and the high cost of these implicit methods might become prohibitive compared to explicit methods.

9.3.2.2 *Nonlinear systems and approximate Jacobian linearizations*

Considering scheme (9.3.13), $H^{n+1} = H(U^{n+1})$ has to be evaluated numerically in some way, since U^{n+1} is unknown.

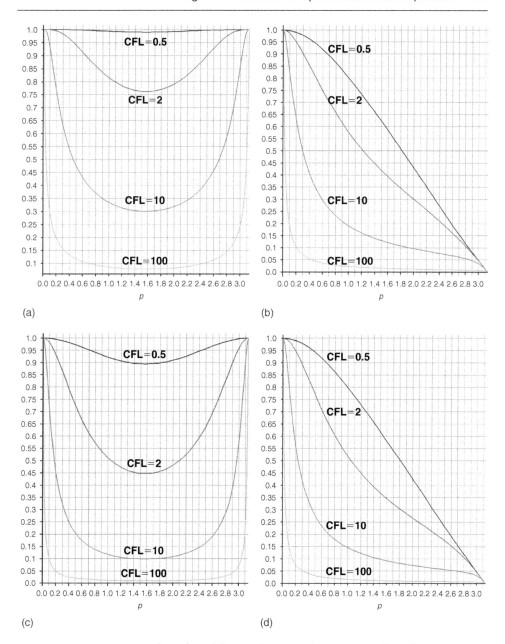

Figure 9.3.4   *Plots of amplification factors and phase errors for different implicit Beam and Warming algorithms at CFL = 0.5, 2, 10 and 100, in function of phase angle in radians, from 0 to π (a) Amplitude error for θ = 1, ξ = 1/2 backward differencing scheme. (b) Phase error for θ = 1, ξ = 1/2 backward differencing scheme. (c) Amplitude error for θ = 1, ξ = 0 backward Euler (implicit) scheme. (d) Phase error for θ = 1, ξ = 0 backward Euler (implicit) scheme.*

A straightforward approach is to apply the Jacobian linearization (9.3.2), written here between the time levels n and $(n+1)$

$$H^{n+1} = H(U^{n+1}) = H^n + \left(\frac{\partial H}{\partial U}\right) \cdot (U^{n+1} - U^n) + O(\Delta U^2) \qquad (9.3.23)$$

which can be written, up to second order

$$H^{n+1} = H^n + J(U^n) \cdot \Delta U^n$$

$$\Delta U^n \overset{\Delta}{=} U^{n+1} - U^n \qquad (9.3.24)$$

The scheme (9.3.12) becomes, with this approach

$$\begin{aligned}
[(1+\xi) - \theta \Delta t J(U^n)] \cdot \Delta U^n &= \Delta t H^n + \phi \Delta t (H^n - H^{n-1}) + \xi \Delta U^{n-1} \\
&= \Delta t H^n + \phi \Delta t J(U^{n-1}) \Delta U^{n-1} + \xi \Delta U^{n-1} \\
&= \Delta t H^n + [\xi + \phi \Delta t J(U^{n-1})] \Delta U^{n-1} \quad (9.3.25)
\end{aligned}$$

This formulation, with $\phi = 0$, is known as the Beam and Warming scheme (1976, 1978) and is often called the **Δ-*form* (*delta-form*)**, although the first application of local linearization to flow problems was developed by Briley and Mc Donald (1975).

In the modal expansion, Ω represents the eigenvalues of the Jacobian matrix $J(U)$, as seen from equation (9.3.4), and the amplification factor of the time integration method $z = z(\Omega \Delta t)$ is solution of equation (9.3.14). However, in practical applications, $H(U)$ will generally be a second or higher order discretization and its Jacobian leads to matrices considered as too complex for practical implementations.

*Therefore it is a general rule to apply an **approximate Jacobian** in the linearization, for instance by selecting the Jacobian associated to a first order upwind discretization.*

This has as consequence a loss in consistency between the space-discretized operators and the applied approximate Jacobian, with severe consequences on reduction of stability range and reduced accuracy. The approach developed in this chapter provides an elegant way of analyzing these effects and evaluate their impact on the scheme properties. If we designate the approximate Jacobian by

$$J^{(1)}(U) = \frac{\partial H^{(1)}}{\partial U} \qquad (9.3.26)$$

where $H^{(1)}$ represents a first order space-discretization of the flux terms, its eigenvalues will be designated by $\Omega^{(1)}$. The linearized scheme (9.3.25) is then replaced by

$$[(1+\xi) - \theta \Delta t J^{(1)}(U^n)] \cdot \Delta U^n = \Delta t H^n + [\xi + \phi \Delta t J(U^{n-1})] \Delta U^{n-1} \quad (9.3.27)$$

and the characteristic polynomial (9.3.7) is replaced by

$$\begin{aligned}
P_1(z) &= [(1+\xi) - \theta(\Omega^{(1)} \Delta t)](z-1) - (\Omega \Delta t) - [\xi + \phi(\Omega \Delta t)](1 - 1/z) \\
&= (1+\xi)z^2 - (1+2\xi)z + \xi - \theta(\Omega^{(1)} \Delta t) z(z-1) \\
&\quad - (\Omega \Delta t)z - \phi(\Omega \Delta t)(z-1) \qquad (9.3.28)
\end{aligned}$$

Its solutions, and the scheme properties will depend on the two eigenvalues Ω and $\Omega^{(1)}$

Actually, the Beam and Warming scheme (9.3.25) is applied in practice under the form (9.3.27) with $\phi = 0$, where the approximate Jacobian is derived from a first order upwind discretization of the convective fluxes.

This can be extended to other time discretization methods and can be applied in a straightforward way when associated to the eigenvalues obtained from the Fourier modes. See Section 9.3.5 for a detailed application.

9.3.3 Predictor–Corrector Methods

When $H(U)$ is nonlinear in U we can avoid the evaluation of Jacobians, by considering an iterative approach, whereby a first guess $\overline{U}^{n+1}$ of U^{n+1} is obtained, for instance by applying an explicit scheme, eventually of lower order, called a **predictor step**, followed by a second step, using the 'predicted' value as a basis for improving the solution. This is called a **predictor–corrector method** and many variants have been studied and applied in practice.

We will present here the most widely known of these methods.

Applied for instance to the scheme (9.3.13), the **corrector step** would become

$$(1+\xi)U^{n+1} - (1+2\xi)U^n = \Delta t[\theta\overline{H^{n+1}} + (1-\theta)H^n] - \xi U^{n-1} \qquad (9.3.29)$$

This defines the **predictor–corrector** sequence, with $\overline{H^{n+1}} = H(\overline{U}^{n+1})$.

This approach can be pursued at the same time level, by repeating equation (9.3.29) for s steps, until some form of convergence between two consecutive estimations of the corrector step solutions $\overline{\overline{U}}^{n+1}$ is achieved. This implies an evaluation of H at each 'local' iteration step and this procedure is generally not recommended, since the evaluation of H is often the most costly operation in the numerical simulation. Repeating the corrector sequence implies that the solution of equation (9.3.29), now designated by $\hat{U}^{n+1}$, is obtained after the second corrector step as

$$(1+\xi)\hat{U}^{n+1} - (1+2\xi)U^n = \Delta t[\theta\overline{\overline{H^{n+1}}} + (1-\theta)H^n] - \xi U^{n-1} \qquad (9.3.30)$$

where $\hat{U}^{n+1}$ is the new value for U^{n+1} and $\overline{\overline{H^{n+1}}} = H(\overline{\overline{U}}^{n+1})$.

A detailed analysis of various options and of the influence of the number of 'local' iterations, can be found in Lambert (1973). One of the essential conclusions, from a comparison of the effect of different numbers of corrector steps coupled to the same predictor, is that the unique sequence of a predictor step followed by a single corrector step appears as being optimal, in view of the reduction of the number of evaluations of the right-hand side.

In order to analyze the order of accuracy and the stability of a predictor–corrector sequence, we can combine the predictor $\overline{U}^{n+1}$ and the known value U^n in a single vector and write the sequence as a system of two equations. The characteristic polynomial is obtained by setting the determinant of the matrix to zero, for the modal equation $H = \Omega U$.

Let us illustrate this for the two-level schemes (9.3.13). By definition, we require the predictor to be explicit; hence we set $\theta = \xi = 0$ and obtain the first order Euler method as predictor (see, however, Problem P.9.23 for an implicit predictor–corrector approach):

$$\overline{U}^{n+1} = U^n + \Delta t H^n \tag{9.3.31}$$

The corrector is equation (9.3.29) written in operator form as

$$[(1 + \xi)\overline{E} - (1 + 2\xi)]U^n = \Delta t[\theta \overline{H}^{n+1} + (1 - \theta)H^n] - \xi U^{n-1} \tag{9.3.32}$$

and can be treated as an explicit equation.

Inserting the modal equation $H = \Omega U$, we obtain the system

$$\begin{vmatrix} \overline{E} & -(1 + \Omega \Delta t) \\ -\theta(\Omega \Delta t)\overline{E} & \xi \overline{E}^{-1} + (1 + \xi)\overline{E} - (1 + 2\xi) - \Omega \Delta t(1 - \theta) \end{vmatrix} \begin{vmatrix} \overline{U}^n \\ U^n \end{vmatrix} = \begin{vmatrix} 0 \\ 0 \end{vmatrix} \tag{9.3.33}$$

leading to the polynomial equation obtained from the determinant of the system set to zero:

$$P_1(z) = (1 + \xi)z^2 - (1 + 2\xi)z + \xi - (\Omega \Delta t)z(1 + \theta \Omega \Delta t) = 0 \tag{9.3.34}$$

It is interesting to compare this equation with (9.3.14) derived for the implicit system, in particular when $\xi = 0$. The above equation becomes, removing a trivial root $z = 0$:

$$z = 1 + (\Omega \Delta t) + \theta(\Omega \Delta t)^2 \tag{9.3.35}$$

This solution is typical of explicit schemes, and the predictor–corrector sequence has become explicit for all values of θ. The sequence is only first order accurate in time, excepted for $\theta = 1/2$ where it is second order, since the quadratic term is equal to the corresponding term in the Taylor development of $\exp(\Omega \Delta t)$ as seen from equation (9.2.13).

Figure 9.3.5 shows the stability regions for the cases $\theta = 0$, $\theta = 1/2$ and $\theta = 1$. The case $\theta = 0$ is the explicit Euler method. Only for $\theta = 1$, does the stability domain include part of the imaginary axis, which indicates that it can be applied with central discretizations of the convection terms. This is not the case anymore when $\theta = 1/2$.

The choice $\theta = 1/2$ corresponds to a second order accurate sequence and is known as **Henn's method**. It forms the basis of the **McCormack scheme** and can be written as

$$\overline{U}^{n+1} = U^n + \Delta t H^n$$
$$U^{n+1} = U^n + \frac{1}{2}\Delta t(\overline{H^{n+1}} + H^n) \tag{9.3.36}$$
$$= \frac{1}{2}(\overline{U}^{n+1} + U^n) + \frac{1}{2}\Delta t \overline{H}^{n+1}$$

This scheme is **unstable** for the convection equation, when central differences are applied, as can be seen from Figure 9.3.4. On the other hand, for $\theta = 1$ we obtain

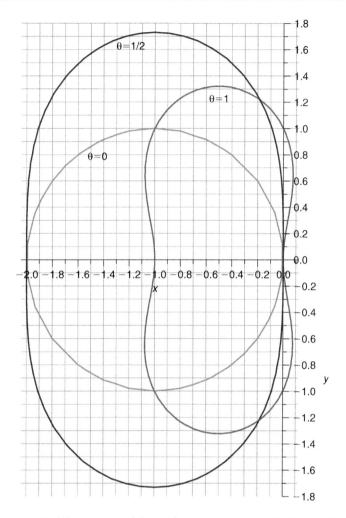

Figure 9.3.5   *Stability regions of the predictor–corrector methods (9.3.34), with*
$\xi = 0$ *and for* $\theta = 0$, $\theta = 1/2$ *and* $\theta = 1$.

the following scheme

$$\overline{U}^{n+1} = U^n + \Delta t H^n$$
$$U^{n+1} = U^n + \Delta t \overline{H^{n+1}}$$

(9.3.37)

which is stable for the convection equation with central differences under the CFL
condition $|\sigma| \leq 1$.

Variants of this scheme have been applied by Brailowskaya (1965) to the Navier–
Stokes equations by keeping the viscous diffusion terms in the corrector step at
level n.

In particular, for the linear convection equation, the method (9.3.37) becomes with a central discretization:

$$\overline{u_i^{n+1}} = u_i^n - \frac{\sigma}{2}(u_{i+1}^n - u_{i-1}^n)$$
$$u_i^{n+1} = u_i^n - \frac{\sigma}{2}(\overline{u_{i+1}^{n+1}} - \overline{u_{i-1}^{n+1}})$$
(9.3.38)

Combining the two steps leads to a scheme of second accuracy in space and first order in time, which involves the points $i - 2$ and $i + 2$ and is stable under the CFL condition $|\sigma| \leq 1$ (see Problem P.9.15).

If scheme (9.3.37) is applied with upwind differences, we obtain the algorithm

$$u_i^{n+1} = u_i^n - \sigma(u_i^n - u_{i-1}^n) + \sigma^2(u_i^n - 2u_{i-1}^n + u_{i-2}^n)$$
(9.3.39)

This scheme is only first order accurate, both in space and time. Comparing with the unique upwind second order accurate scheme of Warming and Beam, equation (7.4.35) or (8.2.30), this scheme differs by the coefficient in front of the last term. The stability conditions can be obtained by plotting the spectrum of eigenvalues of the first order upwind space discretization on the stability curves of Figure 9.3.5. This is shown in Figure 9.3.6, where you can see that the stability condition for $\theta = 1$ is restricted to $0 < \sigma \leq 1/2$ (see also Problem P.9.16). Observe also that the case $\theta = 1/2$ has a stability limit of $0 < \sigma \leq 1$, while the three predictor–corrector schemes are all unstable for CFL $= 1.1$, as some of the eigenvalues are outside the stability limits shown by the full lines.

When an upwind differencing is applied to the convection terms, Henn's method (9.3.36) takes the following form for the linear convection equation

$$\overline{u_i^{n+1}} = u_i^n - \sigma(u_i^n - u_{i-1}^n)$$
$$u_i^{n+1} = u_i^n - \frac{\sigma}{2}\left(\overline{u_i^{n+1}} - \overline{u_{i-1}^{n+1}} + u_i^n - u_{i-1}^n\right)$$
(9.3.40)

or

$$u_i^{n+1} = u_i^n - \sigma(u_i^n - u_{i-1}^n) + \frac{\sigma^2}{2}(u_i^n - 2u_{i-1}^n + u_{i-2}^n)$$
(9.3.41)

This scheme differs also by the coefficient in front of the last term from the second order Warming and Beam scheme and is also only first order accurate. A stability analysis leads to the CFL condition $0 < \sigma \leq 1$ (see Problem P.9.17).

McCormack's scheme

In order to obtain a second order accurate scheme with Henn's predictor–corrector sequence, and a first order space differencing for H^n, we could attempt to compensate the truncation errors in the combined sequence by applying a different space operator in the corrector step.

Considering scheme (9.3.36) and a space truncation error $a_{p+1}\Delta x^p$ in the predictor. We could obtain a higher global accuracy if the corrector would generate an equal,

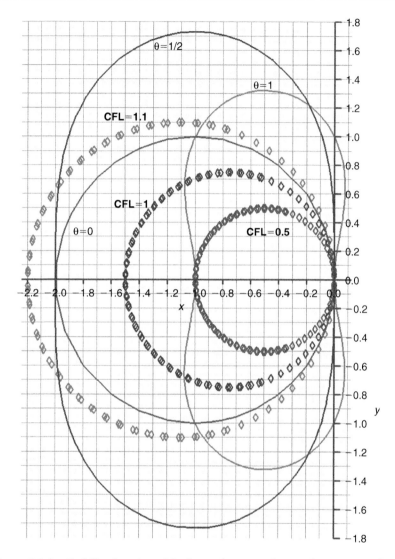

Figure 9.3.6 *Stability domains of the first order upwind space discretization for CFL = 0.5, 1 and 1.1 (symbols) in the $\Omega\Delta t$-plane, together with the predictor–corrector methods (9.3.31) and (9.3.32) for $\xi = 0$ and $\theta = 0, 1/2, 1$.*

but opposite in sign, truncation error. That is, we would have

$$\overline{U^{n+1}} = U^n + \Delta t H^n + \Delta x^p(a_{p+1} + \Delta x^{p+1} a_{p+2}) \tag{9.3.42a}$$

and

$$U^{n+1} = \frac{1}{2}(\overline{U^{n+1}} + U^n) + \frac{1}{2}\Delta t[\overline{H_1^{n+1}} + (\Delta x^p \overline{a_{p+1}} + \Delta x^{p+1} \overline{a_{p+2}})] \tag{9.3.42b}$$

where H_1 is a different space operator from H. If $\overline{a_{p+1}} = -a_{p+1}$, the overall accuracy is increased by one unit, and becomes of order $(p+1)$. For the convection equation, this would be realized if the corrector step would contain a forward space difference, when the predictor contains a backward difference, or vice-versa. This leads to the second order accurate (in space and in time) **McCormack** scheme, which is widely applied for resolution of Euler and Navier–Stokes equations (McCormack, 1969, 1971).

$$\overline{u_i^{n+1}} = u_i^n - \sigma(u_i^n - u_{i-1}^n)$$

$$u_i^{n+1} = \frac{1}{2}(\overline{u_i^{n+1}} + u_i^n) - \frac{\sigma}{2}(\overline{u_{i+1}^{n+1}} - \overline{u_i^{n+1}}) \tag{9.3.43}$$

Since the unique second order scheme for the linear wave equation on the support $(i-1, i, i+1)$ is the Lax–Wendroff scheme, they should be identical (see Problem P.9.18).

However, this is not the case anymore for nonlinear problems. Many variants can be derived if we allow for the freedom of different space operators in the two steps of the predictor–corrector sequence.

For instance, McCormack's scheme becomes for a general flux function f, after one-sided space discretizations of $\partial f / \partial x$

$$\overline{u_i^{n+1}} = u_i^n - \frac{\Delta t}{\Delta x}(f_i^n - f_{i-1}^n)$$

$$u_i^{n+1} = \frac{1}{2}(\overline{u_i^{n+1}} + u_i^n) - \frac{\Delta t}{2\Delta x}(\overline{f_{i+1}^{n+1}} - \overline{f_i^{n+1}}) \tag{9.3.44}$$

9.3.4 The Runge–Kutta Methods

An important family of explicit time integration techniques, of high order of accuracy, limited to two time levels, is provided by the family of **Runge–Kutta methods**. Compared with the linear multistep method, the Runge–Kutta schemes achieve high orders of accuracy by introducing multiple stages, while the former achieve high accuracy by involving multiple time steps.

A detailed description of Runge–Kutta methods can be found in the books of Gear (1971), Lambert (1973, 1991), Van der Houwen (1977).

These methods have been applied to the solution of Euler equations by Jameson et al. (1981), and further developed to highly efficient operational codes, Jameson and Baker (1983, 1984). They are also widely used in high order methods for Computational Aero-Acoustics (CAA), see for instance Hu et al. (1996) and the reviews by Tam (2004), Kurbatskii and Mankbadi (2004).

The basic idea of Runge–Kutta (RK) methods is to evaluate the right-hand side of the differential system (9.3.1) at several values of U in the interval, between $n\Delta t$ and $(n+1)\Delta t$, and to combine them in order to obtain a high order approximation of U^{n+1}. The number of intermediate values is referred to as the **Runge–Kutta stages**.

The most general form of a K-stage Runge–Kutta method is as follows:

$$U^{(1)} = U^n$$
$$U^{(2)} = U^n + \Delta t \alpha_{12} H^{(1)}$$
$$\cdot$$
$$U^{(j)} = U^n + \Delta t \sum_{k=1}^{j-1} \alpha_{kj} H^{(k)} \tag{9.3.45}$$
$$\cdot$$
$$U^{(K)} = U^n + \Delta t \sum_{k=1}^{K-1} \alpha_{kj} H^{(k)}$$
$$U^{n+1} = U^n + \Delta t \sum_{k=1}^{K} \beta_k H^{(k)}$$

The notation $H^{(k)}$ implies

$$H^{(k)} = H(U^{(k)}) \tag{9.3.46}$$

written here for the case where H is explicitly independent of time which is generally the case in fluid mechanical problems.

This formulation requires a large memory storage, since at each stage j all intermediate $H^{(k)}$ of the previous stages need to be stored. Therefore, a *low-storage Runge–Kutta method* is usually applied, given by

$$U^{(1)} = U^n$$
$$U^{(2)} = U^n + \Delta t \alpha_2 H^{(1)}$$
$$U^{(3)} = U^n + \Delta t \alpha_3 H^{(2)}$$
$$\cdot$$
$$\cdot \tag{9.3.47}$$
$$U^{(K)} = U^n + \Delta t \alpha_K H^{(K-1)}$$
$$U^{n+1} = U^n + \Delta t \sum_{k=1}^{K} \beta_k H^{(k)}$$

where for consistency

$$\sum_{k=1}^{K} \beta_k = 1 \tag{9.3.48}$$

A particular choice, often applied, is

$$\beta_i = 0 \quad \text{for } i = 1, \ldots, K-1 \quad \text{and} \quad \beta_K = 1 \tag{9.3.49}$$

leading to

$$U^{n+1} = U^n + \Delta t H^{(K)} \tag{9.3.50}$$

For each number of stages K, an infinite number of Runge–Kutta schemes can be defined, with maximum order of accuracy, or when the requirement on the order of accuracy is relaxed. Various conditions can be imposed on the coefficients of the RK scheme, for instance to minimize dispersion and diffusion errors. See for instance Hu et al. (1996) for a representative example, where several Runge–Kutta schemes are optimized for their dispersion and dissipation behavior and known as *low-dissipation and low-dispersion Runge–Kutta methods* (LDDRK). Another family of optimized RK schemes has been proposed by Ramboer et al. (2006).

A popular version is the fourth order Runge–Kutta method, defined by the coefficients

$$\alpha_2 = \frac{1}{2} \quad \alpha_3 = \frac{1}{2} \quad \alpha_4 = 1$$
$$\beta_1 = \frac{1}{6} \quad \beta_2 = \beta_3 = \frac{1}{3} \quad \beta_4 = \frac{1}{6} \tag{9.3.51}$$

leading to

$$U^{(1)} = U^n$$
$$U^{(2)} = U^n + \frac{1}{2}\Delta t H^{(1)}$$
$$U^{(3)} = U^n + \frac{1}{2}\Delta t H^{(2)} \tag{9.3.52}$$
$$U^{(4)} = U^n + \Delta t H^{(3)}$$

$$U^{n+1} = U^n + \frac{\Delta t}{6}(H^n + 2H^{(2)} + 2H^{(3)} + H^{(4)}) \tag{9.3.53}$$

where $H^{(1)}$ has been written as H^n.

The first order Runge–Kutta method is the Euler explicit scheme.

A well-known two-step Runge–Kutta method, *Henn's method*, is defined by the predictor–corrector scheme (9.3.36). With the restriction to order two, there exist an infinite number of two-stage Runge–Kutta methods with order two, but none with order higher than two. They all can be considered as predictor–corrector schemes. Another popular, second order scheme, of this family is defined by

$$\overline{U^{n+1}} = U^n + \frac{1}{2}\Delta t H^n$$
$$U^{n+1} = U^n + \Delta t \overline{H^{n+1}} \tag{9.3.54}$$

9.3.4.1 *Stability analysis for the Runge–Kutta method*

The properties of the Runge–Kutta methods can be analyzed by the general methodology described in the previous sections.

Introducing the modal equation $H = \Omega U$, in (9.3.47), we obtain, for instance with the assumption (9.3.49)

$$z = 1 + \Omega \Delta t (1 + \alpha_K \Omega \Delta t (1 + \alpha_{K-1} \Omega \Delta t (1 + \cdots (1 + \alpha_2 \Omega \Delta t))) \cdots) \quad (9.3.55)$$

leading to

$$z = 1 + \sum_{j=1}^{K} a_j (\Omega \Delta t)^j \qquad (9.3.56)$$

with

$$a_1 = 0, \quad a_2 = \alpha_K, \quad a_3 = \alpha_K \alpha_{K-1}, \ldots \quad a_K = \alpha_K \alpha_{K-1} \alpha_{K-2}, \ldots, \alpha_2$$

$$(9.3.57)$$

This should be compared with the exact amplification factor (9.2.13):

$$z_P(\Omega \Delta t) \approx e^{\Omega \Delta t} = 1 + \Omega \Delta t + \frac{(\Omega \Delta t)^2}{2} + \frac{(\Omega \Delta t)^3}{3!} + \frac{(\Omega \Delta t)^4}{4!} + \cdots$$

The first term of (9.3.56) that deviates from the exact development determines the order of accuracy in time.

We notice that the maximum order of accuracy in time is K, but various conditions can be imposed on the coefficients, to minimize the numerical errors with a reduction of the order of accuracy in time.

For instance, for the variant (9.3.52), we obtain

$$U^{(2)} = \left(1 + \frac{1}{2} \Omega \Delta t\right) U^n$$

$$U^{(3)} = \left[1 + \frac{1}{2} \Omega \Delta t + \frac{1}{4} (\Omega \Delta t)^2\right] U^n \qquad (9.3.58)$$

$$U^{(4)} = \left[1 + \Omega \Delta t + \frac{1}{2} (\Omega \Delta t)^2 + \frac{1}{4} (\Omega \Delta t)^3\right] U^n$$

and

$$U^{n+1} = \left[1 + \Omega \Delta t + \frac{1}{2} (\Omega \Delta t)^2 + \frac{1}{6} (\Omega \Delta t)^3 + \frac{1}{24} (\Omega \Delta t)^4\right] U^n \equiv z U^n$$

$$(9.3.59)$$

showing that the scheme is fourth order accurate since z_P is the Taylor expansion of the exact amplification $\exp(\Omega \Delta t)$ up to fourth order. The stability region in the $\Omega \Delta t$-plane is shown on Figure 9.3.7, together with the stability regions for the second order method (9.3.36) and for the third to fifth stage Runge–Kutta methods of maximum time accuracy.

Actually all methods with K stages and order K have the same domain of stability.

A very important property of the third and higher order RK methods is that they contain a segment of the imaginary axis of the $\Omega\Delta t$-plane in their stability region. The length of this segment $(-I\beta, +I\beta)$ is increasing with the stage number K of the method, depending on the order of accuracy which is not necessarily equal to its maximum value. *Hence, the higher order RK methods are very well adapted for centrally discretized convection, with a conditional stability condition of $|\sigma| \leq \beta$.* This explains their widespread use for the Euler and Navier–Stokes equations, as initially introduced by Jameson et al. (1981). For the methods with maximum order of accuracy, the imaginary segment inside the stability region is limited by $\beta = \sqrt{3} = 1.73$, $\beta = 2\sqrt{2} = 2.828$, $\beta = 3.44$ for the third, fourth and fifth order methods, respectively.

Note that this does not apply to first or second order RK methods, such as (9.3.36) or (9.3.54), which do not include any portion of the imaginary axis. Hence, they are not adapted to central discretizations of the convection terms. On the other hand, all RK methods are adapted to upwind discretized convection, since the corresponding eigenvalue spectrum is located in the left-hand side of the $\Omega\Delta t$-plane, as seen from Figure 9.1.3.

For second order time accurate methods, the limit value $\beta = K - 1$ can be obtained for K-odd and certain combination of coefficients, Van der Houwen (1977). For even number of stages, for instance $K = 4$, Vichnevetsky (1983) has shown that the value $(K - 1)$ is also an upper limit and Sonneveld and Van Leer (1984) showed that this limit can indeed be reached for first order accurate methods and $K > 2$.

On the other hand, the limits of the RK stability domain on the real negative axis will provide a stability limit on the dissipation terms, this dissipation being either numerical, as obtained from upwind-discretized convection operators, or physical, from centrally discretized diffusion equations.

The limits $(-B)$ on the real axis are -2, -2.51, -2.78, -3.2 for the second, third, fourth and fifth, order respectively, as can be seen from Figure 9.3.5 (see also Lambert, 1973, 1991).

For stationary problems it is important to be able to allow the highest possible time steps and therefore the extension of the stability region is more important than their order. By relaxing the order, one can obtain Runge–Kutta methods of a given stage number, with higher stability regions; for instance a third order Runge–Kutta method with second order accuracy can be defined which cuts the real axis at $(-4.52, 0)$, Lambert (1973).

Other applications of the large number of degrees of freedom available in the selection of the α_k and β_k coefficients of equation (9.3.47) can be directed toward the selective damping of high frequency error components, suitable for integration into multigrid iterative methods (Jameson and Baker, 1983) or for aero-acoustic applications (Hu et al., 1996; Ramboer et al., 2006).

Example E.9.3.1 Diffusion equation $u_t = au_{xx}$ with second order central space differences

If the system

$$\frac{du_i}{dt} = \frac{\alpha}{\Delta x^2}(u_{i+1} - 2u_i + u_{i-1}) \tag{E.9.3.1}$$

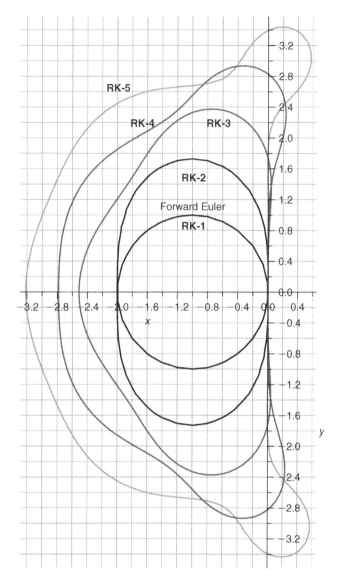

Figure 9.3.7   *Stability limits for Runge–Kutta methods in the complex $\Omega\Delta t$-plane, from 1 to 5 stages (RK-1 to RK-5).*

is solved by a Runge–Kutta method whose stability domain cuts the real axis at $(-B, 0)$, then the method will be stable for

$$0 \le \frac{\alpha\Delta t}{\Delta x^2} \le \frac{B}{4} \qquad\qquad (E.9.3.2)$$

For instance, for the fourth order method the limit is 2.78/4, which is only marginally larger than the single step, first Euler explicit method.

Example E.9.3.2 Convection equation $u_t + au_x = 0$ with second order central space difference

The system

$$\frac{du_i}{dt} = -\frac{a}{2\Delta x}(u_{i+1} - u_{i-1}) \qquad\qquad (E.9.3.3)$$

solved by a fourth order Runge–Kutta method has a stability range on the imaginary axis limited to $\pm I2 \cdot \sqrt{2}$. Hence, with

$$\Omega \Delta t = -I\sigma \sin\phi \qquad\qquad (E.9.3.4)$$

the stability condition becomes

$$|\sigma| < 2\sqrt{2} \qquad\qquad (E.9.3.5)$$

This scheme therefore allows a Courant number close to three times the usual CFL limit of one. However, for $\phi = \pi$, that is for high frequencies, $\Omega = 0$ and the amplification factor z becomes equal to one. Therefore, it is to be expected that this scheme will become unstable when applied to nonlinear hyperbolic problems with central differences, if some dissipation is not added to the scheme. See Section 9.3.6, where the notion of **artificial dissipation** is introduced.

These two examples show that there is little to be gained in applying high order Runge–Kutta methods to pure diffusion problems, with regard to maximum allowable time step, when compared to the one-step explicit Euler method. However, this is not the case for convection equations where high stage Runge–Kutta methods have an increasing segment of the imaginary axis of the $\Omega \Delta t$-plane in their stability region.

Example E.9.3.3 Convection equation $u_t + au_x = 0$ with Fromm's second order upwind biased scheme

We consider here the system

$$\frac{du_i}{dt} = -\frac{a}{4\Delta x}(u_{i+1} + 3u_i - 5u_{i-1} + u_{i-2}) \qquad\qquad (E.9.3.6)$$

and envisage to solve it with a Runge–Kutta method. Plotting the stability domains of the space discretization for CFL $= 0.5$, 1, 1.25 and 1.5 in the $\Omega \Delta t$-plane, together with the RK methods of order 1 to 4 on Figure 9.3.8, we observe that for the one-stage RK-1 method, that is the Euler explicit method, the resulting scheme will be unstable. For the two-stage RK-2 method, we would have a stability condition of CFL ≤ 1. The three-stage RK-3 method has clearly a somewhat higher CFL limit of stability, although you clearly can see that the eigenvalues of the space discretization are not totally inside the RK-3 stability curve. The fourth order RK-4 method is still stable at CFL $= 1.35$, as can be seen from Figure 9.3.9, but a plot for CFL $= 1.4$ shows that the combination is unstable. Applying the definitions seen earlier, we can now determine the accuracy of the resulting scheme by considering the amplification factor $z_P(\Omega(\phi))$ in the complex z-plane (Figure 9.3.9a). From there we derive the dissipation and dispersion errors in function of the phase angle, at constant CFL (Figure 9.3.9b and c).

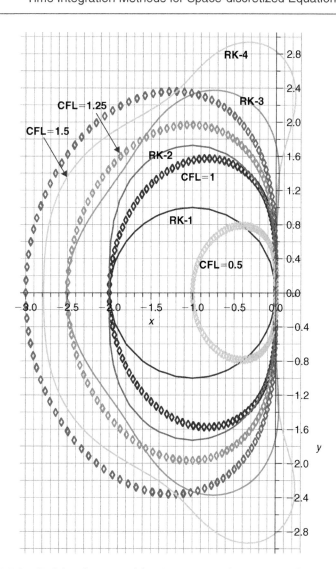

Figure 9.3.8   *Stability domains of the Fromm space discretization for CFL = 0.5, 1, 1.25 and 1.5 (symbols) in the $\Omega \Delta t$-plane, together with the RK methods of order 1 to 4 (continuous lines).*

9.3.5   Application of the Methodology and Implicit Methods

This subsection will now guide you toward the practical workout of the methodology for a few representative examples. We will also consider the effect of an approximate Jacobian and derive some guidelines related to applying implicit methods.

As a first example, we consider the trapezium method for the time integration of the second order upwind space discretization for the linear convection equation. Based

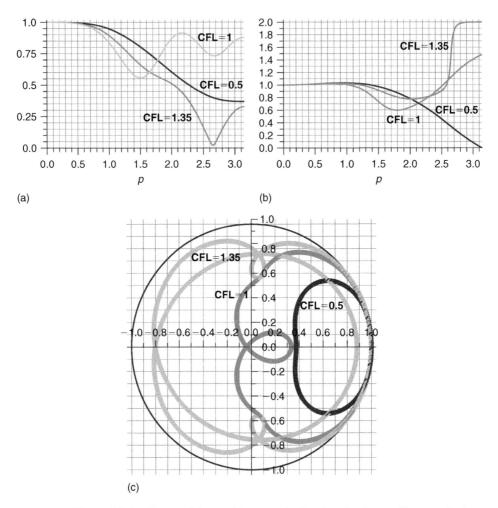

Figure 9.3.9 *Errors of the combination of a fourth order Runge–Kutta method with Fromm's scheme at constant CFL, in function of the phase angle for the same parameters as in Figure 9.3.8: (a) dissipation error; (b) dispersion error; (c) error in the complex z-plane; at CFL = 0.5, 1 and 1.35.*

on Table 9.1.1, we have the system of ODE's in time

$$\frac{du_i}{dt} = -\frac{a}{2\Delta x}(3u_i - 4u_{i-1} + u_{i-2}) \tag{9.3.60}$$

leading to the scheme, based on (9.3.16) (see Problem P.9.8)

$$u_i^{n+1} = u_i^n - \frac{\sigma}{4}[(3u_i - 4u_{i-1} + u_{i-2})^{n+1} + (3u_i - 4u_{i-1} + u_{i-2})^n] \tag{9.3.61}$$

We will analyze this scheme with exact and approximate upwind Jacobians, following step by step the methodology outlined in the previous sections:

- First we determine the eigenvalues of the selected second order upwind scheme, following Table 9.1.1,

$$\Omega \Delta t = -\sigma[(1 - \cos \phi)^2 + I(2 - \cos \phi) \sin \phi] \qquad (9.3.62)$$

- A first source of information is to display the eigenvalue trajectories, in the range $\phi = -\pi, \ldots, \pi$ in the complex $\Omega \Delta t$-plane. This is shown on Figure 9.3.10a for the CFL values 0.5, 1, 2 and 5.
- The characteristic root of the trapezium method z, which is its **amplification factor**, is given by equation (9.3.18), as

$$z = \frac{1 + (\Omega \Delta t)/2}{1 - (\Omega \Delta t)/2} \quad \text{or} \quad \Omega \Delta t = 2\frac{z - 1}{z + 1} \qquad (9.3.63)$$

- The stability domain can be represented either in the $\Omega \Delta t$-plane or in the complex z-plane, by representing the stability limit $|z| \leq 1$. In the $\Omega \Delta t$-plane, we write $z = \exp(I\phi)$ in the second equation (9.3.63). The vertical imaginary axis is the limit of the stability domain that encompasses the whole left side. Hence the scheme is stable for all values of the CFL number. In the z-plane, the stability limit is simply the circle of radius 1, centered at the origin.
- We now represent the eigenvalues of the space discretization in the z-plane, by introducing the eigenvalues (9.3.62) in the first of the equations (9.3.63). This defines the amplification factor as a function of CFL and the phase angle, $z = z(\phi, \text{CFL})$. This is shown in Figure 9.3.10b for the CFL numbers 0.5, 1, 2 and 10. Again you can see that all the eigenvalues are inside the circle of radius 1, indicated as a continuous line, confirming the unconditional stability of the scheme.
- The next step is to analyze the dissipation and dispersion errors. They are defined according to equation (9.2.23) and can be obtained in a straightforward way from the modulus (the dissipation error) and the phase (the dispersion error) of the several curves of Figure 9.3.10b. They are plotted as Figure 9.3.10c and d, over the range $\phi = 0, \ldots, \pi$.
- Notice the strong deterioration of the dispersion errors at high CFL-numbers, for CFL $> \sim 5$. Remember that the dispersion error is the ratio of the numerical wave speed to the physical one. Hence, a dispersion error value of 0.5 for instance indicates that the associated wave travels at half of the correct speed. This will distort significantly the waveforms and puts severe restrictions for unsteady flow simulations with implicit schemes at high CFL values. An upper limit of CFL ~ 2 could be acceptable for unsteady flows. *For steady state computations, on the other hand, this is not a drawback.*
- The diffusion errors decrease (i.e. they become closer to one) when the CFL number increases, indicating diminishing numerical dissipation, which will slow down the convergence at the highest CFL-numbers, for steady state simulations.

We now wish to analyze the behavior of the scheme if we apply an approximate Jacobian, based on the first order upwind discretization, to solve the implicit algebraic

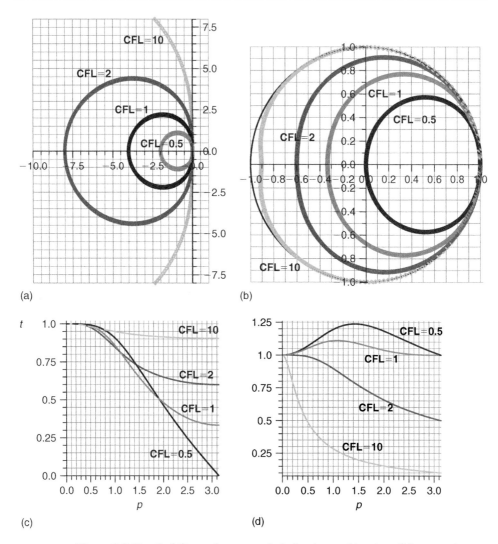

Figure 9.3.10 *Stability and error analysis for the combination of the trapezium implicit time integration applied to the second order upwind space discretization for the linear convection equation. CFL values: 0.5, 1, 2, 10. (a) Eigenvalues of the second order upwind scheme in the $\Omega \Delta t$-plane; (b) Amplification factors for the second order upwind scheme with the trapezium method in the z-plane. (c) Diffusion error. (d) Dispersion error.*

system. This leads to

$$\Delta u_i^n + \sigma(\Delta u_i - \Delta u_{i-1})^n = -\frac{\sigma}{2}(3u_i - 4u_{i-1} + u_{i-2})^n \qquad (9.3.64)$$

Hence, we apply the same steps, but with the modified characteristic polynomial (9.3.28), whose roots are the modified amplification factors. For the trapezium method, $\xi = \phi = 0$, $\theta = 1/2$, we obtain the solution, after removal of the trivial root $z = 0$:

$$z_{mod} = \frac{2 - (\Omega^{(1)}\Delta t) + 2(\Omega\Delta t)}{2 - (\Omega^{(1)}\Delta t)}$$ (9.3.65)

- Introduce the first upwind eigenvalues from Table 9.1.1:

$$\Omega^{(1)}\Delta t = -\sigma[(1 - \cos\phi) + I\sin\phi]$$ (9.3.66)

- We can now represent the modified amplification factors for various CFL numbers in the range $\phi = -\pi, \ldots, \pi$ in the complex z-plane and the associated errors. This is shown on Figure 9.3.11.
- You immediately notice that the modified scheme, with the first order Jacobian, has lost its unconditional stability, since the curves for CFL > 1 all move partly outside the stability domain of the unit circle. This can also be seen on the diffusion error curves, which become larger than one for these CFL values.
- This is a very severe penalty, since the stability limit is now restricted to an explicit type limit of CFL ≤ 1. Hence, this approach becomes useless, as the advantages of running an implicit method with large time steps is totally lost. *This explains why in practice, trapezium methods are never used with upwind space discretizations and an approximate Jacobian.*

You can repeat these developments with a central discretization and you will then observe that the application of the approximate Jacobian does *not* restrict the unconditional stability and has only a marginal impact on the accuracy. *This explains the successful application of the Beam and Warming schemes (9.3.27), coupled to a central space discretization of the convective flux terms.*

A very similar behavior is observed when the trapezium method is applied to other upwind options of Table 9.1.1. Therefore, other A-stable methods have to be considered for applications with upwind space discretizations; for instance a backward differencing, or Gear's method, with second order upwind discretization leads to an unconditionally stable scheme, even with the approximate first order Jacobian (see Problem P.9.9).

9.3.6 The Importance of Artificial Dissipation with Central Schemes

We have seen that central discretizations of the convection terms lead to purely imaginary eigenvalues with the consequence that the high frequencies, corresponding to the $2\Delta x$ shortest waves on the mesh, or $\phi = \pi$, are not damped by the scheme, as can be seen on Figure 9.3.4. This can also lead to marginal stability conditions with implicit time integration methods, since the stability condition reduces to the modulus of the amplification factor equal to one for all values of the phase angle. This creates a risk for weak instabilities generated by the nonlinearity of the convection terms in the conservation laws. We have already encountered this problem with the leapfrog method in Chapter 7.

One way of introducing some high frequency dissipative component, that is providing some real negative part to the eigenvalues, is the selection of an upwind discretization for the convection terms.

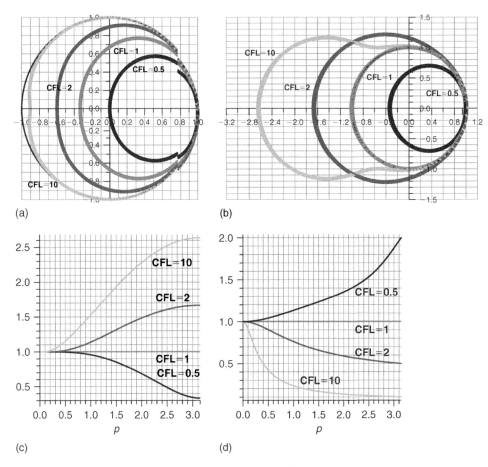

Figure 9.3.11 *Stability and error analysis for the combination of the trapezium implicit time integration applied to the second order upwind space discretization for the linear convection equation, with an approximate, first order upwind, Jacobian. CFL values: 0.5, 1, 2 and 10. (a) Amplification factors for the second order upwind scheme with the trapezium method in the z-plane, based on exact Jacobian. (b) Amplification factors for the second order upwind scheme with the trapezium method in the z-plane with an approximate first order upwind Jacobian. (c) Diffusion error with approximate Jacobian. (d) Dispersion error with approximate Jacobian.*

The other alternative is to add *artificial dissipation* terms, also referred to as *artificial viscosity* terms (AVT) to the centrally discretized convection operator.[1] The objective is to add a dissipative contribution to the scheme, that is an *even order derivative*,

---

[1] There is some terminology confusion in the literature between *numerical dissipation* and *artificial dissipation*. Take good notice that numerical dissipation refers to the dissipation generated by the selected discretization, while artificial dissipation refers to the contributions from terms *added* to the selected scheme, in a somewhat '*artificial*' way, as opposed to the '*natural*' dissipation present in the selected discretization of the model derivatives.

with a truncation error lower than the truncation error of the selected discretization. For instance, with a second order discretization, the artificial dissipation term could be of order four, while selecting a fourth order discretization of the convection term, would require at least a sixth order derivative artificial dissipation term.

Let us consider the standard second order discretization of the convection term

$$\left(a \frac{\partial u}{\partial x} \right)_i = -\frac{a}{2\Delta x}(u_{i+1} - u_{i-1}) \tag{9.3.67}$$

to which we add a fourth order artificial dissipation term (AVT4) of the form:

$$\gamma \Delta x^3 \left(\frac{\partial^4 u}{\partial x^4} \right)_i \cong \frac{\gamma}{\Delta x}(u_{i+2} - 4u_{i+1} + 6u_i - 4u_{i-1} + u_{i-2}) \tag{9.3.68}$$

As the truncation error of the central discretization is of second order, the added fourth order term should not affect the global accuracy of the simulation. Actually, we solve the model

$$u_t + au_x = \gamma \Delta x^3 \frac{\partial^4 u}{\partial x^4} \tag{9.3.69}$$

instead of the homogenous convection equation.

Let us investigate the influence of this term and the eventual stability condition on the dissipation coefficient γ, for several time integration methods. We proceed according to the steps outlined in the previous section:

- We first determine the eigenvalues of the combined space discretization, that is the Fourier symbol, leading to, with $\varepsilon = \gamma \Delta t / \Delta x$

$$\Omega \Delta t = -I\sigma \sin \phi - 4\varepsilon(1 - \cos \phi)^2 \tag{9.3.70}$$

- The eigenvalues deviate now from the imaginary axis, due to the real, negative contribution of the added dissipation terms. They are represented in the $\Omega \Delta t$-plane on Figure 9.3.12a, for CFL $=2$ and values of $\varepsilon = 1/20$, $1/8$, $1/5$ and $1/2$.
- Selecting the trapezium method and applying equation (9.3.63), we obtain the scheme:

$$u_i^{n+1} = u_i^n - \frac{\sigma}{4}[(u_{i+1} - u_{i-1})^{n+1} + (u_{i+1} - u_{i-1})^n]$$

$$+ \frac{\varepsilon}{2}[(u_{i+2} - 4u_{i+1} + 6u_i - 4u_{i-1} + u_{i-2})^{n+1}$$

$$+ (u_{i+2} - 4u_{i+1} + 6u_i - 4u_{i-1} + u_{i-2})^n] \tag{9.3.71}$$

- Deriving the stability curves, you can observe that they deviate from the unit circle when $\varepsilon = 0$, as shown in Figure 9.3.12b, for CFL $=2$ and values of $\varepsilon = 1/20$, $1/8$, $1/5$ and $1/2$, but they still remain fully inside the stability circle for all values of ε. This can also be seen from the stability condition $|z| \leq 1$, in particular for $\phi = \pi$, where the artificial dissipation reaches its maximum effect, leading to

$$z(\phi = \pi) = \frac{2 - 16\varepsilon}{2 + 16\varepsilon} \tag{9.3.72}$$

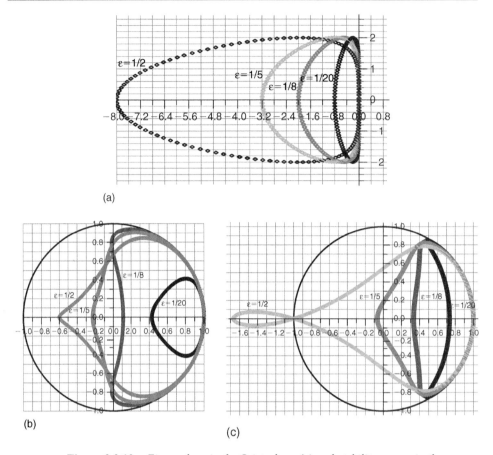

Figure 9.3.12 *Eigenvalues in the $\Omega\Delta t$-plane (a) and stability curves in the z-plane for the exact (b) and modified Jacobians (c) for CFL $= 2$; $\varepsilon = 1/20$, $1/8$, $1/5$ and $1/2$, for the central scheme with fourth order artificial dissipation and the trapezium method. This is the Beam and Warming scheme.*

- It is easily shown that the stability condition is satisfied for all positive values of the dissipation coefficient. *Hence, the trapezium scheme with central discretization and AVT remains unconditionally stable.*
- When the approximate, first order Jacobian is applied, the scheme is modified to

$$\Delta u_i^n + \sigma(\Delta u_i - \Delta u_{i-1})^n = -\frac{\sigma}{2}(3u_i - 4u_{i-1} + u_{i-2})^n$$

$$+ \varepsilon(u_{i+2} - 4u_{i+1} + 6u_i - 4u_{i-1} + u_{i-2})^n$$

(9.3.73)

- Applying now equation (9.3.65), for $\phi = \pi$, we obtain

$$z_{mod}(\phi = \pi) = \frac{1 + \sigma - 16\varepsilon}{1 + \sigma}$$

(9.3.74)

which leads to the stability condition on the dissipation coefficient

$$0 \leq \varepsilon \leq \frac{1 + \sigma}{8} \tag{9.3.75}$$

- Hence, the dissipation coefficient has to satisfy the condition $0 < \varepsilon < 1/8$, in order to satisfy the stability condition for all values of the CFL number. This is not a very severe condition, since the ε-parameter has to remain small for reasons of accuracy.
- The corresponding stability diagram is shown on Figure 9.3.12c, for the same values of the parameters, where the instability for $\varepsilon = 1/2$ can clearly be seen, as the corresponding curve is partly outside the stability circle.

The combination of space centered discretizations with fourth order Runge–Kutta methods is a very widely applied scheme for the convection terms for general Euler and Navier–Stokes equations, following its introduction by Jameson et al. (1981). It forms the basis for many CFD codes and the addition of a fourth order dissipation plays an important role in the elimination of high frequency errors.

Figure 9.3.13a shows the stability domains in the $\Omega\Delta t$-plane, for the values CFL $= 2$ and $\varepsilon = 1/20$; $1/8$; $1/5$, superimposed on the stability domains of the second, third and fourth order Runge–Kutta methods. You can clearly see that the value $\varepsilon = 1/5$ leads to an unstable scheme, as the value on the negative real axis, for $\phi = \pi$, equal to -16ε, is outside the stability region of the fourth order RK method. The stability condition is easily derived, if we denote by $-B$ the intersection of the RK stability curve with the real axis, leading to

$$\varepsilon \leq \frac{B}{16} \tag{9.3.76}$$

for stability. For the fourth order RK method, we have $B \approx 2.78$, giving the condition $\varepsilon < 0.174$.

Figure 9.3.13b and c show in addition the effect of the artificial dissipation terms on the diffusion error of the scheme, in particular on the high frequency behavior, the scheme being now dissipative. The dissipation error larger than one for the highest ε-value reflects the instability of the scheme for this value.

HANDS-ON TASK 6

Apply the methodology described above to analyze combinations of a set of time integration methods, with various discretizations of the convection and diffusion model equations, along the lines of Section 9.3.5.

Follow the steps as outlined with the objective to understand the properties in terms of stability limits and accuracy in function of the scheme parameters:

- Comment and analyze these results.

Apply symbolic mathematical software tools, such as MAPLE, Mathematica or MATLAB.

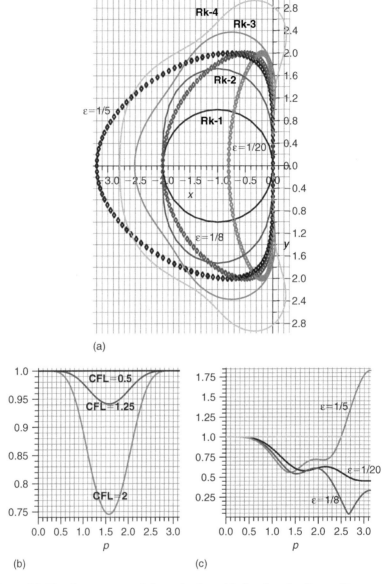

(a)

(b)

(c)

Figure 9.3.13 *Space centered discretization with fourth order Runge–Kutta method and fourth order artificial dissipation. (a) Eigenvalues in the $\Omega \Delta t$-plane for CFL = 2, $\varepsilon = 1/20$, 1/8 and 1/5; (b) dissipation error curves for $\varepsilon = 0$, for CFL = 0.5, 1.25 and 2, compared to the curves (c) at CFL = 2 for dissipation coefficients of 1/20, 1/8 and 1/5. The last value leads to an unstable scheme.*

HANDS-ON TASK 7

Experiment with several of the methods discussed in this chapter by applying them to solve the test cases of Chapter 7, as described in the Hands-On Tasks 2 and 3, as well as the cases of Problems P.7.5:

- Refer also to Problems P.9.13, P.9.14, P.9.19, P.9.20 as part of this task.
- Experiment by yourself with various combinations of time integration methods and space discretizations for the convection and diffusion equations.
- Write a general program to solve the mentioned problems.

A9.4 IMPLICIT SCHEMES FOR MULTIDIMENSIONAL PROBLEMS: APPROXIMATE FACTORIZATION METHODS[*]

When implicit schemes are applied to multidimensional problems, the resulting implicit matrix system is not tridiagonal anymore as for three point discretizations on one-dimensional equations. For instance, the two-dimensional parabolic diffusion equation, with constant diffusion coefficient α,

$$u_t = \alpha(u_{xx} + u_{yy}) \tag{9.4.1}$$

discretized with a five point finite difference Laplace scheme, leads to the system ($\Delta x = \Delta y$):

$$\frac{du_{ij}}{dt} = \frac{\alpha}{\Delta x^2}(u_{i+1,j} + u_{i-1,j} + u_{i,j+1} + u_{i,j-1} - 4u_{ij}) \tag{9.4.2}$$

With a backward Euler scheme for instance, we obtain a pentadiagonal matrix system

$$u_{ij}^{n+1} - u_{ij}^n = \frac{\alpha\Delta t}{\Delta x^2}(u_{i+1,j}^{n+1} + u_{i-1,j}^{n+1} + u_{i,j+1}^{n+1} + u_{i,j-1}^{n+1} - 4u_{ij}^{n+1}) \tag{9.4.3}$$

On an arbitrary mesh, or with higher order discretizations, or with finite elements, more mesh points appear in the Laplace discretization at point i, j, and the implicit matrix will have a more complicated structure than the pentadiagonal one.

The principle of the *Approximate Factorization* or *Alternating Direction Implicit* method (*ADI*), is to separate the operators into one-dimensional components and split the scheme into two (or three, for three-dimensional problems) steps, each one involving only the implicit operations originating from a single coordinate. This method has been introduced by Peaceman and Rachford (1955), Douglas and Rachford (1956) and generalized by Douglas and Gunn (1964). Many developments and extensions have been brought to this approach by Russian authors and given different names, *Fractional step method* by Yanenko (1971) or *Splitting method* by Marchuk (1975). An excellent description of ADI methods can be found in Mitchell (1969), Mitchell and Griffiths (1980).

If the matrix operator S on the right-hand side of

$$\frac{dU}{dt} = S \cdot U + Q \tag{9.4.4}$$

is separated into submatrices acting on the U components in a single direction, that is $S = S_x + S_y + S_z$, each operator in the right-hand side acting on the variable indicated as subscript, then equation (9.4.4) becomes

$$\frac{dU}{dt} = (S_x + S_y + S_z) \cdot U + Q \tag{9.4.5}$$

In equation (9.4.2), S_x and S_y represent the second order derivatives in the x and y direction, respectively, with the shift operators E_x and E_y,

$$S_x \cdot U = \frac{\alpha}{\Delta x^2}(E_x - 2 + E_x^{-1})U \tag{9.4.6}$$

$$S_y \cdot U = \frac{\alpha}{\Delta y^2}(E_y - 2 + E_y^{-1})U \tag{9.4.7}$$

or explicitly

$$S_x \cdot u_{ij} = \frac{\alpha}{\Delta x^2}(u_{i+1,j} - 2u_{ij} + u_{i-1,j}) \tag{9.4.8}$$

$$S_y \cdot u_{ij} = \frac{\alpha}{\Delta y^2}(u_{i,j+1} - 2u_{ij} + u_{i,j-1}) \tag{9.4.9}$$

With the implicit scheme (9.3.13), with $\xi = 0$, $\theta = 1$, defining the first order in time, backward Euler method, we obtain

$$U^{n+1} - U^n = \Delta t(S_x + S_y + S_z) \cdot U^{n+1} + Q\Delta t \tag{9.4.10}$$

The implicit operators appear from

$$[1 - \Delta t(S_x + S_y + S_z)] \cdot U^{n+1} = U^n + Q\Delta t \tag{9.4.11}$$

The basic idea behind the ADI method consists in a factorization of the right-hand side operator in a product of one-dimensional operators. This equation is replaced by

$$[(1 - \tau\Delta t S_x)(1 - \tau\Delta t S_y)(1 - \tau\Delta t S_z)] \cdot U^{n+1} = U^n + Q\Delta t \tag{9.4.12}$$

where τ is a free parameter. Developing equation (9.4.12), leads to

$$U^{n+1} - U^n = \tau\Delta t(S_x + S_y + S_z) \cdot U^{n+1} + \tau Q\Delta t$$
$$- \tau^2 \Delta t^2(S_x S_y + S_y S_z + S_z S_x) \cdot U^{n+1} + \tau^3 \Delta t^3(S_x S_y S_z) \cdot U^{n+1} \tag{9.4.13}$$

to be compared with equation (9.4.10).

The factorization (9.4.12) has introduced two additional terms, which represent errors with respect to the original scheme to be solved. However, these are higher order errors, proportional to Δt^2 and Δt^3, and since the backward Euler method is first order in time, these error terms are of the same order as the truncation error and do not affect the overall accuracy of the scheme.

The parameter τ appears as a relaxation parameter and has to be taken equal to one, if this scheme is to be used for time-dependent simulations.

However, the ADI technique is mostly of application for stationary problems, whereby one attempts to reach the convergence limit as fast as possible. In this connection, the relaxation parameter can be chosen to accelerate this convergence process, since it represents a scaling of the time step (see Chapter 10 for more details).

The factorized scheme is then solved in three steps:

$$(1 - \tau \Delta t S_x) \cdot \overline{U^{n+1}} = [1 + \tau \Delta t (S_y + S_z)] \cdot U^n + \tau \Delta t Q$$

$$(1 - \tau \Delta t S_y) \cdot \overline{\overline{U^{n+1}}} = \overline{U^{n+1}} - \tau \Delta t S_y U^n \qquad (9.4.14)$$

$$(1 - \tau \Delta t S_z) \cdot U^{n+1} = \overline{\overline{U^{n+1}}} - \tau \Delta t S_z U^n$$

Introducing the variations

$$\Delta U^n = U^{n+1} - U^n \quad \overline{\Delta U^n} = \overline{U^{n+1}} - U^n \quad \overline{\overline{\Delta U^n}} = \overline{\overline{U^{n+1}}} - U^n$$

the ADI scheme can be rewritten as

$$(1 - \tau \Delta t S_x) \cdot \overline{\Delta U^n} = \tau \Delta t (S \cdot U^n + Q)$$

$$(1 - \tau \Delta t S_y) \cdot \overline{\overline{\Delta U^n}} = \overline{\Delta U^n} \qquad (9.4.15)$$

$$(1 - \tau \Delta t S_z) \cdot \Delta U^n = \overline{\overline{\Delta U^n}}$$

This is sometimes called the **Δ-formulation**.

By recombining the factors, the ADI approximation can be redefined by the formulation:

$$[(1 - \tau \Delta t S_x)(1 - \tau \Delta t S_y)(1 - \tau \Delta t S_z)] \cdot \Delta U^n = \tau \Delta t (S \cdot U^n + Q) \qquad (9.4.16)$$

For $\tau = 1$, we obtain the **Douglas–Rachford scheme**.

The **Splitting or fractional step method**, leads to another ADI formulation, based on a factorization of the Crank–Nicholson scheme and is therefore second order in time. Equation (9.4.10) is replaced by

$$U^{n+1} - U^n = \Delta t (S_x + S_y + S_z) \cdot \frac{U^{n+1} + U^n}{2} + Q \Delta t \qquad (9.4.17)$$

or

$$\left[1 - \frac{\Delta t}{2} (S_x + S_y + S_z) \right] \cdot U^{n+1} = \left[1 + \frac{\Delta t}{2} (S_x + S_y + S_z) \right] \cdot U^n + Q \Delta t \qquad (9.4.18)$$

This equation is factorized as follows:

$$\left[\left(1 - \frac{\Delta t}{2} S_x \right) \left(1 - \frac{\Delta t}{2} S_y \right) \left(1 - \frac{\Delta t}{2} S_z \right) \right] \cdot U^{n+1}$$

$$= \left[\left(1 + \frac{\Delta t}{2} S_x \right) \left(1 + \frac{\Delta t}{2} S_y \right) \left(1 + \frac{\Delta t}{2} S_z \right) \right] \cdot U^n + Q \Delta t \qquad (9.4.19)$$

and represents an approximation to equation (9.4.17) of second order accuracy, since equation (9.4.19) is equal to

$$\left[1 - \frac{\Delta t}{2}(S_x + S_y + S_z)\right] \cdot U^{n+1} = \left[1 + \frac{\Delta t}{2}(S_x + S_y + S_z)\right] \cdot U^n + Q\Delta t$$

$$- \frac{\Delta t^2}{4}(S_x S_y + S_y S_z + S_z S_x) \cdot (U^{n+1} - U^n) + \frac{\Delta t^3}{8}(S_x S_y S_z) \cdot (U^{n+1} - U^n)$$

$$(9.4.20)$$

Since $(U^{n+1} - U^n)$ is of order Δt, the error terms are $0(\Delta t^3)$. Equation (9.4.19) is then solved as a succession of one-dimensional Crank–Nicholson schemes

$$\left(1 - \frac{\Delta t}{2}S_x\right) \cdot \overline{U^{n+1}} = \left(1 + \frac{\Delta t}{2}S_x\right) \cdot U^n + \Delta t Q$$

$$\left(1 - \frac{\Delta t}{2}S_y\right) \cdot \overline{\overline{U^{n+1}}} = \left(1 + \frac{\Delta t}{2}S_y\right) \cdot \overline{U^{n+1}} \qquad (9.4.21)$$

$$\left(1 - \frac{\Delta t}{2}S_z\right) \cdot U^{n+1} = \left(1 + \frac{\Delta t}{2}S_z\right) \cdot \overline{\overline{U^{n+1}}}$$

When the S_i commute we recover (9.4.19) by elimination of $\overline{\overline{U^{n+1}}}$ and $\overline{U^{n+1}}$. If the S_i operators do not commute, the approximation (9.4.21) is still valid, but is reduced to first order in time.

A9.4.1 Two-Dimensional Diffusion Equation

Considering equation (9.4.1), with the central space discretizations, we can write the factorized ADI scheme (9.4.14) as

$$(1 - \tau \Delta t S_x) \cdot \overline{U^{n+1}} = (1 + \tau \Delta t S_y) \cdot U^n$$

$$(1 - \tau \Delta t S_y) \cdot U^{n+1} = \overline{U^{n+1}} - \tau \Delta t S_y U^n \qquad (9.4.22)$$

Written out explicitly, we obtain the following tridiagonal systems, with $\Delta x = \Delta y$ and $\beta = \alpha \Delta t / \Delta x^2$,

$$\overline{u_{ij}^{n+1}} - \tau\beta(\overline{u_{i+1,j}^{n+1}} - 2\overline{u_{ij}^{n+1}} + \overline{u_{i-1,j}^{n+1}}) = u_{ij}^n + \tau\beta(u_{i,j+1}^n - 2u_{ij}^n + u_{i,j-1}^n)$$

$$u_{ij}^{n+1} - \tau\beta(u_{i,j+1}^{n+1} - 2u_{ij}^{n+1} + u_{i,j-1}^{n+1}) = \overline{u_{ij}^{n+1}} - \tau\beta(u_{i,j+1}^n - 2u_{ij}^n + u_{i,j-1}^n)$$

$$(9.4.23)$$

The first equation is solved as a tridiagonal system along all the horizontal j-lines, sweeping the mesh from $j = 1$ to $j = j_{max}$. After this first step, the intermediate solution $\overline{u_{ij}^{n+1}}$ is obtained at all mesh points. The second equation represents a succession of tridiagonal systems along the i-columns and the solution u_{ij}^{n+1} is obtained after having swept through the mesh along the vertical lines, from $i = 1$ to $i = i_{max}$.

A Von Neumann stability analysis can be performed for this system by defining an intermediate amplification factor $\overline{G}$, by

$$\overline{U^{n+1}} = \overline{G} \cdot U^n \tag{9.4.24}$$

For the first step, with ϕ_x and ϕ_y being the Fourier variables in the x- and y-directions, respectively, we have

$$\overline{G} = \frac{1 - 4\tau\beta \sin^2 \phi_y/2}{1 + 4\tau\beta \sin^2 \phi_x/2} \tag{9.4.25}$$

$$G = \frac{\overline{G} + 4\tau\beta \sin^2 \phi_y/2}{1 + 4\tau\beta \sin^2 \phi_y/2} \tag{9.4.26}$$

and by combining these two equations

$$G = \frac{1 + 16\tau^2\beta^2 \sin^2 \phi_x/2 \cdot \sin^2 \phi_y/2}{(1 + 4\tau\beta \sin^2 \phi_x/2)(1 + 4\tau\beta \sin^2 \phi_y/2)} \tag{9.4.27}$$

The scheme is clearly unconditionally stable since $|G| < 1$. A similar calculation for three dimensions confirms this property.

Another version of the ADI technique is the *Peaceman–Rachford* method, based on the following formulation, with $\tau = 1/2$, in two dimensions

$$\left(1 - \frac{\Delta t}{2} S_x\right) \cdot \overline{U^{n+1}} = (1 + \tau\Delta t S_y) \cdot U^n$$

$$\left(1 - \frac{\Delta t}{2} S_y\right) \cdot U^{n+1} = \left(1 + \frac{\Delta t}{2} S_x\right) \cdot \overline{U^{n+1}} \tag{9.4.28}$$

We leave it to you to show that this scheme is also unconditionally stable for the diffusion equation of the previous example.

In addition, this form of the ADI method is second order accurate in time, as can be seen by eliminating the intermediate solution $\overline{U^{n+1}}$. We obtain

$$U^{n+1} - U^n = \Delta t(S_x + S_y) \cdot \frac{U^{n+1} + U^n}{2} - \frac{\Delta t^2}{4} S_x S_y \cdot (U^{n+1} - U^n) \tag{9.4.29}$$

and since $U^{n+1} - U^n$ is of order Δt, the last term is $0(\Delta t^3)$. The intermediate values $\overline{\Delta U}$ and $\overline{\overline{\Delta U}}$ have not necessarily a physical meaning and boundary conditions have to be defined for these variables in accordance with the physical boundary conditions of U.

Boundary conditions for the intermediate steps can be obtained from the structure of the system (9.4.13). In particular for Dirichlet conditions, we would write for the boundary values $(\ )_B$:

$$\overline{\overline{\Delta U}}\Big|_B = (1 - \tau\Delta t S_z) \cdot (U^{n+1} - U^n)_B$$

$$\overline{\Delta U}\Big|_B = (1 - \tau\Delta t S_y)(1 - \tau\Delta t S_z) \cdot (U^{n+1} - U^n)_B \tag{9.4.30}$$

More details can be found in Mitchell and Griffiths (1980).

A9.4.2 ADI Method for the Convection Equation

Considering the three-dimensional convection equation

$$u_t + au_x + bu_y + cu_z = 0 \tag{9.4.31}$$

the operators S_x, S_y and S_z can be written, for a second order central difference of the space derivatives, as

$$\Delta t S_x u_{ijk} = -\frac{a\Delta t}{2\Delta x}(u_{i+1,jk} - u_{i-1,jk}) \equiv -\frac{\sigma_x}{2}(u_{i+1,jk} - u_{i-1,jk})$$

$$\Delta t S_y u_{ijk} = -\frac{b\Delta t}{2\Delta y}(u_{i,j+1,k} - u_{i,j-1,k}) \equiv -\frac{\sigma_y}{2}(u_{i,j+1,k} - u_{i,j-1,k}) \tag{9.4.32}$$

$$\Delta t S_z u_{ijk} = -\frac{c\Delta t}{2\Delta z}(u_{ij,k+1} - u_{ij,k-1}) \equiv -\frac{\sigma_z}{2}(u_{ij,k+1} - u_{ij,k-1})$$

Considering first the *two-dimensional* case, the ADI scheme (9.4.14) can be written

$$\overline{u_{ij}^{n+1}} + \tau\frac{\sigma_x}{2}(\overline{u_{i+1,j}^{n+1}} - \overline{u_{i-1,j}^{n+1}}) = u_{ij}^n - \tau\frac{\sigma_y}{2}(u_{i,j+1} - u_{i,j-1})$$

$$u_{ij}^{n+1} + \tau\frac{\sigma_y}{2}(u_{i,j+1}^{n+1} - u_{i,j-1}^{n+1}) = \overline{u_{ij}^{n+1}} + \tau\frac{\sigma_y}{2}(u_{i,j+1}^n - u_{i,j-1}^n) \tag{9.4.33}$$

The amplification factors are given by

$$\overline{G} = \frac{1 - I\tau\sigma_y \sin\phi_y}{1 + I\tau\sigma_x \sin\phi_x} \tag{9.4.34}$$

$$G = \frac{\overline{G} + I\tau\sigma_y \sin\phi_y}{1 + I\tau\sigma_y \sin\phi_y} = \frac{1 - \tau^2\sigma_x\sigma_y \sin\phi_x \sin\phi_y}{(1 + I\tau\sigma_x \sin\phi_x)(1 + I\tau\sigma_y \sin\phi_y)} \tag{9.4.35}$$

Since G is of the form

$$G = \frac{1 - \tau^2\sigma_x\sigma_y \sin\phi_x \sin\phi_y}{1 - \tau^2\sigma_x\sigma_y \sin\phi_x \sin\phi_y + I\tau(\sigma_x \sin\phi_x + \sigma_y \sin\phi_y)} \tag{9.4.36}$$

we have always $|G| \leq 1$ and the scheme is unconditionally stable.

However, in three dimensions, this is not the case anymore; the ADI scheme applied to the three-dimensional convection equation, centrally discretized (without dissipation contributions) is unconditionally unstable as shown by Abarbanel et al. (1982).

In three dimensions, the above procedure leads to the following amplification matrix writing $s_x = \sigma_x \sin\phi_x$ and similarly for the y and z components,

$$G = \frac{1 - \tau^2(s_x s_y + s_y s_z + s_z s_x) - I\tau^3 s_x s_y s_z}{(1 + I\tau s_x)(1 + I\tau s_y)(1 + I\tau s_z)} \tag{9.4.37}$$

It can be shown that there are always values of s_x, s_y, s_z such that $|G| > 1$. The following proof is due to Saul Abarbanel and Eli Turkel (private communication).

The amplification matrix G is written as

$$G = \frac{\alpha_1 + I\beta_1}{\alpha_1 + I\beta_2} \tag{9.4.38}$$

and since the real parts are equal, stability is obtained if

$$|\beta_1| \leq |\beta_2| \tag{9.4.39}$$

With

$$\beta_2 = \beta_1 + \tau(s_x + s_y + s_z) \equiv \beta_1 + \gamma$$

the condition (11.4.39) implies

$$(\beta_1 + \gamma)^2 \geq \beta_1^2$$

or

$$\gamma(2\beta_1 + \gamma) \geq 0$$

If we select values of ϕ_x, ϕ_y, ϕ_z such that $\gamma > 0$, assuming $\tau > 0$, the stability condition (9.4.39) becomes

$$2\tau^2 s_x s_y s_z \leq (s_x + s_y + s_z) = \gamma/\tau \tag{9.4.40}$$

However, we can always find values of these variables which do not satisfy this inequality. For instance, take $\gamma = \varepsilon$ through

$$s_x = -1/4 \quad s_y = -1/4 \quad s_z = \varepsilon + 1/2 \tag{9.4.41}$$

the above condition becomes

$$\varepsilon \geq \frac{\tau^2}{16}(1 + 2\varepsilon)$$

or

$$\tau^2 \geq \frac{16\varepsilon}{1 + 2\varepsilon} \tag{9.4.42}$$

We can always select a value of ε sufficiently small, such that this condition is never satisfied for any fixed finite value of τ, since the right-hand side goes to zero with ε. For instance, for $\tau = 1$, $\varepsilon < 1/14$ leads to an unstable scheme.

This instability is associated with low frequencies, since equation (9.4.41) implies that $s_x + s_y + s_z = \varepsilon/\tau$ is a small quantity and can only be satisfied by small values of the wave numbers ϕ_x, ϕ_y, ϕ_z. In addition it can be considered as a weak instability since, from equation (9.4.38), for ε small

$$|G|^2 = \frac{\alpha_1^2 + \beta_1^2}{\alpha_1^2 + (\beta_1 + \varepsilon)^2} \approx 1 - \frac{2\varepsilon\beta_1}{\alpha_1^2 + \beta_1^2} \tag{9.4.43}$$

where β_1 is defined, in the assumption (9.4.41), by

$$\beta_1 = -\frac{\tau^3}{16}\left(\frac{1}{2}+\varepsilon\right) < 0$$

and is a small negative quantity. Hence, $|G|$ is higher than one by an amount proportional to ε and therefore the amplification of errors should remain limited. This is confirmed by computations performed by Compton and Whitesides (1983) with the full system of Euler equations. However, the addition of adequate damping terms can remove the instability as shown by Abarbanel et al. (1982).

Actually, the three-dimensional fractional step method (9.4.21) is stable in three dimensions for the convection equation, although neutrally stable since it is a product of one-dimensional Crank–Nicholson schemes.

An essential difference between the fractional step method and the ADI method in its form (9.4.15), is connected to their behavior for stationary problems. In this case, convergence toward steady state $\Delta U^n = 0$ is sought implying $(SU^n + Q) = 0$, according to equation (9.4.16), when this limit is reached. Hence, the steady state limit resulting from the computation will be independent of the time step Δt, since S is a pure space discretization. Equation (9.4.20), on the other hand, shows that in the limit $\Delta U^n \to 0$, we solve, with $(U^{n+1} + U^n)/2 \approx U^n$

$$\left[1 - \frac{\Delta t^2}{4}(S_xS_y + S_yS_z + S_zS_x)\right]\cdot \Delta U^n = \Delta t(S\cdot U^n + Q) + \frac{\Delta t^3}{4}S_xS_yS_z\cdot U^n$$

(9.4.44)

which produces a stationary solution with a vanishing right-hand side. That is, the obtained stationary solution satisfies

$$S\cdot U^n + Q = -\frac{\Delta t^2}{4}S_xS_yS_z\cdot U^n \qquad (9.4.45)$$

and is function of the time step. In addition, for large time steps, which is what we aim at with implicit methods, the right hand side might become unacceptably high.

CONCLUSIONS AND MAIN TOPICS TO REMEMBER

This chapter generalizes the approach of the previous Chapter 7, by extending the range of schemes toward the modern approach of separate space and time discretizations, as opposed to the Lax–Wendroff family where the time and space discretizations are intertwined.

The essential outcome of this chapter is the methodology enabling you to analyze the association of a time integration method to a pre-selected space discretization. The developed methodology establishes the stability conditions and the accuracy, in terms of dispersion and diffusion errors. We summarize the methodology as follows:

- Select a space discretization method for the flux terms and define the semi-discretized system of ODEs in time.

- Apply a Fourier mode analysis to the space discretization to derive its spectrum of eigenvalues (its Fourier symbol) in function of the phase angle $\phi = k\Delta x$, covering the range $(-\pi, +\pi)$.
- Verify that the real part of the eigenvalues is in the negative part of the $\Omega\Delta t$-plane, for the space discretization to be acceptable.
- Select a time integration method for the system of semi-discretized system of ordinary differential equations (ODEs) in time.
- Apply the selected method to the canonical form of the system of ODE's in time $dw/dt = \Omega w$ and derive the stability domain of the time integration method in the complex $\Omega\Delta t$-plane.
- Verify now the compatibility between the space and the time discretizations, by plotting the trajectories of the eigenvalues of the space discretization in the $\Omega\Delta t$-plane, or in the z-plane. The conditions ensuring that these eigenvalues remain fully within the time integration stability domain will determine the stability conditions of the combined scheme.
- Investigate the dissipation and dispersion errors of the combined scheme, to determine the accuracy level of the selected combination, by analyzing the amplification factor behavior in function of the phase angle.
- A most important topic to remember is related to the amount of dissipation of the space discretization scheme, with the necessity of adding artificial dissipation terms to centrally discretized convection terms.

A distinction has to be made between steady and unsteady simulations:

- For steady problems, the transient is not physical and should decrease as fast as possible. Hence, large dissipation errors are favorable for convergence and dispersion errors are not significant.
- Unsteady simulations are much more demanding in terms of accuracy as they require very low dispersion and diffusion errors. Refer you to the relevant discussion of Section 7.4 and the final conclusions of this chapter.
- Implicit methods, at high CFL, have very strong numerical dissipation, which is excellent for steady problems, since all modes will be rapidly damped, leading to fast convergence. For unsteady problems, this is unacceptable and a severe restriction of the maximum CFL might be required to maintain the desired accuracy.

REFERENCES

Abarbanel, S., Dwoyer, D.D. and Gottlieb, D. (1982). Stable implicit finite difference methods for three dimensional hyperbolic systems. ICASE Report 82-39. ICASE, NASA Langley RC, Hampton, Virginia, USA.

Beam, R.M. and Warming, R.F. (1976). An implicit finite-difference algorithm for hyperbolic systems in conservation law form. *J. Comput. Phys.*, 22, 87–109.

Beam, R.M. and Warming, R.F. (1978). An implicit factored scheme for the compressible Navier–Stokes equations. *AIAA J.*, 16, 393–402.

Beam, R.M. and Warming, R.F. (1980). Alternating direction implicit methods for parabolic equations with a mixed derivative. *SIAM J. Sci. Stat. Comp.*, 1, 131–159.

Beam, R.M. and Warming, R.F. (1982). Implicit Numerical Methods for the Compressible Navier–Stokes and Euler Equations. *Von Karman Institute Lecture Series 1982-04*, Rhode Saint Genese, Belgium.

Brailovskaya, I. (1965). A difference scheme for numerical solutions of the two-dimensional nonstationary Navier–Stokes equations for a compressible gas. *Sov. Phys. Dokl.*, 10, 107–110.

Briley, W.R. and Mc Donald, H. (1975). Solution of the three-dimensional compressible Navier–Stokes equations by an implicit technique. *Proceedings of the Fourth International Conference on Numerical Methods in Fluid Dynamics, Lecture Notes in Physics*, Vol. 35, Springer, Berlin.

Briley, W.R. and Mc Donald, H. (1980). On the structure and use of linearized block implicit schemes. *J. Comp. Phys.*, 6, 428–453.

Compton II, W.B. and Whitesides, J.L. (1983). Three-dimensional Euler solutions for long duct nacelles. *AIAA 21st Aerospace Sciences Meeting*, AIAA Paper 83-0089.

Dahlquist, G. (1963). *A Special Stability Problem for Linear Multistep Methods*. BIT, Copenhagen, Vol. 3, 27–43.

Dahlquist, G. and Bjorck, A. (1974). *Numerical Methods*. Prentice Hall, New Jersey.

Douglas, J. and Gunn, J.E. (1964). A general formulation of alternating direction methods: parabolic and hyperbolic problems. *Numerische Mathematik*, 6, 428–453.

Douglas J. and Rachford, H.H. (1956). On the numerical solution of heat conduction problems in two and three space variables. *Trans. Am. Math. Soc.*, Vol. 82, 421–439.

Gear, G.W. (1971). *Numerical Initial Value Problems in Ordinary Differential Equations*. Prentice-Hall, New Jersey.

Gustafsson, B. and Kreiss, H.O. (1979). Boundary conditions for time dependent problems with an artificial boundary. *J. Comput. Phys.*, 30, 333–351.

Gustafsson, B., Kreiss, H.O. and Sundstrsm, A. (1972). Stability theory of difference approximations for mixed initial boundary value problems. *Math. Comput.*, 26, 649–686.

Hu, F.Q., Hussaini, M.Y. and Manthey, J.L. (1996). Low-dissipation and low-dispersion Runge–Kutta schemes for computational acoustics. *J. Comput. Phys.*, 124, 177–191.

Hairer, E., Wanner, G. (1996). *Solving Ordinary Differential Equations II – Stiff and Differential-Algebraic Problems*. Springer Verlag, Berlin.

Hairer, E., Norsett, S.P. and Wanner, G. (1993). *Solving Ordinary Differential Equations I – Non-stiff Problems*. Springer Verlag, Berlin.

Hundsdorfer, W. and Verwer, J.G. (2003). *Numerical Solution of Time-Dependent Advection–Diffusion-Reaction Equations*. Springer Verlag, Berlin.

Jameson, A. and Baker, T.J. (1983). Solutions of the Euler equations for complex configurations. *AIAA 6th Computational Fluid Dynamics Conference*. AIAA Paper 83-1929.

Jameson, A. and Baker, T.J. (1984). Multigrid solution of the Euler equations for aircraft configurations. AIAA Paper 84-0093, *AIAA 22nd Aerospace Sciences Meeting*.

Jameson, A., Schmidt, W., and Turkel, E. (1981). *Numerical simulation of the Euler equations by finite volume methods using Runge–Kutta time stepping schemes. AIAA 5th Computational Fluid Dynamics Conference*, AIAA Paper 81-1259.

Kreiss, H.O. (1968). Stability theory for difference approximations of mixed initial boundary value problems, I. Math. *Comput.*, 22, 703–714.

Kreiss, H.O. (1970). Initial boundary value problem for hyperbolic systems. *Comm. Pure and Appl. Math.*, 23, 273–298.

Kurbatskii, K.A. and Mankbadi, R.R. (2004). Review of Computational Aeroacoustics Algorithms. *Int. J. Comput. Fluid Dyn.*, 18(6), 533–546.

Lambert, J.D. (1973). *Computational Methods in Ordinary Differential Equations.* J. Wiley & Sons, New York.

Lambert, J.D. (1991). *Numerical Methods for Ordinary Differential Equations.* J. Wiley & Sons, New York.

Mc Cormack, R.W. (1969). The effect of viscosity in hypervelocity impact cratering. *AIAA Paper 69-354.*

Mc Cormack, R.W. (1971). Numerical solution of the interaction of a shock wave with a laminar boundary layer. *Proceedings of the Second International Conference on Numerical Methods in Fluid Dynamics,* Lecture Notes in Physics, Vol. 8, Springer, Berlin, pp. 151–163.

Mitchell, A.R. (1969). *Computational Methods in Partial Differential Equations.* J. Wiley & Sons, New York.

Mitchell, A.R. and Griffiths, D.F. (1980). *The Finite Difference Method in Partial Differential Equations.* J. Wiley & Sons, New York.

Peaceman, D.W. and Rachford, H.H. (1955). The numerical solution of parabolic and elliptic differential equations. *SIAM Journal,* 3, 28–41.

Pulliam, T.H. (1984). Euler and thin layer Navier–Stokes codes: ARC2D, ARC3D. *Proceedings of the Computational Fluid Dynamics User's Workshop.* The University of Tennessee Space Institute Tullahoma, Tennessee.

Ramboer, J., Smirnov, S., Broeckhoven, T. and Lacor, C. (2006). Optimization of time integration schemes coupled to spatial discretization for use in CAA applications. *J. Comput. Phys.,* 213(2), 777–802.

Sonneveld, P. and Van Leer, B. (1984). A minimax problem along the imaginary axis. Nieuw Archief voor Wiskunde, Vol. 3, pp. 19–22.

Tam, C. (2004). Computational aeroacoustics: an overview of computational challenges and applications. *Int. J. Comput. Fluid Dynamics,* 18(6), 547–567.

Van der Houwen, P.J. (1977). *Construction of Integration Formulas for Initial Value Problems.* North Holland Publ. Co, Amsterdam.

Vichnevetsky, R. (1983). New Stability Theorems Concerning One-Step Numerical Methods for Ordinary Differential Equations. *Math. Comput. Simul.,* 25, 199–211.

Yanenko, N.N. (1979). *The Method of Fractional Steps.* Springer Verlag, New York.

Yee, H.C. (1981). Numerical approximation of boundary conditions with applications to inviscid gas dynamics. *NASA Report TM-81265.*

PROBLEMS

P.9.1 Derive the discretization matrix S for the diffusion equation $u_t = au_{xx}$ with the conditions (9.1.9), by applying a central difference at $x = 0$, between $i = -1$ and $i = 1$. Use this equation to eliminate u_{-1} in the equation written for u_0. Obtain the matrix equation for the vector $U^T = \{u_0, u_1, \ldots, u_{N-1}\}$.

P.9.2 Analyze the stability of the two-step (three-level) method, corresponding to the multistep parameters of Table 9.3.1, $\theta = 1/2$; $\xi = -5/6$; $\phi = -1/2$, leading to

$$U^{n+1} + 4U^n - 5U^{n-1} = 3\Delt(2H^{n+1} + H^n)$$

Calculate the characteristic polynomial, its roots and display the stability boundary $|z| = 1$. Plot also the lines of constant amplification factor with values lower (stable) and higher (unstable) than 1. Show that the scheme is always unstable, as the instability region of the time integration method covers part of the negative Ω-domain. Verify also by a Taylor expansion of the roots $z(\Omega)$ that the spurious root is not contained in the domain $|z| < 1$.

P.9.3 Consider the two-step (three-level) Adams–Bashworth method

$$U^{n+1} - U^n = \Delta t(3H^n - H^{n-1})/2$$

corresponding to the multistep parameters of Table 9.3.1, $\theta = 0$; $\xi = 0$; $\phi = 1/2$, applied to the diffusion equation with a central second order discretization as defined by equation (9.1.4), with periodic boundary conditions. Apply the procedures outlined in Section 9.3.5. Show that the resulting scheme is stable under the condition $\beta = \alpha \Delta t / \Delta x^2 \leq 1/4$. Write out the schemes in full.

P.9.4 Repeat the analysis of Problem P.9.3 by combining the Adams–Bashworth method with first order upwind (FOU) and with central discretizations of the convection terms.

Show that the resulting scheme is unstable for the central discretization of the convection terms and conditionally stable for the FOU discretization, under the condition CFL ≤ 0.5. Write out the schemes in full.

P.9.5 Consider the space operator of the Lax–Friedrichs scheme for the convection equation and the associated semi-discretized system of ODE's in time:

$$\frac{du_i}{dt} = -\frac{a}{2\Delta x}(u_{i+1} - u_{i-1}) + \frac{1}{2}(u_{i+1} - 2u_i + u_{i-1})$$

Determine the Fourier eigenvalues and represent them in the complex $\Omega \Delta t$-plane. Show that with an Euler explicit method, the scheme is stable under the CFL condition.

P.9.6 Repeat the analysis of Problem P.9.5 with the four stages Runge–Kutta method (9.3.51) and show that the resulting scheme is stable for a CFL limit close to 2.4. Determine the dispersion and diffusion errors for several CFL values.

Derive the equivalent differential equation, by identifying its dominating terms with the coefficients of the series expansion of the dispersion and diffusion errors in powers of ϕ. (refer to Section 8.1.3).

Consider the combination of the Runge–Kutta method with the first order upwind (FOU) scheme and repeat the analysis. Show that this combination is stable for CFL $< B$, where $-B$ is the abscissa where the stability domain cuts the real negative axis. Determine the dispersion and diffusion errors for several CFL values, for the fourth order Runge–Kutta mehtod.

P.9.7 Consider the general multistep method in the formulation of Beam and Warming, given by equation (9.3.6). Find the roots of the characteristic polynomial (9.3.7) and select the physical root. Apply a Taylor series development in powers of $\Omega \Delta t$ to this root and compare with the exact solution (9.2.13) in order to derive the relations (9.3.8)–(9.3.10).

P.9.8 Consider the scheme (9.3.61) and derive the Jacobian matrix based on the definition (9.3.3) and the general formulation (9.3.25). Write the Jacobian in operator form with the space shift operators $E^k u_i = u_{i+k}$.

Consider the approximate Jacobian (9.3.26), based on the first order upwind space discretization and write the corresponding scheme, based on the formulation (9.3.27)

Hint: Obtain

$$J(U) = -\frac{a}{2\Delta x}(3 - 4E^{-1} + E^{-2})u_i$$

$$J^{(1)}(U) = -\frac{a}{\Delta x}(1 - E^{-1})u_i$$

and the approximate Jacobian scheme

$$\Delta u_i^{n+1} + \sigma\Delta(u_i - u_{i-1})^{n+1} = -\frac{\sigma}{2}(3u_i - 4u_{i-1} + u_{i-2})^n$$

or

$$u_i^{n+1} + \sigma(u_i - u_{i-1})^{n+1} = u_i^n - \sigma(u_i - u_{i-1})^n - \frac{\sigma}{2}(3u_i - 4u_{i-1} + u_{i-2})^n$$

P.9.9 Consider the Gear backward differencing method with the Fromm second order upwind biased scheme for the convection equation.

Write down the complete form of the scheme.

Develop the analysis of the scheme and plot the stability curves for various values of the CFL number.

Show that the scheme remains unconditionally stable even with the approximate first order Jacobian. Compare the behavior of the diffusion and dispersion errors for the exact and approximate Jacobians at all values of the CFL numbers and notice particularly the behavior at large CFL.

P.9.10 Apply the Crank–Nicholson scheme (9.3.16) to the diffusion equation $u_t = \alpha u_{xx}$ with central second order space discretization. Show that the resulting scheme is unconditional stable for $\alpha > 0$. Draw plots of the dissipation error and compare with the dissipation error generated by the explicit forward Euler scheme. Observe the absence of dispersion errors.

Hint: Obtain with $\beta = \alpha\Delta t/\Delta x^2$ the equivalent differential equation

$$u_t - \alpha u_{xx} = \frac{\alpha\Delta x^2}{12}\left(\frac{\partial^4 u}{\partial x^4}\right) + \frac{\alpha\Delta x^4}{12}\left(\beta^2 + \frac{1}{30}\right)\left(\frac{\partial^6 u}{\partial x^6}\right)$$

P.9.11 Repeat Problem P.9.10 for the convection–diffusion equation $u_t + au_x = \alpha u_{xx}$ with a central difference discretization. Apply the global methodology for stability and error analysis.

Calculate the amplitude and phase errors in function of the scheme parameters, and draw plots of both quantities, in function of the phase angle. Observe that the scheme is not dissipative and therefore oscillations might appear in nonlinear problems. Derive the equivalent differential equation by identification with the series development of the dispersion and diffusion errors.

P.9.12 Apply the Euler implicit scheme to the second order space discretized diffusion equation. Determine the root of the characteristic polynomial and obtain the amplification factor. Draw plots of the dissipation error and compare with the results of Problem P.9.10. Determine the equivalent differential equation by identification with the series development of the dispersion and diffusion errors.

P.9.13 Solve the Problem P.8.6 with the upwind implicit Euler backward scheme for $\sigma = 0.5, 1, 2$, after 10, 20, 50, 150 time steps.

Generate a plot of the numerical solution and compare with the exact solution.

P.9.14 Solve the moving shock Problem of P.8.7 with the upwind implicit Euler scheme for $\sigma = 0.5, 1, 2$, after 10, 20, 50, 150 time steps.

Generate a plot of the numerical solution and compare with the exact solution.

P.9.15 Consider the predictor–corrector scheme (9.3.38) and show that is equivalent to the following one-step scheme with five-point support $i, i \pm 1, i \pm 2$:

$$u_i^{n+1} = u_i^n - \frac{\sigma}{2}(u_{i+1}^n - u_{i-1}^n) + \frac{\sigma^2}{4}(u_{i+2}^n - 2u_i^n + u_{i-2}^n)$$

Derive the stability CFL condition $|\sigma| < 1$ from the general methodology and analyze the dispersion and diffusion errors.

P.9.16 Show that equation (9.3.39) is obtained from the first order predictor–corrector method (9.3.37) when a first order upwind difference is applied to the linear convection equation.

Obtain from the general stability analysis the stability condition $0 < \sigma \leq 1/2$. Obtain and plot the dissipation and dispersion errors in function of phase angle. Determine also the equivalent differential equation and compare with the second order Warming and Beam scheme (8.2.30).

P.9.17 Repeat Problem P.9.16 for the scheme (9.3.41), derived from Henn's method applied with first order upwind space discretization.

P.9.18 Show that the Mc Cormack scheme (9.3.43) is identical to the Lax–Wendroff scheme. Write the scheme by taking a forward space difference in the predictor and a backward difference in the corrector. Analyze the stability by the method of the characteristic polynomial, and obtain the CFL condition for stability.

P.9.19 Solve Burgers equation with Mc Cormack's scheme (9.3.44), where $f_i = (u_i)^2/2$, for a stationary discontinuity $(+1, -1)$ and a moving discontinuity $(1, -0.5)$. Consider the values $CFL = 0.25, 0.5$ and 0.75 and analyze the solutions after 10, 20, 30 and 50 time steps.

P.9.20 Repeat Problem P.9.19 with the Euler implicit scheme and with the trapezoidal scheme. Observe the appearance of strong oscillations and explain their origin by the analysis of the amplification function and by the truncation error structure.

P.9.21 Solve the Problems P.8.6 and P.8.7 with Mc Cormack's scheme (9.3.44). Compare with the previous results.

P.9.22 Analyze the stability conditions for the convection–diffusion equation $u_t + au_x = \alpha u_{xx}$ discretized with central differences, when solved with a fourth order Runge–Kutta method. Determine the eigenvalues in the $\Omega \Delta t$-plane, for different CFL numbers σ and for constant values of $\beta = \alpha \Delta t / \Delta x^2$. Show that the stability condition on the diffusion coefficient is defined by $4\beta < B$, where $(-B)$ is the intersection of the RK stability curve with the real axis.

P.9.23 Consider a predictor–corrector method with an implicit corrector step, applied to the one-dimensional diffusion equation $u_t = \alpha u_{xx}$. The general form of the method is defined by

$$\overline{U}^{n+1} = U^n + \Delta t H^n$$

$$L \cdot (U^{n+1} - U^n) = \overline{U}^{n+1} - U^n$$

Take $L = 1 - \varepsilon \Delta x^2 \frac{\partial^2}{\partial x^2}$ as implicit operator, where ε is a free parameter and perform a central discretization.

Determine the stability properties of the resulting scheme, by applying the general methodology and show that the scheme is unconditional stable for all $\varepsilon > 0$. Show that the implicit corrector step increases the time step limit of the explicit corrector step, without modification of the spatial order of accuracy.

Hint: Show that the amplification factor is

$$z = 1 - \frac{4\beta \sin^2 \phi/2}{1 + 4\varepsilon \sin^2 \phi/2}$$

Chapter 10

Iterative Methods for the Resolution of Algebraic Systems

OBJECTIVES AND GUIDELINES

The outcome of any discretization method is an algebraic system for the flow variables associated to the mesh points. In CFD this system can be very large, with a number of unknowns equal to the product of the number of flow variables by the number of mesh points. For a 3D turbulent RANS flow simulation with one million points, and a two-equation turbulence model, we will have 7 flow variables, leading to an algebraic system with 7 million equations for 7 million unknowns. Note that nowadays, one million mesh point for a 3D simulation is considered as a rather coarse grid.

For explicit schemes, the solution of the algebraic system is trivial, as there are no matrices to invert. Full algebraic systems of equations are obtained either as a result of the application of implicit time integration schemes to time-dependent formulations as shown in the previous chapter, or from space discretization of steady state formulations.

Two large families of methods are available for the resolution of a linear algebraic system: the direct and the iterative methods. Direct methods are based on a finite number of arithmetic operations leading to the exact solution of a linear algebraic system.

Iterative methods, on the other hand, are based on a succession of approximate solutions, leading to the exact solution after, in theory, an infinite number of steps. In practice however, the number of arithmetic operations of a direct method can be very high, as they can increase up to N^3, where N is the number of unknowns.

In nonlinear problems, we have to set up an iterative scheme to solve for the nonlinearity, even when the algebraic system, at each iteration step, is solved by a direct method. In these cases, it is often more economical, for large-sized problems, to insert the nonlinear iterations into an overall iterative method for the algebraic system.

The basis of iterative methods is to perform a small number of operations on the matrix elements of the algebraic system in an iterative way, with the aim of approaching the exact solution, within a preset level of accuracy, in a finite, and hopefully small, number of iterations.

Since fluid mechanical problems mostly require fine meshes in order to obtain sufficient resolution, direct methods are seldom applied, with the exception of tridiagonal systems. For these often occurring systems, a very efficient direct solver, known as Thomas algorithm, is applied and because of its importance, the method and FORTRAN subroutines are presented in Appendix A for the scalar case with

Dirichlet or Neumann boundary conditions as well as for the case of periodic boundary conditions.

Resolution methods for algebraic systems have been treated extensively in the literature and are still a subject of much research with the aim of improving the algorithms and reducing the total number of operations. Some excellent expositions of direct and iterative methods can be found in the following books: Varga (1962), Wachspress (1966), Young (1971), Dahlquist and Bjork (1974), Marchuk (1975), Hageman and Young (1981), Meis and Morcowitz (1981), Golub and Meurant (1983), Barrett et al. (1994) and Saad (2003).

A general presentation of matrix properties can be found, for instance in Strang (1976), Berman and Plemmons (1979), Saad (1993). An updated version of the book by Saad is available on http://www-users.cs.umn.edu/~saad/books.html.

With the development of Internet, several texts can be downloaded and various software systems for efficient solutions of algebraic systems are available as freeware. A list of freely available software for linear algebra on the web can be found on http://www.netlib.org/utk/people/JackDongarra/la-sw.html. A highly developed set of routines for sparse iterative solvers is the system known as PETSc (Portable, Extensible Toolkit for Scientific Computation) a suite of data structures and routines for the scalable (parallel) solution of scientific applications modeled by partial differential equations, employing the MPI standard for all message-passing communication. It can be freely downloaded from http://www-unix.mcs.anl.gov/petsc/petsc-as/ and is gaining widespread popularity due to its excellent documentation and reliability.

A very large number of methods are available with various rates of convergence and levels of complexity, but it is not our intention of this introductory text to cover all of them. We wish to introduce you to some basic concepts and to the simplest of the methods, as a guideline to the methodology behind iterative techniques.

Section 10.1 will introduce the iterative methods of *Jacobi and Gauss–Seidel*, firstly on the standard test case of a Poisson equation on a regular grid, followed by a generalization to arbitrary algebraic systems. This will lead us to a general analysis of the stability and convergence rate of an iterative method.

The critical issue with iterative methods is their rate of convergence, as function of the number of unknowns, and the objective is to reach the lowest possible rate. The following sections cover some of the current methods for the acceleration of the convergence, such as *overrelaxation* in Section 10.2 and the more general approach of *preconditioning* in Section 10.3, including the *Alternating Direction Implicit* (ADI) methods which, when optimized, have excellent convergence properties.

Due to the nonlinearity of the flow equations, the algebraic systems are generally nonlinear, and a proper linearization procedure is required. Section 10.4 introduces the treatment of nonlinear algebraic systems, based essentially on a Newton method and some of its approximations.

The most efficient of the iterative techniques, although delicate to program, is the *Multigrid Method*. It is the most general method available to obtain fast convergence rates, as it leads theoretically to the same asymptotic convergence rate, or operation count, as the best Fast Poisson Solvers, without being limited by the same restricted conditions of applicability. It will be introduced in Section 10.5. The roadmap to this chapter is summarized in Figure 10.0.1.

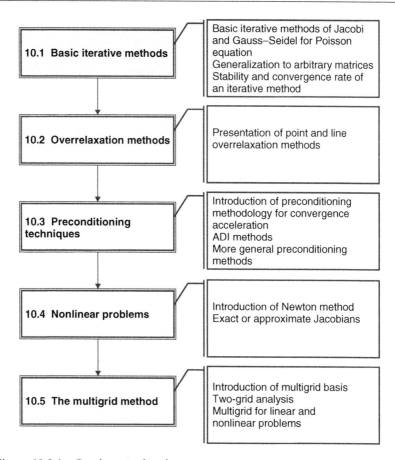

Figure 10.0.1 *Roadmap to this chapter.*

10.1 BASIC ITERATIVE METHODS

The classical example on which the iterative techniques are applied and evaluated is Poisson's equation on a uniform Cartesian mesh. We will follow here this tradition, which allows to clearly demonstrate the different approaches, their properties, structure and limitations.

10.1.1 Poisson's Equation on a Cartesian, Two-Dimensional Mesh

Let us consider the Poisson equation

$$\Delta u = f \quad 0 \le x, y \le L \tag{10.1.1}$$

with

$$u = g \tag{10.1.2}$$

on the boundaries of the rectangular domain and discretized with a second order, five-point scheme as follows with $\Delta x = \Delta y = L/M$, $i, j = 1, \ldots, M$ (see Figure 10.1.1)

$$(u_{i+1,j} - 2u_{ij} + u_{i-1,j}) + (u_{i,j+1} - 2u_{ij} + u_{i,j-1}) = f_{ij} \Delta x^2 \tag{10.1.3}$$

If the vector U is set up with the u_{ij} classified line by line, i.e.

$$U^{\mathsf{T}} \overset{\triangle}{=} (u_{11}, u_{21}, \ldots, u_{M-1,1}, u_{M1}, u_{12}, u_{22}, \ldots, u_{M-1,2}, u_{M2}, \ldots, u_{1j},$$
$$u_{2j}, \ldots, u_{M-1,j}, u_{Mj}, \ldots) \tag{10.1.4}$$

The matrix system obtained from (10.1.3) can be written as, with S being a $(M^2 \times M^2)$ matrix

$$S \cdot U = F \Delta x^2 + G \equiv -Q \tag{10.1.5}$$

$$
\begin{pmatrix}
-4 & \overbrace{1 \cdots 1}^{M\text{-spaces}} & & & & & & \\
\vdots & \vdots & \vdots & \vdots & \vdots & \vdots & \vdots & \vdots \\
& & \overbrace{1 \cdots 1}^{M\text{-spaces}} & -4 & \overbrace{1 \cdots 1}^{M\text{-spaces}} & & & \\
& & & \overbrace{1 \cdots 1}^{M\text{-spaces}} & -4 & \overbrace{1 \cdots 1}^{M\text{-spaces}} & & \\
\vdots & \vdots & \vdots & & & & \overbrace{}^{M\text{-spaces}} & \vdots \\
& & & & & & 1 \cdots 1 & -4
\end{pmatrix}
$$

$$
\times
\begin{pmatrix}
u_{11} \\
\vdots \\
u_{i1} \\
\vdots \\
u_{ij} \\
\vdots \\
u_{i,j+1} \\
\vdots \\
u_{M-1,M-1}
\end{pmatrix}
= \Delta x^2
\begin{pmatrix}
f_{11} \\
\vdots \\
f_{i1} \\
\vdots \\
f_{ij} \\
\vdots \\
f_{i,j+1} \\
\vdots \\
f_{M-1,M-1}
\end{pmatrix}
-
\begin{pmatrix}
g_{10} + g_{01} \\
\vdots \\
g_{i0} \\
\vdots \\
0 \\
\vdots \\
0 \\
\vdots \\
g_{M,M-1} + g_{M-1,M}
\end{pmatrix}
\tag{10.1.6}
$$

The first equation for $i = j = 1$, becomes with the Dirichlet conditions (10.1.2).

$$(u_{21} - 4u_{11} + u_{12}) = f_{11} \Delta x^2 - g_{10} - g_{01} \equiv -q_{11} \tag{10.1.7}$$

Along $j = 1$, we have the equation

$$(u_{i+1,1} - 4u_{i1} + u_{i-1,1} + u_{i2}) = f_{11} \Delta x^2 - g_{i0} \equiv -q_{i1} \tag{10.1.8}$$

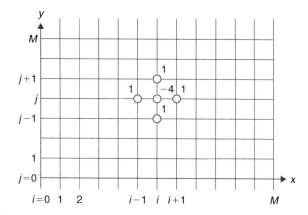

Figure 10.1.1   *Cartesian mesh for Laplace operator.*

Similar equations can be written for the other boundaries (see Problem P.10.1). The vectors F, G are defined by equation (10.1.5). The vector Q will represent the sum of the right-hand side vectors.

By inspection, it is seen that S has the following properties:

(i) S is irreducible diagonal dominant (see the first line).
(ii) S is symmetric and hence non-singular and positive definite.

In order to relate the notation in the system (10.1.6) to the classical notations of linear algebra, we consider the vector U, with elements u_I, $I = 1, \ldots, N = M^2$, where $N = M^2$ is the total number of mesh points. Hence, to each component u_{ij} we associate the component u_I, where I is for instance defined by $I = i + (j-1)M$ as would be the case in a finite element discretization.

Equation (10.1.5) will be written as

$$\sum_{J=1}^{N} s_{IJ} u_J = -Q_I \quad I = 1, \ldots, N \tag{10.1.9}$$

The coefficients s_{IJ} represent the space discretization of the Laplace operator.

10.1.2   Point Jacobi Method/Point Gauss–Seidel Method

In order to solve for the unknowns in system (10.1.3), we could define an initial approximation to the vector U, and attempt to correct this approximation by solving equation (10.1.3) sequentially, sweeping through the mesh, point by point, starting at $i = j = 1$, following the mesh line by line or column per column.

If we indicate by u_{ij}^n (or u_I^n) the assumed approximation (n will be an iteration index); the corrected approximation is u_{ij}^{n+1} (or u_I^{n+1}) and can be obtained (see Figure 10.1.2a) as

$$u_{ij}^{n+1} = \frac{1}{4}(u_{i+1,j}^n + u_{i-1,j}^n + u_{i,j+1}^n + u_{i,j-1}^n) + \frac{1}{4}q_{ij}^n \tag{10.1.10}$$

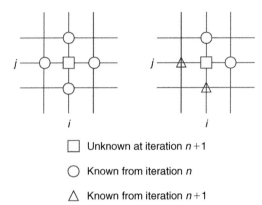

☐ Unknown at iteration $n+1$

◯ Known from iteration n

△ Known from iteration $n+1$

Figure 10.1.2 *Point relaxation method: (a) Jacobi method and (b) Gauss–Seidel method.*

where q_{ij} represents the right-hand side.

In a general formulation, for the system (10.1.9), the point Jacobi method is defined by the algorithm

$$u_I^{n+1} = \frac{1}{s_{II}}\left(-q_I^n - \sum_{\substack{J=1 \\ J\neq I}}^{N} s_{IJ} u_J^n\right) \tag{10.1.11}$$

The general formulation is best represented in matrix form, if we decompose S in a sum of three matrices containing the main diagonal D, the upper triangular part F and the lower triangular part E. That is we write

$$S = D + E + F \tag{10.1.12}$$

with

$$D = \begin{pmatrix} s_{11} & & & \\ & \cdot & & \\ & & s_{II} & \\ & & & \cdot \\ & & & & s_{NN} \end{pmatrix} \qquad E = \begin{pmatrix} 0 & & & & & 0 \\ s_{21} & \cdot & & & & \\ \cdot & \cdot & 0 & & & \\ s_{I1} & s_{I2} & \cdot & 0 & & \\ \cdot & \cdot & \cdot & \cdot & 0 & 0 \\ s_{N1} & s_{N2} & \cdot & & s_{N,N-1} & 0 \end{pmatrix}$$

$$F = \begin{pmatrix} 0 & s_{12} & \cdot & s_{1J} & \cdot & s_{1N} \\ & \cdot & & s_{2J} & \cdot & s_{2N} \\ & & 0 & \cdot & \cdot & \cdot \\ & & & 0 & \cdot & \\ & & & & 0 & s_{N-1,N} \\ 0 & & & & & 0 \end{pmatrix} \tag{10.1.13}$$

For the Laplace operator, this splitting is obvious from the form of the matrix (10.1.6).

Equation (10.1.11) can be written as

$$D \cdot U^{n+1} = -Q^n - (E+F) \cdot U^n \qquad (10.1.14)$$

defining the iterative point Jacobi method for the system (10.1.9) written as

$$(D+E+F) \cdot U = -Q \qquad (10.1.15)$$

The iterative Jacobi scheme can also be written in **Δ-form (delta-form)** by introducing the **Residual R^n** at iteration n

$$R^n \equiv R(U^n) \stackrel{\Delta}{=} S \cdot U^n + Q^n = (D+E+F) \cdot U^n + Q^n \qquad (10.1.16)$$

Equation (10.1.14) can be rewritten, after subtraction of DU^n on both sides,

$$D \cdot \Delta U^n = -R^n \qquad (10.1.17)$$

where

$$\Delta U^n \stackrel{\Delta}{=} U^{n+1} - U^n \qquad (10.1.18)$$

The residual is an important quantity, since it is a measure of the error at a given iteration number n.

Obviously, the iterative method will have to generate a sequence of decreasing values of ΔU^n and R^n when the number of iterations increases, since the converged solution corresponds to a vanishing residual.

For equation (10.1.10), the residual form (10.1.17) is obtained by subtracting u_{ij}^n from both sides, leading to

$$4(u_{ij}^{n+1} - u_{ij}^n) = (u_{i+1,j}^n + u_{i1,j}^n + u_{i,j+1}^n + u_{i,j-1}^n - 4u_{ij}^n) + q_{ij}^n \qquad (10.1.19)$$

Point Gauss–Seidel method

In Figure 10.1.2a, we observe that the points $(i, j-1)$ and $(i-1, j)$ have already been updated at iteration $(n+1)$ when u_{ij} is calculated. We are therefore tempted to use these new values in the estimation of u_{ij}^{n+1} as soon as they have been calculated. We can expect thereby to obtain a higher convergence rate since the influence of a perturbation on u^n is transmitted more rapidly. With the Jacobi method, a perturbation of $u_{i-1,j}^{n+1}$ will be felt on u_{ij} only after the whole mesh is swept since it will occur for the first time at the next iteration through the equation for u_{ij}^{n+2}. With the Gauss–Seidel method, this influence already appears at the current iteration since u_{ij}^{n+1} is immediately affected by $u_{i-1,j}^{n+1}$ (see Figure 10.1.2b). As a by-product, it can be observed that as soon as a new value $u_{i,j}^{n+1}$ is calculated the 'old' value $u_{i,j}^n$ is not needed anymore. Hence, the new value can be stored in the same location and overwrite, in the coding, the local value $u_{i,j}^n$. Therefore only one vector U of length N has to be stored, while two vectors U^{n+1} and U^n have to be saved in the Jacobi method.

The Gauss–Seidel method is, from every point of view more advantageous than the Jacobi method, and is defined by the iterative scheme

$$u_{ij}^{n+1} = \frac{1}{4}(u_{i+1,j}^n + u_{i-1,j}^{n+1} + u_{i,j+1}^n + u_{i,j-1}^{n+1}) + \frac{1}{4}q_{ij}^n \tag{10.1.20}$$

For the general system (10.1.9), the Gauss–Seidel method takes all the variables associated to the lower diagonal of the matrix S, at the new level $(n+1)$.

Hence, instead of equation (10.1.11), we have

$$u_I^{n+1} = \frac{1}{s_{II}}\left(-q_I^n - \sum_{J=1}^{I-1} s_{IJ}u_J^{n+1} - \sum_{J=I+1}^{N} s_{IJ}u_J^n\right) \tag{10.1.21}$$

In operator form, this scheme can be written as

$$(D + E) \cdot U^{n+1} = -Q^n - F \cdot U^n \tag{10.1.22}$$

and in residual form

$$(D + E) \cdot \Delta U^n = -R^n \tag{10.1.23}$$

By comparing this equation with (10.1.17) it can be seen that the Gauss–Seidel method corresponds to another choice for the matrix which 'drives' ΔU^n to zero, that is to convergence, or equivalently, to another splitting of the matrix S.

This observation leads us to the following general analysis of an iterative method.

10.1.3 Convergence Analysis of Iterative Schemes

An iterative method is said to be convergent if the error tends to zero as the number of iterations goes to infinity. If $\overline{U}$ is the exact solution of the system (10.1.5), the error at iteration n is

$$e^n = U^n - \overline{U} \tag{10.1.24}$$

Let us consider an arbitrary splitting of the matrix S

$$S = P + A \tag{10.1.25}$$

and an iterative scheme

$$P \cdot U^{n+1} = -Q^n - A \cdot U^n \tag{10.1.26}$$

or equivalently

$$P \cdot \Delta U^n = -R^n \tag{10.1.27}$$

The matrix P will be called a ***convergence or (pre)conditioning matrix (operator)*** and is selected such that the system (10.1.26) is *easily* solvable. It is important to observe here that the iterative scheme (10.1.27) replaces the full iterative scheme, corresponding to a direct method,

$$S \cdot U^{n+1} = -Q^n \quad \text{or} \quad S \cdot \Delta U^n = -R^n \tag{10.1.28}$$

Subtracting from equation (10.1.26), the relation defining the exact solution $\overline{U}$

$$S \cdot \overline{U} \equiv (P + A) \cdot \overline{U} = -Q \tag{10.1.29}$$

we obtain, assuming that Q is independent of U,

$$P \cdot e^{n+1} = -A \cdot e^n \tag{10.1.30}$$

or

$$e^{n+1} = -(P^{-1} \cdot A) \cdot e^n = -(P^{-1} \cdot A)^n \cdot e^1 = (1 - P^{-1} \cdot S)^n \cdot e^1 \tag{10.1.31}$$

where $(P^{-1}A)^n$ is the matrix $P^{-1}A$ to the power n.

If $\|e^{n+1}\|$ is to go to zero for increasing n, the matrix $P^{-1}A$ should be a convergent matrix and its spectral radius has to be lower than one.

The iterative scheme will be convergent, if and only if the matrix G, called the *iteration* or *amplification* matrix (operator)

$$G = 1 - P^{-1} \cdot S \tag{10.1.32}$$

is a convergent matrix, i.e., satisfying the condition on the spectral radius

$$\rho(G) \leq 1 \tag{10.1.33}$$

or

$$|\lambda_J(G)| \leq 1 \quad \text{for all } J \tag{10.1.34}$$

All the eigenvalues $\lambda_J(G)$ of $G = (1 - P^{-1}S)$ have to be lower than one in modulus for the iterative scheme to converge.

Hence, it is seen that we can replace equation (10.1.28) and the matrix S acting on the corrections ΔU^n, by another operator P, which is expected to be easier to invert, provided the above conditions are satisfied.

What do we mean by 'easy to invert'? The main conditions on the convergence matrix P are:

- The number of arithmetic operations required for the computation of P^{-1} should not be higher than $O(N)$, with a proportional coefficient as low as possible.
- P should be as close as possible to S.

Except for these conditions, the choice of P is arbitrary. If $P = S$ we obtain the exact solution but this corresponds to a direct method, since we have to invert the matrix S of the system.

The residual R^n is connected to the error vector e^n, by the following relation obtained by subtracting R^n from $\overline{R} = S\overline{U} + Q \equiv 0$, assuming Q to be independent of U,

$$R^n = S \cdot e^n \tag{10.1.35}$$

This shows the quantitative relation between the error and the residual and proofs that the residual will go to zero when the error tends to zero.

In practical computations, the error is not accessible since the exact solution is not known. Therefore we generally use the norm of the residual $||R^n||$ as a measure of the evolution toward convergence and practical convergence criteria will be set by requiring that the residual drops by a predetermined number of orders of magnitude. Observe that even when the residual is reduced to machine accuracy (machine zero), it does *not* mean that the solution U^n is within machine accuracy of the exact solution. When SU results from a space discretization, achieving machine zero on the residual will produce a solution U^n which differs from the exact solution $\overline{U}$ of the differential problem by the amount of the truncation error. For example, a second order accurate space discretization, with $\Delta x = 10^{-2}$, will produce an error of the order of 10^{-4} on the solution, which cannot be reduced further even if the residual equals 10^{-14}.

Equation (10.1.26) can also be written as

$$\begin{aligned} U^{n+1} &= (1 - P^{-1} \cdot S) \cdot U^n - P^{-1} \cdot Q \\ &= G \cdot U^n - P^{-1} \cdot Q \end{aligned} \tag{10.1.36}$$

Comparing equation (10.1.31), written here as

$$e^{n+1} = G \cdot e^n = (G)^{n+1} \cdot e^0 \tag{10.1.37}$$

with similar relations in Chapter 7 or 9, interpreting the index n as a pseudo-time index, the similitude between the formulation of an iterative scheme of a stationary problem, and a time-dependent formulation of the same problem clearly appears. The matrix G, being the amplification matrix of the iterative method can be analyzed along the lines developed in Chapters 7 and 9. We will return to this very important link below.

Convergence conditions for the Jacobi and Gauss–Seidel methods

For the Jacobi method $G = -D^{-1} \cdot (E + F) = 1 - D^{-1}S$ and for the Gauss–Seidel method $G = -(D + E)^{-1} \cdot F = 1 - (D + E)^{-1}S$. From the properties of matrices, it can be shown that the Jacobi and Gauss–Seidel methods will converge if S is irreducible diagonal dominant. When S is symmetric, it is sufficient for convergence that S, or $(-S)$ is positive definite for both methods and in addition for the Jacobi method, $(2D - S)$, or $(S - 2D)$ in our case, has also to be positive definite.

Estimation of the convergence rate

An important issue of iterative methods is the estimation of the rate of reduction of the error with increasing number of iterations. The average rate of error reduction over n iterations can be defined by

$$\overline{\Delta e_n} \stackrel{\Delta}{=} \left(\frac{\|e^n\|}{\|e^0\|} \right)^{1/n}$$

(10.1.38)

From equation (10.1.37) we have

$$\overline{\Delta e_n} \le \|(G)^n\|^{1/n}$$

(10.1.39)

Asymptotically it can be shown, Varga (1962), that this quantity tends to the spectral radius of G for $n \to \infty$:

$$s \equiv \lim_{n \to \infty} \|(G)^n\|^{1/n} = \rho(G)$$

(10.1.40)

The logarithm of s is the *average convergence rate*, measuring the number of iterations needed to decrease the error by a given factor.

If we consider s to be a valid measure of $\overline{\Delta e_n}$, the norm of the error will be reduced by an order of magnitude (factor 10), in a number of iterations n, such that

$$\left(\frac{1}{10} \right)^{1/n} \le s = \rho(G)$$

(10.1.41)

or

$$n \ge -1/\log \rho(G)$$

(10.1.42)

Remember that the spectral radius $\rho(G) < 1$ is always lower than one for stability which explains the negative sign of this relation.

This relation is valid asymptotically for large n, and is generally not to be relied on at small values of n. In the early stages of the iterative process it is not uncommon, particularly with nonlinear problems to observe even an increase in the residual before it starts to decrease at larger n. Also, we can often observe different rates of residual reduction, usually an initial reduction rate much higher than s, slowing down gradually to the asymptotic value. The explanation for this behavior is to be found in the frequency distribution of the error and in the way a given iterative method treats the different frequencies in the spectrum. The residual history of curve a of Figure 10.1.3 is typical for a method which damps rapidly the high frequency components but damps poorly the low frequency errors while the opposite is true for curve b. The reason for this behavior is explained from an eigenvalue analysis.

10.1.4 Eigenvalue Analysis of an Iterative Method

The frequency response, as well as the asymptotic convergence rate of an iterative method, are defined by the eigenvalue spectrum of the amplification matrix G, and by the way the matrix P^{-1} treats the eigenvalues of the matrix S. Referring to Section 9.1.3, the solution U^{n+1} can be written as a linear combination of the eigenmodes $V^{(J)}$

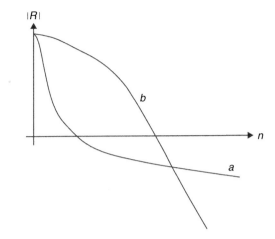

Figure 10.1.3   *Typical residual history of a relaxation method.*

of the matrix S (which represents the space discretization of the differential operator). Each term represents the contribution from the initial component U_{0J} damped by the factor $(\lambda_J)^n$, where (λ_J) is the eigenvalue of the matrix G, associated to the eigenvalue (Ω_J) of the matrix S. We suppose here that S *and* G *commute and have the same set of eigenfunctions*, although this will not always be the true, even in simple cases. For the present example of Poisson's equation on a Cartesian mesh, this property is satisfied for the Jacobi iteration but not for the Gauss–Seidel method, since the matrices $(D + E)^{-1}$ and S do not commute. However, the conclusions of this section with regard to the convergence properties of iterative methods and their relation to the eigenvalues of the space operator will remain largely valid, independently of this hypothesis, as can be seen from the analysis of Section 9.1.3. Hence, we assume

$$U^n = \sum_{J=1} (\lambda_J)^n U_{0J} \cdot V^{(J)} - \sum_{J=1} \frac{Q_J}{\Omega_J} [1 - (\lambda_J)^n] \cdot V^{(J)} \tag{10.1.43}$$

Note that $(\lambda_J)^n$ is (λ_J) to the power n.

The first term is the 'transient' behavior of the homogeneous solution while the last term is a transient contribution from the source term. Since, for convergence, all the (λ_J) are in modulus lower than one, after a sufficiently large number of iterations, we are left with the 'converged' solution

$$\lim_{n \to \infty} U^n = -\sum_{J=1} \frac{Q_J}{\Omega_J} \cdot V^{(J)} \tag{10.1.44}$$

which is the eigenmode expansion of the solution $\overline{U} = -S^{-1} \cdot Q$. Indeed, by writing

$$\overline{U} = \sum_{J=1} U_J \cdot V^{(J)} \tag{10.1.45}$$

we obtain

$$S \cdot \overline{U} = \sum_{J=1} \Omega_J U_J \cdot V^{(J)} = -\sum_{J=1} Q_J \cdot V^{(J)} \tag{10.1.46}$$

Since the U^n form a basis of the U-space, we have

$$U_J = -\frac{Q_J}{\Omega_J} \tag{10.1.47}$$

which shows that equation (10.1.44) is indeed the solution. Equation (10.1.43) shows that the transient will, for large n, be dominated by the largest eigenvalue (λ_J); this explains the relation (10.1.42).

It is seen from equation (10.1.31) that the error e^n behaves like the homogeneous part of (10.1.43). Therefore, if e_J^0 is the initial error distribution, after n iterations the error is reduced to

$$e^n = \sum_{J=1} (\lambda_J)^n e_{0J} \cdot V^{(J)} \tag{10.1.48}$$

The residual R^n can also be expanded in a series of eigenfunctions $V(J)$ since from equation (10.1.35) we can write

$$R^n = \sum_{J=1} (\lambda_J)^n e_{0J} S \cdot V^{(J)} = \sum_{J=1} (\lambda_J)^n \Omega_J e_{0J} \cdot V^{(J)} \tag{10.1.49}$$

If we assume the eigenvectors to be orthonormal, then the L_2-norm of the residual, will be

$$\|R^n\|_{L_2} = \sum_{J=1} [(\lambda_J)^n \Omega_J e_{0J}]^2 \tag{10.1.50}$$

When the iteration number n increases, the low values of (λ_J) will be damped more rapidly, and at large values of n we will be left with the highest (λ_J), in particular those who are the closest to the upper limit of one. Hence, if the iterative scheme is such that the highest eigenvalues $\lambda_J = \lambda(\Omega_J)$ are close to one, the asymptotic convergence rate will be poor. Since generally, the high frequencies of the operator S generate the lowest values of (λ_J), we will tend to have a behavior such as shown on curve a of Figure 10.1.3, where the initial rapid decrease of the residual is due to the damping of the high frequencies (low $\hat{A}$), and where one is left, at large n, with the low frequency end (high(λ_J)) of the error spectrum.

On the other hand, if such an iterative method is combined with an algorithm which damps effectively the low frequencies of S, that is the large (λ_J) region, then we will obtain a behavior of the type shown on curve b of Figure 10.1.3. This is the principle at the basis of the multigrid method, to be described in Section 10.5.

The relation $\lambda_J = \lambda(\Omega_J)$ can be computed explicitly only in simple cases. For instance, for the Jacobi method we have, representing by G the amplification matrix,

$$G_J = 1 - D^{-1} S \tag{10.1.51}$$

and since D is a diagonal matrix

$$\lambda_J = 1 - \frac{1}{d_J}\Omega_J(S) \tag{10.1.52}$$

Note that we define S as the matrix representation of the discretized differential operator, which has to satisfy the condition of stability of the time-dependent counterpart, and therefore have eigenvalues with negative real parts, as seen in Section 9.1. Hence, S will be negative definite when all the eigenvalues are real.

10.1.5 Fourier Analysis of an Iterative Method

The Fourier analysis allows a simple estimation of eigenvalues, when periodic conditions are assumed, such that the Fourier modes are eigenvectors of the S matrix and of the amplification matrix G. This is the case for the Jacobi method, but not for the Gauss–Seidel iterations.

For the discretization (10.1.3) with Dirichlet boundary conditions, the eigenfunctions of S reduce to $\sin(i\phi_x) \cdot \sin(j\phi_y)$ and the eigenvalues are equal to, dropping the index J,

$$\Omega = -4(\sin^2\phi_x/2 + \sin^2\phi_y/2)$$
$$\phi_x = l\pi/M \quad \phi_y = m\pi/M \quad J = l + (m-1)M \quad l, m = 1, \ldots, M \tag{10.1.53}$$

The eigenvalues of G_J are therefore, with $d = -4$

$$\lambda(G_J) = 1 - (\sin^2\phi_x/2 + \sin^2\phi_y/2) = \frac{1}{2}(\cos\phi_x + \cos\phi_y) \tag{10.1.54}$$

At the low frequency end, around $\phi_x = \phi_y = 0$, the damping rate is very poor, since $\lambda = 1$. The intermediate frequency range is strongly damped, that is for the frequencies corresponding to high values of Ω. The spectral radius $\rho(G_J)$ is defined by the highest eigenvalue, which corresponds to the lowest frequencies of the Ω-spectrum. These are obtained for $\phi_x = \phi_y = \pi/M$

$$\rho(G_J) = \cos\pi/M \tag{10.1.55}$$

Hence the convergence rate becomes, for $\Delta x = \Delta y$ that is, for large number of mesh points

$$\rho(G_J) \approx 1 - \frac{\pi^2}{2M^2} = 1 - \frac{\pi^2}{2N} = 1 - O(\Delta x^2) \tag{10.1.56}$$

and, from equation (10.1.42), the number of iterations, needed to reduce the error by one order of magnitude is, asymptotically $(2N/0.43\pi^2)$. Since each Jacobi iteration requires $5N$ operations, the Jacobi method will require roughly $2kN^2$ operations for a reduction of the error by k orders of magnitude.

Gauss–Seidel method

Applying a straightforward Neumann analysis to the homogeneous part of equation (10.1.20), leads to

$$(4 - e^{-I\phi_x} - e^{-I\phi_y})\lambda_{GS} = e^{I\phi_x} + e^{I\phi_y} \tag{10.1.57}$$

However, this does not give the eigenvalues of the matrix G_{GS}, since this matrix does not commute with S. It can be shown, Young (1971), that the corresponding eigenvalues of the amplification matrix of the Gauss–Seidel method are equal to the square of the eigenvalues of the Jacobi method.

$$\lambda(G_{GS}) = \frac{1}{4}(\cos \phi_x + \cos \phi_y)^2 = \lambda(G_J)^2 \tag{10.1.58}$$

Hence, the spectral radius is

$$\rho(G_{GS}) = \rho(G_J)^2 = \cos^2 \pi/M \tag{10.1.59}$$

and for large M,

$$\rho(G_{GS}) \approx 1 - \frac{\pi^2}{M^2} = 1 - \frac{\pi^2}{N} \tag{10.1.60}$$

and the method converges *twice* as fast as the Jacobi method, since the number of iterations for one order of magnitude reduction in error, is $N/(0.43\pi^2)$.

For a more general case, it can be shown, Varga (1962), that the Gauss–Seidel method converges if S is an irreducible diagonal dominant matrix, or if $(-S)$ is symmetric, positive definite. In this latter case we have $\lambda(G_{GS}) = \lambda(G_J)^2$. Interesting considerations with regard to the application of Fourier analysis to iterative methods can be found in LeVeque and Trefethen (1986).

10.2 OVERRELAXATION METHODS

We could be tempted to increase the convergence rate of an iterative method, by 'propagating' the corrections $\Delta U^n = U^{n+1} - U^n$ faster through the mesh. This idea is the basis of the overrelaxation method, introduced independently by Frankel and Young in 1950; see Young (1971) and Hageman and Young (1981).

The overrelaxation method is formulated as follows: if $\overline{U^{n+1}}$ is the value obtained from the basic iterative scheme, the value introduced at the next level U^{n+1} is defined by

$$U^{n+1} = \omega\overline{U^{n+1}} + (1 - \omega)U^n \tag{10.2.1}$$

where ω is the **overrelaxation coefficient**. Alternatively, a new correction is defined

$$\Delta U^n = U^{n+1} - U^n \quad \text{with} \quad \Delta U^n = \omega\overline{\Delta U^n} \tag{10.2.2}$$

When appropriately optimized for maximum convergence rate, a considerable gain can be achieved as will be seen next.

10.2.1 Jacobi Overrelaxation

The Jacobi overrelaxation method becomes for the Laplace operator,

$$u_{ij}^{n+1} = \frac{\omega}{4}(u_{i+1,j}^n + u_{i-1,j}^n + u_{i,j+1}^n + u_{i,j-1}^n + q_{ij}^n) + (1 - \omega)u_{ij}^n \tag{10.2.3}$$

In operator form, equation (12.2.3) becomes

$$D \cdot U^{n+1} = -\omega Q^n - \omega(E + F) \cdot U^n + (1 - \omega)D \cdot U^n$$
$$= -\omega(S \cdot U^n + Q^n) + D \cdot U^n \tag{10.2.4}$$

or

$$D \cdot \Delta U^n = -\omega R^n \tag{10.2.5}$$

instead of equation (10.1.17). The new amplification matrix of the iterative scheme $G_J(\omega)$ is from equation (10.2.4)

$$G_J(\omega) = (1 - \omega)I + \omega G_J = 1 - \omega D^{-1} \cdot S \tag{10.2.6}$$

The convergence will be ensured if the spectral radius of $G_J(\omega)$ is lower than one, i.e. if

$$\rho(G_J(\omega)) \leq |1 - \omega| + \omega\rho(G_J) < 1$$

This will be satisfied if

$$0 < \omega < \frac{2}{1 + \rho(G_J)} \tag{10.2.7}$$

The eigenvalues are related by

$$\lambda(G_J(\omega)) = (1 - \omega) + \omega\lambda(G_J) \tag{10.2.8}$$

and this allows to select optimum values of ω for maximal damping in a given frequency range. In function of ω, the eigenvalue $\lambda(G_J(\omega))$ will vanish for the particular choice $\omega = 1/(1 - \lambda(G_J))$ and the corresponding frequency component in the development (10.1.48) will not contribute. Hence, if we attempt to damp selectively the high frequencies, corresponding to the values of λ toward the end-point of -1, we could select ω values lower than one. On the other hand the low frequencies generally span the λ-region close to the other end point of the permissible λ-range, namely $\lambda \leq 1$, and we would select values of ω toward the higher end of its range if the goal would be to damp more selectively the low frequencies.

We can define also an optimum value of ω, ω_{opt} which averages an optimal damping over the whole range. From Figure 10.2.1, where $|\lambda(G_J(\omega))|$ is plotted against ω, at constant values of $\lambda(G_J)$ the intersection of the curves associated respectively to the minimum value of $\lambda(G_J)$, λ_{min}, and the maximum eigenvalue λ_{max}, appears to improve best the damping over the whole range. Hence, ω_{opt} is defined by

$$-1 + \omega_{opt}(1 - \lambda_{min}) = 1 - \omega_{opt}(1 - \lambda_{max})$$

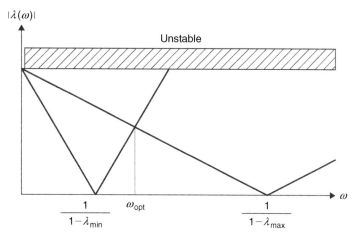

Figure 10.2.1 *Optimal relaxation for Jacobi method.*

or

$$\omega_{\text{opt}} = \frac{2}{2 - (\lambda_{\min} + \lambda_{\max})} \qquad (10.2.9)$$

For the Laplace operator, from equation (10.1.54)

$$\lambda_{\max} = -\lambda_{\min} = \cos \pi/M \qquad (10.2.10)$$

and $\omega_{\text{opt}} = 1$. Hence the Jacobi method is optimal by this definition.

10.2.2 Gauss–Seidel Overrelaxation: Successive Overrelaxation (SOR)

The benefits of the overrelaxation concept take on their full range when applied to the Gauss–Seidel method, and in this case the methods are usually called *successive overrelaxation* methods, indicated by the symbol SOR. Applying the definition (10.2.1) to the Gauss–Seidel method for the Laplace operator, leads to the following iterative scheme:

$$\overline{u_{ij}^{n+1}} = \frac{1}{4}(u_{i+1,j}^{n} + u_{i-1,j}^{n+1} + u_{i,j+1}^{n} + u_{i,j-1}^{n+1}) + \frac{1}{4}q_{ij}^{n}$$

$$u_{ij}^{n+1} = \omega \overline{u_{ij}^{n+1}} + (1 - \omega)u_{ij}^{n} \qquad (10.2.11)$$

In operator form, this corresponds to the general residual formulation

$$D \cdot \overline{\Delta U^{n}} = -R^{n} - E \cdot \Delta U^{n} \qquad (10.2.12)$$

and the overrelaxation method is defined by the iteration operator $(D + \omega E)$, since equation (10.2.12) reads

$$(D + \omega E) \cdot \Delta U^{n} = -\omega R^{n} \qquad (10.2.13)$$

The amplification matrix $G_{SOR}(\omega)$ is now

$$G_{SOR}(\omega) = 1 - \omega(D + \omega E)^{-1} \cdot S$$
$$= (D + \omega E)^{-1} \cdot [(1 - \omega)D - \omega F] \tag{10.2.14}$$

Condition on the relaxation parameter z

Since $G_{SOR}(\omega)$ is a product of triangular matrices, its determinant is the product of its diagonal elements. Hence,

$$\det G_{SOR}(\omega) = \det(I + \omega D^{-1}E)^{-1} \cdot \det[(1 - \omega)I - \omega D^{-1}F]$$
$$= I \cdot \det[(1 - \omega)I - \omega D^{-1}F] = (1 - \omega)^N \tag{10.2.15}$$

On the other hand, the determinant of a matrix is equal to the product of its eigenvalues, and hence

$$\rho(G_{SOR}(\omega))^N \geq |\det G_{SOR}(\omega)| = (1 - \omega)^N \tag{10.2.16}$$

The spectral radius will be lower than one if

$$|(1 - \omega)| \leq \rho(G_{SOR}(\omega)) < 1 \tag{10.2.17}$$

or

$$0 < \omega < 2 \tag{10.2.18}$$

For irreducible diagonal dominant matrices, it can be shown (Young, 1971) that the SOR method will converge, if

$$0 < \omega < \frac{2}{1 + \rho(G_{GS})} \tag{10.2.19}$$

More information can be obtained for symmetrical matrices. The eigenvalues of $G_{SOR}(\omega)$ satisfy the following relations, for symmetrical matrices of the form

$$S = \begin{pmatrix} D_1 & F \\ F^T & D_2 \end{pmatrix} \tag{10.2.20}$$

where D_1 and D_2 are block diagonal, writing $\lambda(\omega)$ for $\lambda(G_{SOR}(\omega))$:

(i) The eigenvalues of the SOR amplification factor satisfy the relation

$$\lambda(\omega) = 1 - \omega + \omega\lambda^{1/2}(\omega) \cdot \lambda(G_J) \tag{10.2.21}$$

(ii) For $\omega = 1$, we recover relation (10.1.58)

$$\lambda(\omega = 1) \equiv \lambda(G_{GS}) = \lambda^2(G_J) \tag{10.2.22}$$

(iii) an optimal relaxation coefficient can be defined, rendering $\rho(G_{SOR}(\omega))$ minimum, as

$$\omega_{opt} = \frac{2}{1 + \sqrt{1 - \rho^2(G_J)}} \qquad (10.2.23)$$

with

$$\rho(G_{SOR}(\omega_{opt})) = \omega_{opt} - 1 \qquad (10.2.24)$$

Note the role played by the eigenvalues of the Jacobi point iteration matrix G_J. For Laplace's operator, we have

$$\rho(G_J) = \cos \pi/M \qquad (10.2.25)$$

$$\omega_{opt} = \frac{2}{1 + \sin \pi/M} \approx 2 \left(1 - \frac{\pi}{M} + \frac{\pi^2}{M^2} \right) \qquad (10.2.26)$$

and the spectral radius at optimum relaxation coefficient is

$$\rho(G_{SOR}(\omega_{opt})) = \frac{1 - \sin \pi/M}{1 + \sin \pi/M} \approx 1 - \frac{2\pi}{M} + O \left(\frac{1}{M^2} \right) \qquad (10.2.27)$$

The number of iterations for a one order of magnitude reduction in error norm is of the order of $2.3M/\pi \approx \sqrt{N}$. Compared to the corresponding values for Jacobi and Gauss–Seidel iterations, the optimal SOR will require $kN \cdot \sqrt{N}$ operations, for a reduction of the error of k orders of magnitude. Hence, this represents a considerable gain in convergence rate.

For non-optimal relaxation parameters however, this rate can seriously deteriorate. Since, for more general problems, the eigenvalues are not easily found, one can attempt to find numerical estimates of the optimal relaxation parameter. For instance, for large n, the error being dominated by the largest eigenvalue, we have from equations (10.1.36) and (10.1.48)

$$\rho(G) = \lim_{n \to \infty} \frac{\|e^{n+1}\|}{\|e^n\|} \qquad (10.2.28)$$

and when this ratio stabilizes, we can deduct $\rho(G_J)$ from equation (10.2.21) by introducing this estimation of $\rho(G)$ for $\lambda(\omega)$ and find ω_{opt} from equation (10.2.23). Other strategies can be found in Wachspress (1966), Hageman and Young (1981).

10.2.3 Symmetric Successive Overrelaxation (SSOR)

The SOR method, as described by equation (10.2.11), sweeps through the mesh from the lower left corner to the right upper corner. This gives a bias to the iteration scheme, which might lead to some error accumulations, for instance, if the physics of the system to be solved is not compatible with this sweep direction. This can be avoided by alternating SOR sweeps in both directions. We obtain a two-step iterative method (a form of predictor–corrector approach), whereby a predictor $U^{n+1/2}$ is obtained from a SOR sweep (10.2.12), followed by a sweep in the reverse direction starting at the end-point of the first step.

The SSOR can be described in residual form by the equations

$$(D + \omega E) \cdot \overline{\Delta U^{n+1/2}} = -\omega R^n \tag{10.2.29}$$

$$(D + \omega F) \cdot \overline{\Delta U^n} = -\omega R(\overline{U^{n+1/2}}) \tag{10.2.30}$$

with

$$\overline{U^{n+1/2}} = U^n + \overline{\Delta U^{n+1/2}}$$
$$U^{n+1} = \overline{U^{n+1/2}} + \overline{\Delta U^n} = U^n + \overline{\Delta U^{n+1/2}} + \overline{\Delta U^n} \tag{10.2.31}$$

This method converges under the same conditions as the SOR method and an optimal relaxation parameter can only be estimated approximately as

$$\omega_{\text{opt}} \approx \frac{2}{1 + \sqrt{2(1 - \rho^2(G_J))}} \tag{10.2.32}$$

Hence, the SSOR method converges twice as fast as SOR, but since it requires twice as much work, there is not much to be gained in convergence rate. However, SSOR might lead to better error distributions and is a good candidate for preconditioning matrices. Also, when applied in conjunction with non-stationary relaxation, significant improvements in convergence rate over SOR can be achieved (Young, 1971).

10.2.4   Successive Line Overrelaxation Methods (SLOR)

We could still accelerate the convergence of the SOR method by increasing the way the variations ΔU^n are transmitted throughout the field, if we are willing to increase the workload per iteration by solving small systems of equations.

Referring to Figure 10.2.2, for the Laplace operator, we could consider the three points on a column i, or on a line j, as simultaneous unknowns and solve a tridiagonal system along each vertical line, sweeping through the mesh column by column from $i = 1$ to $i = M$ (Figure 10.2.2a). or along horizontal lines as in Figure 10.2.2b, from $j = 1$ to $j = M$. This is called line Gauss–Seidel ($\omega = 1$) or line successive overrelaxation, generally abbreviated as SLOR, when $\omega \neq 1$.

For the Laplace operator, the SLOR iterative scheme along vertical lines (VLOR) is defined by

$$\overline{u_{ij}^{n+1}} = \frac{1}{4}(u_{i+1,j}^n + \overline{u_{i-1,j}^{n+1}} + \overline{u_{i,j+1}^{n+1}} + \overline{u_{i,j-1}^{n+1}}) + \frac{1}{4}q_{ij}^n$$
$$u_{ij}^{n+1} = \omega \overline{u_{ij}^{n+1}} + (1 - \omega)u_{ij}^n \tag{10.2.33}$$

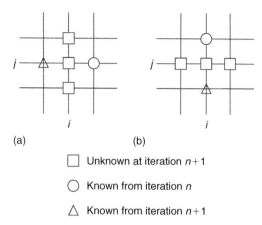

☐ Unknown at iteration $n+1$

◯ Known from iteration n

△ Known from iteration $n+1$

Figure 10.2.2 *Line relaxation method.*

or in incremental form

$$4\Delta U_{ij}^n - \Delta U_{i,j-1}^n - \Delta U_{i,j+1}^n - \omega\Delta U_{i-1,j}^n = \omega R_{ij}^n \tag{10.2.34}$$

which is solved by the tridiagonal Thomas algorithm in $5M$ operations for each line. Hence $5M^2 = 5N$ operations are required for all the tridiagonal systems. This number is to be roughly doubled to take into account the computation of the right-hand side terms.

A Jacobi line relaxation method can also be defined and written as

$$\overline{u_{ij}^{n+1}} = \frac{1}{4}(u_{i+1,j}^n + u_{i-1,j}^n + \overline{u_{i,j+1}^{n+1}} + \overline{u_{i,j-1}^{n+1}}) + \frac{1}{4}q_{ij}^n$$

$$u_{ij}^{n+1} = \omega\overline{u_{ij}^{n+1}} + (1 - \omega)u_{ij}^n \tag{10.2.35}$$

or in incremental form

$$4\Delta U_{ij}^n - \Delta U_{i,j-1}^n - \Delta U_{i,j+1}^n = \omega R_{ij}^n \tag{10.2.36}$$

and corresponds in Figure 10.2.2 to a replacement of the triangles by a circle symbol.

Many variants of the relaxation method can be derived by combining point or line relaxations on alternate positions, or on alternate sweep directions. It can be shown that the line relaxations for the Laplace operator have a convergence rate twice faster than the point relaxations. The overrelaxation method SLOR, is $\sqrt{2}$ times faster than SOR at the same value of the optimal ω.

(a) Red–Black point relaxation

In this approach, the relaxation method is applied to two distinct series of mesh points, called red and black points, where the red points can be considered as having an even number $I = i + (j - 1)M$ and the black points an odd number (see Figure 10.2.3).

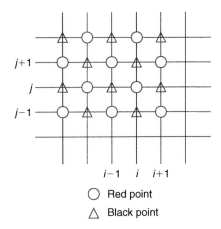

Figure 10.2.3 *Red–Black point relaxation method.*

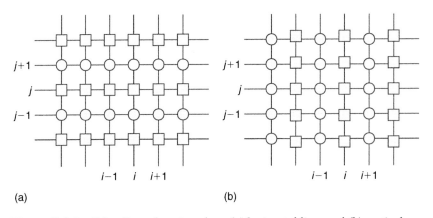

(a)   (b)

Figure 10.2.4 *Zebra line relaxation along (a) horizontal lines and (b) vertical lines.*

(b) Zebra line relaxation

Here line relaxations are performed on alternating lines, on the horizontal lines, or on vertical lines (Figure 10.2.4). Various combinations can be defined such as interchanging horizontal and vertical lines, alternating the sweeping directions or considering multiple zebra's with different families of lines. It is to be noted that these schemes can be used by themselves and also as 'smoothers' in multigrid iterations, as presented in Section 10.5.

10.3 PRECONDITIONING TECHNIQUES

All the iterative schemes of the previous sections can be put under the residual form (10.1.27), with various forms for the convergence or conditioning operator P. As it appeared already in Section 10.1.3, this brings forward a clear connection between the behavior of iterative schemes with the iteration number n, and the behavior of time-dependent formulations where n is a time step index. Hence if we view n in this way, we can consider the general iterative scheme

$$P\frac{\Delta U^n}{\tau} = -\omega(S \cdot U^n + Q^n) = -\omega R^n \qquad (10.3.1)$$

as the explicit Euler integration, with time step τ, of the differential system

$$P\frac{dU}{dt} = -\omega(S \cdot U + Q) \qquad (10.3.2)$$

where we are only interested in the asymptotic steady state. The operator P can be chosen in an arbitrary way, with the only restriction that the amplification matrix G

$$G = 1 - \omega\tau P^{-1} \cdot S \qquad (10.3.3)$$

should have all its eigenvalues lower than one, i.e.

$$\rho(G) \leq 1 \qquad (10.3.4)$$

In addition, we know that the operator $(-\omega\tau P^{-1} \cdot S)$ must have eigenvalues with non-positive real parts for stability. The parameter τ appears as a pseudo-time step, which can eventually be absorbed by ω or left as an additional optimization parameter, or simply set equal to one.

It is seen from equation (10.3.3) that for $P = (\omega\tau \cdot S)$, $G = 0$ implying that we have the exact solution in one iteration. But of course, this amounts to the solution of the stationary problem. Therefore, the closer $P/\omega\tau$ approximates S, the better the iterative scheme, but on the other hand P has to be easily solvable. Let us consider a few well-known examples.

10.3.1   Richardson Method

This is the simplest case, where $P/\tau = -1$. The choice (-1) is necessary because S has eigenvalues with real negative parts. For the Laplace equation all the eigenvalues of S are negative. The iterative scheme is

$$\begin{aligned} U^{n+1} &= U^n + \omega(S \cdot U^n + Q^n) = (1 + \omega S) \cdot U^n + \omega Q^n \\ &\equiv G_R \cdot U^n + \omega Q^n \end{aligned} \qquad (10.3.5)$$

defining the iteration or amplification operator G_R as $G_R = 1 + \omega S$. The eigenvalues of G_R satisfy the relations (where the eigenvalues Ω_J of S, have non-positive real parts),

$$\lambda_J(G_R) = 1 + \omega\Omega_J \qquad (10.3.6)$$

The parameter ω can be chosen to damp selectively certain frequency bands of the Ω_J-spectrum, but has to be limited for stability by the following condition,

$$0 < \omega < \frac{2}{|\Omega_J|_{max}} = \frac{2}{\rho(S)} \qquad (10.3.7)$$

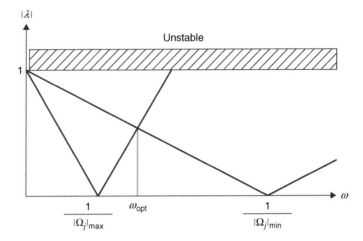

Figure 10.3.1 *Spectrum of Richardson relaxation operator for real negative Ω.*

An optimal relaxation parameter can be selected, following Figure 10.3.1 as

$$\omega_{opt} = \frac{2}{|\Omega_J|_{max} + |\Omega_J|_{min}} \tag{10.3.8}$$

The corresponding spectral radius of the Richardson iteration operator is

$$\rho(G_R) = 1 - \frac{2|\Omega_J|_{max}}{|\Omega_J|_{max} + |\Omega_J|_{min}} = \frac{\kappa(S) - 1}{\kappa(S) + 1} \tag{10.3.9}$$

where the condition number of the S matrix has been introduced

$$\kappa(S) = \frac{|\Omega_J|_{max}}{|\Omega_J|_{min}} \tag{10.3.10}$$

For the Laplace operator, the eigenvalue (10.1.53) leads to

$$|\Omega_J|_{max} = 8 \quad |\Omega_J|_{min} = 8 \sin^2 \frac{\pi}{2M} \approx \frac{2\pi^2}{N}$$

and the condition number is

$$\kappa(S) = \frac{1}{\sin^2 \frac{\pi}{2M}} \approx \frac{4N}{\pi^2} \tag{10.3.11}$$

The approximate equality refers to the case of large values of M. On a mesh 50×50 points $\kappa(S) \approx 1000$.

The convergence rate (10.1.40) becomes, for large M

$$\rho(G_R) = 1 - \frac{2\pi^2}{N} = 1 - \frac{2}{\kappa(S)} \tag{10.3.12}$$

giving the same (poor) convergence rate as the point Jacobi method, see equation (10.1.56).

We could improve the convergence properties by adopting a series of different values of ω, such as to cover the whole range from $1/|\Omega_J|_{\max}$ to $1/|\Omega_J|_{\min}$. Such a process, whereby ω changes from one iteration to another, is called ***non-stationary relaxation***. When ω is constant, we deal with a ***stationary*** relaxation.

The selection of ω in a sequence of p-values covering this range can be optimized to yield the maximal convergence rate. This is realized when the intermediate values of ω are distributed between the two extreme values $1/|\Omega_J|_{\max}$ to $1/|\Omega_J|_{\min}$, as the zeros of the Chebyshev polynomials, Young (1971) and Marchuk (1975).

$$\frac{1}{\omega_k} = \frac{|\Omega_J|_{\max} + |\Omega_J|_{\min}}{2} + \frac{|\Omega_J|_{\max} - |\Omega_J|_{\min}}{2} \cos\left(\frac{2k-1}{p}\frac{\pi}{2}\right)$$

$$k = 1, 2, \ldots, p \qquad\qquad (10.3.13)$$

Observe, that for antisymmetric S matrices $(S = -S^T)$, the eigenvalues are purely imaginary, for instance with the centrally discretized first space derivative in the convection equation. In this case, the eigenvalues of the Richardson iteration matrix become equal to $\lambda(G_R) = 1 + I\omega\Omega$, and the iterative method will diverge since $|\lambda(G_R)| > 1$. This is in full agreement with the instability of the Euler explicit scheme for the central differenced convection equation.

10.3.2 Alternating Direction Implicit Method (ADI)

An effective iterative method is obtained by using as conditioning operator P, the ADI-space factorized operators described in Section A9.4. Referring to equation (9.4.16), the ADI iterative method is defined by

$$(1 - \tau S_x)(1 - \tau S_y)(1 - \tau S_z) = \tau\omega(S \cdot U^n + Q^n) \qquad\qquad (10.3.14)$$

and solved in the sequence

$$
\begin{aligned}
(1 - \tau S_x)\overline{\Delta U} &= \tau\omega(S \cdot U^n + Q^n) \\
(1 - \tau S_y)\overline{\overline{\Delta U}} &= \overline{\Delta U} \\
(1 - \tau S_z)\Delta U^n &= \overline{\overline{\Delta U}}
\end{aligned}
\qquad\qquad (10.3.15)
$$

The parameters ω and τ have to be optimized in order to make the ADI iterations efficient. For optimized parameters, the convergence rate can be of the order of $N \log N$ and is an order of magnitude faster than SOR. However, the ADI method is difficult to optimize for general problems. A discussion of this topic is to be found in the books of Wachspress (1966), Mitchell (1969) and Mitchell and Griffiths (1980).

It can be shown that an optimal value of ω is $\omega_{\mathrm{opt}} \approx 2$. A guideline to selections of τ is

$$\tau_{\mathrm{opt}} = \frac{1}{\sqrt{|\Omega_J|_{\min}|\Omega_J|_{\max}}}$$

or to distribute τ in a sequence of p-values covering the range of frequencies in a selected direction. If the S matrix has the Fourier harmonics as eigenvectors (for periodic conditions), then the eigenvalues of the different submatrices can be obtained by the Von Neumann analysis. For the Laplace equation, the eigenvalues $\lambda(G_{ADI})$ are given by equation (9.4.27) for $\omega = 1$, but with $\omega = 2$, we obtain, with $\tau_1 = \tau/\Delta x^2$

$$\lambda(G_{ADI}) = \frac{(1 - 4\tau_1 \sin^2 \phi_x/2)(1 - 4\tau_1 \sin^2 \phi_y/2)}{(1 + 4\tau_1 \sin^2 \phi_x/2)(1 + 4\tau_1 \sin^2 \phi_y/2)} \tag{10.3.16}$$

The ADI iterative method is clearly unconditionally stable for positive values of τ. If the frequencies in a given direction, say y, are selected to be damped we can distribute τ between the minimum and maximum values associated respectively with the high frequency $\phi_y = \pi$ and the lowest frequency $\phi_y = \pi/M$

$$(\tau_1)_{min} = \frac{1}{4} \quad (\tau_1)_{max} = \frac{1}{4\sin^2(\pi/2M)} \approx \frac{M^2}{\pi^2} \tag{10.3.17}$$

The sequence of τ-values could be distributed by a Chebyshev optimization, for instance following equation (10.3.13). Another distribution law has been applied by Ballhaus et al. (1978) to the computation of solutions to the transonic potential flow equation

$$\tau = \tau_{min} \left(\frac{\tau_{max}}{\tau_{min}} \right)^{(k-1)/(p-1)} \tag{10.3.18}$$

where p is taken between 5 and 10. Even with an approximate optimization, the ADI method is very efficient and generally requires less computational time than point or line overrelaxation.

10.3.3 Other Preconditioning and Relaxation Methods

All the matrices representing the relaxation methods can be considered as preconditioning matrices. An essential guideline toward the selection of the operator P is to be obtained from the generalization of equation (10.3.1) defining the Richardson iteration. If we write the general form of the iteration scheme as

$$U^{n+1} = U^n - \tau \omega P^{-1}(S \cdot U^n + Q^n) \tag{10.3.19}$$

we can consider this method as a Richardson iteration for the operator $B = \tau P^{-1}S$, which is positive definite by definition of P. Hence, the eigenvalues of the iteration matrix $G = 1 - \tau P^{-1}S$, satisfy the relation

$$\lambda(G) = 1 - \omega\lambda(\tau P^{-1} \cdot S) \tag{10.3.20}$$

An optimal relaxation parameter can be defined by

$$\omega_{opt} = \frac{2}{\lambda_{min}(\tau P^{-1} \cdot S) + \lambda_{max}(\tau P^{-1} \cdot S)} \tag{10.3.21}$$

and the spectral radius of G, becomes

$$\rho(G) = \frac{\kappa(\tau P^{-1} S) - 1}{\kappa(\tau P^{-1} S) + 1} \tag{10.3.22}$$

The Richardson iteration has a poor convergence rate, because of the very high value of the condition number. In practical problems, the mesh is often dense in certain regions and more importantly, not uniform. A large variation in mesh spacing contributes to an increase in the condition number, and values of 10^5–10^6 are not uncommon in large-size problems. On the other hand, a condition number close to one leads to an excellent convergence rate. Therefore, various methods have been developed which aim at choosing P such that the eigenvalues of $\tau P^{-1} S$ are much more concentrated. We mention here a few of them, referring the reader to the literature cited for more details.

Incomplete Choleski factorization, Meijerink and Van Der Vorst (1977, 1981), Kershaw (1978)

During an LU factorization of the banded matrix S, elements appear in L and U, at positions where the S matrix has zero-elements. If an approximate matrix P is defined formed by $\overline{L}$ and $\overline{U}$, $P = \overline{L}\,\overline{U}$ where $\overline{L}$ and $\overline{U}$ are derived from L and U by maintaining only the elements which correspond to the location of non-zero elements of S, we can expect P to be a good approximation to S and that $\rho(P^{-1} S)$ will be close to one. Hence, $\overline{L}$ is defined by

$$\begin{aligned}\overline{l}_{IJ} &= l_{IJ} \quad \text{if } s_{IJ} \neq 0 \\ &= 0 \quad\ \ \text{if } s_{IJ} = 0\end{aligned} \tag{10.3.23}$$

and similarly for U. Since P is very close to S, the matrix $P^{-1} S = (\overline{LU})^{-1} LU$ will be close to the unit matrix and the condition number of $P^{-1} S$ will be strongly reduced compared to the condition of the S matrix. Hence, $\lambda(G) = 1 - P^{-1} S$ will be small and high convergence rates can be expected. When used with a conjugate gradient method, very remarkable convergence accelerations on symmetric, positive definite matrices, have been obtained, Kershaw (1978). The coupling with the conjugate gradient method is essential, since the P matrix obtained from the incomplete Choleski decomposition although close to unity, still has a few extreme eigenvalues, which will be dominant at large iteration numbers. The conjugate gradient method, when applied to this preconditioned system, rapidly eliminates the extreme eigenvalues and corresponding eigenmodes. We are then left with a system where all the eigenvalues are close to one and the amplification matrix becomes very small.

Strongly Implicit Procedure (SIP), Stone (1968), Schneider and Zedan (1981)

This is also an incomplete factorization technique, based on the matrix structure arising from the discretization of elliptic operators. The obtained matrix is block tridiagonal or block pentadiagonal for the most current discretizations. The SIP method defines a conditioning matrix P as $P = L \cdot U$, where $U = L^{T}$ for symmetrical matrices such that L and U have the non-zero elements in the same location as the S matrix. A procedure is derived allowing the determination of L and L^{T} and the resulting system is solved by a forward and backward substitution as

$$L\overline{\Delta U} = -R^n$$
$$L^T \Delta U^n = \overline{\Delta U}$$

(10.3.24)

The L and L^T matrices are obtained by a totally different procedure from the Choleski factorization.

Conjugate Gradient Method, Reid (1971), Concus et al. (1976), Kershaw (1978)

This method, originally developed by Hestenes and Stiefel is actually a direct method, which did not produce the expected results with regard to convergence rates. However as an acceleration technique, coupled to preconditioning matrices, it can produce extremely efficient convergence accelerations. Coupled to the incomplete Choleski factorization, the conjugate gradient method converges extremely rapidly, see Golub and Meurant (1983) and Hageman and Young (1981) for some comparisons on elliptic test problems. However they require more storage than the relaxation methods and this might limit their application for three-dimensional problems.

Generalized Minimum Residual (GMRES), Saad and Schultz (1985), Saad (2003)

The Generalized Minimum Residual (GMRES) method can be considered as an extension of the family of conjugate gradient methods, and has been developed for non-symmetric linear systems. It can be very efficient, once the iterative solution is close enough to the converged solution. However, memory requirements could be prohibitive for very large-scale problems.

We refer you the recent literature, in particular Barrett et al. (2000), Saad (2003) for more information on these methods, as well as to the PETSc library for applications to complex problems.

For simplified problems, as will be presented in Chapters 11 and 12, we suggest you program the simplest iterative methods, described in Sections 10.1 and 10.2 yourself, in order to develop a 'feeling' for their properties.

10.4   NONLINEAR PROBLEMS

In a nonlinear problem, we define a Newton linearization in order to enter an iterative approximation sequence. If the nonlinear system is of the form

$$S(U) = -Q$$

(10.4.1)

the Newton iteration for U^{n+1} is

$$S(U^{n+1}) = S(U^n + \Delta U) = S(U^n) + \left(\frac{\partial S}{\partial U}\right) \cdot \Delta U = -Q$$

$$J(U) \stackrel{\Delta}{=} \left(\frac{\partial S}{\partial U}\right)$$

(10.4.2)

where higher order terms in ΔU have been neglected. The Jacobian of S with respect to U, also called tangential stiffness matrix in the finite element literature, defines the iterative system,

$$J(U) \cdot \Delta U^n = -R^n \tag{10.4.3}$$

In the linear case $J = S$ and (10.4.3) is identical to the basic system (10.1.28). In the nonlinear case, the Newton formulation actually replaces the basic system (10.4.1). Indeed, close enough to the converged solution, the higher order terms can be made as small as we wish and the algebraic system (10.4.3) can be considered as an 'exact' equation.

If this equation would be solved by a direct method, we would obtain the exact solution $\overline{U}$ as

$$\overline{U} = U^n - J(U)^{-1} \cdot R^n \tag{10.4.4}$$

The error $e^n = U^n - \overline{U}$, with respect to this exact solution satisfies the relation

$$J(U)e^n = R^n \tag{10.4.5}$$

If the system (10.4.3) is solved by an iterative method, represented by the conditioning operator P/τ,

$$\frac{P}{\tau} \cdot \Delta U^n = -R^n \tag{10.4.6}$$

the error e^n will be amplified by the operator G, such that

$$\begin{aligned} e^{n+1} &= U^{n+1} - \overline{U} = e^n + \Delta U = e^n - \tau P^{-1} R^n \\ &= (1 - \tau P^{-1} J)e^n \equiv Ge^n \end{aligned} \tag{10.4.7}$$

Hence, for a nonlinear system of equations, we can consider that the conditioning operator should be an approximation to the jacobian matrix of the system (10.4.1).

The iterative scheme (10.4.6) will converge if the spectral radius of

$$G = 1 - \tau P^{-1} J \tag{10.4.8}$$

is lower or equal one.

Equation (10.4.8) replaces therefore (10.3.3) for a nonlinear system. Actually, since $J = S$ for linear equations, (10.4.8) is the most general form of the amplification matrix of an iterative scheme with operator P/τ, applied to linear or nonlinear equations.

The definition (10.4.8) allows therefore the analysis of convergence conditions for nonlinear systems when the jacobian operator $J(U)$ can be determined.

Constant and secant stiffness method

This denomination originates from finite element applications, where the following conditioning matrices are often used, essentially for elliptic equations.

Constant stiffness

$$P = S(U^0) \tag{10.4.9}$$

The convergence matrix P is taken as the operator S at a previous iteration and kept fixed.

Secant stiffness

$$P = S(U^n) \tag{10.4.10}$$

The systems obtained can be solved by any of the mentioned methods, direct or iterative. In the former case we have a semi-direct method in the sense that the iterations treat only the nonlinearity.

10.5 THE MULTIGRID METHOD

The multigrid method is the most efficient and general iterative technique known today. Although originally developed for elliptic equations, such as the discretized Poisson equation, Fedorenko (1962, 1964), the full potential of the multigrid approach was put forward and systemized by Brandt (1972, 1977, 1982).

It has since then been applied to a variety of problems with great success, ranging from potential equations to Euler and Navier–Stokes discretizations on structured as well as unstructured grids, and is still subject to considerable research and development (see for instance Thomas et al., 2003).

The following books provide more information on the basics of multigrid methods with selected applications: Hackbusch and Trottenberg (1982), Briggs et al. (2000), Trottenberg et al. (2000) and Wesseling (2004).

The multigrid method finds its origin in the properties of conventional iterative techniques, discussed in the previous sections. Their asymptotic slow convergence, consequence of the poor damping of the low frequency errors (long wavelengths) is a major problem of most iterative techniques. Indeed, as can be seen from equations (10.1.48) and (10.1.50) the asymptotic behavior of the error (or of the residual) is dominated by the eigenvalues of the amplification matrix close to one (in absolute value). These are associated to the eigenvalues of the space discretization operator S with the lowest absolute value, i.e. to the lowest frequencies. For the Laplace operator for instance, the lowest eigenvalue $|\Omega_J|_{min}$ is obtained from equation (10.1.53) for the low frequency component $\phi_x = \phi_y = \pi/M$.

We can therefore consider, on a fairly general basis, that the error components situated in the low frequency range of the spectrum of the space discretization operator S, are the slowest to be damped in the iterative process. On the other hand, the higher frequencies are very effectively damped and after a few iterations, a large part of the high frequency error spectrum will be significantly reduced. Remember that each mode of the Fourier spectrum is reduced by a factor $(\lambda_J)^n$, where λ_J is the eigenvalue of the amplification matrix, as seen from equation (10.1.50). In practical terms, if $\lambda_J = 0.5$, the associated mode will have decreased by a factor 2^n after n iterations, while when $\lambda_J = 0.9999$, it requires 23,000 iterations to reduce the associated error by a factor 10 (see equation (10.1.42)).

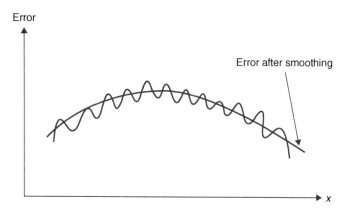

Figure 10.5.1   *Smoothing of the error by relaxation sweeps, damping its high frequency content.*

The multigrid method is entirely based on these properties: if we perform a few iteration sweeps through the mesh (also called relaxation sweeps), we will have damped most of the high frequency part of the residual or error spectrum. As a consequence, we are left with the 'smooth' part of the residual, so that it can be adequately represented on a coarser mesh. We can say that the iterative method acts as a smoother of the error.

See Figure 10.5.1 for a graphical illustration of these properties.

Since the low convergence rate of an iterative method is defined by the lowest frequencies of the space operator S, *the objective of the multigrid method is to transfer these low frequencies to successive coarser grids, where they progressively become part of the high frequency range, which are then very effectively damped by the relaxation sweeps*.

Let us look at how this fundamental property is exploited on a one-dimensional problem.

We consider a one-dimensional space of length L, divided into M cells, with $\Delta x = L/M$, on which we wish to obtain a numerical solution to our problem. We consider this mesh as the 'fine' mesh. The spectrum of the Fourier modes ranges from $\phi = 2\pi \Delta x / \Lambda = k\Delta x = \pi/M$ to $\phi = \pi$, where Λ is the wavelength as seen in previous chapters, covering the wavelengths $2\Delta x$ to $2L$ (Figure 10.5.2).

We will define the 'high frequency' region as contained in the range $\pi/2 < \phi < \pi$; while the low frequency range is defined by the subdomain $0 < \phi < \pi/2$.

If we create now a 'coarse' mesh (mesh C1) by removing every second point, the mesh spacing is equal to $2\Delta x$ and the shortest wave that can be represented on this mesh, equal to twice the mesh size, is now $\Lambda = 4\Delta x$, which corresponds to a highest frequency defined by the phase angle $\phi = \pi/2$; Figure 10.5.2.

Hence the whole range of wavenumber variables is situated between $\phi = 0$ and $\phi = \pi/2$. On this coarser mesh, the high frequency region is defined by the range $[\pi/4, \pi/2]$, while the low frequency range is $[0, \pi/4]$. A few relaxation sweeps on this coarser mesh will damp this high frequency part of the spectrum $[\pi/4, \pi/2]$, *which is part of the low frequency spectrum of the fine mesh*.

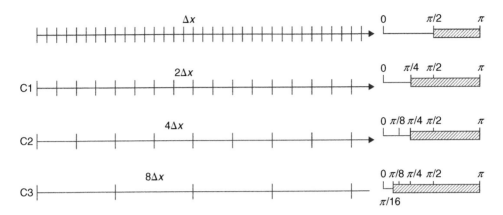

Figure 10.5.2 *Fine and three coarse grid levels with associated parts of the spectrum damped by the relaxation sweeps. The shaded area indicates the damped regions in the phase variable ϕ.*

We can now repeat this process on the next coarser grid (mesh C2) by removing again every second point; we obtain a grid with mesh size $4\Delta x$ and the Fourier spectrum covers the range $[0,\pi/4]$, with a high frequency range $[\pi/8, \pi/4]$. Performing again a few relaxation sweeps will damp the spectrum in this range. Hence, after this action on the second coarse grid, we have damped the fine grid spectrum from π to $\pi/8$. Note that these coarse grid actions have very reduced costs, since the number of unknowns has been divided by 2 (for grid C1) and by 4 (for grid C2). For a Gauss–Seidel method, for which the cost per iteration is proportional to N^2, this means that the cost is reduced respectively by 4 and 16, compared to a fine grid iteration.

For each additional coarser grid, this strategy will damp in a very effective way additional parts of the low frequency fine grid spectrum. On the last of the successive coarser grids, we perform a larger number of sweeps to damp all the remaining frequency. If this is the fourth grid level, e.g. the cost per iteration will divided by a factor 256!

As a consequence, in a single multigrid sequence, the complete error spectrum has been damped. This is a remarkable property and explains the high performance level of the multigrid methods.

The general multigrid methodology is defined by the following succession of steps:

(i) Define a series of successive coarser grids by removing every second line in each direction.
(ii) Transfer the system to be solved from the fine grid to the coarser grids.
(iii) Apply one or more sweeps of an iterative method with good smoothing properties of the higher frequency components, on each of the coarse grids.
(iv) Transfer the corrected solution back to the fine grid in order to generate a new approximation of the solution.

These different steps will be defined more precisely in the next sections, but it should be clear that the error smoothing on the fine mesh is the essential property of multigrid methods.

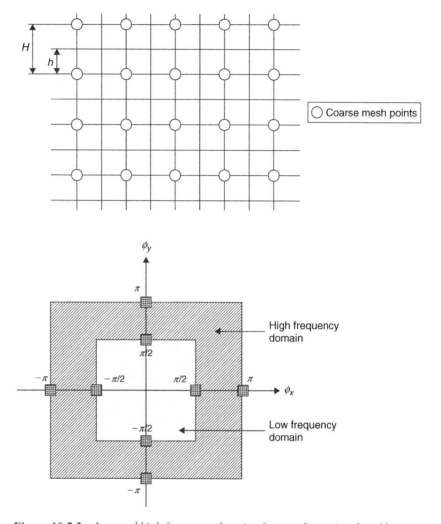

Figure 10.5.3 *Low and high frequency domains for two-dimensional problems.*

10.5.1 Smoothing Properties

The smoothing qualities of an iterative (relaxation) method can be estimated by evaluating the maximum eigenvalues of the amplification matrix in the high frequency domain of the associated wave number variable $\phi = [\pi/2, \pi]$.

In two dimensions, the high frequency domain is defined by the region $\pi/2 \leq \phi_x$, $\phi_y \leq \pi$ (Figure 10.5.3), when the Fourier modes are considered, with a coarsening whereby every second point is removed along the coordinate lines. This is the so-called **standard coarsening**. Hence, the smoothing factor μ is defined by

$$\mu = \underset{\frac{\pi}{2} \leq \phi \leq \pi}{\text{Max}} |\lambda(G)| \tag{10.5.1}$$

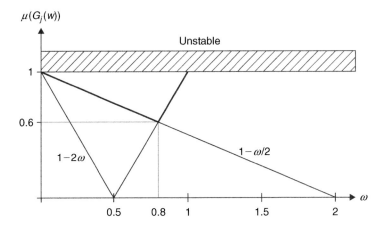

Figure 10.5.4 *Smoothing factor for Jacobi overrelaxation.*

For instance, the Jacobi method applied to the Laplace operator has the eigenvalues (10.1.54), which take on the value -1 for $\phi_x = \phi_y = \pi$.

Therefore the smoothing factor (10.5.1) is equal to one, indicating that some high frequency error components are not damped. As a consequence, the Jacobi method is not a valid smoother. However, with relaxation methods, we can select the relaxation coefficient ω, in order to optimize the high frequency damping properties. With equation (10.2.8), the smoothing factor for Jacobi overrelaxation becomes, introducing the extreme eigenvalues $\lambda(G_J) = -1$ and $+1/2$ in the high frequency domain.

$$\mu(G_J(\omega)) = \text{Max}[|1 - 2\omega|, |1 - \omega/2|] \tag{10.5.2}$$

This is illustrated in Figure 10.5.4, where the optimum relaxation coefficient $\omega = 4/5$ gives the lowest smoothing factor of 0.6, showing that after four iterations the high frequency components will be damped by nearly one order of magnitude.

The Gauss–Seidel iterative method, leads to a still better smoothing factor for the Laplace operator, since we obtain, Hackbusch and Trottenberg (1982), $\mu_{GS} = 0.5$. Here, an order of magnitude reduction in high frequency error components is achieved after three iterations. All the iterative methods can be analyzed for their smoothing properties, and the interested reader will find more information in the mentioned literature. In general, the smoothing factor may depend on the number of relaxation sweeps performed and also on the definition of the coarse mesh points with respect to the fine mesh.

For the Laplace operator on a uniform mesh, one can compare various line relaxation methods for Jacobi or SOR, as well as Red–Black, Zebra, and other variants. The following values are obtained (Hackbusch and Trottenberg, 1982):

Line Jacobi relaxation	0.6
Line Gauss–Seidel	0.447
Red–Black point relaxation	0.25
Zebra line relaxation	0.25
Alternating Zebra	0.048

Red–Black relaxation has a smoothing factor that increases with the number of iterations. This is due to the coupling between high and low frequencies introduced by the red–black sequences and consequently, there is no gain in increasing the number of relaxation sweeps in this case. Note the excellent smoothing properties of alternating zebra line relaxation.

10.5.2   The Coarse Grid Correction Method (CGC) for Linear Problems

After a few relaxation sweeps, the remaining error is expected to be sufficiently smooth to be approximated on a coarser grid, with $2\Delta x$ for instance as basic step size. *The procedure for generating this coarse grid correction is the building block of the multigrid method.*

The fine mesh, on which the solution is sought, will be represented by the symbol h, which will also be used as subscript or superscript in order to indicate the mesh on which the operators or the solutions are defined. Similarly, H will designate the coarse mesh. Both symbols h, H designate also representative mesh sizes, with $H > h$, for instance $H = 2h$ when every second mesh point is removed in the standard coarsening, as shown on Figure 10.5.3.

We consider the linear problem on the fine mesh h,

$$S_h U_h = -Q_h \tag{10.5.3}$$

and an iterative scheme under the residual form (10.1.27)

$$P \cdot \Delta U_h = -R_h \tag{10.5.4}$$

Actually, the coarse grid correction method consists in selecting P proportional to the basic operator S *defined on the coarse mesh.* This process involves three steps:

1. The transfer from the fine to the coarse grid, characterized by a **restriction operator** I_h^H. This operator defines the way the mesh values on the coarse grid are derived from the surrounding fine mesh values. In particular, the residual on the coarse grid, R_H, is obtained by

   $$R_H = I_h^H R_h \tag{10.5.5}$$

2. The solution of the problem on the coarse mesh

   $$S_H \cdot U_H = -Q_H \tag{10.5.6}$$

   or

   $$S_H \cdot \Delta U_H = -R_H \tag{10.5.7}$$

3. The transfer of the corrections ΔU_H from the coarse mesh to the fine mesh. This defines an interpolation or **prolongation operator** I_H^h.

   $$\Delta U_h = I_H^h \Delta U_H \tag{10.5.8}$$

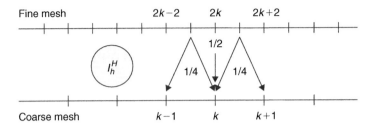

Figure 10.5.5 *Restriction operator for one-dimensional problem.*

Restriction operator

Let us consider a one-dimensional mesh and a second order derivative discretized by the central difference operator, $h = \Delta x$,

$$R_h = \frac{1}{h^2}(u_{i+1} - 2u_i + u_{i-1}) + q_i \tag{10.5.9}$$

where i represents the fine mesh points (Figure 10.5.5). In order to relate i to the coarse mesh points, obtained by removing every second point, we will set $i = 2k$. The coarse mesh points are then designated by $k - 1, k, k + 1, \ldots$. Hence, equation (10.5.9) becomes, for constant q

$$R_h = \frac{1}{h^2}(u_{2k+1} - 2u_{2k} + u_{2k-1}) + q \tag{10.5.10}$$

The full weighting restriction operator is defined by the distribution of Figure 10.5.5, such that each coarse mesh point receives a contribution from itself and the two neighbouring fine mesh points, weighted respectively by the factors 1/2, 1/4, 1/4. The sum of the weighting factors should always be equal to one for consistency.

The coarse mesh residual R_H is obtained by

$$R_H = I_h^H R_h \tag{10.5.11}$$

leading to

$$R_H = \frac{1}{4h^2}(u_{k+1} - 2u_k + u_{k-1}) + q \tag{10.5.12}$$

which is the discretization of the second derivative on the coarse mesh.

This restriction operator is represented here by a matrix $M/2 \times M$, where M is the number of cells in the fine mesh.

$$
I_h^H = \begin{array}{c} k-1 \\ k \\ k+1 \end{array}
\begin{array}{ccccccc}
2k-1 & 2k & 2k+1 & & & \\
\end{array}
\left|
\begin{array}{ccccccc}
\cdot & \cdot & \cdot & \cdot & \cdot & \cdot & \cdot \\
1/4 & 1/2 & 1/4 & \cdot & \cdot & \cdot & \cdot \\
\cdot & \cdot & 1/4 & 1/2 & 1/4 & \cdot & \cdot \\
 & & & & 1/4 & 1/2 & 1/4 \\
\end{array}
\right| \tag{10.5.13}
$$

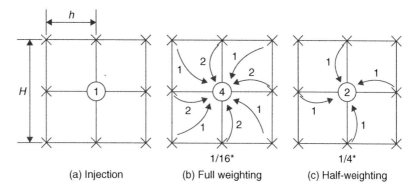

(a) Injection (b) Full weighting (c) Half-weighting

Figure 10.5.6 *Restriction operator in two-dimensions on a uniform mesh.*

Another choice is referred to as **Simple Injection**, whereby the coarse mesh functions are taken equal to their fine mesh value H.

$$I_h^H = 1 \quad \text{for all } 2k \text{ fine mesh points}$$

$$I_h^H = \begin{vmatrix} . & . & . & . & . & . & . & . \\ 0 & 1 & 0 & & & & & \\ & 0 & 1 & 0 & & & & \\ & & 0 & 1 & 0 & & & \\ . & . & . & . & . & . & . & . \end{vmatrix} \tag{10.5.14}$$

In two dimensions, the simple injection is represented by the molecule on Figure 10.5.6a, where the central coarse mesh point is shown surrounded by its fine mesh neighbors. The number inside the circle shows the local weight factor between the fine and coarse mesh functions in this point. Full weighting restriction is shown by the molecule of Figure 10.5.6b, while a half-weighting restriction operator is indicated on Figure 10.5.6c.

Prolongation operator

Prolongation or interpolation operators are generally defined by tensor products of one-dimensional interpolation polynomials of odd order, such as the linear (or cubic) finite element interpolation functions. For the previous one-dimensional example of Figure 10.5.5, a linear interpolation would lead to the representation shown on Figure 10.5.7.

This is represented by a matrix $M \times M/2$

$$I_H^h = \begin{matrix} & \\ 2k-1 \\ 2k \\ 2k+1 \\ & \end{matrix} \begin{vmatrix} & & k-1 & k & k+1 & & & \\ . & . & . & . & . & . & . \\ 0 & 1/2 & 1/2 & 0 & . & . & . \\ 0 & 0 & 1 & 0 & 0 & & \\ 0 & 0 & 1/2 & 1/2 & 0 & . & . \\ & & 0 & 1 & 0 & & \\ & & 1/2 & 1/2 & & & \end{vmatrix} \tag{10.5.15}$$

Generalized to a uniform two-dimensional mesh, we obtain the molecule shown on Figure 10.5.8.

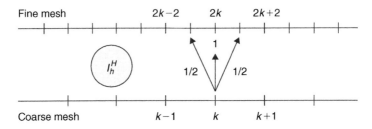

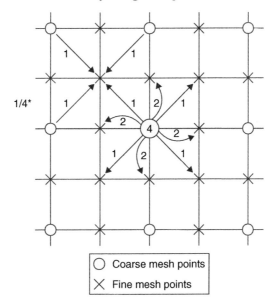

Figure 10.5.7 *One-dimensional prolongation operator.*

Figure 10.5.8 *Two-dimensional prolongation operator.*

Coarse grid correction operator

Combining the three steps mentioned above, the conditioning operator for the coarse grid correction method P_{CGC}

$$P_{CGC} \cdot \Delta U_h = -R_h \tag{10.5.16}$$

is obtained by combining equations (10.5.5), (10.5.7) and (10.5.8), leading to

$$\Delta U_h = I_H^h \Delta U_H$$
$$= -I_H^h S_H^{-1} \cdot R_H \tag{10.5.17}$$
$$= -I_H^h S_H^{-1} \cdot I_h^H R_h$$

Hence

$$P_{CGC}^{-1} = I_H^h S_H^{-1} \cdot I_h^H \tag{10.5.18}$$

The amplification matrix associated to equation (10.5.16), is defined by

$$U_h^{n+1} = G_{CGC} U_h^n - P_{CGC}^{-1} Q \tag{10.5.19}$$

with

$$G_{\text{CGC}} = 1 - P_{\text{CGC}}^{-1} S_h$$
$$= 1 - I_H^h S_H^{-1} \cdot I_h^H S_h \qquad (10.5.20)$$

It is assumed here that the problem is solved exactly on the coarse grid H, i.e. that S can be evaluated.

10.5.3 The Two-Grid Iteration Method for Linear Problems

The multigrid method on the two grids h, H is now completely defined when the smoothing is combined with the coarse grid correction. Hence, the two-grid iteration method is obtained by the following sequence:

1. Perform n_1 relaxation sweeps with smoother S_1 on the fine mesh solution U_h^n
2. Perform the coarse grid correction; obtaining $U_h^{n+1} = U_h^n + \Delta U_h$
3. Perform n_2 relaxation sweeps with a smoother S_2 on the fine mesh solution U_h^{n+1}

Denoting by $G_{h,H}$ the amplification matrix of the two-grid iteration method, by G_S the amplification matrix of the relaxation method selected as smoothing operator, we obtain

$$G_{h,H} = G_{S_2}^{n_2} \cdot G_{\text{CGC}} \cdot G_{S_1}^{n_1} = G_{S_2}^{n_2} \cdot [1 - I_H^h S_H^{-1} \cdot I_h^H S_h] \cdot G_{S_1}^{n_1} \qquad (10.5.21)$$

Typical values for n_1, n_2 are one or two, and the smoothers S_1 and S_2 may be different. For instance when an alternating line relaxation technique is applied (one of the most efficient smoothers for elliptic equations), S_1 could be connected to the horizontal lines and S_2 to the vertical lines. Another option is to modify the value of some relaxation coefficients between S_1 and S_2.

Convergence properties
The two-grid method will converge, if the spectral radius of $G_{h,H}$ is lower than one

$$\rho(G_{h,H}) \leq 1 \qquad (10.5.22)$$

The convergence properties of the two-grid method can be analyzed on simple problems, such as the Poisson equation and uniform meshes, by a Fourier–Von Neumann method, whereby the eigenvalues of the different operators in (10.5.21) are determined. Examples of this approach can be found in Hackbusch amd Trottenberg (1982), where the spectral radius $\rho(G_{h,H}) \leq 1$ is obtained for various combinations and choices of the operators in (10.5.21).

The two-grid method has remarkable convergence properties, which have been proved by W. Hackbusch (see Hackbusch and Trottenberg (1982)), for a rather large class of problems. *Namely, the asymptotic convergence rate is independent of the mesh size h, that is independent of the number of mesh points.*

Hence the computational work, in a two-grid method is determined essentially by the work required to perform the smoothing steps. Since the relaxation sweeps require a number of operations proportional to the number of fine mesh points N, the total work will vary linearly with N.

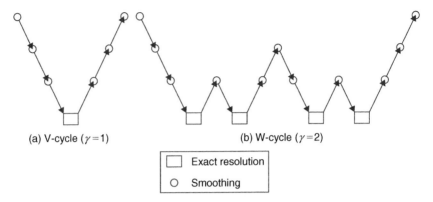

(a) V-cycle ($\gamma = 1$)   (b) W-cycle ($\gamma = 2$)

☐ Exact resolution

○ Smoothing

Figure 10.5.9 *Multigrid strategies on a four-grid method.*

10.5.4 The Multigrid Method for Linear Problems

In the two-grid method, the solution step on the coarser grid, that is obtaining ΔU_H from (10.5.7), is supposed to have been performed accurately. In the multigrid approach, this solution is obtained by application of another two-grid iteration on a coarser grid, for instance a grid of size $2H = 4h$. This can be repeated for the solution of (10.5.7) on the grid $2H$, and so on. The multigrid method is therefore defined on a succession of coarser and coarser grids, applying recursively the two-grid iteration, whereby the 'exact' solution of (10.5.7) has only to be obtained on the last, very coarse grid.

Usually, four to five grids are used and various strategies can be chosen in the sequences of transfer and smoothing between successive grids. The V-cycle, shown on Figure 10.5.9a, consists in a succession of smoothing (symbol 0) and transfer to the next coarser grid, with a unique exact solution (symbol ☐) on the coarsest grid, followed by a unique sequence of transfer and smoothing, back to the finest grid. In the W-cycles intermediate V-cycles are performed on the coarser grids, as shown on Figure 10.5.4b, where (γ) indicates the number of internal V-cycles.

The amplification matrix of the multigrid operator is obtained from equation (10.5.21) by replacing S_H^{-1} in $G_{h,H}$ by

$$S_H^{-1} \Rightarrow (1 - G_{H,2H})S_H^{-1} \tag{10.5.23}$$

and performing the same replacement for the operator S_{2H}^{-1} appearing in $G_{H,2H}$ until the coarsest grid is reached.

It can be shown, that the convergence properties of the two-grid method apply also to the multigrid approach, under fairly general conditions, namely the mesh-size independence of the convergence rate. Therefore the computational work of a multigrid cycle is essentially dominated by the number of operations required for the two-grid computation from the finest to the next grid. Hence, the total work will remain proportional to the number of mesh points N of the fine grid. Note that W-cycles are roughly 50% more expensive than V-cycles, but are generally more robust and should be preferred, particularly for sensitive problems.

Other iterative techniques, require a number of operations for an error reduction of one order of magnitude, of the order of $N^{3/2}$ (optimal SOR) or $N^{5/4}$ for preconditioned

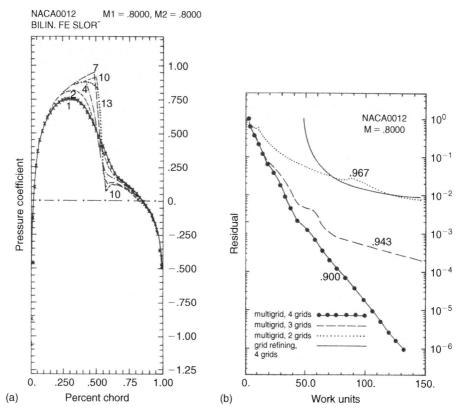

Figure 10.5.10   *Multigrid solutions for transonic flow over a NACA 0012 airfoil at Mach number 0.8 (a) Convergence history of the solution (b) Convergence history of the residuals. The numbers indicate the average convergence rate.*

conjugate gradient methods, against N for the multigrid approach. Hence, multigrid methods are optimal in convergence rates and operation count as well as having general validity.

An additional advantage of multigrid methods, which appears in practice, is the extremely rapid way the solution approaches its final shape, after only a very few multigrid cycles. This is actually a consequence of the fact that the errors are damped over the whole frequency range in every cycle, when passing from the finest to the coarsest grid and back. With other methods, the high frequencies are generally first damped, then progressively as the number of iterations increases, more and more frequencies in the medium and lower range of the spectrum are treated and removed. With the multigrid method, the whole spectrum is scanned during each cycle. An example of this behavior is shown on Figure 10.5.10, from Deconinck and Hirsch (1982), for a potential flow computation over an airfoil in the transonic regime, with multigrid cycles of four grids.

Figure 10.5.10a shows the evolution of the surface Mach number, containing a shock, from the first to the 13th multigrid cycle after which it cannot be distinguished from the final converged solution. After one single cycle, we are already very close to the overall shape of the converged solution, and basically 10–15 multigrid cycles are sufficient for a fully converged solution. Figure 10.5.10b compares the convergence rates of the residual for a line relaxation SLOR iterative method and the multigrid with two, three and four grids. The numbers indicated represent the average convergence rate of the residuals and are an approximation of the spectral radius of the multigrid amplification matrix. One work unit on the horizontal scale corresponds to one SLOR iteration on the fine grid. The improvement with respect to SLOR is spectacular and we can also notice the influence of the number of grids on the overall convergence rates. Note that these computations have been performed for a nonlinear problem by the method presented in the following section.

10.5.5 The Multigrid Method for Nonlinear Problems

For nonlinear problems, the multigrid method can be applied in association with a linearization process such as the Newton linearization (10.4.3). However, we can also adapt the multigrid approach directly to the nonlinear equations by operating simultaneously on the residuals R_h and on the solution U_h itself instead of the corrections ΔU_h. This is known as the **Full Approximation Scheme (FAS)** (Brandt, 1977).

Considering the nonlinear problem on mesh h,

$$S_h(U_h) = -Q_h \tag{10.5.24}$$

the two-grid nonlinear FAS solves for $U^{\text{new}} = (U_h + \Delta U_h)$, solution of

$$S_h(U_h + \Delta U_h) - S_h(U_h) = -R_h \tag{10.5.25}$$

by a transfer to a coarser grid H.

After n_1 relaxation sweeps on U_h, a transfer of residuals and of the solution to the coarse grid is defined by two restriction operators I_h^H and $\hat{I}_h^H$

$$R_H = I_h^H R_h \tag{10.5.26}$$

$$U_H = \hat{I}_h^H U_h \tag{10.5.27}$$

The two restriction operators may be different, since they are operating on different spaces. Hence, on the coarse grid H, equation (10.5.25) becomes

$$S_H(U_H + \Delta U_H) - S_H(U_H) = -R_H \tag{10.5.28}$$

or

$$\begin{aligned} S_H(U_H + \Delta U_H) &= -I_h^H R_h + S_H(\hat{I}_h^H U_h) \\ &= -I_h^H Q_h - I_h^H S_h(U_h) + S_H(\hat{I}_h^H U_h) \end{aligned} \tag{10.5.29}$$

In the multigrid method, equation (10.5.29) is solved by subsequent transfer to a next coarser grid. On the final grid, this nonlinear equation has to be solved exactly. *The key motivation behind this formulation is to ensure that the coarse grid solutions maintain the truncation errors of the fine grid.*

When transferring from the coarse to the fine grid, a prolongation operator is applied to the *corrections* ΔU_H defined by

$$\Delta U_H = U_H^{\text{new}} - U_H \tag{10.5.30}$$

leading to

$$\Delta U_h = I_H^h \Delta U_H \tag{10.5.31}$$

and to the fine grid solution

$$U_h^{\text{new}} = U_h + I_H^h \Delta U_H \tag{10.5.32}$$

The multigrid cycle is closed after performing n_2 smoothing relaxation sweeps on the new solution U_h^{new}. In the full approximation scheme no global linearization is required, except on the last coarsest mesh and for the local linearizations of the relaxation sweeps.

CONCLUSIONS AND MAIN TOPICS TO REMEMBER

Iterative methods for the solution of algebraic systems are an important component of many CFD methods. We have provided in this chapter a brief introduction in order to allow you to grasp the basic components and some of the essential properties of iterative methods. In particular, the following properties are to be kept in mind:

- Each iterative method, also called relaxation method, can be defined by a convergence, or preconditioning, matrix P.
- Iterative methods are very effective in damping effectively the lowest eigenvalues of the amplification matrix $G = 1 - P^{-1}S$.
- The low eigenvalues of G are associated with the high frequencies of the space discretization matrix S, represented by their Fourier modes, and inversely, the high eigenvalues of G are associated with the low frequencies of the matrix S.
- Many variants of preconditioning are available, such as overrelaxation methods. The field of preconditioned methods is very wide and we recommend to the interested reader to consult the cited references for in-depth information.
- From all the current methods, multigrid is by far the most effective and the most general of all convergence acceleration techniques. It has the remarkable property of a convergence rate independent of the number of mesh points; i.e. it reaches the ideal situation of a cost of the order of $O(N)$. Hence, application of multigrid is highly recommended.

REFERENCES

Ballhaus, W.F., Jameson, A. and Albert, T.J. (1978). Implicit approximate factorization schemes for steady transonic flow problems. *AIAA J.*, 16, 573–579.

Barrett, R. et al. (10 authors) (1994). *Templates for the Solution of Linear Systems: Building Blocks for Iterative Methods*. SIAM, Philadelphia, Can be downloaded from http://netlib2.cs.utk.edu/linalg/html_templates/Templates.html

Berman, A. and Plemmons, R.J. (1979). *Non-Negative Matrices in the Mathematical Sciences*. Academic Press, New York.

Brandt, A. (1972). Multilevel adaptive technique for fast numerical solution to boundary value problems. *Proceedings of the 3rd International Conference on Numerical Methods in Fluid Dynamics. Lecture Notes in Physics*, Vol. 18. Springer Verlag, New York, pp. 82–89.

Brandt, A. (1977). Multilevel adaptive solutions to boundary value problems. *Math. Comput.*, 31, 333–390.

Brandt, A. (1982). Guide to multigrid development. In *Multigrid Methods, Lecture Notes in Mathematics*, Vol. 960. Springer Verlag, New York.

Briggs, W., Van Emden, H. and McCormick, H. (2000). A multigrid tutorial. *SIAM – Society for Industrial and Applied Mathematics*, 2nd edn., Philadelphia

Concus, P., Golub, G.H. and O'Leary, D.P. (1976). A generalized conjugate gradient method for the numerical solution of elliptic partial differential equations. In *Space Matrix Computations*, J.R. Bunch and D.J. Rose (Eds). Academic Press, New York.

Dahlquist, G. and Bjork, A. (1974). *Numerical Methods*. Prentice Hall, New Jersey.

Deconinck, H. and Hirsch, Ch. (1982). A multigrid method for the transonic full potential equation discretized with finite elements on an arbitrary body fitted mesh. *J. Comput. Phys.*, 48, 344–365.

Fedorenko, R.P. (1962). A relaxation method for solving elliptic differential equations. *USSR Comput. Math. and Math. Phys.*, 1, 1092–1096.

Fedorenko, R.P. (1964). The speed of convergence of an iterative process. *USSR Comput. Math. and Math. Phys.*, 4, 227–235.

Golub, G.H. and Meurant, G.A. (1983). *Resolution Numerique des Grands Systemes Lineaires*. Eyrolles, Paris.

Hackbusch, W. and Trottenberg, U. (Eds) (1982). *Multigrid Methods. Lecture Notes in Mathematics*, Vol. 960. Springer Verlag, New York.

Hageman, L. and Young, D.M. (1981). *Applied Iterative Methods*. Academic Press, New York.

Kershaw, D.S. (1978). The incomplete Cholesky conjugate gradient method for the iterative solution of linear equations. *J. Comput. Phys.*, 26, 43–65.

LeVeque, R.A. and Trefethen, L.N. (1986). Fourier analysis of the SOR iteration. NASA-CR 178191, ICASE Report 86-63.

Marchuk, G.I. (1975). *Method of Numerical Mathematics*. Springer Verlag, Berlin.

Meijerink, J.A. and Van Der Vorst, H.A. (1977). An iterative solution method for linear systems of which the coefficient matrix is a M-matrix. *Math. Comput.*, 31, 148–162.

Meijerink, J.A. and Van Der Vorst, H.A. (1981). Guidelines for the usage of incomplete decompositions in solving sets of linear equations as they occur in practice. *J. Comput. Phys.*, 44, 134–155.

Meis, T. and Marcowitz, U. (1981). *Numerical Solution of Partial Differential Equations*. Springer Verlag, New York.

Mitchell, A.R. (1969). *Computational Methods in Partial Differential Equations*. J. Wiley & Sons, New York.

Mitchell, A.R. and Griffiths, D.F. (1980). *The Finite Difference Method in Partial Differential Equations*. J. Wiley & Sons, New York.

Reid, J.K. (1971). On the method of conjugate gradients for the solution of large sparse systems of linear equations. In *Large Sparce Sets of Linear Equations*, J.K. Reid (Ed.). Academic Press, New York

Saad, Y. (1993). *Numerical Methods for Large Eigenvalue Problems*, Manchester University Press Series. Out of print. An updated version is available on: http://www-users.cs.umn.edu/~saad/books.html.

Saad, Y. (2003). *Iterative Methods for Sparse Linear Systems*, 2nd edn. SIAM. An updated first edition, originally published in 1996, can be downloaded from http://www-users.cs.umn.edu/~saad/books.html.

Saad, Y. and Schultz, M. (1985). Conjugate gradient-like algorithms for solving non-symmetric linear systems. *Math. Comput.*, 44, 417–424.

Schneider, G.E. and Zedan, M. (1981). A modified strongly implicit procedure for the numerical solution of field problems. *Num. Heat Transfer*, 4, 1–19.

Stone, H.L. (1968). Iterative solution of implicit approximations of multidimensional partial differential equations. *SIAM J. Num. Anal.*, 5, 530–558.

Strang, G. (1976). *Linear Algebra and Its Applications*. Academic Press, New York.

Thomas, J.L., Diskin, B. and Brandt, A. (2003). Textbook multigrid efficiency for fluid simulations. *Ann. Rev. Fluid Mech.*, 35, 317–340.

Trottenberg, U., Oosterlee, C.W. and Schuller, A. (2000). *Multigrid*. Academic Press, New York

Varga, R.S. (1962). *Matrix Iterative Analysis*. Prentice Hall, Inc., New Jersey.

Wachspress, E.L. (1966). *Iterative Solution of Elliptic Systems*. Prentice Hall, New Jersey.

Wesseling, P. (2004). *An Introduction to Multigrid Methods*. R.T. Edwards, Inc, Philadelphia

Young, D.M. (1971). *Iterative Solution of Large Linear Systems*. Academic Press, New York.

PROBLEMS

P.10.1 Write the discretized equations for the nodes close to all the boundaries for Poisson's equation (10.1.1), (10.1.2) based on Figure 10.1.1. Repeat the same exercise for Neumann conditions along the boundary $j = 0$ ($y = 0$), of the form

$$\frac{\partial u}{\partial y} = g \quad y = 0$$

by applying a one sided difference.

P.10.2 Consider the stationary diffusion equation $\alpha u_{xx} = q$ in the domain $0 < x < 1$ and the boundary conditions

$$x = 0 \ u(0) = 0$$
$$x = 1 \ u(1) = 0$$

with $q/\alpha = -4$. Apply a central second order difference and solve the scheme with the tridiagonal algorithm for $\Delta x = 0.1$ and $\Delta x = 0.02$ and compare with the analytical solution.

P.10.3 Solve Problem P.10.2 with the Jacobi method. Write the matrices D, E, F as tridiagonal matrices $B(a, b, c)$.

P.10.4 Solve Problem P.10.2 with the Gauss–Seidel method and compare the convergence rate with the Jacobi method.

P.10.5 Obtain the eigenvalue $\lambda(G_{ADI})$ of equation (10.3.16), for the ADI preconditioning operator.

P.10.6 Consider the steady state convection–diffusion equation $au_x = \alpha u_{xx}$ discretized with second order central differences. Write explicitly the tridiagonal system obtained and apply a Gauss–Seidel method.

Repeat and compare to a SOR method.

P.10.7 Solve the Poisson equation

$$\Delta u = -2\pi^2 \sin \pi x \cdot \sin \pi y$$

on a square $0 < x < 1$, $0 < y < 1$, with the homogeneous Dirichlet boundary conditions, $u = 0$ on the four sides. Select a five-point discretization on a rectangular mesh.

Consider a 11×11 mesh and solve with a Jacobi iteration and compare the convergence rate with the same computation on a 21×21 mesh. Compare the results with the exact solution $u = \sin \pi x \sin \pi y$.

P.10.8 Repeat Problem P.10.7 with a Gauss–Seidel, SOR, SLOR and ADI methods. Try different relaxation parameters and compare convergence rates and computational times.

P.10.9 Consider the alternative definition of a line overrelaxation method, where the intermediate values are obtained with the fully updated values in the right-hand side. Instead of (10.2.33), define

$$\overline{u_{ij}^{n+1}} = \frac{1}{4}(u_{i+1,j}^n + u_{i-1,j}^{n+1} + u_{i,j+1}^{n+1} + u_{i,j-1}^{n+1}) + \frac{1}{4}q_{ij}^n$$

$$u_{ij}^{n+1} = \omega \overline{u_{ij}^{n+1}} + (1 - \omega)u_{ij}^n$$

and obtain the incremental form

$$4\Delta U_{ij}^n - \omega\Delta U_{i,j-1}^n - \omega\Delta U_{i,j+1}^n - \omega\Delta U_{i-1,j}^n = \omega R_{ij}^n$$

Apply to Problems P.10.7, P.10.8 and compare with the other methods.

APPENDIX A: Thomas Algorithm For Tridiagonal Systems

A.1 Scalar Tridiagonal Systems

For tridiagonal systems, the LU decomposition method leads to an efficient algorithm, known as Thomas' algorithm. For a system of the form

$$a_k x_{k-1} + b_k x_k + c_k x_{k+1} = f_k \quad k = 1, \dots, N \tag{A.1}$$

with

$$a_1 = c_N = 0 \tag{A.2}$$

the following algorithm is obtained

Forward step
Calculate successively

$$\beta_1 = b_1 \quad \beta_k = b_k - a_k \frac{c_{k-1}}{\beta_{k-1}} \quad k = 2, \dots, N$$

$$\gamma_1 = \frac{f_1}{\beta_1} \quad \gamma_k = \frac{-a_k \gamma_{k-1} + f_k}{\beta_k} \quad k = 2, \dots, N \tag{A.3}$$

Backward step

$$x_N = \gamma_N$$

$$x_k = \gamma_k - x_{k+1}\frac{c_k}{\beta_k} \quad k = N-1, \ldots, 1 \tag{A.4}$$

This requires in total $5N$ operations.

It can be shown that the above algorithm will always converge if the tridiagonal system is diagonal dominant, i.e. if

$$|b_k| \geq |a_k| + |c_k| \quad k = 2, \ldots, N-1$$
$$|b_1| \geq |c_1| \quad \text{and} \quad |b_N| \geq |a_N| \tag{A.5}$$

If a, b, c are matrices, we have a ***block-tridiagonal system***, and the same algorithm can be applied. Due to the importance of tridiagonal system, we present here a subroutine, which can be used for an arbitrary scalar tridiagonal system.

Subroutine TRIDAG

```
      SUBROUTINE TRIDAG(AA,BB,CC,FF,N1,N)
C
C********************************************************************
C      SOLUTION OF A TRIDIAGONAL SYSTEM OF N-N1+1 EQUATIONS OF THE FORM
C
C      AA(K)*X(K-1) + BB(K)*X(K) + CC(K)*X(K+1) = FF(K)      K=N1,...,N
C
C      K RANGING FROM N1 TO N
C      THE SOLUTION X(K) IS STORED IN FF(K)
C      AA(N1) AND CC(N) ARE NOT USED
C      AA,BB,CC,FF ARE VECTORS WITH DIMENSION N, TO BE SPECIFIED IN THE
C      CALLING SEQUENCE
C
C
C********************************************************************
C
      DIMENSION AA(1),BB(1),CC(1),FF(1)
      BB(N1)=1./BB(N1)
      AA(N1)=FF(N1)*BB(N1)
      N2=N1+1
      N1N=N1+N
      DO 10 K=N2,N
      K1=K-1
      CC(K1)=CC(K1)*BB(K1)
      BB(K) =BB(K)-AA(K)*CC(K1)
      BB(K) =1./BB(K)
      AA(K) =(FF(K)-AA(K)*AA(K1))*BB(K)
   10 CONTINUE
C
C      BACK SUBSTTUTION
C
      FF(N) =AA(N)
      DO 20 K1=N2,N
      K=N1N-K1
```

```
      FF(K)=AA(K)-CC(K)*FF(K+1)
  20 CONTINUE
      RETURN
      END
```

A.2 Periodic Tridiagonal Systems

For periodic boundary conditions, and a tridiagonal matrix with one in the extreme corners as in equation (9.1.16), the above method does not apply. The following approach leads to an algorithm whereby two tridiagonal systems have to be solved.

If the periodic matrix $B_p(\vec{a}, \vec{b}, \vec{c})$ has $(N+1)$ lines and columns resulting from a periodicity between points 1 and $N+2$, the solution X is written as a linear combination

$$X = X^{(1)} + x_{N+1}X^{(2)} \quad \text{or} \quad x_k = x_k^{(1)} + x_{N+1} \cdot x_k^{(2)} \tag{A.6}$$

where the vectors $X^{(1)}$ and $X^{(2)}$ are solutions of tridiagonal systems obtained by removing the last line and last column of B_p, containing the periodic elements. If this modified matrix is called $B^{(N)}(\vec{a}, \vec{b}, \vec{c})$, we solve successively, where the right-hand side terms f_k are put in a column vector F.

$$B^{(N)}(\vec{a}, \vec{b}, \vec{c})X^{(1)} = F \tag{A.7}$$

followed by

$$B^{(N)}(\vec{a}, \vec{b}, \vec{c})X^{(2)} = G \tag{A.8}$$

with

$$G^T = (-a_1, \ldots, 0, -c_N) \tag{A.9}$$

The last unknown x_{N+1} is obtained from the last equation by back-substitution

$$x_{N+1} = \frac{f_{N+1} - c_{N+1}x_1^{(1)} - a_{N+1}x_N^{(1)}}{b_{N+1} + c_{N+1}x_1^{(2)} + a_{N+1}x_N^{(2)}} \tag{A.10}$$

The periodicity condition, determines x as

$$x_{N+2} = x_1 \tag{A.11}$$

The subroutine TRIPER is included here, based on this algorithm. Note that if the periodicity condition is

$$x_{N+2} = x_1 + C \tag{A.12}$$

then the periodicity constant C has to be added to the right-hand side of the last instruction, defining $FF(N+2)$.

Subroutine TRIPER

```
      SUBROUTINE TRIPER(AA,BB,CC,FF,N1,N,GAM2)
C
C************************************************************************
C       SOLUTION OF A TRIDIAGONAL SYSTEM OF EQUATIONS WITH PERIODICITY
C       BETWEEN THE POINTS K=N1 AND K=N+2
```

```
C
C          AA(K)*X(K-1) + BB(K)*X(K) + CC(K)*X(K+1) = FF(K)     K=N1,...,N+1
C
C          THE ELEMENT IN THE UPPER RIGHT CORNER IS STORED IN AA(N1)
C          THE ELEMENT IN THE LOWER LEFT  CORNER IS STORED IN CC(N+1)
C          AA,BB,CC,FF,GAM2 ARE VECTORS WITH DIMENSION N+2, TO BE SPECIFIED
C          IN THE CALLING SEQUENCE
C          GAM2 IS AN AUXILIARY VECTOR NEEDED FOR STORAGE
C          THE SOLUTION X(K) IS STORED IN FF(K)
C
C*********************************************************************
C
           DIMENSION AA(1),BB(1),CC(1),FF(1), GAM2(1)
           BB(N1)=1./BB(N1)
           GAM2(N1)=-AA(N1)*BB(N1)
           AA(N1)=FF(N1)*BB(N1)
           N2=N1+1
           N1N=N1+N
           DO 10 K=N2,N
           K1=K-1
           CC(K1)=CC(K1)*BB(K1)
           BB(K) =BB(K)-AA(K)*CC(K1)
           BB(K) =1./BB(K)
           GAM2(K)=-AA(K)*GAM2(K1)*BB(K)
           AA(K) = (FF(K)-AA(K)*AA(K1))*BB(K)
        10 CONTINUE
           GAM2(N)=GAM2(N)-CC(N)*BB(N)
C
C          BACK SUBSTTUTION
C
           FF(N)=AA(N)
           BB(N)=GAM2(N)
           DO 20 K1=N2,N
           K=N1N-K1
           K2=K+1
           FF(K)=AA(K)-CC(K)*FF(K2)
           NN(K)=GAM2(K)-CC(K)*BB(K2)
        20 CONTINUE
C
           K1=N+1
           ZAA=FF(K1)-CC(K1)*FF(N1)-AA(K1)*FF(N)
           ZAA=ZAA/(BB(K1)+AA(K1)*BB(N)+CC(K1)*BB(N1))
           FF(K1)=ZAA
           DO 30 K=N1,N
           FF(K)=FF(K)+BB(K)*ZAA
        30 CONTINUE
C
           FF(N+2)=FF(N1)
           RETURN
           END
```

Part V

Applications to Inviscid and Viscous Flows

We have now reached the stage where you can start applying the acquired methodology to compute realistic flows.

You certainly have gained the awareness from the previous chapters, that for any mathematical model an unlimited number of options are available to set up the numerical model, although practical experience reduces the choice to a more restricted range. In the previous chapters, we have focused essentially on numerical schemes of second order accuracy in space, which is generally considered as providing the best compromise in terms of cost to accuracy ratio. This still leaves many options open, as different schemes have different dissipation and dispersion error properties.

Writing CFD codes is a learning process where many components have to be taken into account, step by step. Whatever mathematical model you select, from the simplest potential flow to Euler equations, up to laminar and turbulent full Navier–Stokes models, you have to make a choice on each of the following topics and to evaluate their impact on the solution accuracy, on convergence behavior and on computational time:

- The type of grid and its mesh point density, cell-centered or cell-vertex configurations.
- The numerical scheme, defined by the selected time and space discretization. For steady state simulations, you need to select only the space discretization.
- The boundary conditions and their numerical implementation.
- The resolution method of the obtained algebraic system and the treatment of the nonlinearities.

Once your code runs without bugs, you should give great attention to **verification and validation**. This is a fundamental step in assuring that your code not only works properly, but even more importantly, in assessing the accuracy dependence with varying grid density and with parameters of the numerical scheme and its implementation.

As you will experience when writing your own code, '*the devil is in the details*', since as soon as you deviate from the ideal configuration of a uniform Cartesian grid, you have to make choices at any step of the discretization on non-uniform grids and with the nonlinear components. We refer here to issues such as the selection of the points where face quantities are evaluated, at mid-point, or by taking averages of neighboring points; the evaluation of gradients; the discretization of boundary conditions and the choice of interpolation between near-boundary points; and many other details that you will face with increasing complexity of the CFD models.

Verification and validation are critical issues of any simulation and refer to two different steps in code development and code assessment. Formal definitions of these concepts are subject to much debate among experts, and the definitions agreed upon in the AIAA[1] Committee on Standards in CFD, AIAA (1998), or the ASME[2] PTC 60 Committee on Verification and Validation in Computational Solid Mechanics, ASME (2006), are not identical. Our own definition of verification is a synthesis, aiming at clarity and simplicity, compatible with the more advanced options from these Committees.

Verification: The process of determining that a model implementation accurately represents the underlying mathematical model and its solutions.

Verification is thus defined as the steps by which you verify that the mathematical model is correctly programmed. It is based on comparing the numerical results with exact, analytical solutions, and should help you assess the correctness and the accuracy of the numerical discretization, as well as its grid dependence.

Validation: The process of determining the degree to which a model is an accurate representation of the real world from the perspective of the intended uses of the model.

Validation refers to comparisons with real world experimental data and will allow you to assess the accuracy of the physical model assumptions entering in your mathematical model. For instance, validating turbulence models, two-phase models, empirical data related to real gas effects, temperature dependence of viscosity, etc, can only be achieved by comparing with experimental data, when and if available.

The relation between the real world, the conceptual mathematical models selected to describe it and the numerical simulation tool, is summarized in Figure V.1, from Oberkampf et al. (2004), where more details on the implications of verification and validation can be found.

The issue of validation of CFD codes for industrial relevant applications is an extremely complex process as it is hardly possible to generate experimental data at the same level of details as provided by the CFD simulation. In addition, both experimental and numerical data are subjected to many error and uncertainty sources, with the consequence that an *absolute* validation of simulation results remains an unreachable objective. It requires, at the end, to take into account the uncertainties and to call upon good judgment and expert knowledge, both of the physical properties of the investigated system and of the numerical properties of the selected scheme, to assess the validity range and the confidence level that we can attach to a numerical result. These uncertainties associated to real life problems, make the issue of verification even more essential, with the objective to ensure that at least the numerical discretization is full proof and has a controllable accuracy.

Verification is therefore a prerequisite, upstream of the validation process and requires a dedicated effort and an adequate and systematic methodology, to cover all the numerical aspects of a CFD code.

[1] AIAA: American Institute for Aeronautics and Astronautics.
[2] ASME: American Society of Mechanical Engineers.

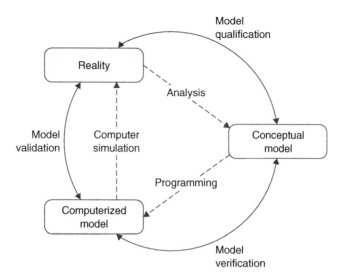

Figure V.1   *Conceptual relation between the real world, the conceptual models and the computer model.*

Consequently, we will restrict your first steps in the world of CFD code development to this verification phase and we will select flow configurations having analytical or well-established reference solutions (obtained for instance on very fine meshes), enabling you an in-depth verification of all the steps of the numerical implementation.

Keep in mind that when comparing with exact solutions, as opposed to experimental results, you are considering only the numerical error sources and it is essential to develop a good understanding of their impact on the solution accuracy. In particular, the verification process should lead to an evaluation of grid dependence, in terms of effects of grid quality and grid density on accuracy. In practical terms, you should be able to establish guidelines on grid density for a pre-selected level of accuracy, as you will wish to know for instance how many mesh points you need for an accuracy of 1% or for an accuracy of 0.1% on a selected quantity. Note also that the response to this question will vary according to the selected quantity. Requiring an accuracy of 0.1% on drag or on a heat transfer coefficient will lead to more severe requirements on the grid density compared to a similar level of accuracy on less sensitive quantities, such as lift or pressure coefficients, since the latter are basically of inviscid origin.

This part contains two chapters, separated into inviscid flow models and laminar flows. Chapter 11 will guide you through the steps for writing a program for inviscid flows, modeled by the potential equation and the Euler equations. We will first select a flow case with an exact solution, such as the incompressible potential flow around a circular cylinder in 2D, which will help us illustrate some major differences between potential and Euler equation models. We will handle next a representative internal compressible flow case of the flow on a circular bump on a lower wall, with a flat upper wall. This flow has no analytical solution, but a reference solution can be obtained by selecting a very fine grid. As third case, a supersonic flow over a wedge, generating an oblique shock with an analytical solution, will be treated, with the objective to introduce you to some aspects of transonic and supersonic flows.

Chapter 12 will concentrate on viscous, laminar flows in two dimensions and some well-known laminar flow cases, such as the Couette flow between two plates, including thermal effects and the flat plate laminar boundary layer, for which exact numerical solutions are known. This chapter will allow us to introduce an alternative solution methods, particularly well adapted to incompressible flows, namely the pressure correction method, which you will be able to compare with the compressible based method developed in Chapter 11. It will be applied to another standard test case, namely the lid driven cavity, for which reference solutions on fine grids are available.

Our main objective with these two chapters is not to show nice or perfect solutions, but on the contrary, to draw your attention to all the potential error sources that are hidden behind a CFD code, and which can affect the accuracy of your solution. This applies in very much the same way to your own code or to a third party code that you would apply as a 'user'. By putting the emphasis on the details of the algorithm implementation, the associated options and eventual user dependent parameters, such as artificial viscosity coefficients, relaxation parameters, boundary conditions, convergence levels and behavior, . . . , we wish to raise your awareness so as to be able to exercise your critical judgment when evaluating a CFD application. We also expect it will guide you in your readiness to ask the right questions to CFD developers or CFD code providers.

The sections on the applications in these chapters have been written with the active participation of Dr. Benoit Tartinville, from Numeca International, who produced also the results of the three inviscid test cases of Chapter 11 and the two viscous cases of Section 12.3, and of Dr. Sergey Smirnov, at the Vrije Universiteit Brussel, who contributed to the sections on the pressure correction method, Sections 12.4 and 12.5, and produced the program and the results for the lid driven cavity of Section 12.5. Their contribution is gratefully acknowledged.

Chapter 11

Numerical Simulation of Inviscid Flows

OBJECTIVES AND GUIDELINES

In this first chapter dedicated to the practical implementation of the simplest of inviscid flows, we will guide you through the various steps required to write a CFD code.

Inviscid flows are modeled in general by the system of Euler equations. Considering flow conditions with uniform inflow, that is an irrotational flow far upstream, we know from inviscid flow theory and Helmholtz theorem, that the flow will remain irrotational everywhere. In other words, it is equivalent to a potential flow. The latter is the highest level of simplification of a flow description, as all the flow variables can be obtained from the single scalar potential function.

The main difference between the potential and the time-dependent Euler equation models, as identified in Chapter 3, lies in their mathematical properties. We have seen, and we refer you to Chapter 3, Sections 3.2.2 and 3.4.1, that the steady potential equation is elliptic at subsonic speeds and hyperbolic in the supersonic range; while the unsteady Euler equations are always hyperbolic in space and time, independently of the flow regime. This has major consequences on the discretization approach, as the schemes for Laplace-like equations of subsonic potential flows, fail when applied to supersonic conditions; while this will not be the case with the Euler equations. Hence, we will focus on time-dependent discretization methods for the Euler equations, looking for the numerical steady solution, as advocated in the previous chapters.

The other important difference is connected to the possible generation of numerical entropy with the Euler equation model, as a consequence of the numerical dissipation of the selected scheme. ***This will allow you to identify one of the most important properties of numerical CFD solutions, namely the 'visible' effects of the numerical dissipation***. We will show that numerical dissipation generates vorticity and entropy and as a consequence regions of entropy increase provide a picture of the influence of numerical dissipation. Since potential flows are by definition irrotational, this *marker* of numerical dissipation is not present with the potential model.

Another important property of flows around solid bodies is the generation of lift and drag, resulting from the balance of pressure and shear stresses on the surface. With inviscid flows, there is no viscous drag and the numerical 'production' of drag is also a global marker of the influence of numerical viscosity.

As the Euler equations form the basis for the full Navier–Stokes solutions, particularly at high Reynolds numbers where the flow properties are dominated by convection, the observed properties will be of critical importance when the same discretization of the convective terms is applied to viscous flows. Indeed, when the numerical viscosity is too high, for instance when applying first order schemes on coarse grids, or when the grid resolution or quality is not sufficient with second order schemes, the effects of this numerical dissipation on the flow behavior can overshadow the effects of the molecular or turbulent viscosity. Therefore, an accurate identification of the level and of the flow areas where this is likely to happen is essential in order to establish the reliability of your numerical results.

It is therefore advocated, when you run a Navier–Stokes simulation, to first perform an Euler simulation of the same test case and on the same grid, looking for the regions influenced by numerical dissipation, by tracking the growth of entropy.

Before starting, we also wish to provide you with some general guidelines on how to structure and organize your program. The key issue is to be able to verify, at each step, the correctness of groups of instructions, as well as your input and output sections. This requires a systematic and modular construction of your program, based on 'building bocks' formed by separate subroutines or modules, each one of them being verified separately.

We recommend you to proceed as follows:

- Start with a main program whose function is exclusively to control all the steps of your algorithm, as well as the input and output modules. Do not introduce algorithmic elements in the main program, concentrating them in separate subroutines.
- Define subroutines with a single objective, i.e. avoid different functionalities in the same subroutine. For instance, if you need to solve algebraic systems, consider a routine that fulfills this objective without adding other functionalities, such as preparing output plots for instance.
- Verify each of these subroutines as an isolated subprogram. For the example mentioned, create algebraic systems with known exact solutions and apply your subprogram to verify that it operates correctly. Once this is achieved, and if errors appear in your program, you will be assured that it does not come form the verified subroutine, but most probably from incorrect or invalid input data.
- During the debugging phase, introduce print instructions at all steps of your program, before and after the instructions, which you will remove when all the bugs are fixed. For instance, when you read in geometrical or flow data from a file, we recommend printing immediately these data, to make sure that they are correctly read, in the proper format and in the expected units. Similarly, when the main program calls a subroutine, put print instructions before and after the call, to verify that the input data as well as the output data are as expected.

The potential and Euler models and some of their essential properties are presented in Sections 11.1 and 11.2. The steady potential flow solutions are treated in Section 11.3. The flow case suggested for the development of your first CFD program is the potential flow around a cylinder, for which we know an exact solution in the limit of incompressible flows. This will allow you to verify the accuracy of your numerical solution, in function of discretization options and grid density. The extension to compressible potential flows will be included, and when keeping the Mach number low enough, you will also be able to verify your numerical solution by comparison with the exact incompressible solution.

Section 11.4 focuses on the application of the Finite Volume Method (FVM) to the system of Euler equations, on a cell-centered grid. We will select a central space discretization, requiring the introduction of ***artificial dissipation***, which we will couple to a Runge–Kutta time integration method. A most critical issue with the Euler equations is the definition and the number of physical boundary conditions we are allowed to impose. This will be related to the hyperbolic properties and the associated characteristic speeds of propagation.

Section 11.5 will guide you to the applications of your FVM code to three test cases. The first one is the flow around the cylinder, as a first verification case. The second case is the compressible internal flow between two solid walls, with a circular bump on the lower wall. Although this flow has no known exact solution, a numerical solution obtained on a fine grid will serve as a reference solution for verification. A third case, with an exact solution, is the supersonic flow over a wedge, generating an oblique shock and will put you in a first contact with some important issues related to the presence of shock discontinuities.

Figure 11.0.1 summarizes the guide through this chapter.

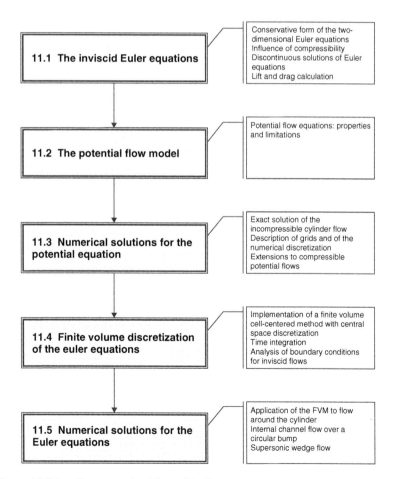

Figure 11.0.1   *Content and guide to this chapter.*

11.1 THE INVISCID EULER EQUATIONS

The system of Euler equations describes flows where the influence of viscous shear stresses and heat conduction effects can be neglected.

Referring to Section 2.7, the system of Euler equations, is written in a compact, conservative form as

$$\frac{\partial U}{\partial t} + \vec{\nabla} \cdot \vec{F} = 0 \tag{11.1.1}$$

This system of first order partial differential equations is hyperbolic in time and space, and in two dimensions the flux vector $\vec{F}$ has the Cartesian components (f, g) given by

$$f = \begin{vmatrix} \rho u \\ \rho u^2 + p \\ \rho u v \\ \rho u H \end{vmatrix} \quad g = \begin{vmatrix} \rho v \\ \rho v u \\ \rho v^2 + p \\ \rho v H \end{vmatrix} \quad U = \begin{vmatrix} \rho \\ \rho u \\ \rho v \\ \rho E \end{vmatrix} \tag{11.1.2}$$

Note that when combining the continuity equation (first equation) with the energy conservation equation (fourth equation), we obtain

$$\frac{\partial H}{\partial t} + u \frac{\partial H}{\partial x} + v \frac{\partial H}{\partial y} = \frac{1}{\rho} \frac{\partial p}{\partial t} \tag{11.1.3}$$

which reduces, for steady flows, to the constancy of total enthalpy H:

$$H = H_{\text{inlet}} \quad \text{along each streamline} \tag{11.1.4}$$

It is important to notice the properties of the entropy variations in an inviscid flow. From equation (2.7.3) and in absence of heat sources, the entropy equation for continuous flow variations reduces to

$$T \left(\frac{\partial s}{\partial t} + \vec{v} \cdot \vec{\nabla} s \right) = 0 \tag{11.1.5}$$

expressing that entropy is constant along a flow path. For steady flows, we have

$$s = s_{\text{inlet}} \quad \text{along each streamline} \tag{11.1.6}$$

The value of the entropy can however vary from one flow path to another. This is best seen from Crocco's form of the momentum equation (1.5.13), which reduces, for a stationary inviscid flow in absence of external forces, to

$$-(\vec{v} \times \vec{\zeta}) = T \vec{\nabla} s - \vec{\nabla} H \tag{11.1.7}$$

In an intrinsic coordinate system with unit vectors $(\vec{e}_l, \vec{e}_n, \vec{e}_b)$, where l is directed along the velocity and b is the binormal direction, this equation becomes, when projected in the normal direction n, for a uniform total enthalpy,

$$|\vec{v}| \zeta_b = T \frac{\partial s}{\partial n} \tag{11.1.8}$$

This relation shows that entropy variations in the direction normal to the local velocity direction are connected to vorticity. ***Hence entropy variations will generate vorticity and inversely, vorticity will create entropy gradients.***

11.1.1   Steady Compressible Flows

The constancy of the total enthalpy, identical to the constancy of the total temperature T_0, takes the following form, assuming perfect gas conditions, where A is a point on the inlet surface:

$$H = c_p T_0 = H_A = c_p T_{0A}$$

$$T_0 = T + \frac{\vec{v}^2}{2c_p} = T\left(1 + \frac{\gamma - 1}{2} M^2\right) \tag{11.1.9}$$

The isentropic relations between pressure, temperature and density, are

$$\frac{\rho}{\rho_0} = \left[\frac{T}{T_0}\right]^{1/(\gamma-1)} = \left[\frac{p}{p_0}\right]^{1/\gamma} \tag{11.1.10}$$

Important property
Combining the isentropic condition s = const., with equation (11.1.9), expressing the constancy of total temperature or total enthalpy, and taking into account equation (1.4.36), repeated here for convenience

$$s - s_A = -r \ln \frac{p_0/p_{0A}}{(H/H_A)^{\gamma/(\gamma-1)}} \tag{11.1.11}$$

we see immediately that the ***total pressure*** has to be constant

$$p_0 = p\left[1 + \frac{\gamma - 1}{2} M^2\right]^{\gamma/(\gamma-1)} = p_{0A} \tag{11.1.12}$$

Hence, all stagnation conditions are constant along streamlines. Note that these constant values can change from one streamline to the other. However, if the incoming flow is uniform, leading to a potential flow, then there is only one constant value over the whole flow field.

11.1.2   The Influence of Compressibility

An important question in practical applications is related to the choice between a compressible flow model and a purely incompressible model, for low velocity flows of gases, such as air.

In other words, below which velocity, or Mach number levels, can we consider a flow of air as incompressible?

The answer to this question can be obtained from the above relations, for instance by evaluating the numerical influence of Mach number on total pressure, equation (11.1.12).

Table 11.1.1 *Mach number influence on compressibility effects.*

M	$\gamma M^4/8$	$M^2/4$
0.01	1.75×10^{-9}	0.000025
0.05	1.09375×10^{-6}	0.000625
0.1	1.75×10^{-5}	0.0025
0.2	0.000280	0.01
0.3	0.00141750	0.0225

It is known from basic fluid mechanics and Bernoulli equation in particular, that in an incompressible flow with constant density, the stagnation pressure is defined by

$$p_0 = p + \rho \frac{\vec{v}^2}{2} \tag{11.1.13}$$

Referring to the general definition (11.1.12), we can expand this equation in power series of Mach number M, in order to compare with equation (11.1.13).
We obtain

$$\frac{p_0}{p} = \left[1 + \frac{\gamma - 1}{2} M^2 \right]^{\gamma/(\gamma-1)} \cong 1 + \frac{\gamma}{2} M^2 + \frac{\gamma}{8} M^4 + \frac{2 - \gamma}{48} \gamma M^6 + O(M^8) \tag{11.1.14}$$

With the definition of Mach number as

$$M^2 = \frac{\vec{v}^2}{\gamma r T} \tag{11.1.15}$$

where the speed of sound is defined as $c = \sqrt{\gamma r T}$, and with the perfect gas law $p = \rho r T$, we obtain

$$p_0 \cong p + \rho \frac{\vec{v}^2}{2} + \frac{\gamma}{8} p M^4 + p \frac{2 - \gamma}{48} \gamma M^6 + O(M^8) \tag{11.1.16}$$

The dominating correction factor for compressibility influence is the third term of this equation $\gamma p M^4/8$, but we prefer to evaluate the ratio of this correction to the second term, ratio equal to $M^2/4$, as seen from equation (11.1.14). For the standard value of $\gamma = 1.4$, valid in particular for air, these factors take the following values for various Mach numbers (Table 11.1.1).

The second column is the ratio between the third term and the static pressure p (first term), while the third column is the ratio between the third term and the **dynamic pressure** (second term). At $M = 0.1$, the compressibility effect represents a correction to the dynamic pressure of 0.25% and reaches 1% at $M = 0.2$.

For practical reasons we may consider a 1% error as acceptable and conclude that for $M \leq 0.2$, the compressibility effects can be neglected and the gas flow can be considered as incompressible. For a flow of air at atmospheric conditions ($T = 288$ K), the speed of sound is $c \approx 340$ m/s and the limit $M = 0.2$ corresponds to a velocity of 68 m/s or 250 km/h. This can be considered as a serious storm if we think about it as an atmospheric wind speed.

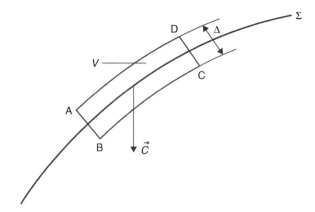

Figure 11.1.1   *Control volume around a moving discontinuity surface Σ (Δ is an infinitesimal distance normal to Σ).*

11.1.3 The Properties of Discontinuous Solutions

As is known, the set of Euler equations allows also discontinuous solutions in certain cases, namely **vortex sheets, contact discontinuities** or **shock waves** occurring in supersonic flows. The properties of these discontinuous solutions can only be obtained from the integral form of the conservation equations, since the gradients of the fluxes are not defined at discontinuity surfaces.

For a discontinuity surface Σ, moving with velocity $\vec{C}$, the integral conservation laws are applied to the infinitesimal volume V of Figure 11.1.1.

Referring to equation (1.1.2) in absence of source terms, the integral form of the Euler equations, takes the following form:

$$\frac{\partial}{\partial t}\int_V U \, d\Omega + \oint_S \vec{F}\cdot d\vec{S} = 0 \tag{11.1.17}$$

The time derivative of the volume integral has to take into account the motion of the surface Σ and hence of the control volume V, through

$$\frac{\partial}{\partial t}\int_V U \, d\Omega = \int_V \frac{\partial U}{\partial t} d\Omega + \int_V U \frac{\partial}{\partial t}(d\Omega)$$

$$= \int_V \frac{\partial U}{\partial t} d\Omega - \oint_S U\vec{C}\cdot d\vec{S} \tag{11.1.18}$$

expressing the conservation of the volume V in the translation with velocity $\vec{C}$.

The flux term in equation (11.1.17) can be rewritten for vanishing volumes $V(\Delta \to 0)$ as

$$\oint_S \vec{F}\cdot d\vec{S} = \int_\Sigma (\vec{F}_2 - \vec{F}_1)\cdot d\vec{\Sigma} \equiv \int_\Sigma [\vec{F}\cdot\vec{e}_n]d\Sigma \tag{11.1.19}$$

where $d\vec{\Sigma}$ is normal to the discontinuity surface Σ and where the notation

$$[A] \equiv A_2 - A_1 \tag{11.1.20}$$

denotes the jump in the variable A when crossing the discontinuity.

Combining (11.1.18) and (11.1.19) we obtain, for vanishing volumes V,

$$\int_{\Sigma} ([\vec{F}] - \vec{C}[U]) \cdot d\vec{\Sigma} = 0 \tag{11.1.21}$$

leading to the local form of the conservation laws over a discontinuity, called the **Rankine–Hugoniot relations**

$$[\vec{F}] \cdot \vec{e}_n - \vec{C}[U] \cdot \vec{e}_n = 0 \tag{11.1.22}$$

If $\Sigma(\vec{x}, t) = 0$ is the discontinuity surface, then we have

$$\frac{d\Sigma}{dt} \equiv \frac{\partial \Sigma}{\partial t} + (\vec{C} \cdot \vec{\nabla})\Sigma = 0 \tag{11.1.23}$$

With the unit vector along the normal $\vec{e}_n$ defined by

$$\vec{e}_n = \frac{\vec{\nabla}\Sigma}{|\vec{\nabla}\Sigma|} \tag{11.1.24}$$

equation (11.1.22) takes the form

$$[\vec{F}] \cdot \vec{\nabla}\Sigma + [U]\frac{\partial \Sigma}{\partial t} = 0 \tag{11.1.25}$$

Various forms of discontinuities are physically possible:

- **Shocks** where all flow variables undergo a discontinuous variation.
- **Contact discontinuities and vortex sheets**, also called **slip lines**, across which no mass transfer takes place but where density, as well as the tangential velocity, maybe discontinuous, although pressure and normal velocity remain continuous.

The properties of these discontinuous solutions can best be seen from a reference system moving with the discontinuity. In this system the discontinuity surface is stationary, $C = 0$, and the **Rankine–Hugoniot relations** for the Euler equations become

$$[\rho v_n] = 0 \tag{11.1.26a}$$

$$[\rho v_n \vec{v}] + [p]\vec{e}_n = 0 \tag{11.1.26b}$$

$$\rho v_n[H] = 0 \tag{11.1.26c}$$

where v_n is the normal component of the velocity vector $v_n = \vec{v} \cdot \vec{e}_n$.

The third equation shows that total enthalpy always remains constant through the discontinuity.

This system admits solutions with the following properties.

11.1.3.1   *Contact discontinuities*

They are defined by the condition of no mass flow through the discontinuity:

$$v_{n_1} = v_{n_2} = 0 \tag{11.1.27}$$

and, following equation (11.1.26), by continuity of pressure

$$[p] = 0 \tag{11.1.28}$$

allowing non-zero values for the jump in specific mass, as seen from equation (11.1.26a)

$$[\rho] \neq 0 \tag{11.1.29}$$

The tangential velocity variation over the discontinuity could be continuous or not, as seen from the tangential projection of equation (11.1.26b).

When the tangential velocity is continuous, we have a ***contact discontinuity***

$$[v_t] = 0 \tag{11.1.30}$$

11.1.3.2   *Vortex sheets or slip lines*

They are also defined by the conditions of no mass flow through the discontinuity, continuous pressure and discontinuous density, as for the contact discontinuity, but with a jump in tangential velocity:

$$v_{n_1} = v_{n_2} = 0$$
$$[p] = 0$$
$$[\rho] \neq 0$$
$$[v_t] \neq 0 \tag{11.1.31}$$

11.1.3.3   *Shock surfaces*

Shocks are solutions of the Rankine–Hugoniot relations with *non-zero mass flow* through the discontinuity, which appear with **supersonic flows**. Consequently, pressure and normal velocity undergo discontinuous variations, while the tangential velocity remains continuous. Hence, shocks satisfy the following properties:

$$[v_n] \neq 0$$
$$[p] \neq 0$$
$$[\rho] \neq 0$$
$$[v_t] = 0 \tag{11.1.32}$$

Note that since the stagnation pressure p_0 is *not constant* across the shock, the inviscid shock relations imply a ***discontinuous entropy variation*** through the shock. This variation has to be positive, corresponding to compression shocks and excluding hereby expansion shocks, for physical reasons connected to the second principle of thermodynamics, (Shapiro, (1953); Zucrow and Hoffmann, (1976)).

It has to be added that expansion shocks, whereby the entropy jump is negative, are also valid solutions of the inviscid equations. Hence, there is no mechanism allowing to distinguish between discontinuities with entropy increase (positive entropy jump) or entropy decrease (negative entropy variation). An additional condition, called the ***entropy condition*** has therefore to be added to the inviscid equation in order to exclude these non-physical solutions, (Lax, 1973). This is necessary for all inviscid flow models and a more detailed discussion of the entropy condition is presented in Volume II.

The mathematical formulation of the second principle of thermodynamics can be expressed, for an adiabatic flow without heat conduction nor heat sources $q_H = 0$, following equation (1.4.18)

$$\rho T \left(\frac{\partial s}{\partial t} + \vec{v} \cdot \vec{\nabla} s \right) = \varepsilon_v \qquad (11.1.33)$$

Since ε_v is the viscous dissipation and always positive, this equation states that any solution of the Euler equations which has a physical sense as a limit, for vanishing viscosity, of real fluid flow phenomena, has to satisfy the following entropy inequality:

$$\left(\frac{\partial s}{\partial t} + \vec{v} \cdot \vec{\nabla} s \right) \geq 0 \qquad (11.1.34)$$

In addition, a non-uniform discontinuity such as a shock with varying intensity will generate a non-uniform entropy field in the direction normal to the velocity. Equation (11.1.8) then shows that as a consequence, vorticity will be generated downstream of the shock. ***Hence, even for irrotational flow conditions upstream of the shock, a rotational flow will be created by a non-uniform shock intensity.***

11.1.4 Lift and Drag on Solid Bodies

The lift and drag resulting forces exerted by the flow on a solid body can be obtained by an extension of the momentum conservation law in integral form, equation (1.3.9). See also the 'Advanced' section A1.6.1 in Chapter 1.

If the control volume Ω contains a solid body, then an additional force $(-\vec{R})$ has to be added to the right-hand side of equation (1.3.9), where $\vec{R}$ is the total force exerted ***by*** the fluid on the body. The total force $\vec{R}$ is composed of the ***lift force*** $\vec{L}$, defined as the component normal to the incoming velocity, and the ***drag force*** $\vec{D}$ defined as the component parallel and opposed to the incoming velocity direction.

We obtain, in absence of external forces, for stationary flows:

$$\oint_S \rho \vec{v}(\vec{v} \cdot d\vec{S}) = -\oint_S p \, d\vec{S} + \oint_S \bar{\bar{\tau}} \cdot d\vec{S} - \vec{R} \qquad (11.1.35)$$

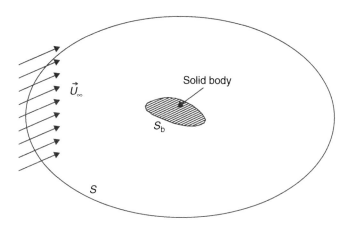

Figure 11.1.2   *Far-field control surface for lift and drag determination on enclosed solid body.*

Based on this equation, two methods can be applied to calculate lift and drag forces:

(i) For a surface S located in the far field (Figure 11.1.2), where the viscous shear stresses can be considered as negligible, the sum of the stationary lift and drag forces are given by the following relation, for stationary flows:

$$-\vec{R} = -(\vec{L} + \vec{D}) = \oint_S \rho\vec{v}(\vec{v} \cdot d\vec{S}) + \oint_S p\,d\vec{S} \tag{11.1.36}$$

(ii) On the other hand, if the control surface S is taken **on the solid body surface** S_b, where the velocity field is either zero due to the non-slip condition of viscous flows or having zero normal velocity with inviscid flows, then the lift and drag forces are also defined by the following relation, since the left-hand side term of equation (11.1.35) is zero

$$\vec{R} = \vec{L} + \vec{D} = -\oint_{S_b} p\,d\vec{S} + \oint_{S_b} \bar{\bar{\tau}} \cdot d\vec{S} \tag{11.1.37}$$

This is an important relation, which is currently applied to determine lift and drag forces from computed flow fields.

For inviscid flows, there are no shear stresses and the lift force over the body is the resultant of the pressure forces:

$$\vec{L} = -\oint_{S_b} p\,d\vec{S} \tag{11.1.38}$$

Note that, for aeronautical applications, the lift force is often considered as defined by the **vertical component of the pressure forces** and its horizontal component is called the **pressure drag**, as it will act alongside the drag force, although it results from the inviscid pressure forces.

11.2 THE POTENTIAL FLOW MODEL

The most impressive simplification of the mathematical description of a flow system is obtained with the approximation of a non-viscous, *irrotational* flow.

The condition of zero vorticity

$$\vec{\zeta} = \vec{\nabla} \times \vec{v} = 0 \tag{11.2.1}$$

will be automatically satisfied if the three-dimensional velocity field can be described by a single scalar potential function ϕ, defined by

$$\vec{v} = \vec{\nabla}\phi \tag{11.2.2}$$

since the rotation of a gradient is identical zero, for any value of the function ϕ.

This reduces the knowledge of the three velocity components to the determination of a single scalar function ϕ.

It is known from the theory of inviscid flows, that the vorticity remains constant in any streamtube. Hence, in a flow where the rotation free condition (11.2.1) is satisfied at the inlet boundary, it will remain so everywhere in the flow domain. In particular, since a uniform flow is rotation free, any subsonic inviscid flow with uniform inlet conditions will remain rotation free everywhere.

This has far-reaching consequences, namely that all the flow variables of a 3D inviscid potential flow are completely defined by the single potential function. Hence we are left with one unknown instead of five, which represents a considerable simplification.

The applications handled in this chapter will be restricted to steady potential flows, which offers an additional simplification.

It is seen from equation (11.1.8) that a potential flow is always isentropic, and with the isentropic relation (11.1.10), the density itself is completely defined by the potential function, as shown by equation (2.8.6), assuming perfect gas conditions:

$$\frac{\rho}{\rho_0} = \left[1 - \frac{\vec{v}^2}{2H}\right]^{1/(\gamma-1)} = \left[1 - \frac{(\vec{\nabla}\phi)^2}{2H}\right]^{1/(\gamma-1)} \tag{11.2.3}$$

where ρ_0 is the stagnation density, constant throughout the whole flow field and γ is the specific heat ratio, equal to $\gamma = 1.4$ for air.

Since all stagnation conditions are constant throughout the whole flow field all thermodynamic properties are known as soon as we know the potential function. This demonstrates the simplification introduced by potential flows, where the knowledge of the single scalar potential function determines all the five flow variables in 3D.

From equation (2.8.5) we obtain the steady potential equation:

$$\vec{\nabla} \cdot (\rho\vec{\nabla}\phi) = 0 \tag{11.2.4}$$

Both for steady and unsteady flows, the inviscid boundary condition along a solid boundary is zero normal relative velocity between flow and solid boundary

$$v_n = \frac{\partial \phi}{\partial n} = \vec{u}_w \cdot \vec{e}_n = 0 \tag{11.2.5}$$

where $\vec{u}_w$ is the velocity of the solid boundary with respect to the considered system of reference and n is the direction normal to the solid wall.

11.2.1 The Limitations of the Potential Flow Model for Transonic Flows

If we consider the steady state potential model for continuous flows, the constancy of entropy and total enthalpy, coupled to irrotationality, form a set of conditions fully consistent with the system of Euler equations. Hence, the model defined by

$$s = s_0 = \text{const.}$$

$$H = H_0 = \text{const.} \tag{11.2.6}$$

and $\vec{v} = \vec{\nabla}\phi$ or $\vec{\nabla} \times \vec{v} = 0$, where ϕ is solution of the mass conservation equation, ensures that the momentum and energy conservation laws are also satisfied. *Therefore, it can be considered that an inviscid continuous flow, with initial conditions satisfying the condition (11.2.1) , will be exactly described by the potential flow model*.

However, in presence of discontinuities such as shock waves, this will not be the case anymore since the Rankine–Hugoniot relations lead to an entropy increase through a shock. If the shock intensity is uniform, then the entropy will remain uniform downstream of the shock, but at another value than the initial constant value. In this case, according to equation (11.1.8), the flow remains irrotational. However, if the shock intensity is not constant, which is most likely to occur in practice, for instance for curved shocks, then equation (11.1.8) shows that the flow is not irrotational anymore and hence the mere existence of a potential downstream of the discontinuity cannot be justified rigorously. Therefore, the *potential flow model in presence of shock discontinuities cannot be made fully compatible with the system of Euler equations*, since the potential model implies constant entropy and has therefore no mechanisms to generate entropy variations over discontinuities.

11.2.2 Incompressible Potential Flows

When the density is constant the potential equation (11.2.4) reduces to the simplest Laplace equation:

$$\Delta \phi = 0 \tag{11.2.7}$$

with the boundary condition of zero normal velocity for a fixed cylinder, expressed by the Neumann boundary condition:

$$\frac{\partial \phi}{\partial n} = 0 \quad \text{on the solid walls} \tag{11.2.8}$$

In the far field, where the inflow velocity $\vec{U}$ is constant, we have from the definition of the potential function:

$$\phi - \phi_0 = \vec{U} \cdot \vec{x} \tag{11.2.9}$$

11.3 NUMERICAL SOLUTIONS FOR THE POTENTIAL EQUATION

We will deal here with a single case, defined by the flow around a cylinder, for which an exact solution is available in the incompressible limit. It will offer a first opportunity to illustrate several aspects of a CFD code related to grid properties and to the accuracy dependence with grid density.

Although applied to the simplest of the models, the conclusions will nevertheless be of more general validity.

11.3.1 Incompressible Flow Around a Circular Cylinder

We start with one of the simplest potential flows, the 2D **incompressible** flow around a circular cylinder of radius a, with uniform inlet velocity U, as illustrated in Figure 11.3.1.

An exact solution is known from classic fluid dynamics, defined by the complex potential function $\zeta = \phi + I\psi$ ($I = \sqrt{-1}$), where ψ is the streamfunction

$$\zeta(z) = \phi(x, y) + I\psi(x, y) \quad z = x + Iy \tag{11.3.1}$$

The complex velocity is defined by

$$u - Iv = \frac{d\zeta}{dz} \tag{11.3.2}$$

leading to the relations

$$u = \frac{\partial \phi}{\partial x} = \frac{\partial \psi}{\partial y}$$

$$v = \frac{\partial \phi}{\partial y} = -\frac{\partial \psi}{\partial x} \tag{11.3.3}$$

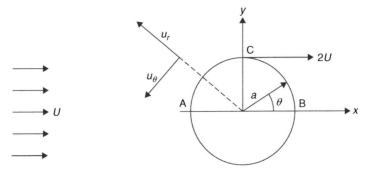

Figure 11.3.1 *Two-dimensional, incompressible potential flow around a circular cylinder for a uniform incident velocity field.*

The exact incompressible potential solution for a cylinder of radius a, is defined by

$$\zeta(z) = \phi(x, y) + I\psi(x, y) = U\left(z + \frac{a^2}{z}\right)$$ (11.3.4)

leading to

$$\phi(x, y) = Ux\frac{x^2 + y^2 + a^2}{x^2 + y^2} \qquad \psi(x, y) = Uy\frac{x^2 + y^2 - a^2}{x^2 + y^2}$$ (11.3.5)

and

$$u - Iv = U\left(1 - \frac{a^2}{z^2}\right)$$ (11.3.6)

or

$$u = U\left[1 - \frac{a^2(x^2 - y^2)}{(x^2 + y^2)^2}\right] \qquad v = -2Ua^2\frac{xy}{(x^2 + y^2)^2}$$ (11.3.7)

On the cylinder surface, that is for $x^2 + y^2 = a^2$, we have

$$\phi = 2Ux \qquad\qquad \psi = 0$$
$$u = 2U\left(1 - \frac{x^2}{a^2}\right) \qquad v = -2U\frac{xy}{a^2}$$ (11.3.8)

Another representation is obtained when replacing the complex position variable z by its polar, instead of its Cartesian, form

$$z = x + Iy = re^{I\theta} = r(\cos\theta + I\sin\theta)$$ (11.3.9)

The exact solution to this potential flow becomes, in cylindrical coordinates

$$\zeta(z) = \phi(r, \theta) + I\psi(r, \theta) = U\left(re^{I\theta} + \frac{a^2}{r}e^{-I\theta}\right)$$
$$\phi(r, \theta) = U\left(r + \frac{a^2}{r}\right)\cos\theta \qquad \psi(r, \theta) = U\left(r - \frac{a^2}{r}\right)\sin\theta$$ (11.3.10)

The polar velocity components are defined by

$$u_r = \frac{\partial\phi}{\partial r} = \frac{1}{r}\frac{\partial\psi}{\partial\theta} = U\left(1 - \frac{a^2}{r^2}\right)\cos\theta$$
$$u_\theta = \frac{1}{r}\frac{\partial\phi}{\partial\theta} = -\frac{\partial\psi}{\partial r} = -U\left(1 + \frac{a^2}{r^2}\right)\sin\theta$$ (11.3.11)

In particular on the cylinder surface, defined by

$$z = ae^{I\theta} = a(\cos\theta + I\sin\theta)$$ (11.3.12)

the complex potential function on the surface reduces to

$$\phi(r,\theta) = 2Ua\cos\theta \quad \psi(r,\theta) = 0 \tag{11.3.13}$$

and the velocity components take the following values:

$$u_r = 0 \quad u_\theta = -2U\sin\theta \tag{11.3.14}$$

This confirms that the cylinder surface is a streamline where the velocity is along the tangential direction. The negative sign results from the definition of the tangential velocity component as being positive in the anti-clockwise direction, while the incoming velocity is in the positive x-direction, as shown on Figure 11.3.1. Note that the velocity is zero at the stagnation points A, B, but reaches a value equal to $2U$, that is twice the incoming velocity at the top of the cylinder in point C, at $\theta = 90°$.

The pressure field is obtained from the constancy of the stagnation pressure (11.1.13), written here for incompressible flows

$$p_0 = p + \rho\frac{\vec{v}^2}{2} = p_\infty + \rho\frac{U^2}{2} \tag{11.3.15}$$

and is best expressed by a non-dimensional pressure coefficient C_p, which is *independent of the inlet velocity*:

$$C_p = \frac{p - p_\infty}{\rho\dfrac{U^2}{2}} = 1 - \frac{\vec{v}^2}{U^2} = 1 - \frac{|\vec{\nabla}\phi|^2}{U^2} \tag{11.3.16}$$

The pressure coefficient on the surface (indicated by a subscript S) is generally plotted in function of the solid wall arc length, and becomes here, with (11.3.14)

$$C_p|_S = \frac{p_S - p_\infty}{\rho\dfrac{U^2}{2}} = 1 - \frac{\vec{v}_S^2}{U^2} = 1 - 4\sin^2\theta \tag{11.3.17}$$

Figure 11.3.2 displays the streamlines and potential lines, as well as the surface pressure coefficient.

We now proceed by following the steps of the previous chapters, as developed in Parts I–IV:

(a) Select the mathematical model.
(b) Define the grid.
(c) Define the numerical scheme.
(d) Establish the stability and accuracy properties of the scheme, based on the material of the previous chapters. If necessary, perform a new analysis.
(e) Solve the algebraic system.
(f) Analyze the results and evaluate the grid dependence and overall accuracy.

These steps should be translated into a flowchart, establishing the structure of your main program, as seen on Figure 11.3.3. Each box should represent a call to one or

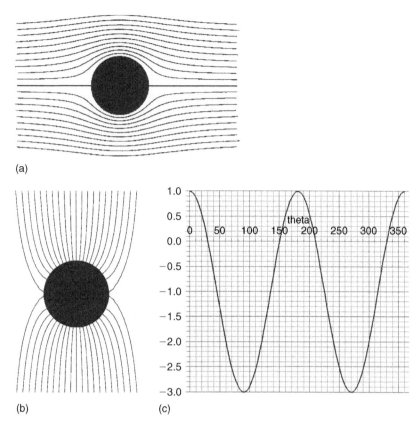

Figure 11.3.2 *Potential flow on a circular cylinder: (a) streamlines, (b) potential lines and (c) surface pressure coefficient C_{pS} in function of angular position θ.*

more separate subroutines, each one having a single objective. It is recommended during the debugging phase, to add print instructions before and after each call to a subroutine, to verify that the intended operations are correctly executed.

Let us now apply this to the mathematical model of the incompressible potential flow equation (11.2.17) with the boundary conditions (11.2.18) and (11.2.9).

11.3.1.1 *Define the grid*

We will select a straightforward analytical grid in polar coordinates, formed by circles and radial lines, allowing you to have full control of mesh density and mesh spacing.

You are now faced with your first decision, namely you have to fix the outer boundary of your computational domain for this external flow problem. This is an important decision, since we apply free undisturbed flow conditions on this boundary and therefore it should be located far enough from the solid body in order to ensure that its influence is negligible. Keep in mind that in subsonic flows, all the points in the flow domain influence each other, as is typical for elliptic equations. Hence, in theory

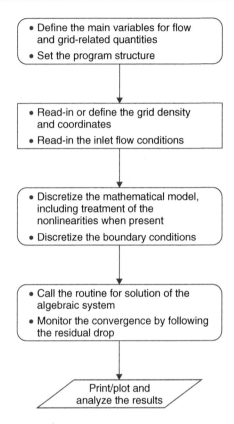

Figure 11.3.3 *Structure of main program.*

we will always have some disturbance from uniform flow conditions on the outer boundary, but if far enough it could be neglected.

In practical terms, a distance of the order of 40–50 times the radius should be recommended for the outer boundary, at least for non-lifting bodies.

For lift generating airfoils, the far field is influenced by a free vortex singularity defined by the circulation around the airfoil, which tends to zero like the inverse of the distance. Either this correction is introduced in the far field, or we have to increase the distance of the outer boundary to values closer to 100 chords.

This initial decision can be a first source of errors, and in case of doubt, you should apply your code with several positions of the downstream boundary, and verify its influence on the solution.

How to choose the grid spacing?

Option 1: Select circles equally spaced in radius value (index i) and radial lines equally spaced in angular position (index j), as shown on Figure 11.3.4. If the outer boundary is fixed at 40 times the cylinder radius a, then if we consider a mesh with Ni points in the radial direction, the radial spacing between the circles should be equal to $40a/Ni$.

Option 2: The first option might seem straightforward, but at second thought it is not such a good idea. Indeed, an important guideline to a good grid is to concentrate more

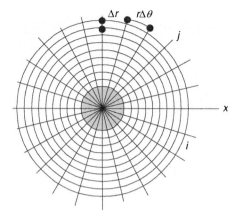

Figure 11.3.4 *Equidistant mesh for circular cylinder.*

grid points in regions of stronger flow variations that is in regions of strong gradients of the flow variables. If we denote du/dn a gradient of a variable u in the n-direction, a good criterion is to aim at a grid where the numerical variation $\Delta u \simeq (\partial u/\partial n) \, \Delta n$ over the cells remains of the same order over the flow domain. This implies that the grid spacing Δn in the n-direction should be inversely proportional to the local gradient intensity. *This guideline leads to smaller grid spacing near the solid boundaries, where the flow gradients will always be significant and larger grid spacing when approaching the far field, where the flow approaches uniform conditions.* Hence, we should select a grid where the radial spacing increases from the solid body surface to the far field with a clustering factor R, as defined for instance by equation (4.3.1), written here as

$$r_{i+1} = aR^i \quad \text{or} \quad \Delta r_i \overset{\Delta}{=} r_{i+1} - r_i = r_i(R - 1) \tag{11.3.18}$$

The factor R is defined by the position of the outer boundary and the number of mesh points Ni in the radial direction:

$$r_{Ni} = aR^{Ni-1} \tag{11.3.19}$$

For instance, selecting $Ni = 33$ points in the radial direction and an outer boundary at $40a$, we obtain

$$40a = aR^{32} \quad \text{or} \quad R = 40^{1/32} = 1.12218478 \tag{11.3.20}$$

The resulting radial coordinates are listed in Table 11.3.1, as the second column, applying equation (11.3.18), for $a = 1$.

Option 3: This grid can still be improved in terms of quality, if we look at the curvilinear polar grid (r, θ) as a transformation from the grid in the physical space. For the considered polar grid, we have

$$x = r \cos \theta \quad y = r \sin \theta \tag{11.3.21}$$

The arc length in the radial direction is Δr and the arc length of a cell side along the circular mesh lines is $r\Delta\theta$. Since the Cartesian grid is considered as the ideal grid, having equal spacing in both directions, we can improve the generated grid by requiring that the polar grid also satisfies the condition of equal arc lengths in the radial and circular directions: $\Delta r_i = r_i\Delta\theta$.

This condition can be satisfied on a grid defined by equation (11.3.18), leading to $(R - 1) = \Delta\theta$, and on a mesh with 128 points in the circular direction, we would have $R = 1 + 2\pi/128 = 1.04908738522$. This value does not allow controlling the

Table 11.3.1 *Radial coordinates for grid options around circular cylinder of radius 1, based on equations (11.3.18) and (11.3.24).*

Radial index i	Radius number $i+1 = a \cdot R^i$ equation (11.3.18)	Radius number $i+1$ equation (11.3.24)
1	1	1
2	1.122185	0.3454
3	1.259299	0.402619
4	1.413166	0.469318
5	1.585833	0.547065
6	1.779598	0.637692
7	1.997038	0.743333
8	2.241045	0.866474
9	2.514867	1.010015
10	2.822145	1.177334
11	3.166968	1.372373
12	3.553924	1.599721
13	3.988159	1.864732
14	4.475452	2.173645
15	5.022284	2.533732
16	5.63593	2.953472
17	6.324555	3.442747
18	7.09732	4.013075
19	7.964504	4.677884
20	8.937645	5.452826
21	10.02969	6.356146
22	11.25516	7.409109
23	12.63037	8.636508
24	14.17361	10.06724
25	15.90541	11.73499
26	17.84881	13.67901
27	20.02967	15.94509
28	22.47699	18.58656
29	25.22333	21.66563
30	28.30524	25.25477
31	31.76371	29.4385
32	35.64475	34.3153
33	40	40

outer boundary position with a user defined number of mesh points in the radial direction. To reach an outer boundary of $40a$, this value of R would require 77 radial mesh points. Inversely, if we keep the value of R defined by condition (11.3.20), we would be constrained in the number of points Nj in the circular direction by $(R-1) = \Delta\theta = 2\pi/Nj$, leading to $Nj \sim 51$.

To keep full control of the number of mesh points in both directions, we should introduce an additional mesh scaling parameter. The straightforward attempt to rescale the clustered grid defined by equation (11.3.18), with a factor k.

$$r_{i+1} = kaR^i \quad \text{or} \quad \Delta r_i \overset{\Delta}{=} r_{i+1} - r_i = kr_i(R-1) \tag{11.3.22}$$

leads to the values $R = 1.165660769$ and $k = 0.296312673$. Because of the low value of k, the first few circular mesh lines are at a radius below the cylinder radius. Hence, this option is not acceptable.

Instead, we could require a mesh clustering, defined by

$$r_{i+1} = r_i + kaR^i \tag{11.3.23}$$

leading to

$$r_{i+1} = a\left(1 + k\sum_{m=0}^{i} R^m\right) \tag{11.3.24}$$

The condition of equal arc lengths of the curvilinear cells, $\Delta r_i = r_i \Delta\theta$ implies

$$\Delta r_i = kaR^i = r_i \Delta\theta = r_i \frac{2\pi}{Nj} \tag{11.3.25}$$

which clearly cannot be satisfied for all cells, as it would require a relation such as (11.3.22), which is different from the choice (11.3.24). However, you could satisfy this condition in an approximate way, by applying it at one point, for instance at the center of the computational domain, for a certain value of the mesh index i. Another simple option is to approach this condition at the level of the cylinder, for $i = 1$, with the choice $k = 2\pi/Nj$, neglecting the effect of the factor R. For $Nj = 128$ with $k = 2\pi/128$, applying (11.3.24) at a distance of $40a$ with 33 mesh points in the radial direction, using symbolic algebra software tools, such as MAPLE or Mathematica, you can obtain the value of the clustering factor R, as $R = 1.1580372$. The resulting radial coordinates are given in the third column of Table 11.3.1 for $a = 1$.

The main message at this initial stage of the mesh generation of your first CFD code is that each step requires sound judgment and a readiness to make choices and approximations.

It is up to you to choose, either to satisfy the equality of arc lengths everywhere, giving away the full control of the grid density and number of mesh points, or to fully control the number of mesh points in both directions, giving away the arc length uniformity.

And of course many other options for generation of grids around the circular cylinder are possible, as presented in Chapter 6.

We consider in the following table that you have selected one of the meshes just described, summarized in Table 11.3.1.

11.3.1.2   *Define the numerical scheme*

Here again, you are faced with many choices for the discretization of the Laplace equation (11.2.7) and its associated boundary conditions. We can write the potential equation in Cartesian coordinates or in cylindrical coordinates and apply a finite difference method (FDM). Alternatively, we can write the equation in integral form and apply a finite volume method (FVM), after having made a selection between the large numbers of possible choices for the control volumes. Finally, you could also apply a finite element method.

We will focus here on the FDM, while the FVM will be applied for the Euler equations.

Finite difference method in cylindrical coordinates

The Laplace equation for the potential function (11.2.7) in cylindrical coordinates is written as

$$
\frac{\partial}{\partial r}\left(r\frac{\partial \phi}{\partial r}\right) + \frac{\partial}{\partial \theta}\left(\frac{1}{r}\frac{\partial \phi}{\partial \theta}\right) = 0
\tag{11.3.26}
$$

and we wish to discretize this equation directly in the (r, θ) space, based on the mesh formed by circles and radial lines.

We refer you to Chapter 4 and equation (4.2.17) for a second order central discretization of this equation. Applied to the configuration of Figure 11.3.5, we obtain the scheme

$$
\frac{1}{(r_{i+1/2,j} - r_{i-1/2,j})}\left(r_{i+1/2,j}\frac{\phi_{i+1,j} - \phi_{i,j}}{\Delta r_i} - r_{i-1/2,j}\frac{\phi_{i,j} - \phi_{i-1,j}}{\Delta r_{i-1}}\right)
$$
$$
+ \frac{1}{\Delta\theta}\left(\frac{1}{r_{i,j+1/2}}\frac{\phi_{i,j+1} - \phi_{i,j}}{\Delta\theta} - \frac{1}{r_{i,j-1/2}}\frac{\phi_{i,j} - \phi_{i,j-1}}{\Delta\theta}\right) = 0
\tag{11.3.27}
$$

The mid-point radii are defined as follows:

$$
r_{i\pm1/2,j} = \frac{1}{2}(r_{i\pm1,j} + r_{i,j}) \quad r_{i,j\pm1/2} = r_{i,j}
\tag{11.3.28}
$$

where $r_{i,j}$ is independent of j, as the mesh line i is of constant radius and the sum

$$
r_{i+1/2,j} - r_{i-1/2,j} = \frac{1}{2}(r_{i+1,j} - r_{i-1,j})
\tag{11.3.29}
$$

Also

$$
\Delta r_i = r_{i+1} - r_i \quad \Delta r_{i-1} = r_i - r_{i-1}
\tag{11.3.30}
$$

All these quantities are defined by the selected grid point distribution.

Refer here to Section 4.3.1.1 in Chapter 4 for an evaluation of the truncation errors associated to these conservative finite difference formulas, in particular

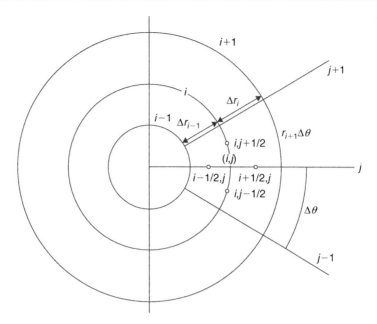

Figure 11.3.5   *Flow around a cylinder: mesh formed by circular and radial lines. The i-index refers to constant radii mesh lines and the j-index to constant angular positions.*

formula (4.3.9). As point (i, j) is not at the center between the points $(i - 1/2, j)$ and $(i + 1/2, j)$, the formulas are first order accurate in general, but in the present case, with the smooth grid variations defined by (11.3.18) or (11.3.24) the factor $(\Delta r_{i+1} - \Delta r_i) \sim O(\Delta r^2)$ and the scheme remains of second order accuracy.

Boundary conditions

Equation (11.3.27) can be applied from the value $i = 2$ on, while for $i = 1$, which is the surface of the cylinder, we have to apply the Neumann boundary condition (11.2.8). This condition can be discretized by a forward difference along the normal direction to connect the points at $i = 2$ to the points at $i = 1$. With the selected mesh this is straightforward since the radial mesh lines are the normals to the cylinder surface.

This is a clear example of the importance of a grid selection adapted to the geometry, since in more general cases the FDM discretization of the normal derivative is slightly more complicated. We will come back to this important issue when applying the finite volume method.

Along a constant radial j-line, the Neumann boundary condition can be approximated as

$$\phi_{1,j} = \phi_{2,j} \tag{11.3.31}$$

When introduced in equation (11.3.20), written for $i = 2$, we obtain

$$\frac{2}{(r_{3,j} - r_{1,j})} \left(r_{5/2,j} \frac{\phi_{3,j} - \phi_{2,j}}{r_{3,j} - r_{2,j}} \right)$$

$$+ \frac{1}{\Delta\theta} \left(\frac{1}{r_{2,j+1/2}} \frac{\phi_{2,j+1} - \phi_{2,j}}{\Delta\theta} - \frac{1}{r_{2,j-1/2}} \frac{\phi_{2,j} - \phi_{2,j-1}}{\Delta\theta} \right) = 0 \quad (11.3.32)$$

once $\phi_{2,j}$ is obtained from the solution of the algebraic system, the wall value is known from equation (11.3.31).

You will notice that the accuracy of the forward difference (11.3.31) will be enhanced if the distance between the two first radii at $i = 2$ and $i = 1$ is small. This adds to the requirements for a higher grid density close to the solid boundaries.

On the far-field side, we have to apply the Dirichlet condition (11.2.9), where we put $\phi_0 = 0$, leading to, if Ni is the index of the outer boundary circle

$$\phi_{Ni,j} = \vec{U} \cdot \vec{x}_{Ni,j} = U r_{Ni,j} \cos\theta_{Ni,j} \quad (11.3.33)$$

This value will be inserted in the system equation (11.3.27), written for $i = Ni - 1$.

We have hereby completed the algebraic system of unknowns for the potential function by combining equations (11.3.32) for $i = 2$, (11.3.27) from $i = 3$ to $i = Ni - 1$, while equation (11.3.33) fixes the values at the outer boundary $i = Ni$.

This system can be written in condensed form, defining the coefficients $a(i, j)$, $b(i, j), c(i, j), d(i, j), e(i, j)$ and the right-hand side $f(i, j)$, by

$$a(i, j)\phi_{i+1,j} + b(i, j)\phi_{i-1,j} + c(i, j)\phi_{i,j+1} + d(i, j)\phi_{i,j-1} - e(i, j)\phi_{i,j} = f(i, j)$$
$$(11.3.34)$$

with the consistency condition

$$e(i, j) = a(i, j) + b(i, j) + c(i, j) + d(i, j) - f(i, j) \quad (11.3.35)$$

For $i = 2$, we have

$$\frac{1}{(r_{3,j} - r_{1,j})} \left((r_{3,j} + r_{2,j}) \frac{\phi_{3,j} - \phi_{2,j}}{r_{3,j} - r_{2,j}} \right)$$

$$+ \frac{1}{\Delta\theta} \left(\frac{1}{r_{2,j}} \frac{\phi_{2,j+1} - \phi_{2,j}}{\Delta\theta} - \frac{1}{r_{2,j}} \frac{\phi_{2,j} - \phi_{2,j-1}}{\Delta\theta} \right) = 0$$

$$a(2, j) = \frac{1}{(r_{3,j} - r_{1,j})} \frac{r_{3,j} + r_{2,j}}{r_{3,j} - r_{2,j}} \quad b(2, j) = 0$$

$$c(2, j) = \frac{1}{r_{2,j}(\Delta\theta)^2} \quad d(2, j) = \frac{1}{r_{2,j}(\Delta\theta)^2} \quad f(2, j) = 0$$

$$e(2, j) = a(2, j) + c(2, j) + d(2, j) \quad (11.3.36)$$

For $i = 3$ to $Ni - 2$

$$\frac{1}{(r_{i+1,j} - r_{i-1,j})} \left((r_{i+1,j} + r_{i,j}) \frac{\phi_{i+1,j} - \phi_{i,j}}{r_{i+1,j} - r_{i,j}} - (r_{i-1,j} + r_{i,j}) \frac{\phi_{i,j} - \phi_{i-1,j}}{r_{i,j} - r_{i-1,j}} \right)$$

$$+ \frac{1}{\Delta\theta} \left(\frac{1}{r_{i,j}} \frac{\phi_{i,j+1} - \phi_{i,j}}{\Delta\theta} - \frac{1}{r_{i,j}} \frac{\phi_{i,j} - \phi_{i,j-1}}{\Delta\theta} \right) = 0$$

$$a(i,j) = \frac{1}{(r_{i+1,j} - r_{i-1,j})} \frac{r_{i+1,j} + r_{i,j}}{r_{i+1,j} - r_{i,j}} \quad b(i,j) = \frac{1}{(r_{i+1,j} - r_{i-1,j})} \frac{r_{i,j} + r_{i-1,j}}{r_{i,j} - r_{i-1,j}}$$

$$c(i,j) = \frac{1}{r_{i,j}(\Delta\theta)^2} \quad d(i,j) = \frac{1}{r_{i,j}(\Delta\theta)^2} \quad f(i,j) = 0 \tag{11.3.37}$$

For $i = Ni - 1$, the system generates a right-hand side as a consequence of the boundary condition (11.3.33), leading to

$$\frac{1}{(r_{Ni,j} - r_{Ni-2,j})} \left((r_{Ni,j} + r_{Ni-1,j}) \frac{\phi_{Ni,j} - \phi_{Ni-1,j}}{r_{Ni,j} - r_{Ni-1,j}} \right.$$

$$\left. - (r_{Ni-2,j} + r_{Ni-1,j}) \frac{\phi_{Ni-1,j} - \phi_{Ni-2,j}}{r_{Ni-1,j} - r_{Ni-2,j}} \right)$$

$$+ \frac{1}{\Delta\theta} \left(\frac{1}{r_{Ni-1,j}} \frac{\phi_{Ni-1,j+1} - \phi_{Ni-1,j}}{\Delta\theta} - \frac{1}{r_{Ni-1,j}} \frac{\phi_{Ni-1,j} - \phi_{Ni-1,j-1}}{\Delta\theta} \right) = 0$$

$$a(Ni-1,j) = 0 \quad b(Ni-1,j) = \frac{1}{(r_{Ni,j} - r_{Ni-2,j})} \frac{r_{Ni-1,j} + r_{Ni-2,j}}{r_{Ni-1,j} - r_{Ni-2,j}}$$

$$c(Ni-1,j) = \frac{1}{r_{Ni-1,j}(\Delta\theta)^2} \quad d(Ni-1,j) = \frac{1}{r_{Ni-1,j}(\Delta\theta)^2}$$

$$f(Ni-1,j) = -\frac{1}{(r_{Ni,j} - r_{Ni-2,j})} \frac{r_{Ni,j} + r_{Ni-1,j}}{r_{Ni,j} - r_{Ni-1,j}} U r_{Ni,j} \cos\theta_{Ni,j}$$

$$e(i,j) = \frac{1}{(r_{Ni,j} - r_{Ni-2,j})} \frac{r_{Ni,j} + r_{Ni-1,j}}{r_{Ni,j} - r_{Ni-1,j}} + b(Ni-1,j)$$

$$+ c(Ni-1,j) + d(Ni-1,j) \tag{11.3.38}$$

We are now ready to move to the next step, the resolution of the algebraic system.

11.3.1.3 *Solve the algebraic system*

To solve the system (11.3.34), we refer you to Chapter 10 and you can apply any of the presented methods.

We suggest that you start with the Jacobi method, which can be applied in a straightforward way.

Next you can program the Gauss–Seidel method, taking into account that the coefficients $b(i, j)$ and $d(i, j)$ coefficients form the lower diagonal matrix E of equation (10.1.13).

11.3.1.4 *Analyze the results and evaluate the accuracy*

Select a grid among the three options described above, selecting a distance of 40 times the radius $a = 1$, for the outer boundary of the computational domain. Define a series of grids $Ni * Nj$ with $33 * 128$, $17 * 64$, $9 * 32$, $5 * 16$ mesh points. The first number Ni refers to the circular mesh lines and the second number Nj to the radial lines.

Perform the following tests:

- Monitor the convergence rate by plotting the residual in function of the iteration number.
- Compare the convergence rates of Jacobi and Gauss–Seidel methods.
- Plot the wall pressure coefficients C_{pS} and compare with the exact solution for different grid densities. Compare in particular the uniform grid option of Figure 11.3.4 with the other two options, where the grid density in clustered near the solid surface.
- Compare the velocity components with the exact values by applying appropriate finite difference formulas to the potential mesh point values, based on the definitions (11.3.11), since the mesh lines follow the cylindrical coordinates.

  Here again you are faced with various options, as you can choose to derive the velocity components in the mesh points (i, j), or at the mid-face values $(i \pm 1/2, j)$ and $(i, j \pm 1/2)$. Referring to Figure 11.3.5, you can apply the following formulas of nominally second order accuracy.

  At mid-points:

  $$(u_r)_{i \pm 1/2, j} = \frac{\phi_{i \pm 1, j} - \phi_{i, j}}{r_{i \pm 1, j} - r_{i, j}}$$

  $$(u_\theta)_{i, j \pm 1/2} = \pm \frac{\phi_{i, j \pm 1} - \phi_{i, j}}{r_{i, j} \Delta \theta} \qquad (11.3.39)$$

  At the mesh points, a good approximation is provided by

  $$(u_r)_{i, j} = \frac{1}{2}[(u_r)_{i+1/2, j} + (u_r)_{i-1/2, j}]$$

  $$(u_\theta)_{i, j} = \frac{1}{2}[(u_\theta)_{i, j+1/2} + (u_\theta)_{i, j-1/2}] \qquad (11.3.40)$$

  You can also apply other formulas based on Chapter 4, Section 4.3.

  Compare the numerical values of the velocity components with their exact values.

- Calculate lift and drag by applying equation (11.1.38). In the present case, this can be calculated as follows:

  $$\vec{L} = -\oint_{S_b} p \, d\vec{S} = -\oint_{S_b} p \, dx \, \vec{e}_y + \oint_{S_b} p \, dy \, \vec{e}_x$$

  $$= \oint_{S_b} pr \, d\theta \, \vec{e}_r$$

  $$= \oint_{S_b} pr \cos \theta \, d\theta \, \vec{e}_x + \oint_{S_b} pr \sin \theta \, d\theta \, \vec{e}_y \qquad (11.3.41)$$

based on the relations between the cylindrical and Cartesian coordinates, on the cylinder surface of radius $r = a$

$$
\begin{aligned}
\mathrm{d}\vec{S} &= -\mathrm{d}y\,\vec{e}_x + \mathrm{d}x\,\vec{e}_y = -r\,\mathrm{d}\theta\,\vec{e}_r = -r\,\mathrm{d}\theta(\cos\theta\vec{e}_x + \sin\theta\vec{e}_y) \\
&= -r\cos\theta\,\mathrm{d}\theta\,\vec{e}_x - r\sin\theta\,\mathrm{d}\theta\,\vec{e}_y
\end{aligned}
\tag{11.3.42}
$$

The integrals in the last line of equation (11.3.41) are easily evaluated numerically by applying a trapezium formula, over the Nj points in the circular direction.

For the circular cylinder, apply

$$
L_y = a \sum_{j=1}^{Nj-1} \frac{1}{2}[p_{1,j}\sin\theta_{1,j} + p_{1,j+1}\sin\theta_{1,j+1}]\Delta\theta
$$

$$
L_x = a \sum_{j=1}^{Nj-1} \frac{1}{2}[p_{1,j}\cos\theta_{1,j} + p_{1,j+1}\cos\theta_{1,j+1}]\Delta\theta
\tag{11.3.43}
$$

- Check the influence of the far-field boundary position by comparing the wall pressure for ratios of distance to radius, ranging from 20 to 80.
- Apply a series of grids: 33 * 128, 17 * 64, 9 * 32, 5 * 16 and plot the error (for instance the L_2-norm of the surface pressure) in function of number of mesh points in log scale. The slope should be close to 2 for second order accuracy.

11.3.2 Compressible Potential Flow Around the Circular Cylinder

Extend your program to handle compressibility by introducing the density based on equation (11.2.4).

The modification to the numerical scheme (11.3.27) is straightforward and leads to the scheme:

$$
\begin{aligned}
\frac{1}{(r_{i+1/2,j} - r_{i-1/2,j})} &\left(\rho_{i+1/2,j} r_{i+1/2,j} \frac{\phi_{i+1,j} - \phi_{i,j}}{\Delta r_i} \right. \\
&\left. - \rho_{i-1/2,j} r_{i-1/2,j} \frac{\phi_{i,j} - \phi_{i-1,j}}{\Delta r_{i-1}} \right) \\
+ \frac{1}{\Delta\theta} &\left(\frac{\rho_{i,j+1/2}}{r_{i,j+1/2}} \frac{\phi_{i,j+1} - \phi_{i,j}}{\Delta\theta} - \frac{\rho_{i,j-1/2}}{r_{i,j-1/2}} \frac{\phi_{i,j} - \phi_{i,j-1}}{\Delta\theta} \right) = 0
\end{aligned}
\tag{11.3.44}
$$

However, two additional complications arise now, since the density is not constant and secondly its dependence on the potential function is not linear. Hence, you have to make choices on these two issues.

11.3.2.1 *Numerical estimation of the density and its nonlinearity*

The density is function of the velocity, following equation (11.2.3) and you can apply directly formulas (11.3.39) to obtain the densities at these mid-point values.

However, you need the two velocity components in the same mesh points:

$$\rho_{i\pm1/2,j} = \rho_0 \left[1 - \frac{(\vec{v}^2)_{i\pm1/2,j}}{2H} \right]^{1/(\gamma-1)}$$

$$(\vec{v}^2)_{i\pm1/2,j} = (u_r^2)_{i\pm1/2,j} + (u_\theta^2)_{i\pm1/2,j} \tag{11.3.45}$$

We leave it to you now to apply various formulas for the u_θ components at the points $(i\pm1/2, j)$, and similarly for the u_r components at the points $(i, j\pm1/2)$.

In equation (11.3.44), the density depends on the potential values and you have to linearize the system by either an explicit or an implicit method. The simplest method is obviously the explicit option, whereby you estimate the density based on the velocities of the previous iteration.

You can also consider improving this approximation by performing a few local iterations on this nonlinear treatment, for each of the relaxation iterations.

Solve the compressible potential flow for a low Mach number value, for instance, $M = 0.05$, and compare with the incompressible exact solution. As seen in Table 11.1.1, at this low value of Mach number the compressibility effect is of the order of 0.0625% of the dynamic pressure, which should correspond to a deviation on the velocity of the order of 0.03125%. Even at $M = 0.1$, the compressibility effect on the dynamic pressure is of 0.25% or 0.125% on the velocity.

11.3.2.2 *Transonic potential flow*

If you push your curiosity to increase the incident Mach number for the compressible version of your program, say at $M = 0.6$ or higher, a shock will appear on the upper surface of the cylinder in the region around $\theta = 90°$ and your program will 'blow up' and no convergent solution will be possible.

This is due to the fact that the potential equation becomes hyperbolic in the variables (x, y) in the supersonic regions, as seen in Chapter 3. The method to cure this problem is to take into account the physical properties of hyperbolic supersonic flows at the level of the discretization.

The first successful computation of a steady transonic potential flow was obtained by Murman and Cole (1971) for the small disturbance equation in two dimensions. This basic work marked a breakthrough that initiated considerable activity in this field, giving rise to an extremely rapid development which led, in about 10 years time, to the situation where the computation of transonic potential flows could be considered as a practically solved problem. A large number of operational codes exist by now, which compute three-dimensional transonic potential flows in a few seconds of CPU time on the most advanced computers, (Holst and Thomas, 1983).

The original idea of Murman and Cole consisted of using different finite difference formulas in the supersonic and subsonic regions. In the subsonic, elliptic, region a central difference is adequate and compatible with the physics of diffusion; while in the supersonic, hyperbolic, region the direction of propagation has to be respected, which leads to the choice of an upwind difference. As with many ideas which appear simple afterwards, the original development required deep understanding of the underlying problems both numerical and physical.

This work was a major landmark in the history of CFD. It is a beautiful example of how new discoveries and progress are made in science in general, and CFD in particular. It is fascinating and instructive to read the historical account of the genesis of these ideas, as reported by Hall (1981) and we hope that you will share our pleasure in quoting this account:

> *Earll Murman had been working for a year or so at Boeing on finite difference methods for integrating the compressible Navier–Stokes equations when, in 1968, Julian Cole arrived on a 1-year visit.*
>
> **Cole writes**: *'It was Goldberg who suggested that transonic flow was a timely subject. I decided on a joint analytical and numerical approach and he said that Earll and I could work together (since my programming was feeble)'. Our approach was founded on several bits of previous experience.*

> *(iii) The fact that (the) Lax–Wendroff (scheme) could give the correct shock jumps (had) made a deep impression and I (had) learned about artificial viscosity, diffusion and dispersion of difference schemes. Yosh (Yoshihara) was convinced that steady flows could not be calculated directly but I decided while at Boeing to try using a conservative scheme (a la Lax) in order to catch shocks.*
> *(iv) I was aware of Howard Emmons very early 'successful' relaxation calculations of mixed flows in nozzles and decided to try a relaxation method.*
> *(v) I had studied the fundamentals of small disturbance theory ... rather extensively earlier. I knew it had all the essential difficulties and could even be a good approximation. It was clear that it would make the numerical work easier.*

> **Murman writes** *that Cole 'spent several months systematically deriving a (transonic) small disturbance (TSP) theory from the complete Euler equations. It laid the theoretical groundwork for our later developments. In January 1969 we started some computations solving Laplace equations and then the TSP using centered finite differences. By April we found that we could not get the calculations to converge for supercritical flow. It was in the following several months that we hit upon the idea of switching and type dependent schemes. I believe that the idea grew out of an afternoon brainstorming session when we were discussing finite difference methods for elliptic and hyperbolic problems and how the two were basically different. Julian, I believe, threw out a comment that maybe we could combine them somehow.*
>
> *I have often reflected back on that event to realize how important it is in research to be open-minded, imaginative, and receptive to unconventional suggestions'.*
>
> **Cole adds**, *'I knew enough numerical analysis to know that hyperbolic schemes were unstable if the domain of dependence was incorrect. Even though the time-like direction was unclear I thought that perhaps we should have only downstream influence. So we decided to switch schemes: explicit hyperbolic was ruled out by the CFL condition near the sonic line'.*
>
> **Murman continues** *'My experience the previous year on the Navier–Stokes computations allowed us to make rapid progress. It was clear that we should*

maintain conservation form to calculate shock waves. Unfortunately we missed the essential point of the shock point operator. For stability reasons, the hyperbolic operator had to be implicit. This naturally led to a line relaxation algorithm so that the method would work in the limits of both purely supersonic and purely subsonic flow. In July we programmed up the first code and it worked almost immediately'.

After this initial work, Murman and Cole's procedure was extended to three dimensions by Ballhaus and Bailey (1972); to the non-conservative full potential equation for two dimensions by Steger and Lomax (1972), Garabedian and Korn (1972), and three dimensions by Jameson (1974). The conservative full potential equation was solved initially by Jameson (1975) for two-dimensional flows and extended to three-dimensional configurations by Jameson and Caughey (1977).

These developments will be dealt with in Volume II.

11.3.3 Additional Optional Tasks

If you have successfully achieved the tasks suggested in the previous section, you have now a good basis to go a step further and extend your experience to other options. We list here a few of them for your consideration, and of course you can think of many others, based on your personal interest and curiosity:

- Apply other iterative methods, such as point and line relaxation methods, as described in Chapter 10.
- The streamfunction ψ is also a solution of the Laplace equation:

$$\Delta \psi = 0 \tag{11.3.46}$$

with the boundary conditions

$$\begin{aligned} \psi &= 0 \quad \text{on the solid surface} \\ \psi &= Uy \quad \text{in the far field} \end{aligned} \tag{11.3.47}$$

You can apply the same discretization as defined by equation (11.3.27), the only change will occur for the values $i = 2$ and $i = Ni - 1$, due to the change in boundary condition.

11.4 FINITE VOLUME DISCRETIZATION OF THE EULER EQUATIONS

This section introduces you to the numerical solutions for the system of Euler equations in two dimensions, and we will guide you through the steps required to solve the same inviscid flow around the cylinder as treated in the previous section, as well as two new test cases. One is typical of internal flows and is represented by the flow on a bump placed on a flat plate, while the second case is a supersonic flow over a wedge with an oblique shock, introducing you to some of the issues related to the numerical simulation of shocks.

In the subsonic range, the solutions of the Euler equations for uniform inflow conditions should be identical to the potential flow solutions, although we have to solve four differential equations, instead of a single one for the potential equation. But the main difference, as stated in the introduction to this chapter, is that Euler equations do not guarantee that the calculated flow remains irrotational or isentropic. Although equation (11.1.5) indicates that in an inviscid subsonic flow with uniform inlet conditions, entropy remains constant and uniform in the whole flow domain, you should be aware that the numerical dissipation generated by the numerical scheme will 'mimic' in some way the physical dissipation of viscous flows. Seen from the point of view of the discretized system, we have to realize indeed that the computer cannot distinguish physical dissipation from numerical dissipation. *Therefore, the numerical system will obey equation (11.1.33) instead, where the right-hand side represents the global dissipation of the numerical model. As a consequence, entropy will not remain constant and we can consider it as a unique indicator of the presence of numerical dissipation in the scheme. In regions where the calculated entropy increases, we can be assured that these regions are influenced by numerical dissipation.* This is a very important property, which provides a direct measure of the quality of the numerical scheme on the selected grid, by monitoring and post-processing the evolution of entropy. Since the numerical dissipation is proportional to a power of the mesh size, the only way to reduce the dissipation of the selected scheme is to refine the grid in the regions where an excessive entropy generation would occur. This will be the case in regions with high-velocity gradients, such as leading edges, trailing edges of airfoils, or regions with abrupt geometry changes, such as sharp corners, and these regions will require higher mesh densities to keep the numerical dissipation influence below an acceptable level.

The system of time-dependent Euler equations is hyperbolic in space and time and we will select the time-dependent numerical formulation, as recommended in the previous chapters, to find the steady state solution.

There is however a problem with incompressible flows, since taking constant density removes the time-dependent term in the continuity equation and this requires special methods to overcome this absence, namely methods of *artificial compressibility* (Chorin, 1967) also called *preconditioning methods* (Merkle and Choi, 1985). These methods extend the time-dependent approach to cover all speed regimes, including the incompressible limit, but their presentation is outside the scope of this introductory text. It will be treated in more details in Volume II.

Hence, we will handle the cylinder case in the domain of small Mach numbers, for instance at values below $M = 0.1$, where compressibility is present but remains small, so that a comparison with the incompressible solution remains meaningful.

We will select a most representative scheme of the family of separate space and time discretizations, combining a centered space discretization with an explicit Runge–Kutta method, as described in Sections 9.3.4 and 9.3.6.

Also, we will apply here a *finite volume method*, in order to generate a program that you can apply to any grid configuration.

11.4.1   Finite Volume Method for Euler Equations

We refer you to Chapter 5 for the general formulation of finite volume methods (FVM) and their practical discretization. If needed, we suggest you study again this chapter to refresh your memory with the main properties of finite volume methods. In particular, the most general formulation is provided by equation (5.2.7), repeated here, for the case of the homogenous Euler equations without source terms, as

$$\frac{d}{dt}\left[\overline{U}_{i,j}\Omega_{i,j}\right] = -\sum_{\text{faces}} \vec{F}^* \cdot \Delta\vec{S} \equiv -R_{i,j} \tag{11.4.1}$$

The right-hand side defines the **residual** $R_{i,j}$ as the balance of fluxes over all the faces forming the cell (i,j).

Some of the essential properties of the FVM are:

- The solution $\overline{U}_{i,j}$ of the system (11.4.1) is the **cell-averaged value** of the conservative variable U over the cell (i, j).
- When the results of the simulation are post-processed, we will assign the cell-averaged values to the center of the cell. **This introduces an error, generally of second order**, which is part of the discretization error.
- The numerical flux $\vec{F}^*$ represents the discretization of the physical fluxes, as defined by the selected numerical scheme.

11.4.1.1 *Space discretization*

For two-dimensional structured grids, we can select either a cell-centered or a cell-vertex approach. We will select here the cell-centered method, which is also a most current approach as illustrated in Figure 11.4.1.

In the cell-centered approach the grid coordinates, as read-in from your mesh input file, are the points such as A, B, C, D, while we attribute the numerical coordinates (i,j) to a cell-center point such as $P(i,j)$. Its coordinates are obtained by the arithmetic average of the four corner cells A, B, C, D. For instance, point $P(i,j)$ is defined by its coordinates

$$\vec{x}_{i,j} = \frac{1}{4}(\vec{x}_A + \vec{x}_B + \vec{x}_C + \vec{x}_D) \tag{11.4.2}$$

and similarly for the other centers of the control volumes.

Figure 11.4.1 shows the relation between a curvilinear grid, formed for instance by the circular arcs of the cylinder mesh applied for the potential solution, and the associated quadrilateral grid that is seen by the program, when each cell side is considered as formed by a straight line. This is typical for second order approximations and is compatible with second order numerical schemes. If higher order schemes would be considered, then a more accurate representation of the cell sides, as curved boundaries, would have to be considered.

To construct your main program, proceed as follows with each step forming the subject of a separate subroutine:

- Read the grid coordinates.
- Define the cell-center coordinates, the cell areas and the face normals, based on the formulas defined in equations (5.3.2):

$$\Delta\vec{S}_{i+1/2,j} = \Delta\vec{S}_{AB} = \Delta y_{AB}\vec{1}_x - \Delta x_{AB}\vec{1}_y = (y_B - y_A)\vec{1}_x - (x_B - x_A)\vec{1}_y$$

$$\tag{11.4.3}$$

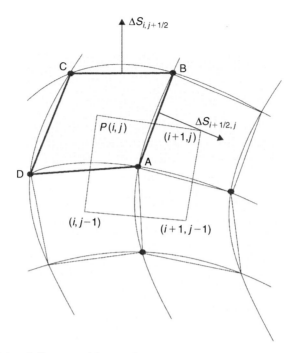

Figure 11.4.1 *Cell-centered finite volume space discretization on a structured grid.*

and the area, following equation (5.3.4)

$$\Omega_{i,j} = \Omega_{ABCD} = \frac{1}{2}|\vec{x}_{AC} \times \vec{x}_{BD}|$$

$$= \frac{1}{2}[(x_C - x_A)(y_D - y_B) - (x_D - x_B)(y_C - y_A)]$$

$$= \frac{1}{2}(\Delta x_{AC}\Delta y_{BD} - \Delta x_{BD}\Delta y_{AC}) \tag{11.4.4}$$

- The numerical flux $\vec{F}^*$ is written as the addition of a central flux, representing the central scheme, plus a dissipation term, under the following form:

$$(\vec{F}^* \cdot \Delta\vec{S})_{i+1/2,j} \equiv \left[\frac{1}{2}(\vec{F}_{i,j} + \vec{F}_{i+1,j}) - D_{i+1/2,j}\right] \cdot \Delta\vec{S}_{i+1/2,j} \tag{11.4.5}$$

where the dissipation term $d_{i+1/2,j}$ is either part of the numerical dissipation or/and contains artificial dissipation terms.

- For the central scheme, an artificial viscosity term is added, following the formulation of Section 9.3.6. It is written here as a difference of third order derivatives, with

$$D_{i+1/2,j} = \gamma_{i+1/2,j}(U_{i+2,j} - 3U_{i+1,j} + 3U_{i,j} - U_{i-1,j}) \tag{11.4.6}$$

- Organize the subroutine for the calculation of this artificial dissipation term, as it requires mesh point values from two additional rows of cells. The value of the γ-coefficient is based on the formulation of Jameson et al. (1981), with the spectral radius of the Jacobian as coefficient

$$\gamma_{i+1/2,j} = \frac{1}{2}\kappa^{(4)}[\vec{v} \cdot \Delta\vec{S} + c|\Delta\vec{S}|]_{i+1/2,j} \qquad (11.4.7)$$

  where c is the speed of sound and $\kappa^{(4)}$ is a non-dimensional coefficient of dissipation, with the minimum value of $\kappa^{(4)} = 1/256$.

- Calculate the cell residuals by *programming a loop over all the cell faces*. For each face, calculate the associated flux, following formula:

$$(\vec{F}^* \cdot \Delta\vec{S})_{i+1/2,j} = (\vec{F}^* \cdot \Delta\vec{S})_{AB} = f^*_{AB}(y_B - y_A) - g^*_{AB}(x_B - x_A) \qquad (11.4.8)$$

- Send the contribution to the right cell and its negative value to the left cell, based on an anti-clockwise positive orientation.

- For the faces situated on the boundaries, take into account the particular boundary conditions, as defined hereafter. In particular on a solid boundary face, all the convective fluxes are zero, since the normal velocity v_n has to be zero and the dissipation term is set to zero.

11.4.1.2 *Time integration*

Apply the fourth stage Runge–Kutta time integration method, following Section 9.3.4, equations (9.3.47–9.3.50), with the following coefficient:

$$U^{(1)}_{i,j} = U^n_{i,j} - \frac{\Delta t}{\Omega_{i,j}}\alpha_1 R^n_{i,j}$$

$$U^{(2)}_{i,j} = U^n_{i,j} - \frac{\Delta t}{\Omega_{i,j}}\alpha_2 R^{(1)}_{i,j}$$

$$U^{(3)}_{i,j} = U^n_{i,j} - \frac{\Delta t}{\Omega_{i,j}}\alpha_3 R^{(2)}_{i,j}$$

$$U^{n+1}_{i,j} = U^n_{i,j} - \frac{\Delta t}{\Omega_{i,j}}\alpha_4 R^{(3)}_{i,j} \qquad (11.4.9)$$

with

$$\alpha_1 = \frac{1}{4} \quad \alpha_2 = \frac{1}{3} \quad \alpha_3 = \frac{1}{2} \quad \alpha_4 = 1 \qquad (11.4.10)$$

Another option, which optimizes the dissipation of the scheme, is given by

$$\alpha_1 = \frac{1}{8} \quad \alpha_2 = 0.306 \quad \alpha_3 = 0.587 \quad \alpha_4 = 1 \qquad (11.4.11)$$

- Choose the CFL number, under the stability condition CFL < 2.8.
- The time step has to be evaluated based on the sufficient condition expressing that the physical domain of dependence should be completely contained in the

numerical domain of dependence. However, as we are not interested in the transient behavior of the solution, we can choose a *local time step*, whereby each cell progresses at its maximum possible time step $\Delta t_{i,j}$. This looses the time consistency of the transient since each cell has its own time step, but provides significant convergence acceleration. Hence, we make $\Delta t_{i,j}$ proportional to the local cell size $\Omega_{i,j}$ and is calculated as follows:

$$\Delta t_{i,j} \le \mathrm{CFL} \frac{\Omega_{i,j}}{|(\vec{v} + c)_{i,j} \cdot \Delta \vec{S}_i| + |(\vec{v} + c)_{i,j} \cdot \Delta \vec{S}_{j+1/2}|}$$

$$\Delta \vec{S}_i = \frac{1}{2}(\Delta \vec{S}_{i+1/2,j} + \Delta \vec{S}_{i-1/2,j}) \quad \Delta \vec{S}_j = \frac{1}{2}(\Delta \vec{S}_{i,j+1/2} + \Delta \vec{S}_{i,j-1/2})$$

$$(11.4.12)$$

- To start the calculation you need to define an initial solution. By lack of knowledge it is customary to take a uniform initial solution corresponding to the inlet condition, distributed uniformly over the mesh. Note that the convergence behavior can be very sensitive to the initial solution: the closer this initial guess is to the final solution, the faster the convergence.

Observe here that once again many 'local' decisions have to be taken as to where to evaluate the velocities and speed of sound. We have in these simulations, where all quantities vary from point to point, to continuously decide if we evaluate the values at the mesh points or at the cell faces, and how we connect one to the other.

11.4.1.3   *Boundary conditions for the Euler equations*

The last item of importance in the generation of your program for the Euler equations is the definition of the boundary conditions. This is a most critical component of any CFD code, and has to be compatible with both physical and numerical properties of the problem to be solved.

The time-dependent hyperbolic system of Euler equations contains four unknown dependent variables, and we have to determine how to handle these variables at the boundaries of the computational domain. These boundaries are of three types: solid walls, inlet and outlet boundaries, and each one of them will require a dedicated treatment.

We know from Chapter 3 that the system of time-dependent Euler equations is propagation dominated and we are faced therefore with the following questions:

(i)   How many conditions of physical origin are to be imposed at a given boundary?
(ii)   What physical quantities are to be imposed at a boundary?
(iii)   How are the remaining variables to be defined at the boundaries?

Go back to Chapter 3, Section 3.4.1 and to the Example E.3.4.3 dealing with the isentropic form of the Euler equation, where the energy equation has been removed. The three eigenvalues of the system, which correspond to the speed of propagation of three characteristic quantities, are given by

$$\lambda_1 = \vec{v} \cdot \vec{\kappa}/\kappa$$

$$\lambda_2 = \vec{v} \cdot \vec{\kappa}/\kappa + c$$

$$\lambda_3 = \vec{v} \cdot \vec{\kappa}/\kappa - c \qquad (11.4.13)$$

where $\vec{\kappa}$ is the wave number vector. The last two correspond to the speeds of the acoustic waves.

It can be shown, that by adding the energy equation a fourth eigenvalue appears equal to the first one, which becomes double valued.

Since, the transport properties at a surface are determined by the normal components of the fluxes, the number and type of conditions at a boundary of a multi-dimensional domain will be defined by the eigenvalue spectrum of the Jacobian matrices associated to the normal to the boundary. Hence, at a boundary surface, the behavior of the Euler system will be determined by the propagation of waves with the following speeds:

$$\lambda_1 = \vec{v} \cdot \vec{e}_n = v_n$$

$$\lambda_2 = \vec{v} \cdot \vec{e}_n = v_n$$

$$\lambda_3 = \vec{v} \cdot \vec{e}_n + c = v_n + c$$

$$\lambda_4 = \vec{v} \cdot \vec{e}_n - c = v_n - c \qquad (11.4.14)$$

where v_n is the normal velocity component at the considered surface.

This defines locally *quasi-one-dimensional propagation properties* and we can therefore look at how the propagation of information behaves at a boundary, in function of the sign of these quantities. The first two eigenvalues are equal to the normal component of the velocity and correspond to the entropy and vorticity waves, while the two remaining eigenvalues, are associated to the acoustic waves. Hence, the sign of these eigenvalues will be determined by the velocity components normal to the boundary surfaces.

The key to the understanding of the issue of the number of boundary conditions is indeed the awareness that the characteristics convey information in the (n, t) space formed by the local normal direction and time. If λ represents the propagation speed, the trajectory of the corresponding information path is given by $dn/dt = \lambda$. When information is propagated from outside toward the inside of the computational domain, it means that this information has to be defined from outside and represents a *physical boundary condition*. When λ is positive, the information carried by the associated characteristics, propagates from the boundary toward the interior of the flow domain and a *physical boundary condition* has to be imposed. On the other hand, when the eigenvalue λ is negative and the propagation occurs from the interior of the domain toward the boundary, it means that the related information is determined at the boundary by the interior flow and *cannot* be imposed from the outside. It will have to be expressed numerically, through *numerical boundary conditions*.

In summary, the number of physical conditions to be imposed at a boundary with normal vector $\vec{n}$, pointing toward the flow domain, is defined by the number of characteristics entering the domain.

Referring to Figure 11.4.2 for an inlet boundary, if the inlet flow is subsonic in the direction normal to the inlet surface, three eigenvalues are positive and one, $\lambda_4 = v_n - c$ is negative. *Therefore, three quantities will have to be fixed by the physical flow conditions at the inlet of the flow domain, while the remaining one will be determined by the interior conditions, through a numerical boundary condition.*

At an outlet boundary (Figure 11.4.3) with subsonic normal velocity, three eigenvalues are negative, since the normals are defined as pointing toward the interior flow domain. Three numerical boundary conditions have therefore to be set, while the fourth condition, associated to the positive eigenvalue $(-|v_n| + c)$, propagates

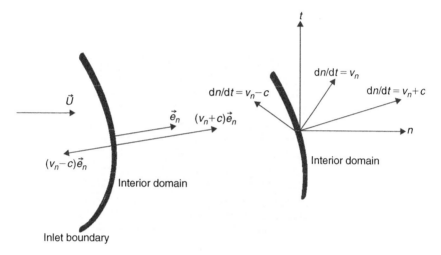

Figure 11.4.2   *Characteristic propagation properties at an inlet boundary with subsonic conditions.*

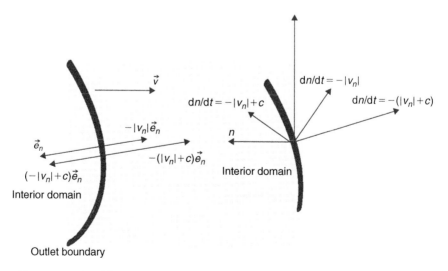

Figure 11.4.3   *Characteristic propagation properties at an outlet boundary with subsonic conditions.*

information from the boundary toward the flow region. It is consequently associated to a physical boundary condition.

This is a very important result, as it indicates that we cannot fix all four quantities at a subsonic inlet, but only three of them, neither can we know all quantities at a subsonic outlet, where we have to impose one variable, the other three being defined from the inlet flow properties.

If the flow is supersonic normal to the inlet surface, all boundary conditions are physical. With the same circumstances at outlet, all eigenvalues are of negative sign and no physical conditions have to be given. All the boundary variables are defined by the interior flow, for instance via extrapolation formulas.

The next question to answer is: *What quantities should be fixed as physical boundary conditions?*

This question has no unique answer and forms a vast complex subject, which is outside the scope of this introductory text. It is treated more in details in Volume II. For practical applications, you can proceed as follows:

Inlet boundary: You can select to fix the inlet velocity and inlet temperature for an external flow problem.

Another option, often applied to internal flows such as channels or cascade computations, is to specify two thermodynamic variables such as the upstream stagnation pressure and temperature, and an inlet Mach number or velocity magnitude, and have the inlet flow angle defined by the computed flow, or inversely, fix the incident flow angle, determining inlet Mach number from the computed flow. This has as consequence that the mass flow is not defined, but is a result from the computation.

Outlet boundary: The most appropriate physical condition, particularly for internal flows and corresponding to most experimental situations, consists in fixing the downstream static pressure. This can also be applied for external flow problems. However in this latter case, free stream velocity could also be imposed.

At a solid wall boundary: The normal velocity is zero, since no mass or other convective flux can penetrate the solid body. Hence, only one eigenvalue is positive and only one physical condition can be imposed, namely $v_n = 0$. The other variables at the wall, in particular tangential velocity components and pressure have to be determined by extrapolation from the interior to the boundary (Figure 11.4.4).

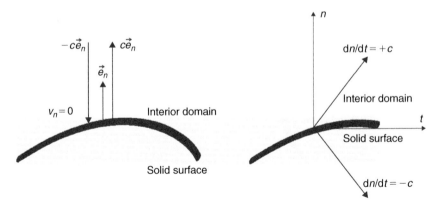

Figure 11.4.4 *Characteristic propagation properties at a solid surface boundary.*

Numerical boundary conditions: The other issue to be fixed in your program is to implement the numerical boundary conditions, in order to obtain the numerical values of the remaining variables at the boundaries. This is particularly important for solid walls, where we want to determine the pressure variations.

We suggest you apply a simple extrapolation from the inside point to the next surface point along the mesh lines. This is strictly valid if they are orthogonal to the surface, but should be acceptable for regular grids of the type suggested here. An even simpler way is to take the value at the boundary equal to the value at the cell center of the associated cell. This assumes the quantity to be piecewise constant in the considered cell and is a form of zero order extrapolation.

A more accurate way, but less robust, is to perform a linear extrapolation from two cells in the normal direction.

11.5 NUMERICAL SOLUTIONS FOR THE EULER EQUATIONS[1]

We will treat here in detail several test cases, in order to apply the numerical methods to a variety of flow conditions and identify the associated numerical effects. We will handle first the cylinder flow, in order to compare with the potential flow model and illustrate in particular the effects of numerical dissipation though the entropy field.

The next case will be typical of internal channel flows for several Mach number flow conditions; while the last case will be representative of supersonic flows with an oblique shock. All these cases will be treated assuming *perfect gas* relations for the considered fluid.

11.5.1 Application to the Flow Around a Cylinder

Proceed as follows:

- Consider the grids applied in the previous section, following equation (11.3.18) with 33 points in the radial direction and 128 points in the circumferential direction (see Figure 11.5.1). By removing every second point in both directions, new grids with a reduced density can be constructed, defining hereby a fine mesh (33×128), an intermediate mesh (17×64) and a coarse mesh (9×32).
- Assume atmospheric conditions, for pressure and temperature with $p_a = 101300$ Pa, $T_a = 288$ K, and take $M = 0.1$ as incoming flow conditions.
- In order to match these inlet conditions, the total pressure and total temperature are fixed at the inlet to 102010 Pa and 288.6 K, respectively, by following equations (11.1.9) and (11.1.12). The inlet velocity direction is imposed along the x-direction. The atmospheric pressure is also imposed at the outlet. Since the computational domain is circular, we will consider that the left half circle of the outer boundary forms the inlet section, while the right half circle will form the outlet section of the computational domain.
- Apply the Runge–Kutta method (11.4.9) with the coefficients (11.4.11) and take $\text{CFL} = 2$. You can choose here between a local time step based on equations

---

[1] This section has been written with the active participation of Dr. Benoit Tartinville, who produced also the results of the three test cases.

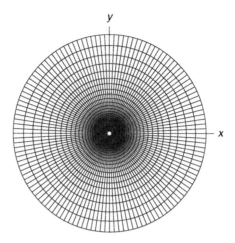

Figure 11.5.1 *Fine mesh used for the circular cylinder flow.*

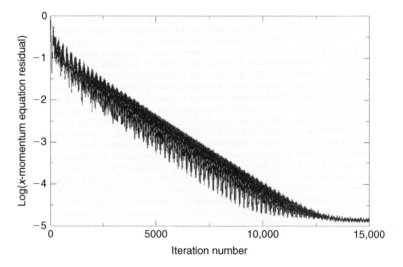

Figure 11.5.2 *Convergence history of the normalized L_2-norm of the momentum residual for the calculation on the fine mesh.*

(11.4.12), or a global time step by taking the minimum value of all the local time steps. We choose here the first alternative, but you can experience with both options.

- To start the calculation take a uniform initial solution equal to the values defined by the boundary conditions.
- Solve the Euler equations for the different grid densities considered, until a steady state is reached, by selecting a value of the artificial dissipation coefficient $\kappa^{(4)} = 1/50$, in equation (11.4.7).
- Monitor and plot the L_2-norm of the residuals in function of number of time steps. Figure 11.5.2 shows the evolution of the axial velocity residual, normalized by

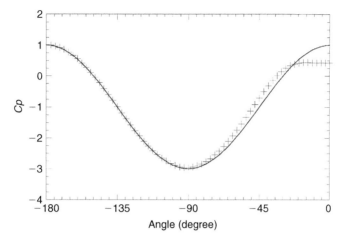

Figure 11.5.3 *Circumferential distribution of pressure coefficient from the numerical result (plus signs), compared to the exact solution (continuous line).*

the residual of the initial solution for the finest mesh, with the ordinate plotted on a logarithmic scale. The residual has dropped by nearly 5 orders of magnitude, which is close to single precision machine accuracy.

- If you cannot reach machine accuracy, if the residual drops for instance by 3 orders of magnitude only, it will not necessarily mean that your solution is not valid, but it indicates that somewhere in the flow domain, often around the boundaries, there is a local error source. This error could result from an inaccuracy in the implementation or from some small local oscillation between values at neighboring points. This is called a '*limit cycle*' in the convergence behavior. You can sometimes detect it by looking at the flow regions with the highest values of the residual.

- Observe that the convergence requires many time steps before reaching machine accuracy. To accelerate the convergence, you would need to apply the multigrid method, which would reduce the number of iterations easily by a factor of the order of 10 or more. Alternatively you could apply an implicit method, which is however also more complicated to implement.

- Note also the oscillatory behavior of the residual. This is generally an indication that the acoustic waves, which transport part of the initial errors, are reflected at the boundaries prior to their damping by the numerical scheme. This occurs when the boundary condition implementation does not allow these error waves to leave the computational domain without being reflection. The imposition of the outlet pressure is typical of what is called a ***reflecting boundary condition***. To avoid these effects, we have to impose ***non-reflecting boundary conditions***, based on the characteristic properties associated to the reflected waves. This is outside the scope of this introductory text and will be dealt with in more details in Volume II.

- Plot the wall surface pressure coefficient and compare with the exact incompressible solution, equation (11.3.17), as shown on Figure 11.5.3.

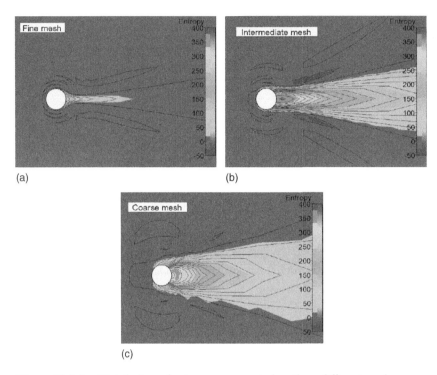

(a) (b)

(c)

Figure 11.5.4 *Distribution of entropy as computed on three different meshes:*
(a) fine mesh, (b) intermediate mesh and (c) coarse mesh (for color image refer
Plate 11.5.4).

- In the front part of the cylinder, between the points A and C of Figure 11.3.1, the
  solution is nearly perfect. However in the back part, between C and B a growing
  discrepancy can be noticed, with a significant deviation in the wake region around
  point B. You can observe that the left–right symmetry of the exact solution is lost.
  This is a major property of Euler calculations and is due, as explained above,
  to the numerical dissipation and the related numerical generation of entropy, as
  seen from Figures 11.5.4.
- Plot the entropy isolines and color maps, by monitoring the quantity:

$$s - s_\infty = \frac{p}{\rho^\gamma} - \left(\frac{p}{\rho^\gamma} \right)_\infty \qquad (11.5.1)$$

  where the subscript ∞ refers to the far field upstream conditions.
- For an inviscid flow, the entropy has to remain constant and equal to its upstream
  value, which is clearly not the case for the results of the Euler calculations on
  the three grids, as you see from Figure 11.5.4. Looking at the entropy evolution
  along the axis of symmetry and the cylinder surface, you observe that the entropy
  increases already from the upstream stagnation point (point A of Figure 11.3.1)
  along the solid surface, which appears as a source of entropy, like in a viscous
  boundary layer and keeps increasing further downstream, forming a wake.

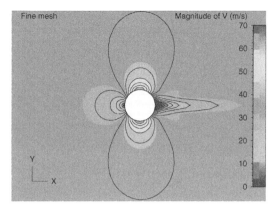

Figure 11.5.5 *Distribution of velocity magnitude and iso-velocity lines, as computed on the finest mesh (for color image refer Plate 11.5.8).*

- Observe also that the level of entropy decreases when the mesh is refined. It would require finer grids to reduce this effect to an acceptable low level and we recommend that you further refine the grid by dividing the cell sizes by two in each direction, until you reduce the entropy to a sufficiently low level, reached when you recover a nearly symmetrical solution.
- This effect can be seen also on the iso-velocity plot of Figure 11.5.5, where the 'numerical' wake is clearly visible. The behavior of the downstream region mimics viscous flows, and is the direct consequence of the presence of the numerical dissipation.

    This is a major requirement on accuracy. You should always attempt to refine the grids up to a level where the numerical dissipation is sufficiently low as to minimize the numerical entropy generation.

- Calculate the lift and drag components following equation (11.3.43). Both quantities should be zero and their non-zero values are a direct measure of the numerical errors of the simulation. Since the mesh is symmetrical in the present test case, the calculated lift should reflect the machine accuracy and only the drag coefficient is a measure of the numerical dissipation. The drag coefficient is defined by

$$C_D = \frac{L_x}{\frac{1}{2}\rho U^2 S} \quad \text{with } S = 2\pi a \tag{11.5.2}$$

It should tend to zero, as the mesh is refined as seen from Table 11.5.1.

11.5.2 Application to the Internal Flow in a Channel with a Circular Bump

This is a representative example of an internal flow configuration. It consists of a channel of height L and length $3L$ with, along the bottom wall, a circular arc of length L and thickness equal to $0.1L$. We have no analytical exact solution for this case, but a reference solution can be obtained from a fine mesh simulation.

Table 11.5.1 *Drag coefficient on the cylinder, computed on three different meshes.*

	Drag coefficient
Exact	**0**
128×32	0.0094
65×17	0.0784
33×9	0.1867

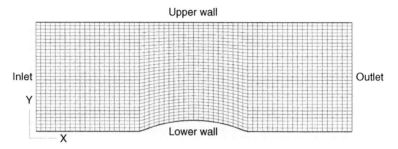

Figure 11.5.6 *Mesh used for the channel flow test case.*

Proceed now as follows:

- An H-grid can be constructed using 65 points in the axial direction with equal spacing and 33 points in-between walls. A quasi-uniform grid distribution is selected by taking $\Delta x = 3L/64$ and dividing each vertical mesh line in 32 uniform steps (see Figure 11.5.6).
- Assume atmospheric conditions, for pressure and temperature with $p_a = 101300$ Pa and $T_a = 288$ K, and $M = 0.1$ as inlet flow condition.
- In order to match these flow conditions, the total pressure and temperature are fixed at the inlet to 102010 Pa and 288.6 K, respectively, by following equations (11.1.9) and (11.1.12). The inlet velocity vector is imposed in the axial direction. The atmospheric pressure is also imposed at the outlet.
- At the solid walls, obtain the pressure and tangential velocity by taking them equal to the values of the corresponding cell centers.
- Apply the Runge–Kutta method (11.4.9) with the coefficients (11.4.11) and take CFL $= 2$. Select a local time step, based on equations (11.4.12).
- To start the calculation take a uniform initial solution equal to the values defined by the boundary conditions.
- Solve the Euler equations until a steady state is reached, with CFL $= 2$ and $\kappa^{(4)} = 1/256$.
- For internal flows, an important consistency check is the satisfaction of mass flow conservation, that is the mass flow rate through any section should be equal, up to an acceptably level of accuracy, to the inlet mass flow. The mass flow rate through a vertical mesh line i is equal to

$$\dot{m} = \int_{section} \rho \vec{v} \cdot d\vec{S} = \sum_{j} \left[\frac{(\rho u)_{i,j} + (\rho u)_{i,j+1}}{2} (y_{i,j+1} - y_{i,j}) \right] \qquad (11.5.3)$$

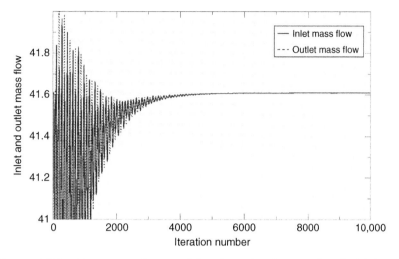

Figure 11.5.7 *Convergence history of the inlet and outlet mass flow (kg/s/m).*

referring to Figure 11.4.1, and applying a second order trapezium formula for the integration,

- Typically you should aim at mass flow errors between inlet and exit, below 0.1%. Note that this can only be achieved when the steady state is reached, since during the compressible transient the time derivative of density is not zero and hence the mass flow rate, as defined by equation (11.5.3), will not be constant.
- This is clearly seen on Figure 11.5.7, which shows the evolution of the inlet and exit mass flows (in kg/s/m) during the iterative process. You observe that after 5000 time steps, we have reached a satisfactory level of mass conservation, with an error below 0.01%.
- The Mach number isolines and color maps are shown on Figure 11.5.8 for the (65×33) mesh superimposed on the color map of a reference solution on a (225×113) mesh. One isoline has been drawn on the color map every 0.001 ranging from 0.07 to 0.13. The bottom figure compares the Mach number isolines of the two solutions, where the darkest line is the reference solution. You can observe here again the loss of symmetry between the upstream and downstream parts of the circular bump. The reference solution, obtained on a grid of 225×113 mesh points, has still a small deviation from symmetry, while the numerical solution on the selected grid shows a larger loss of symmetry in the wake region.
- If we compare the distribution of the wall surface pressure coefficient with the reference solution, shown on Figure 11.5.9, you observe that we cannot distinguish them. ***This indicates that the pressure is less sensitive to the numerical dissipation than entropy or drag.***
- Compare the solutions obtained at lower and higher Mach numbers (for instance 0.01, 0.05, 0.1 and 0.7) to investigate the effects of compressibility. This can be seen on Figure 11.5.10. You can observe that within the low Mach number range, the solution does not greatly depend on Mach number and is practically identical

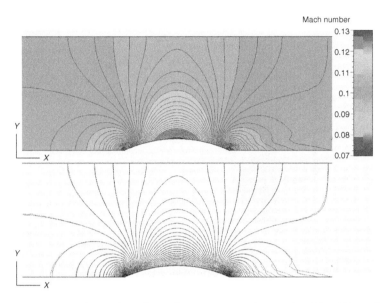

Figure 11.5.8 *Isolines of Mach number as computed using the (65 × 33) mesh superimposed on the color map of a reference solution on a (225 × 113) mesh. One isoline has been drawn every 0.001 ranging from 0.07 to 0.13. The bottom figure compares the Mach number isolines of the two solutions, where the darkest line is the reference solution (for color image refer Plate 11.5.8).*

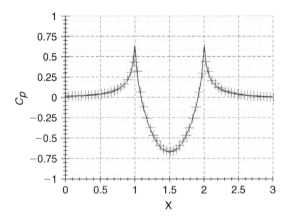

Figure 11.5.9 *Axial distribution of pressure coefficient as computed using the (65 × 33) mesh (plus signs) and compared to the (225 × 113) mesh (continuous line).*

to the incompressible flow. When increasing the Mach number, compressibility effects become more important and for an inlet Mach number slightly above 0.6, the sonic speed of $M = 1$ is reached on the bump surface, followed by a supersonic region terminated by a shock. This can be seen for the case of an incident velocity of $M = 0.7$.

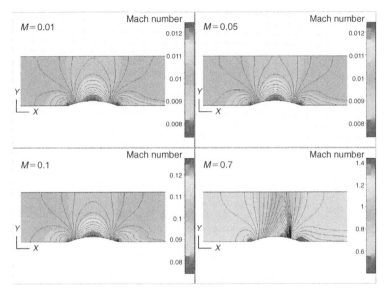

Figure 11.5.10   *Distribution of velocity as computed on the (65 × 33) mesh, for different values of the incident Mach number, from 0.01 to 0.7. Observe the shock appearing at M = 0.7 (for color image refer Plate 11.5.10).*

11.5.3   Application to the Supersonic Flow on a Wedge at *M* = 2.5

This example of a supersonic flow over a wedge of angle 15° generates an oblique shock. The Mach number upstream of the shock is fixed to 2.5. This test case has an exact analytical solution, satisfying the Rankine–Hugoniot relations (11.1.26); see for instance Anderson (1982), formed by two regions of constant states, separated by the oblique shock.

The downstream flow conditions are listed in Table 11.5.2.

Table 11.5.2   *Analytical solution for the supersonic flow on a 15° wedge at M = 2.5.*

Downstream Mach number (M_2)	1.87
Pressure ratio (p_2/p_1)	2.47
Entropy ratio (s_2/s_1)	1.03

This problem cannot be solved with the implementation of the Euler solver as applied for the two previous test cases. Indeed, in presence of shocks, we have to add an additional dissipation term to control the numerical oscillations appearing in presence of discontinuities, as shown for linear convection in Chapters 7 and 8. To this end, a second order artificial dissipation term is added to capture numerically strong gradients such as shock waves. Therefore, equation (11.4.6) is reformulated by adding a nonlinear dissipation term of second order, that reduces locally to first order at shock positions, in order to achieve a nearly monotone behavior, following

(Jameson et al. 1981):

$$D_{i+1/2,j} = \eta_{i+1/2,j}(U_{i+1,j} - U_{i,j}) + \gamma_{i+1/2,j}(U_{i+2,j} - 3U_{i+1,j}$$
$$+ 3U_{i,j} - U_{i-1,j}) \tag{11.5.4}$$

The coefficient η and γ are based on the following relations:

$$\eta_{i+1/2,j} = \frac{1}{2}\kappa^{(2)}[\vec{v} \cdot \Delta\vec{S} + c|\Delta\vec{S}|]_{i+1/2,j} \max(v_{i-1}, v_i, v_{i+1}, v_{i+2})$$

$$\gamma_{i+1/2,j} = \max\left(0, \frac{1}{2}\kappa^{(4)}[\vec{v} \cdot \Delta\vec{S} + c|\Delta\vec{S}|]_{i+1/2,j} - \eta_{i+1/2,j}\right) \tag{11.5.5}$$

where the variable v_i are sensors that activate the second order dissipation in regions of strong gradients. They are based on the pressure variations and are defined as

$$v_i = \left|\frac{p_{i+1,j} - 2p_{i,j} + p_{i-1,j}}{p_{i+1,j} + 2p_{i,j} + p_{i-1,j}}\right| \tag{11.5.6}$$

Observe that the numerator is proportional to the second derivative of the pressure and is a second order discretization of $\Delta x^2(\partial^2 p/\partial x^2)$ when the pressure variations are smooth. In the shock region, however, this term becomes close to 1 and the first term of equation (11.5.4) reduces to first order.

$\kappa^{(2)}$ is a non-dimensional coefficient of the order of unity.

You can notice that if the pressure variations are linear, the sensor v_i vanishes, the second order dissipation is not activated and the dissipation term is identical to the expression (11.4.6) defined for the subsonic applications.

With the addition of the second order dissipation to the inviscid numerical fluxes, we can proceed as follows:

- A uniform H-grid can be constructed using 97 points in the axial direction and 65 points in-between walls. The distribution of grid points is set to uniform along the x-direction and in-between walls (see Figure 11.5.11).
- In order to match the flow conditions, the static pressure and temperature are fixed at the inlet to 101353 Pa and 288.9 K, respectively. The inlet velocity vector is fixed to axial with a magnitude of 852.4 m/s.
- Since the outlet is supersonic, none of the characteristics enter the domain through this boundary. Therefore, all the variables are extrapolated at this boundary.
- Apply the Runge–Kutta method (11.4.9) with the coefficients (11.4.11) and take CFL $= 2$. Select here a local time step, based on equations (11.4.12).
- To start the calculation take a uniform initial solution equal to the values defined by the inlet boundary conditions.
- Solve the Euler equations until a steady state is reached, with CFL $= 2$, $\kappa^{(2)} = 1$ and $\kappa^{(4)} = 1/10$.
- Monitor the L_2-norm of the axial momentum residual in function of number of time steps (see Figure 11.5.12). The residual has decreased by more than five orders of magnitude, which is in accordance with single precision machine

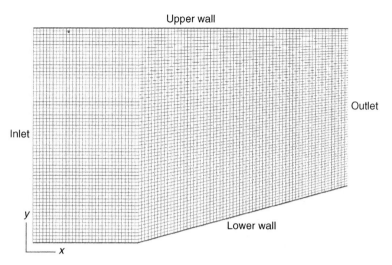

Figure 11.5.11   *Mesh used for the wedge test case.*

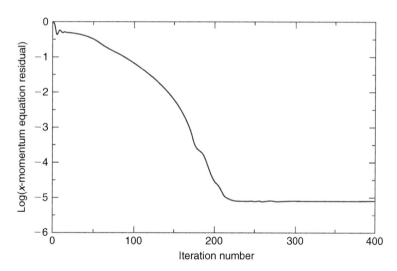

Figure 11.5.12   *Convergence history of the normalized L_2-norm of the x-momentum residual.*

accuracy. You can reproduce the same calculation by using double precision machine accuracy and observe the drop in residual to much lower levels.

- If you compare this convergence behavior with the two previous test cases, you can notice how much faster the supersonic case converges, requiring only 150 iterations, compared to thousands for the subsonic cases. This is very typical and is largely connected to the boundary conditions and to the intrinsic properties of supersonic flows, which do not allow for upstream wave propagation of any quantity, including numerical errors.

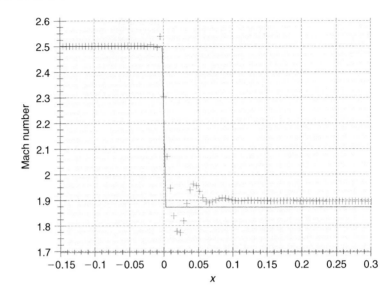

Figure 11.5.13   *Axial distribution of Mach number along the bottom wall for the exact solution (continuous line) and from the numerical result (plus signs).*

- Plot the Mach number variations along the bottom wall, and compare with the exact solution (see Figure 11.5.13). Upstream of the wall corner the solution is nearly perfect. Downstream of the shock, along the inclined wall, the numerical solution shows an oscillatory behavior of the Mach number. This illustrates the non-monotonic behavior of the second order central scheme with the dissipation term (11.5.4), which does not take into account the monotonicity requirements, discussed in Chapter 8.
- In addition, the first undershoot of the Mach number after the shock is due to the zero order interpolation of the pressure from the cell center to the wall, which is incorrect when the cell center is located upstream of the shock in the cells near the wedge corner. This illustrates the difficulties in imposing numerical boundary conditions, in presence of sharp local changes in geometry and flow conditions.
- Plot the iso-Mach numbers and iso-entropy lines, and observe the non-uniform behavior upstream and downstream of the shock wave, resulting from the non-monotone behavior of the scheme (see Figures 11.5.14 and 11.5.15). An important numerical error source is generated at the wedge corner, as a result of the local treatment of the wall boundary condition.
- Increase the mesh size in both directions and observe the progressive reduction of the errors.

This example clearly shows that, although a representative solution can be obtained for this simple supersonic flow, additional work has still to be done to eliminate the numerical oscillations, by imposing more severe monotone schemes with appropriate limiters, as discussed in Chapter 8.

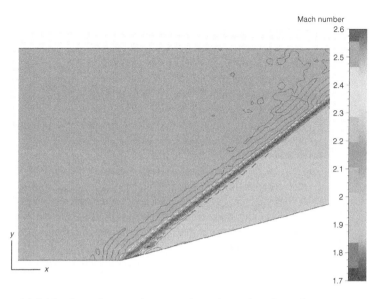

Figure 11.5.14   *Distribution of computed Mach number (for color image refer Plate 11.5.14).*

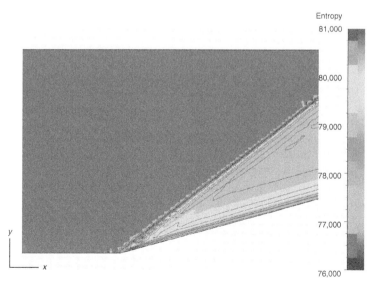

Figure 11.5.15   *Distribution of computed entropy (for color image refer Plate 11.5.15).*

11.5.4   Additional Hands-On Suggestions

With the developed programs, you can now extend the range of applications and experience with many other set-ups. In particular, you can investigate the effects of many parameters, such as inflow conditions; grid densities and qualities, numerical algorithms. Some suggested actions are:

- Exercise with different grid types and grid densities to develop a 'feeling' and experience with ill-conditioned grids. In particular, introduce grid distortions and grid discontinuities close to the solid walls to observe their effects on the quality of the numerical solution.
- Look at the influence of the artificial dissipation coefficients $\kappa^{(4)}$ and $\kappa^{(2)}$ on convergence rate and calculated solution accuracy.
- Investigate the effects of compressibility by applying your potential or Euler code to the considered flow cases, with increasing incident Mach number. Observe what is happening to your code when shocks appear and in particular the non-monotone behavior over the shock with the Euler equations.
- Observe the sensitivity of the Mach number distribution downstream of the shock wave to the second and fourth order dissipation coefficients $\kappa^{(2)}$ and $\kappa^{(4)}$, for the wedge flow test case.
- Try other time integration methods, such as the Beam and Warming schemes introduced in Chapter 9.

Other options are left to your imagination and creativity.

CONCLUSIONS AND MAIN TOPICS TO REMEMBER

The main objective of this chapter is to guide you in writing your first CFD codes for inviscid potential and Euler flows on simple configurations, for which we have either an analytic solution or a reference numerical solution obtained on fine grids.

In running these test cases, we have insistently drawn you attention to the many error sources associated to the discretization of the inviscid Euler equations, at the level of implementation issues, boundary conditions as well as grid sensitivity.

On all the three test cases, you will have observed that the obtained results are not perfect and the errors shown are highly representative of 'real life' CFD simulations of industrial relevant systems. When you do not have a reference solution to compare with, the internal consistency checks, such as mass flow conservation, and the monitoring of entropy will provide you with a clear picture of the error levels of your simulation.

Main topics to remember are:

- Great care has to be exercised at all levels of the detailed implementation and programming choices.
- Give great attention to the implementation of boundary conditions.
- Perform as many 'consistency' checks as possible, by monitoring global conserved quantities, such as mass flow with internal flows, or total enthalpy. The variations of these quantities are indicators of local error sources.
- The issue of the entropy generated by the numerical dissipation is a most critical issue in CFD. Applying the Euler equation model is a unique method to identify the regions where this dissipation has a significant influence on the numerical solution. This should be your guide toward the required mesh refinement, in

order to bring this effect below an acceptable level. The same is true for the numerical calculated drag forces, which should be zero for inviscid flows.

- Grid quality and grid density have a major influence on the accuracy of the numerical simulation.

REFERENCES

AIAA (1998). *Guide for the Verification and Validation of Computational Fluid Dynamics Simulations*. American Institute of Aeronautics and Astronautics, AIAA-G-077-1998, Reston, VA.

Anderson, J.D. (1982). *Modern Compressible Flow.* McGraw Hill Inc., New York.

ASME (2006). *V&V 10-2006 Guide for Verification and Validation in Computational Solid Mechanics*. ASME Publications; http://catalog.asme.org/Codes/PrintBook.

Ballhaus, W.F. and Bailey, F.R. (1972). *Numerical Calculation of Transonic Flow about Swept Wings*. AIAA Paper 72-677.

Chorin, A.J. (1967). A numerical method for solving incompressible viscous flow problems. *J. Computational Physics* 2, 12.

Garabedian, P.R. and Korn, D. (1972). Analysis of transonic airfoils. *Commun. Pure Appl. Math* 24, 841–851.

Hall, M.G. (1981). Computational Fluid Dynamics – A Revolutionary Force in Aerodynamics. AIAA Paper 81-1014 – Proc. Fifth AIAA Computational Fluid Dynamics Conference , pp.176–188.

Holst, T. and Thomas, S. (1983). Numerical solution of transonic wing flow fields. *AIAA Journal* 21, 863–870.

Jameson, A. (1974). Iterative solutions of transonic flows over airfoils and wings, including flows at Mach 1. *Commun. Pure Appl. Math*, 27, 283–309.

Jameson, A. (1975). Transonic potential flow calculations using conservative form. *Proceedings of the AIAA Second Computational Fluid Dynamics Conference*, Hartford, pp.148–161.

Jameson, A. and Caughey, D.E. (1977). A Finite Volume Method for Transonic Potential Flow Calculations. Proc. AIAA 3rd Computational Fluid Dynamics Conference, Albuquerque, pp. 35–54.

Jameson, A., Schmidt, W. and Turkel, E. (1981). Numerical Simulation of the Euler Equations by Finite Volume Methods using Runge–Kutta Time Stepping Schemes. AIAA Paper 81-1259, AIAA 5th Computational Fluid Dynamics Conference.

Lax, P.D. (1973). Hyperbolic Systems of Conservation Laws and the Mathematical Theory of Shock Waves, SIAM Publications, Philadelphia, USA.

Merkle, C.L. and Choi, Y.-H. (1985). Computation of low speed compressible flows with time-marching methods. *Int. J. Numer. Method. Eng.*, 25, 293–311.

Murman, E.M. and Cole, J.D. (1971). Calculation of plane steady transonic flows. *AIAA J.* 9, 114–121.

Oberkampf, W.L., Trucano, T.G. and Hirsch, Ch. (2004). Verification, validation and predictive capability in computational engineering and physics. *Appl. Mech. Rev.*, 57(5), pp. 345–384.

Shapiro, A.H. (1953). The Dynamics and Thermodynamics of Compressible Fluid Flow. Ronald Press, New York.

Steger, J.L. and Lomax, H. (1972). Transonic flow about two-dimensional airfoils by relaxation procedures. *AIAA J.* 10, 49–54.

Zucrow, M.J., Hoffman, J.D. (1976). Gas Dynamics Vol. I & II. J. Wiley & Sons, New York.

Chapter 12

Numerical Solutions of Viscous Laminar Flows

OBJECTIVES AND GUIDELINES

This chapter will guide you into the programming of CFD codes for viscous laminar flows, modeled by the Navier–Stokes equations, where the main additions to the inviscid model are the viscous internal friction stresses and the thermal conduction, both being diffusion type terms, represented by second order derivatives. The basic equations are summarized in Section 12.1.

This is a major change from inviscid flows, as it affects the solid wall boundary conditions, where all the velocity components have to vanish, following the non-slip boundary condition for viscous flows. The main consequence of it is the presence of viscous and thermal boundary layers near solid walls, where the velocity and temperature profiles can vary extremely rapidly over a short distance, requiring hereby high density grids in these near-wall regions.

We will consider essentially low speed flows and this will provide us with the opportunity to introduce an alternative numerical approach for the solution of the Navier–Stokes equations, namely the pressure correction method. The method is widely used, also in several commercial codes, and was initiated many years ago, for low speed industrial flows.

This particular numerical approach is to be put in relation with the method presented in the previous chapter for the Euler equations, which has been developed within the aeronautical community and oriented at high speed flows where the variations of density can become dominant. Hence, they are considered as density-based methods. Today, both families of methods have been extended to handle flows at all speed regimes, although the pressure correction methods might display some difficulties for high supersonic flow conditions.

Very few viscous test cases are known having analytical solutions. Among the best known is the Couette flow, generated by the uniform motion of a flat plate at a certain distance from a fixed plate, leading to a linear velocity distribution as a result of the viscous effects. When a temperature difference is imposed between the plates, the exact parabolic solution for the temperature profile allows the verification of thermal effects.

Another fundamental test case is the laminar boundary layer flow over a flat plate, with the well-known Blasius profile, considered as an exact solution. Despite its simple geometry, it displays most of the flow features encountered in viscous external or internal flows, such as the flow over a wing or a turbomachinery blade row, with the presence of leading edge, although of zero thickness, a boundary layer flow and a wake.

Section 12.2 will first guide you to the extension toward viscous flows of the program developed for the Euler equations in Chapter 11. This is not so straightforward

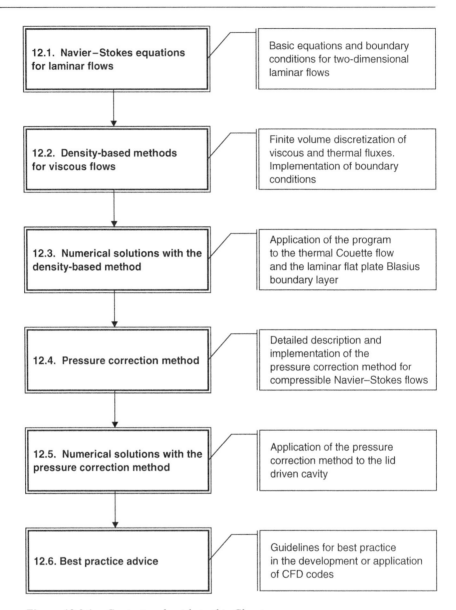

Figure 12.0.1   *Content and guide to this Chapter.*

extension, particularly on a general finite volume structured grid, as it requires the addition of the flux terms associated to the central discretized viscous and heat conduction fluxes, and a modification of the wall boundary conditions. You will be able to apply your program to the thermal Couette flow and to the laminar flat plate boundary layer flow, following Section 12.3.

Section 12.4 will introduce you to the pressure correction method and will guide you to its programming, with an application to the well-known test case of the lid driven cavity flow in Section 12.5. The lid driven cavity case has no analytical solution, but is representative of a complex internal flow structure and is considered as a standard test case.

Finally the last section 12.6 will provide some Best Practice Advice for either the development of a CFD code, or the use of existing, commercial or research, CFD codes.

Figure 12.0.1 summarizes the roadmap of this chapter.

Sections 12.2 and 12.3 have been written with the active participation of Dr. Benoit Tartinville, from Numeca International, who produced also the results of the two viscous cases of Section 12.3. Dr. Sergey Smirnov, from the Vrije Universiteit Brussel, contributed significantly to Section 12.4 on the pressure correction method, and produced the program and the results for the lid driven cavity of Section 12.5. Their contribution is gratefully acknowledged.

12.1 NAVIER–STOKES EQUATIONS FOR LAMINAR FLOWS

The system of Navier–Stokes equations can be written in various ways based on the derivations in Chapter 1. We refer you to Table 1 of Chapter 1 for a summary of the conservation laws in different forms. A general differential form is provided by the following equations, in absence of external forces and heat fluxes:

$$\frac{\partial \rho}{\partial t} + \vec{\nabla} \cdot (\rho \vec{v}) = 0$$

$$\frac{\partial \rho \vec{v}}{\partial t} + \vec{\nabla} \cdot (\rho \vec{v} \otimes \vec{v} + p\bar{\bar{I}} - \bar{\bar{\tau}}) = 0 \qquad (12.1.1)$$

$$\frac{\partial \rho E}{\partial t} + \vec{\nabla} \cdot (\rho \vec{v} H - k\vec{\nabla} T - \bar{\bar{\tau}} \cdot \vec{v}) = 0$$

Assuming Newtonian fluids, the shear stress tensor has the following Cartesian components, based on equation (1.3.6):

$$\tau_{ij} = \mu \left[\left(\frac{\partial v_j}{\partial x_i} + \frac{\partial v_i}{\partial x_j} \right) - \frac{2}{3} (\vec{\nabla} \cdot \vec{v}) \delta_{ij} \right] \qquad (12.1.2)$$

In two-dimensions and Cartesian coordinates, we obtain the system:

$$\frac{\partial \rho}{\partial t} + \frac{\partial \rho u}{\partial x} + \frac{\partial \rho v}{\partial y} = 0$$

$$\frac{\partial \rho u}{\partial t} + \frac{\partial (\rho u^2 + p)}{\partial x} + \frac{\partial (\rho uv)}{\partial y} = \frac{\partial \tau_{xx}}{\partial x} + \frac{\partial \tau_{yx}}{\partial y}$$

$$\frac{\partial \rho v}{\partial t} + \frac{\partial (\rho uv)}{\partial x} + \frac{\partial (\rho v^2 + p)}{\partial y} = \frac{\partial \tau_{xy}}{\partial x} + \frac{\partial \tau_{yy}}{\partial y}$$

$$\frac{\partial \rho E}{\partial t} + \frac{\partial (\rho u H)}{\partial x} + \frac{\partial (\rho v H)}{\partial y} = \frac{\partial (\tau_{xx} u + \tau_{xy} v)}{\partial x} + \frac{\partial (\tau_{yx} u + \tau_{yy} v)}{\partial y}$$

$$+ \frac{\partial}{\partial x} \left(k \frac{\partial T}{\partial x} \right) + \frac{\partial}{\partial y} \left(k \frac{\partial T}{\partial y} \right) \qquad (12.1.3)$$

The right-hand side collects the terms to be added to the inviscid fluxes.

The energy equation can be written in many equivalent forms, for instance equation (1.4.15) for the internal energy:

$$\rho\frac{de}{dt} \equiv \rho\left[\frac{\partial e}{\partial t} + (\vec{v}\cdot\vec{\nabla})e\right] = -p(\vec{\nabla}\cdot\vec{v}) + \vec{\nabla}\cdot(k\vec{\nabla}T) + (\bar{\bar{\tau}}\cdot\vec{\nabla})\cdot\vec{v} \qquad (12.1.4)$$

For incompressible flows, the above equations simplify further, since the continuity equation reduces to the condition of divergence free velocity:

$$\vec{\nabla}\cdot\vec{v} = 0 \qquad (12.1.5)$$

and the shear stress definition becomes

$$\tau_{ij} = \mu\left(\frac{\partial v_j}{\partial x_i} + \frac{\partial v_i}{\partial x_j}\right) \qquad (12.1.6)$$

leading to the following non-conservative momentum equation (1.4.40)

$$\frac{\partial\vec{v}}{\partial t} + (\vec{v}\cdot\vec{\nabla})\vec{v} = -\frac{1}{\rho}\vec{\nabla}p + \nu\Delta\vec{v} \qquad (12.1.7)$$

and the energy equation for the temperature, writing $e = c_v T$

$$\rho c_v\left[\frac{\partial T}{\partial t} + (\vec{v}\cdot\vec{\nabla})T\right] = \vec{\nabla}\cdot(k\vec{\nabla}T) + (\bar{\bar{\tau}}\cdot\vec{\nabla})\cdot\vec{v} \qquad (12.1.8)$$

Note that for liquids the specific heat coefficients at constant volume and constant pressure are equal, i.e. $c_v = c_p$ and either values can be used in equation (12.1.8), when applied to incompressible liquids.

You can choose to discretize directly this system by applying finite differences, if you can generate a Cartesian grid for your problem. For a general curvilinear grid, we recommend that you apply the finite volume method, by extending directly the program you have developed for the Euler equations.

With reference to equation (11.4.1), we have to add the contributions from the viscous and thermal fluxes to the momentum and energy conservation equations. This gives the following finite volume discretized equation:

$$\frac{d}{dt}[\bar{U}_{i,j}\Omega_{i,j}] = -\sum_{\text{faces}}[\vec{F}^*\cdot\Delta\vec{S} - \vec{F}_v\cdot\Delta\vec{S}] \equiv -R_{i,j} \qquad (12.1.9)$$

where $\vec{F}_V$ represents the viscous and thermal fluxes, with components f_v and g_v:

$$f_v = \begin{vmatrix} 0 \\ \tau_{xx} \\ \tau_{xy} \\ \tau_{xx}u + \tau_{xy}v + k\dfrac{\partial T}{\partial x} \end{vmatrix} \qquad g_v = \begin{vmatrix} 0 \\ \tau_{yx} \\ \tau_{yy} \\ \tau_{yx}u + \tau_{yy}v + k\dfrac{\partial T}{\partial y} \end{vmatrix} \qquad (12.1.10)$$

The estimation of the viscous and thermal fluxes requires the calculation of the velocity and temperature gradients. *Since these terms describe diffusion effects, they have to be discretized by central formulas.*

12.1.1 Boundary Conditions for Viscous Flows

The presence of the viscous and thermal fluxes makes the time-dependent Navier–Stokes equations of the mixed parabolic–hyperbolic type, as seen in Chapter 3, and this has an impact on the number and nature of boundary conditions to be imposed, in particular at solid walls.

We consider here high Reynolds number flows, which is the most often occurring situation in both industrial as environmental systems. If you recall the significance of the Reynolds number as the ratio of convective to viscous effects, a high Reynolds number means that the flow system will be dominated by the convective terms, that is will be dominated by its inviscid properties. This can be clearly seen when analyzing viscous flows, with the important exception of the near-wall regions, where the viscous effects, leading to a boundary layer configuration, dominate the flow behavior.

The main consequence is that at inlet and outlet boundaries, we can keep the same boundary conditions as for the inviscid computations, but the wall boundary conditions will change drastically.

At solid walls, the velocity relative to the wall has to vanish. This is the *no-slip boundary condition*, whereby the three velocity components are zero at the body surface.

We have to add a boundary condition for the temperature status of the solid wall. You can encounter several situations; the most current being:

- *Adiabatic walls*: Whereby we express that there is no heat flux through the solid surface. This is expressed by the Neumann condition

$$\frac{\partial T}{\partial n} = 0 \quad \text{at the solid wall} \tag{12.1.11}$$

- *Constant temperature wall*: Here we assume that the solid surface is kept at a fixed temperature T_w, leading to a Dirichlet type boundary condition

$$T = T_w \quad \text{at the solid wall} \tag{12.1.12}$$

- *Imposed heat flux*: In this case, the solid wall is the source of a fixed heat flux q_e to or from the fluid flow, for instance when the solid surface is part of a heat exchanger system. This flux will be positive for a heated wall or negative for a cooled wall. The boundary condition generalizes the adiabatic condition (12.1.11) to

$$\frac{\partial T}{\partial n} = q_e \quad \text{at the solid wall} \tag{12.1.13}$$

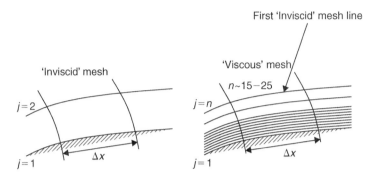

Figure 12.1.1 *Difference between an inviscid grid and a boundary layer oriented grid.*

12.1.2 Grids for Boundary Layer Flows

The presence of boundary layers has dramatic consequences on the grid requirements.

If you consider a flat plate, the developing boundary layer, as already discussed in Section 4.3, when we introduced the discretization for non-uniform grids, has a thickness to length ratio of the order of $1/\sqrt{Re}$. Hence, for a Reynolds number of 10^6 and a plate with a length of 1 m, the boundary layer thickness will be of the order of 1 mm at the end of the plate. For an incoming moderate velocity of say 30 m/s, the velocity will vary from zero at the wall to 30 m/s over a distance of 1mm, which is a very strong variation. As a consequence you need to arrange for a minimum number of grid points, of the order of 20–25 mesh points over this short distance, as illustrated on Figure 12.1.1. You can consider that the boundary layer mesh is inserted between the wall and the fist grid line of the inviscid mesh ($j = 2$). Moreover, since the velocity gradients take their highest values near the wall and decrease progressively toward the edge of the boundary layer, it is recommended to cluster the grid points close to the solid surface with a progressive coarsening when moving away from the wall. Refer to Section 4.3 for a discussion of this important issue.

Hence, viscous grids must be highly concentrated near solid walls.

12.2 DENSITY-BASED METHODS FOR VISCOUS FLOWS

To extend your inviscid finite volume program developed in Chapter 11, proceed as follows:

- Add an option in the main program to distinguish between inviscid or viscous simulations.
- Add subroutines to calculate the velocity and temperature gradients.
- Add subroutines to calculate the viscous stresses and thermal fluxes.
- Add subroutines for the viscous solid wall boundary conditions.

Referring to the cell-centered configuration of Figure 11.4.1, the viscous and thermal fluxes are calculated directly for each face, in a finite volume formulation, as described hereafter.

12.2.1   Discretization of Viscous and Thermal Fluxes

All the points and grid coordinates used in the following equations refer to Figure 11.4.1. Independently of the choice made for the inviscid part of the solver – either using an upwind or a central scheme – *the viscous and thermal fluxes should always be discretized using a central scheme*.

Following equations (12.1.9) and (12.1.10), the viscous and thermal fluxes at the cell face $(i + 1/2, j)$ can be expressed as

$$(\vec{F}_v \cdot \Delta \vec{S})_{i+1/2, j} = (\vec{F}_v \cdot \Delta \vec{S})_{AB} = f_{v, AB}(y_B - y_A) - g_{v, AB}(x_B - x_A) \quad (12.2.1)$$

The main difficulty resides in the dependence of the flux components f_v and g_v on the velocity and temperature gradients, and we have to decide how to estimate these gradients at the cell faces. Note that all quantities appearing in the flux terms are to be considered as *face-averaged values*.

Again we are confronted with a 'local' decision, as several options can be selected to evaluate the gradients.

They can be computed directly at a mid-point of the cell face, or evaluated at the cell corners and averaged to obtain an average cell-face value.

Though the first approach is more direct, it is more expensive than the second one. Indeed for a two-dimensional calculation, the total number of cell faces is about two times the number of corners. Nonetheless, the first approach will be retained here, as it is more robust.

Thus, we can proceed as follows:

- Calculate the face velocity and temperature gradients by *programming a loop over all the cell faces*.
- Compute the face-averaged gradients by using the Gauss divergence theorem over a selected control volume, following equation (5.3.16), where the overbar indicates the cell average value:

$$\int_{\Omega} (\vec{\nabla} U) d\Omega \stackrel{\Delta}{=} \overline{\vec{\nabla} U} \Omega = \oint_S U d\vec{S} \quad (12.2.2)$$

This equation is interpreted here in a two-dimensional space, where Ω is the face area and the contour integral is performed over the face boundaries.

- In order to obtain the gradients on face AB of Figure 12.2.1, we consider the control volume 1234 around AB, indicated as the shaded area. As faces 1 and 3 pass through the cell centers of cells (i, j) and $(i + 1, j)$, the values at these points can be used directly. However, for the faces 2 and 4, an arithmetic average over four cells is required in order to compute the variables on these faces. Thus, the computation of the gradients at a cell face requires an access to the quantities at six different points. For instance, the gradient on face AB, labeled $(i + 1/2, j)$, requires the quantities at points $(i, j), (i, j - 1), (i, j + 1), (i + 1, j), (i + 1, j - 1),$

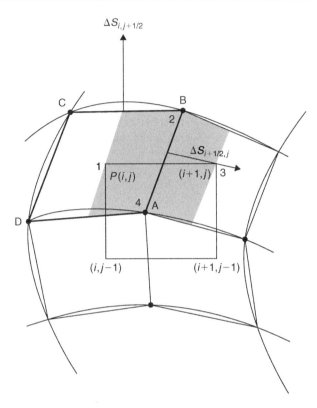

Figure 12.2.1 *Control volume used for computing the velocity and temperature gradients at face $i + 1/2, j$ (dashed area). The faces around this volume are numbered 1234.*

and $(i + 1, j + 1)$ following:

$$
\begin{cases}
U_{\text{face1}} = U_{i,j} \\
U_{\text{face2}} = \dfrac{1}{4}(U_{i,j} + U_{i+1,j} + U_{i,j+1} + U_{i+1,j+1}) \\
U_{\text{face3}} = U_{i+1,j} \\
U_{\text{face4}} = \dfrac{1}{4}(U_{i,j} + U_{i+1,j} + U_{i,j-1} + U_{i+1,j-1})
\end{cases}
\tag{12.2.3}
$$

$$
(\overline{\vec{\nabla} U})_{i+1/2,j} = \frac{1}{\Omega_{1234}}(U_{\text{face1}}\Delta\vec{S}_{\text{face1}} + U_{\text{face2}}\Delta\vec{S}_{\text{face2}} + U_{\text{face3}}\Delta\vec{S}_{\text{face3}} \\
+ U_{\text{face4}}\Delta\vec{S}_{\text{face4}})
$$

The x- and y-derivatives are obtained by projecting the average gradients in the corresponding direction. For instance, for the temperature gradients:

$$
\begin{aligned}
\left.\frac{\partial T}{\partial x}\right|_{i+1/2,j} &= (\overline{\vec{\nabla} T})_{i+1/2,j} \cdot \vec{e}_x \\
\left.\frac{\partial T}{\partial y}\right|_{i+1/2,j} &= (\overline{\vec{\nabla} T})_{i+1/2,j} \cdot \vec{e}_y
\end{aligned}
\tag{12.2.4}
$$

- In case the viscosity is not constant, for instance when the temperature is variable, the dynamic viscosity at cell faces is estimated by using arithmetic averaging of cell-center values.
- Compute the shear stresses at cell faces by using equation (12.1.2), following:

$$(\tau_{xx})_{i+1/2,j} = \mu_{i+1/2,j} \left[2 \left. \frac{\partial u}{\partial x} \right|_{i+1/2,j} - \frac{2}{3} \left(\left. \frac{\partial u}{\partial x} \right|_{i+1/2,j} + \left. \frac{\partial v}{\partial y} \right|_{i+1/2,j} \right) \right]$$

$$(\tau_{xy})_{i+1/2,j} = \mu_{i+1/2,j} \left[\left. \frac{\partial u}{\partial y} \right|_{i+1/2,j} = \left. \frac{\partial v}{\partial x} \right|_{i+1/2,j} \right] \tag{12.2.5}$$

- For faces located on boundaries, take into account the particular boundary conditions, as defined hereafter.
- Compute the fluxes using equation (12.2.1) and send the contribution to the right cell and its negative value to the left cell, in a counter-clockwise sense.

12.2.2   Boundary Conditions

As you certainly have experienced by now, if you have followed the code the development in Chapter 11, the numerical translation of the boundary conditions is one of the most critical issues in CFD.

12.2.2.1   *Physical boundary conditions*

As already mentioned above the boundary treatment described in Chapter 11 can still be applied for inlet and outlet boundaries. But the boundary conditions on walls have to be adapted to the viscous flow solver. Therefore, it is recommended to make two independent subroutines for inviscid and viscous wall boundary conditions.

As mentioned above, the major change compared to the inviscid code is the solid wall no-slip boundary condition, which imposes that all the velocity components vanish at the solid body surface. This is expressed by the Dirichlet type condition:

$$\vec{v} = 0 \quad \text{at the solid walls.} \tag{12.2.6}$$

The computation of the cell-face gradients, following equation (12.2.3), requires an access to all the neighboring cell values, which are not available for faces on the solid boundaries. Hence, a special treatment has to be implemented for these cells, by selecting control volumes entirely inside the computational domain. The control volume 1234, defined to compute the gradients at wall faces, is modified according to Figure 12.2.2, and the formulas (12.2.3) are replaced by

$$\begin{cases} U_{\text{face}1} = U_{i,j} \\ U_{\text{face}2} = \frac{1}{4}(U_{i,j} + U_{i+1/2,j} + U_{i,j+1} + U_{i+1/2,j+1}) \\ U_{\text{face}3} = U_{i+1/2,j} \\ U_{\text{face}4} = \frac{1}{4}(U_{i,j} + U_{i+1/2,j} + U_{i,j-1} + U_{i+1/2,j-1}) \end{cases}$$

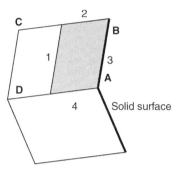

Figure 12.2.2 *Control volume used for computing the velocity and temperature gradients at face $i + 1/2, j$ which corresponds to a viscous wall (dashed area). The faces around this volume are numbered 1234.*

$$(\overline{\overline{\nabla U}})_{i+1/2,j} = \frac{1}{\Omega_{1234}}(U_{\text{face1}}\Delta\vec{S}_{\text{face1}} + U_{\text{face2}}\Delta\vec{S}_{\text{face2}} + U_{\text{face3}}\Delta\vec{S}_{\text{face3}}$$

$$+ U_{\text{face4}}\Delta\vec{S}_{\text{face4}}) \qquad (12.2.7)$$

where the values at locations $i + 1/2$ are the values imposed on the solid boundaries.

This requires an additional loop on all the cells close to the solid body.

Another method, which does not require a special loop over the boundary cells, consists in adding a row of additional cells outside of the computational domain. The flow variables are imposed in these 'ghost' cells in order to match the required boundary condition at the corresponding cell faces. The external loop on all the faces of the computational domain can be directly used. Though this method is less time consuming it requires to store variables in 'ghost' cells and therefore requires more memory.

You are here confronted with a representative choice between time and memory consumption.

For the energy equations three boundary conditions can be used: adiabatic, isothermal and with imposed heat flux. Depending on the type of boundary condition specified, i.e. Dirichlet or Neumann, either the temperature or its gradient has to be specified on the wall cell face.

If a Neumann type condition is applied the temperature gradient is directly defined and does not need to be calculated. The wall temperature can be inferred by writing the normal temperature gradient as a function of the temperature derivatives along grid lines:

$$(\vec{n}\cdot\vec{\nabla}T)_{i+1/2,j} = \frac{\Delta\vec{S}_{i+1/2,j}}{|\Delta\vec{S}_{i+1/2,j}|\Omega_{i,j}}\cdot\left[\Delta\vec{S}_{i+1/2,j}\frac{\partial T}{\partial\xi}\bigg|_{\text{wall}} + \Delta\vec{S}_j\frac{\partial T}{\partial\eta}\bigg|_{\text{wall}}\right]$$

$$(12.2.8)$$

where ξ, η represent the coordinate in the i,j directions, and assuming i as being the direction away from the wall. The surface $\Delta\vec{S}_j$ is defined as in equation (11.4.12).

This allows to estimate the gradients along the mesh line direction:

$$\left.\frac{\partial T}{\partial \xi}\right|_{\text{wall}} = -\frac{1}{|\Delta \vec{S}_{i+1/2,j}|^2}\left[\Delta \vec{S}_{i+1/2,j} \cdot \Delta \vec{S}_j \left.\frac{\partial T}{\partial \eta}\right|_{\text{wall}}\right]$$

$$+\frac{\Omega_{i,j}}{|\Delta \vec{S}_{i+1/2,j}|}(\vec{n} \cdot \vec{\nabla} T)_{i+1/2,j} \qquad (12.2.9)$$

The temperature difference along the j-direction can be deduced following:

$$\left.\frac{\partial T}{\partial \eta}\right|_{\text{wall}} = \frac{1}{2}(T_{i,j+1} - T_{1,j-1}) \qquad (12.2.10)$$

and the temperature at the cell face can be computed as

$$T_{i+1/2,j} = T_{i,j} + \frac{1}{2}\left.\frac{\partial T}{\partial \xi}\right|_{\text{wall}} \qquad (12.2.11)$$

If a Dirichlet type condition is imposed, the same procedure used for the momentum equations should be used. Either the subroutines that compute the gradients are adapted to this boundary condition, or the values inside the 'ghost' cells have to be adapted to the imposed condition.

12.2.2.2 *Numerical boundary conditions*

As for the inviscid flow solver, other variables have to be imposed on the solid wall boundaries. Since we have already imposed the condition on the velocity vector and on the temperature, only the pressure remains. A common assumption is that the pressure gradient normal to the wall vanishes. Since this gradient can be expressed as function of the pressure derivative along grid lines, following equation (12.2.8), we can write

$$(\vec{n} \cdot \vec{\nabla} p)_{i+1/2,j} = \frac{\Delta \vec{S}_{i+1/2,j}}{|\Delta \vec{S}_{i+1/2,j}|\,\Omega_{i,j}} \cdot \left[\Delta \vec{S}_{i+1/2,j}\left.\frac{\partial p}{\partial \xi}\right|_{\text{wall}} + \Delta \vec{S}_j\left.\frac{\partial p}{\partial \eta}\right|_{\text{wall}}\right]$$

$$(12.2.12)$$

Therefore, following the same procedure as for the temperature, the pressure difference along the i-direction can be obtained on the boundary cell faces, and the pressure at the wall can be computed.

The density follows from the knowledge of pressure and temperature at the wall surface.

12.2.2.3 *Periodic boundary conditions*

Periodic boundary conditions are often used when a periodicity is observed in the geometry of the domain to be meshed or in the flow motion to be represented. This is often the case in problems where the flow does not vary in a given direction, such as axisymmetric problems, or for turbomachinery applications, where the geometry has a periodicity, allowing only one blade passage to be meshed.

Another application of a periodic boundary condition occurs for simple one-dimensional problems, for which the streamwise direction is infinite without any flow variation, and we need to handle this problem with a two-dimensional code. In this case, we define a two-dimensional mesh with a few cells in the streamwise direction, selecting 2 or 3 cells in that direction. The constancy of all variables in that direction will be modeled by imposing a periodicity condition to connect the upstream and downstream boundaries in the streamwise direction, expressing equality of all the flow variables between these two boundaries.

Such a boundary condition is used hereafter for the simulation of the one-dimensional Couette flow problem with your 2D code.

12.2.3 Estimation of Viscous Time Step and CFL Conditions

As for the inviscid flow solver, since we are not interested in the transient behavior, a local time step could be used. Therefore, each cell will progress at its maximum time step. Following equation (E.9.3.1) the local viscous time step is defined as

$$\Delta t_{i,j} \leq \text{VNN} \frac{\Omega_{i,j}^2}{8\mu(|\Delta \vec{S}_i|^2 + |\Delta \vec{S}_j|^2 + 2|\Delta \vec{S}_i \Delta \vec{S}_j|)}$$

$$\Delta \vec{S}_i = \frac{1}{2}(\Delta \vec{S}_{i+1/2,j} + \Delta \vec{S}_{i-1/2,j}) \quad \Delta \vec{S}_j = \frac{1}{2}(\Delta \vec{S}_{i,j+1/2} + \Delta \vec{S}_{i,j-1/2})$$

$$(12.2.13)$$

where VNN is the von Neumann number, defined by $\alpha \Delta t/\Delta x^2$, for the one-dimensional diffusion equation. Here the diffusion coefficient is replaced by the kinematic viscosity for the momentum equation and by the thermal diffusivity for the energy equation.

For the complete Navier–Stokes equations the local time step to be applied is the minimum between the inviscid and viscous time steps, along the lines of equation (8.2.15).

12.3 NUMERICAL SOLUTIONS WITH THE DENSITY-BASED METHOD

We will guide you now to the application of the general finite volume code, developed along the guidelines just described in Section 12.2, to two problems with a simple geometry, where in fact Cartesian grids can be used. However, once you have acquired the first experience in solving viscous and thermal problems, you will also be able to apply your code to more general cases, such as the cylinder flow, the internal bump case of Chapter 11, and to many other cases you might be interested in, requiring more general curvilinear grids.

The first test case of the Couette flow is rather simple, as it does not have a streamwise variation and is essentially one-dimensional. Nevertheless, the incorporation of thermal effects makes the case of interest, as it will help us illustrate some of the requirements associated with thermal problems.

The second test case is the laminar flat plate boundary layer, with the Blasius profile as reference solution.

The exact solutions of these two test cases are for incompressible conditions, but will be treated here in the low Mach number compressible mode.

12.3.1   Couette Thermal Flow

One of the simplest cases to verify the discretization of viscous effects is the laminar flow between two parallel walls, a fixed wall and a moving one at a distance L with velocity U. In order to include the computation of the thermal fluxes both walls are considered as isothermal, at different temperatures. The moving upper wall has a higher fixed temperature than the static wall (see Figure 12.3.1).

The analytical solution for incompressible flow conditions is easily derived. Since the plates have infinite lengths, there is no physically relevant length scale in the streamwise direction and therefore all x-derivatives have to vanish. In addition, since the flow has only a streamwise velocity component, the normal velocity component is zero everywhere, i.e. $v = 0$, reducing the momentum equation to its streamwise component. The shear stress tensor (12.1.6) is also reduced to a single component τ_{12}.

$$\tau_{12} = \mu \frac{\partial u}{\partial y} \quad \text{all other } \tau_{ij} = 0 \tag{12.3.1}$$

Consequently, equations (12.1.7) and (12.1.8) simplify considerably to the one-dimensional system:

$$\frac{\partial u}{\partial t} = v \frac{\partial u^2}{\partial y^2}$$

$$\rho c_v \frac{\partial T}{\partial t} = k \frac{\partial^2 T}{\partial y^2} + \mu \left(\frac{\partial u}{\partial y} \right)^2 \tag{12.3.2}$$

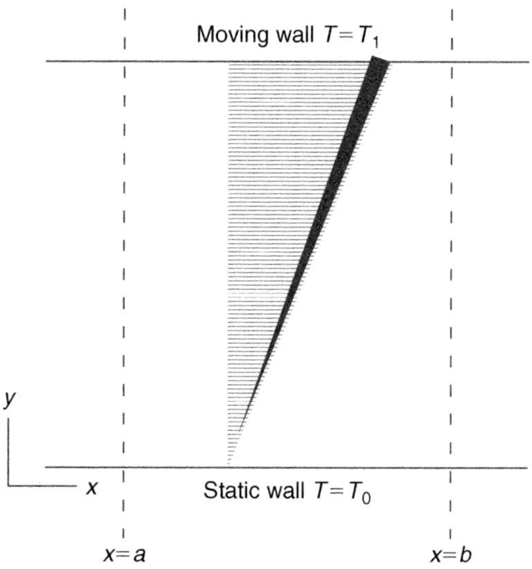

Figure 12.3.1   *Representation of the Couette flow test case.*

Note that these equations are parabolic in space and time, since all the convection terms have vanished.

They are to be solved with the Dirichlet boundary conditions for velocity and temperature:

$$
\begin{array}{llll}
y = 0 & u = 0 & T = T_0 \\
y = L & u = U & T = T_1
\end{array}
\tag{12.3.3}
$$

The steady state solution is now easily obtained, by setting the time derivatives to zero. The momentum equation reduces to

$$
\frac{\partial^2 u}{\partial y^2} = 0
\tag{12.3.4}
$$

leading to a linear velocity profile

$$
u(y) = \frac{y}{L} U
\tag{12.3.5}
$$

where L denotes the distance between the two plates and U is the velocity of the moving wall.

Introducing this solution in the steady temperature equation, leads to

$$
\frac{\partial^2 T}{\partial y^2} = -\frac{\mu}{k} \left(\frac{\partial u}{\partial y} \right)^2 = -\frac{\mu U^2}{kL^2}
\tag{12.3.6}
$$

This generates a parabolic temperature profile, depending on the parameter in the right-hand side.

The velocity scale is fixed by the upper wall velocity U; the temperature scale is determined by the temperature difference $\Delta T = T_1 - T_0$ and the length scale by the distance L between the plates. Hence, we define the non-dimensional variables

$$
\tilde{U} = u/U \quad \tilde{T} \triangleq \frac{T(y) - T_0}{T_1 - T_0} \quad Y = y/L
\tag{12.3.7}
$$

and the above temperature equation becomes

$$
\frac{\partial^2 \tilde{T}}{\partial Y^2} = -\frac{\mu U^2}{k \Delta T}
\tag{12.3.8}
$$

The right-hand side coefficient can be written as the product of the Prandtl and Eckert non-dimension numbers:

$$
P_r = \frac{\mu c_\mathrm{p}}{k} \quad E_c = \frac{U^2}{c_\mathrm{p} \Delta T} \quad P_r E_c = \frac{\mu U^2}{k \Delta T}
\tag{12.3.9}
$$

The Eckert number is the ratio of the dynamic temperature induced by fluid motion to the characteristics temperature difference in the fluid.

The analytical velocity and temperature profiles are then easily obtained as

$$\tilde{U}(Y) = Y$$

$$\tilde{T}(Y) = \frac{T(y) - T_0}{T_1 - T_0} = \left[1 + \frac{1}{2}P_rE_c(1 - Y)\right]Y \qquad (12.3.10)$$

The wall heat transfer coefficient is an important quantity in engineering applications and is generally expressed by the non-dimensional Nusselt number, defined here as a measure of the intensity of the heat flux via

$$Nu \triangleq \frac{\partial \tilde{T}}{\partial Y} = \frac{L}{\Delta T}\frac{\partial T}{\partial y} \qquad (12.3.11)$$

For the Couette flow, it takes the following values at the wall

$$Nu = 1 + \frac{P_rE_c}{2} \quad \text{at } y = 0$$

$$Nu = 1 - \frac{P_rE_c}{2} \quad \text{at } y = L \qquad (12.3.12)$$

An interesting property of this solution is that the fluid maximum temperature is greater than the upper wall temperature, when the product P_rE_c is greater than two.

12.3.1.1 *Numerical simulation conditions*

If you exercise your critical judgment, referring to the scheme properties developed in Chapter 9, you might recognize that we have here a purely parabolic problem and wonder as to the adequacy of the application of the explicit Runge–Kutta time integration method to this diffusion dominated test case. It is true indeed that this option is not optimal for the Couette flow test case, but we wish to guide you here in the practice of a general finite volume code, valid for low and high speeds, so that you can verify for yourself the range of applications you can cover. As seen in Chapter 9, the domain of stability of the Runge–Kutta method includes a part of the negative real axis of the eigenvalue spectrum, and therefore it remains valid for pure diffusion problems.

A first issue is the treatment of a one-dimensional flow case, where nothing is happening in the x-direction. This can be treated by generating a two-dimensional mesh with a limited number of mesh points in the x-direction, and applying periodic boundary conditions at the two ends of the domain, between $x = a$ and $x = b$, in Figure 12.3.1. The periodic boundary conditions express that all quantities at $x = b$ are equal to their corresponding values at the same ordinate at $x = a$. In principle two or three mesh points should be sufficient in the x-direction.

Another issue is connected to the numerical values of the flow variables. Although the non-dimensional solution (12.3.10) is independent of the levels of temperature differences and physical distances between the two walls, your code is written for the physical variables and consequently their numerical values can influence the overall accuracy of the computed results. Moreover, we apply here a density-based code, for which we consider low compressible conditions, with a Mach number around 0.1.

As seen in Chapter 11, this is totally acceptable for the analysis of incompressible flows, assuming perfect gas relations for the considered fluid. Because of the thermal effects, we have to ensure that numerically the perfect gas relations still remain close to constant density conditions and therefore we have to limit the absolute values of the temperature difference between the two endplates.

Since the non-dimensional solution only depends on the product $P_r E_c$, this parameter defines completely the numerical solution, when solved in the non-dimensional form. As this is not the case here, we have to select all the physical quantities of the fluid and the physical set-up, in order to fully define the dimensional form of the solution.

We select here the following values:

- The fluid is a perfect gas with the following properties:
  Specific Heat: $c_p = 1006$ J/kg/K
  Gamma: $\gamma = 1.4$
  Kinematic viscosity: $\nu = 1.57 \times 10^{-5}$ m^2/s
  Prandtl number: $P_r = 0.708$.
- The Reynolds numbers based on the velocity of the moving wall is 4000.
- The physical conditions of this Couette flow are chosen as $P_r E_c = 4$, with the following variables set according to:
  $L = 0.83$ mm
  $T_0 = 293°$K
  $T_1 = 294°$K
  $U = 75.4$ m/s
- Note the very small temperature difference selected of 1°K, which requires double precision arithmetic.

Since this problem is slightly more complex than the simple Couette flow without any thermal effect, it is recommended that you verify first the implementation of the viscous effects, without taking into account the thermal fluxes, to check the obtained linear velocity distribution.

12.3.1.2 *Grid definition*

A regular grid has been set-up for this case with 65 points in-between walls and 3 points in the axial direction, over a length L.

The two lateral sides of the computational domain are connected assuming a periodic repetition of the channel.

In order to investigate the sensitivity of the numerical results to the grid density, coarser meshes are constructed by simply removing each second point in the vertical the wall-to-wall direction. Hereby we generate four different grids: $(65 \times 3), (33 \times 3), (17 \times 3)$ and (9×3).

12.3.1.3 *Results*

The Navier–Stokes equations are solved using the cell-centered approach until a steady state is reached, with CFL and Von Neumann numbers put to 1.8 and with the dissipation coefficient $\kappa^{(4)} = 1/100$.

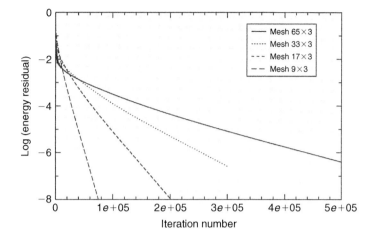

Figure 12.3.2 *Convergence history of the normalized L₂-norm of the energy residual on four different meshes.*

A Runge–Kutta time integration method is also applied with the coefficients provided by equation (11.4.11). The computations are stopped when the residuals have decreased by more than 6 orders of magnitude.

The convergence history of the energy equation is provided in Figure 12.3.2. The energy residual is decreased by about 8 orders of magnitude. Such a convergence level can only be obtained by using a double precision version of your code. It is seen that the rate of convergence strongly depends on the mesh used. A coarser mesh is associated to a more rapid convergence, whereas the convergence rate decreases if the grid density is refined.

The very slow convergence results largely from the application of the general code you have developed, which is oriented at 2D convection dominated inviscid or viscous flows. The Couette flow, on the other hand, is a pure one-dimensional parabolic problem since convection does not play any role in this particular case, and the applied algorithms are therefore not optimal.

Note also that in a more advanced code, multigrid acceleration will normally be available, reducing considerably the required number of iterations.

In practice, 3–4 orders of magnitude might be sufficient for an 'engineering solution', but is it important that you verify that your code can reach machine accuracy, to ensure that the algorithm is correctly programmed, in all its details.

The accuracy of the solution is also strongly influenced by the grid density used. On the one hand, all the meshes used are able to reproduce the linear distribution of axial velocity in-between walls (see Figure 12.3.3) demonstrating the second order accuracy of the scheme, *for which a linear variation has to be exactly reproduced*. On the other hand, the quadratic distribution of temperature is not accurately captured when using too coarse meshes (see Figure 12.3.4). The meshes with 33 and 65 points from wall to wall are able to accurately predict the analytical solution. This finding illustrates that a sufficient refinement is necessary in order to capture the flow features present in laminar boundary layers.

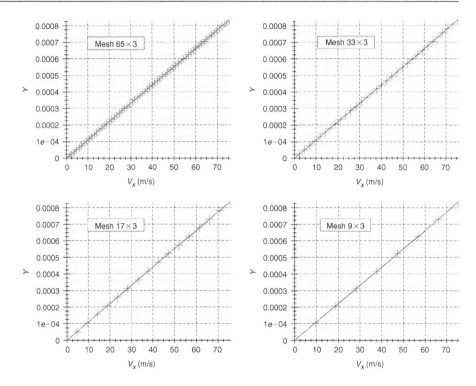

Figure 12.3.3 *Wall-to-wall distribution of axial velocity as deduced from analytical result (continuous line) and from the numerical result (plus signs) (for color image refer Plate 12.3.3).*

An important aspect of many engineering flow problems is the ability to predict local extrema of important flow variables. Since thermal stresses have an impact on solid structure lifetime, it is of importance to accurately predict the maximum temperature inside a flow. Therefore, we will perform a grid convergence study based on the prediction of the maximum temperature between the solid walls. According to equation (12.3.10) the maximum temperature inside the flow is obtained at a position $y/L = 3/4$ from the static wall and its value is 294.125 K. The computed maximum temperatures are reported in Table 12.3.1, including the relative error in % of the wall temperature difference. As expected, the maximum temperature error is reduced if the mesh is refined, and an error lower than 1% of the temperature variation can be obtained with a mesh having at least 33 grid cells in the wall normal direction. This number has to be doubled if we need a precision of less than 0.1%.

Another critical quantity in presence of thermal effects is the heat flux through a boundary, as expressed by the Nusselt number, defined here by equation (12.3.11). This is generally highly grid dependent, although in the present case, as the temperature gradient is linear, it will be less sensitive to the grid, as seen from Table 12.3.2.

12.3.1.4 *Other options for solving the Couette flow*

The method applied here, based on a more general finite volume formulation, is of course not optimal for this simple Couette flow. Since the problem is actually mono-dimensional, you can solve equations (12.3.2) and (12.3.3) much easier and

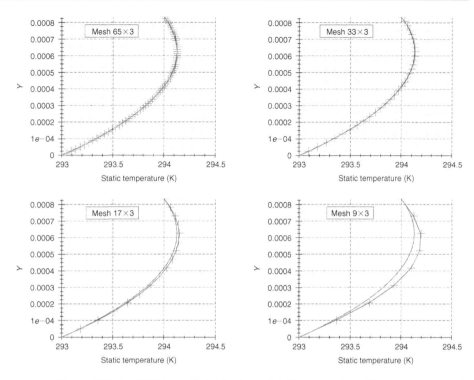

Figure 12.3.4   *Wall-to-wall distribution of static temperature (line with plus signs), compared to the analytical solution (continuous line), for the four different grids (for color image refer Plate 12.3.4).*

Table 12.3.1   *Maximum temperature computed using the different meshes.*

	Temperature (K)	Error in temperature (K)	Relative error in temperature (%)
Analytical	294.1250	–	–
65 × 3	294.1255	5×10^{-4}	0.05
33 × 3	294.1284	3.4×10^{-3}	0.34
17 × 3	294.1462	2.12×10^{-2}	2.12
9 × 3	294.1898	6.48×10^{-2}	6.48

Table 12.3.2   *Estimated values of the Nusselt number on several grids.*

	Nusselt number at lower wall	Nusselt number at upper wall	Relative error at lower wall (%)	Relative error at upper wall (%)
Analytical	3	−1	–	–
65 × 3	3.0072	−1.0034	0.24	0.34
33 × 3	3.0075	−1.0030	0.25	0.30
17 × 3	3.0070	−1.0025	0.23	0.25
9 × 3	3.0008	−0.9993	0.17	−0.07

faster, by applying a dedicated algorithm for parabolic problems. For instance an implicit Crank–Nicholson scheme to the centrally discretized diffusion terms, will lead to a much faster code. We recommend that you program equations (9.3.16), referring also to Problem P.9.10.

12.3.2   Flat Plate

One of the most popular applications of laminar viscous flows is the boundary layer development along a flat plate. The main advantages of this case are its relevance for a number of practical flow problems and the availability of an exact solution, obtained by solving the Blasius equation.

As for the Couette flow problem, we consider here weak compressible conditions, with a free stream Mach number around 0.2, which is still acceptable for the analysis of incompressible flows, assuming perfect gas relations for the considered fluid.

The fluid is a perfect gas fluid with the following properties:

Specific Heat: $c_p = 1006$ J/kg/K
Gamma: $\gamma = 1.4$
Kinematic viscosity: $\nu = 1.57 \times 10^{-5}$ m^2/s
Prandtl number: $P_r = 0.708$

We select the length of the plate equal to 0.2 m and the free stream velocity is fixed to match a Mach number of 0.2, leading to $U = 68.3$ m/s.

The Reynolds number based on the free stream velocity and the plate length is 8.7×10^5.

12.3.2.1   *Exact solution*

The exact solution of the development of a laminar boundary layer over a flat plate with an incoming uniform velocity U in the axial plat direction has been solved by Blasius nearly a century ago. It is based on the self-similarity of the velocity profiles along the plate for an incompressible fluid. A detailed description of the Blasius solution can be found in Schlichting (1979).

After a few simplifications of the Navier–Stokes equations, and assuming the boundary layer approximations, we obtain to the Blasius equation:

$$f\frac{d^2f}{d\eta^2} + 2\frac{d^3f}{d\eta^3} = 0 \tag{12.3.13}$$

with the following boundary conditions

$$\begin{aligned} f(0) &= 0 \\ f'(0) &= 0 \\ f'(\eta) &\to 1 \quad \text{if } \eta \to \infty \end{aligned} \tag{12.3.14}$$

where f is the function solution of the Blasius equation. η is a non-dimensional coordinate normal to the plate

$$\eta = \frac{y}{\left(\dfrac{\nu x}{U}\right)^{1/2}} \tag{12.3.15}$$

Equation (12.3.13) is an ordinary differential equation and can be solved numerically with an arbitrary accuracy. The two components of the velocity vector can be inferred from the function f

$$u = U\frac{df}{d\eta}$$

$$v = \frac{1}{2}\left(\frac{U\nu}{x}\right)^{1/2}\left(\eta\frac{df}{d\eta} - f\right) \tag{12.3.16}$$

where U is the free stream velocity.

From this solution we can deduced the distribution of the friction coefficient along the single sided plate

$$C_f = \frac{0.664}{\sqrt{Re_x}} \qquad Re_x = \frac{Ux}{\nu} \tag{12.3.17}$$

The Blasius solution is tabulated in Table 12.3.3.

12.3.2.2 *Grid definition*

A first important decision is the selection of the computational domain. We have actually two options, each with its specific problems:

Option 1: Locate the computational domain boundaries upstream and downstream of the leading and trailing edges of the plate. This is the most realistic choice, with the advantage of allowing the simulation of the approach of the flow toward the leading edge and the downstream wake. However, it requires a very dense mesh around the leading and trailing edges, and the associated range of boundary conditions.

Option 2: Select the computational domain between the leading edge and the trailing edge of the plate. This avoids the grid concentrations of option 1, but creates a non-realistic flow at the leading edge, since a uniform flow is assumed at the leading edge.

Due to its simplicity we choose here the second option.

To close the computational domain, an outlet boundary has to be defined at a certain distance parallel to the plate. This boundary may not influence the development of the boundary layer and it should be placed sufficiently far from the plate. According to the Blasius solution, the thickness of the boundary layer at a distance x from its leading edge is of the order:

$$\delta_\infty(x) \cong 5\frac{x}{\sqrt{Re_x}} \tag{12.3.18}$$

For the above-mentioned conditions, the boundary layer thickness at the end of the plate $x = 0.2$ m, is of the order of 1 mm. Therefore, the outlet boundary parallel to the

Table 12.3.3 *Blasius solution for the flat plate boundary layer.*

η	$\mathrm{d}f/\mathrm{d}\eta$	$\eta\,\mathrm{d}f/\mathrm{d}\eta - f$
0.00000E+00	0.00000E+00	0.00000E+00
0.14142E+00	0.46960E−01	0.33177E−02
0.28284E+00	0.93910E−01	0.13282E−01
0.42426E+00	0.14081E+00	0.29858E−01
0.56569E+00	0.18761E+00	0.53025E−01
0.70711E+00	0.23423E+00	0.82696E−01
0.84853E+00	0.28058E+00	0.11873E+00
0.98995E+00	0.32653E+00	0.16098E+00
0.11314E+01	0.37196E+00	0.20916E+00
0.12728E+01	0.41672E+00	0.26296E+00
0.14142E+01	0.46063E+00	0.32193E+00
0.15556E+01	0.50354E+00	0.38563E+00
0.16971E+01	0.54525E+00	0.45345E+00
0.18385E+01	0.58559E+00	0.52475E+00
0.19799E+01	0.62439E+00	0.59881E+00
0.21213E+01	0.66147E+00	0.67483E+00
0.22627E+01	0.69670E+00	0.75202E+00
0.24042E+01	0.72993E+00	0.82955E+00
0.25456E+01	0.76106E+00	0.90656E+00
0.26870E+01	0.79000E+00	0.98226E+00
0.28284E+01	0.81669E+00	0.10558E+01
0.31113E+01	0.86330E+00	0.11940E+01
0.33941E+01	0.90107E+00	0.13167E+01
0.36770E+01	0.93060E+00	0.14209E+01
0.39598E+01	0.95288E+00	0.15051E+01
0.42426E+01	0.96905E+00	0.15720E+01
0.45255E+01	0.98037E+00	0.16215E+01
0.48083E+01	0.98797E+00	0.16569E+01
0.50912E+01	0.99289E+00	0.16812E+01
0.53740E+01	0.99594E+00	0.16972E+01
0.56569E+01	0.99777E+00	0.17072E+01
0.59397E+01	0.99882E+00	0.17133E+01
0.62225E+01	0.99940E+00	0.17168E+01
0.65054E+01	0.99970E+00	0.17187E+01
0.67882E+01	0.99986E+00	0.17198E+01
0.70711E+01	0.99994E+00	0.17203E+01
0.73539E+01	0.99997E+00	0.17206E+01
0.76368E+01	0.99999E+00	0.17207E+01
0.79196E+01	0.99999E+00	0.17207E+01
0.82024E+01	0.10000E+01	0.17208E+01
0.84853E+01	0.10000E+01	0.17208E+01

wall is fixed at a distance of about 0.02 m from the plate, i.e. 20 times the boundary layer thickness.

This decision results from a compromise between accuracy and computational cost. Indeed, the location of the lateral boundary can be a source of errors, but putting this limit too far from the plate will result in a higher number of points in the free stream region where the flow is almost uniform.

Based on the above definition of the computational domain, a rectangular grid is set-up, with an adequate refinement in the regions of strong flow variations, i.e. in the leading edge region and close to the plate. Far from the plate the flow is almost uniform and allows coarser grid cells.

An exponential stretching technique is applied for which the ratio of two subsequent grid cells is constant:

$$\frac{\Delta y_i}{\Delta y_{i-1}} = a \tag{12.3.19}$$

and the coordinates of the grid points are given by

$$y_i = y_1 + \frac{a^{i-1} - 1}{a - 1} \Delta y_1 \tag{12.3.20}$$

Since in our application the plate is located at $y_1 = 0$, the grid distribution normal to the plate depends on the cell width at the first inner cell and on the number of cells. Since we want enough points in the laminar boundary layer, the cell spacing at the first inner cell will be fixed to 10^{-5} m, and we select 65 points in the wall normal direction. In order to have the outer point located at a distance 0.02 m from the plate, the factor a should be equal to $a = 1.083317311$, leading to a distribution of 28 cells located in the 1 mm thick boundary layer at $x = 0.2$ m.

Since the thickness of the boundary layer evolves as the square root of the distance to the leading edge, the mesh should also be refined in the streamwise direction. We select a streamwise distribution of grid points following also an exponential stretching as described in equation (12.3.20), with 65 points along the 0.2 long plate and the thickness of the first grid cell in the streamwise direction is fixed to 0.1 mm. Therefore, the parameter a is fixed to $a = 1.083317311$.

With these distributions in the streamwise and wall normal directions we can construct the regular mesh displayed in Figure 12.3.5.

The boundary in the far field *must be treated as an outlet boundary* through which mass flow can escape, since due to the boundary layer growth, the flow field has a non-zero vertical component at any distance from the wall. If this boundary is considered as an inviscid boundary on which the free stream velocity is imposed, this will impose a pressure force on the flat plate in contradiction with the physics of the problem as it will not allow the vertical velocity component to expand as it should from the theory.

In order to investigate the sensitivity of the numerical results to the grid density, coarser meshes are constructed by simply removing every second point in the wall normal direction, defining five different grids: (65×65), (65×33), (65×17), (65×9) and (65×5).

This will also allow you to evaluate the influence of the number of boundary layer grid points on the numerical accuracy. This is an important issue, as in practical

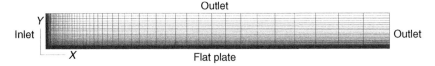

Figure 12.3.5 *Mesh defined for the laminar flat plate test case.*

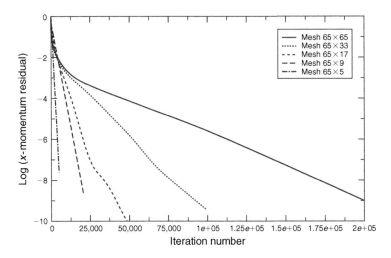

Figure 12.3.6 *Convergence history of the normalized L_2-norm of the x-momentum residual on five different meshes.*

industrial case, it is quite difficult and costly to ensure 25–30 points in boundary layers of complex 3D geometries.

Boundary conditions are specified at the inlet and outlet of the computational domain. Assuming atmospheric condition for the static temperature and pressure and an inlet Mach number of 0.2, the inlet total pressure and total temperature are fixed to 104165 Pa and 290.304 K, respectively (following equations (11.1.9) and (11.1.12)). The inlet velocity vector is imposed in the x-direction. At the outlet boundaries, as shown on Figure 12.3.5, the static pressure is fixed to the atmospheric value, i.e. 101300 Pa.

12.3.2.3 *Results*

The Navier–Stokes equations are solved using the cell-centered approach until a steady state is reached, with CFL and Von Neumann numbers put to 1.5 and the dissipation coefficient $\kappa^{(4)} = 1/100$.

A Runge–Kutta time integration method is also used with the coefficients provided by equation (11.4.11). The computations are stopped when the residuals have decreased by more than 6 orders of magnitude.

The convergence history of the x-momentum equation is provided in Figure 12.3.6. This residual is decreased by about 8 orders of magnitude. Such a convergence level

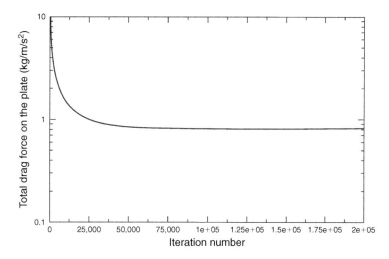

Figure 12.3.7 *Convergence history of the drag force along the plate for the finest mesh.*

can only be obtained by using a double precision version of your code. As for the Couette flow problem, it is seen that the rate of convergence strongly depends on the mesh used. A coarser mesh is associated to a more rapid convergence, whereas the convergence rate decreases if the grid density is refined.

You can see that after about 25,000 iterations the x-momentum residual on the finest mesh calculation has decreased by more than 3 orders of magnitude. From this observation, we could be led to judge that the simulation has reached a sufficient level of convergence and that it can be stopped. However, by monitoring the total drag force acting on the plate, it can be seen on Figure 12.3.7 that it we have to wait for more than 50,000 iterations, for a fully converged solution. Indeed, small cells located all along the plate require more iterations to reach a fully converged solution. Therefore, it is suggested to monitor not only the global residuals but also some key parameters in order to identify if a converged solution has been reached. This is particularly important for viscous calculations where numerous small cells should be located in the boundary layer.

Note that this high number of iterations would be considerably reduced by the addition of the multigrid technique.

The solution obtained on the finest mesh is displayed on Figure 12.3.8. It appears that, with a sufficient number of points in the boundary layer the solver is able to reproduce the analytical distribution of friction coefficient along the plate and the mainstream flow component.

Note however that in the region of the largest curvature of the axial velocity profile, the numerical results show a lower velocity compared to the exact solution, indicating the influence of numerical dissipation. This can be improved by more sophisticated dissipation terms, as obtained from formulations based on matrix dissipation, where the coefficients of the artificial dissipation terms (11.4.7) are replaced by terms containing the full Jacobian matrix, as opposed to its spectral radius (see for instance Jameson, 1995a, b). Alternatively, the second order upwind schemes, as will be seen

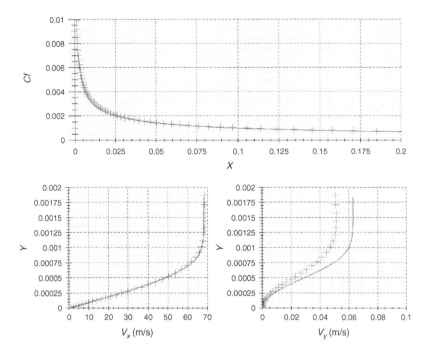

Figure 12.3.8   *Distribution of friction coefficient (upper panel), axial and wall normal velocities at x = 0.2 m (lower left and right panels) as obtained from analytical result (continuous line) and from the numerical result on the finest mesh (plus signs). The vertical axis is the distance in meters (for color image refer Plate 12.3.8).*

in Volume II, have also a reduced dependency to the number of boundary layer grid points.

The velocity in the wall normal direction differs significantly from the Blasius profiles, since relatively small errors in the axial velocity have a great impact on the wall normal velocity, as the latter is about three orders of magnitude lower than the former. This is also related to the large dissipation associated to the selected formulation of the central schemes.

An important outcome of viscous flow computations is the prediction of the friction coefficient along the solid surfaces. According to equation (12.3.17) the analytical drag coefficient at the end of the plate ($x = 0.2$ m) is equal to 0.00713. The computed drag coefficients are reported in Table 12.3.4 together with the relative error. As can be seen, the error in the friction coefficient is reduced if the mesh is refined, but is still at high values with the meshes used in the present calculations. The second column displays an interesting information, namely the number of mesh points in the inner part of the boundary layer, over the first one millimeter.

Accurate predictions of drag coefficients form a challenging and difficult issue in CFD, and represent one of the most sensitive criteria for accuracy assessment.

The interested reader might consult with interest the summary papers of recent workshops held on drag prediction evaluations of aeronautical relevant configurations,

Table 12.3.4   *Friction coefficient computed using the different meshes.*

	Number of cells in the boundary layer (first one millimeter)	Cf	Relative error in Cf (%)
Analytical		0.000713	–
65 × 65	28	0.00075	5.2
65 × 33	14	0.000803	12.7
65 × 17	7	0.000917	28.7
65 × 9	4	0.001371	92.3
65 × 5	2	0.002196	208.0

in Hemsch and Morrison (2004), also referred to in the general introduction to this book.

The errors on the friction coefficient are a consequence of the reduced accuracy on the velocity profiles, when the mesh is coarsened. This can be seen from Figure 12.3.9, where the distributions of skin friction and velocity are displayed on the two successive coarser grids, namely 65 × 33 and 65 × 17, having respectively, 14 and 7 points in the boundary layer. You will notice that 14 points might still be acceptable, for 'engineering' accuracy, but the error on the skin friction is of 12%, as seen from Table 12.3.4.

12.4   PRESSURE CORRECTION METHOD

The methods known as **pressure correction** are among the first developed for the numerical solutions of the full Navier–Stokes equations for incompressible flows. The method was originally applied by Harlow and Welch (1965) in the MAC, Marker-and-Cell, method for the computation of free surface incompressible flows. It is closely related to the **fractional step method**, also called **projection method**, developed independently by Chorin (1967), (1968) and Temam (1969); see also Temam (1977).

It has been adapted to industrial flow simulations by Patankar and Spalding (1972) and described in details in several books (Patankar (1980); Anderson (1995); Ferziger and Peric (1997); Wesseling (2001)).

The methods falling in this class can be applied to the stationary as well as to the time-dependent incompressible flow equations. They consist of a basic iterative procedure between the velocity and the pressure fields. For an initial approximation of the pressure, the momentum equation can be solved to determine the velocity field. The obtained velocity field does not satisfy the divergence free, continuity equation and has therefore to be corrected. Since this correction has an impact on the pressure field, a related pressure correction is defined, obtained by expressing that the corrected velocity satisfies the continuity equation. This leads to a Poisson equation for the pressure correction.

The pressure correction methods have been extended to compressible flows, and various approaches can be defined. Weak compressibility can be handled through a simplified system of equations obtained by developing a low Mach number expansion of the Navier–Stokes equations, omitting then the terms that are of higher order in

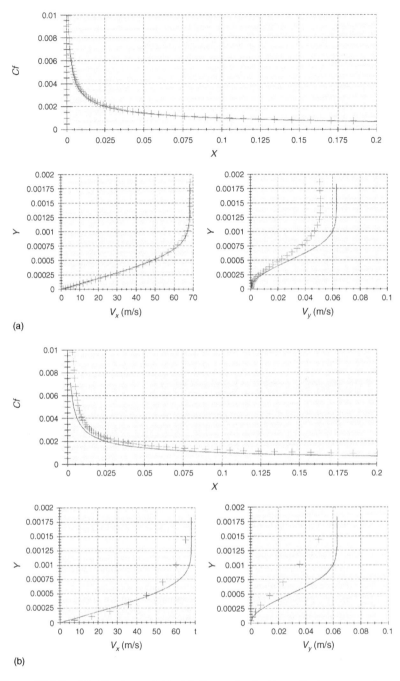

Figure 12.3.9 *Distributions of skin friction and velocity on the two successive coarser grids of 65 × 33 and 65 × 17, having respectively, 14 and 7 points in the boundary layer. (a) Numerical solution on 65 × 33 mesh with 14 points in the boundary layer and (b) numerical solution on 65 × 17 mesh with 7 points in the boundary layer.*

the Mach number (see, for instance, Majda and Sethian (1985)). A full compressible extension can be found in the book of Ferziger and Peric (1997).

12.4.1   Basic Approach of Pressure Correction Methods

We restrict the presentation here to strictly incompressible, isothermal flows, defined in Section 1.4.4 by the system formed by the continuity and momentum equations, written here in non-conservative form.

The mass conservation equation reduces in the case of incompressible flows to

$$\vec{\nabla} \cdot \vec{v} = 0 \tag{12.4.1}$$

which appears as a constraint to the general time-dependent equation of motion, written here in absence of external volume forces

$$\frac{\partial \vec{v}}{\partial t} + (\vec{v} \cdot \vec{\nabla})\vec{v} = -\frac{1}{\rho}\vec{\nabla}p + \nu\Delta\vec{v} \tag{12.4.2}$$

The only unknowns are velocity and pressure.

An equation for the pressure can be obtained by taking the divergence of the momentum equation (12.4.2), and introducing the divergence free velocity condition (12.4.1), leading to

$$\frac{1}{\rho}\Delta p = -\vec{\nabla} \cdot (\vec{v} \cdot \vec{\nabla})\vec{v} \tag{12.4.3}$$

which can be considered as a Poisson equation for the pressure for a given velocity field. Note that the right-hand side contains only products of first order velocity derivatives, because of the incompressibility condition (12.4.1). Indeed, in tensor notations, the velocity term in the right-hand side term is equal to $(\partial_j v_i) \cdot (\partial_i v_j)$.

Before describing the pressure correction method we have to select a time integration scheme for the momentum equations, considering the pressure gradient as known. For reasons of simplicity and in order to point out the essential properties of the pressure correction approach, we will select an explicit method of first order accuracy in time, although it is not recommended in practice. Even for time-dependent problems the time step restriction imposed by stability conditions for the parabolic, convection–diffusion momentum equations is generally smaller than the physical time constant of the flow. Hence, the time steps allowed by the requirements of physical accuracy are large enough to allow the larger numerical time steps of implicit schemes. Typically, semi-implicit time integration schemes are recommended. For instance, the viscous fluxes can be treated implicitly by means of the Crank–Nicholson formulation, while convective fluxes can be handled with an Adams–Bashworth second order method. This provides a higher accuracy for time-dependent simulations and allows for large time steps leading to more efficient calculations in terms of computational costs.

The fundamental approach of pressure correction methods is the decoupling of the pressure field from the velocity field. This is expressed by solving the momentum equation with a known pressure field, for instance with the pressure obtained at the previous iteration.

A variety of methods have been developed and applied in practice, based on various decoupling approaches. In its simplest form, with an explicit time discretization, we solve for an intermediate velocity field $\vec{v}^*$, solution of

$$\frac{\vec{v}^* - \vec{v}^n}{\Delta t} = -\vec{\nabla} \cdot (\vec{v} \otimes \vec{v})^n - \frac{1}{\rho}\vec{\nabla}p^n + \nu\Delta\vec{v}^n \tag{12.4.4}$$

The solution $\vec{v}^*$ of this equation does not satisfy the continuity equation. Hence the final values are defined by adding corrections to the intermediate values

$$\vec{v}^{n+1} = \vec{v}^* + \vec{v}' \quad p^{n+1} = p^n + p' \tag{12.4.5}$$

where the final values with superscript $n + 1$ have to be solutions of

$$\frac{\vec{v}^{n+1} - \vec{v}^n}{\Delta t} = -\vec{\nabla} \cdot (\vec{v} \otimes \vec{v})^n - \frac{1}{\rho}\vec{\nabla}p^{n+1} + \nu\Delta\vec{v}^n$$
$$\vec{\nabla} \cdot \vec{v}^{n+1} = 0 \tag{12.4.6}$$

Introducing (12.4.5) in the above equation and subtracting (12.4.4), leads to the following relation between the pressure and velocity corrections:

$$\vec{v}' = -\frac{\Delta t}{\rho}\vec{\nabla}p' \tag{12.4.7}$$

Note that expressing the velocity correction as a gradient of a scalar function conserves the vorticity of the intermediate velocity field. That is, the correction field is a potential flow.

Taking the divergence of the first of the equations (12.4.6) gives the Poisson equation for the pressure correction:

$$\Delta p' = \frac{\rho}{\Delta t}\vec{\nabla} \cdot \vec{v}^* \tag{12.4.8}$$

Equation (12.4.3) assumes that the solution at time level n satisfies exactly the divergence-free condition. In the numerical process, the velocity at level n might not satisfy exactly this condition. In this case, the non-zero value of $D^n \stackrel{\Delta}{=} \vec{\nabla} \cdot \vec{v}^n$ should be introduced in the pressure Poisson equation. This situation is more likely to occur in stationary computations where n represents an iteration count. With time-dependent calculations, it is recommended to satisfy accurately mass conservation at each time step, in particular by discretizing the integral form of the mass conservation law on a finite volume mesh.

The Poisson equation for the pressure is solved with Neumann boundary conditions, on the normal pressure gradient, obtained by taking the normal component of equations (12.4.6). The details of the implementation depend on the selected space discretization and on the mesh.

An alternative approach is the ***fractional step, or projection***, method based on a slightly different definition of the intermediate pressure, whereby the pressure term is simply omitted, leading to the complete decoupling of the intermediate velocity field

$$\frac{\vec{v}^* - \vec{v}^n}{\Delta t} = -\vec{\nabla} \cdot (\vec{v} \otimes \vec{v})^n + \nu \Delta \vec{v}^n \tag{12.4.9}$$

followed by the pressure equation

$$\frac{\vec{v}^{n+1} - \vec{v}^*}{\Delta t} = -\frac{1}{\rho} \vec{\nabla} p^{n+1} \tag{12.4.10}$$

The final value of velocity field is obtained from equation (12.4.10). The pressure is calculated in such a way that the velocity field at level $(n + 1)$ satisfies the divergence free condition:

$$\frac{\rho}{\Delta t} [\vec{\nabla} \cdot \vec{v}^{n+1} - \vec{\nabla} \cdot \vec{v}^*] = -\Delta p^{n+1} \tag{12.4.11}$$

leading to the an equation similar to (12.4.8)

$$\Delta p^{n+1} = \frac{\rho}{\Delta t} \vec{\nabla} \cdot \vec{v}^* \tag{12.4.12}$$

The final value of the velocity field $\vec{v}^{n+1}$ velocities are updated from (12.4.10).

The numerical resolution of the pressure Poisson equation is a crucial step of the whole approach, since the overall efficiency of the code will depend on its performance. Hence all possible convergence optimization and acceleration techniques should be applied. In particular preconditioning and multigrid techniques are strongly recommended for this step of the computation, and eventually for other steps.

12.4.2 The Issue of Staggered Versus Collocated Grids

The choice of a space discretization is, as for compressible flows, between centered or upwind methods, at least for the convection terms, since the diffusive contributions are always centrally discretized.

The most current choice is the central discretization of the convection terms, which raises a particular problem with the pressure correction approach.

The central discretization for the convection terms requires the addition of higher order artificial dissipation terms to create the required damping of high frequency errors, as introduced in the previous examples. However the absence of the time derivative of the density in the continuity equation creates an additional decoupling in the centrally discretized equations, with pressure correction methods. This is best illustrated on the one-dimensional system of incompressible Navier–Stokes equations. The conservation equations take the form

$$\frac{\partial u}{\partial x} = 0$$

$$\frac{\partial u}{\partial t} + \frac{\partial u^2}{\partial x} = -\frac{1}{\rho} \frac{\partial p}{\partial x} + \nu \frac{\partial^2 u}{\partial x^2} \tag{12.4.13}$$

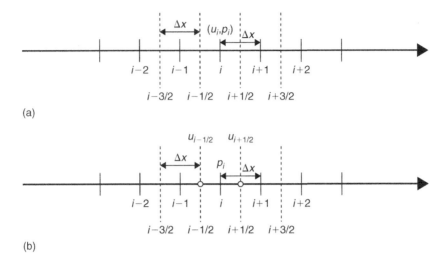

(a)

(b)

Figure 12.4.1 *Standard 'collocated' and staggered grids: (a) standard collocated grid and (b) staggered grid.*

We consider a central finite difference discretization of the above equations on a standard uniform grid, where all the variables are defined on the same mesh points. This is called a ***collocated mesh***, in the context of pressure correction methods (see Figure 12.4.1a). The centrally discretized equations become

$$\frac{u_{i+1}^{n+1} - u_{i-1}^{n+1}}{2\Delta x} = 0 \tag{12.4.14}$$

$$\frac{u_i^{n+1} - u_i^n}{\Delta t} + \frac{(u_{i+1}^n)^2 - (u_{i-1}^n)^2}{2\Delta x} = -\frac{1}{\rho}\frac{p_{i+1} - p_{i-1}}{2\Delta x} + v\frac{u_{i+1}^n - 2u_i^n + u_{i-1}^n}{\Delta x^2} \tag{12.4.15}$$

In the framework of the pressure correction method, the momentum equation is split into two parts by introducing the intermediate velocity u^*, based on the fractional step method, solution of

$$\frac{u_i^* - u_i^n}{\Delta t} + \frac{(u_{i+1}^n)^2 - (u_{i-1}^n)^2}{2\Delta x} = v\frac{u_{i+1}^n - 2u_i^n + u_{i-1}^n}{\Delta x^2} \tag{12.4.16}$$

$$\frac{u_i^{n+1} - u_i^*}{\Delta t} = -\frac{1}{\rho}\frac{p_{i+1} - p_{i-1}}{2\Delta x} \tag{12.4.17}$$

The expression for u_i^{n+1}, obtained from equation (12.4.17), can be substituted into (12.4.14), resulting in the following 1D equation:

$$\frac{p_{i+2} - 2p_i + p_{i-2}}{4\Delta x^2} = \frac{\rho}{\Delta t}\frac{u_{i+1}^* - u_{i-1}^*}{2\Delta x} \tag{12.4.18}$$

This is a Poisson equation for the pressure, which ensures that the continuity equation (12.4.14) is satisfied for the newly updated velocity u_i^{n+1}. The stencil on which the Laplace operator is discretized $(i + 2, i, i - 2)$ contains, however, only odd or even indices, which leads to a decoupling of the discrete pressure field and often results in high frequency oscillations of pressure.

As can be seen indeed, the pressure at point i is not influenced by the velocity component u_i^n of the same point and in return u_i^n is not affected by p_i. Hence velocity and pressure are decoupled on even and odd points; see also Section 4.2 and the discussion around the Lax–Friedrichs scheme in Chapter 7, for an illustration of analog cases. This decoupling is not present with compressible flows due to the density–velocity coupling in the continuity equation. It will generate additional high frequency oscillations, requesting the introduction of artificial dissipation terms.

A solution to the odd–even decoupling problem, has been introduced by Harlow and Welch (1965), by defining a ***staggered mesh***, where the velocity and pressure are not defined in the same mesh points. As seen in Figure 12.4.1b, the velocity is directly defined at the half mesh points, while the pressure remains defined at the central mesh point. The central discretization of the continuity equation of (12.4.13) now becomes

$$\frac{u_{i+1/2}^{n+1} - u_{i-1/2}^{n+1}}{\Delta x} = 0 \tag{12.4.19}$$

With the fractional step method, equation (12.4.17) becomes on the staggered mesh

$$\frac{u_{i+1/2}^{n+1} - u_{i+1/2}^{*}}{\Delta t} = -\frac{1}{\rho}\frac{p_{i+1} - p_i}{\Delta x} \tag{12.4.20}$$

By substituting expressions for $u_{i+1/2}^{n+1}$ and $u_{i-1/2}^{n+1}$ derived from (12.4.20) into this equation, we obtain

$$\frac{p_{i+1} - 2p_i + p_{i-1}}{\Delta x^2} = \frac{\rho}{\Delta t}\frac{u_{i+1/2}^{*} - u_{i-1/2}^{*}}{\Delta x} \tag{12.4.21}$$

In this discrete Poisson equation the pressure and velocity in all nodes are fully coupled, which completely eliminates the problem of odd–even decoupling.

Staggered meshes are currently applied with central discretization and the most popular two-dimensional arrangement is shown in Figure 12.4.2, where the u and v velocity components are located on different cell faces. The equations are discretized in conservation form, the control volumes depending on the considered equations. The mass equation is discretized on the volume centered on the point (i, j), while the x-momentum conservation is expressed on the volume centered for the location of u, i.e. $(i + 1/2, j)$. Similarly, the y-momentum conservation is expressed on the volume centered on the location of v, i.e. $(i, j + 1/2)$.

The Poisson equation for the pressure is obtained from the divergence of the discretized momentum equation. *This step should be performed by exactly the same discrete operations as applied to express mass conservation.* This is required for global consistency and conservation. It is fairly straightforward on a Cartesian mesh, but becomes essential on arbitrary meshes.

The Poisson equation for the pressure is solved with Neumann boundary conditions on the normal pressure gradient on the walls and at the inlet, while at the outlet

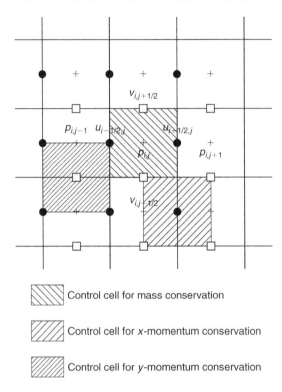

Control cell for mass conservation

Control cell for x-momentum conservation

Control cell for y-momentum conservation

Figure 12.4.2 *Staggered, two-dimensional finite difference mesh for centrally discretized pressure correction methods.*

the pressure is set to a certain value (e.g. atmospheric pressure). The details of the implementation depend on the selected space discretization and on the mesh.

An additional condition is essential for the numerical accuracy of the resolution of the pressure equation, namely that the compatibility condition, obtained from Green's theorem applied to the Poisson equation, should be identically satisfied by the space discretization. Applied to equation (12.4.12), we should have identically, for the integral of the normal pressure gradient on boundary $\vec{S}$ of the computational domain Ω:

$$\int_{\Omega} \Delta p^{n+1} d\Omega = \oint_S \vec{\nabla} p \cdot d\vec{S} = \oint_S \frac{\partial p}{\partial n} dS = \frac{\rho}{\Delta t} \int_{\Omega} \nabla \cdot \vec{v}^* d\Omega = \frac{\rho}{\Delta t} \oint_S \vec{v}^* \cdot d\vec{S}$$

$$(12.4.22)$$

12.4.3 Implementation of a Pressure Correction Method

We consider the application of the fractional step method to the simulation of incompressible flows. The method is described for a cell-centered finite volume method on a Cartesian mesh, with all flow variables defined in the same points at the centers of the computational cells, although the staggered grid approach is applied to connect values at the cell centers with face defined values, as seen on Figure 12.4.3, where different control volumes are shown. Boundaries of the computational domain are

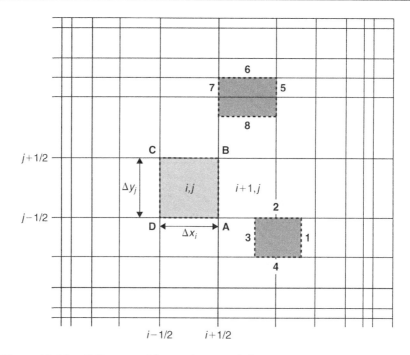

Figure 12.4.3   *Cell-centered finite volume mesh for pressure correction method on non-uniform Cartesian grid.*

located on cell faces and you can follow the treatment of the boundary conditions as described in Section 12.2.

Another way to impose boundary conditions in such a configuration can be considered through introduction of ghost or dummy cells, i.e. rows of cells neighboring the computational domain and having the same size as the first row of cells inside the domain.

You can extend the generality of your code by allowing for non-uniform Cartesian grids to account for clustered grids near solid walls. We consider a rectangular domain with side lengths L_x and L_y with N_x and N_y mesh points in both directions and variable mesh sizes. The Cartesian mesh has $N_x \times N_y$ nodes, dividing the domain into $(N_x - 1)(N_y - 1)$ rectangular cells, the sides of varying sizes: $\Delta x_i = x_{i+1} - x_i$, $\Delta y_i = y_{i+1} - y_i$.

The two-dimensional incompressible Navier–Stokes equations are discretized on the non-uniform Cartesian grid of Figure 12.4.3. A cell-centered second order finite volume discretization is selected, with an explicit first order time integration.

$$\frac{\partial u}{\partial x} + \frac{\partial v}{\partial y} = 0$$

$$\frac{\partial u}{\partial t} + \frac{\partial u^2}{\partial x} + \frac{\partial uv}{\partial y} = -\frac{1}{\rho}\frac{\partial p}{\partial x} + \frac{1}{\rho}\frac{\partial \tau_{xx}}{\partial x} + \frac{1}{\rho}\frac{\partial \tau_{xy}}{\partial y} = -\frac{1}{\rho}\frac{\partial p}{\partial x} + v\frac{\partial^2 u}{\partial x^2} + v\frac{\partial^2 u}{\partial y^2}$$

$$\frac{\partial v}{\partial t} + \frac{\partial uv}{\partial x} + \frac{\partial v^2}{\partial y} = -\frac{1}{\rho}\frac{\partial p}{\partial y} + \frac{1}{\rho}\frac{\partial \tau_{xy}}{\partial x} + \frac{1}{\rho}\frac{\partial \tau_{yy}}{\partial y} = -\frac{1}{\rho}\frac{\partial p}{\partial y} + v\frac{\partial^2 v}{\partial x^2} + v\frac{\partial^2 v}{\partial y^2}$$

$$(12.4.23)$$

12.4.3.1 *Numerical discretization*

In the pressure correction method, the velocities are updated from the momentum equations, while pressure is obtained by solving the Poisson equation derived by taking a divergence of the momentum equations and taking conservation of mass into account.

We refer to Figure 12.4.3 and apply a finite volume formulation on the contour ABCD for the velocity components u and v at the cell centers.

Applying the fractional step formulation of equations (12.4.9) to (12.4.12), we write the scheme as follows, for the intermediate velocity components u^* and v^*

$$\frac{u_{i,j}^* - u_{i,j}^n}{\Delta t} + \frac{(u^2)_{i+1/2,j}^n - (u^2)_{i-1/2,j}^n}{\Delta x_i} + \frac{(uv)_{i,j+1/2}^n - (uv)_{i,j-1/2}^n}{\Delta y_j}$$

$$= \frac{1}{\rho}\frac{(\tau_{xx})_{i+1/2,j}^n - (\tau_{xx})_{i-1/2,j}^n}{\Delta x_i} + \frac{1}{\rho}\frac{(\tau_{xy})_{i,j+1/2}^n - (\tau_{xy})_{i,j-1/2}^n}{\Delta y_j} \quad (12.4.24)$$

$$\frac{v_{i,j}^* - v_{i,j}^n}{\Delta t} + \frac{(uv)_{i+1/2,j}^n - (uv)_{i-1/2,j}^n}{\Delta x_i} + \frac{(v^2)_{i,j+1/2}^n - (v^2)_{i,j-1/2}^n}{\Delta y_j}$$

$$= \frac{1}{\rho}\frac{(\tau_{yx})_{i+1/2,j}^n - (\tau_{yx})_{i-1/2,j}^n}{\Delta x_i} + \frac{1}{\rho}\frac{(\tau_{yy})_{i,j+1/2}^n - (\tau_{yy})_{i,j-1/2}^n}{\Delta y_j} \quad (12.4.25)$$

where the interface values are obtained by the weighted averages

$$u_{i+1/2,j} = \frac{\Delta x_i u_{i+1,j} + \Delta x_{i+1} u_{i,j}}{\Delta x_i + \Delta x_{i+1}}$$

$$v_{i,j+1/2} = \frac{\Delta y_j v_{i,j+1} + \Delta y_{j+1} v_{i,j}}{\Delta y_j + \Delta y_{j+1}} \quad (12.4.26)$$

and similarly for the other variables.

You can calculate the shear stress components as follows:

$$(\tau_{xx})_{i+1/2,j}^n = 2\mu_{i+1/2,j}^n (u_x)_{i+1/2,j}^n$$

$$(\tau_{xy})_{i+1/2,j}^n = \mu_{i+1/2,j}^n ((u_y)_{i+1/2,j}^n + (v_x)_{i+1/2,j}^n)$$

$$(u_x)_{i+1/2,j}^n = \frac{u_{i+1,j}^n - u_{i,j}^n}{(\Delta x_{i+1} + \Delta x_i)/2} \quad (v_x)_{i+1/2,j}^n = \frac{v_{i+1,j}^n - v_{i,j}^n}{(\Delta x_{i+1} + \Delta x_i)/2}$$

$$(u_y)_{i+1/2,j}^n = \frac{1}{2}\left[\frac{u_{i+1,j+1}^n - u_{i+1,j-1}^n}{(\Delta y_{j+1} + 2\Delta y_j + \Delta y_{j-1})/2} + \frac{u_{i,j+1}^n - u_{i,j-1}^n}{(\Delta y_{j+1} + 2\Delta y_j + \Delta y_{j-1})/2}\right]$$

$$(v_y)_{i+1/2,j}^n = \frac{1}{2}\left[\frac{v_{i+1,j+1}^n - v_{i+1,j-1}^n}{(\Delta y_{j+1} + 2\Delta y_j + \Delta y_{j-1})/2} + \frac{v_{i,j+1}^n - v_{i,j-1}^n}{(\Delta y_{j+1} + 2\Delta y_j + \Delta y_{j-1})/2}\right]$$

$$(12.4.27)$$

$$(\tau_{yy})_{i,j+1/2}^{n} = 2\mu_{i,j+1/2}^{n}(v_y)_{i+1/2,j}^{n}$$

$$(\tau_{yx})_{i,j+1/2}^{n} = \mu_{i,j+1/2}^{n}((u_y)_{i,j+1/2}^{n} + (v_x)_{i,j+1/2}^{n})$$

$$(u_y)_{i,j+1/2}^{n} = \frac{u_{i,j+1}^{n} - u_{i,j}^{n}}{(\Delta y_{j+1} + \Delta y_j)/2} \qquad (v_y)_{i,j+1/2}^{n} = \frac{v_{i,j+1}^{n} - v_{i,j}^{n}}{(\Delta y_{j+1} + \Delta y_j)/2}$$

$$(u_x)_{i,j+1/2}^{n} = \frac{1}{2}\left[\frac{u_{i+1,j+1}^{n} - u_{i-1,j+1}^{n}}{(\Delta x_{i+1} + 2\Delta x_i + \Delta x_{i-1})/2} + \frac{u_{i+1,j}^{n} - u_{i-1,j}^{n}}{(\Delta x_{i+1} + 2\Delta x_i + \Delta x_{i-1})/2}\right]$$

$$(v_x)_{i,j+1/2}^{n} = \frac{1}{2}\left[\frac{v_{i+1,j+1}^{n} - v_{i-1,j+1}^{n}}{(\Delta x_{i+1} + 2\Delta x_i + \Delta x_{i-1})/2} + \frac{v_{i+1,j}^{n} - v_{i-1,j}^{n}}{(\Delta x_{i+1} + 2\Delta x_i + \Delta x_{i-1})/2}\right]$$

$$(12.4.28)$$

Once the intermediate velocity components are estimated, you can obtain the values of the updated velocity and pressure variables, satisfying the continuity equation, following (12.4.10). Applying a finite volume formulation on the contour ABCD to this equation and projecting in the x- and y-directions, we obtain

$$\frac{u_{i,j}^{n+1} - u_{i,j}^{*}}{\Delta t} = -\frac{1}{\rho}\frac{p_{i+1/2,j} - p_{i-1/2,j}}{\Delta x_i}$$

$$\frac{v_{i,j}^{n+1} - v_{i,j}^{*}}{\Delta t} = -\frac{1}{\rho}\frac{p_{i,j+1/2} - p_{i,j+1/2}}{\Delta y_j} \qquad (12.4.29)$$

The pressure is obtained by expressing that the velocity components at level $(n+1)$ satisfy the divergence free continuity equation, as in (12.4.6). Discretized on the finite volume mesh ABCD we obtain

$$\frac{u_{i+1/2,j}^{n+1} - u_{i-1/2,j}^{n+1}}{\Delta x_i} + \frac{v_{i,j+1/2}^{n+1} - v_{i,j-1/2}^{n+1}}{\Delta y_j} = 0 \qquad (12.4.30)$$

The interface velocities at the $(n+1)$ level are obtained by applying once again the finite volume formulation of equation (12.4.10), but this time on a staggered control volume such as 1234, Figure 12.4.3, centered on face AB, for the x component equation

$$\rho\frac{u_{i+1/2,j}^{n+1} - u_{i+1/2,j}^{*}}{\Delta t}\Delta y_j\frac{\Delta x_{i+1} + \Delta x_i}{2} = (p_{i+1,j}^{n+1} - p_{i,j}^{n+1})\Delta y_j \qquad (12.4.31)$$

and on the control volume 5678, centered around BC, for the vertical component, leading to

$$\rho\frac{v_{i,j+1/2}^{n+1} - v_{i,j+1/2}^{*}}{\Delta t}\Delta x_i\frac{\Delta y_{j+1} + \Delta y_j}{2} = (p_{i,j+1}^{n+1} - p_{i,j}^{n+1})\Delta x_i \qquad (12.4.32)$$

The values on faces CD and DA are obtained similarly.

The intermediate values u^* and v^* on the cell faces are obtained from the relations (12.4.26).

Substituting these relations in equation (12.4.30), leads to the pressure Poisson equation:

$$\frac{1}{\Delta x_i}\left[\frac{p_{i+1,j}^{n+1} - p_{i,j}^{n+1}}{(\Delta x_{i+1} + \Delta x_i)/2} - \frac{p_{i,j}^{n+1} - p_{i-1,j}^{n+1}}{(\Delta x_i + \Delta x_{i-1})/2}\right]$$

$$+\frac{1}{\Delta y_j}\left[\frac{p_{i,j+1}^{n+1} - p_{i,j}^{n+1}}{(\Delta y_{j+1} + \Delta y_j)/2} - \frac{p_{i,j}^{n+1} - p_{i,j-1}^{n+1}}{(\Delta y_j + \Delta y_{j-1})/2}\right]$$

$$=\frac{\rho}{\Delta t}\left[\frac{u_{i+1/2,j}^* - u_{i-1/2,j}^*}{\Delta x_i} + \frac{v_{i,j+1/2}^* - v_{i,j-1/2}^*}{\Delta y_j}\right] \tag{12.4.33}$$

The whole procedure of updating the solution is the following:

- Calculate the intermediate velocity field $u_{i,j}^*, v_{i,j}^*$ from (12.4.24) and (12.4.25).
- Obtain the pressure by solving the Poisson equation (12.4.33).
- Obtain the solution at the next time step $u_{i,j}^{n+1}, v_{i,j}^{n+1}$ from (12.4.29), where the pressure at the cell faces is obtained by applying relations (12.4.26).

We have now to focus on the most critical issue of pressure correction methods, namely the efficient resolution of the pressure Poisson equation. This is a crucial step of the whole approach, since the overall efficiency of your code will depend on its performance.

12.4.3.2 *Algorithm for the pressure Poisson equation*

The pressure Poisson equation (12.4.33) is a standard elliptic equation and you can call upon the various methods introduced in Chapter 10. In advanced codes, various convergence optimization and acceleration techniques are applied, in particular preconditioning and multigrid techniques are strongly recommended for this step of the computation, and many of these techniques are described in the literature on pressure correction methods.

Here, we suggest you to choose a simple line Gauss–Seidel method along a vertical line. As a first approximation of the pressure, its discrete values obtained on the previous time level are used. Given an approximation of the pressure p^k, the next one is obtained from the following relation:

$$\frac{1}{\Delta x_i}\left[\frac{p_{i+1,j}^{k} - p_{i,j}^{k+1}}{(\Delta x_{i+1} + \Delta x_i)/2} - \frac{p_{i,j}^{k+1} - p_{i-1,j}^{k+1}}{(\Delta x_i + \Delta x_{i-1})/2}\right]$$

$$+\frac{1}{\Delta y_j}\left[\frac{p_{i,j+1}^{k+1} - p_{i,j}^{k+1}}{(\Delta y_{j+1} + \Delta y_j)/2} - \frac{p_{i,j}^{k+1} - p_{i,j-1}^{k+1}}{(\Delta y_j + \Delta y_{j-1})/2}\right] = Q_{i,j} \tag{12.4.34}$$

where $Q_{i,j}$ is the right-hand side of (12.4.33). Not that the k index denotes an iteration number and not a time level (n in the previous section). Equation (12.4.34) can be rewritten as follows:

$$a_{i,j}p_{i,j-1}^{k+1} + b_{i,j}p_{i,j}^{k+1} + c_{i,j}p_{i,j+1}^{k+1}$$

$$= Q_{i,j} - \frac{1}{\Delta x_i}\left[\frac{p_{i+1,j}^{k}}{(\Delta x_{i+1} + \Delta x_i)/2} + \frac{p_{i-1,j}^{k+1}}{(\Delta x_i + \Delta x_{i-1})/2}\right] \qquad (12.4.35)$$

where

$$a_{i,j} = \frac{2}{\Delta y_j}\frac{1}{\Delta y_j + \Delta y_{j-1}}$$

$$c_{i,j} = \frac{2}{\Delta y_j}\frac{1}{\Delta y_{j+1} + \Delta y_j} \qquad (12.4.36)$$

$$b_{i,j} = -\frac{2}{\Delta x_i}\left[\frac{1}{\Delta x_{i+1} + \Delta x_i} + \frac{1}{\Delta x_i + \Delta x_{i-1}}\right] - (a_{i,j} + c_{i,j})$$

We have to add the boundary conditions, for instance for $i = 1$ and a Neumann boundary condition (12.4.35) can be rewritten taking into account the boundary condition at $i = 1/2$:

$$a_{1,j}p_{1,j-1}^{k+1} + \left(b_{1,j} + \frac{1}{\Delta x_1}\frac{2}{\Delta x_1 + \Delta x_0}\right)p_{1,j}^{k+1} + c_{1,j}p_{1,j+1}^{k+1}$$

$$= Q_{1,j} - \frac{1}{\Delta x_1}\frac{p_{2,j}^{k}}{(\Delta x_2 + \Delta x_1)/2} \qquad (12.4.37)$$

where the Neumann boundary condition is expressed as $p_0^{k+1} = p_1^{k+1}$.

You can apply this similarly for the other boundary conditions. If the pressure is imposed at certain boundaries, as a Dirichlet condition, then you can introduce this value directly in the corresponding equation.

The algebraic system can be efficiently solved with the Thomas Algorithm (see Appendix A in Chapter 10).

The iterations are to be repeated until a prescribed convergence criterion is satisfied (e.g. $\max_{i,j} |p_{i,j}^{k+1} - p_{i,j}^{k}| < \varepsilon$).

For reasons of accuracy, it is recommended to alternate this algorithm with a line Gauss–Seidel method in the horizontal direction (at $j = $ constant), applying the Thomas algorithm in two different mesh directions. Equation (12.4.34) is replaced by

$$\frac{1}{\Delta x_i}\left[\frac{p_{i+1,j}^{k+1} - p_{i,j}^{k+1}}{(\Delta x_{i+1} + \Delta x_i)/2} - \frac{p_{i,j}^{k+1} - p_{i-1,j}^{k+1}}{(\Delta x_i + \Delta x_{i-1})/2}\right]$$

$$+ \frac{1}{\Delta y_j}\left[\frac{p_{i,j+1}^{k} - p_{i,j}^{k+1}}{(\Delta y_{j+1} + \Delta y_j)/2} - \frac{p_{i,j}^{k+1} - p_{i,j-1}^{k+1}}{(\Delta y_j + \Delta y_{j-1})/2}\right] = Q_{i,j} \qquad (12.4.38)$$

We leave it to you as an exercise to work out the details of its implementation.

12.5 NUMERICAL SOLUTIONS WITH THE PRESSURE CORRECTION METHOD

We apply now the developed method to the incompressible lid driven cavity flow. This well-known flow configuration results from the uniform motion of the upper wall of a square box, wherein the flow is induced by the viscous stresses, similarly to the Couette flow. The flow within the lid driven rectangular two-dimensional cavity is maintained by the continuous diffusion of kinetic energy injected by the moving wall. This energy is initially confined to a thin viscous layer of fluid next to the moving boundary. After a period of time, which depends on the Reynolds number, the redistribution of energy reaches an equilibrium leading to a steady state laminar flow. In case of high Reynolds number flows, this steady state solution is never reached, due to instabilities leading to transition to turbulence.

It can actually be considered as the two-dimensional extension of the Couette flow.

12.5.1 Lid Driven Cavity

We consider the domain included in a square of unit length, with $0 \leq x, y \leq 1$, where the upper boundary at $y = 1$, moves with a constant velocity $U = 1$.

The Reynolds number based on the size of the domain, the velocity of the moving wall, density $\rho = 1$ and viscosity $\mu = 0.01$ is Re $= 100$.

The boundary conditions are set as follows

$$
\begin{aligned}
u(x,0) &= 0 & v(x,0) &= 0 \\
u(x,1) &= 1 & v(x,1) &= 0 \\
u(0,y) &= 0 & v(0,y) &= 0 \\
u(1,y) &= 0 & v(1,y) &= 0
\end{aligned}
\tag{12.5.1}
$$

The calculation is performed on a 41×41 Cartesian uniform mesh, which divides the computational domain in 1600 equidistant cells, with $\Delta x = \Delta y = 0.025$. In the solution of the pressure Poisson equation a Neumann boundary condition is imposed on the boundaries. The time step is taken as $\Delta t = 0.01$. A line Gauss–Seidel method is used to solve the pressure equation iteratively, iterations repeated till the maximum absolute value of the residual is smaller that 10^{-3}. In the selected time-dependent approach, the Poisson equation is converged for each time step. This option is selected to enable you also to handle unsteady flows or to detect spontaneous unsteadiness when they occur.

Figure 12.5.1 shows the convergence history of the pressure equation plotted for the first 10 time steps. As can be seen from the graph, the residual value for the first iteration of the Poisson solver at each time step decreases as the calculation proceeds, which means that less and less iterations are needed to reach the convergence criterion. This is typical for pressure correction methods when applied to unsteady flows with a steady state limit, at which the Poisson equation is satisfied automatically. Each jump in this figure represents the passage at the next time step, while the residual reduction in between represents the convergence behavior of the Gauss–Seidel relaxation method.

The diagrams on Figure 12.5.2 display the streamlines of the solution at different transient stages, at $t = 0.5$, 1, 10 and 30. The flow undergoes a recirculation motion

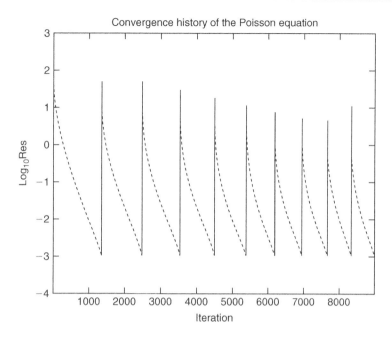

Figure 12.5.1 *Convergence history of the Poisson equation for the pressure.*

imposed by the viscous effects. The solution at $t = 0.5$ and $t = 1$ represent a transient stage of the solution, where the circular motion of the fluid is still developing. At $t = 10$ and $t = 30$ the flow in the cavity is fully developed and has reached its steady state.

On a quantitative basis, Figure 12.5.3 compares the velocity distributions along the centerlines $x = 0.5$ and $y = 0.5$, with a reference solution obtained by Dr. Sergey Smirnov, at the Vrije Universiteit Brussel, Belgium, on a fine grid of 161×161 with a fourth order accurate compact scheme for the space discretization.

The maximum error on the velocity distribution is of the order of 10% on this 41×41 mesh, indicating that a finer resolution is required.

12.5.2 Additional Suggestions

You can now run your code on many other cases, such as:

- The lid driven cavity by increasing the Reynolds number of your simulation, until you start detecting the initial process toward transition. This will require you to increase the grid resolution.
- The flat plate problem.
- Other cases with a Cartesian grid, such as the backward facing step.
- You could also extend now your code to more general grids and run the cylinder case in laminar mode.

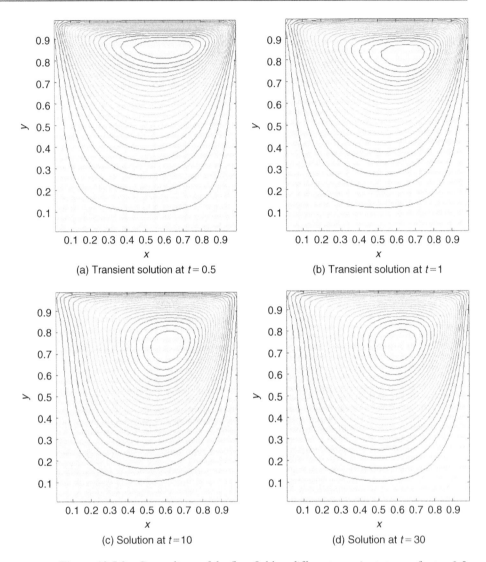

Figure 12.5.2   *Streamlines of the flow field at different transient stages, for t = 0.5, 1, 10 and 30 (for color image refer Plate 12.5.2).*

12.6   BEST PRACTICE ADVICE

CFD software systems form today an essential part of the world of Computer Aided Engineering (CAE), supporting the design and analysis of industrial products involving fluid flows. Many design decisions of systems, whose performance depends on their internal or external flow behavior, are based on the results of CFD simulations, either with in-house or commercial CFD codes. This raises the question of

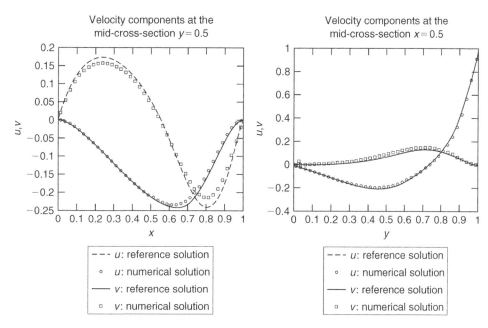

Figure 12.5.3 *Comparison of the computed velocity distributions along the centerlines x = 0.5 and y = 0.5, with a reference solution.*

the reliability and the confidence we can attach to the results of CFD, in presence of numerous sources of errors and of uncertainties.

The overwhelming majority of industrial and environmental fluid flow systems are turbulent and the modeling of turbulence remains a dominant factor of uncertainty, as none of the available models today are fully satisfactory in their prediction capability of the complex phenomena of turbulence.

We have not dealt with turbulence, in this introductory text on the basics of CFD, and left this important issue to the Volume II, although it has a considerable effect on the level of uncertainty of CFD results.

In order to respond to the needs of the increasing number of CFD users in industry, a demand has arisen for recommendations of best practices in the application of CFD codes, in regard of the complexity of industrial flow systems.

One of the first efforts toward the establishment of best practice guidelines (BPG) for CFD has been generated by the ERCOFTAC (European Research Community on Flow, Turbulence And Combustion) association; http://www.ercoftac.org. This effort has led to a document, Casey and Wintergerste (2000), providing an extensive set of recommendations, and is available from this organization.

As we have considered here only the numerical issues and since we attempted to develop your awareness of the various pitfalls and error sources, particularly in the last two chapters, we will summarize here some of the basic guidelines and recommendations in applying CFD codes, based essentially on the content of this ERCOFTAC document.

12.6.1 List of Possible Error Sources

The first issue is to attempt to summarize all the possible error sources. We can distribute them as follows:

- *Discretization or numerical error*: These errors are due to the approximations resulting from the space and time numerical discretization, and have been analyzed in details in Parts III and IV of this book.
- *Iteration or convergence error*: These errors occur due to the difference between a fully converged solution on a finite number of grid points and a solution that is not fully converged. Ideally, each calculation should be run up to the reduction of the residuals to machine accuracy. Although this can be performed on simple cases as illustrated in the last two chapters, it is hardly ever possible on industrial simulations with million of points.
- *Round-off errors*: These errors are due to the fact that the difference between two values of a parameter is below the machine accuracy of the computer. This is caused by the limited number of computer digits available for storage of a given physical value. It might require to shift to double precision arithmetic, when dealing with very small variations between flow variables, or when very short distances between mesh points are introduced.
- *Application uncertainties*: Many variables defining the flow conditions, such as operational and/or geometrical data are often not precisely defined or not well known. Examples of this are uncertainties in the precise geometry due to manufacturing tolerances, uncertain inflow data or models, such as turbulence properties or fluid properties.
- *User errors*: Errors can arise from mistakes introduced by the user. They can cover various aspects, such as inadequate or poor grid generation; incorrect boundary condition; incorrect choice of numerical parameters, such as time step or relaxation coefficients; post-processing errors. Experience and great care are required to minimize their risk of occurrence.
- *Code errors*: Errors due to bugs in the software cannot be excluded, despite all verification efforts, as it is humanly impossible to cover all possible combinations of code parameters with a finite number of verification tests.
- *Model uncertainties*: It refers to the physical models that have to be introduced to describe complex flow properties, such as turbulence, multiphase flows, combustion, real Newtonian or non-Newtonian fluids.

To minimize the effects of these error sources, a series of recommendations can be collected, that we recommend to your attention, as a kind of checklist when running a CFD code.

12.6.2 Best Practice Recommendations

As seen in the previous chapters, an essential component of a CFD simulation and a major potential source of errors is the choice of the grid and the resulting grid quality.

RECOMMENDATIONS ON GRIDS

The key recommendation is to ensure smooth grids, avoiding abrupt changes in grid size or shape, as this can lead to a significant loss of accuracy. Hence take good care to:

- Define the computational domain, in order to minimize the influence and interactions between the flow and the far-field conditions. In particular,
  - Place inlet and outlet boundaries as far away as possible from the region of interest. In particular, if uniform far-field conditions are imposed, you should ensure that the boundary is not in a region where the flow may still vary significantly.
  - Avoid inlet or outlet boundaries in regions of strong geometrical changes or in regions of recirculation.
- Avoid jumps in grid density or in grid size.
- Avoid highly distorted cells or small grid angles.
- Ensure that the grid stretching is continuous.
- Avoid unstructured tetrahedral meshes in boundary layer regions.
- Refine the grids in regions with high gradients, such as boundary layers, leading edges of airfoils and any region where large changes in flow properties might occur.
- Make sure that the number of points in the boundary layers is sufficient for the expected accuracy. Avoid less than 10 points over the inner part of the boundary layer thickness.
- Monitor the grid quality by adequate mesh parameters, available in most of the grid generators, such as aspect ratio, internal angle, concavity, skewness, negative volume.

RECOMMENDATIONS ON SOLUTION ASSESSMENT

Once you run your code, the following recommendations will be useful to enhance your confidence in the results obtained:

- Check very carefully the selected boundary conditions for correctness and compatibility with the physics of the flow you are modeling.
- Verify all the numerical settings and parameters, before launching the CFD run.
- Verify that your initial solution is acceptable for the problem to be solved.
- Monitor the convergence to ensure that you reach machine accuracy. It is recommended to monitor, in addition to the residuals, the convergence of representative quantities of your problem, such as a drag force or coefficient, a velocity, temperature or pressure at selected points in the flow domain.
- Look carefully at the behavior of the residual convergence curve in function of number of iterations. If the behavior is oscillatory, or if the residual does not converge to machine accuracy by showing a limit cycle at a certain level of residual reduction, it tells you that some inaccuracy affects your solution process.

- Apply internal consistency and accuracy criteria, by verifying:
  - Conservation of global quantities such as total enthalpy and mass flow in steady flow calculations.
  - The entropy production and drag coefficients with inviscid flows, which are strong indicators of the influence of numerical dissipation, as they should be zero.
- Check, whenever possible, the grid dependence of the solution by comparing the results obtained on different grid sizes.
- Some quantities are more sensitive than others to error sources. Pressure curves are less sensitive than shear stresses, which in turn are less sensitive than temperature gradients or heat fluxes, which require finer grids for a given accuracy level.
- If your calculation appears difficult to converge, you can
  - Look at the residual distribution and associated flow field for possible hints, e.g. regions with large residuals or unrealistic levels of the relevant flow parameters.
  - Reduce the values of parameters controlling convergence, such as the CFL number or some under-relaxation parameter, when available.
  - Consider the effects of different initial flow conditions.
  - Check the effect of the grid quality on the convergence rate.
  - Use a more robust numerical scheme, such as a first order scheme, during the initial steps of the convergence and switch to more accurate numerical schemes as the convergence improves.

RECOMMENDATIONS ON EVALUATION OF UNCERTAINTIES

This is a very difficult issue, as the application uncertainties are generally not well defined and require a sound judgment about the physics of the considered flow problem. Some recommendations can be offered:

- Attempt to list the most important uncertainties, such as
  - Geometrical simplifications and manufacturing tolerances around the CAD definition.
  - Operational conditions, such as inlet velocity or inlet flow angle.
  - Physical approximations, such as handling an incompressible flow as a low Mach number compressible flow. This type of uncertainty is manageable, as it can more easily be quantified.
  - Uncertainties related to turbulence or other physical models.
- Perform a sensitivity analysis of the relevant uncertainty to investigate its influence.

CONCLUSIONS AND MAIN TOPICS TO REMEMBER

If you have followed closely the guidelines of this chapter, you have now available a general 2D finite volume density-based code, which allows you to handle practically any flow configuration, from low to supersonic speed. You can even simulate incompressible flow conditions, by considering low Mach numbers, say below 0.2, for which the numerical solution is an excellent approximation of incompressible fluid flows.

You also have available another option, with a code based on the pressure correction method, suitable for compressible and incompressible flows, although it is restricted to the subsonic range.

You have certainly experienced, by following the steps of the last two chapters in running the various proposed test cases, that the way to achieve high accuracy and reliability of the CFD results on general grids is a difficult process, requiring a close attention to all the details of the implementation of a selected scheme.

Our main ambition with these two chapters was to introduce you to this awareness and to guide you in your ability to ask the 'right questions' when faced with the development of a CFD code or when using a third party code.

The main topics to remember are summarized in the best practice guidelines of Section 12.6. The main message being that you have to exercise critical judgment at all stages of the code development. If you apply a third party code, your critical judgment should apply to your assessment of all aspects of the schemes and its implementation as proposed by the options you select. Make sure that you have enough information on:

- Formal order of the scheme, but also on its behavior on a non-uniform grid.
- The level of numerical dissipation generated on your grid. This can be obtained by running the same case as an inviscid problem, monitoring the entropy distribution.
- The details of the boundary condition implementation and their effect on the accuracy and convergence.
- Convergence levels of the solution, in terms of residuals, but also by monitoring some of the quantities relevant for the problem you are interested in.

We also hope that these exercises will have stimulated your interest and enthusiasm for the beautiful world of numerical flow simulations.

REFERENCES

Anderson, J.D. (1995). *Computational Fluid Dynamics. The Basics with Applications*. McGraw-Hill, New York.

Casey, M. and Wintergerste, T. (2000). Best Practice Guidelines. *ERCOFTAC Special Interest Group on Quality and Trust in Industrial CFD*. ERCOFTAC, http://www. ercoftac.org.

Chorin, A.J. (1967). A numerical method for solving incompressible viscous flow problems. *J. Comput. Phys.*, 2, 12–26.

Chorin, A.J. (1968). Numerical solution of the Navier–Stokes equations. *Math. comput.*, 23, 341–54.

Ferziger, J.H. and Peric, M. (1997). *Computational Methods for Fluid Dynamics*. Springer Verlag, Berlin.

Harlow, F.H. and Welch, J.E. (1965). Numerical calculation of time-dependent viscous incompressible flow of fluid with free surface. *Phys. Fluid.*, 8, 2182–2189.

Hemsch, M.J. and Morrison, J.H. (2004). Statistical analysis of CFD solutions from 2nd drag prediction workshop. *42nd AIAA Aerospace Sciences Meeting*, Reno, AIAA Paper 2004-556.

Jameson, A. (1995a). Analysis and design of numerical schemes for gas dynamics. 1. Artificial diffusion, upwind biasing, limiters and their effect on multigrid convergence. *Int. J. Comp. Fluid Dyn.*, 4, 171–218.

Jameson, A. (1995b). Analysis and design of numerical schemes for gas dynamics. 2. Artificial diffusion and discrete shock structure. *Int. J. Comp. Fluid Dyn.*, 5, 1–38.

Majda, A. and Sethian, J. (1985). The derivation and numerical solution of the equations for zero Mach number combustion. *Combust. Sci. Technol.*, 42, 185.

Patankar, S.V. (1980). *Numerical Heat Transfer and Fluid Flow*. Hemisphere Publ. Co., New York.

Patankar, S.V. and Spalding, D.B. (1972). A calculation procedure for heat, mass and momentum transfer in three-dimensional parabolic flows. *Int. J. Heat Mass Transfer*, 15, 1787–1806.

Schlichting, H. (1979). *Boundary Layer Theory*. McGraw-Hill, New York.

Temam, R. (1969). Sur l'approximation de la solution de equations de Navier–Stokes par la methode des pas fractionnnaires. *Arch. Rational Mech. Anal.*, 32, 135–153; (II): *Arch. Rational Mech. Anal.*, 33, 377–385.

Temam, R. (1977). *Navier–Stokes Equations*. North-Holland, Amsterdam.

Wesseling, P. (2001). *Principles of Computational Fluid Dynamics*. Springer Verlag, Berlin.

Index

Printed and bound by CPI Group (UK) Ltd, Croydon, CR0 4YY

03/10/2024

01040332-0007